Role of Green Chemistry in Ecosystem Restoration to Achieve Environmental Sustainability

Woodhead Publishing Series in Advances in Pollution Research

Role of Green Chemistry in Ecosystem Restoration to Achieve Environmental Sustainability

Edited by

Arun Lal Srivastav
Chitkara University School of Engineering and Technology, Chitkara University, Kallujhanda, Himachal Pradesh, India

Ajmer Singh Grewal
Department of Pharmaceutical Chemistry, Guru Gobind Singh College of Pharmacy, Yamunanagar, Haryana, India

Markandeya
Department of Civil Engineering, IIT (BHU) Varanasi, Varanasi, Uttar Pradesh, India

Tien Duc Pham
University of Science, Vietnam National University, Hanoi, Vietnam

ELSEVIER

WP WOODHEAD PUBLISHING
An imprint of Elsevier

Woodhead Publishing is an imprint of Elsevier
50 Hampshire Street, 5th Floor, Cambridge, MA 02139, United States
125 London Wall, London EC2Y 5AS, United Kingdom

ISBN: 978-0-443-15291-7 (print)
ISBN: 978-0-443-15292-4 (online)

For Information on all Woodhead Publishing publications
visit our website at https://www.elsevier.com/books-and-journals

Publisher: Candice G. Janco
Acquisitions Editor: Jessica Mack
Editorial Project Manager: Rupinder K. Heron
Production Project Manager: R. Vijay Bharath
Cover Designer: Vicky Pearson Esser

Typeset by MPS Limited, Chennai, India

Contents

Section B
Water sustainability

List of contributors

Md Altamash Ahmad Chitkara College of Pharmacy, Chitkara University, Rajpura, Punjab, India

Bharat Kumar Allam Department of Chemistry, Rajiv Gandhi University (A Central University), Doimukh, Arunachal Pradesh, India

Aryan Anand Department of Environmental Sciences and Technology, Central University of Punjab, Bathinda, Punjab, India

Ponnusamy Thillai Arasu College of Natural and Computational Sciences, Wollega University, Nekemte, Oromiya, Ethiopia

Vimal Arora University Institute of Pharma Sciences, Chandigarh University, Mohali, Punjab, India

Arunima Department of Chemistry, M. L. K. (P. G.) College, Balrampur, Uttar Pradesh, India

Segni Asafa Ashana College of Natural and Computational Sciences, Wollega University, Nekemte, Oromiya, Ethiopia

Himani Bajaj Adarsh Vijendra Institute of Pharmaceutical Sciences, Shobhit University, Gangoh, Uttar Pradesh, India

Sushmita Banerjee Department of Environmental Sciences, Sharda University, Greater Noida, Uttar Pradesh, India

Bhaskar Behera P.G. Department of Bioscience and Biotechnology, Fakir Mohan University, Balasore, Odisha, India

Bijay Kumar Behera Aquatic Environmental Biotechnology and Nanotechnology Division, ICAR-Central Inland Fisheries Research Institute, Kolkata, West Bengal, India

Jiban Kumar Behera P.G. Department of Zoology, Fakir Mohan University, Balasore, Odisha, India

Manojit Bhattacharya P.G. Department of Zoology, Fakir Mohan University, Balasore, Odisha, India

Gargi Biswas Waste Management (I) Division, Central Pollution Control Board, Delhi, India

Paromita Chakraborty Center for Research in Environment, Sustainability Advocacy and Climate Change (REACH), SRM Institute of Science and Technology, Kattankulathur, Tamil Nadu, India; Faculty of Biology and Environmental Protection, University of Lodz, Lodz, Poland

Balakumar Chandrasekaran Faculty of Pharmacy, Philadelphia University, Amman, Jordan

Bhawna Chopra Guru Gobind Singh College of Pharmacy, Yamunanagar, Haryana, India

Ratnakar Dash Department of Chemistry, Mahima Mahavidyalaya Joranda, Atinda, Odisha, India

Subhakanta Dash Department of Chemistry, Synergy Institute of Engineering and Technology, Dhenkanal, Odisha, India

Jyotsna Dei Aquatic Environmental Biotechnology and Nanotechnology Division, ICAR-Central Inland Fisheries Research Institute, Kolkata, West Bengal, India

Geeta Deswal Guru Gobind Singh College of Pharmacy, Yamunanagar, Haryana, India

Mohini Devi Institute of Pharmaceutical Sciences, Kurukshetra University, Kurukshetra, Haryana, India

Priyanka Devi Department of Agronomy, School of Agriculture, Lovely Professional University, Phagwara, Punjab, India

Shipa Rani Dey Department of Agronomy, School of Agriculture, Lovely Professional University, Phagwara, Punjab, India

Sujata Dey Aquatic Environmental Biotechnology and Nanotechnology Division, ICAR-Central Inland Fisheries Research Institute, Kolkata, West Bengal, India

Ashwani K. Dhingra Guru Gobind Singh College of Pharmacy, Yamunanagar, Haryana, India

Nadjib Drouiche CRTSE-Division CCPM- N°2, Alger, Algeria

Dibyangana Ganguly Foundation for Innovative New Diagnostics (FIND), New Delhi, India

Chinky Gangwar Department of Chemistry, Lucknow University, Lucknow, Uttar Pradesh, India

Vinod K. Garg Department of Environmental Sciences and Technology, Central University of Punjab, Bathinda, Punjab, India

Rupesh Kumar Gautam Department of Pharmacology, Indore Institute of Pharmacy, Indore, Madhya Pradesh, India

Lameesa Gazi-Khan North South University, Dhaka, Bangladesh

Arpita Ghosh Indian Institute of Management Sirmaur, Paonta Sahib, Himachal Pradesh, India

Dipankar Ghosh Microbial Engineering & Algal Biotechnology Laboratory, Department of Biosciences, JIS University, Agarpara, Kolkata, India

Divya Goel Department of Environmental Science, School of Vocational Studies and Applied Sciences (SoVSAS), Gautam Buddha University (GBU), Government of Uttar Pradesh, Greater Noida, Uttar Pradesh, India

Alyson D. Gonzalez-Torres Department of Biology, College of Arts and Sciences, Appalachian State University, Boone, NC, United States

Rajat Goyal MM College of Pharmacy, Maharishi Markandeshwar (Deemed to be University), Mullana-Ambala, Haryana, India

Ajmer Singh Grewal Guru Gobind Singh College of Pharmacy, Yamunanagar, Haryana, India

Aadil Gulzar Department of Environmental Sciences, University of Kashmir, Srinagar, Jammu and Kashmir, India

Manjul Gupta PARAMARSH Consultancy Services, Ghaziabad, Uttar Pradesh, India

Piyush Gupta Department of Chemistry, Faculty of Engineering and Technology, SRM Institute of Science and Technology, NCR Campus, Delhi-NCR Campus, Ghaziabad, Uttar Pradesh, India

Sumeet Gupta MM College of Pharmacy, Maharishi Markandeshwar (Deemed to be University), Mullana-Ambala, Haryana, India

Shama E. Haque North South University, Dhaka, Bangladesh

Vijaya Ilango Department of Chemistry, Birla Institute of Technology & Science Pilani, Dubai Campus, Dubai, United Arab Emirates

Sylvester C. Izah Department of Microbiology, Faculty of Science, Bayelsa Medical University, Yenagoa, Bayelsa State, Nigeria

Sarita Joshi Department of Environmental Science, School of Vocational Studies and Applied Sciences (SoVSAS), Gautam Buddha University (GBU), Government of Uttar Pradesh, Greater Noida, Uttar Pradesh, India

Atul Kabra University Institute of Pharma Sciences, Chandigarh University, Mohali, Punjab, India

Azra N. Kamili Department of Botany, Central University of Kashmir, Ganderbal, Jammu and Kashmir, India

Biswabandita Kar Department of Chemistry, School of Applied Sciences, KIIT Deemed to be University, Bhubaneswar, Odisha, India

Amardeep Kaur Himalayan Institute of Pharmacy, Kala Amb, Himachal Pradesh, India

Gagandeep Kaur Chitkata School of Pharmacy, Chitkara University, Himachal Pradesh, India

Rajwinder Kaur Chitkara College of Pharmacy, Chitkara University, Rajpura, Punjab, India

Jyotsna Kaushal Centre for Water Sciences, Chitkara University Institute of Engineering & Technology, Chitkara University, Rajpura, Punjab, India

Peeyush Kaushik Guru Gobind Singh College of Pharmacy, Yamunanagar, Haryana, India

Priyanka Kriplani Guru Gobind Singh College of Pharmacy, Yamunanagar, Haryana, India

Deepak Kumar SJVN Limited, Shimla, Himachal Pradesh, India

Indresh Kumar Department of Chemistry, Lucknow University, Lucknow, Uttar Pradesh, India

Jitendra Kumar Department of Chemistry, M. L. K. (P. G.) College, Balrampur, Uttar Pradesh, India

Maneesh Kumar Department of Biotechnology, Magadh University, Bodh Gaya, Bihar, India

Manoj Kumar Department of Hydro and Renewable Energy, Indian Institute of Technology, Roorkee, Uttrakhand, India

Nitin Kumar Department of Environmental Sciences and Technology, Central University of Punjab, Bathinda, Punjab, India

Prasann Kumar Department of Agronomy, School of Agriculture, Lovely Professional University, Phagwara, Punjab, India

Pratik Kumar Clinton Health Access Initiative (CHAI), New Delhi, India

Rahul Kumar Chair of Hydrobiology and Fishery, Institute of Agricultural and Environmental Sciences, Estonian University of Life Sciences (EMÜ), Fr. R. Kreutzwaldi 1, Tartu, Estonia

Arti Kumari Department of Biotechnology, Patna Women's College, Patna, Bihar, India

Neeta Kumari Department of Civil and Environmental Engineering, Birla Institute of Technology, Mesra, Ranchi, Jharkhand, India

Manzari Kushwaha Department of Applied Chemistry, School of Vocational Studies and Applied Sciences (SoVSAS), Gautam Buddha University (GBU), Government of Uttar Pradesh, Greater Noida, Uttar Pradesh, India

Damanpreet Kaur Lang Chitkara College of Pharmacy, Chitkara University, Rajpura, Punjab, India

Jitendra Maharana Department of Bioinformatics, Odisha University of Agriculture & Technology, Bhubaneswar, Odisha, India

Lone Rafiya Majeed Department of Environmental Science, Vivekananda Global University, Jaipur, Rajasthan, India

Cherubala N. Malikia Department of Geography and Regional Planning, College of Arts and Sciences, Appalachian State University, Boone, NC, United States

Arumugam Manohar Department of Chemistry, Periyar Maniammai Institute of Science and Technology, Thanjavur, Tamil Nadu, India

Markandeya Ex-Department of Civil Engineering, Indian Institute of Technology (Banaras Hindu University), Varanasi, Uttar Pradesh, India

Mark J. McCarthy Chair of Hydrobiology and Fishery, Institute of Agricultural and Environmental Sciences, Estonian University of Life Sciences (EMÜ), Fr. R. Kreutzwaldi 1, Tartu, Estonia; Department of Earth and Environmental Sciences, Wright State University, Dayton, OH, United States

Anuradha Mishra Department of Applied Chemistry, School of Vocational Studies and Applied Sciences (SoVSAS), Gautam Buddha University (GBU), Government of Uttar Pradesh, Greater Noida, Uttar Pradesh, India

Brijesh Kumar Mishra Department of Environmental Science and Engineering, Indian Institute of Technology (Indian School of Mines), Dhanbad, Jharkhand, India

Nishish Mishra Department of Biotechnology, National Institute of Technology Durgapur, Durgapur, West Bengal, India

Pabitra Mishra P.G. Department of Zoology, Fakir Mohan University, Balasore, Odisha, India

Devendra Mohan Department of Civil Engineering, Indian Institute of Technology (Banaras Hindu University), Varanasi, Uttar Pradesh, India

Itishree Mohanty Department of Chemistry, Synergy Institute of Engineering and Technology, Dhenkanal, Odisha, India

Arumugam Murugan Department of Chemistry, North Eastern Regional Institute of Science and Technology, Nirjuli, Itanagar, Arunachal Pradesh, India

Radhey Mohan Naik Department of Chemistry, Lucknow University, Lucknow, Uttar Pradesh, India

Ujjval Nautiyal Himalayan Institute of Pharmacy, Kala Amb, Himachal Pradesh, India

Rashmi Nayak CSIR-National Botanical Research Institute, Lucknow, Uttar Pradesh, India

Rudra Prasanna Nayak Department of Chemistry, Synergy Institute of Engineering and Technology, Dhenkanal, Odisha, India

Thuong Tuan Nguyen Institute of Life Sciences, Thai Nguyen University, Quyet Thang Commune, Thai Nguyen City, Thai Nguyen, Vietnam

Bisma Nisar Department of Environmental Sciences, University of Kashmir, Srinagar, Jammu and Kashmir, India

Peter Nõges Chair of Hydrobiology and Fishery, Institute of Agricultural and Environmental Sciences, Estonian University of Life Sciences (EMÜ), Fr. R. Kreutzwaldi 1, Tartu, Estonia

Tiina Nõges Chair of Hydrobiology and Fishery, Institute of Agricultural and Environmental Sciences, Estonian University of Life Sciences (EMÜ), Fr. R. Kreutzwaldi 1, Tartu, Estonia

Matthew C. Ogwu Goodnight Family Department of Sustainable Development, Appalachian State University, Boone, NC, United States

Heena Nisar Pahalvi Department of Environmental Sciences, University of Kashmir, Srinagar, Jammu and Kashmir, India

Laxmidhar Panda Department of Chemistry, School of Applied Sciences, KIIT Deemed to be University, Bhubaneswar, Odisha, India

Amit Ranjan Pandey Department of Applied Science and Humanities (APSH), Rajkiya Engineering College, Ambedkar Nagar, Uttar Pradesh, India

Pradeep Kumar Pandey Department of Chemistry, Lucknow University, Lucknow, Uttar Pradesh, India

Rishi Ranjan Pandey Department of Chemistry, M. L. K. (P. G.) College, Balrampur, Uttar Pradesh, India

Soumya Pandey Department of Civil and Environmental Engineering, Birla Institute of Technology, Mesra, Ranchi, Jharkhand, India

Satya Narayan Parida Aquatic Environmental Biotechnology and Nanotechnology Division, ICAR-Central Inland Fisheries Research Institute, Kolkata, West Bengal, India

Bidyutlata Patra Kuntala Kumari Sabat Women's College, Fakir Mohan University, Balasore, Odisha, India; P.G. Department of Zoology, Fakir Mohan University, Balasore, Odisha, India

Jyotirmayee Pradhan Kuntala Kumari Sabat Women's College, Fakir Mohan University, Balasore, Odisha, India

Smruti Priyambada Pradhan Department of Community Medicine, IMS & SUM HOSPITAL, Siksha o' Anusandhan University (Deemed to be University), Bhubaneswar, Odisha, India

Sukanta Kumar Pradhan Department of Bioinformatics, Odisha University of Agriculture & Technology, Bhubaneswar, Odisha, India

Siya Ram Life Sciences Discipline, School of Sciences, Indira Gandhi National Open University (IGNOU), Maidan Garhi, Delhi, India

Lata Rani Chitkara School of Pharmacy, Chitkara University, Baddi, Himachal Pradesh, India; Centre for Water Sciences, Chitkara University Institute of Engineering & Technology, Chitkara University, Rajpura, Punjab, India

Habiba Rashid North South University, Dhaka, Bangladesh

Sumaira Rashid Department of Environmental Sciences, University of Kashmir, Srinagar, Jammu and Kashmir, India

C.R. Ravikumar Research Centre, Department of Chemistry, East West Institute of Technology, Bangalore, Karnataka, India

Ajaya Kumar Rout P.G. Department of Bioscience and Biotechnology, Fakir Mohan University, Balasore, Odisha, India; Aquatic Environmental Biotechnology and Nanotechnology Division, ICAR-Central Inland Fisheries Research Institute, Kolkata, West Bengal, India

Alka Sahu Department of Environmental Science and Engineering, Indian Institute of Technology (Indian School of Mine), Dhanbad, Jharkhand, India

Balraj Saini Chitkara College of Pharmacy, Chitkara University, Rajpura, Punjab, India

Joy Sarkar Department of Chemistry, Lucknow University, Lucknow, Uttar Pradesh, India

Devi Selvaraj PG and Research Department of Chemistry, Cauvery College for Women (Autonomous), Tiruchirappalli, Tamil Nadu, India

Mitun Sen Microbial Engineering & Algal Biotechnology Laboratory, Department of Biosciences, JIS University, Agarpara, Kolkata, India

S. Senthilkumar Department of Chemistry, Ayya Nadar Janaki Ammal College, Sivakasi, Tamil Nadu, India

Shiv Shankar Department of Environmental Science, School of Vocational Studies and Applied Sciences (SoVSAS), Gautam Buddha University (GBU), Government of Uttar Pradesh, Greater Noida, Uttar Pradesh, India

Anjali Sharma Guru Gobind Singh College of Pharmacy, Yamunanagar, Haryana, India

Devkant Sharma Ch. Devi Lal College of Pharmacy, Jagadhri, Haryana, India

Khushbu Sharma Department of Agronomy, School of Agriculture, Lovely Professional University, Phagwara, Punjab, India

Prateek Sharma Government Pharmacy College, Nagrota Bagwan, Kangra, Himachal Pradesh, India

Ram Babu Sharma Himalayan Institute of Pharmacy, Kala Amb, Himachal Pradesh, India

Dericks Praise Shukla School of Engineering, Indian Institute of Technology, Mandi, Himachal Pradesh, India

Sheo Prasad Shukla Rajkiya Engineering College, Banda, Uttar Pradesh, India

Swayam Siddha Department of Environmental Studies, The MS University of Baroda, Vadodara, Gujrat, India

Aakansha Singh Department of Environmental Science and Engineering, Indian Institute of Technology (Indian School of Mines), Dhanbad, Jharkhand, India

Indu Singh Department of Chemistry, Lucknow University, Lucknow, Uttar Pradesh, India

Nakshatra Bahadur Singh Department of Chemistry and Biochemistry, Sharda University, Greater Noida, Uttar Pradesh, India

Neeraj Kumar Singh Central Mine Planning and Design Institute Limited (CMPDIL), Regional Institute-7, Bhubaneswar, Odisha, India

Varinder Singh Chitkara College of Pharmacy, Chitkara University, Rajpura, Punjab, India

Nilothpal Sinha Microbial Engineering & Algal Biotechnology Laboratory, Department of Biosciences, JIS University, Agarpara, Kolkata, India

V. Siva Department of Physics, Karpagam Academy of Higher Education, Coimbatore, Tamil Nadu, India

Siyaram School of Sciences, Indira Gandhi Open National University, New Delhi, India

Richa Soni Physics Department, Lancaster University, Lancaster, United Kingdom

Parul Sood Chitkata School of Pharmacy, Chitkara University, Himachal Pradesh, India

Arun Lal Srivastav Chitkara University School of Engineering and Technology, Chitkara University, Baddi, Himachal Pradesh, India

Aidan Stansfield Goodnight Family Department of Sustainable Development, Appalachian State University, Boone, NC, United States

Avnish Kumar Tripathi Department of Chemistry, M. L. K. (P. G.) College, Balrampur, Uttar Pradesh, India

Thi Thao Truong TNU-University of Sciences, Thai Nguyen University, Tan Thinh Ward, Thai Nguyen City, Thai Nguyen, Vietnam

Usman Lawal Usman Department of Environmental Sciences, Sharda University, Greater Noida, Uttar Pradesh, India; Department of Biology, Umaru Musa Yar'adua University, Katsina, Nigeria

Tharmaraj Vairaperumal Institute of Clinical Medicine, College of Medicine, National Cheng Kung University, Tainan, Taiwan ROC; Center for Research in Environment, Sustainability Advocacy and Climate Change (REACH), SRM Institute of Science and Technology, Kattankulathur, Tamil Nadu, India

Truong Xuan Vuong TNU-University of Sciences, Thai Nguyen University, Tan Thinh Ward, Thai Nguyen City, Thai Nguyen, Vietnam

Manish Yadav Central Mine Planning and Design Institute Limited (CMPDIL), Regional Institute-7, Bhubaneswar, Odisha, India

Mayank Yadav Adarsh Vijendra Institute of Pharmaceutical Sciences, Shobhit University, Gangoh, Uttar Pradesh, India

Bushra Yaseen Department of Chemistry, Lucknow University, Lucknow, Uttar Pradesh, India

About the editors

Dr. Arun Lal Srivastav works as an assistant professor at Chitkara University, Himachal Pradesh (India). He obtained his PhD from the Indian Institute of Technology (BHU), Varanasi, India on the adsorption of nitrate and fluoride from water. He has also carried out postdoctoral research at the National Chung Hsing University, Taiwan. He is currently involved in the teaching of environmental science, environmental engineering, disaster management, and design thinking to undergraduate engineering students. His research interests include water quality surveillance, climate change, water treatment, river ecosystem, soil health maintenance, engineering education, phytoremediation, and waste management. He has published around 70 research papers in prestigious journals. He has filed 19 patents (including 5 granted). He is involved in editing 12 books with Elsevier, Springer, Wiley, etc. He is one of the series book editors of NOVA and Elsevier on e-waste management and intelligent energy systems. He is working on research projects (worth around $\sim$16 million INR) funded by the Government of India and the State government of Himachal Pradesh.

Dr. Ajmer Singh Grewal works as an associate professor at the Department of Pharmaceutical Chemistry, at Guru Gobind Singh College of Pharmacy, Yamuna Nagar, Haryana (India). He obtained his PhD in pharmaceutical sciences on "Design, synthesis and biological evaluation of newer glucokinase activators" from the I.K. Gujral Punjab Technical University, Jalandhar, Punjab (India). His present areas of research include structure-based drug design and medicinal chemistry of matrix metalloproteinase inhibitors as anticancer agents, antibacterial drug resistance, glucokinase activators as antidiabetic agents, aldose reductase inhibitors for diabetic complications, PPAR agonists for treatment of metabolic syndrome, multipotent anti-Alzheimer's agents, pesticides, and environmental chemistry. He has a total of 60 publications in national and international highly reputable journals and has contributed to two book chapters, one book, and two patents.

Dr. (Er.) Markandeya has worked as a visiting faculty at the Department of Civil Engineering, Indian Institute of Technology (BHU), Varanasi. He has also worked as an assistant environmental engineer at the Spectro Analytical Labs Ltd., New Delhi and Green Circle, Inc., Vadodara as an environmental engineer. He availed life membership of the Indian Association for Air Pollution Control, New Delhi, the Institution of Engineers (India), Kolkata, and the National Institute of Hydrology, Roorkee, India. He works in various fields related to air, water, and wastewater and is also an editor and reviewer of various peer-reviewed national and international journals. He has published more than 75 articles including an international patent (01), international peer-reviewed journals (34), national journal (10), book chapters (07), international conferences (12), and other national conferences.

Dr. Tien Duc Pham is a lecturer at the University of Science, Vietnam National University, Hanoi. He obtained his PhD from the University of Tsukuba, Japan in 2014. His research interests focus on adsorption characteristics, mechanisms of surfactants, polyelectrolytes onto metal oxides and soils. He studies the adsorptive removal of organic pollutants such as antibiotics, dyes, and pesticides as well as inorganic contaminants including heavy metal and ammonium ions using novel modified nanomaterials. He also investigates the new green analytical methods based on spectroscopy and separation methods for environmental and food analysis. He is currently teaching analytical chemistry, instrumental analysis, and advanced spectroscopic methods to both undergraduate and graduate students. He has published about 50 ISI papers in prestigious journals, 35 national papers, and numerous conference papers. He currently works as a lead guest editor, topic editor, guest editor, and editorial board member of 5 ISI journals on nanomaterials with various applications.

Preface

Environmental deterioration around the globe has become a global problem for human civilization. Moreover, the solutions to these environmental problems require huge finance. In developing countries, apart from environmental problems, many other socioeconomic problems exist. These problems include the scarcity of clean water, degradation of air & soil quality, exponential population growth, unplanned urbanization, unemployment, and depletion of natural resources. Modern lifestyle has also created many hurdles in achieving environmental sustainability. Embracing green/ environmental chemistry and sustainable approaches can significantly contribute to achieving environmental sustainability and creating a better world.

In the book *Role of Green Chemistry in Ecosystem Restoration to Achieve Environmental Sustainability*, attempts have been made to deal with current environmental problems using the principles of green chemistry and also to maintain harmony between society, environment, and economy which are considered basic pillars of sustainability. Moreover, by embracing green chemistry, environmental chemistry, and sustainable approaches, policymakers and researchers can gain valuable insights to develop effective strategies for achieving sustainable development goals. This comprehensive book is divided into four thought-provoking sections, each dedicated to essential aspects of environmental sustainability: environmental management, water sustainability, waste management & environment protection, and agricultural sustainability.

In the first section, that is, environmental management, valuable scientific contributions have been conveyed by the researchers on the topics, prospects of green chemistry and nanotechnology in environmental sustainability, the significance of microbial consortium and green chemistry for xenobiotics removal from the environment, the role of environmental sustainability for climate change adaptations, the importance of indigenous knowledge in achieving environmental sustainability, recent advances in sustainability science for environmental conservation, microplastic in the environment: analytical techniques and health impacts of human, etc.

In the second section, that is, water sustainability, the significant contributions are the green synthesis of nanomaterials for the removal of emerging water pollutants, application of pyrolysis techniques to produce bio-sorbents for water treatment, mycogenic synthesis of nanoparticles and their application in dye degradation, heavy metals in water: challenges and remediation, hazardous consequences of pharmaceutical wastes to groundwater, remediation strategies for the removal of microplastics from the water, application of graphene oxide for wastewater treatment, antibiotic waste in water: impact and remediation strategies, etc.

In the third section, that is, waste management and environment protection, application of agriculture waste biopolymers for soil strengthening and improvement in erosion-prone areas, bio-plastics: solution to a green environment and sustainability, resource recovery from the e-wastes through bioleaching, hydrochar from agro-wastes: a low-cost adsorbent for environmental application and sustainable food processing waste management for environmental protection are the important scientific contributions from the valued researchers.

In the fourth section, that is, agricultural sustainability, major contributions are sequestration of pesticide residues using bio-fabricated nanomaterials: challenges and prospects, nitrogen cycle and its effect on phytoplankton community structure, mercury phytovolatilization: an overview of the mechanism and mitigation, plant growth–promoting rhizobacteria for sustainable agriculture: recent progress and challenges and perspectives of crop improvement techniques and green chemistry toward sustainable agriculture.

Based on the top-class research contributions received from worldwide researchers, it can be inferred that the present book has the latest practical and theoretical aspects of the environmental crisis of the world along with recent research progresses that can reduce the environmental threats and promote sustainable management strategies. These research/review articles may also serve as the baseline information for environmental issues and their solutions. This book can be an asset for undergraduate or university students, teachers, and researchers, mostly working in areas of environmental conservationists, water science researchers, and institutes working toward sustainable development etc. Primary audiences are graduate & postgraduate students, environmental engineers, chemists, industrialists, scientists, and research scholars working on the issues related to environmental issues and their solutions for society. This book is also useful to the person working in the field of nature conservation and health sectors. This book is also helpful for people working in NGOs.

Section A

Environmental management

Chapter 1

Insights into the prospects of green chemistry and nanotechnology in environmental sustainability

Rajat Goyal[1], Mohini Devi[2], Rupesh Kumar Gautam[3] and Sumeet Gupta[1]
[1]MM College of Pharmacy, Maharishi Markandeshwar (Deemed to be University), Mullana-Ambala, Haryana, India, [2]Institute of Pharmaceutical Sciences, Kurukshetra University, Kurukshetra, Haryana, India, [3]Department of Pharmacology, Indore Institute of Pharmacy, Indore, Madhya Pradesh, India

Introduction

One of the major tasks currently is to deliver sustainable development for future generations via the utilization of the green chemistry principles and concepts of green engineering to create nanomaterials and nano-products without harmful by-products, and to incorporate lifecycle thinking throughout all phases of engineering and design. Nanotechnology has a great potential to utilize novel nanomaterials for the rehabilitation of wastewater, groundwater, and surface water that are polluted by organic or inorganic solutes, ions of heavy metal, and a variety of microorganisms. The most effective strategy to diminish and abolish the release of harmful chemicals into the air, water, and soil environments is through green manufacturing (Khan, 2020). The necessary substance can be obtained by using "green" chemistry in the safest way (Soltys et al., 2021). Green chemistry is characterized as a contemporary branch of chemistry that develops very quickly nowadays. By employing green methodologies, we can not only avoid the use of noxious and harmful chemicals but also avoid the production of by-products (Goyal et al., 2021; El Azab et al., 2014). The usage of environmentally friendly solvents and catalysts in pharmaceutical chemistry is one area of intense interest. The utilization of nonvolatile solvents and/or green catalysts supports a wide range of organic transformations with added value, both economically and environmentally (Bhosle et al., 2012). Green chemistry intends to reimagine the production and application of chemicals used in our society to make them intrinsically safer and more effective. Green chemistry delivers a rational and intellectual framework that develops both trans-materialization and dematerialization approaches within the biochemical enterprise. Trans-materialization is the process of moving away from risky, nonrenewable resources and toward secure, reusable, or renewable materials. Dematerialization seeks to reduce the number of resources and energy used by society while preserving its wealth. The green chemistry principles and other sustainability indicators aid in identifying the prospects for modernization (Mulvihill et al., 2011; Bardos et al., 2015; Khan, 2020). Nanoscience, considered a developing field, offers a useful foundation for investigating the uses and effects of green chemistry more thoroughly and securely. In contrast to earlier remediation techniques, nano-remediation entails an overall decrease in contaminations (Bardos et al., 2015). The conception of indicators for the environment and efficiency of economic resources, environmental health hazards, natural resource basis, as well as amenities and services, economic prospects, and strategies in regulation and socio-economic framework, is all part of monitoring green growth strategies (Basiuk and Basiuk, 2015).

Basic principles of green chemistry

"Green chemistry" refers to the development of chemical products and procedures that reduce or eliminate the use and manufacturing of hazardous materials. The advancements in green chemistry address risks that are evident and connected to a variety of comprehensive challenges, that is, food production, energy release, climate variation, and accessibility of a safe and sufficient water supply. The formulation of the principles was motivated by the need to address

Role of Green Chemistry in Ecosystem Restoration to Achieve Environmental Sustainability. DOI: https://doi.org/10.1016/B978-0-443-15291-7.00030-4

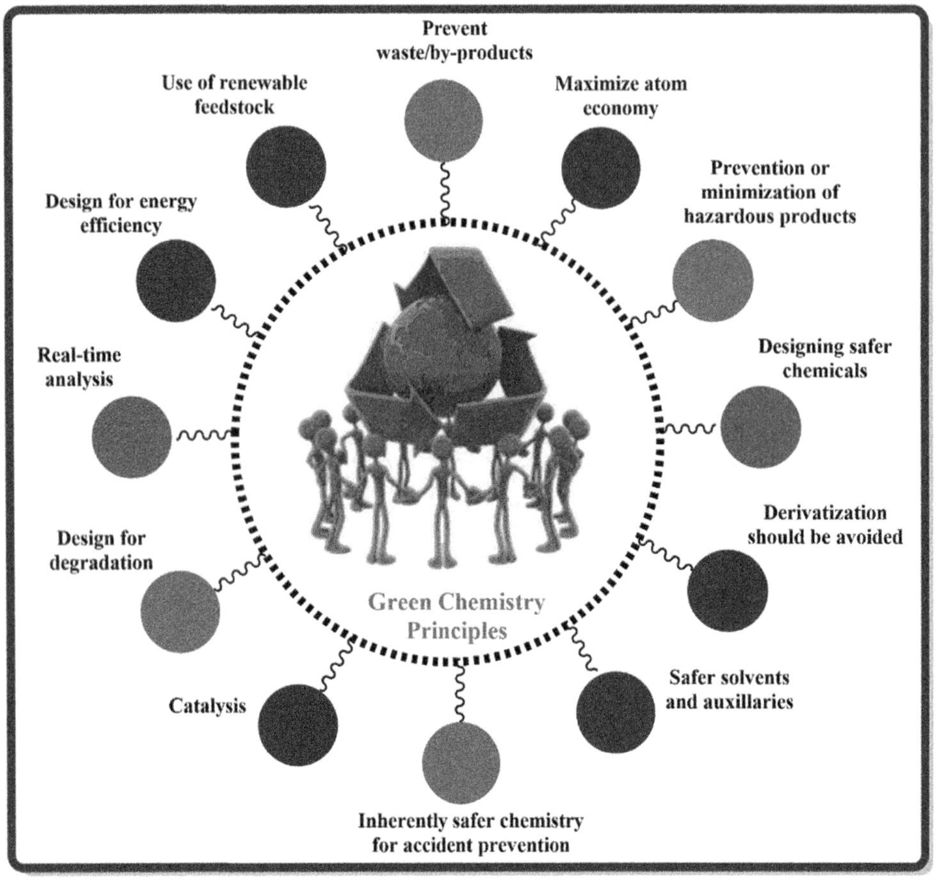

FIGURE 1.1 The pictorial representation of twelve principles of green chemistry.

decades of unintentional environmental contamination and public health effects from the production and usage of harmful chemicals (Jessop et al., 2009). The basic principles of green chemistry cover a broad area of man-made organic synthesis: designing organic synthesis processes to minimize the use of harmful chemicals/raw materials, reduce waste generation/by-products, and use solvents that are more environmentally friendly, (bio)catalysts, renewable raw materials, and methods to increase energy efficiency (Ivanković et al., 2017). These principles of green chemistry (Sharma et al., 2021; Dhage, 2013) are described in Fig. 1.1, and these condensed principles are written in the mnemonic pattern: "PRODUCTIVELY" (Tang et al., 2005) and are illustrated in Fig. 1.2.

Properties and principles of nanotechnology

Nanotechnology imparts a potential part in the improvement of the ecosystem healthcare system through the direct usage of nanomaterials in the detection, prevention, and removal of pollutants, and indirect usage through improved industrial design processes and the manufacture of environmental-friendly products (Taran et al., 2021). Nanoparticles have a dimension of at least 1–100 nm; measured within the scale of a nanometer (nm) or whose functional unit is within 3D space (Sharma et al., 2020). The usage of nanomaterials is rising day by day in cosmetics and food industries to upgrade, bioavailability, shelf life, packaging, and production (Hulla et al., 2015).

There is a constant quest for consistent and eco-friendly ways to create metallic and, metal oxide nanoparticles while reducing or even doing away with the need for dangerous compounds. To accomplish the efficient and validated performance of products and processes, nano-manufacturing standards must be established to appropriately synthesize the nanodevices and nanomaterials. These challenges must trail in the following directions:

1. Atomic-scale synthesis: To create and accumulate the processes that necessitate atomically precise standards. These comprise the work engaged in solving object integrity, dimensional metrology, precision placement, and problems during manufacturing.

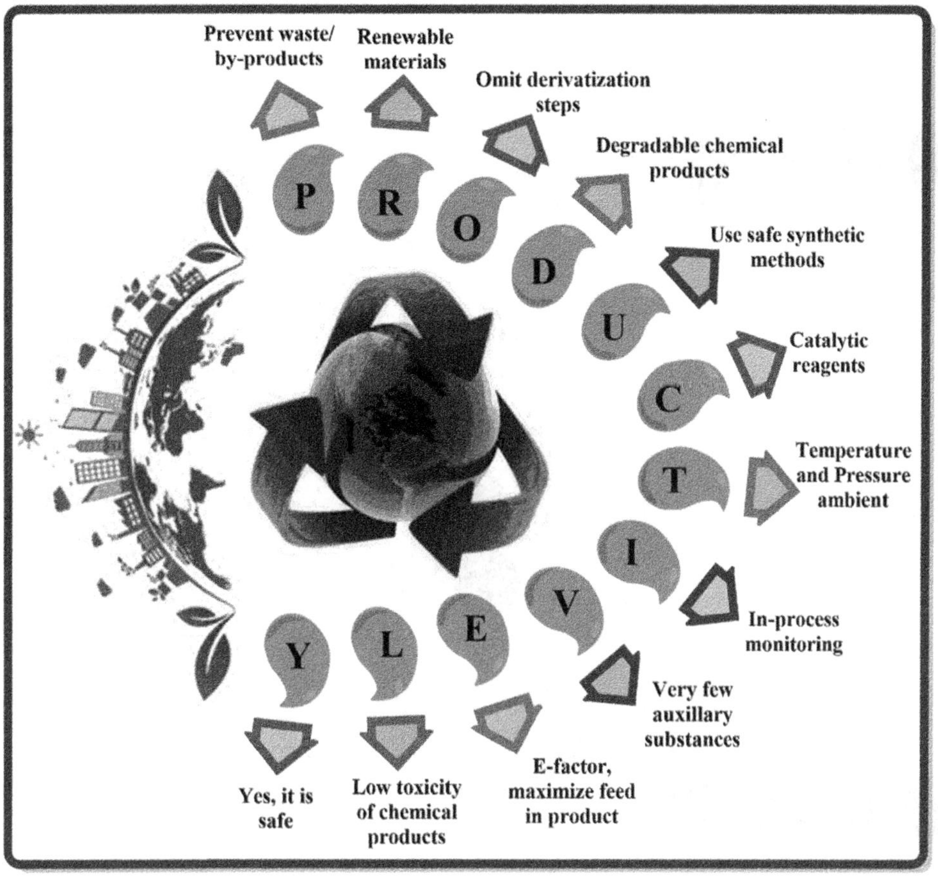

FIGURE 1.2 Principles of green chemistry written in the mnemonic pattern: "PRODUCTIVELY."

2. Molecular-scale manipulation and assembly: To recognize and discourse the basic control, dimensions, and standard problems associated with the manipulation and assembling of nanomaterials by using physicochemical and optical approaches.
3. Micro-to-millimeter-scale synthesis technologies: To create the systems that need to be positioned, assembled, manipulated, and manufactured on scales ranging from nanometers to millimeters (Virkutyte and Varma, 2013).

Green nanotechnology and its applications

Nanotechnology has captivated researchers' interest because of the high surface-to-volume ratio of nanoparticles and microscopic size. Because of these properties, nanomaterials have a broad range of applications in agricultural, environmental, and biomedical fields (Khan et al., 2022). Technologies for preventing pollution could benefit greatly from nanotechnology. For instance, house lighting based on nanotechnology could decrease energy use by 10% United States, saving \$100 billion yearly and reducing carbon emissions by 200 million tonnes. (Masciangioli and Zhang, 2003). The focus of green nanotechnology is on applying nanotechnology to the creation of safer and environmentally friendly procedures and products (Dornfeld et al., 2013). Applications based on nanotechnology may have advantages including more efficiency, lower cost, lesser toxicity, complexity, and reliability. Additionally, it provides the prospect of avoiding negative consequences even before they occur (Owen and Depledge, 2005). The role of green nanotechnologies (Bharti et al., 2022) is described in Fig. 1.3.

Green nanotechnology seeks to exploit the appealing physicochemical features of nanomaterials in a variety of green innovative applications that are both economically, and energy-effective and environmentally sustainable, with the potential to have a significant impact on a wide variety of public areas. These outcomes may provide opportunities to decrease pressure on raw materials trading on renewable energy, improve power delivery systems to be more efficient, safe, and reliable, and usage of nano-enabled construction products or unconventional water sources, resulting in healthier livelihood conditions and better ecosystem (Iavicoli et al., 2014). Green nanotechnology has two main components. The first

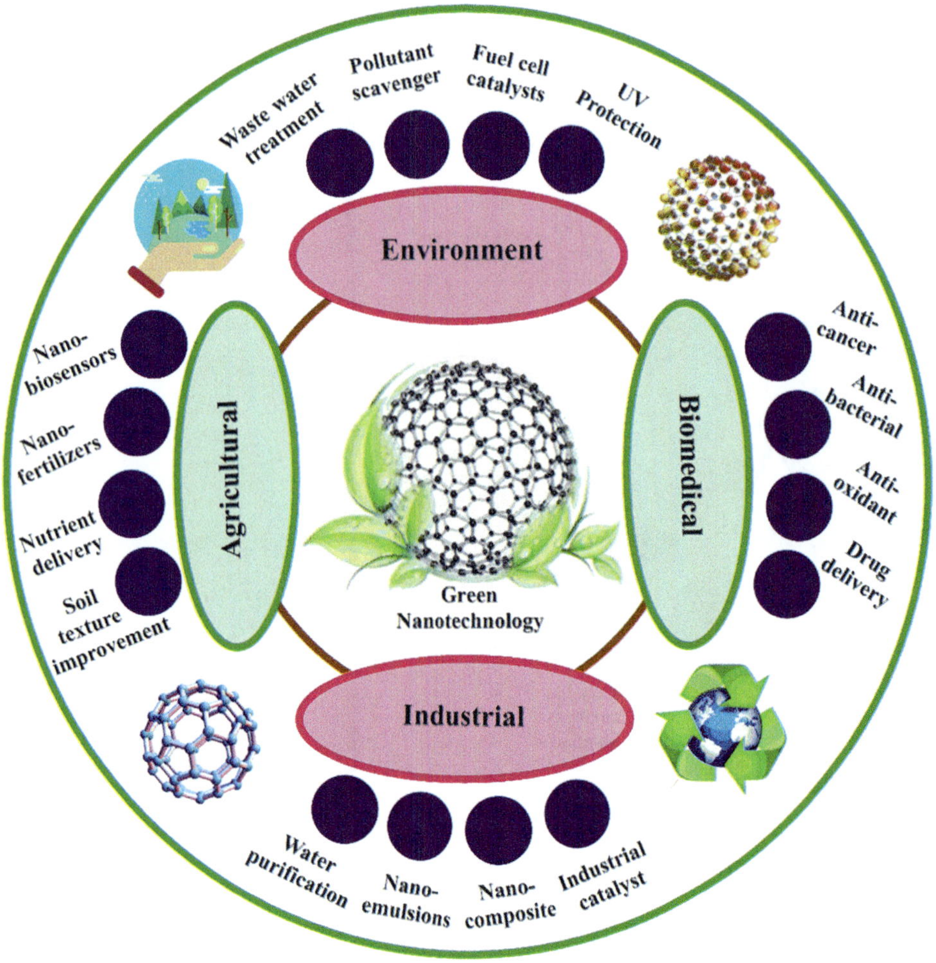

FIGURE 1.3 Role of green nanotechnologies.

entails nano-products that address environmental problems. In addition to being utilized in environmental technologies to clean up dirty streams, repair hazardous waste sites, and desalinate water, these green nano-products are employed to stop harm from recognized toxins. Nanomaterials provide clean and safe drinking water through nano-based processes or membrane technologies that eliminate pathogens and hazardous compounds. Also, green nano-products allow erudite sensing and monitoring tools that may detect harmful pollutants, plant diseases, and associated toxins. The creation of nanoparticles and products containing them to minimize damage to the ecosystem healthcare system constitutes the second component of green nanotechnology. The majority of nanomaterials are created by using chemical processes, which may or may not produce waste products, waste energy, or waste materials (Karn and Bergeson, 2009). Various applications of green nanotechnologies in different fields (Iavicoli et al., 2014) are elaborated in Table 1.1.

Role in environmental sustainability

The phrase "sustainable development" has gained a lot of attention. The common area between three spheres, that is, market, society, and environment are the subject of sustainable development (Tobiszewski et al., 2010). A new approach to green chemistry implementation pays proactive attention to the eradication of chemistry's harmful environmental impacts to support environmental sustainability (Marques and Machado, 2014). The "green chemistry" as a concept, which has become a fundamental developing theme, endeavors to address the scientific challenges of protecting both public health and the environment (Varma et al., 2014). The revolutionary nature of nanotechnology will have an impact on economies through newer consumer goods, production techniques, and material utilization. The creation of renewable energy sources like fuel cells and solar energy is another aspect of nanotechnology (Albrecht et al., 2006). Nanomaterials are almost everywhere because of their multifunctionality uses and characteristics. The usage of nanotechnology is growing day by day, with promising advantages to the environment, including materials for cooling down

TABLE 1.1 Various applications of green nanotechnologies in different fields.

Sr. no.	Field	Applications
1.	Energy conversion and storage	• Improvement of the power delivery systems to be more effective, safe, and consistent by using smart energy nanotechnology. • Nanodevices may craft renewable energy released from naturally replenished possessions, such as wind and sunlight, which may diminish the fossils as resources of energy.
2.	Water clean-up technologies	• Designing nano-enabled infrastructures that are obligatory to clean the water and preserve it for better livelihood circumstances and economic development. • Speculation in the infrastructures and substantial rejuvenation of water treatment policies are essential to decrease the price to face water shortages.
3.	Construction industry	• Nanotechnology intends to upsurge the productivity of a building's usage of resources, that is, water, energy, and materials, while minimizing the building impacts on the human well-being and environment through better designing, siting, and construction. • Nanomaterials applied to the shells of a building's structural elements may subsidize environmental cleaning via photo-catalytic reactions.
4.	Other applications	• The applications of nanotechnology deliver a moderate proclamation and dose of fertilizers, and an effective water reservoir for plants that subsidize the superior productivity of agriculture, particularly in areas with protracted dry spells. • Nano-packaging with amended mechanical, and barrier assets may permit extensive safe food storage, exclusively in provinces, where refrigeration is not easily accessible. • Nano-remediation of environmental contamination may exceed the traditional approaches in speed and effectiveness, • Nanosensors may augment the quality and lessen the cost of environmental intensive care.

without using refrigerant, nanocatalysts for environmental remediation, thermoelectric effective photovoltaic, the lightweight composition of nanomaterials materials for vehicles, nanosensors, and mini devices to decrease the material consumption, which abolishes the prerequisite of wet chemical analysis (Khan, 2020).

Nanotechnology in water treatment

One of the most important issues of the twenty-first century is water scarcity, due to accelerated industrialization development and global population growth. An increased amount of contaminated water is produced annually, which harms the ecosystem and spreads deadly diseases to both humans and animals. There is a substantial need for clean and safe water in developing countries, particularly in rural areas. Thus it is essential and crucial to develop high-performance contaminant detection and water treatment procedures. To improve performance and efficiency, green nanotechnology has recently been used for impurity detection and water treatment. Several nanomaterials including nanotubes, quantum dots, nanosheets, and nanoparticles, have been utilized and fabricated into sensors, photocatalysts, sorbents, and membranes (Li et al., 2022). Adsorbents, catalysts, and membranes based on nanotechnology can produce environmentally beneficial methods of wastewater treatment (Tiwari et al., 2008). Particulates can be removed from contaminated water using nano absorbents like nanoclays, metal oxide nanoparticles, zeolites, polymeric adsorbents, and nanoporous carbon fibers. Nanocatalysts and redox-active nanoparticles can convert harmful organic solutions into harmless by-products. (Shapira and Youtie, 2015). Various approaches, including photocatalysis, adsorption, and nanofiltration, by using polymer membranes, nanowire membranes, ceramic membranes, zinc oxide (ZnO), magnetic nanoparticles, titanium dioxide (TiO_2), and carbon nanotubes, are employed to resolve the problems involving wastewater treatment (Saha et al., 2013), for example, TiO_2 nanoparticles are one of the most promising and emerging photocatalysts for the filtration of water (Amin et al., 2014).

Nanotechnology for the remediation of plastic wastes

Plastic waste has long been a critical problem on a global scale. In recent years, microplastics and nanoplastics, which are degraded from huge plastic wastes, pose a serious hazard to the ecosystem and public health. Due to their potential for biodegradability and safety, biodegradable plastics have provided an excellent solution to the growing environmental problems caused by plastic waste and ensured sustainable development. Additionally, benefiting from the amazing advancements in nanotechnology, a variety of nanomaterials with superior physicochemical properties have greatly enhanced the performance of polymers, and offer a viable technique for cleaning up plastic debris (Zheng et al., 2022; Chellasamy et al., 2022).

Renewable energy generation

One of the best ways to generate renewable energy is through the sun, that is, solar energy. The ecosystem benefits greatly from solar energy. For instance, if a distributed solar system can supply 1% of the world's electricity needs, it might prevent the emissions of almost 40 million tonnes of carbon dioxide per year. The greater area-to-volume ratio of nanoparticles should boost sun energy efficacy by revealing greater conducting surfaces to the light of the sun. By employing substances like lead-selenide, nanotechnology will also improve the efficiency of solar cells. Quantum dots, titanium dioxide, silver, and cadmium telluride are among the best nanomaterials used in cells, together with the polymer which can effectively soak up significant amounts of solar energy (Wang et al., 2008). Nanotechnology may offer attractive ways to lower the cost of solar cells made by novel processes that produce thin-film solar cells (Baxter et al., 2009; Hussein, 2015).

Conclusion

Green chemistry, over the past few decades, has demonstrated how fundamental scientific approaches can protect the environment and human well-being in an economically valuable means. Significant advancements are being achieved in various imperative scientific fields, including catalysis, the creation of safe chemicals or solvents, and growth of the renewable feedstocks. The concept of nanotechnology provides the prospect of treasuring the resolution to universal issues that affect our society. The emphasis has switched to nano-based applications after analyzing the advancements and applications of the field since nanotechnology offers a framework to research the features and characteristics of green chemistry. Green nanotechnology has the substantial prospective to contribute to resolving environmental issues and promoting sustainable development.

References

Albrecht, M.A., Evans, C.W., Raston, C.L., 2006. Green chemistry and the health implications of nanoparticles. Green Chemistry 8 (5), 417–432. Available from: https://doi.org/10.1039/B517131H.

Amin, M.T., Alazba, A.A., Manzoor, U., 2014. A review of removal of pollutants from water/wastewater using different types of nanomaterials. Advances in Materials Science and Engineering 2014. Available from: https://doi.org/10.1155/2014/825910.

Bardos, P., Bone, B., Černík, M., Elliott, D.W., Jones, S., Merly, C., 2015. Nanoremediation and international environmental restoration markets. Remediation Journal 25 (2), 83–94. Available from: https://doi.org/10.1002/rem.21426.

Basiuk, V.A., Basiuk, E.V., 2015. Green Processes for Nanotechnology. Springer, p. 446. Available from: https://doi.org/10.1007/978-3-319-15461-9.15.

Baxter, J., Bian, Z., Chen, G., Danielson, D., Dresselhaus, M.S., Fedorov, A.G., et al., 2009. Nanoscale design to enable the revolution in renewable energy. Energy & Environmental Science 2 (6), 559–588. Available from: https://doi.org/10.1039/B821698C.

Bharti, B., Kumar, R., Kumar, H., Li, H., Zha, X., Ouyang, F., 2022. Advanced applications and current status of green nanotechnology in the environmental industry. In: Shanker, U., Hussain, C.M., Rani, M. (Eds.), Green Functionalized Nanomaterials for Environmental Applications. Elsevier, pp. 303–340. Available from: http://doi.org/10.1016/B978-0-12-823137-1.00012-9.

Bhosle, M.R., Mali, J.R., Mulay, A.A., Mane, R.A., 2012. Polyethylene glycol mediated one-pot three-component synthesis of new 4-thiazolidinones. Heteroatom Chemistry 23 (2), 166–170. Available from: https://doi.org/10.1002/hc.20766.

Chellasamy, G., Kiriyanthan, R.M., Maharajan, T., Radha, A., Yun, K., 2022. Remediation of microplastics using bionanomaterials: a review. Environmental Research 208, 112724. Available from: https://doi.org/10.1016/j.envres.2022.112724.

Dhage, S.D., 2013. Applications of green chemistry principles in everyday life. International Journal of Research in Pharmacy and Chemistry 3 (3), 501–518.

Dornfeld, D., Yuan, C., Diaz, N., Zhang, T., Vijayaraghavan, A., 2013. Introduction to green manufacturing. In: Dornfeld, D.A. (Ed.), Green Manufacturing. Springer, Boston, MA, pp. 1–23. Available from: https://doi.org/10.1007/978-1-4419-6016-0_1.

El Azab, I.H., Youssef, M.M., Amin, M.A., 2014. Microwave-assisted synthesis of novel 2 H-chromene derivatives bearing phenylthiazolidinones and their biological activity assessment. Molecules (Basel, Switzerland) 19 (12), 19648–19664. Available from: https://doi.org/10.3390/molecules191219648.

Goyal, R., Sharma, A., Thakur, V.K., Ola, M., Sharma, P.C., 2021. Green chemistry approaches towards the design and synthesis of anti-infective fluoroquinolone derivatives. Current Research in Green and Sustainable Chemistry 4, 100044. Available from: https://doi.org/10.1016/j.crgsc.2020.100044.

Hulla, J.E., Sahu, S.C., Hayes, A.W., 2015. Nanotechnology: History and future. Human & Experimental Toxicology 34 (12), 1318–1321. Available from: https://doi.org/10.1177/0960327115603588.

Hussein, A.K., 2015. Applications of nanotechnology in renewable energies—a comprehensive overview and understanding. Renewable and Sustainable Energy Reviews 42, 460–476. Available from: https://doi.org/10.1016/j.rser.2014.10.027.

Iavicoli, I., Leso, V., Ricciardi, W., Hodson, L.L., Hoover, M.D., 2014. Opportunities and challenges of nanotechnology in the green economy. Environmental Health 13 (1), 1–11. Available from: https://doi.org/10.1186/1476-069X-13-78.

Ivanković, A., Dronjić, A., Bevanda, A.M., Talić, S., 2017. Review of 12 principles of green chemistry in practice. International Journal of Sustainable and Green Energy 6 (3), 39–48. Available from: https://doi.org/10.11648/j.ijrse.20170603.12.

Jessop, P.G., Trakhtenberg, S., Warner, J., 2009. The twelve principles of green chemistry. In: Flank, W.H., Abraham, M.A., Matthews, M.A. (Eds.), Innovations in Industrial and Engineering Chemistry. pp. 401–436. Available from: https://doi.org/10.1021/bk-2009-1000.ch012.

Karn, B.P., Bergeson, L.L., 2009. Green nanotechnology: straddling promise and uncertainty. Nat. Resources & Env't 24, 9.

Khan, S.H., 2020. Green nanotechnology for the environment and sustainable development. In: Naushad, M., Lichtfouse, E. (Eds.), Green Materials for Wastewater Treatment. Springer, Cham, pp. 13–46. Available from: https://doi.org/10.1007/978-3-030-17724-9_2.

Khan, F., Shariq, M., Asif, M., Siddiqui, M.A., Malan, P., Ahmad, F., 2022. Green nanotechnology: plant-mediated nanoparticle synthesis and application. Nanomaterials 12 (4), 673. Available from: https://doi.org/10.3390/nano12040673.

Li, B., Wang, N., Zheng, J., Zhu, H., Wang, S., Ye, E., et al., 2022. Green nanotechnology for high-performance impurity detection and water treatment. In: Li, Z., Zheng, J., Ye, E. (Eds.), Sustainable Nanotechnology. RSC, pp. 33–64. Available from: https://doi.org/10.1039/9781839165771-00033.

Marques, C.A., Machado, A.A., 2014. Environmental Sustainability: implications and limitations to Green Chemistry. Foundations of Chemistry 16 (2), 125–147. Available from: https://doi.org/10.1007/s10698-013-9189-x.

Masciangioli, T., Zhang, W.X., 2003. Peer reviewed: environmental technologies at the nanoscale. Environmental Science & Technology 37 (5), 102A–108A.

Mulvihill, M.J., Beach, E.S., Zimmerman, J.B., Anastas, P.T., 2011. Green chemistry and green engineering: a framework for sustainable technology development. Annual Review of Environment and Resources 36 (1), 271–293. Available from: https://doi.org/10.1146/annurev-environ-032009-095500.

Owen, R., Depledge, M., 2005. Nanotechnology and the environment: risks and rewards. Marine Pollution Bulletin 50 (6), 609–612. Available from: https://doi.org/10.1016/j.marpolbul.2005.05.001.

Saha, I., Bhattacharya, S., Mukhopadhyay, A., Chattopadhyay, D., Ghosh, U., Chatterjee, D., 2013. Role of nanotechnology in water treatment and purification: potential applications and implications. International Journal of Chemical Science and Technology 3 (3), 59–64.

Shapira, P., Youtie, J., 2015. The economic contributions of nanotechnology to green and sustainable growth. In: Basiuk, V., Basiuk, E. (Eds.), Green Processes for Nanotechnology. Springer, Cham, pp. 409–434. Available from: http://doi.org/10.1007/978-3-319-15461-9_15.

Sharma, P.C., Sharma, D., Sharma, A., Saini, N., Goyal, R., Ola, M., et al., 2020. Hydrazone comprising compounds as promising anti-infective agents: chemistry and structure-property relationship. Materials Today Chemistry 18, 100349. Available from: https://doi.org/10.1016/j.mtchem.2020.100349.

Sharma, D., Sharma, V., Sharma, A., Goyal, R., Tonk, R.K., Thakur, V.K., et al., 2021. Green chemistry approaches for thiazole containing compounds as a potential scaffold for cancer therapy. Sustainable Chemistry and Pharmacy 23, 100496. Available from: https://doi.org/10.1016/j.scp.2021.100496.

Soltys, L., Olkhovyy, O., Tatarchuk, T., Naushad, M., 2021. Green synthesis of metal and metal oxide nanoparticles: Principles of green chemistry and raw materials. Magnetochemistry 7 (11), 145. Available from: https://doi.org/10.3390/magnetochemistry7110145.

Tang, S.L., Smith, R.L., Poliakoff, M., 2005. Principles of green chemistry: PRODUCTIVELY. Green Chemistry 7 (11), 761–762. Available from: https://doi.org/10.1039/B513020B.

Taran, M., Safaei, M., Karimi, N., Almasi, A., 2021. Benefits and application of nanotechnology in environmental science: an overview. Biointerface Research in Applied Chemistry 11 (1), 7860–7870. Available from: https://doi.org/10.33263/BRIAC111.78607870.

Tiwari, D.K., Behari, J., Sen, P., 2008. Application of nanoparticles in waste water treatment. The Pennsylvania State University, p. 1. Available from: http://citeseerx.ist.psu.edu/viewdoc/summary?doi = 10.1.1.388.2845.

Tobiszewski, M., Mechlińska, A., Namieśnik, J., 2010. Green analytical chemistry—theory and practice. Chemical Society Reviews 39 (8), 2869–2878. Available from: https://doi.org/10.1039/B926439F.

Varma, R.S., 2014. Journey on greener pathways: from the use of alternate energy inputs and benign reaction media to sustainable applications of nano-catalysts in synthesis and environmental remediation. Green Chemistry 16 (4), 2027–2041. Available from: https://doi.org/10.1039/C3GC42640H.

Virkutyte, J., Varma, R.S., 2013. Green synthesis of nanomaterials: environmental aspects. In: Shamim, N., Sharma, V.K. (Eds.), Sustainable Nanotechnology and the Environment: Advances and Achievements. American Chemical Society, pp. 11–39. Available from: http://doi.org/10.1021/bk-2013-1124.ch002.

Wang, X., Zhi, L., Müllen, K., 2008. Transparent, conductive graphene electrodes for dye-sensitized solar cells. Nano Letters 8 (1), 323–327. Available from: https://doi.org/10.1021/nl072838r.

Zheng, Y., Mao, C., Zhang, Z., Liu, R., Yan, B., 2022. Nanotechnology for the remediation of plastic wastes. In: Li;, Z., Zheng;, J., Ye, E. (Eds.), Sustainable Nanotechnology. Royal Society of Chemistry, pp. 117–143.

Chapter 2

Importance of microbial consortia and green chemistry in the removal of xenobiotics from the environment

Dibyangana Ganguly[1], Pratik Kumar[2], Arti Kumari[3] and Maneesh Kumar[4]

[1]Foundation for Innovative New Diagnostics (FIND), New Delhi, India, [2]Clinton Health Access Initiative (CHAI), New Delhi, India, [3]Department of Biotechnology, Patna Women's College, Patna, Bihar, India, [4]Department of Biotechnology, Magadh University, Bodh Gaya, Bihar, India

Introduction

Numerous industrial and nonindustrial processes, growing urbanization, and careless waste disposal contribute to the introduction of a diverse array of hydrocarbon molecules (Devi et al., 2022). These new pollutants have indefinite impacts and represent grave dangers to the ecosystem, as well as to the environmental balance of groundwater and agricultural soils. Such pollutants are predominantly present in air, soils, aquatic habitats, and dust particles (Louati et al., 2019). They are known as xenobiotic agents because they function as foreign substances in biological systems (Ramaka et al., 2020; Iqbal et al., 2022). The word "xenobiotic" is mostly explained concerning the milieu of environmental toxins, consisting of enormous by-products or end products as synthetic chemicals resulting from domestic, agricultural, and industrial usage (Embrandiri et al., 2016; Atashgahi et al., 2018; Dinka, 2018). Some of the common xenobiotics include PhACs (pharmaceutical active compounds), chlorinated substances, personal care products, PAHs (polycyclic aromatic hydrocarbons), phenolics, and various other industrial compounds (Idle and Gonzalez, 2007; Zinedine et al., 2007). The major concern among environmentalists at the moment is the vast range of dangerous chemical compounds, which may include cosmetics, dyes, pesticides, food preservatives, artificial flavors, food additives, fragrances, toxic phthalates, phenols, pigments, and other environmental contaminants causing significant harm (Štefanac et al., 2021; Devi et al., 2022). Many international policies from various countries have expressed concern about health risks and xenobiotic compound analysis. With these xenobiotic compounds, the integrated approach and interdependencies between human, animal, and plant health would increase transdisciplinary cooperation in the environment (Bronzwaer et al., 2021; Buschhardt et al., 2021). To degrade and detoxify xenobiotic compounds, many physical and chemical methods have been adopted such as adsorption, filtration, electrolysis ozonation, coagulation, and chemical precipitation (Ravindra and Haq, 2019). However, none of them proved to be effective enough to eradicate xenobiotic exposure. These treatments have often proven to be expensive (Perelo, 2010; Paul et al., 2005). This chapter focuses on the proper utilization of microbes through green chemistry in the elimination of xenobiotic compounds to minimize their negative effects on the ecosystem.

Adverse impacts of xenobiotic compounds

Xenobiotic substances are responsible for countless biological disorders. These chemical compounds accumulate as pollutants in the environment regularly and pose a significant threat to food safety and human health (Kumari and Chaudhary, 2020). Obesity, type-2 diabetes, and other endocrine disorders are all linked to hormonal imbalances (Louati et al., 2019). Some of the common outcomes of xenobiotic exposure can be impaired immunological functions, dysfunctional nervous and endocrine systems, pulmonary bronchitis, cognitive impairment disrupting behavior and development, and mutagenic impacts that may eventually lead to carcinogenesis (Bertotto et al., 2020; Dinka, 2018; Mishra et al., 2019a,b).

Role of Green Chemistry in Ecosystem Restoration to Achieve Environmental Sustainability. DOI: https://doi.org/10.1016/B978-0-443-15291-7.00032-8

Bioremediation of xenobiotic compounds

Bioremediation involves the stimulation of metabolic activities in certain microorganisms to remove contaminants (oil spills, soil contamination, etc.) from the environment, by converting them into solvents like water or gases like carbon dioxide (Azubuike et al., 2016; Gilliespie and Philp, 2013). Microbial remediation has come out to be powerful and cost-effective in proficiently working against xenobiotic stress, resulting in minimal to zero generation of harmful by-products. During the degradation process, microorganisms convert harmful organic contaminants into various nitrogen and carbon compounds for consumption and sustenance of their metabolic and growth activities (Arora et al., 2018; Chen et al., 2013; Zhan et al., 2018; Mishra et al., 2021). When utilized properly, several genetically modified organisms can decrease a variety of polycyclic hydrocarbons, providing another means of safeguarding the environment from pollutants (Kang et al., 2016; Luo et al., 2022). Various ecological and physiological factors influence the rate of biodegradability by microbes, such as carbon and nitrogen sources, humidity percentage, salinity content, soil nutrients, temperature, pH level, concentration of inoculums, and so on (Bhatt et al., 2019; Wu et al., 2014). The intensity of the bioremediation process can be extensively improved by modifications in the microbial genome via genetic engineering, genome editing, and other biochemical methods. Such genomic alterations lead to the development of "super bacteria" or highly upgraded microbial strains, with a greater impact on xenobiotic detoxification than the original existing strains (Janssen and Stucki, 2020; Hussain et al., 2018). The basic criteria behind which the chosen bioremediation technique involves microbes is because of their diverse genetic constitution and functionality (Chen et al., 2015; Dangi et al., 2018). As per recent studies, it has been noted that microbial consortia have a better hold over xenobiotic stress reduction than pure monocultures of microbes.

Purpose of microbial consortium in bioremediation

The active degradation of toxic and xenobiotic substances has become one of the major environmental contamination issues globally. Microbial consortia are effectively involved in biodegrading hydrocarbons and xenobiotics. Microbes potentially make use of xenobiotic compounds to support their growth and various metabolic activities (Kumar et al., 2017; Sandhya et al., 2021). Many investigations have demonstrated that a single strain cannot completely degrade contaminants. Since different strains have diverse metabolic routes, bacteria with varying removal abilities are mixed, and the microbial consortium can combine the benefits of each strain to accomplish efficient pollutant degradation. Mixed microbial consortia have shown impressive substrate tolerance and contaminant degradation. Some of the existing microbial strains isolated from intestinal and natural floras include *Lactobacilli, Actinobacteria, Pseudomonas, Clostridium, Salmonella*, and *Escherichia coli*. These strains have the potential to break down toxins on their own and are appropriate for bioremediation (Bhatt et al., 2021). Members of the consortium can also indirectly improve biodegradation by secreting metabolites that induce co-metabolic degradation or metabolic cross-feeding (Kumar et al., 2022). Programmed communication between populations is a crucial prerequisite for engineering microbial consortia. Small diffusible molecules, such as quorum sensing (QS) signals, critical metabolites, or secreted enzymes, can be used to communicate. Bacteria employ QS to determine the density of their population in the wild. Each cell in the population synthesizes and secretes a tiny diffusible molecule (the QS signal), the concentration of which increases with cell density. The QS signal, at high enough concentrations, can stimulate the expression of target genes, allowing the population to respond to increased cell density (Chan et al., 2022). Industrial processes can be improved and made more regulated by using microbial consortia that are less complicated but equally effective. For instance, a significant portion of functional genes was drastically improved, and by lowering the biodiversity of a microbial population from diesel-contaminated soils, the efficiency of diesel microbial degradation was enhanced (Jung et al., 2016). Therefore, it is essential to develop trustworthy methods for reducing diversity to get optimum microbial consortia from environmental samples. Treatment of mixed pollutants has shown to benefit greatly from microbe-assisted degradation of xenobiotic compounds. Numerous biotic or abiotic elements may impact the degradation pathway; however various in situ or ex-situ bioremediation strategies are utilized to break down toxins all over the world (Kumar and Akhtar, 2019; Prosekov et al., 2015; Tak et al., 2022). The majority of bioremediation activities occur in aerobic settings; however anaerobic conditions are appropriate for resistant molecules. The most capable and well-known bacteria that contribute substantial biodegradation activities are *Aeromonas, Alcaligenes, Aspergillus, Aureobasidium, Candida, Cellulosimicrobium, Flavobacterium, Methanospirillum, Microbacterium, Micrococcus, Penicillium, Trichoderma, Sphingobium, Streptomyces, Rhodococcus*, and *Rhodotorula*. From xenobiotic pollutants present in the soil, they extract carbon or nitrogen as nutrients (Kumar et al., 2018a,b; Sandhya et al., 2021). Some of the microbes have been listed (Table 2.1), with their active involvement in the degradation of harmful xenobiotics.

TABLE 2.1 Mechanism of removals of xenobiotics.

Microorganisms	Xenobiotics	Sources	References
Bacteria			
Achromobacter xylosoxidans, PY4 strain	Polycyclic aromatic hydrocarbons (PAHs) like salicylic acid, catechol, naphthalene, anthracene and phenanthrene	Contaminated soil	Nzila et al. (2018)
Acinetobacter baumannii ZH-14	Permethrin	Contaminated soil	Zhan et al. (2018)
Actinomycetes	Organochlorine	Organochlorine contaminated soil	Fuentes et al. (2010)
Arthrobacter phenanthrenivorans Sphe3	Phenanthrene, phthalate, and glucose	Creosote contaminated soil	Vandera et al. (2015)
Bacillus aryabhattai (VITNNDJ5) strain	Monocrotophos	Contaminated agricultural soil	Dash and Osborne (2020)
Bacillus cereus S13, *B. cereus* 12–2, *B. cereus* strain JMG-01, *B. cereus* strain WD-2	Anthracene, Lead (II), Prochloraz-manganese	Hydrocarbon contaminated soil	Chen et al. (2016); Das et al. (2017); Bibi et al. (2018); Jiang et al. (2019)
Bacillus drentensis strain S1	Acetaminophen	Contaminated water	Chopra and Kumar (2020)
Bacillus firmus	Reactive Blue 160	Contaminated soil from the textile dye industry, Noyyal River basin, Tamil Nadu, India	Barathi et al. (2020)
Bacillus megaterium, B. megaterium STB1	Azo dye, Remazol Blue	Contaminated water from textile industries	Joshi et al. (2013), Shah et al. (2013)
Bacillus pumilus B12	Polylactic acid	Contaminated water from agricultural region	Bonifer et al. (2019)
Bacillus subtilis RKJ 700	4-Chloro-2-nitrophenol; Cypermethrin	Contaminated soil	Arora (2012); Gangola et al. (2018)
Bacillus thuringiensis strain SG4, *B. thuringiensis* strain ZS-19	Cypermethrin, Cyhalothrin	Contaminated soil slurry	Bhatt et al. (2020a,b); Chen et al. (2015)
Cycloclasticus sp. P1	Polycyclic aromatic hydrocarbons (PAHs)	Deep sea sediments	Wang et al. (2018)
Enterobacter spp., *Bacillus* spp., *Providencia* spp., and *Pseudomonas* spp.	Atrazine	Contaminated Agricultural soil	El-Bestawy et al. (2014)
Gordonia sp. 1D	Naphthalene	Oil-refinery soil	Delegan et al. (2019)
Halomonas sp. ATBC28 and *Mycobacterium* sp. DBP42	Phthalic acid ester plasticizers: bis(2-ethyl hexyl) phthalate (DEHP) and dibutyl phthalate (DBP)	Seawater	Wright et al. (2020)
Microbacterium Y2	Decarbodimethyl ether (BDE-209)	Contaminated water-sediment system	Yu et al. (2020)
Lactobacillus plantarum	Phorate	Contaminated soil	Li et al. (2019)

(*Continued*)

TABLE 2.1 (Continued)

Microorganisms	Xenobiotics	Sources	References
Photobacterium ganghwense strain 6046	Cyfluthrin	Seawater	Wang et al. (2019)
Pseudomonas sp., *Rhodococcus* sp., *Bordetella* sp., *Achromobacter* sp., and *Varivorex* sp.	Biphenyl	Contaminated soil	Garrido-Sanz et al. (2018)
Pseudomonas nitroreducens strain CW7	Allethrin	Soil	Bhatt et al. (2020a,b)
Pseudomonas putida, Pseudomonas aeruginosa, and *Acinetobacter faecalis*	Pesticides: Glyphosate	Contaminated Agricultural soil	Olawale and, Akintobi (2011)
Sphingobacterium multivorum	Hexaconazole,	sludge	Yao et al. (2020)
Sphingobium chungbukense DJ77	Phenanthrene, Naphthalene, and Biphenyls (PNB)	Contaminated soil	Lee et al. (2016)
Staphylococcus succinus HLJ-10	D-cyphenothrin	Contaminated soil	Huang et al. (2020)
Stenotrophomonas malthophilia, Proteus vulgaris, Vibrio metschinkouii, Serratia ficaria, Serratia spp., and *Yersinia enterocolitica*	Tetrachlorvinphos	Agricultural Soil	Ortiz-Hernández and Sánchez-Salinas (2010)
Fungi			
Drechslera sp.	Methyl tertiary-butyl ether (MtBE)	Soil	d'Errico et al. (2021)
Metarhizium brunneum ARSEF 2017	s-Triazene		
Metarhizium robertsii	4-n-Nonylphenol (4-n-NP)	Contaminated soil	Szewczyk et al. (2014)
Paecilomyces sp. strain SF-8	Polycyclic aromatic hydrocarbons	Marine water	Velmurugan et al. (2017)
Phanerochaete chrysosporium	Terabromobisphenol A	Contaminated soil	Chen et al. (2019)

Concept of green chemistry

Green chemistry emphasizes the idea that contributors must adopt an approach to decrease environmental effects through the conscious design of chemical products, which not only includes the procurement of raw materials and how industrial and consumer goods are manufactured and utilized, but also how these materials and goods may be reused, repurposed, or up-cycled. It can help advance closed-loop economies while also serving as a role model for innovation in their respective chemistry sub-disciplines on a basic level (Ganesh et al., 2021). The 12 principles of green chemistry are "design guidelines" that might assist chemists in achieving the deliberate objective of sustainability as focused in Fig. 2.1.

Careful chemical synthesis planning and molecular design are hallmarks of green chemistry, which aim to minimize unfavorable effects. Synergies, not trade-offs, may be achieved via good design (Anastas and Eghbali, 2010).

Green chemistry and the role of microbial consortia

In comparison to chemical methods, bioconversion can achieve ecological purity of output since the raw materials for its implementation are renewable materials of animal origin, and the product itself is entirely used in the process of consumption (d'Errico et al., 2020; Liang et al., 2022). Modular coculture engineering could increase the capacity of microbial organisms to generate industrial chemicals from renewable resources. This could reduce metabolic burden through the division of labor, improve enzyme expression and catalysis through compartmentalization, and streamline pathway optimization by using complicated substrates (Mishra et al., 2019a,b; Li et al., 2022).

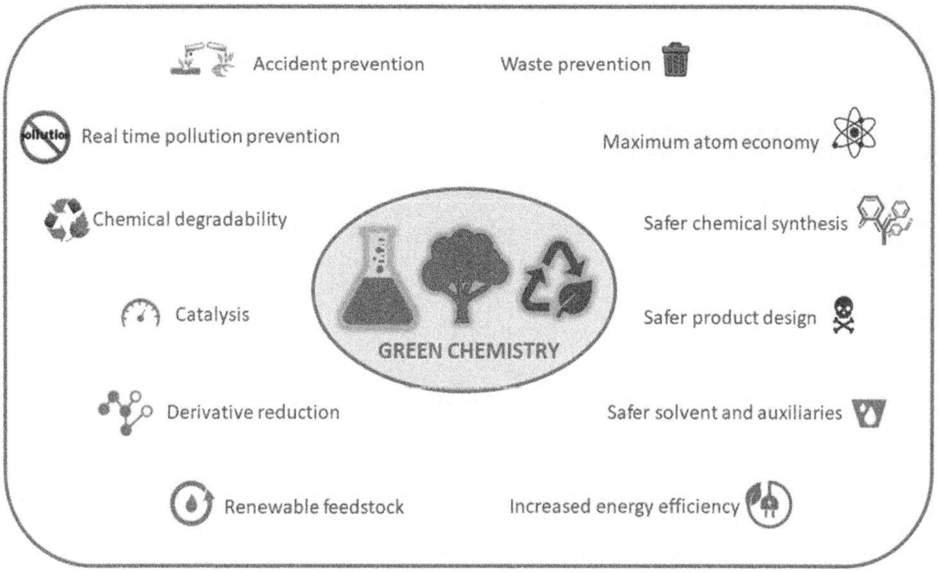

FIGURE 2.1 Twelve principles of green chemistry.

Role of green chemistry and microbial consortium in xenobiotics removal

The high toxicity, lengthy persistence, and low biodegradability of many xenobiotic chemicals have serious negative effects on the ecosystem. Degradation of xenobiotics with the help of microorganisms is the preferred method (Jeffries et al., 2019). The xenobiotic pollutants provide carbon or nitrogen for microorganisms' development and metabolism. Many different kinds of microorganisms can live in polluted places, and some of these microbes are very good at breaking down waste (Yan et al., 2018; Mishra et al., 2021). The development of genetic modification technologies has led to more accurate data and research into how highly competent bacteria might be able to absorb xenobiotics in the future (Mishra et al., 2021). These technologies, with the help of the Green chemistry approach, enable extensive help in solving biodegradability problems caused and vividly understand the mechanism behind a microorganism's ability to break down waste. The microbial consortium engaged in exploring this approach is explained in Fig. 2.2.

Significance of omics technology in green chemistry

Omics technology in bioremediation focuses on the study of genetic manipulation, molecular level alterations, metabolic activity, and stimulation of enzymes and proteins in microbes. The capability to outgrow the existing microbial community centered around recent developments in omics technology surely helps build a stronger strike in detoxifying the environment by targeting potential xenobiotic breakdown. It provides a platform for a variety of current technologies to consider the integrated approach of omics in dealing with complicated biological challenges, as suggested in Fig. 2.3. With time, omics have significantly improved agricultural practices. NGS technologies have emerged for various sequencing uses, which are both rapid and cost-effective.

By recovering spatial and temporal statistics from NGS-acquired metagenome analysis, microbial community interactions can be vividly studied in alleviating environmental distress (Hidalgo et al., 2020). Studies reveal the metagenomic analysis of soil contaminated with harmful pesticides to investigate the degree of degradation in xenobiotic compounds containing organophosphorus (Jeffries et al., 2019). As per the reports, the richness of metabolic pathways targeting carbohydrates and fatty acids along with virulence genes is higher in slowly degrading soil. On the contrary, the presence of transposable elements, nutrient cycling, and membrane transport has been seen much higher in soils with a fast degradation rate (Kumar et al., 2018c). Transcriptomics serves as an extraordinary tool for the detection and analysis of genome and proteome bridges in microbial consortia, along with phenotypic alterations by exploring transcriptional mRNA profiles (Bashiardes et al., 2016; McGrath et al., 2008; Singh and Nagaraj, 2006). Advances in the field of metatranscriptomics, RNA seq, transcriptomics, microarray, and similar technologies have displayed a widely diverse range of microbiota in both land and aquatic environments in synergic interactions with flora, fauna, and other microbes in the given surrounding towards bioremediation (White et al., 2016; Crump et al., 2018; Berg et al., 2018; Perez-Losada et al., 2015). A diversified taxonomic phylum of the dominant bacterial consortium has been identified in the rhizosphere of wheat with the help of metatranscriptomic analysis. Owing to playing key roles in degrading

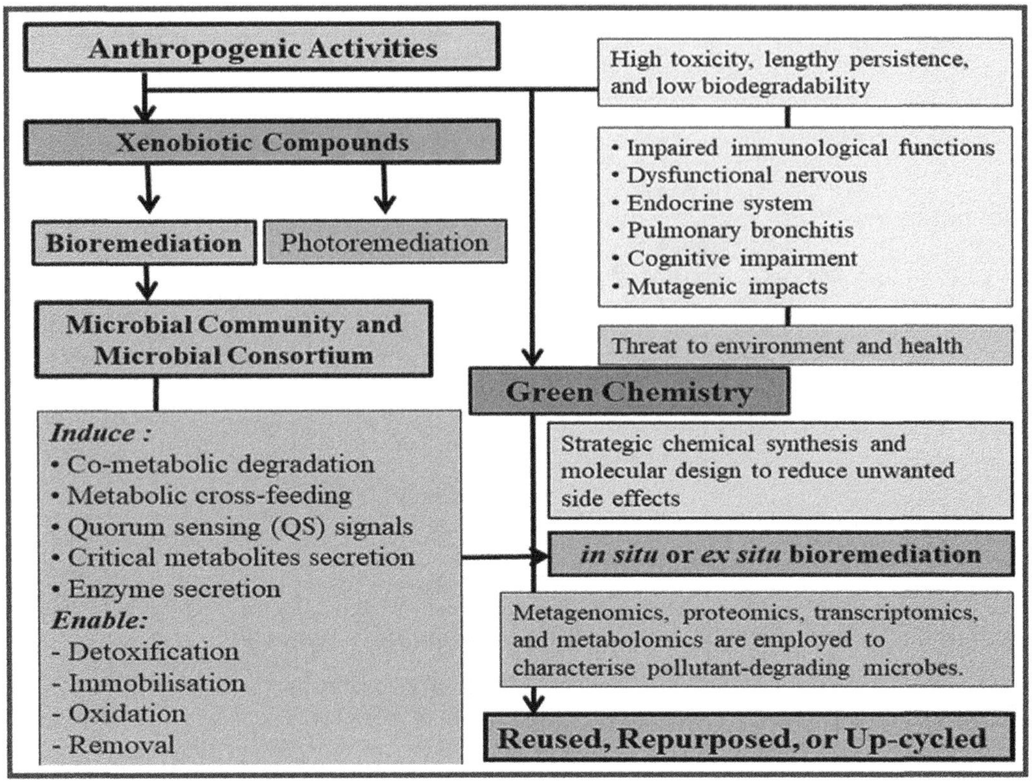

FIGURE 2.2 Mechanism of xenobiotics removal.

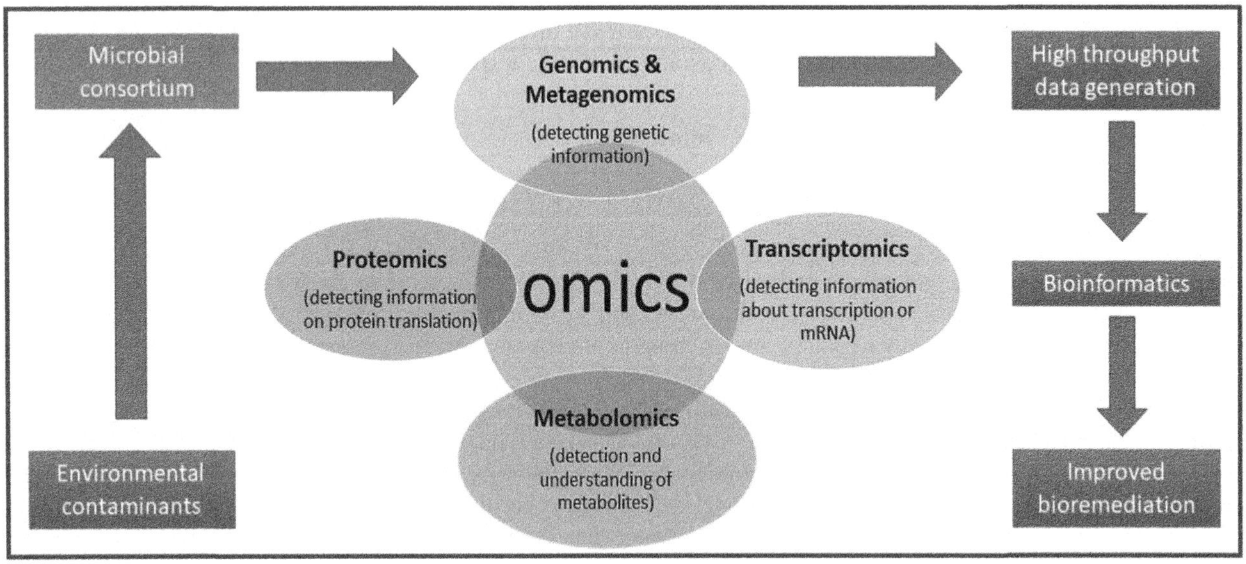

FIGURE 2.3 Omics technology in green chemistry.

xenobiotic chemicals and pollutants, the bacterial population includes Cyanobacteria, Steptophyta, Acidobacteria, Firmicutes, Bacteroidetes, and Ascomycota (Singh et al., 2018). One of the studies explicates the transcriptome analysis for *Sphingobacterium multivorum* involved in the breakdown of hexaconazole present in activated sludge. *S. multivorum* is observed to be capable of breaking down hexaconazole by 85.6% in merely six days, and subsequently producing 3 distinct metabolites- M1 or (2-(2,4-dichlorophenyl)-1-(1H-1,2,4-triazol-1-yl) hexane-2,5diol), M2 or (2-(2,4-dichlorophenyl) hexane-1,2-diol), and M3 or (1H-1,2,4-triazol). The production of M1 metabolite is possible because of the involvement of monooxygenase gene regulation in *S. multivorum* (Singh et al., 2016; An et al., 2020). A combined methodology involving both metatranscriptomics and metagenomics uncovers the synergistic and metabolic activities between *Pseudomonas* sp., *Pusillimonas* sp., and *Sphingomonas* sp. in breaking down bisphenol A in wastewater treatment and other aquatic effluents (Yu et al., 2019). To better understand and investigate associated enzymatic and degradative pathways in the bioremediation of various xenobiotics from contaminated areas, proteomics serves as a potential approach (Wei et al., 2017; Liu et al., 2017; Kim et al., 2004). Metabolomics involves a scientific and extensive study of all the biochemical processes comprising metabolites, postulating chemical fingerprints characteristically specific to defined cellular productivity by referring to metabolite profiles of micro-molecules (Withers et al., 2020; Malla et al., 2018). A collective methodology involving metabolomics and reliable profiles of chemical data can aid in mediating resource exchanges between the host and its community. The science behind metabolomics finds its widespread use in environmental sustainability studies; to understand alterations in metabolite concentrations on exposure of organism models to toxic chemicals like xenobiotics (Mallick et al., 2019; Wang et al., 2018; Seo et al., 2009). Metabolomics studies show a comparative analysis of 156 metabolites in the biotransformation and biodegradation pathways of cyfluthrin (Wang et al., 2019). Sequence analysis and enzymatic pathways are collectively explored via metabolomics profiling, which demonstrates significant inhibition in the metabolic processes in phorate/PAH-polluted soil associated with amino acids, fatty acids, and carbohydrates due to altered pesticide-degradative metabolic pathway of *Lactobacillus plantarum* P9 strain (Li et al., 2018). In another recent study, two bacterial isolates—*Halomonas* sp. ATBC28 and *Mycobacterium* sp. DBP42—present in the marine environment has been found to exhibit degradative properties against plasticizers and phthalates like DEHP, ATBC, and DBP (Wright et al., 2020). Metabolic analysis of *Metarhizium brunneum* ARSEF 2017, an entomopathogenic and cosmopolitan fungal strain interacting with ametryn and resulting in bioremediation of the herbicide. Triazine herbicides like ametryn are found mostly in water bodies and are extremely toxic to aquatic fauna, due to high bioaccumulate properties and degradability resistance. *M. brunneum* ARSEF 2017 showcases promising results as a xenobiotic degrader in eliminating ametryn via metabolic alterations in ametryn stressed environment.

Challenges and future scopes of xenobiotics removal using microorganisms

The presence of microorganisms can substantially aid in the biodegradation of xenobiotic-contaminated soil and water environments. However, a lot of microorganisms associated with bioremediation remain unidentified since not all naturally occurring microbes can be cultured under in vitro conditions. Instead, they reside in viable-but-nonculturable (VBNC) settings. To discover the mysteries behind these VBNC organisms, researchers need hi-tech tools and strategies for studying the genetic and molecular biology of microbes. The complete molecular and genetic characterization of microbial communities can be more effectively approached with the use of modern molecular tools. Omics technologies like metagenomics, transcriptomics, and proteomics can be utilized to provide insight into biodegradation by-products including nucleic acids, enzymes, catabolic genes, plasmids, and metabolic pathways and metabolites. The system-biology perspective shows how genes, proteins, and environmental factors work together to control the whole process of microbial degradation.

Conclusion

A growing number of approaches, such as local similarity analysis and similarity-based methodologies, have been created to decipher microbial interactions in microbial consortia. An efficient network is required to look for potential biotic connections, to strengthen the relationship between species and degradative ability. As anticipated, there is a strong positive association between degradation efficiency and both cell density and enzyme activity. By redesigning chemical products and processes, we can likely reduce or completely eradicate the generation of hazardous waste using the "green chemistry" concept. Additionally, it guarantees a cost-efficient and "ecologically benign" technique. Green chemistry is progressing beyond laboratories into industrial settings. Sustainable practices include a substitution of volatile organic solvents that make up the majority of a reaction mixture, the development of recyclable catalysts, the development of energy-efficient synthesis, and the promotion of renewable raw materials. Many businesses have greatly improved their turnover by adhering to the green chemistry principles, which subsequently has incurred positive impacts on the environment and the economy.

References

An, X., Tian, C., Xu, J., Dong, F., Liu, X., Wu, X., et al., 2020. Characterization of hexaconazole-degrading strain *Sphingobacterium multivorum* and analysis of transcriptome for biodegradation mechanism. The Science of the Total Environment 722, 37171.

Anastas, P., Eghbali, N., 2010. Green chemistry: principles and practice. Chemical Society Reviews 39 (1), 301–312.

Arora, P.K., 2012. Decolourization of 4-chloro-2-nitrophenol by a soil bacterium, *Bacillus subtilis* RKJ 700. PLOS ONE 7 (12), e52012.

Arora, P.K., Srivastava, A., Garg, S.K., Singh, V.P., 2018. Recent advances in degradation of chloronitrophenols. Bioresource Technology 250, 902–909.

Atashgahi, S., Shetty, S.A., Hauke, S., de Vos, W.M., 2018. Flux, impact and fate of halogenated xenobiotic compounds in the Gut. Frontiers in Physiology 9, 888.

Azubuike, C.C., Chikere, C.B., Okpokwasili, G.C., 2016. Bioremediation techniques – classification based on site of application: principle, advantages, limitations, and prospects. World Journal of Microbiology and Biotechnology 32, 180.

Barathi, S., Karthik, C., Nadanasabapathi, S., Padikasan, I.A., 2020. Biodegradation of textile dye Reactive Blue 160 by *Bacillus firmus* (Bacillaceae: Bacillales) and non-target toxicity screening of their degraded products. Toxicology Reports 7, 16–22.

Bashiardes, S., Zilberman-Schapira, G., Elinav, E., 2016. Use of metatranscriptomics in microbe research. Bioinformatics and Biology Insights 10, 19–25.

Berg, C., Dupont, C.L., Asplund-Samuelsson, J., Celepli, N.A., Eiler, A., Allen, A.E., 2018. Dissection of microbial community functions during a cyanobacterial bloom in the Baltic Sea via metatranscriptomic. Frontiers in Marine Science 5, 55.

Bertotto, L.B., Catron, T.R., Tal, T., 2020. Exploring interactions between xenobiotics, microbiota and neurotoxicity in zebrafish. NeuroToxicology 76, 235–244.

Bhatt, P., Huang, Y., Hui, Z., Chen, S., 2019. Insights into microbial applications for the biodegradation of pyrethroid insecticides. Frontiers in Microbiology 10, 177.

Bhatt, P., Huang, Y., Zhang, W., Sharma, A., Chen, S., 2020a. Enhanced cypermethrin degradation kinetics and metabolic pathway in *Bacillus thuringiensis* strain SG4. Microorganisms. 8 (2), 223.

Bhatt, P., Rene, E.R., Kumar, A.J., Zhang, W., Chen, S., 2020b. Binding interaction of allethrin with esterase: Bioremediation potential and mechanism. Bioresource Technology 315, 123845.

Bhatt, P., Gangola, S., Bhandari, G., Zhang, W., Maithani, D., Mishra, S., et al., 2021. New insights into the degradation of synthetic pollutants in contaminated environments. Chemosphere 268, 128827.

Bibi, N., Hamayun, M., Khan, S.A., Iqbal, A., Islam, B., Shah, F., et al., 2018. Anthracene biodegradation capacity of newly isolated rhizospheric bacteria *Bacillus cereus* S13. PLOS ONE 13 (8), e0201620.

Bonifer, K.S., Wen, X., Hasim, S., Phillips, E.K., Dunlap, R.N., Gann, E.R., 2019. *Bacillus pumilus* B12 degrades polylactic acid and degradation is affected by changing nutrient conditions. Frontiers in Microbiology 10, 2548.

Bronzwaer, S., Geervliet, M., Hugas, M., Url, B., 2021. EFSA's expertise supports one health policy needs. EFSA Journal e190501.

Buschhardt, T., Günther, T., Skjerdal, T., Torpdahl, M., Gethmann, J., Filippitzi, M.E., et al., 2021. One health glossary to support communication and information exchange between the human health, animal health and food safety sectors. One Health 13, 100263.

Chan, S.S., Khoo, K.S., Chew, K.W., Ling, T.C., Show, P.L., 2022. Recent advances biodegradation and biosorption of organic compounds from wastewater: microalgae-bacteria consortium—a review. Bioresource Technology 344, 126159.

Chen, S., Dong, Y.H., Chang, C., Deng, Y., Zhang, X.F., Zhong, G., et al., 2013. Characterization of a novel cyfluthrin-degrading bacterial strain *Brevibacterium aureum* and its biochemical degradation pathway. Bioresource Technology 132, 16–23.

Chen, S., Deng, Y., Chang, C., Lee, J., Cheng, Y., Cui, Z., et al., 2015. Pathway and kinetics of cyhalothrin biodegradation by *Bacillus thuringiensis* strain ZS-19. Scientific Reports 5, 8784.

Chen, Z., Pan, X., Chen, H., Guan, X., Lin, Z., 2016. Biomineralization of Pb (II) into Pb-hydroxyapatite induced by *Bacillus cereus* 12-2 isolated from lead–zinc mine tailings. Journal of Hazardous Materials 301, 531–537.

Chen, Z., Yin, H., Peng, H., Lu, G., Liu, Z., Dang, Z., 2019. Identification of novel pathways for biotransformation of tetrabromobisphenol A by *Phanerochaete chrysosporium*, combined with mechanism analysis at proteome level. The Science of the Total Environment 659, 1352–1361.

Chopra, S., Kumar, D., 2020. Characterization, optimization and kinetics study of acetaminophen degradation by *Bacillus drentensis* strain S1 and waste water degradation analysis. Bioresources and Bioprocessing 7 (1), 1–8.

Crump, B.C., Wojahn, J.M., Tomas, F., Mueller, R.S., 2018. Metatranscriptomics and amplicon sequencing reveal mutualism in seagrass microiome. Frontiers in Microbiology 9, 388.

d'Errico, G., Aloj, V., Flematti, G.R., Sivasithamparam, K., Worth, C.M., 2020. Metabolites of *Drechslera* sp. endophyte with potential as biocontrol and bioremediation agent. Natural Product Research 11, 1–9.

d'Errico, G., Aloj, V., Flematti, G.R., Sivasithamparam, K., Worth, C.M., Lombardi, N., et al., 2021. Metabolites of a *Drechslera* sp. endophyte with potential as biocontrol and bioremediation agent. Natural Product Research 35 (22), 4508–4516.

Dangi, A.K., Sharma, B., Hill, R.T., Shukla, P., 2018. Bioremediation through microbes: systems biology and metabolic engineering approach. Critical Reviews in Biotechnology 39, 79–98.

Das, M., Bhattacharya, A., Banu, S., Kotoky, J., 2017. Enhanced biodegradation of anthracene by *Bacillus cereus* strain JMG-01 isolated from hydrocarbon contaminated soils. Soil and Sediment Contamination: An International Journal 26 (5), 510–525.

Dash, D.M., Osborne, J.W., 2020. Biodegradation of monocrotophos by a plant growth promoting *Bacillus aryabhattai* (VITNNDJ5) strain in artificially contaminated soil. International Journal of Environmental Science and Technology 17 (3), 1475–1490.

Delegan, Y.A., Valentovich, L.N., Shafieva, S.M., Ganbarov, K.G., Filonov, A.E., Vainstein, M.B., 2019. Characterization and genomic analysis of highly efficient thermotolerant oil-degrading bacterium *Gordonia* sp. 1D. Folia Microbiologica 64 (1), 41–48.

Devi, S.P., Jani, K., Sharma, A., Jha, D.K., 2022. Bacterial communities and their bioremediation capabilities in oil-contaminated agricultural soils. Environmental Monitoring and Assessment 194 (1), 1–3.

Dinka, D.D., 2018. Environmental xenobiotics and their adverse health impacts-a general review. Journal of Environment Pollution and Human Health 6, 77–88.

El-Bestawy, E., Sabir, J., Mansy, A.H., Zabermawi, N., 2014. Comparison among the efficiency of different bioremediation technologies of Atrazine-contaminated soils. Journal of Bioremediation and Biodegradation 5 (5), 237.

Embrandiri, A., Kiyasudeen, S.K., Rupani, P.F., Ibrahim, M.H., 2016. Environmental xenobiotics and its effect on natural ecosystem. In: Singh, A., Prasad, S.M., Singh, R.P. (Eds.), Plant Response to Xenobiotic. Springer, pp. 1–18.

Fuentes, M.S., Benimeli, C.S., Cuozzo, S.A., Saez, J.M., Amoroso, M.J., 2010. Microorganisms capable to degrade organochlorine pesticides, Current Research, Technology and Education Topics in Applied Microbiology and Microbial Biotechnology, 2. Formatex Research Centre, pp. 1255–1264.

Ganesh, K.N., Zhang, D., Miller, S.J., Rossen, K., Chirik, P.J., Kozlowski, M.C., et al., 2021. Green chemistry: a framework for a sustainable future. Environmental Science & Technology 55 (13), 8459–8463.

Gangola, S., Sharma, A., Bhatt, P., Khati, P., Chaudhary, P., 2018. Presence of esterase and laccase in *Bacillus subtilis* facilitates biodegradation and detoxification of cypermethrin. Sci. rep. 8 (1), 1.

Garrido-Sanz, D., Manzano, J., Martín, M., Redondo-Nieto, M., Rivilla, R., 2018. Metagenomic analysis of a biphenyl-degrading soil bacterial consortium reveals the metabolic roles of specific populations. Frontiers in Microbiology 9, 232.

Gilliespie, I.M.M., Philp, J.C., 2013. Bioremediation, an environmental remediation technology for the bioeconomy. Trends in Biotechnology 31, 329–332.

Hidalgo, K.J., Teramoto, E.H., Soriano, A.U., Valoni, E., Baessa, M.P., Richnow, H.H., et al., 2020. Taxonomic and functional diversity of the microbiome in a jet fuel contaminated site as revealed by combined application of *in situ* microcosms with metagenomic analysis. The Science of the Total Environment 708, 135152.

Huang, Y., Lin, Z., Zhang, W., Pang, S., Bhatt, P., Rene, E.R., et al., 2020. New insights into the microbial degradation of D-cyphenothrin in contaminated water/soil environments. Microorganisms. 8 (4), 473.

Hussain, I., Aleti, G., Naidu, R., Puschenreiter, M., Mahmood, Q., Rahman, M.M., 2018. Microbe and plant assisted remediation of organic xenobiotics and its enhancement by genetically modified organisms and recombinant technology: a review. The Science of the Total Environment 628, 1582–1599.

Idle, J.R., Gonzalez, F.J., 2007. Metabolomics. Cell Metabolism 6, 348–351.

Iqbal, H.M., Bilal, M., Zdarta, J., 2022. Strategies for environmental contaminants monitoring and remediation. Environmental Technology & Innovation 28, 102641.

Janssen, D.B., Stucki, G., 2020. Perspectives of genetically engineered microbes for groundwater bioremediation. Environmental Science: Processes & Impacts 22, 487.

Jeffries, T.C., Rayu, S., Nielsen, U.N., Lai, K., Ijaz, A., Nazaries, L., et al., 2019. Metagenomics function potential predicts functional rates of a model organophosphorus xenobiotic in pesticide-contaminated soil. Frontiers in Microbiology 9, 147.

Jiang, J., Tang, M., Chen, J., Yang, Y., 2019. Identification and degradation characteristics of Bacillus cereus strain WD-2 isolated from prochloraz-manganese-contaminated soils. PLOS ONE 14 (8), e0220975.

Joshi, B., Kabariya, K., Nakrani, S., Khan, A., Parabia, F.M., Doshi, H.V., 2013. Biodegradation of turquoise blue dye by *Bacillus megaterium* isolated from industrial effluent. American Journal of Environmental Protection 1 (2), 41–46.

Jung, J., Philippot, L., Park, W., 2016. Metagenomic and functional analyses of the consequences of reduction of bacterial diversity on soil functions and bioremediation in diesel-contaminated microcosms. Scientific Reports 6 (1), 23012.

Kang, C.H., Kwon, Y.J., So, J.S., 2016. Bioremediation of heavy metals by using bacterial mixtures. Ecological Engineering 89, 64–69.

Kim, S.J., Jones, R.C., Cha, C.J., 2004. Identification of proteins induced by polycyclic aromatic hydrocarbon in *Mycobacterium vanbaalenii* PYR-1 using two-dimensional polyacrylamide gel electrophoresis and denovo sequencing methods. Proteomics 4, 3899–3908.

Kumar, M., Akhtar, M.S., 2019. Application of microbial biotechnology in improving salt stress and crop productivity. Salt Stress, Microbes, and Plant Interactions: Mechanisms and Molecular Approaches. Springer, Singapore, pp. 133–159. Available from: http://doi.org/10.1007/978-981-13-8805-7_7.

Kumar, M., Rana, S., Kumar, H., Kumar, J., Mansuri, R., Chandra Sahoo, G., 2017. Inhibition of product template (PT) domain of aflatoxin producing polyketide synthase enzyme of *Aspergillus parasiticus*. Letters in Drug Design & Discovery 14 (7), 811–818.

Kumar, P., Kumar, M., Kumar, H., Kumar, J., 2018a. In vitro effects of rosuvastatin on *Mycobacterium tuberculosis*. International Journal of Current Research in Life Sciences 7 (4), 1885–1887.

Kumar, M., Sahoo, G., Topno, R., Pandey, K., Ravidas, V., Dikhit, M., et al., 2018c. Molecular interaction of antifungal compounds with fungal type I Polyketide synthase and preventing hepatocarcinogenesis. International Journal of Infectious Diseases: IJID: Official Publication of the International Society for Infectious Diseases 73, 285.

Kumar, M., Topno, R.K., Kumar, H., Kumar, J., 2018b. Biological conversion of Carcinogenic Aflatoxin B1 to AFB1–8, 9-exo-epoxide and its potential prevention. International Journal of Research and Analytical Reviews 5 (4), 935–946.

Kumar, M., Kochhar, N., Kavya, I.K., Shrivastava, S., Ghosh, A., Rawat, V.S., 2022. Perspectives on the microorganism of extreme environments and their applications. Current Research in Microbial Sciences. 100134.

Kumari, A., Chaudhary, D.R., 2020. Engineered microbes and evolving plastic bioremediation technology. Bioremediation of Pollutants. Elsevier, pp. 417–443.

Lee, S.Y., Sekhon, S.S., Ban, Y.H., Ahn, J.Y., Ko, J.H., Lee, L., et al., 2016. Proteomic analysis of polycyclic aromatic hydrocarbons (PAHs) degradation and detoxification in *Sphingobium chungbukense* DJ77. Journal of Microbiology and Biotechnology 26 (11), 1943–1950.

Li, C., Ma, Y., Mi, Z., Huo, R., Zhou, T., Hai, H., et al., 2018. Screening for *Lactobacillus plantarum* strains that posses organophosphorus pesticide-degrading activity and metabolomic analysis of phorate degradation. Frontiers in Microbiology 9, 2048.

Li, X., Qu, C., Bian, Y., Gu, C., Jiang, X., Song, Y., 2019. New insights into the responses of soil microorganisms to polycyclic aromatic hydrocarbon stress by combining enzyme activity and sequencing analysis with metabolomics. Environmental Pollution 255, 113312.

Li, X., Zhou, Z., Li, W., Yan, Y., Shen, X., Wang, J., et al., 2022. Design of stable and self-regulated microbial consortia for chemical synthesis. Nature Communications 13 (1), 1–9.

Liang, Y., Ma, A., Zhuang, G., 2022. Construction of environmental synthetic microbial consortia: based on engineering and ecological principles. Frontiers in Microbiology 437.

Liu, S., Gu, C., Dang, Z., Liang, X., 2017. Comparative proteomics reveal the mechanism of Tween 80 enhanced phenantherene biodegradation by *Sphingomonas* sp. GY2B. Ecotoxicology and Environmental Safety 137, 256–264.

Louati, I., Dammak, M., Nasri, R., Belbahri, L., Nasri, M., Abdelkafi, S., et al., 2019. Biodegradation and detoxification of bisphenol A by bacteria isolated from desert soils. 3 Biotech 9 (6), 1.

Luo, H., Liu, C., He, D., Xu, J., Sun, J., Li, J., et al., 2022. Environmental behaviors of microplastics in aquatic systems: a systematic review on degradation, adsorption, toxicity and biofilm under aging conditions. Journal of Hazardous Materials 423, 126915.

Malla, M.A., Dubey, A., Yadav, S., Kumar, A., Hashem, A., Abd-Allah, E.F., 2018. Understanding and designing the strategies for the microbe-mediated remediation of environmental contaminants using omics approaches. Frontiers in Microbiology 9, 1132.

Mallick, H., Franzosa, E.A., Mclver, L.J., Banerjee, S., Sirota-Madi, A., Kostic, A.D., et al., 2019. Predictive metabolomic profiling of microbial communities using amplicon or metagenomic sequences. Nature Communications 10, 3136.

McGrath, K.C., Thomas-Hall, S.R., Cheng, C.T., Leo, L., Alexa, A., Schmidt, S., et al., 2008. Isolation and analysis of mRNA from environmental microbial communities. Journal of Microbiological Methods 75, 172–176.

Mishra, J., Kumar, M., Sharan, A.K., Kumar, J., 2019a. Detoxification assessment of Aflatoxin in *Aspergillus flavus* under the effect of *Calendula officinalis* Linn's methanolic extract. Agricultural Science Digest-A Research Journal 39 (1), 21–25.

Mishra, V.K., Singh, G., Shukla, R., 2019b. Impact of xenobiotics under a changing climate scenario, Climate Change and Agricultural Ecosystems, 2. Elsevier, pp. 133–151.

Mishra, S., Lin, Z., Pang, S., Zhang, W., Bhatt, P., Chen, S., 2021. Recent advanced technologies for the characterization of xenobiotic-degrading microorganisms and microbial communities. Frontiers in Bioengineering and Biotechnology 9, 632059.

Nzila, A., Ramirez, C.O., Musa, M.M., Sankara, S., Basheer, C., Li, Q.X., 2018. Pyrene biodegradation and proteomic analysis in *Achromobacter xylosoxidans*, PY4 strain. International Biodeterioration & Biodegradation 130, 40–47.

Olawale, A.K., Akintobi, O.A., 2011. Biodegradation of glyphosate pesticide by bacteria isolated from agricultural soil. Report and Opinion 3 (1), 124–128.

Ortiz-Hernández, M., Sánchez-Salinas, E., 2010. Biodegradation of the organophosphate pesticide tetrachlorvinphos by bacteria isolated from agricultural soils in México. Revista Internacional de Contaminación Ambiental 26 (1), 27–38.

Paul, D., Pandey, G., Pandey, J., Jain, R.K., 2005. Accessing microbial diversity for bioremediation and environmental pollutant restoration. Trends in Biotechnology 23, 135–142.

Perelo, L.W., 2010. In situ and bioremediation of organic pollutants in aquatic sediments. Journal of Hazardous Materials 177, 81–89.

Perez-Losada, M., Castro-Nallar, E., Bendall, M.L., Freishtat, R.J., Crandall, K.A., 2015. Dual transcriptomic profiling of host and microbiota during health and disease in pediatric asthma. PLOS ONE 10, e0131819.

Prosekov, A., Milenteva, I., Sukhikh, S., Dyshlyuk, L., Babich, O., Mitrokhin, P., et al., 2015. Optimization of conditions for biodegradation of poultry industry wastes by microbial consortium. Asian Journal of Microbiology, Biotechnology & Environmental Sciences 17 (3), 515–519.

Ramaka, S., Srinivas, R., Vasudeva, M.S., Raghuram Rao, A., 2020. Xenobiotics in health and disease: the two sides of a coin: a clinician's perspective. Open Access Journal of Toxicology 4, 555641.

Ravindra, A.P., Haq, S.A., 2019. Effects of xenobiotics and their biodegradation in marine life. In: Bhatt, P. (Ed.), Smart Bioremediation Technologies: Microbial Enzymes. Academic Press, pp. 63–81.

Sandhya, M., Ziqui, L., Shimei, P., Wenping, Z., Pankaj, B., Shaohua, C., 2021. Recent advanced technologies for the characterization of xenobiotic-degrading microorganisms and microbial communities. Frontiers in Bioengineering and Biotechnology 9, 632059.

Seo, J.S., Keum, Y.S., Li, Q.X., 2009. Bacterial degradation of aromatic compounds. International Journal of Environmental Research and Public Health 6, 278–309.

Shah, M.P., Patel, K.A., Nair, S.S., Darji, A.M., 2013. Potential effect of two *Bacillus* spp. on decolorization of azo dye. Journal of Bioremediation and Biodegradation 4 (7).

Singh, O.V., Nagaraj, N.S., 2006. Transcriptomics, proteomics and interactions: unique approaches to track the insights of bioremediation. Briefings in Functional Genomics 4, 355–362.

Singh, A., Chaudhary, S., Dubey, B., Prasad, V., 2016. Microbial-mediated management of organic xenobiotic pollutants in agricultural lands. In: Singh, A., Prasad, S.M., Singh, R.P. (Eds.), Plant Response to Xenobiotics. Springer, Singapore, pp. 211–230.

Singh, D.P., Prabha, R., Gupta, V.K., Verma, M.K., 2018. Metatranscriptomic analysis deciphers multifunctional genes and enzymes linked with the degradation of aromatic compounds and pesticide in the wheat rhizosphere. Frontiers in Microbiology 9, 1331.

Štefanac, T., Grgas, D., Landeka Dragičević, T., 2021. Xenobiotics—division and methods of detection: a review. Journal of Xenobiotics 11, 130–141.

Szewczyk, R., Sobon, A., Sylwia, R., Dzitko, K., Waidelich, D., Dlugonski, J., 2014. Intracellular proteome expression during 4-n-nonylphenol biodegradation by filamentous fungus *Metarhizium robertsii*. International Biodeterioration and Biodegradation 93, 44–53.

Tak, Y., Kaur, M., Tilgam, J., Kaur, H., Kumar, R., Gautam, C., 2022. Microbes assisted bioremediation: a green technology to remediate pollutants. Bioremediation of Environmental Pollutants. Springer, Cham, pp. 25–52.

Vandera, E., Samiotaki, M., Parapouli, M., Panayotou, G., Koukkou, A.I., 2015. Comparative proteomic analysis of *Arthrobacter phenanthrenivorans* Sphe3 on phenanthrene, phthalate and glucose. Journal of Proteome Research 113, 73–89.

Velmurugan, N., Lee, H.M., Cha, H.J., Lee, Y.S., 2017. Proteomic analysis of the marine-derived fungus Paecilomyces sp. strain SF-8 in response to polycyclic aromatic hydrocarbons. Botanica Marina 60 (4), 381–392.

Wang, W., Wang, L., Shao, Z., 2018. Polycyclic aromatic hydrocarbon (PAH) degradation pathways of the obligate marine PAH degrader *Cycloclasticus* sp. strain P1. Applied and Environmental Microbiology 84 (21), e01261-18.

Wang, Y.F., Zhu, H.W., Wang, Y., Zhang, X.L., Tam, N.F.Y., 2018. Diversity and dynamics of microbial community structure in different mangrove, marine and freshwater sediments during anaerobic debromination of PBDEs. Frontiers in Microbiology 9, 952.

Wang, T., Hu, C., Zhang, R., Sun, A., Li, D., Shi, X., 2019. Mechanism study of cyfluthrin biodegradation by *Photobacterium ganghwense* with comparative metabolomics. Applied Microbiology and Biotechnology 103, 473–488.

Wei, K., Yin, H., Peng, H., Liu, Z., Lu, G., Dang, Z., 2017. Characteristics and proteomic analysis of pyrene degradation by *Brevibacillus brevis* in liquid medium. Chemosphere 178, 80–87.

White, R.A., Bottos, E.M., Roy, C.T., Zucker, J.D., Brislawn, C.J., Nicora, C.D., et al., 2016. Moleculo long-read sequencing facilitates assembly and genomic binning from complex soil metagenomes. mSystems 1, e00045-16.

Withers, E., Hill, P.W., Chadwick, D.R., Jones, D.L., 2020. Use of untargeted metabolomics for assessing soil quality and microbial function. Soil Biology & Biochemistry 143, 107758.

Wright, R., Bosch, R., Gibson, M.I., Christie-Oleza, J., 2020. Plasticizer degradation by a marine bacterial isolates: a proteogenomic and metabolomic characterization. Environmental Science & Technology 54, 2244–2256.

Wu, X., Yin, Y., Wang, S., Yu, Y., 2014. Accumulation of chlorothalonil and its metabolite, 4-hydroxychlorothalonil in soil after repeated applications and its effects on soil microbial activities underground greenhouse conditions. Environmental Science and Pollution Research 21, 3452–3459.

Yan, X., Jin, W., Wu, G., Jiang, W., Yang, Z., Ji, J., et al., 2018. Hydrolase CehA and monooxygenase CfdC are responsible for carbofuran degradation in *Sphingomonas* sp. strain CDS-1. Applied and Environmental Microbiology 84 (16), e00805–e00818.

Yao, Y., Pan, J., Wang, W., Liu, Z., Kan, H., Qiu, Y., et al., 2020. Association of particulate matter pollution and case fatality rate of COVID-19 in 49 Chinese cities. The Science of the Total Environment 741, 140396.

Yu, K., Yi, S., Li, B., Guo, F., Peng, X., Wang, Z., et al., 2019. An integrated meta-omics approach reveals substrates involved in synergistic interactions in a bisphenol A (BPA)-degrading microbial community. Microbiome 7, 16.

Yu, Y., Yin, H., Peng, H., Lu, G., Dang, Z., 2020. Proteomic mechanism of decabromodiphenyl ether (BDE-209) biodegradation by *Microbacterium* Y2 and its potential in remediation of BDE-209 contaminated water-sediment system. Journal of Hazardous Materials 387, 121708.

Zhan, H., Wang, H., Liao, L., Feng, Y., Fan, X., Zhang, L., et al., 2018. Kinetics and novel degradation pathway of cypermethrin in *Acinetobacter baumannii* ZH-14. Frontiers in Microbiology 9, 98.

Zinedine, A., Soriano, J.M., Molto, J.C., Manes, J., 2007. Review on the toxicity, occurrence, metabolism, detoxification, regulations and intake of zearalenone: an oestrogenic mycotoxin. Food and Chemical Toxicology: an International Journal Published for the British Industrial Biological Research Association 45, 1–18.

Chapter 3

Role of environmental sustainability for climate change adaptations

Vijaya Ilango
Department of Chemistry, Birla Institute of Technology & Science Pilani, Dubai Campus, Dubai, United Arab Emirates

Introduction

Climate variation is one of the most complicated problems that we face today. It has numerous facets—society, politics, economics, science, and ethical and moral problems. It is a global issue that will be sensed on a local scale for decades or eras. The Earth (mostly the seas) takes a long time to adjust to warming because CO_2, the heat-trapping greenhouse gas that has driven recent global warming, can linger in the atmosphere for hundreds of years. As a consequence, climate change and global warming would continue to disturb forthcoming generations, even if we completely stopped producing greenhouse gases today. As a result, mankind has "committed" to climate change to a certain degree. The position of adaptation was strengthened by the Delhi Declaration from the 8th Conference of the Parties in November 2002. The Delhi Declaration effectively linked action and financing for adaptation to the effects of climate change to the developing world's commitment to defending emissions (Adger et al., 2003). Since the apparent taboo on adaptation to climate change has been raised, it has been diplomatically acknowledged and established at many levels of power (Pielke et al., 2007) for instance, through the development of financial tools at the global level of the United National Framework Convention on Climate Change (UNFCCC), the Climate Change Adaptation Strategy of the European Union, the increasing number of National Climate Change Adaptation plans and strategies and the ample local and regional initiatives to prepare for future climate change risks (Biesbroek et al., 2010; Dreyfus and Patt, 2012).

Adaptation to climate change

Adaptation

Alteration in natural or anthropological systems in response to authentic or predictable climatic provocations or their consequences. This harms and exploits the desirable prospects in a reasonable manner (Levina and Tirpak, 2006). The concept of "adaptation" encompasses both the act of adapting as well as the state of having adopted. In ecology, adaptation is the process by which an organism or species adjusts to its environment (Abercrombie et al., 2017). In the social sciences, adaptation is the collective behavior of societal systems and individual changes.

Types of adaptation

Adaptation categories (i.e., in what way adaptation happens) have been classified based on several aspects. There are several types of adaptation:

1. Anticipatory and reactive adaptation, which refers to activities that occur in response or before to the effects of climate change
2. Private along with the public adaptation
3. Spontaneous or autonomous adaptations
4. Planned adaptation, which contrasts public policy decisions

Role of Green Chemistry in Ecosystem Restoration to Achieve Environmental Sustainability. DOI: https://doi.org/10.1016/B978-0-443-15291-7.00024-9

Adaptation can be long-term or short-term confined or extensive and can serve several purposes and take frequent forms (Smit et al., 2001). Uncertainty about future climate circumstances makes it difficult to make decisions and establish adaptation (and mitigation) strategies (Adger et al., 2007). Adapting to climate change is now widely recognized as a critical component of our climate change response strategy. This is because, unfortunately, climate change mitigation methods will not be enough to prevent the negative consequences of current climate change (Bowen and Friel, 2012; Mertz et al., 2009; Lwasa, 2014). Even though there is evidence of adaptation practice, there is little agreement on whether it is climate adaptation.

Climate change adaptation steps

To determine whether long-term concerns are taken into account in climate adaptation policies and theoretical conceptions, first assess the effectiveness of current adaptation policy methods and the presence of characteristics like reflexivity, resilience, scalability, flexibility, responsiveness, and uncertainty. These qualities are necessary to adjust scientifically to future conditions (Haasnoot et al., 2013; Vij et al., 2017). Climate change adaptation steps are an organized technique to promote the expansion of multi-dimensional adaptation policies. A study (Tung et al., 2019; Creighton, 2010; Haasnoot et al., 2018) involving climate adaptation algorithms to detail the steps involved is considered and explained below. The steps are depicted in Fig. 3.1.

Pillars of climate action

Adaptation and mitigation are considered the two pillars of climate action. Addressing the global common tragedy is necessary for mitigation. Many international accords have worked to convert the atmosphere from an open-access resource to a common pool system since the 1992 United Nations Framework Convention on Climate Change (UNFCCC). By establishing national-level emissions objectives, every country is supposed to incorporate its domestic policies for climatic changes. The second pillar of climate policy, after mitigation is adaptation climate adaptation refers to reactive or proactive programs that aim to lessen a specific area's, organization, population group, or individual social, biophysical, and economic susceptibility to climate change (or increase flexibility) (Santos et al., 2022). Comparatively, adaptation generates local benefits and ought to have fewer problems with collective action (Dolšak and Prakash, 2018).

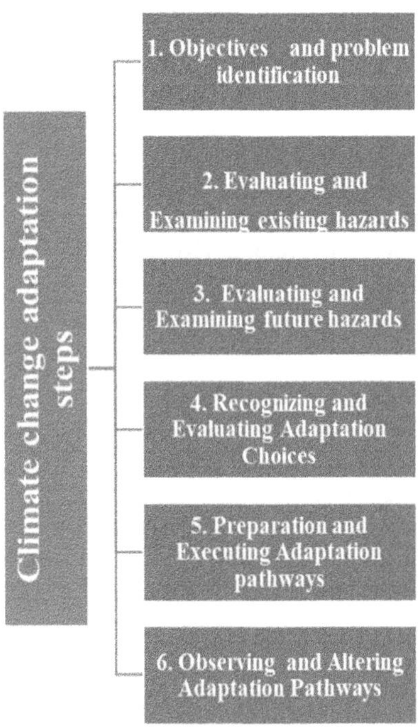

FIGURE 3.1 Steps involved in the climate change adaptation.

Strategies for climate change adaptation

When it comes to addressing climate change, efforts have typically focused on one of two strategies: reducing or eliminating greenhouse gas emissions from the atmosphere, or adapting systems and civilizations to resist the consequences of climate change. Approaches and skills that together limit climate change and manage its effects that previously exist. The strategies that can simultaneously slow climate change and assist us in coping with its effects include (Suarez, 2020):

Defend coastal marshlands: Seagrass, salt marshes, and mangroves are characteristic coastal ecosystems that act as marine habitats and natural water filters. By absorbing storm surges and floodwaters, they protect coastlines from sea level rise. They also store massive amounts of carbon in the soil and roots of their plants. Rising protected coastal wetlands and recuperating by 2050, roughly 40% of the world's ecosystems could reduce one Giga ton of CO_2 annually, or more than three years' worth of emissions (Caldeira et al., 2019)

Encourage the advantages of sustainable agroforestry: It is obvious that the current land management practices need to alter, given that agriculture and forestry contribute to almost 25% of human-caused greenhouse gas emissions (Kumar and Aravindakshan, 2022).

Reorganize energy supply: Examples of centralized energy systems, which are more susceptible to climate change due to the possibility of network-wide interruptions at a single point in the system. Decentralized systems are more robust to climate change because they regularly use renewable energy, and have smaller dispersal areas and shorter transmission lines (Suarez, 2020).

Protected Native peoples' property rights: About 50% of the land on the earth is managed by indigenous and local populations (Veit and Reytar, 2017), on whom around 2.5 billion people depend. Over many generations, these people have lived by the values of adaptation on their lands. A substantial body of conventional knowledge has been produced that can aid others in understanding how to adjust to a changing environment.

Progress mass transportation: Road transport is responsible for 72% of all emissions connected to transportation worldwide and this number will rise until more affordable and reachable low-carbon transportation solutions are produced. Storms and extreme heat are the two climate change influences that will affect the transportation infrastructure (Sims et al., 2014).

Environmental sustainability

According to the Sustainability philosophy, complete sustainability would be solved by the three pillars namely human sustainability, environmental sustainability, and economic sustainability. The goal of human sustainability is to protect and develop society's human capital. Under the general heading of "human sustainability," projects include funds in the education and health systems, access to amenities, nourishment, acquaintance, and skills. The ultimate objective of sustainable development is to promote equitable access to healthcare and education while simultaneously reducing hunger and poverty globally. Environmental sustainability is possibly defined as a state of equilibrium, adaptability, and interrelatedness that enables humanity to meet its requirements without going beyond the size of the ecosystems (Morelli., 2011).

The sense of environmental sustainability was created primarily to assist environmental professionals and other investors in functioning, a portion of the idea of sustainable development as stated in our common future. They are sorted out into the following categories concerning the environment professionals (Morelli, 2011):

Societal needs: Not to produce anything that will impose continuous consideration on the part of future generations. Deliver products and design services that subsidize to a more justifiable budget.

Conservation of biodiversity: Select raw materials that preserve the biodiversity of the natural resources. Use ecologically accountable and sustainable energy origins and capitalize on refining energy productivity.

Reformative capacity: Keep yield rates of renewable resource inputs within the reformative capacities of the natural system that produces them (Goodland, 1995).

Recycle and reuse: Design manufacturing and business processes as closed-loop systems to eliminate emissions and waste (or modify them, as necessary). Design with recyclability and re-usability (Robinson, 2004).

Restraints of nonrenewable resources and waste generation: The size of the human economic subsystem should be kept to a minimum at a level that, if not ideal, is at least sustainable and within carrying capacity.

Environmental sustainability challenges

The new approach of adaptation to climate variation allows us to quicken and measure up our planning. Vice-President of the European Commission, Frans Timmermans, stresses that we still have a chance to develop climate change

TABLE 3.1 Environmental sustainability challenges.

Global environmental issues	Threats posed	Recommendations
Global warming due to the emissions of carbon dioxide	Severe climatic disasters like droughts, fires, and floods	Sturdily directed to mitigate the climate change
Air pollution and contamination of water	90% of people breathe dirty air, due to which seven million fatalities annually. Pose major health hazards.	Air pollution reduction, restriction of the use of chemicals, treating more wastewater.
Inland water, sea, and ocean pollution	Dumping of plastic and synthetic materials, fuel spills, and wastewater	Use of plastic-free alternatives, to reduce carbon footprints.
Slow energy conversion and inadequate share of renewable energy	Lacks access to electricity and dependence on fossil fuels	Sustainable community with cleaner, more accessible, and competent energy transition models.
Detrimental food production	Overutilization of natural resources, soil depletion, and damage to marine ecosystems.	Focus on a plant-based diet to save energy and reduce gas emissions.
Loss of animal species and natural habits	Threats to forest and natural heritage	Protection of biodiversity
Urban strategy/population/waste management	Big environmental challenge, impact on health and environment	Sustainable urban development, prevention, reduction, reuse, and recycling of waste.
Water scarcity and hydric stress	More than 70% of water is used for agriculture affecting the hydrological resources	Protection of the water ecosystem and slow climate change

resilience if we act now and get ready. Numerous global environmental sustainability challenges that need to be resolved by 2030 can readily be listed, calculated, and prioritized (Table 3.1; Nowodziński, 2021).

Advantages of environmental sustainability

The impact of sustainable living on health is one of its key benefits. Other benefits also support living a healthy lifestyle. Advantages and several simple ways (Mollenkamp, 2022) for leading a more sustainable lifestyle in our day-to-day life are schematically represented in Fig. 3.2.

Role of environmental sustainability for climate change adaptations

Protection, adaptation to climate change, and resource management are all necessary for sustainable development. Planning for adaptation to climatic change should identify people or sectors that are at risk, and provide the essential environmental management actions or set of tools for the community to familiarize and scrutinize options to build resilience (Berkhout et al., 2006; Janssen, 2006; Smit and Wandel, 2006).

Sustainable development goals

To encourage regulators to make decisions that will make the world a living and sustainable place for people, many environmental targets and indicators are present in the regulatory framework. They offer a viewpoint from a human-centered society that values people. Such goals and metrics prosper at many levels of decision-making. The SDGs are the most current ones to be announced by the UN among them. They are part of a strategy to encourage all countries to "protect and repair our planet" and "move the globe onto a sustainable and resilient path" (United Nations, 2015). The SDGs call for the achievement of 17 objectives and 169 targets by the year 2030. In addition to placing a strong emphasis on achieving environmental sustainability, the SDGs have emphasized strengthening technology transfer, capacity-building in developing countries, and raising local public awareness to make it easier to achieve these targets (Dong and Hauschild, 2017).

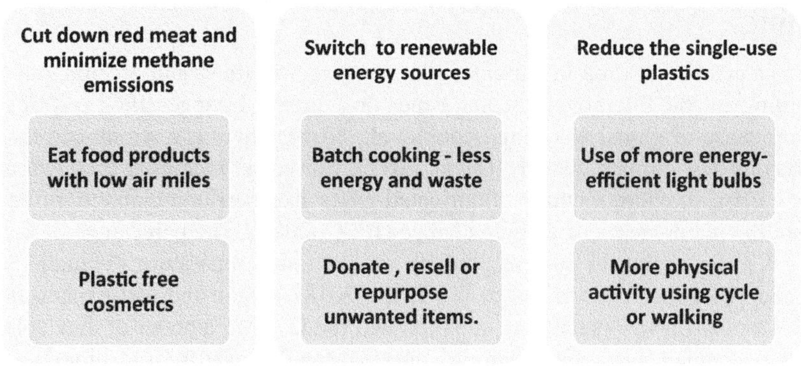

FIGURE 3.2 Sustainable lifestyle in our daily life.

Sustainable energy supply choices

Superior energy competence and an added justifiable supply of inexpensive, environmentally conscious energy that minimizes emissions that contribute to global warming could assist both urban and rural communities in improving the quality of life and combating climate change. Sustainable and inexpensive energy can compromise national safety for some nations (Farah, 2015). Global warming is largely caused by the production of electricity using fossil fuels. Although there has been improvement in energy efficacy and the switch to low-carbon energy sources, these developments are still insufficient to meet the 2015 Paris Agreement targets (UN Environment, 2019). Hence, organizations are trying to attempt to create a market for e-vehicles to overcome challenges with cost, accessibility, awareness, and infrastructure (Sanguesa et al., 2021).

Prevention of pollution

A key issue for a sustainable environment is environmental pollution along with its health impacts (Robert et al., 2005). The pollution prevention approach, on the other hand, aims to improve a process's efficiency to reduce the amount of pollution generated at its source (Sherman and McGain, 2016). Pollution-control strategies typically aim to manage a pollutant after it is emitted and lessen its impact on the environment. For effective, economically feasible production, service delivery, and resolving many environmental concerns, fundamental concepts of preventing pollution rather than fixing problems are crucial (Jorgenson and Wilcoxen 1990). An environmental management hierarchy that encompasses prevention, recycling, treatment, and disposal or discharge has prevention, as its top priority (Elleuch et al., 2018).

The impact of digital transformation on environmental sustainability

Internet of Things (IoT), environmental sustainability, and digital transformation artificial intelligence (AI), social media platforms, mobile technologies, and big data analytics are examples of digital technologies that have a positive impact on society and business (Vial, 2019). Digital technologies are being used more to enhance eco-friendly sustainability and leveraged by businesses to develop new platforms and products that will improve environmental sustainability. The novel application of AI to detect waterborne diseases that have infected waters was highlighted (Goralski and Tan, 2020) in the discussion of AI-based solutions such as Smart Water Management Systems and Plant Village. Nine cases from nine different countries were used in the study to address environmental sustainability concerns and enhance the environment. Big data analytics can be used to create a method that improves the traceability of the food chain and the certification of products based on their direct environmental performance (i.e., carbon footprint) or the procedures employed during their production processes (Weersink et al., 2018). In a similar vein, blockchain is viewed as a tool with immense potential for achieving sustainability in industrial and business activities (Leng et al., 2020). Blockchain has the potential to increase sustainability by extending product life cycles, maximizing resource use, and lowering carbon emissions (Esmaeilian et al., 2020). Digitalization and environmental sustainability are increasingly intertwined at all levels of organizations, industries, and nations (Feroz et al., 2021; Esmaeilian et al., 2020).

Climate change and global warming

Typically, the term "climate change" refers to changes over time in variables like wind, temperature, and precipitation. Since the middle of the 20th century, according to the Intergovernmental Panel on Climate Change (IPCC, 2007), anthropogenic emissions have been the primary cause of global warming (Cook et al., 2016). There is a strong correlation between the rate of carbon dioxide emissions and climate change (Davis, 2017). Currently, increased production and buildup of GHGs such as nitrous oxide, carbon dioxide, methane, fluorinated gases like perfluorocarbons, sulfur hexafluoride (SF_6), and hydrofluorocarbons are the main causes of climate change (IPCC, 2013). The percentage of the individual greenhouse gases is given in Fig. 3.3. According to current predictions, extreme temperature fluctuations over many parts of the world are likely to happen sooner than expected (Li et al., 2018). Ocean thermal expansion brought on by warming oceans due to global warming has worsened the situation and reduced the amount of available land (Cazenave and Llovel, 2010; Khan, 2019). Climate change mitigation can also greatly benefit from biological alternatives and solutions.

Garden cities

Future city plans have been proposed in a variety of forms, some more daring than others. Some are being constructed. All strive to be environmentally friendly and sustainable, utilizing the most recent innovations in building, recycling, renewable energy, and transportation. The building of new societies according to local requirements was emphasized in the British Government's announcement of a proposal for innovative garden cities in the United Kingdom. Although maybe evident, the aspirational "wish list" refers to Howard (Howard, 2003) and outlines the desired British garden city (Riffat et al., 2016):

- Strong visualization, guidance, and communal engagement.
- Long-term benefit governance and communal-owned land ownership.
- Land rate detention for the good of the public.
- Mixed-tenancy dwellings and housing options that are reasonable for the regular public.
- A thriving local labor market in Garden City itself, with a selection of work opportunities conveniently situated close to homes.
- Beautiful and imaginatively designed homes with gardens combine the best aspects of city and country living to create healthy residences in vibrant neighborhoods.
- A lot of green space that is connected to the natural surroundings, such as a countryside buffer zone to avoid sprawl, well-connected public parks with a variety of species, and a mix of private and public networks of well-managed, tree-lined streets, open spaces, and top-notch gardens.

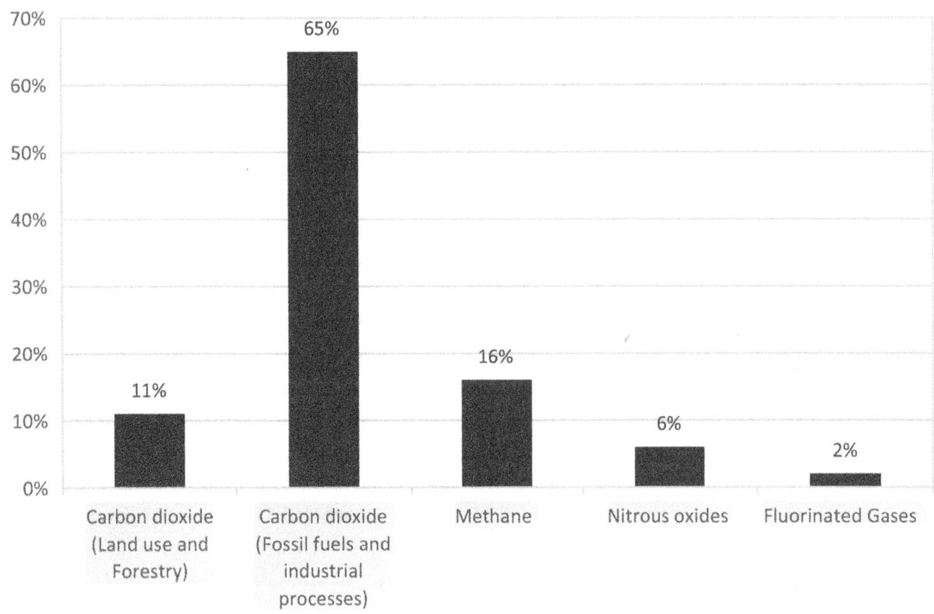

FIGURE 3.3 Percentage of the individual greenhouse gases (Arora et al., 2018).

- Numerous, quality local amenities for shopping, recreation, and culture in walkable neighborhoods.
- Comprehensive low-carbon employment prospects are provided through integrated and accessible rapid transit systems connecting several settlements.

These kinds of garden cities will mostly make use of current technology, a strategy that has been used elsewhere. In the center of Kowloon Bay, Hong Kong's developing and thriving primary commercial zone, is where you'll find the Zero Carbon Building. Its 14,700 m^2 total area, which includes a three-story Zero Carbon Building and a landscape area, not only displays cutting-edge eco-building design and technologies to the local and international construction industry but also raises public alertness of low-carbon living in Hong Kong. In addition to sustainability, the "visuality" and "functionality" of buildings should also be taken into account. The 15-story Jockey Club Innovation Tower (Hong Kong Polytechnic University) is an intriguing illustration (Zaha Hadid Architects, 2009).

Waste management analysis

In all low- and middle-income nations, municipal solid waste management is a glaringly underappreciated aspect of environmental management (Rana, 2007). The informal waste management industry may aid in resource recycling and create income and job opportunities. Poorly managed waste streams have negative impacts on the environment and may pose health risks, which will make things more difficult for low-income families in particular (Misra and Pandey, 2005; Sandhu et al., 2017). In a case study (Sahar and Ahmad, 2019) done in Faisalabad (in Pakistan) one of the top 10 polluted cities in the world the following conclusions were drawn from the perception of the general public and informal waste collectors. It is listed below:

- For both the present and the future, the community has to be more aware of environmental sustainability.
- If officials ensure that locals are involved, the amount of money the government invests can be lowered.
- The public should be made more aware of the improved management may help the environment, according to policymakers.
- A healthy environment can be created by making a small investment in the interests, abilities, education, and health of informal garbage collectors.

Water crisis management

Another area of possibility is the water. The availability of clean, usable water in proportion to the world's expanding population is a major issue. Numerous freshwater ecosystems are deteriorating. India and the surrounding countries' history and development have benefited greatly from the use of water (Amrith, 2018). With population expansion, freshwater availability per person is declining, which also mobilizes and intensifies environmental and human health problems associated with human activity (UN Environment, 2019) as water is becoming increasingly scarce in many places and most of what is available is contaminated. While it is particularly severe in urban areas like Bengaluru, some farmers are committing suicide as a result of it in rural areas (Auzanneau et al., 2018).

Anthropogenic activities and global warming

The average surface temperature of the earth has risen by more than 2°C for the past 100 years. During this time, sea level rise, ocean acidification, and global surface temperatures have increased in particular (King and Karoly, 2017). The combustion of fossil fuels is the primary cause of the atmosphere's rising greenhouse gases. Fossil fuels have been burned by mankind for more than 2000 years. The naturally occurring greenhouse gases in our atmosphere have been greatly augmented over time, causing the greenhouse effect and global warming to increase. However, warming is related to an increase in CO_2 emissions that can alter the climate, which is also related to the combustion of fossil fuels (Ekwurzel et al., 2017). However, the impact of man on the current cycle of climate change is far bigger. Because of this, our current inactivity on climate change and global warming may be originated by natural phenomena rather than human behavior (King and Karoly, 2017). The reality remains that the combustion of fossil fuels by man has resulted in significant CO_2 emissions, which is the primary reason of current global warming. It is now obvious that heat-trapping greenhouse gases, such as methane and CO_2 have been given out into the atmosphere by human beings (Baldocchi and Penuelas, 2019).

Conclusions

Environmental sustainability is the capacity to exist and advance without consuming current or future natural resources. In the framework of internet governance and digital policy, greater attention needs to be paid to advancing environmental education and raising awareness of environmental sustainability. Limiting emissions of greenhouse gases like carbon dioxide is typically emphasized in most "solutions" while addressing the effects of climate change. The strategies envisioned to combat climate change through restrictions on greenhouse gases are nearly without a doubt unsustainable because they permit significant costs, have little effect on the climate, and will undoubtedly increase poverty, making it harder for the poor to adapt to climate change. Care must be taken in materializing them to the real world, since a lack of understanding regarding the well-being of forthcoming generations hinders confident use of any of the proposed techniques, aside from simplifying assumptions. It had been intended to be reviewed for easy reference to discuss the environmental sustainability issues faced and to adapt to the tasks posed by climate change.

References

Abercrombie, M., Hickman, C.J., Johnson, M.L., 2017. A Dictionary of Biology. Routledge. Available from: https://doi.org/10.4324/9781315083636.

Adger, W.N., Huq, S., Brown, K., Conway, D., Hulme, M., 2003. Adaptation to climate change in the developing world. Progress in Development Studies 3 (3), 179–195. Available from: https://doi.org/10.1191/1464993403ps060oa.

Adger, W.N., Agrawala, S., Mirza, M.M.Q., Conde, C., O'Brien, K., Pulhin, J., et al., 2007. Assessment of adaptation practices, options, constraints and capacity. In: Parry, M.L., Canziani, O.F., Palutikof, J.P., van der Linden, P.J., Hanson, C.E. (Eds.), Climate Change 2007: Impacts, Adaptation and Vulnerability. Contribution of Working Group II to the Fourth Assessment Report of the Intergovernmental Panel on Climate Change. Cambridge University Press, Cambridge, pp. 717–774. Available at: https://www.ipcc.ch/site/assets/uploads/2018/02/ar4-wg2-chapter17-1.pdf.

Amrith, S., 2018. Unruly Waters: How Rains, Rivers, Coasts, and Seas Have Shaped Asia's History. New York: Basic Books. Available at: http://www.tinyurl.com/y2yojteg (Accessed 30 October 2023).

Arora, N.K., Fatima, T., Mishra, I., Verma, M., Mishra, J., Mishra, V., 2018. Environmental sustainability: challenges and viable solutions. Environmental Sustainability 1 (4), 309–340.

Auzanneau, M., Reynolds, J.F., Heinberg, R., 2018. Oil, Power, and War: A Dark History. Chelsea Green Publishing, White River Junction, Vermont [online] Greenwich Library. Available at: https://greenwichlibrary.bibliocommons.com/v2/record/S86C1250887.

Baldocchi, D., Penuelas, J., 2019. The physics and ecology of mining carbon dioxide from the atmosphere by ecosystems. Global Change Biology 25 (4), 1191–1197.

Berkhout, F., Hertin, J., Gann, D.M., 2006. Learning to adapt: organisational adaptation to climate change impacts. Climatic Change 78 (1), 135–156. Available from: https://doi.org/10.1007/s10584-006-9089-3, https://link.springer.com/article/10.1007/s10584-006-9089-3.

Biesbroek, G.R., Swart, R.J., Carter, T.R., Cowan, C., Henrichs, T., Mela, H., et al., 2010. Europe adapts to climate change: comparing national adaptation strategies. Global Environmental Change 20 (3), 440–450 [online].

Bowen, K.J., Friel, S., 2012. Climate change adaptation: where does global health fit in the agenda? Globalization and Health 8 (1), 10. Available from: https://doi.org/10.1186/1744-8603-8-10, https://globalizationandhealth.biomedcentral.com/articles/10.1186/1744-8603-8-10#citeas.

Caldeira, K., Chopin, T., Gaines, S., Haugan, P., Hemer, M., Howard, J., et al., 2019. The Ocean as a Solution to Climate Change Five Opportunities for Action Ove Hoegh-Guldberg EXPERT AUTHORS. Available at: https://khrono.no/files/2019/09/25/19_HLP_Report_Ocean_Solution_Climate_Change_final.pdf (Accessed 30 October 2023).

Cazenave, A., Llovel, W., 2010. Contemporary sea level rise. Annual Review of Marine Science 2 (1), 145–173. Available from: https://doi.org/10.1146/annurev-marine-120308-081105.

Cook, J., Oreskes, N., Doran, P.T., Anderegg, W.R.L., Verheggen, B., Maibach, E.W., et al., 2016. Consensus on consensus: a synthesis of consensus estimates on human-caused global warming. Environmental Research Letters 11 (4), 048002. Available from: https://doi.org/10.1088/1748-9326/11/4/048002 [online].

Creighton, J., 2010. How to conduct a shared vision planning process. In: IWR Report 10-R-6. US Army Institute for Water Resources, Alexandria, VA. Available at: https://www.iwr.usace.army.mil/Portals/70/docs/iwrreports/10-R-6.pdf.

Davis, W., 2017. The relationship between atmospheric carbon dioxide concentration and global temperature for the last 425 million years. Climate 5 (4), 76. Available from: https://doi.org/10.3390/cli5040076.

Dolšak, N., Prakash, A., 2018. The politics of climate change adaptation. Annual Review of Environment and Resources 43 (1), 317–341. Available from: https://doi.org/10.1146/annurev-environ-102017-025739.

Dong, Y., Hauschild, M.Z., 2017. Indicators for Environmental Sustainability. Procedia CIRP 61, 697–702 [online].

Dreyfus, M., Patt, A., 2012. The European Commission White Paper on adaptation: appraising its strategic success as an instrument of soft law. Mitigation and Adaptation Strategies for Global Change 17 (8), 849–863 [online].

Ekwurzel, B., Boneham, J., Dalton, M.W., Heede, R., Mera, R.J., Allen, M.R., et al., 2017. The rise in global atmospheric CO2, surface temperature, and sea level from emissions traced to major carbon producers. Climatic Change 144 (4), 579–590.

Elleuch, B., Bouhamed, F., Elloussaief, M., Jaghbir, M., 2018. Environmental sustainability and pollution prevention. Environmental Science and Pollution Research 25 (19), 18223–18225.

Esmaeilian, B., Sarkis, J., Lewis, K., Behdad, S., 2020. Blockchain for the future of sustainable supply chain management in Industry 4.0. Resources, Conservation and Recycling 163, 105064.

Farah, P.D., 2015. Special issue on 'sustainable energy investments and national security: arbitration and negotiation issues' published by the SSCI - Journal of World Energy Law and Business (JWELB), Oxford University Press, Volume 8, Issue 6, December 2015. Journal of World Energy Law and Business (JWELB). Available at: https://www.academia.edu/19391577/Special_Issue_on_Sustainable_Energy_Investments_and_National_Security_Arbitration_and_Negotiation_Issues_published_by_the_SSCI_Journal_of_World_Energy_Law_and_Business_JWELB_Oxford_University_Press_Volume_8_Issue_6_December_2015.

Feroz, A.K., Zo, H., Chiravuri, A., 2021. Digital transformation and environmental sustainability: a review and research agenda. Sustainability 13 (3), 1530.

Goodland, R., 1995. The concept of environmental sustainability. Annual Review of Ecology and Systematics 26 (1), 1–24.

Goralski, M.A., Tan, T.K., 2020. Artificial intelligence and sustainable development. The International Journal of Management Education 18 (1), 100330 [online].

Haasnoot, M., Kwakkel, J.H., Walker, W.E., ter Maat, J., 2013. Dynamic adaptive policy pathways: A method for crafting robust decisions for a deeply uncertain world. Global Environmental Change 23 (2), 485–498.

Haasnoot, M., van 't Klooster, S., van Alphen, J., 2018. Designing a monitoring system to detect signals to adapt to uncertain climate change. Global Environmental Change 52, 273–285.

Howard, E., 2003. Garden Cities Of To-Morrow. Organization & Environment 16 (1), 98–107. Available from: https://doi.org/10.1177/1086026602250259.

IPCC, 2007. In: Parry, M.L., Canziani, O.F., Palutikof, J.P., van der Linden P.J., Hanson, C.E. (Eds.), Climate Change 2007: Impacts, Adaptation and Vulnerability. Contribution of Working Group II to the Fourth Assessment Report of the Intergovernmental Panel on Climate Change. Cambridge University Press, Cambridge, 976 pp. Available at: https://www.ipcc.ch/site/assets/uploads/2018/03/ar4_wg2_full_report.pdf.

IPCC, 2013. In: Stocker, T.F., Qin, D., Plattner, G.-K., Tignor, M., Allen, S.K., Boschung, J., Nauels, A., Xia, Y., Bex, V., Midgley, P.M. (Eds.), Climate Change 2013: The Physical Science Basis. Contribution of Working Group I to the Fifth Assessment Report of the Intergovernmental Panel on Climate Change. Cambridge University Press, Cambridge, New York, NY, 1535 pp. Available at: https://www.ipcc.ch/report/ar5/wg1/.

Janssen, M.A., 2006. Historical institutional analysis of social-ecological systems. Journal of Institutional Economics 2 (2), 127–131 [online].

Jorgenson, D.W., Wilcoxen, P.J., 1990. Environmental Regulation and U.S. Economic Growth. The Rand Journal of Economics 21 (2), 314.

Khan, A.A., 2019. Why would sea-level rise for global warming and polar ice-melt? Geoscience Frontiers 10 (2), 481–494.

King, A.D., Karoly, D.J., 2017. Climate extremes in Europe at 1.5 and 2 degrees of global warming. Environmental Research Letters 12 (11), 114031.

Kumar, B.M., Aravindakshan, S., 2022. Carbon footprints of the Indian AFOLU (Agriculture, Forestry, and Other Land Use) sector: a review. Carbon Footprints 1, 7. Available from: https://doi.org/10.20517/cf.2022.04.

Leng, J., Ruan, G., Jiang, P., Xu, K., Liu, Q., Zhou, X., et al., 2020. Blockchain-empowered sustainable manufacturing and product lifecycle management in industry 4.0: a survey. Renewable and Sustainable Energy Reviews 132 (C). Available at: https://www.researchgate.net/publication/343276482_Blockchain-empowered_sustainable_manufacturing_and_product_lifecycle_management_in_industry_40_A_survey.

Levina, E., Tirpak, D., 2006. Adaptation to Climate Change: Key Terms. Available at: https://www.oecd.org/environment/cc/36736773.pdf.

Li, C., Fang, Y., Caldeira, K., Zhang, X., Diffenbaugh, N.S., Michalak, A.M., 2018. Widespread persistent changes to temperature extremes occurred earlier than predicted. Scientific Reports 8 (1). Available from: https://doi.org/10.1038/s41598-018-19288-z, https://www.nature.com/articles/s41598-018-19288-z.

Lwasa, S., 2014. A systematic review of research on climate change adaptation policy and practice in Africa and South Asia deltas. Regional Environmental Change 15 (5), 815–824 [online].

Mertz, O., Halsnæs, K., Olesen, J.E., Rasmussen, K., 2009. Adaptation to climate change in developing countries. Environmental Management 43 (5), 743–752 [online].

Misra, V., Pandey, S.D., 2005. Hazardous waste, impact on health and environment for development of better waste management strategies in future in India. Environment International 31 (3), 417–431 [online].

Mollenkamp, D., 2022. What Is Sustainability? [online]. Investopedia. Available at: https://www.investopedia.com/terms/s/sustainability.asp.

Morelli, J., 2011. Environmental sustainability: a definition for environmental professionals. Journal of Environmental Sustainability 1 (1), . Available from: https://doi.org/10.14448/jes.01.0002Article 2, https://scholarworks.rit.edu/jes/vol1/iss1/2.

Nowodziński, P., 2021. Sustainability Challenges. [online]. MIT Enterprise Forum CEE. Available at: https://mitefcee.org/sustainability-challenges/.

Pielke, R., Prins, G., Rayner, S., Sarewitz, D., 2007. Lifting the taboo on adaptation. Nature 445 (7128), 597–598 [online].

Rana, T.M., 2007. Illuminating the silence: understanding the structure and function of small RNAs. Nature Reviews. Molecular Cell Biology 8 (1), 23–36.

Riffat, S., Powell, R., Aydin, D., 2016. Future cities and environmental sustainability. Fut Cit & Env 2, 1.

Robert, K.W., Parris, T.M., Leiserowitz, A.A., 2005. What is sustainable development? Goals, indicators, values, and practice. Environment: Science and Policy for Sustainable Development 47 (3), 8–21 [online].

Robinson, J., 2004. Squaring the circle? Some thoughts on the idea of sustainable development. Ecological Economics 48 (4), 369–384.

Sahar, Ahmad, I., 2019. Waste management analysis from economic-environment sustainability perspective. International Journal of Scientific & Technology Research 8 (12), 1540–1543 [online].

Sandhu, K., Burton, P., Dedekorkut-Howes, A., 2017. Between hype and veracity; privatization of municipal solid waste management and its impacts on the informal waste sector. Waste Management 59, 545–556.

Sanguesa, J.A., Torres-Sanz, V., Garrido, P., Martinez, F.J., Marquez-Barja, J.M., 2021. A review on electric vehicles: technologies and challenges. Smart Cities 4 (1), 372–404 [online].

Santos, F.D., Ferreira, P.L., Pedersen, J.S.T., 2022. The climate change challenge: a review of the barriers and solutions to deliver a Paris solution. Climate 10 (5), 75.

Sherman, J., McGain, F., 2016. Environmental sustainability in anesthesia. Advances in Anesthesia 34 (1), 47–61.

Smit, B., Wandel, J., 2006. Adaptation, adaptive capacity and vulnerability. Global Environmental Change 16 (3), 282–292. Available from: https://doi.org/10.1016/j.gloenvcha.2006.03.008 [online].

Sims, R., Schaeffer, R., Creutzig, F., Cruz-Núñez, X., D'Agosto, M., Brazil, A., et al., 2014. 8 Transport Coordinating Lead Authors: Review Editors: Chapter Science Assistant. Available at: https://www.ipcc.ch/site/assets/uploads/2018/02/ipcc_wg3_ar5_chapter8.pdf.

Smit, B., Pilifosova, O., Burton, I., Challenger, B., Huq, S., Klein, R.J.T., et al., 2001. Adaptation to climate change in the context of sustainable development and equity. Available at: https://www.semanticscholar.org/paper/Adaptation-to-climate-change-in-the-context-of-and-Smit-Pilifosova/63b5c9371ac8dc511e109a02947ffdff6438c179.

Suarez, I., 2020. 5 Strategies that Achieve Climate Mitigation and Adaptation Simultaneously. Available at: https://www.wri.org/insights/5-strategies-achieve-climate-mitigation-and-adaptation-simultaneously.

Tung, C.-P., Tsao, J.-H., Tien, Y.-C., Lin, C.-Y., Jhong, B.-C., 2019. Development of a novel climate adaptation algorithm for climate risk assessment. Water 11 (3), 497. Available from: https://doi.org/10.3390/w11030497.

UN Environment, 2019. Global Environment Outlook – GEO-6: Healthy Planet, Healthy People. Cambridge University Press.

United Nations, 2015. Transforming Our World: The 2030 Agenda for Sustainable Development. [online]. Un.org. Available at: https://sustainabledevelopment.un.org/post2015/transformingourworld/publication.

Veit, P., Reytar, K., 2017. By the Numbers: Indigenous and Community Land Rights. Available at: https://www.wri.org/insights/numbers-indigenous-and-community-land-rights.

Vial, G., 2019. Understanding digital transformation: a review and a research agenda. The Journal of Strategic Information Systems 28 (2), 118–144 [online].

Vij, S., Moors, E., Ahmad, B., Arfanuzzaman, Md, Bhadwal, S., Biesbroek, R., et al., 2017. Climate adaptation approaches and key policy characteristics: cases from South Asia. Environmental Science & Policy 78, 58–65.

Weersink, A., Fraser, E., Pannell, D., Duncan, E., Rotz, S., 2018. Opportunities and challenges for big data in agricultural and environmental analysis. Annual Review of Resource Economics 10 (1), 19–37. Available from: https://doi.org/10.1146/annurev-resource-100516-053654.

Zaha Hadid Architects, 2009. Jockey Club Innovation Tower. Zaha Hadid Architects. [online]. Available at: https://www.zaha-hadid.com/architecture/jockey-club-innovation-tower/.

Chapter 4

The role of green chemistry and nanotechnology in developing environmental sustainability

Ram Babu Sharma[1], Amardeep Kaur[1], Ujjval Nautiyal[1], Atul Kabra[2] and Vimal Arora[2]
[1]Himalayan Institute of Pharmacy, Kala Amb, Himachal Pradesh, India, [2]University Institute of Pharma Sciences, Chandigarh University, Mohali, Punjab, India

Introduction

According to the United Nations (UN) World Commission on Environment and Development, environmental sustainability means acting in a way that ensures the conservation of natural resources for future generations for their better existence. Whereas, the UN's definition is fairly common and has been broadened throughout time to include non-economic elements, such as education and health, clean air and water, and the preservation of natural beauty.

- Alternate definition: The meaning of environmental sustainability is to increase the capabilities of human beings in the context of improving the ecosystem through their constant and sustainable efforts.
- Alternate definition: It can also be defined as environmental sustainability is precisely maintaining a delicate balance between the natural environment and human refinement (International Union for Conservation of Nature, 2020).

Powl Howken, a well-known environmentalist, offers another term that reflects his worry that people use natural resources far faster than they can be regenerated (Rayne, 2020). Carbon emissions in the environment are the major cause of pollution and because of that, carbon dioxide is released into the atmosphere, which is the main cause of global warming. Global warming indicates that the earth's temperature is rising daily, causing glaciers to melt and sea levels to rise, posing a significant challenge to the survival of coastal cities. All countries have their constraints in terms of social and economic fabrication. Therefore, policies cannot be implemented uniformly. To fight against environmental issues, each country has to form its policy as per its feasibility and implement it properly, like, in the United States, the EPA is the main policy-forming organization that deals with controlling causes that pose an environmental threat (EPA, 2020).

Environmental sustainability issues

In today's era, increasing industrialization, urbanization, and the use of automobiles create enormous problems worldwide. Every individual must promise to protect the natural heritage of the earth to address the problems facing the planet. Technology, innovation, and teamwork should be used by conscious civilizations to change the current developing trend toward ecologically sustainable growth. Sustainability is concerned not just with environmental challenges but includes social science, environmental science, and future technologies. Sustainability defends our natural environment and human and ecological health and promotes innovation without jeopardizing our way of life (Ajmal et al., 2018). Some challenges of sustainability are discussed in the following subsections.

Climate change

This is the most crucial area of interest for all scientists, which is that drastic changes are coming into the environment for several reasons, and one of the most inevitable reasons is "global warming," which is the emission of carbon dioxide into

Role of Green Chemistry in Ecosystem Restoration to Achieve Environmental Sustainability. DOI: https://doi.org/10.1016/B978-0-443-15291-7.00009-2

the environment. According to the UN, CO_2 emissions have risen about 50% since 1990, which is a deep threat to the environment and puts the lives of millions of people in trouble. Therefore we must minimize the global warming effect by reducing the emission of gases that are the main causes of global warming, mainly carbon dioxide (Ajmal et al., 2018).

Natural resource use

Optimum use of natural resources and their management is an art, but most countries have failed to use them appropriately. Today's era is based upon economics and those countries that are abundant with natural resources are recognized as rich and influential countries across the world.

Waste production

Recycling waste is also an essential aspect of environmental management. Plastic waste management is the most difficult task for everyone. Electronic devices, particularly mobile phones, are another major source of concern, and their widespread usage is creating a severe hazard. Another concern is food waste management, which is not well practiced in wealthy nations but causes large losses early in the development cycle. It has been suggested to stop the misuse of natural resources and try to adapt the reducing, reusing, and recycling of waste materials up to a maximum extent (Lang and Murphy, 2014).

Water pollution

As per the data released, water pollution is a big problem in today's times across the world. It is estimated that >80% of polluted water will be discharged into the environment without proper treatment. The untreated discharge water further deteriorates the groundwater while its seepage into the groundwater results in degrading the quality of the soil surface, turning it into a wilderness.

Deforestation

According to NASA, the global forest covers up to 33% of the total land area. It explains the vital role of upholding the ecosystem of the earth. It downplays the role of carbon dioxide in the ecosystem, which is the main cause of global warming on earth. It also helps with erosion, fluid control in the rainy season, promoting biodiversity, etc. Uncontrolled deforestation, blazing, and cutting down of trees in the name of development, is the main cause of concern. The resultant forest area is shrinking day by day.

Overfishing

Overfishing also leads to disbalance in the environment and has long-term consequences. A human chain system may get disturbed and turbulence in the environment may be generated, which causes a food chain system to completely collapse.

Ocean acidification

The oceans have turned into massive trash dumps. The most recent research has discovered that the ecosystem's CO_2 output degrades ocean water quality and causes it to become acidic (pH change). This changes the fundamental characteristics of seawater, endangering the lives of marine species. Additionally, it has been noted that seawater absorbs more than 30% of the CO_2 created in the atmosphere. As a result, it contributes significantly to global warming and aids in keeping the earth's temperature stable (Lang and Murphy, 2014).

Air pollution

According to data published by the World Health Organization, 90% of the world's population breathes polluted air, resulting in seven million deaths each year. The main reason for deaths is that breathing of fine particles penetrates deep areas of the lungs, and heart due to severe illnesses like asthma, respiratory infection, lung cancer, cardiovascular disease, obstructive pulmonary disease, etc. As per the recommendation of the UN, the use of natural gas, green energy, tidal energy, wind energy, and solar energy, instead of petroleum products as fuel in automobiles, controls air pollution to an extent (Rogers, 2019).

Environmental sustainability in pharmaceutical manufacturing

The governments of every nation place the utmost focus on pharmaceutical product quality since it affects billions of lives. On the other hand, the control and sustainability measures are relatively little from an environmental standpoint. Product and process quality and safety are ensured by cGMP rules; however, pollution is not taken into account . In 2020 the CPhI urged pharmaceutical companies to put environmental sustainability first . To ensure the eco-ongoing system's viability, pharmaceutical corporations use cutting-edge technologies. Lately, the energy consumption intensity (EUI) of a pharmaceutical facility is more than 15 times that of an ordinary office building (Caparella, 2013). Recently, various new sources of energy have developed, mainly solar energy and natural gas like CNG, which leave behind no residue, unlike coal, which emits various pollutants like carbon dioxide, carbon monoxide, etc. into the environment. The use of renewable energy is not only eco-friendly but also helps save businesses money. However, waste hierarchy techniques may be used to appropriately manage toxic waste, therefore manufacturers should commit to building a process that reduces the waste output created by it. According to a survey conducted in 2021, environmental sustainability is an area of concern for environmental, social, and governance criteria which are required to be handled on a priority basis. It is also a very crucial part of the pharmaceutical industry to improve sustainability but also to create new strategies that can pave the way for achieving the long-term goal in a very pragmatic manner.

UN sustainability goals

In September 2015 the UN General Assembly approved the 2030 Agenda for Sustainable Development, which consists of 17 sustainable development goals (SDGs). The new agenda emphasizes a comprehensive strategy based on the notion of "leaving no one behind" to achieve sustainable development for all. The SDGs address disability in several areas, such as human settlement accessibility, inequality, growth & employment, education, and data gathering and monitoring .

Green chemistry for environment sustainability

Description of the green chemistry was mainly the utilization of skills in chemistry along with the knowledge to diminish the usage or manufacture of harmful moieties when planning chemicals, manufacturing, and its employment so that it reduces hazards to the operator and environmental health (Anastas, 1999). Paul Anastas and John Warner published the very first Green Chemistry textbook in 1998, which included 12 guiding principles for the field (Anastas and Warner, 1998a,b). Numerous marine and terrestrial species of animals suffered loss, and over 1500 ha of plants were devastated (). Therefore it's possible to recognize the amplitude of climate, economic, and infrastructural damage that such a disaster could cause (). This aspect, termed the "E factor" (efficiency factor) by Roger Sheldon, was invented (Sheldon, 1997). In this sense, "waste" refers to everything that was created during the process, excluding the ultimate result. As a result, the E factor must be the lowest in all processes. It is clear to see that the lower the E factor, the higher the segment's production. By using value, it is entirely feasible to determine that, even as big manufacturing processes are thought to be more harmful to the climate than smaller businesses, they aren't (Lenardao et al., 2003; Sheldon, 1997; Sheldon, 1996; Poliakoff and Licence, 2007). The pharma industry's bad accomplishment in this area, which is also evident within chemical companies, is because of the aspect that its industrial processes were designed to use conventional stoichiometric operations, which generate massive amounts of inorganic compounds as squandering (Lenardao et al., 2003). There seem to be specific references to the carcinogenicity of solvents that are implemented by educational universities and companies. In a study released in 2014, Prat and collaborators conducted a fascinating contrast of several manuals and elucidated a chart with the collation, wherein they categorized the organic solvent into six groups (Fig. 4.1; Prat et al., 2014). Most solvents are very volatile and can quickly pollute the environment. Furthermore, many of them are combustible and poisonous. As a consequence, whenever possible, different solvents must be replaced with water. Nevertheless, since many organic compounds are immiscible in this liquid, this may not always be feasible (Poliakoff and Licence, 2007).

Process

Green chemistry is concerned with synthesis techniques that have no or little negative impact (Marques and Machado, 2014) on both human health and the environment, following its 12 fundamental principles. These would be significant and precise ideas; however, they do not recognize all aspects of chemical methods, especially upgrading (industrial chemical

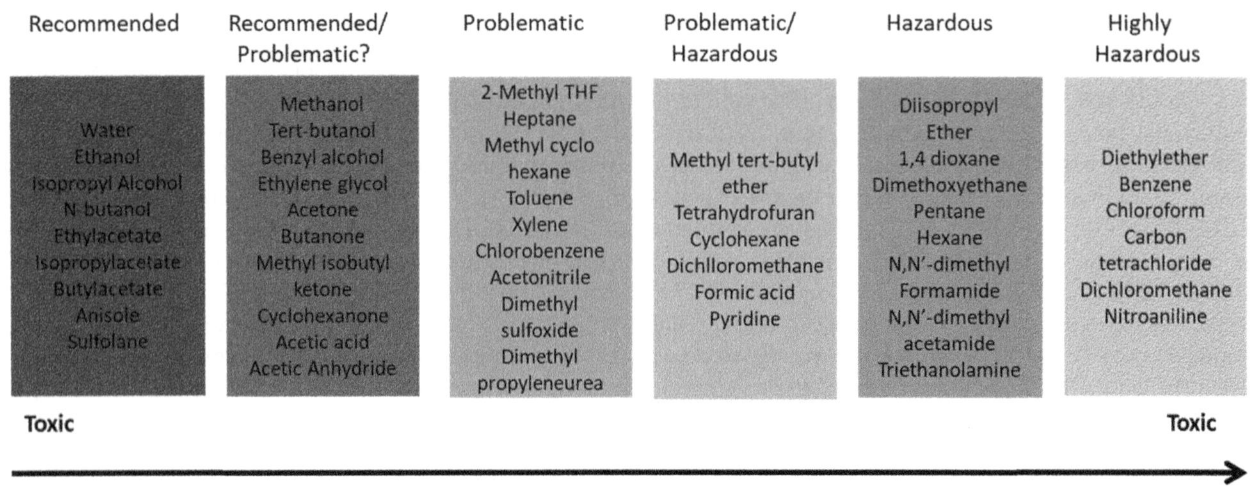

FIGURE 4.1 Classification of solvents in green chemistry.

processes) and cost analysis. These methodologies can assist in minimizing the negative impact on the climate, but they cannot effectively stop dangerous chemicals from interacting with the climate (Parveen et al., 2016). If an organic response permits product expulsion via distillation or excavation, the catalyst might very well remain in such an eluent; in this case, both (the catalyst and liquid) could be reprocessed and reused. Organic responses could also be conducted without the need for organic solvent in a supercritical medium of CO_2 (another option is microwave synthesis); all these methods are regarded as environment acceptable. In several processes, toluene can be substituted for the more hazardous benzene; recyclable solvents are yet another sustainable green alternative (Sheldon and Isabel Arends, 2007).

Chemical and physical methods

Different physical methods being employed in the green chemistry approach, such as magnetic field–assisted synthesis, are uncommon approaches (Do and Friscic, 2017; Dąbrowska et al., 2018; Darr et al., 2017; Ye et al., 2018; Babu et al., 2015; Krajewski, 2017) outline the main recent reviews in these domains.

Ball milling

Ball milling (Rivas, 2018) is a solid-state reaction (complexation) production that is extensively used in chemical synthesis, in addition to therapeutic potential in inorganic solid-state mechanisms (Margetic and Strukil, 2016). Like an illustration of instrumentation, a Retsch planetary-type high-energy ball mill is described. While using milling, we could indeed increase energy efficiency while trying to dodge hazardous toxins and solvents. Ball milling is regarded as a sustainable resource for chemical reactions because it is simple and environmentally friendly, however, it is not widely used by research scientists despite its tremendous ability.

Microwave irradiation

Based on conduction and dipolar polarization, microwave synthesis is one of the major approaches in the green synthesis of compounds. Several organic compounds have been described as having been generated in this cost-effective, efficient, rapid, and clean manner, which is now acknowledged as a standard synthetic chemistry instrument that has made significant contributions to organic synthesis (Ravichandran and Karthikeyan, 2018; Kitchen et al., 2014). To undertake MW synthesis successfully, every one of the reactants could perhaps preferably be an excellent absorber; instead, an even more material acting as a heating element (susceptor) and transforming MW power to heat is considered necessary.

Photocatalysis

Green chemistry reactions are photocatalytic reactions in the presence of ultraviolet light that are premised on stimulated emission, which changes the physicochemical characteristics of reactants in chemical synthesis. Glasing et al. (2016)

explained superoxide anion generation but also its participation in photooxygenation (the inclusion of oxygen molecules into atoms). The amalgamation of photochemical reactions with catalytic reactions, and the utilization of consistent flows or upgradeable to enhance them (Ardila Arias et al., 2016).

Hydro(solvo)thermal synthesis

In this hydro(solvo)thermal methodology, an autoclave holds solvent (water) and dissolved/ suspended reactants. Pressure, heat, diluents, response time, and reactant ratio all may be changed to modify the morphological characteristics and crystalline nature of the products made. This method results in little chemical loss and often better production yields, making it particularly useful for acquiring classic and much fewer nanomaterials with required shape and size control: granules, films, and, in general, and especially, 1- to 3D-microstructures (Abbas et al., 2019; Xu et al., 2018).

Ultrasound-assisted (sonochemical) synthesis

Ultrasonic (sonochemical) production is a solvent-based methodology (even though clogging only takes place in solvents) that appears to work in a solvent phase as a consequence of the hypothesis of cavitation bubbles (bubbles emerging in a solvent grow as well as implosively crumble, causing extreme local stress and pressure of 1000 atm as well as temperature levels of 5000K) without the need for a real impact upon that bond vibrational energy.

Magnetic field–assisted synthesis

Magnetic field–assisted production has been researched as a possible replacement for existing strategies. Some older methodologies necessitate the utilization of toxic solvents and otherwise supplementary phases that consume more energy and create undesirable by-products. These unneeded processes can be avoided by using magnetic properties even during the synthesis of the desired substance. The utilization of electromagnetic fields in synthesis allows for the development of morphological features that distinguish from all those produced by traditional methods. This seems to be possible because of magnetism, which causes the formation of metallic crystals to evolve in a specific orientation during the manufacture of nanorods (Wang et al., 2007). These processes can help with the manufacturing of adjustable electrodes for devices of the future. Magnetic flux was also demonstrated to be aided by a variety of compounds, particularly nickel, cobalt, arsenic, and several of its alloys (Li and Liao, 2009; Wu et al., 2010; Hu and Sugawara, 2009). On the contrary, cobalt and carbide of cobalt made using conventional methods without magnetism frequently display spherical morphology (Farghaly et al., 2012). Nanorods could also be made using the solvothermal method of bismuth production in magnetism (Xu et al., 2008).

Solvents and catalysis in green processes

The 12 regulations of green chemical synthesis can be satisfactorily used throughout organic synthesis, which frequently uses toxic solvents (Anastas and Warner, 1998a,b; Makone and Niwadange, 2016): maximum particle economy (attempting to avoid by-products and waste material, in general, and especially by solvent-less methods, i.e., watery media); non-hazardous and safe compound paths without toxic substances; utilization renewable starting materials (i.e., trees and shrubs rather than fossil fuels); catalyst supports in tiny amounts. To function as green materials for chemical synthesis, solutes would have to be recyclable, minimally toxic, have a significant boiling point, become reusable, as well as be inconsistent with water. These conditions are best served by water, polyethylene glycols, ILs, and different SC liquids. Accessibility, renewability, relatively inexpensive, negligible vapor pressure, as well as a strong boiling point, are just a few of its advantages. There seem to be numerous glycerol-based synthesis schemes provided in the research (Díaz-Álvarez et al., 2014).

Green supercritical solvents

Inflammable, harmless, and environmentally friendly supercritical CO_2 has historically been acknowledged as a workable substitute fluid for synthetic chemistry amongst many other ecological liquids (Wood et al., 2004). Additionally, supercritical water has been employed as an ecological solvent (Adschiri et al., 2011; Verma et al., 2014). Nowadays, techniques rely on supercritical fluids, specifically in the case of water or carbon dioxide, assist to in chemical synthesis throughout the formation of nanomaterials (for instance, in the production of semiconductors utilizing SC CO_2).

These techniques help to address some ecological consequences (Lonkar et al., 2019). To create a homogeneous phase, supercritical water can be mixed with both inorganic and organic compounds at temperatures and pressures higher than 22.1 MPa and 646K.

Biological methods

Considering their sluggish metal-lowering speed, biological techniques might be a more environmentally friendly synthetic pathway than traditional physical and chemical procedures. Although microorganisms might be employed for commercial silver retrieval, their investigations are still very limited, especially those outlining crucial aspects of biosynthesis, hence this technique is currently being investigated in a batch mode (Verma et al., 2014).

Uses of green chemistry in environmental sciences

Chemistry would have been the catalyst for revolution up until the middle of the 20th century. Throughout this time, new medications and antibiotics were found. The availability of food across the globe has also increased significantly as a consequence of the invention of hybrid varieties, better agricultural techniques, improved seeds, and the use of pesticides, insecticides, and nutrients. Soon later, chemistry's development started to show its harmful effects (Rayne, 2020). Lengthy gases including CO_2, SO_2, and oxides of nitrogen interact in seawater, causing acidification. This acidity has lately been recognized as a threat to marine life, especially marine calcifies. It was around this point that green chemistry officially began. Chemistry that does not negatively impact the environment is referred to as "chemical synthesis." Among the areas receiving the most attention right now is green science. Green computing research's main objective is to minimize the production of harmful residues while enhancing the targeted products in an ecologically responsible way. The three major advances in green chemical synthesis are the utilization of supercritical CO_2, freshwater as green organic solvents, aquatic hydrogen peroxide as just an oxidizer, and the utilization of hydrogen for asymmetric synthesis. To reduce carbon impact, microwave irradiation is slowly replacing conventional heating methods. Additionally, it saves a considerable amount of time. These days, consumers favor food with lower carbon footprints (Lang and Murphy, 2014). Green chemistry aims to reduce the negative consequences on the petrochemical just on the environment and human well-being. The quantity of major heavy substances and other dangerous chemicals released into the atmosphere will be reduced as a result of the use of superior fuel and modified green procedures. Chemical operations without solvents or the use of water in place of polar solvents are two ways to reduce the number of volatile compounds (VOC) released into the air. Energy output and reaction duration have both decreased as a result of the usage of radiation in chemical reactions. The green synthesis tenets of minimize, recover, and recycle will lead to a decrease in both marine waste and ecosystem contamination. The most challenging problem is to integrate green chemical synthesis into lab, commercial, and daily activities to prevent environmental deterioration and, consequently, ocean degradation, at the origin (EPA, 2020). There has been a lot of good work accomplished, but there remains more to be completed. A future era of chemists can be educated and trained to achieve this. All grades of student curricula should contain biodegradable materials so that each person is aware of the importance of leading an ecologically responsible lifestyle.

Conclusion

The green chemistry approach has provided various way outs leading to no or insignificant harmful effects as compared to that of the processes following conventional approaches like microwave-assisted synthesis, biocatalysis, microwave-assisted synthesis, Ultrasonic wave-assisted synthesis, etc., with certain benefits like utilization of limited resources, cost-effectiveness, preventing pollution and above all economic or financial benefits. A discussion has also been made in context to the green approaches utilized these days in the pharmaceutical and chemical industry, majorly addressing the type and extent of use of different solvents in these industries. Therefore, we may conclude that the green chemistry approach is the need of the hour; looking into the environmental hazards offered by the conventional approaches or processes and the parallel involvement of the nanotechnological intervention are capable of giving synergistic effects and thus can help in saving the environment for the upcoming generations.

References

Abbas, W., et al., 2019. Facilely green synthesis of 3D nano-pyramids Cu/carbon hybrid sensor electrode materials for simultaneous monitoring of phenolic compounds. SensActuat B. 282, 617–625.

Adschiri, T., Lee, Y.W., Goto, M., Takami, S., 2011. Green materials synthesis with supercritical water. Green Chemistry: An International Journal and Green Chemistry Resource: GC 13 (6), 1380–1390.

Ajmal, M., Khan, M., Hussain, M., Helo, P., 2018. Conceptualizing and incorporating social sustainability in the business world. International Journal of Sustainable Development & World Ecology 25, 327–339.

Anastas, P.T., 1999. Green chemistry and the role of analytical methodology development. Critical Reviews in Analytical Chemistry 29 (3), 167–175.

Anastas, P.T., Warner, J.C., 1998a. Green Chemistry: Theory and Practice. *Oxford University Press, Oxford.*

Anastas, P.T., Warner, J.C., 1998b. Principles of green chemistry. Green Chemistry: Theory and Practice. *Oxford University Press, Oxford*, pp. 29–56.

Ardila Arias, A.N., et al., 2016. Mineralización de etilenglicolporfoto-fentonasistido con ferrioxalato. Revista Internacional de Contaminación Ambiental 32 (2), 213–226.

Babu, S.G., Neppolian, B., Kumar, M., 2015. Ultrasound-Assisted Synthesis of Nanoparticles for Energy and Environmental Applications. *Springer Science + Business Media, Singapore*, pp. 1–34.

Caparella, J., 2013. Energy benchmarking in the pharmaceutical industry. Pharmaceutical Engineering 33 [Online]. Available from: https://ipsdb.com/files/docs/13SO-Capparella.pdf.

Dąbrowska, S., Chudoba, T., Wojnarowicz, J., Łojkowski, W., 2018. Current trends in the development of microwave reactors for the synthesis of nanomaterials in laboratories and industries: a review. Crystals 8 (10), 379.

Darr, J.A., Zhang, J., Makwana, N.M., Weng, X., 2017. Continuous hydrothermal synthesis of inorganic nanoparticles: applications and future directions. Chemical Reviews 117 (17), 11125–11238.

Díaz-Álvarez, A.E., Francos, J., Croche, P., Cadierno, V., 2014. Recent advances in the use of glycerol as Green solvent for synthetic organic chemistry. Current Green Chemistry 1 (1), 51–65.

Do, J.L., Friscic, T., 2017. Mechanochemistry: a force of synthesis. ACS Central Science 3 (1), 13–29.

EPA. 2020. Our Mission and What We Do. Available from: https://www.epa.gov/aboutepa/our-mission-and-what-we-do. Accessed 27.06.2020.

Farghaly, A.A., Huba, Z.J., Carpenter, E.E., 2012. Magnetic field assisted polyol synthesis of cobalt carbide and cobalt microwires. Journal of Nanoparticle Research 14 (10), 1159.

Glasing, J., Champagne, P., Cunningham, M.F., 2016. Graft modification of chitosan, cellulose and alginate using reversible deactivation radical polymerization (RDRP). Current Opinion in Green and Sustainable Chemistry 2, 15–21.

Hu, H., Sugawara, K., 2009. Magnetic-field-assisted synthesis of Ni nanostructures: selective control of particle shape. Chemical Physics Letters 477 (1–3), 184–188.

International Union for Conservation of Nature. 2020. Sustainable Development Goals. Available from: https://www.epa.gov/aboutepa/our-mission-and-what-we-do. Accessed 27.06.2020.

Kitchen, H.J., et al., 2014. Modern microwave methods in solid-state inorganic materials chemistry: from fundamentals to manufacturing. Chemical Reviews 114 (2), 1170–1206.

Krajewski, M., 2017. Magnetic-field-induced synthesis of magnetic wire-like micro- and nanostructures. Nanoscale 9 (43), 16511–16545.

Lang A., Murphy H., 2014. Business and sustainability. In: Top 12 Sustainability Issues in 2020, Network for Business Sustainability. Springer, Cham. nbs.net.

Lenardao, J.E., et al., 2003. Green chemistry: the 12 principles of green chemistry and it insertion in the teach and research activities. Química Nova 26, 123–129.

Li, H., Liao, S., 2009. A magnetic-field-assisted solution-phase route to cobalt thin film composed of cobalt nanosheets. Journal of Materials Chemistry 19 (29), 5207–5211.

Lonkar, S.P., Pillai, V., Abdala, A., 2019. Solvent-free synthesis of ZnO-graphene nanocomposite with superior photocatalytic activity. Applied Surface Science 465, 1107–1113.

Makone, S.S., Niwadange, S.N., 2016. Green chemistry alternatives for sustainable development in organic synthesis. Green Chemistry: An International Journal and Green Chemistry Resource: GC 3, 113–115.

Margetic, D., Strukil, V., 2016. Mechanochemical Organic Synthesis, 1st ed *Elsevier, Amsterdam.*

Marques, C.A., Machado, A.A.S.C., 2014. Environmental sustainability: implications and limitations to green chemistry. Foundations of Chemistry 16 (2), 125–147.

Parveen, K., Banse, V., Ledwani, L., 2016. Green synthesis of nanoparticles: their advantages and disadvantages. AIP Conference Proceedings 16, 1724.

Poliakoff, M., Licence, P., 2007. Sustainable technology: green chemistry. Nature 450 (7171), 810–822.

Prat, D., Hayler, J., Wells, A., 2014. A survey of solvent selection guides. Green Chemistry: an International Journal and Green Chemistry Resource: GC 16 (10), 4546–4551.

Ravichandran, S., Karthikeyan, E., 2018. Microwave synthesis—a potential tool for Green chemistry. Int J ChemTech Res 3, 466–470.

Rayne, L. "Conscious Spending. Conscious Life," Page 241. Available from: https://www.epa.gov/aboutepa/our-mission-and-what-we-do. Accessed 27.06.2020.

Rivas, M.E., 2018. Ball milling towards green synthesis: applications, projects, challenges. Johnson Matthey Technol Rev. 60 (2), 148–150.

Rogers, M., These 9 Technological Innovations Will Shape The Sustainability Agenda In 2019.

Sheldon, R.A., 1996. Selective catalytic synthesis of fine chemicals: opportunities and trends. Journal of Molecular Catalysis A: Chemical 1107 (1–3), 75–83.

Sheldon, R.A., 1997. Catalysis: the key to waste minimization. Journal of Chemical Technology and Biotechnology (Oxford, Oxfordshire: 1986) 68 (4), 381–388.

Sheldon, R.A., Isabel Arends, U.H., 2007. Green Chemistry and Catalysis. *Wiley-VCH Press, Weinheim.*

Verma, R., Kumar, L., Kurba, V.B., 2014. Green chemistry principles in organic compound synthesis and analysis. International Journal of Pharmacology 4, 10–22.

Wang, J., Wu, Y., Zhu, Y., 2007. Fabrication of complex of Fe3O4 nanorods by magnetic-field-assisted solvothermal process. Materials Chemistry and Physics 106 (1), 1–4.

Wood, C.D., Cooper, A.I., Desimone, J.M., 2004. Green synthesis of polymers using supercritical carbon dioxide. Current Opinion in Solid State & Materials Science 8 (5), 325–331.

Wu, M., et al., 2010. Magnetic field-assisted solvothermal assembly of one-dimensional nanostructures of Ni–Co alloy nanoparticles. Journal of Alloys and Compounds 491 (1–2), 689–693.

Xu, Y., et al., 2008. Magnetic-field-assisted solvothermal growth of single-crystalline bismuth nanowires. Nanotechnology 19 (11), 115602.

Xu, Y., et al., 2018. Facile preparation of rod-like MnOnanomixtures via hydrothermal approach and highly efficient removal of methylene blue for waste water treatment. Nanomaterials (Basel) 9 (1), 10.

Ye, N., et al., 2018. A review: conventional and supercritical hydro/solvothermal synthesis of ultra fine particles as cathode in lithium battery. Ceramics International 44 (5), 4521–4537.

Chapter 5

Nanobioremediation: a sustainable approach for environmental monitoring with special reference to the restoration of heavy metal contaminated soil and wastewater treatment

Sujata Dey[1], Satya Narayan Parida[1], Ajaya Kumar Rout[1], Jyotsna Dei[1], Jitendra Maharana[2], Smruti Priyambada Pradhan[3], Sukanta Kumar Pradhan[2] and Bijay Kumar Behera[1]

[1]*Aquatic Environmental Biotechnology and Nanotechnology Division, ICAR-Central Inland Fisheries Research Institute, Kolkata, West Bengal, India,* [2]*Department of Bioinformatics, Odisha University of Agriculture & Technology, Bhubaneswar, Odisha, India,* [3]*Department of Community Medicine, IMS & SUM HOSPITAL, Siksha o' Anusandhan University (Deemed to be University), Bhubaneswar, Odisha, India*

Introduction

Nanotechnology is a potential strategy for eliminating heavy metals from polluted environments. It is a growing industry combined with the most frequent traditional methods of removing HMs from soil and wastewater (Yang et al., 2019). Chemicals and impurities can react with such substances (Chen et al., 2017). Human activities contaminate the soil with heavy metals and organic and inorganic compounds, and exceeding the threshold limit might cause severe damage (Rajput et al., 2022). Soil health management is critical for ensuring food and agricultural output and preserving the environment (Lal, 2015). The significant harm to land and soil has caused worries and highlighted the urgent need for these polluted resources to be remedied. Advancements in nanosciences provide new insights into nearly every aspect of science. Nanobioremediation is a method for clearing environmental pollution by speeding up natural biodegradation processes with nanoparticles (NPs) (Mueller and Nowack, 2010). Sedimentation, screening, oxygenation, filtration, and other physicochemical processes for wastewater treatment have all been widely studied. Various human activities in corporations, markets, agricultural areas, and regular residential behaviours generate massive amounts of wastewater. The world's population is steadily expanding, and cities are becoming congested and exceeding capacity. Enterprises are increasing faster to meet the increased need for social beings, resulting in significantly more effluent than previously, thus exacerbating the problem. Sedimentation, screening, oxygenation, filtration, and other physicochemical processes for wastewater treatment have all been widely studied. Due to limitations such as partial treatment, increased expense, secondary pollutant formation, significant solid deposition, and the usage of numerous chemicals, biological techniques offer a preferable wastewater treatment solution (Das and Kundu, 2019).

Soil remediation

Nanomaterial in soil heavy metal remediation

Heavy metals often injure plants and lower soil microorganisms by producing reactive oxygen species (ROS), which degrade macromolecules like proteins and nucleic acids. Site stabilization is one of the most important ways to eliminate HMs because it immobilizes them in a particular location. *Halomonas* sp. has successfully utilized FeO NPs coated with polyvinylpyrrolidone to enhance the bioremediation of Pb and Cd-contaminated soil. After 24 and 48 hours, this

Role of Green Chemistry in Ecosystem Restoration to Achieve Environmental Sustainability. DOI: https://doi.org/10.1016/B978-0-443-15291-7.00012-2

approach cleared almost 100% of Pb and nearly 100% of Cd respectively. A biosorbent (n-Fe_3O_4-Path-S complex) removed 83.0%–89.5% for Cu^{2+}, 99.4%–100% for Pb^{2+}, and 92.6%–97.5% for Ni^{2+}. A recent study using HM-resistant microbes including *L. macroides* (PMBL-7) and *B. cereus* (PMBL-3) to remove Pb, Cd, Cr, and Cu proved that ZnO NPs synergistically remove Pb by 85%, Cr by 60%, and Cu by 70% when compared to *L. macroides* (55 and 50%) and *B. cereus* (80 and 60%) at neutral pH. Fe and magnetic oxide nanoparticles are designed to eliminate Cr (VI) pollution from the environment (Kumar et al., 2019).

Nanophytoremediation of soil

Nanophytoremediation is the collection of plants (hyperaccumulators) with suitable NPs, which might be a potential strategy for polluted soil repair that is concerned with the removal of pollutants containing synthetic NPs and using hyperaccumulator plants in combination with appropriate adsorbent material as a potential technique for removing HMs from polluted soil (Guerra et al., 2018). Adsorbents such as biochar, activated carbon, and NPs have recently become widely accessible; these adsorbent materials have quick adsorption capability, cover vast surface areas, give additional interaction sites for HMs, and are inexpensive (Ramos-Ruiz et al., 2017). HM-contaminated soil has mostly been treated with carbon or noncarbon NPs (Abbas et al., 2016). Silver NPs, TiO_2 NPs, carbon nanotubes (CNTs), and nZVI (nano zero-valent iron) have all been extensively studied for their soil remediation potential (Mubarak et al., 2016). At 10.0 pH, CNTs (0.05 g) remove 99.9% of Zn^{2+} (Rai et al., 2021), and a large percentage of Cd^{2+} is removed at 3.0 pH (Sun et al., 2016); FeS NPs remove 99.65% of $Cr6^{+}$ at 6.0 pH, but the clearance rate of FeS NPs is lowered at high pH (> 10.0) (Wu et al., 2021). The adsorption of chitosan-alginate NPs at 30 °C remove Hg^{2+} (Dubey et al., 2016) and Pb^{2+} absorption by Fe_3O_4 NPs increases from 298K to 328K (Nassar, 2010). Other NPs, such as modified carbon, significantly boosted Suaeda salsa plant growth while decreasing Cd and Ni absorption compared to the control (Cheng et al., 2019).

Use of nanoparticles in soil remediation

Efficiency of relatively stable Fe–Mn binary oxide NPs for in situ selenite immobilization in groundwater and soil and discovered that starch and carboxymethyl cellulose may both function as excellent stabilizers to aid in the dispersion of Fe–Mn NPs with significantly higher (> 3 times) Se (IV) absorption capacity than all other absorbents revealed (Xie et al., 2015). In this regard, Zhang et al. (2010) demonstrated the efficacy of NPs based on iron, including Fe_3O_4, FeS, and nZVI particles, for immobilizing in contaminated soil. Among them, Fe_3O_4 NPs have been reported to be more effective than other NPs. NPs were also used in certain circumstances to immobilize heavy metals in polluted soil. Nanostructured ZnO, TiO_2, and MgO were discovered to be excellent adsorbents for extracting Cr ions from waste-treated soil in the leather industry (Taghipour and Jalali, 2015). The capacity of the nanoscale amorphous MnO to treat soils polluted with heavy metals (HMs) such as Pb, Cd, As, Cu, and Zn was also investigated (Michálková et al., 2014).

Wastewater treatment

Nanomaterials in the treatment of wastewater

The electron aggregates of nanomaterials collectively produce the quantum phenomena. Their sizes are the principal determinant of distinct and noticeable characteristics. These are used in various domains, including photonics, catalysis, electronics, and biology (Yadav et al., 2017). Compared to bulk materials, nanomaterials can have a wide range of properties, allowing us to create new materials through various applications. By using a specific treatment technique that may remove a variety of contaminants from dirty water, NPs can increase the energy efficiency of wastewater treatment (Kamat et al., 2002). Highly selective nanomaterials (NMs) adsorb and remove HM ions from wastewater due to their high surface-to-volume ratio (Mallikarjunaiah et al., 2020). NPs have easily penetrated deeper, increased reactivity, and more effectively removed hazardous metals (Bystrzejewski et al., 2009). Diffusion on the adsorbent's outer surface occurs after dispersion on the adsorbent's pores (Lata, Samadder, 2016). The surface chemistry, shape, size, fractal dimensions, crystal structure, solubility, and grouping condition of nanoadsorbents has an impact on their properties. NPs, like bulk materials, allow for atomic-level modifications, resulting in a flood of novel features not found in bulk materials (Subramaniam et al., 2019).

Nanomembrane technology in wastewater treatment

To remove HMs from wastewater, evaporation, freezing, membrane ultrafiltration (UF), ion exchange, sedimentation, and electrochemical techniques have been tried. Because of the enormous cost of extracting small amounts of heavy metals and creating sludge, none of them was completely successful. In wastewater treatment operations, nanomembranes constructed of nanofibers offer promising possibilities for eliminating undesired HMs (Khulbe and Matsuura, 2018). These thin-layer permeable membranes resist bacteria, HMs, salt, and other contaminants. Various membranes with surface charges that repel HM ions, membrane distillation, adsorptive membranes, and HM removal by size exclusion can be utilized to extract HM. HM ions flow through UF membranes due to their comparatively large pores. Nanofiltration (NF) membranes are a potential replacement for traditional UF membranes as an HM removal agent. The basic assumption of nanomembranes is that unwanted contaminants are filtered out. The use of nanomembranes speeds up the treatment process considerably. Bacteria, viruses, large multivalent ions, herbicides, dissolved organic matter, insecticides, and other pollutants can all be eliminated with this method. It outperforms microfilters and ultrafilters while consuming less energy. In addition, the cost of running NF is lower, occupies less space, and is easier to build and operate. To treat dye-containing wastewater, a cross-linked polyetherimide-reliant NF was employed (Gunawan et al., 2019) m-phenylenediamine was used to create the NF membrane. The Fe-MMT (modified montmorillonite) nanomembrane was employed to filter wastewater containing Hg combined with cyanide (Lodo and Diaz, 2019). Organic chemical contamination in water has emerged as a key concern that must be addressed as civilization advances. Due to their greater surface area, nanoparticles enhanced the chemical activity and adsorption capacity of the surface metals of the nano adsorbents.

Nanozyme-based treatment of wastewater

Pollution is becoming a global issue that impacts our everyday lives and may endanger our survival in the future. Traditional physicochemical pollutant treatment approaches are ineffective and frequently result in the creation of harmful by-products (Ihsanullah et al., 2020). Due to its high removal effectiveness, moderate reaction conditions, and low equipment needs, the enzymatic method has been explored as a feasible approach for addressing these environmental challenges. However, the fact that enzymes are easy to break down, expensive to make, and can't be recycled has kept them from being widely used to clean up environmental pollution. Nanozymes' absorption ability, for example, eliminates pollutants directly and enhances catalytic degradation by promoting substrate binding to the catalytic site, resulting in more efficient pollution treatment. The process of Fe_3O_4 NPs for pollutant 4-chlorophenol (4-CP) elimination was investigated (Cheng et al., 2015). During the degradation, 10% of 4-CP was absorbed onto the Fe_3O_4 surface and destroyed by Fe_3O_4 NPs-produced •OH. As a result, Fe_3O_4 NPs might be utilized as both catalysts and absorbents in removing 4-CP. Even though many nanozymes have both adsorption and catalytic abilities, nanozymes are primarily used as catalysts to break down contaminants. Nanozymes' adsorption capacity serves as a backup for wastewater treatment. Table 5.1 shows recent advances in treating various contaminants by various nanozymes.

Peroxidase nanozymes

As chemical raw materials or intermediate products in the resin, paint, dyeing, and textile industries are commonly used organic compounds like chloroanilines, phenols, and polycyclic aromatic hydrocarbons. As a result, they are common contaminants in industrial effluent. Solvent extraction, chemical oxidation, and charcoal adsorption are among the physical and chemical processes now utilized to remove phenolic compounds from industrial effluents. However, due to the inadequate removal of phenolic chemicals, these procedures are costly, time-demanding, and inefficient. These chemical molecules, fortunately, are peroxidase substrates. For example, peroxidases in the presence of hydrogen peroxide catalytically convert phenolic substrates to phenoxy radicals, resulting in free radical polymerization. When phenol radicals self-couple, they precipitate polymeric compounds easily removed from the water. Nonetheless, many issues must be addressed, including stability, redox potential, and large-scale manufacturing before natural peroxidases may be used in pollutant transformation. Nanozyme-based wastewater treatment, on the other hand, has the advantages of high performance, simplicity, and stability. Iron oxide NPs with peroxidase activity were utilized to remove pollutants by accelerating the conversion of H_2O_2 into the hydroxyl radical (•OH), which has a high oxidation capacity and can destroy phenols (Zhang et al., 2008). After that, the peroxidase-like catalytic activity of nanozymes has been intensively researched for its potential to catalyze the breakdown of environmental contaminants, in particular, dyes such as methylene blue (Huang et al., 2018), rhodamine B (Ibarbia et al., 2020) and xylenol orange (Jain et al., 2019). In the presence

TABLE 5.1 Nanozyme-based wastewater treatment

Enzymatic activity	Nanozyme	Pollutant	Removal efficiency (%)	Mechanism	References
Oxidase and peroxidase	C-CoM-HNCs	RhB	Over 95	ROS oxidation caused by PMS	Li et al. (2019)
	BSA-$NiCo_2O_4$@ MnO_2/C	TCH	97.60	O_2• − and •OH oxidation	Yang et al. (2021)
Hydrolase	AuNP-Zn(II)	Methyl paraoxon		Phosphate triester bond hydrolysis	Lyu et al. (2021)
	Zr-incorporated CeO_2	Methyl paraoxon	More than 90	Phosphate triester bond hydrolysis	Khulbe et al. (2018)
SOD and catalase	CuTA	O_2• −, H_2O_2, and •OH in the smoke	71.70	Catalytic oxidation and reduction	Lin et al. (2020)
Peroxidase	Fe_3O_4	MB	77	•OH oxidation	Jain et al. (2019)
		MO	60	–	–
		RhB	98	–	–
	Peroxidase-like carbon-doped Fe_3O_4 nanozyme	NOM in the membrane	–	•OH oxidation	Yang et al. (2020)
	Iron oxide particles encased in textile	Crystal violet	–	•OH oxidation	Safarik et al. (2021)
	Au@CuxS/ZnO	MB	>70	•OH oxidation	Kuo et al. (2022)
	Fe_3O_4-TiO_2/rGO	Atrazine	97	•OH oxidation	Boruah et al. (2021)
	DhHP-6-c-ZrMOF	Phenol	98	•OH oxidation	Ding et al. (2020)
Oxidase	Fe-NC	NOM	–	Oxidation by O_2• −	Yang et al. (2020)
	FeBi-NC	RhB	100	ROS oxidation caused by PMS	Chen et al. (2022)
Haloperoxi-dases	CeO_{2-x} nanorods	Biofoul	69	Producing HOBr to impair microorganism quorum sensing	Hu et al. (2018)
Laccase	GNFs	Neutral red	Over 90	Catalytic oxidation	Tran et al. (2021)
	CuNi/$CoMoO_4$	AF	99.45	Catalytic oxidation	Sun et al. (2021)
	CH-Cu	2,4-DP and hydroquinone	>80	Catalytic oxidation	Wang et al. (2019)

2,4-DP, 2,4-Dichlorophenol; *AF*, acid Fuchsin; *MB*, methyl blue; *MO*, methyl orange; *NOM*, natural organic matter; *PMS*, peroxymonosulfate; *RhB*, rhodamine B; *ROS*, reactive oxygen species; *TCH*, tetracycline hydrochloride.

of hydrogen peroxide, haloperoxidase nanozymes can catalyze the conversion of halogen ions into hypohalites XO^- (where X is either Cl, Br, or I) or organic halides, as shown by Eq. (5.1).

$$H_2HO_2 + H^- + X^- \rightarrow H_2O + XOH \tag{5.1}$$

Hypohalous acids impact microbe quorum sensing, which regulates biofilm development. The release of HOBr hence avoids seaweed biofouling (Hu et al., 2018; L. Wang et al., 2019).

Oxidase nanozyme

Reactive oxygen species are largely used in oxidase nanozyme wastewater treatment, much like in peroxidase nanozyme wastewater treatment, to oxidize organic compounds like phenol and organic dyes (Jiang et al., 2018; Safarik et al., 2021; Yang et al., 2020). Oxidases use oxygen as an oxidant rather than H_2O_2 to avoid the entry of external pollutants, which is a superior property of the enzymes. One theory is that the oxidase nanozyme catalyzes the formation of activated oxygen molecules such as superoxide anion free radicals from dissolved oxygen (Liu et al., 2021; Xu et al., 2020). Liu and colleagues created an Au-NiFe-layered double hydroxide/rGO nanocomposite for the breakdown of organic mercury in wastewater as an oxidase nanozyme that produces $O2\bullet^-$. FeBi-NC oxidase nanozymes activate PMS to create $\bullet OH$, $SO_4\bullet^-$ and $^1O_2$ for effective rhodamine B breakdown, according to research (Chen et al., 2022).

Other nanozymes

Other nanozymes, in contrast to oxidase or peroxidase, which catalyze the production of reactive oxygen species, primarily focus on the breakdown of specific contaminants. Catalase nanozymes, for example, are designed to remove H_2O_2 and are frequently employed in conjunction with peroxidase to remove any leftover H_2O_2 after the reaction (Wang et al., 2020). According to Aman's research, CeO_2 NPs have phosphatase-like activity in degrading organophosphate insecticides such as paraoxon (Trenque et al., 2019). Furthermore, laccase nanozyme showed outstanding results in accelerating the breakdown of phenol contaminants (Wang et al., 2019).

Merits and demerits

For the treatment of soil pollutants, field-scale applications of alternative nanomaterials are urgently needed in addition to nZVI. To increase the usefulness of nanoremediation in practice, the following concerns must be considered. (1) The limitation of dose modulation is among the application of NMs greatest issues. Higher discharge levels of NMs in water may be helpful for removing a particular pollutant, but when applied to soil, they can be quite harmful. Traditional manufacturing facilities cannot handle the massive amount of NMs needed to restore damaged soils. Additionally, high levels of NPs, such as nZVI, may be toxic to microbes. (2) There are no standard techniques for using NMs to clean the environment. For field size applications, appropriate application methods for various NMs must be documented and maintained. (3) More research is required to understand how applied NMs recover in soil fully. Once NMs have been introduced or injected into the soil, it is challenging to get them out because the soil is a rather heterogeneous media by nature. Contaminants that have been adsorbent may eventually wash off the nanoadsorbents. (4) Despite being rather expensive, several NMs, including CNTs and GO, and NPs, including BNPs, polymer NPs, Fe and Cu^- based NPs, hold significant promise for the field-scale remediation of organic and inorganic pollutants. Green synthesis methods (like those used for nZVI) should also be considered for these NMs to reduce environmental toxicity. (5) Before using NMs, they should be thoroughly assessed to establish their effects and long-term fate on soil and aquatic life. This will ensure the environmental and ecological safety of NMs (Besha et al., 2020). (6) Standardizing the working parameters (such as pH, temperature, competing ions, and organic matter) of various NMs for cleaning up various contaminants is important due to the variability of field sites.

Environmental significance of nano-based composites

Studies have aimed to comprehend the processes of NM interactions with ambient variables, microbial populations, and target pollutants in response to the potential widespread usage of NMs for remediating contaminated soil and water (Biswas and Sarkar, 2019). Numerous crucial factors (soil organic matter, pH, and texture) significantly impact how NPs behave, particularly in soils. By nZVI, microbial communities may suffer (Lefevre et al., 2016). *Escherichia coli* cells treated with high concentrations of nZVI particles displayed comparable antibacterial activity, resulting in oxidative damage and cell inactivation (Chaithawiwat et al., 2016). The hydroponically grown Arabidopsis thaliana was used to test the effects of excessive nZVI treatment on plants because nZVI is a potent As(V) removal agent in aquatic environments (Zhang et al., 2018). Pi and Mg-ATP2 levels were tracked in vivo using biosensors for inorganic phosphate (Pi) and in the cells. Overexposure to nZVI caused plants to become ravenous for Pi, which was bad for plant growth

(Zhang et al., 2018). According to ecotoxicity testing conducted by Hjorth et al. (2017), most nZVI products, except the ball-milled nZVI particles, would not be classified as environmental risks under European standards. According to Li et al. (2016), Ti-based NPs also negatively affected soil microbes. Therefore, it is advised to replace riskier Ti-based NPs wherever possible with safer metal oxide NPs including Zn, Fe, and Cu. Even after all pollutants have been removed from the water, residual NPs may still exist (Peeters et al., 2016). The iron aggregates settled more slowly after mixing with tetramethylammonium hydroxide to disperse the Fe NPs. Additionally, because NM-based herbicides and fertilizers stimulate crop development while posing little environmental risks, researchers have advocated their use (e.g., Cu-, Fe-, Si-, Mn-, and Mg-based NMs) (Adisa et al., 2019; Kopittke et al., 2019).

Economic factors, challenges, and future recommendations

In addition to social costs and operational details, rehabilitation operations rely on socio-economic ramifications. Because nanotechnology is a relatively new field, it took a lot of work to evaluate the financial effects of nano remediation. Analyzing the procedure's financial implications will aid decision-makers and others in establishing the approach and goods that will be employed (Hussain, 2020). According to official studies, activated carbon is not practical in a marketing setting. Alternatively, we may employ the same substitution technique to make the treatment more cost-effective and expand the range of regenerative behavior. Nanozyme-based disease diagnostics, environmental monitoring, and pollution removal face numerous hurdles:

- Nanozyme activity is currently confined to hydrolases (e.g., phosphatase) and oxidoreductases (e.g., peroxidase) with additional types being utilized only infrequently. Lipase nanozymes.
- The sensitivity of molecular detection is determined by the activity of nanozymes. The therapeutic effect of pollutants is determined by the activity of nanozymes. Although a tiny percentage of nanozymes have better catalytic activity than natural enzymes. Their catalytic activity still has to be improved urgently.
- Regarding nanozyme-based pollution detection, specificity is a crucial factor to consider. Multiple environmental contaminants have been observed to decrease nanozyme activity, causing the detection signal to alter.
- Furthermore, most nanozyme demonstrate significant catalytic activity only in a narrow pH range, restricting their use. The versatility of nanozyme-based technology for pollution remediation and treatment might be increased if nanozymes can work in a wider temperature and pH range.
- Currently, nanozymes are predominantly employed in the water-phase treatment of sewage for ecological reasons.
- As a result, more advances in nanozyme technology are required before they can be used on a large scale in illness detection, pollution monitoring, and environmental remediation.

Conclusions

Compared to other commercialized industries, the environmental nanotechnology sector is still developing. The worldwide market is gradually opening up to this business area more and more. Recently, it was shown that industrial dyes can adsorb and chemically disintegrate, having negative impacts. The removal of organic trace pollutants in the treatment of municipal wastewater is however expected to advance as a result of nanotechnology. Additional purification procedures using, for example, reactive surfaces coated with catalytic NPs may be an option for removing anthropogenic pollutants after biological treatment. Similar improvements and trends are occurring in the drinking water treatment sector. Currently, the main applications of nanotechnologies are the elimination of organic compounds, the adsorption of heavy metals, and the sanitization of microbes. The research on heavy metal adsorption and organic compound pollution appears to still be in its development, even though the application for disinfection is well established. However, New uses for NMs in the treatment of drinking water are more economical and environmentally friendly than traditional treatment methods by various scientific teams. Therefore, higher acceptance and information sharing are needed in this area to enhance the utilization of nanotechnologies.

References

Abbas, A., Al-Amer, A.M., Laoui, T., Al-Marri, M.J., Nasser, M.S., Khraisheh, M., et al., 2016. Heavy metal removal from aqueous solution by advanced carbon nanotubes: critical review of adsorption applications. Separation and Purification Technology 157, 141–161.

Adisa, I.O., Pullagurala, V.L.R., Peralta-Videa, J.R., Dimkpa, C.O., Elmer, W.H., Gardea-Torresdey, J.L., et al., 2019. Recent advances in nano-enabled fertilizers and pesticides: a critical review of mechanisms of action. Environmental Science: Nano 6 (7), 2002–2030.

Besha, A.T., Liu, Y., Bekele, D.N., Dong, Z., Naidu, R., Gebremariam, G.N., 2020. Sustainability and environmental ethics for the application of engineered nanoparticles. Environmental Science & Policy 103, 85–98. Available from: https://doi.org/10.1016/j.envsci.2019.10.013.

Biswas, J.K., Sarkar, D., 2019. Nanopollution in the aquatic environment and ecotoxicity: no nano issue!. Current Pollution Reports 5 (1), 4–7.

Boruah, P.K., Darabdhara, G., Das, M.R., 2021. Polydopamine functionalized graphene sheets decorated with magnetic metal oxide nanoparticles as efficient nanozyme for the detection and degradation of harmful triazine pesticides. Chemosphere 268, 129328.

Bystrzejewski, M., Pyrzyńska, K., Huczko, A., Lange, H., 2009. Carbon-encapsulated magnetic nanoparticles as separable and mobile sorbents of heavy metal ions from aqueous solutions. Carbon 47 (4), 1201–1204.

Chaithawiwat, K., Vangnai, A., McEvoy, J.M., Pruess, B., Krajangpan, S., Khan, E., 2016. Role of oxidative stress in inactivation of Escherichia coli BW25113 by nanoscale zero-valent iron. Science of the Total Environment 565, 857–862.

Cheng, R., Li, G.Q., Cheng, C., Shi, L., Zheng, X., Ma, Z., 2015. Catalytic oxidation of 4-chlorophenol with magnetic Fe3 4 nanoparticles: mechanisms and particle transformation. RSC Advances 5 (82), 66927–66933.

Cheng, J., Sun, Z., Yu, Y., Li, X., Li, T., 2019. Effects of modified carbon black nanoparticles on plant-microbe remediation of petroleum and heavy metal co-contaminated soils. International Journal of Phytoremediation 21 (7), 634–642.

Chen, K., He, J., Li, Y., Cai, X., Zhang, K., Liu, T., et al., 2017. Removal of cadmium and lead ions from water by sulfonated magnetic nanoparticle adsorbents. Journal of Colloid and Interface Science 494, 307–316.

Chen, Q., Liu, Y., Lu, Y., Hou, Y., Zhang, X., Shi, W., et al., 2022. Atomically dispersed Fe/Bi dual active sites single-atom nanozymes for cascade catalysis and peroxymonosulfate activation to degrade dyes. Journal of Hazardous Materials 422, 126929.

Das, A., Kundu, P., 2019. Bioremediation of wastewater using microalgae. Environmental Biotechnology for Soil and Wastewater Implications on Ecosystems. Springer, Singapore, pp. 55–60.

Ding, Y., Li, Z., Jiang, W., Yuan, B., Huang, T., Wang, L., et al., 2020. Degradation of phenol using a peroxidase mimetic catalyst through conjugating deuterohemin-peptide onto metal-organic framework with enhanced catalytic activity. Catalysis Communications 134, 105859.

Dubey, R., Bajpai, J., Bajpai, A.K., 2016. Chitosan-alginate nanoparticles (CANPs) as potential nanosorbent for removal of Hg (II) ions. Environmental Nanotechnology, Monitoring and Management 6, 32–44.

Guerra, F.D., Attia, M.F., Whitehead, D.C., Alexis, F., 2018. Nanotechnology for environmental remediation: materials and applications. Molecules 23 (7), 1760.

Gunawan, F.M., Mangindaan, D., Khoiruddin, K., Wenten, I.G., 2019. Nanofiltration membrane cross-linked by m-phenylenediamine for dye removal from textile wastewater. Polymers for Advanced Technologies 30 (2), 360–367.

Hjorth, R., Coutris, C., Nguyen, N.H., Sevcu, A., Gallego-Urrea, J.A., Baun, A., et al., 2017. Ecotoxicity testing and environmental risk assessment of iron nanomaterials for sub-surface remediation – recommendations from the FP7 project NanoRem. Chemosphere 182, 525–531.

Huang, D., Qin, X., Peng, Z., Liu, Y., Gong, X., Zeng, G., et al., 2018. Nanoscale zero-valent iron assisted phytoremediation of Pb in sediment: impacts on metal accumulation and antioxidative system of Lolium perenne. Ecotoxicology and Environmental Safety 153, 229–237.

Hussain, C.M. (Ed.), 2020. The Handbook of Environmental Remediation: Classic and Modern Techniques. Royal Society of Chemistry.

Hu, M., Korschelt, K., Viel, M., Wiesmann, N., Kappl, M., Brieger, J., et al., 2018. Nanozymes in nanofibrous mats with haloperoxidase-like activity to combat biofouling. ACS Applied Materials & Interfaces 10 (51), 44722–44730.

Ibarbia, A., Sánchez-Abella, L., Lezama, L., Grande, H.J., Ruiz, V., 2020. Graphene quantum dot-based hydrogels for photocatalytic degradation of organic dyes. Applied Surface Science 527, 146937.

Ihsanullah, I., Jamal, A., Ilyas, M., Zubair, M., Khan, G., Atieh, M.A., 2020. Bioremediation of dyes: current status and prospects. Journal of Water Process Engineering 38, 101680.

Jain, S., Panigrahi, A., Sarma, T.K., 2019. Counter anion-directed growth of iron oxide nanorods in a polyol medium with efficient peroxidase-mimicking activity for degradation of dyes in contaminated water. ACS Omega 4 (8), 13153–13164.

Jiang, J., He, C., Wang, S., Jiang, H., Li, J., Li, L., 2018. Recyclable ferromagnetic chitosan nanozyme for decomposing phenol. Carbohydrate Polymers 198, 348–353.

Kamat, P.V., Huehn, R., Nicolaescu, R., 2002. A “sense and shoot” approach for photocatalytic degradation of organic contaminants in water. The Journal of Physical Chemistry B 106 (4), 788–794.

Khulbe, K.C., Matsuura, T., 2018. Removal of heavy metals and pollutants by membrane adsorption techniques. Applied Water Science 8 (1), 1–30.

Khulbe, K., Roy, P., Radhakrishnan, A., Mugesh, G., 2018. An unusual two-step hydrolysis of nerve agents by a nanozyme. ChemCatChem 10 (21), 4826–4831.

Kopittke, P.M., Lombi, E., Wang, P., Schjoerring, J.K., Husted, S., 2019. Nanomaterials as fertilizers for improving plant mineral nutrition and environmental outcomes. Environmental Science: Nano 6 (12), 3513–3524.

Kumar, S., Sinha, S.K., Goud, V.V., Das, S., 2019. Removal of Cr (VI) by magnetic iron oxide nanoparticles synthesized from extracellular polymeric substances of chromium resistant acid-tolerant bacterium Lysinibacillus sphaericus RTA-01. Journal of Environmental Health Science and Engineering 17 (2), 1001–1016. Available from: https://doi.org/10.1007/s40201-019-00415-5.

Kuo, K.Y., Chen, S.H., Hsiao, P.H., Lee, J.T., Chen, C.Y., 2022. Day-night active photocatalysts obtained through effective incorporation of Au@CuxS nanoparticles onto ZnO nanowalls. Journal of Hazardous Materials 421, 126674.

Lal, R., 2015. Restoring soil quality to mitigate soil degradation. Sustainability 7 (5), 5875–5895.

Lata, S., Samadder, S.R., 2016. Removal of arsenic from water using nano adsorbents and challenges: a review. Journal of Environmental Management 166, 387–406.

Lefevre, E., Bossa, N., Wiesner, M.R., Gunsch, C.K., 2016. A review of the environmental implications of in situ remediation by nanoscale zero valent iron (nZVI): behavior, transport and impacts on microbial communities. Science of the Total Environment 565, 889–901.

Lin, S., Cheng, Y., Zhang, H., Wang, X., Zhang, Y., Zhang, Y., et al., 2020. Copper tannic acid coordination nanosheet: a potent nanozyme for scavenging ROS from cigarette smoke. Small 16, 1902123.

Liu, H., Guo, Y., Wang, Y., Zhang, H., Ma, X., Wen, S., et al., 2021. A nanozyme-based enhanced system for total removal of organic mercury and SERS sensing. Journal of Hazardous Materials 405, 124642.

Li, S., Hou, Y., Chen, Q., Zhang, X., Cao, H., Huang, Y., 2019. Promoting active sites in MOF-derived homobimetallic hollow nanocages as a high-performance multifunctional nanozyme catalyst for biosensing and organic pollutant degradation. ACS Applied Materials & Interfaces 12 (2), 2581–2590.

Li, D., Li, B., Wang, Q., Hou, N., Li, C., Cheng, X., 2016. Toxicity of TiO2 nanoparticle to denitrifying strain CFY1 and the impact on microbial community structures in activated sludge. Chemosphere 144, 1334–1341.

Lodo, M.J., Diaz, L.J., 2019. Reusability of Fe-modified MMT nanomembranes and the retrieval of the adsorbed mercury metal. IOP Conference Series: Earth and Environmental Science 345 (1), 012011.

Lyu, Y., Morillas-Becerril, L., Mancin, F., Scrimin, P., 2021. Hydrolytic cleavage of nerve agent simulants by gold nanozymes. Journal of Hazardous Materials 415, 125644.

Mallikarjunaiah, S., Pattabhiramaiah, M., Metikurki, B., 2020. Application of nanotechnology in the bioremediation of heavy metals and wastewater management. Nanotechnology for Food, Agriculture, and Environment. Springer, Cham, pp. 297–321.

Michálková, Z., Komárek, K., Šillerová, H., Della Puppa, L., Joussein, E., Bordas, F., et al., 2014. Evaluating the potential of three Fe-and Mn-(nano) oxides for the stabilization of Cd, Cu and Pb in contaminated soils. Journal of Environmental Management 146, 226–234. Available from: https://doi.org/10.1016/j.jenvman.2014.08.004.

Mubarak, N.M., Sahu, J.N., Abdullah, E.C., Jayakumar, N.S., Ganesan, P., 2016. Microwave-assisted synthesis of multi-walled carbon nanotubes for enhanced removal of Zn (II) from wastewater. Research on Chemical Intermediates 42 (4), 3257–3281.

Mueller, N.C., Nowack, B., 2010. Nanoparticles for remediation: solving big problems with little particles. Elements 6 (6), 395–400.

Nassar, N.N., 2010. Rapid removal and recovery of Pb (II) from wastewater by magnetic nanoadsorbents. Journal of Hazardous Materials 184 (1–3), 538–546.

Peeters, K., Lespes, G., Zuliani, T., Ščančar, J., Milačič, R., 2016. The fate of iron nanoparticles in environmental waters treated with nanoscale zero-valent iron, FeONPs and Fe3O4 NPs. Water Research 94, 315–327.

Rai, G.K., Bhat, B.A., Mushtaq, M., Tariq, L., Rai, P.K., Basu, U., et al., 2021. Insights into decontamination of soils by phytoremediation: a detailed account on heavy metal toxicity and mitigation strategies. Physiologia Plant 173 (1), 287–304.

Rajput, V.D., Minkina, T., Kumari, A., Shende, S.S., Ranjan, A., Faizan, M., et al., 2022. A review on nanobioremediation approaches for restoration of contaminated soil. Eurasian Journal of Soil Science 11 (1), 43–60.

Ramos-Ruiz, A., Sesma-Martin, J., Sierra-Alvarez, R., Field, J.A., 2017. Continuous reduction of tellurite to recoverable tellurium nanoparticles using an upflow anaerobic sludge bed (UASB) reactor. Water Research 108, 189–196.

Safarik, I., Prochazkova, J., Schroer, M.A., Garamus, V.M., Kopcansky, P., Timko, M., et al., 2021. Cotton textile/iron oxide nanozyme composites with peroxidase-like activity: preparation, characterization, and application. ACS Applied Materials & Interfaces 13 (20), 23627–23637.

Subramaniam, M.N., Goh, P.S., Lau, W.J., Ismail, A.F., 2019. The roles of nanomaterials in conventional and emerging technologies for heavy metal removal: a state-of-the-art review. Nanomaterials 9 (4), 625.

Sun, T.Y., Bornhöft, N.A., Hungerbühler, K., Nowack, B., 2016. Dynamic probabilistic modeling of environmental emissions of engineered nanomaterials. Environmental Science & Technology 50 (9), 4701–4711.

Sun, M., Dang, Y., Du, R., Lu, Z., Wang, Y., Wang, X., et al., 2021. Experiment and theoretical insights into CuNi/CoMoO4 multi-functional catalyst with laccase-like: catalysis mechanism, smartphone biosensing and organic pollutant efficient degradation. Chemical Engineering Journal 425, 130586.

Taghipour, M., Jalali, M., 2015. Effect of clay minerals and nanoparticles on chromium fractionation in soil contaminated with leather factory waste. Journal of Hazardous Materials 297, 127–133. Available from: https://doi.org/10.1016/j.jhazmat.2015.04.067.

Tran, T.D., Nguyen, P.T., Le, T.N., Kim, M.I., 2021. DNA-copper hybrid nanoflowers as efficient laccase mimics for colorimetric detection of phenolic compounds in paper microfluidic devices. Biosensors and Bioelectronics 182, 113187.

Trenque, I., Magnano, G.C., Bolzinger, M.A., Roiban, L., Chaput, F., Pitault, I., et al., 2019. Shape-selective synthesis of nanoceria for degradation of paraoxon as a chemical warfare simulant. Physical Chemistry Chemical Physics 21 (10), 5455–5465.

Wang, L., Hou, J., Liu, S., Carrier, A.J., Guo, T., Liang, Q., et al., 2019. CuO nanoparticles as haloperoxidase-mimics: chloride-accelerated heterogeneous Cu-Fenton chemistry for H2O2 and glucose sensing. Sensors and Actuators B: Chemical 287, 180–184.

Wang, J., Huang, R., Qi, W., Su, R., Binks, B.P., He, Z., 2019. Construction of a bioinspired laccase-mimicking nanozyme for the degradation and detection of phenolic pollutants. Applied Catalysis B: Environmental 254, 452–462.

Wang, F., Zhang, Y., Liu, Z., Ren, J., Qu, X., 2020. A mesoporous encapsulated nanozyme for decontaminating two kinds of wastewater and avoiding secondary pollution. Nanoscale 12 (27), 14465–14471.

Wu, X., Hu, J., Wu, F., Zhang, X., Wang, B., Yang, Y., et al., 2021. Application of TiO2 nanoparticles to reduce bioaccumulation of arsenic in rice seedlings (Oryza sativa L.): A mechanistic study. Journal of Hazardous Materials 405, 124047.

Xie, W., Liang, Q., Qian, T., Zhao, D., 2015. Immobilization of selenite in soil and groundwater using stabilized Fe–Mn binary oxide nanoparticles. Water Research 70, 485–494.

Xu, X., Luo, J., Wei, S., Zou, X., Niu, X., Pan, J., 2020. Three-dimensional flower-like multifunctional adsorbents with excellent sorptive removal and colorimetric detection of arsenate. Chemical Engineering Journal 398, 125649.

Yadav, K.K, Singh, J.K., Gupta, N., Kumar, V., 2017. A review of nanobioremediation technologies for environmental cleanup: a novel biological approach. Journal of Materials and Environmental Sciences 8 (2), 740–757.

Yang, J., Hou, B., Wang, J., Tian, B., Bi, J., Wang, N., et al., 2019. Nanomaterials for the removal of heavy metals from wastewater. Nanomaterials 9 (3), 424.

Yang, W., Li, J., Lyu, Y., Yan, X., Yang, P., Zuo, M., 2021. Bioinspired 3D hierarchical BSA-NiCo2O4@ MnO2/C multifunctional micromotors for simultaneous spectrophotometric determination of enzyme activity and pollutant removal. Journal of Cleaner Production 309, 127294.

Yang, H., Wu, X., Su, L., Ma, Y., Graham, N.J., Yu, W., 2020. The Fe–N–C oxidase-like nanozyme used for catalytic oxidation of NOM in surface water. Water Research 171, 115491.

Yang, H., Xue, W., Liu, M., Yu, K., Yu, W., 2020. Carbon doped Fe3O4 peroxidase-like nanozyme for mitigating the membrane fouling by NOM at neutral pH. Water Research 174, 115637.

Zhang, W., Lo, I.M., Hu, L., Voon, C.P., Lim, B.L., Versaw, W.K., 2018. Environmental risks of nano zerovalent iron for arsenate remediation: impacts on cytosolic levels of inorganic phosphate and MgATP2–in Arabidopsis thaliana. Environmental Science & Technology 52 (7), 4385–4392.

Zhang, M., Wang, Y., Zhao, D., Pan, G., 2010. Immobilization of arsenic in soils by stabilized nanoscale zero-valent iron, iron sulfide (FeS), and magnetite (Fe3O4) particles. Chinese Science Bulletin 55 (4), 365–372.

Zhang, J., Zhuang, J., Gao, L., Zhang, Y., Gu, N., Feng, J., et al., 2008. Decomposing phenol by the hidden talent of ferromagnetic nanoparticles. Chemosphere 73 (9), 1524–1528.

Chapter 6

Plants as monitors and managers of pollution

Matthew C. Ogwu
Goodnight Family Department of Sustainable Development, Appalachian State University, Boone, NC, United States

Introduction

Industrialization and urbanization are often tied to globalization and if unchecked might have negative impacts on flora and fauna well-being, as well as the environment that supports them (Vwioko et al., 2018; Ogwu, 2019a). Moreover, the rising number of automobiles and other fossil fuel-powered machines are significant contributors to local, regional, and global atmospheric pollution by increasing the concentrations of CO_x and NO_x in air, water, and soil and turn cause climate change (Skrynetska et al., 2018; Ikhajiagbe and Ogwu, 2020; Ikhajiagbe et al., 2021). Roadside vegetation in urban and suburban centers is particularly vulnerable to environmental pollution with visible effects on plant growth and developmental processes (Osawaru et al., 2013a,b). Environmental pollutants affect not only plant health but also animal development and alterations in environmental status (Ogwu et al., 2014; Ogwu and Oladeji, 2014; Osawaru and Ogwu, 2014). In the case of plants, the symptoms via injuries, disability, discoloration, premature aging, etc. due to environmental pollutants like oxides of nitrogen, sulfur, and carbon (NO_x, SO_x, and CO_x), hydrocarbons, heavy metals, and particulate matter (Vwioko et al., 2018; Sukumaran, 2014). Generally, pollution of the environment from diverse anthropogenic activities is considered the main driver of climate change and biodiversity loss (Kumar and Nandini, 2013; Bellard et al., 2012; Manisalidis et al., 2020; Tran et al., 2020). Plant leaves are their most sensitive organ when it comes to indicating, monitoring, and managing these pollutants compared to their stem and roots. Therefore it is anticipated that pollutants can be detected and possibly remediated by several technological tools with these plant organs as the focus. Biological indicators (i.e., biomonitors) are living organisms exploited in the detection and monitoring of pollutants and their levels in a specific ecosystem (Ogwu, 2019b). Khatri and Tyagi (2015) elaborated on the significance of monitoring the emergent properties of these indicators with factors like moisture, light, temperature, and suspended solids. Such observation can also be tailored to use these indicators as pollution managers to safeguard human and environmental health as well as for climate change mitigation and adaptation. This chapter aims to appraise the concept of using plants as biological monitors and managers of the ecosystem using a literature review approach.

Environmental pollution

The environment, that is, the living and nonliving things and the totality of space, at a particular time, socio-cultural settings, and interactions that exist for their sustenance (Barry, 2007; Hardin, 2009; Osawaru et al., 2016; Ogwu et al., 2018). Pollutants are capable of harming the components of the environment including humans and other living things through direct toxic effects, or indirectly by changing pH, solubilization, and mineralization of certain elements with potential negative mechanical effects (Osawaru and Ogwu, 2020). In the view of Gheorghe and Barbu (2011), pollutants are capable of direct, that is, identifiable impacts on the environment as well as secondary damage mainly via minor perturbations over long periods. Societal industrialization via increasing production of social and environmental goods, services, capital, and waste products coupled with exponential growth and loss in the human population and biodiversity respectively has caused an exponential rise in environmental disasters throughout the world (Ogwu, 2019c, 2020). Policies and government responses often fall short of targets and instead of being proactive are rather reactive with little planning, mitigation, and adaptation strategies (Ikhajiagbe and Ogwu, 2020; Ogwu, 2019b).

Role of Green Chemistry in Ecosystem Restoration to Achieve Environmental Sustainability. DOI: https://doi.org/10.1016/B978-0-443-15291-7.00022-5

Air pollution

Air pollutants are mostly gaseous (mainly SO_2, NO_x, CO, O_3, etc.) but also dust, particulate matter, and some heavy metals. Air pollution is a global challenge with the potential to get worse due to continuous and unpredictable changes in the concentrations of these pollutants, especially from human activities (Mandal and Mukherji, 2000). According to Ram et al. (2015), approximately millions of lives are lost each year, especially in developing countries due to poor indoor and outdoor air quality as they contain one, or more, hazardous substances. The main pollutants are polycyclic aromatic hydrocarbons, heavy metals, ozone, halogens, CO, NO_2, etc. with potential and actual disastrous effects on all environmental components (Ioannis et al., 2020). Wildfires are particularly burdened by toxic levels leading to reproductive failure, birth defects, and death (Zuhara and Isaifan, 2018). The effects of atmospheric pollution are more on plant leaves and epidermal tissues, which are often the first receptors as they pass through the stomata with physiological and biochemical changes coming later on (Randhi and Reddy, 2012; Lakshmi et al., 2009). The overall feedback and emergent properties can be measured using parameters like pH, chlorophyll levels and integrity, phytochemical changes, and relative water content (Lakshmi et al., 2009; Jyothi and Jaya, 2010; Skrynetska, et al., 2018, 2019; Molnár et al., 2020; Sarkar et al., 2021). The importance of this air pollution tolerance index also suggests if the plant may be used as an air pollution indicator cum monitor as well as a potential manager, simply because they reflect the adaptive capacity of the plant.

Water pollution

Water is polluted when it contains a toxic concentration of foreign or native substances that limit the use and ecosystem function of the water body. Owa (2013) defined polluted water or aquatic ecosystem as one that contains excess amounts of confirmed or suspected pollutants that make it unfit for use such as drinking, bathing, cooking, etc., and unable to sustain the normal functioning of aquatic organisms. Death due to water-borne diseases like cholera and diarrhea is also common in developing countries, especially in high-population-density areas (NADC [National Association of Drainage Contractors], 2021; Hasan et al., 2019).

Land pollution

The land is the recipient of a majority of human waste and acts as a sink for many pollutants. Lithospheric degradation is caused directly or indirectly by anthropogenic processes such as the misuse of land, unsustainable agricultural practices, exploration and exploitation activities, indiscriminate dumping of waste, and indiscriminate disposal of urban wastes, which put together, have negative impacts on land and global development targets like food security. The problem of land pollution is much bigger than it appears because humans as Earth's custodians are not taking adequate steps despite a majority of the resulting changes being irreversible (Kaberi, 2012). Three main drivers of land pollution are industrialization, overpopulation, and urbanization, in addition to improper agricultural, domestic, industrial, solid, radioactive, and other waste disposals as well as littering, mining, land misuse, and the application of chemicals for crop and animal production (Savasan, 2017). Research into the sources, fate, and occurrence of land pollution suggests it can also lead to some health problems such as respiratory and skin infections including some forms of cancers (Savasan, 2017; Bhatnagar, 2017; Campanale et al., 2020). A pertinent challenge in this effort is that diverse locations exist alongside the sources where these pollutants can be found in toxic concentrations on land and the requirement of a transdisciplinary approach to device appropriate concepts for land security (Skála et al., 2018). Recently, Khan et al. (2021) suggested that knowledge about the status of pollutants in diverse parts of the land as well as a risk assessment using single and total complex pollution indices on a global and regional scale can contribute to addressing the menace of land pollution sustainably.

Plants as monitors of pollutants

Plants show a quick response to environmental stress which makes them viable indicators (Joanna, 2006). The role of plants in pollution monitoring is to maintain ecological balance, which is mostly an active process often through nutrient cycling, absorption, adsorption, and accumulation of pollutants (Ugwu, 2014; Oertel and Salánki, 2003; Li et al., 2010; Chu and Karr, 2017; Santamaría and Méndez, 2012). Biomonitoring is vital for sustainable human and societal development as well as climate change response through adaptation and mitigation via their assessment of toxicants in the environment. For instance, besides air quality monitoring, bryophytes and lichens have been used to present the spatial distribution of mercury despite its long half-life and low reactivity of mercury (Lodenius, 2013; Ogwu, 2019a). Again, environmental stress may be indicated by the presence, health status, or disappearance of lichen in forest ecosystems as they are sensitive to changes in the level of sulfur and nitrogen oxides (Khatri and Tyagi, 2015). In the same vein, the sensitivity to cadmium by the small edible plant, *Wolffia globose* presents the argument for using plants as pollution monitors in another dimension due to the risk of poisoning and food change contamination. Therefore the basic mechanism of plants as environment lies in the injury caused by the pollutants on the plant and the symptoms expressed by the plants. An example (of injury and symptom) is cell destruction and chlorophyll collapse leading to leaf shrinkage and discoloration. Higher plants are exposed to diverse ecosystem pollutants due to their immovability and lengthened existence and hence often seem to be worthy monitors (Macedo-Miranda et al., 2016; Sadowska et al., 2001; Allan et al., 2011). Phytomonitoring is an essential component of water quality assessment as well as for conducting studies on water pollution. The status of wetlands can be properly monitored using a plant-based index of biological integrity, which is a numerical presentation of human influence on wetlands, due to their seasonal consistency (Wilson et al., 2013). Generally, aquatic macrophytes have a higher rate of pollutant immobility coupled with time and are well-suited for restoration efforts. Marine plants provide invaluable information to assess the oceanic environment because they are immobile and can rapidly attain or exist in equilibrium with their natural surrounding (Jain et al., 2010). The physiological responses of *Amaranthus hybridus, Datura alba, Lycoiperisum esculentus*, and *Ricinus communis* have been successfully used in the biomonitoring and remediation of urban air pollution in Nigeria and Pakistan (Vwioko et al., 2018; Khalid et al., 2019). Plants in urban environments are acutely exposed to toxic agents from major to minor to trace with related human and environmental risks and their leaves and needles bear the legacy like the results of Rucandio et al. (2011) from Madrid Spain show on *Cedrus deodara* (silver, mercury, molybdenum, vanadium), *Cupressus sempervirens* (zirconium), etc. Therefore plants are crucial to environmental assessments and biogeographic changes as biomonitors. Biological monitoring using plant qualitative and quantitative information incorporates various developmental phenotypic and genomic variables (Chakrabortty and Paratkar, 2006; Balduzzi et al., 2017; Godebo et al., 2022; Middleton et al., 2012). Plant species diversity and their response to environmental changes like climate change are known to be vital to biological monitoring (Joanna, 2006). They act as a mirror to the impacts of toxic agents on the natural world with minimum preparation and training (Kumar et al., 2017). Phytoplanktons and similar species are useful for water pollution monitoring depending on the transmission of light, water, temperature, and suspended solids (Singh and Gupta, 2016). Changes in the phytoplankton species diversity like *Euglena clastica*, *Phacus tortus*, and *Trachelon anas* have been documented to suggest a certain degree of phosphorus and nitrogen pollution within marine ecosystems (Jain et al., 2010). Saber et al. (2016) stated that in addition to their roles as a primary food source for many species, phytoplankton is essential in discussing marine ecosystem sustainability. Prasad (2003) provided a list of plants used in monitoring heavy metal levels and pollution in the environment and includes *Anthoxanthum odoratum*, *Agrostis capillaris*, *A. gigantean*, *A. stolonifera*, *A. tenuis*, *Arrhenatherum pratensis*, *Avenella flexuosa*, *Brachypodium sylvaticum*, *Bromus ramosus*, etc. Plant monitors are used in assessing the health of the environment as a tool for monitoring (even slight) positive and negative changes in the environment as well as their subsequent effects at the systems level. With the aid of phytomonitors, the health status of a certain region—humans, biodiversity, and environment, can be predicted (Khatri and Tyagi, 2015). The responses of these phytomonitors—physical, development, or otherwise, are considered representative of the number and concentration of contaminants supplied and detected even though this may be spatially and temporally extensive and depending on the plant species and types of receptors and pollutants (Kosior et al., 2020; Calzoni et al., 2007). Plant monitors have been appraised in the detection of trace metal depositions, accumulation, and distribution in ecosystems (El-Tanbouly et al., 2021; Ghosh and Singh 2005; Islam et al., 2018). They provide a simple and cheap method for monitoring trace amounts of contaminants in the atmosphere (Dogan et al., 2014; Csavina et al., 2012; Öğütücü, et al., 2021; Pogorzelski, et al., 2021). However, morphologically visible symptoms are considered the first site of impact on phytomonitors (Saber et al., 2016).

Plants as managers of pollutants

Plants can manage pollution through diverse processes including the popular phytoremediation, phytoaccumulation and mobilization, phytoextraction, phytostimulation, photoabsorption, or phytosequestration phytofiltration, phytostabilization, phylloremediation, rhizoremediation, phytovolatilization, phytoaugmentation, phytodegradation, etc. However, the plant pollution management process involves addressing environmental problems through documenting plant interactions with pollutants and microorganisms such as *Betula*, and *Populus* and heavy metal pollutants and *Desulfomonile*, *Clostridium*, *Pseudomonas*, and *Actinobacter*. Therefore they manage these pollutants by removing them from the environment, thereby preventing risks to human health as well as reducing climate change effects. Some pollutants that have been successfully managed by plants include O_3, NO_2, SO_2, CO, polycyclic aromatic hydrocarbons, heavy metals, particulate matter, volatile organic compounds, formaldehyde, radon, halogens, tobacco smoke, asbestos, and several pathogens considered harmful to human and environmental health. Fast-growing ones like those classified as ephemerals and high-biomass-producing plants such as *Jatropha*, willow, and poplar are useful for pollution managers such as Fe, Al, Cu, Hg, Zn, Mn, Cr, As, and energy production (Henry, 2010; Teng et al., 2015; Abhilash et al., 2012; Apte and Salvi, 2016). The process of pollution management may also involve the uptake or absorption of pollutants from the air, land (soil), or water by whole plant or plant parts and their translocation to other parts and/or bioaccumulation in other parts of the part, that is, the shoot (Sumiahadi and Acar, 2018; Schmitz et al., 2000). The movement of pollutants like heavy metals from root to shoot is an essential biochemical process that is necessary for effective phytoextraction (Yang et al., 2014; Zacchini et al., 2009; Suman et al., 2018; Tangahu et al., 2011; Mleczek et al., 2017). In phytofiltration, the contaminants are absorbed or adsorbed and thus their movement to underground waters is minimized. Phytostabilization is also known as phytoimmobilization and is the use of certain plants or plant parts for the stabilization of contaminants in contaminated soils and is used to reduce the mobility and bioavailability of pollutants in the environment, thus preventing their transportation to groundwater or their entry into the food chain (Vandecasteele et al., 2005; Kumar et al., 2020; Guo et al., 2019). Phytostabilization can limit the accumulation of heavy metals and limit leaching in significant quantities into underground water systems. However, this may not be a permanent solution because the heavy metals may persist in the soil as this transport process is limited. Phytovolatilization is the uptake of pollutants from the soil by plants, their conversion to volatile form, and subsequent release into the atmosphere. This phytomanagement of pollution approach can be applied to the detoxification of organic pollutants and some heavy metals like Se, Hg, and As (Mahar et al., 2016). The use is also limited by the fact that it does not remove the pollutant completely but transfers from one segment (soil) to another (atmosphere) from where it can be redeposited. Phytovolatilization is considered the most controversial of all phytomanagement of pollution technologies (Padmavathiamma and Li, 2012). Phytodegradation of organic pollutants requires the help of enzymes such as dehalogenase and oxygenase and rhizospheric microorganisms (Vishnoi and Srivastava 2008). Some studies have reported the use of genetically modified plants (e.g., transgenic poplars) for phytodegradation (Doty et al., 2000; 2003; Mukhopadhyay and Maiti, 2010). Rhizodegradation, that is, the breakdown of pollutants in the soil by plant and microbial processes is a process by which plants manage pollution. Phytodesalination, that is, a phytoremediation process that leads to the removal of excess salts from soils using halophytic plants like *Ipomoea aquatic*, *Alternanthera philoxeroides*, and *Ludwigia adscendens* (Manousaki and Kalogerakis, 2009; Zorrig et al., 2019; Islam et al., 2009). Halophytic plants are adapted to manage heavy metals and other halophytic compounds compared to glycophytic plants, for example, *Suaeda maritima* and *Sesuvium portulacastrum* have been documented to accumulate NaCl from highly salinated soils and allow for subsequent crop production on the same land which increases in soil fertility after repeated crop cultivation and harvest (Ravindran et al., 2007). In the case of recalcitrant pollutants, plants recruit microorganisms to assist in the rhizospheric management of the pollutants. Several plant species like *R. communis, Solanum nitidum, Brassica rapa, Fuertesimalva echinata, Urtica urens*, and *Lupinus ballianus*. Various plants, including *Thlaspi caerulescens, Viola calaminaria, Brassica napus, Avena sativa*, and *Hordeum vulgare*, are known to manage pollutants such as selenium, copper, nickel, cadmium, zinc, and other heavy metals due to physiological and molecular adaptation. With technological advances like "-omics" the molecular basis of how plants manage pollution and pathogens in the environment is becoming increasingly clearer. Some plant pollution managers and pollutants that they can manage are presented in Table 6.1.

Limitations associated with using plants as pollution monitors and ecosystem managers

The specificity of the plant to manage and monitor certain types of pollutants is not well understood leading to unplanned use (Jiang et al., 2018). Therefore there is a need to establish a monitoring and managing index for certain plant species' responses to certain groups of pollutants. For as much as phytoremediation and biomonitoring offer a promising alternative to solving pollution problems more needs to be done to elucidate their applications effectively (Ghosh and Singh, 2005; Karami and Shamsuddin, 2010). These include:

- Optimizing the planning, development, execution, and treatment process of using plants as pollution monitors and ecosystem managers in place of business as usual without evaluation.
- Improving the effectiveness and efficiency of every aspect of photomanagement (of plants employed in the process) by improving their genetic materials.

TABLE 6.1 Some higher plants that have been used in managing pollution.

Plant name	Pollutants that can be managed
Alyssum	Ni accumulator
Amaranthus retroflexus	Accumulator of ^{137}Cs
Armoracia rustica	Hairy-root cultures remove heavy metals
Armeria maritima	Lead accumulator
Atriplex prostrata	Removes some salts from the soil
Azolla pinnata	Accumulator of lead, copper, cadmium, and iron
Brassica canola	Remediates ^{137}Cs-contaminated soil
B. juncea, B. rapa, flax, hemp, cotton, *Cannabis sativa, Cardamonopsis hallerii*	Hyperaccumulators of metals like Cr
Ceratophyllum demersum	Metals accumulator
Datura innoxia	Barium accumulator
Eucalyptus sp.	Removes sodium and arsenic
Eichhornia crassipes	Accumulator of Pb, Cu, Cd, Fe, As, Cs, Zn, and inorganic nutrients
Helianthus annus	Accumulator of Pb, Zn, and Cd and uranium. Removes ^{137}Cs and 90 Sr in hydroponic reactors
Hydrocotyle umbellata	Accumulator of lead, copper, cadmium, and iron
Kochia scoparia	Removes ^{137}Cs and other radionuclides
Lemna minor	Accumulator of lead, copper, cadmium, and iron
Phaseolus acutifolius	Accumulator of ^{137}Cs
Pteris vittata	Arsenic hyperaccumulator
Salix sp.	Phytoextraction of heavy metals, wastewater, and leachate
Jatropha curcas, Canola, oat, wheat, King grass (*Pennisetum americanum, Pennisetum purpureum*), *Arachis hypogaea*	Cd Phytoremediation
Ipomoea batatas, Chenopodium album, Vetiveria zizanioides, Hemidesmus indicus	Pb
Willow	A wide array of contaminants
Poplars	Fertilizers, petrochemicals compounds, inorganic metals and metalloids, and soluble radionuclides
Zea mays, Elsholtzia splendens, Festuca rubra, L. minor	Cu, Cu, Zn, Fe
Miscanthus sp. and *Arundo donax*	Zn, Cr, and Pb
Oryza sativa, Vettiveria zizanioides, Solanum nigrum, and *L. minor*	Cd, Cu
Berkheya coddii	Cobalt
Phytolacca americana	Mn
Pistia stratiotes and *Marrubium vulgare*	Hg
Alyssum lesbiacumAgropyron elongatum, L. minor	Ni
Gypophila sphaerocephala	Boron
P. vittata and *Piricum sativum*	Arsenic
Allium schoenoprasum and *Medicago sativa*	Ni, Co, Cd
Cajanus cajan	As, Cd

(*Continued*)

TABLE 6.1 (Continued)

Plant name	Pollutants that can be managed
Cicer aeritinum	Cd, Pb, Cr, Cu
Cucumis sativus	Pb
Lantana camara and *Lens camara*	Pb
Lepidium sativum	As, Cd, Fe, Pb, Hg
Lactuca sativa	Cu, Fe, Mn, Zn, Ni, Cd, Pb, Co, As
P. stratiotes	Cr, Cd, As
Pisum sativum	Pb, Cu, Zn, Fe, Cd, Ni, As, Cr
Rapanus sativus	As, Cd, Fe, Pb, Cu
Spinacia oleracea	Cd, Cu, Fe, Ni, Pb, Zn, Cr
Sorghum bicolor	Cd, Cu, Zn, Fe
Colocasia esculenta, *Alocasia macrorrhiza*, *A. odora*, and *P. stratiotes*	Wastewater and heavy metal management,
Spathiphyllum species, *Deiffenbachia* species, *Philodendron domesticum*, *Epipermum aureum*, *Syngonium podophyllum*, and *Rhapis excelsa*	Managing air pollutants (CO_2, xylene, benzene, and toluene) and pathogens (bacterial, actinomycetes, formaldehyde, and mold)

Source: Adapted from Kumar et al. (2017), Prasad (2003), El-Tanbouly et al. (2021), Sumiahadi and Acar (2018), Jabeen et al. (2009), Sharma et al. (2018), and Aydogan and Cerone (2021).

- Understanding the concentration of macro- and micronutrients in polluted media and the associated molecular and biochemical mechanisms within the plant.
- First elucidating the interactions within sites of non-point source contaminations and designing appropriate plant-based monitoring and management plans for the site.
- Time required to fully clean up the contaminated site versus plant life cycle—development processes.
- Natural limits to the phytoremediation efficiency of plants that have been documented as pollutant hyperaccumulators as well as their growth, development, and biomass should be well understood.
- Associated difficulty in the mobilization, accumulation, and removal of a tightly bound fraction of pollutants from the environment.
- Site management requirements about plant habits.
- The human risk associated with consuming plants used in the process and or the best method of disposal after use to prevent transfer into the human and animal food chain.

Prospects of using plants as pollution monitors and ecosystem managers

Phytoremediation is considered a green remediation technology with immense potential. Cen (2015) posited that plant monitoring and management of pollution is an economic and ecologically viable option that is better than conventional approaches. However, most research to elucidate the process is limited to mesocosms—laboratory and greenhouse scale studies and only a few have been conducted in the field. Also, results from field studies can differ significantly from mesocosm ones that are conducted in a laboratory or greenhouse (Valentín-Vargas et al., 2014; Qiu et al., 2013; Beans, 2017). This may be due to the controlled or uncontrol system in mesocosm compared to the broad observation and inferential power of field studies. Nonetheless, results from mesocosms are widely accepted for understanding plant roles as pollution managers and monitors because of their size and repeatability. Some of the factors that may affect plants' role as pollution managers and monitors include abiotic conditions like temperature, climate, nutrients, humidity, precipitation and moisture, plant pathogens and herbivory, uneven distribution of contaminants, and other conditions specific to the soil type, location, and contaminant (Vangronsveld et al., 2009). The efficiency of the pollution monitoring and management roles of plants for specific groups of heavy metals may show different results under different conditions. Understanding the cellular level metabolic mechanism responsible for the process and how they can be manipulated to improve the process and then commercialized will be key to future societal development. For instance, after the identification of desirable traits in hyperaccumulators like edenfern (*Pteris*), the selection of such traits either by breeding techniques, hybridization (such as protoplast fusion), or manipulation of gene expression in transgenic plants their commercial demand increased significantly (Beans, 2017). Research is in progress to identify more genes coding for the hyperaccumulation of specific heavy metals in different plant species. Transgenic plants may also be developed that can better secrete selective metal ligands into the rhizosphere and solubilize toxic elements of interest (Rathore et al., 2015). Thus a combination of desirable traits can be put together and transferred into a single plant species, which may better serve the intended purpose.

Conclusion

The monitoring and management of pollution is a major challenge in the world especially as they are significantly contributing to climate change and human health issues. Also, the geographic distribution and spread of human diseases, as well as the rate of climate change, correlate with environmental pollution. Higher and lower plant forms are viable solution for the monitoring and management of pollution that is cheap, efficient, effective, and easily standardizable. Moreover, diverse plants are available almost all over the world. However, the use of plants as pollution monitors and ecosystem managers will benefit from more awareness and outreach programs aimed at improving acceptability. This is more sustainable than the use of more advanced and technology-based approaches that may be economically expensive, environmentally unsafe, and socially costly. The approach should be supported by relevant policies including those intended to reduce the cost and acceptability. Also, global strategy, policy, and consensus are needed to reduce the effects of monitoring and managing pollution on plant diversity as well as ecological stability and resilience. Sharing of outstanding plant cultivars, species, subspecies, and varieties for monitoring pollution and managing ecosystems should be encouraged.

References

Abhilash, P.C., Powell, J.R., Singh, H.B., Singh, B.K., 2012. Plant-microbe interactions: novel applications for exploitation in multipurpose remediation technologies. Trends in Biotechnology 30, 416–420. Available from: https://doi.org/10.1016/j.tibtech.2012.04.004.

Allan, E., Weisser, W., Weigelt, A., Roscher, C., Fischer, M., Hillebrand, H., 2011. More diverse plant communities have higher functioning over time due to turnover in complementary dominant species. Proceedings of the National Academy of Sciences of the United States of America 108 (41), 17034–17039.

Apte, K., Salvi, S., 2016. Household air pollution and its effects on health. F1000Research 5, F1000.

Aydogan, A., Cerone, R., 2021. Review of the Effects of Plants on Indoor Environments. Indoor and Built Environment 30 (4), 442–460.

Balduzzi, M., Binder, B.M., Bucksch, A., Chang, C., Hong, L., Iyer-Pascuzzi, A.S., et al., 2017. Reshaping Plant Biology: Qualitative and Quantitative Descriptors for Plant Morphology. Frontiers in Plant Science 8, 117.

Barry, J., 2007. Environment and Social Theory, Second Edition Routledge, UK, p. 345.

Beans, C., 2017. Phytoremediation advances in the lab but lags in the field. PNAS 114 (29), 7475–7477.

Bellard, C., Bertelsmeier, C., Leadley, P., Thuiller, W., Courchamp, F., 2012. Impacts of climate change on the future of biodiversity. Ecology Letters 15 (4), 365–377.

Bhatnagar, A., 2017. Environmental determinants of cardiovascular disease. Circulation Research 121 (2), 162–180.

Calzoni, G.L., Antognoni, F., Pari, E., Fonti, P., Gnes, A., Speranza, A., 2007. Active biomonitoring of heavy metal pollution using Rosa rugosa plants. Environmental Pollution (Barking, Essex: 1987) 149 (2), 239–245.

Campanale, C., Massarelli, C., Savino, I., Locaputo, V., Uricchio, V.F., 2020. A detailed review study on potential effects of microplastics and additives of concern on human health. International Journal of Environmental Research and Public Health 17 (4), 1212.

Cen, S., 2015. Biological monitoring of air pollutants and its influence on human beings. The Open Biomedical Engineering Journal 9, 219–223.

Chakrabortty, S., Paratkar, G.T., 2006. Biomonitoring of trace element air pollution using mosses. Aerosol Air Qual. Res 6, 247–258.

Chu, E.W., Karr, J.R., 2017. Environmental impact: concept, consequences, measurement. Reference Module in Life Sciences. Available from: https://doi.org/10.1016/B978-0-12-809633-8.02380-3.

Csavina, J., Field, J., Taylor, M.P., Gao, S., Landázuri, A., Betterton, E.A., et al., 2012. A review on the importance of metals and metalloids in atmospheric dust and aerosol from mining operations. The Science of the Total Environment 433, 58–73.

Dogan, Y., Unver, M.C., Ugulu, I., Calis, M., Durkan, N., 2014. Heavy metal accumulation in the bark and leaves of *Juglans regia* planted in Artvin City, Turkey. Biotechnology, Biotechnological Equipment 28 (4), 643–649.

Doty, S.L., Shang, Q.T., Wilson, A.M., Moore, A.L., Newman, L.A., Strand, S.E., et al., 2003. Metabolism of the soil and groundwater contaminants, ethylene dibromide and trichloroethylene, by the tropical leguminous tree, *Leucaena leucocephala*. Water Research 37, 441–449.

Doty, S.L., Shang, Q.T., Wilson, A.M., Tangen, J., Westergreen, A., Newman, L.A., et al., 2000. Enhanced metabolism of halogenated hydrocarbons in transgenic plants containing mammalian P450 2E1. Proceedings of the National Academy of Sciences of the United States of America 97, 6287–6291.

El-Tanbouly, R., Hassan, Z., El-Messeiry, S., 2021. The role of indoor plants in air purification and human health in the context of COVID-19 pandemic: a proposal for a novel line of inquiry. Frontiers in Molecular Biosciences 8, 709395.

Gheorghe, I., Barbu, I., 2011. The effects of air pollutants on vegetation and the role of vegetation in reducing atmospheric pollution. The Impact of Air Pollution on Health, Economy, Environment and Agricultural Source 12, 241–280.

Ghosh, M., Singh, S., 2005. A review on phytoremediation of heavy metals and utilization of its by-products. Applied Ecology and Environmental Research 3. Available from: https://doi.org/10.15666/aeer/0301_001018.

Godebo, A.T., Wee, N., Yost, C.K., Walley, F.L., Germida, J.J., 2022. A meta-analysis to determine the state of biological control of Aphanomyces root rot. Frontiers in Molecular Biosciences 8, 777042.

Guo, W., Pan, B., Sakkiah, S., Yavas, G., Ge, W., Zou, W., et al., 2019. Persistent organic pollutants in food: contamination sources, health effects and detection methods. International Journal of Environmental Research and Public Health 16 (22), 4361.

Hardin, G.L., 2009. Environmental determinism: broken paradigm or viable perspective? Electronic Theses and Dissertations 1839.

Hasan, M.K., Shahriar, A., Jim, K.U., 2019. Water pollution in Bangladesh and its impact on public health. Heliyon 5 (8), e02145.

Henry, R.J., 2010. Evaluation of plant biomass resources available for replacement of fossil oil. Plant Biotechnology Journal 8 (3), 288–293.

Ikhajiagbe, B., Igiebor, F.A., Ogwu, M.C., 2021. Growth and yield performances of rice (*Oryza sativa* var. nerica) after exposure to biosynthesized nanoparticles. Bulletin of the National Research Centre 45, Article number 62.

Ikhajiagbe, B., Ogwu, M.C., 2020. Hazard quotient, microbial diversity and plant composition of spent crude oil polluted-soil. Beni-Suef University Journal of Basic and Applied Sciences 9 (26).

Ioannis, M., Elisavet, S., Agathangelos, S., Eugenia, B., 2020. Environmental and health impacts of air pollution: a review. Frontiers in Public Health 8 (14), 1–13.

Islam, M.M., Hoque, M.A., Okuma, E., Murata, Y., 2009. Exogenous proline and glycinebetaine increase antioxidant enzyme activities and confer tolerance to cadmium stress in cultured tobacco cells. Journal of Plant Physiology 166, 1587–1597.

Islam, M.S., Hosen, M.M.L., Uddin, M.N., 2018. Phytodesalination of saline water using *Ipomoea aqautica, Alternanthera philoxeroides* and *Ludwigia adscendens*. International Journal of Environmental Science and Technology 16, 965–972.

Jabeen, R., Ahmad, A., Iqbal, M., 2009. Phytoremediation of Heavy Metals: Physiological and Molecular Mechanisms. The Botanical Review 75, 339–364.

Jain, A., Singh, B.N., Si, S., Singh, H.B., Singh, S., 2010. Exploring biodiversity as biological indicators for aquatic pollution. In: National Conf. on Biodiversity, Development and Poverty Alleviation. Uttar Pradesh State Biodiversity Board, Lucknow, Uttar Pradesh.

Jiang, Y., Fan, M., Hu, R., Zhao, J., Wu, Y., 2018. Mosses are better than leaves of vascular plants in monitoring atmospheric heavy metal pollution in urban areas. International Journal of Environmental Research and Public Health 15 (6), 1105.

Joanna, B., 2006. Bioindicators: types, development, and use in ecological assessment and research. Environmental Bioindicators 1, 22–39.

Jyothi, S.J., Jaya, D.S., 2010. Evaluation of air pollution tolerance index of selected plant species along roadsides in Thiruvananthapuram, Kerala. Journal of Environmental Biology 31 (3), 379–386.

Kaberi, M., 2012. Land Pollution: Causes, Effects and Remedies. EO-GE-03-TH-Environmental Geography. Undergraduate Resources of Raja Narendralal Khan Women's College (an affiliate of Vidyasagar University), Midnapore, West Bengal, India, p. 4. Available from: http://www.rnlkwc.ac.in/pdf/studymaterial/geography/4th_Sem_General_GEO_GE_03_TH_4.%20Land%20Pollution.pdf.

Karami, A., Shamsuddin, Z., 2010. Phytoremediation of heavy metals with several efficiency enhancer methods. African Journal of Biotechnology 9 (25), 3689–3698.

Khalid, N., Masood, A., Noman, A., Aqeel, M., Qasim, M., 2019. Study of the responses of two biomonitor plant species (*Datura alba* & *Ricinus communis*) to roadside air pollution. Chemosphere 235, 832–841. Available from: https://doi.org/10.1016/j.chemosphere.2019.06.143.

Khan, S., Naushad, M., Lima, E.C., Zhang, S., Shaheen, S.M., Rinklebe, J., 2021. Global soil pollution by toxic elements: current status and future perspectives on the risk assessment and remediation strategies – a review. Journal of Hazard Materials 5 (417), 126039.

Khatri, N., Tyagi, S., 2015. Influences of natural and anthropogenic factors on surface and ground aquatic quality in rural and urban areas. Front Life Sci. 8, 23–39.

Kosior, G., Frontasyeva, M., Ziembik, Z., Zincovscaia, I., Dołhańczuk-Śródka, A., Godzik, B., 2020. The moss biomonitoring method and neutron activation analysis in assessing pollution by trace elements in selected polish national parks. Archives of Environmental Contamination and Toxicology 79 (3), 310–320. Available from: https://doi.org/10.1007/s00244-020-00755-6.

Kumar, A., Kumar, A., MMS, C.P., Chaturvedi, A.K., Shabnam, A.A., Subrahmanyam, G., et al., 2020. Lead toxicity: health hazards, influence on food chain, and sustainable remediation approaches. International Journal of Environmental Research and Public Health 17 (7), 2179.

Kumar, M., Nandini, N., 2013. Identification and evaluation of air pollution tolerance index of selected avenue tree species of urban Bangalore, India. International Journal of Emerging Technologies in Computational and Applied Sciences 13, 388–390.

Kumar, S., Singh, R., Kumar, V., Rani, A., Jain, R., 2017. *Cannabis sativa:* a plant suitable for phytoremediation and bioenergy production. In: Bauddh, K., Singh, B., Korstad, J. (Eds.), Phytoremediation Potential of Bioenergy Plants. Springer, Singapore, pp. 269–285. Available from: https://doi.org/10.1007/978-981-10-3084-0_10.

Lakshmi, P.S., Sravanti, K.L., Srinivas, N., 2009. Air pollution tolerance index of various plant species growing in industrial areas. The Ecoscan 2 (2), 203–206.

Li, L., Binghui, Z., Lusan, L., 2010. Biomonitoring and bioindicators used for river ecosystems. Procedia Environmental Sciences 2, 1510–1524.

Lodenius, M., 2013. Use of plants for biomonitoring of airborne mercury in contaminated areas. Environmental Research 125, 113–123. Available from: https://doi.org/10.1016/j.envres.2012.10.014.

Macedo-Miranda, G., Avila-Pérez, P., Gil-Vargas, P., Zarazúa, G., Sánchez-Meza, J.C., Zepeda-Gómez, C., et al., 2016. Accumulation of heavy metals in mosses: a biomonitoring study. SpringerPlus 5 (1), 715.

Mahar, A., Wang, P., Ali, A., Awasthi, M.K., Lahori, A.H., Wang, Q., 2016. Challenges and opportunities in the phytoremediation of heavy metals contaminated soils: a review. Ecotoxicology and Environmental Safety 126, 111–121.

Mandal, M., Mukherji, S., 2000. Changes in chlorophyll context, chlorophyllase activity, Hill reaction, photosynthetic CO2 uptake, sugar and starch contents in five dicotyledonous plants exposed to automobile exhaust pollution. Journal of Environmental Biology 21, 37–41.

Manisalidis, I., Stavropoulou, E., Stavropoulos, A., Bezirtzoglou, E., 2020. Environmental and health impacts of air pollution: a review. Frontiers in Public Health 8, 14.

Manousaki, E., Kalogerakis, N., 2009. Phytoextraction of Pb and Cd by the Mediterranean saltbush (*Atriplex halimus* L.): metal uptake in relation to salinity. Environmental Science and Pollution Research International 16 (7), 844–854. Available from: https://doi.org/10.1007/s11356-009-0224-3.

Middleton, A.M., Farcot, E., Owen, M.R., Vernoux, T., 2012. Modeling regulatory networks to understand plant development: small is beautiful. The Plant Cell 24 (10), 3876–3891.

Mleczek, M., Goliński, P., Krzesłowska, M., Gąsecka, M., Magdziak, Z., Rutkowski, P., et al., 2017. Phytoextraction of potentially toxic elements by six tree species growing on hazardous mining sludge. Environmental Science and Pollution Research International 24 (28), 22183–22195.

Molnár, V.É., Tőzsér, D., Szabó, S., Tóthmérész, B., Simon, E., 2020. Use of leaves as bioindicator to assess air pollution based on composite proxy measure (APTI), dust amount and elemental concentration of metals. Plants (Basel, Switzerland) 9 (12), 1743.

Mukhopadhyay, S., Maiti, S.K., 2010. Phytoremediation of metal mine waste. Appl. Ecol. Environ. Res. 8, 207–222.

NADC (National Association of Drainage Contractors), 2021. Watercourse pollution and your liability. <https://nadc.org.uk/watercourse-pollution-and-your-liability/> (accessed March 06.03. 22).

Oertel, N., Salánki, J., 2003. Biomonitoring and bioindicators in aquatic ecosystems. In: Ambasht, R.S., Ambasht, N.K. (Eds.), Modern Trends in Applied Aquatic Ecology. Kluwer Academic, NY, USA, pp. 209–219.

Öğütücü, G., Özdemir, G., Acararıcın, Z., Aydın, A., 2021. Trend analysis of lead content in roadside plant and soil samples in Turkey. Turkish Journal of Pharmaceutical Sciences 18 (5), 581–588.

Ogwu, M.C., 2019a. Ecologic and economic significance of Bryophyte. In: Rathoure, A.K., Chauhan, P.B. (Eds.), Current State and Future Impacts of Climate Change on Biodiversity. IGI Global, Pennsylvania, PA, pp. 54–78.

Ogwu, M.C., 2019b. Towards sustainable development in Africa: the challenge of urbanization and climate change adaptation. In: Cobbinah, P.B., Addaney, M. (Eds.), The Geography of Climate Change Adaptation in Urban Africa. Springer Nature, Switzerland, pp. 29–55.

Ogwu, M.C., 2019c. Understanding the composition of food waste: an "-omics" approach to food waste management. In: Gunjal, A.P., Waghmode, M.S., Patil, N.N., Bhatt, P. (Eds.), Global Initiatives for Waste Reduction and Cutting Food Loss. IGI Global, Pennsylvania, PA, pp. 212–236.

Ogwu, M.C., 2020. Value of *Amaranthus* [L.] species in Nigeria. In: Waisundara, V. (Ed.), Nutritional Value of Amaranth. IntechOpen, UK, pp. 1–21.

Ogwu, M.C., Oladeji, T.S., 2014. Effect of municipal waste effluent on albino rats [*Rattus norvegicus* (Amori)]. Italian Journal of Occupation and Environmental Hygiene 5 (1), 4–10.

Ogwu, M.C., Osawaru, M.E., Ahana, C.M., 2014. Challenges in conserving and utilizing plant genetic resources (PGR). International Journal of Genetics and Molecular Biology 6 (2), 16–22.

Ogwu, M.C., Chime, A.O., Oseh, O.M., 2018. Ethnobotanical survey of tomato in some cultivated regions in Southern Nigeria. Maldives National Research Journal 6 (1), 19–29.

Osawaru, M.E., Ogwu, M.C., 2014. Conservation and utilization of plant genetic resources. In: Omokhafe, K., Odewale, J. (Eds.), Proceedings of 38th Annual Conference of the Genetics Society of Nigeria. Empress Prints Nigeria Limited, pp. 105–119.

Osawaru, M.E., Ogwu, M.C., 2020. Survey of plant and plant products in local markets within Benin city and environs. In: Filho, L.W., Ogugu, N., Ayal, D., Adelake, L., da Silva, I. (Eds.), African Handbook of Climate Change Adaptation. Springer Nature, Switzerland, pp. 1–24.

Osawaru, M.E., Ogwu, M.C., Braimah, L., 2013a. Growth responses of two cultivated Okra species (*Abelmoschus caillei* (A. Chev.) Stevels and *Abelmoschus esculentus* (Linn.) Moench) in crude oil contaminated soil. Nigerian Journal of Basic and Applied Science 21 (3), 215–226.

Osawaru, M.E., Ogwu, M.C., Chime, A.O., 2013b. Assessment of growth performance of two Okra species (*Abelmoschus esculentus* [L.] Moench and *Abelmoschus caillei* [A. Chev.] Stevels) exposed to crude oil contaminated soil. Nigerian Journal of Biotechnology 26, 11–20.

Osawaru, M.E., Ogwu, M.C., Omoigui, I.D., Aiwansoba, R.O., Kevin, A., 2016. Ethnobotanical survey of vegetables eaten by Akwa Ibom people residing in Benin City, Nigeria. University of Benin Journal of Science and Technology 4 (1), 70–93.

Owa, F.D., 2013. Water pollution: sources, effects, control and management. Mediterranean Journal of Social Sciences 4 (8), 65–68.

Padmavathiamma, P.K., Li, L.Y., 2012. Rhizosphere influence and seasonal impact on phytostabilisation of metals—a field study. Water, Air, and Soil Pollution 223, 107–124.

Pogorzelski, S.J., Rochowski, P., Grzegorczyk, M., Boniewicz-Szmyt, K., 2021. Snowpack-stored atmospheric surface-active contaminants traced with snowmelt water surface film rheology. Environmental Science and Pollution Research International 28 (5), 5443–5454.

Prasad, M.N.V., 2003. Phytoremediation of metal-polluted ecosystems: hype for commercialization. Russian Journal of Plant Physiology 50 (5), 686–700.

Qiu, J.R., Guo, X.F., Cai, Q.Y., Liu, W., Zhang, M.W., Wei, Z.B., et al., 2013. Phytotreatment of sewage sludge contaminated by heavy metals and Pahs by co-planting *Sedum alfredii* and *Alocasia marorrhiza*. Int. J. Phytoremediation 16, 1–13.

Ram, S.S., Majumder, S., Chaudhuri, S., Chanda, S.C., Santra, A., Chakraborty, A., et al., 2015. A review on air pollution monitoring and management using plants with special reference to foliar dust adsorption and physiological stress responses. Critical Reviews in Environ. Sci. and Technol.

Randhi, U., Reddy, A., 2012. Evaluation of tolerant plant species in urban environment: a case study from Hyderabad, India. Universal Journal of Environmental Research and Technology 2, 300–304.

Rathore, K.S., Sunilkumar, G., Campbell, L.M., 2015. Cotton Plant With Seed Specific Reduction in Gossypol. Springer Nature, Germany, p. 17.

Ravindran, K.C., Venkatesan, K., Balakrishnan, V., Chellappan, K.P., Balasubramanian, T., 2007. Restoration of saline land by halophytes for Indian soils. Soil Biology and Biochemistry 39 (10), 2661–2664.

Rucandio, M.I., Petit-Domínguez, M.D., Fidalgo-Hijano, C., García-Giménez, R., 2011. Biomonitoring of chemical elements in an urban environment using arboreal and bush plant species. Environmental Science and Pollution Research International 18 (1), 51–63.

Saber, M., Hoballah, E., Ramadan, R., El-Ashry, S., Zaghloul, A.M., 2016. Kinetic assessment of potential toxic elements desorption from contaminated soil ecosystems irrigated with low quality aquatic. International Journal of Soil Science 11, 71–78.

Sadowska, A., Pluygers, E., Niklińska, W., Maria, M.R., Obidoska, G., 2001. Use of higher plants in the biomonitoring of environmental genotoxic pollution. Folia Histochemica et Cytobiologica 39 (Suppl 2), 52–53.

Santamaría, L., Méndez, P.F., 2012. Evolution in biodiversity policy – current gaps and future needs. Evolutionary Applications 5 (2), 202–218.

Sarkar, S., Mondal, K., Sanyal, S., Chakrabarty, M., 2021. Study of biochemical factors in assessing air pollution tolerance index of selected plant species in and around Durgapur industrial belt, India. Environmental Monitoring and Assessment 193 (8), 474. Available from: https://doi.org/10.1007/s10661-021-09253-1.

Savasan, Z., 2017. Encyclopedia of Big Data. Springer International Publishing AG.

Schmitz, H., Hilgers, U., Weidner, M., 2000. Assimilation and metabolism of formaldehyde by leaves appear unlikely to be of value for indoor air purification. The New Phytologist 147 (2), 307–315.

Sharma, S.D., Jain, P.K., Sharma, V., 2018. Cleanest, greenest solution for maintaining indoor air quality in urban areas: plants. IJTSRD 2 (2), 1442–1450.

Singh, M.R., Gupta, A., 2016. Water pollution-sources, effects and control. Pointer Publishers Jaipur, India 1–16 pp.

Skrynetska, I., Ciepał, R., Kandziora-Ciupa, M., 2018. Ecophysiological responses to The effect of soil on human health: an overview. In: Steffan, J. J., Brevik, E.C., Burgess, L.C., Cerdà, A. (Eds.), European Journal of Soil Science, 69. pp. 159–171. (*1*).

Skrynetska, I., Karcz, J., Barczyk, G., Kandziora-Ciupa, M., Ciepał, R., Nadgórska-Socha, A., 2019. Using Plantago major and Plantago lanceolata in environmental pollution research in an urban area of Southern Poland. Environmental Science and Pollution Research International 26 (23), 23359–23371. Available from: https://doi.org/10.1007/s11356-019-05535-x.

Skála, J., Vácha, R., Čupr, P., 2018. Which compounds contribute most to elevated soil pollution and the corresponding health risks in floodplains in the headwater areas of the central european watershed? International Journal of Environmental Research and Public Health 15 (6), 1146.

Sukumaran, D., 2014. Effect of air pollution on the anatomy some tropical plants. Applied Ecology and Environmental Sciences 2, 32–36.

Suman, J., Uhlik, O., Viktorova, J., Macek, T., 2018. Phytoextraction of heavy metals: a promising tool for clean-up of polluted environment? Frontiers in Plant Science 9, 1476.

Sumiahadi, A., Acar, R., 2018. A review of phytoremediation technology: heavy metals uptake by plants. IOP Conference Series: Earth and Environmental Science 142, 012023.

Tangahu, B.V., Sheikh Abdullah, S.R., Basri, H., Idris, M., Anuar, N., Mukhlisin, M., 2011. A review on heavy metals (As, Pb, and Hg) uptake by plants through phytoremediation. International Journal of Chemical Engineering 2011, 939161.

Teng, Y., Wang, X., Li, L., Li, Z., Luo, Y., 2015. Rhizobia and their bio-partners as novel drivers for functional remediation in contaminated soils. Frontiers in Plant Science 6, 32.

Tran, V.V., Park, D., Lee, Y.C., 2020. Indoor air pollution, related human diseases, and recent trends in the control and improvement of indoor air quality. International Journal of Environmental Research and Public Health 17 (8), 2927.

Ugwu, P.N. (2014). Effects of Heavy Metals on the Air Pollution Tolerance Indices (APTI) of Five Medicinal Plants Growing Within Quarry Site in Ishiagu, Ebonyi State, Nigeria (M.Sc. thesis). Dept. of Chemistry, University of Nigeria, Nsukka, 124 p.

Valentín-Vargas, A., Root, R.A., Neilson, J.W., Chorover, J., Maier, R.M., 2014. Environmental factors influencing the structural dynamics of soil microbial communities during assisted phytostabilization of acid-generating mine tailings: a mesocosm experiment. The Science of the Total Environment 500–501, 314–324.

Vandecasteele, B., Meers, E., Vervaeke, P., De Vos, B., Quataert, P., Tack, F.M.G., 2005. Growth and trace metal accumulation of two Salix clones on sediment-derived soils with increasing contamination levels. Chemosphere 58, 995–1002.

Vangronsveld, J., Herzig, R., Weyens, N., Boulet, J., Adriaensen, K., Ruttens, A., et al., 2009. Phytoremediation of contaminated soils and groundwater: lessons from the field. Environmental Science and Pollution Research International 16 (7), 765–794.

Vishnoi, S.R., Srivastava, P.N., 2008. Phytoremediation-green for environmental clean. In: The 12th World Lake Conference, pp. 1016–1021.

Vwioko, D.E., Okoekhian, I., Ogwu, M.C., 2018. Stress analysis of *Amaranthus hybridus* L. and *Lycopersicon esculentum* Mill. exposed to sulphur and nitrogen dioxide. Pertanika Journal of Tropical Agricultural Sciences 41 (3), 1169–1191.

Wilson, M.J., Bayley, S.E., Rooney, R.C., 2013. A plant-based index of biological integrity in permanent marsh wetlands yields consistent scores in dry and wet years. Aquatic Conservation: Marine and Freshwater Ecosystems 23, 698–709.

Yang, W.D., Wang, Y.Y., Zhao, F.L., Ding, Z.L., Zhang, X.C., Zhu, Z.Q., et al., 2014. Variation in copper and zinc tolerance and accumulation in 12 willow clones: implications for phytoextraction. Journal of Zhejiang University. Science. B 15 (9), 788–800.

Zacchini, M., Pietrini, F., Mugnozza, G.S., 2009. Metal tolerance, accumulation and translocation in poplar and willow clones treated with cadmium in hydroponics. Water, Air, and Soil Pollution 197 (1–4), 23–34. Available from: https://doi.org/10.1007/s11270-008-9788-7.

Zorrig, W., et al., 2019. Genetic analysis of cadmium accumulation in lettuce (*Lactuca sativa*). Plant Physiology and Biochemistry 136, 67–75. Available from: https://doi.org/10.1016/j.plaphy.2019.01.011.

Zuhara, S., Isaifan, R., 2018. The impact of criteria air pollutants on soil and water: a review. Journal of Environmental Science and Pollution Research 4 (2), 278–284.

Chapter 7

Climate change mitigation and adaptation strategies, the environment, and impacts of the COVID-19 pandemic: a review of the literature

Shama E. Haque, Lameesa Gazi-Khan and Habiba Rashid
North South University, Dhaka, Bangladesh

Introduction

The climate is the long-term weather pattern experienced in a specific region, usually over fairly long periods, and climate change refers to long-term alternations in Earth's climate over time scales of decades or longer owing to natural variability or impacts of anthropogenic activities (IPCC, 2007). Global warming is the remarkably rapid increase in global temperatures primarily due to the rising levels of greenhouse gases (GHGs) in our planet's atmosphere (Ramanatham and Feng, 2008). Fossil fuel (e.g., coal, oil, gas) combustion for energy releases heat-trapping GHGs into Earth's atmosphere, which intensifies the greenhouse effect, raising our planet's average air temperatures (Stager, 2012). Over time, GHGs can be removed or sequestered from the atmosphere by chemical reactions or by emissions sinks (e.g., oceans and vegetation). However, since the initiation of the Industrial Revolution and the invention of coal-powered steam engines, human-generated GHGs are entering the planet's atmosphere at a faster pace than they are being removed (IPCC, 2007). As the industrial revolution gained momentum in the 1800s, anthropogenic activities markedly raised the quantities of GHGs emitted into the atmosphere. Further, evidence of human-induced climate change is being observed in extremes, such as intense heatwaves, drier droughts, heavy rainfall events, and tropical cyclones (IPCC, 2014). Additionally, the increased atmospheric GHG concentrations are affecting Earth's radiative balance, which in turn is influencing climate and weather patterns both at global and regional levels (Loubere, 2012; IPCC, 2021). In 2019 global CO_2 levels in Earth's atmosphere were greater than at any time during the past 2 million years, and historical measurements reveal those levels of CH_4 and N_2O were unprecedented compared with the last 800,000 years (very high confidence; IPCC, 2021). Specifically, China, the European Union, and the United States are the top three emitters of CO_2, accounting for more than 50% of the production (Delgado, 2021). Since 1880, the warming has increased by approximately 0.08°C per decade, and during the last four decades, the rate of increase has more than doubled (Lindsey and Dahlman, 2021). The warming trend is expected to continue, and by 2100, global average surface temperatures are projected to increase between 1.1°C and 6.4°C (IPCC, 2007). With the impacts of changing climate, the intensity and frequency of precipitation, and storm patterns are also anticipated to change, although the direction and magnitude of change will likely vary by region and season (Lawler, 2009). Lindsey (2020) reports that since scientific record keeping started in 1880, the global sea level has risen approximately 21–24 cm, and the sea level has risen from 1.4 mm/year during most of the 20th century to 3.6 mm/year between 2006 and 2015. Further, in 2020, the sea level rise reached a new record high of 9.13 cm. Sea level rise, overpumping of groundwater, and a combination of other factors can lead to the encroachment of seawater into coastal aquifers (Carretero et al., 2013). In many developing countries, global warming and changing climate are exacerbating the impacts of high population growth, poverty, and rapid urbanization (Eriksen et al., 2007; Mondal, 2019). Additionally, in recent years, the number and intensity of disasters driven by climate change have increased in many areas of Asia, Europe, Australia, and the United States

Role of Green Chemistry in Ecosystem Restoration to Achieve Environmental Sustainability. DOI: https://doi.org/10.1016/B978-0-443-15291-7.00006-7

(Kundzewicz et al., 2013), and in parts of the world, changing climate is expected to exaggerate existing problems and produce new combinations of risks (IPCC, 2007). According to the IPCC, to keep temperature rise below 2°C, GHG emissions need to be reduced by 25% by 2030, and recent climate change mitigation efforts have focused on reducing atmospheric GHG to slow down the rate of climate change (IPCC, 2022). This book chapter is based on a systematic and well-organized literature search of the previously published work to identify a breadth of good quality references on climate change and adaptation strategies, and environmental sustainability with influences of the COVID-19 pandemic on climate change discussion.

Climate change mitigation measures and adaptation strategies

According to the IPCC (2014), mitigation is the effort to govern the anthropogenic sources of changing climate and their cumulative effects, particularly GHG emissions and other pollutants, which also influence Earth's energy balance. Climate change mitigation means renewable energy sources, retrofitting buildings to improve energy efficiency; using more fuel-efficient vehicles; and reducing emissions from deforestation and forest degradation to reduce or prevent the GHG emissions linked to human activities (USEPA, 2022). It is also imperative to understand that the longer we delay mitigation action or fail to implement mitigation policies, the economic cost of mitigating changing climate will continue to rise (Rosen and Guenther, 2015). Adapting to changing climate refers to modifications in procedures, methods, and structures to protect a community or ecosystem against the consequences of changing climate or to benefit from climate change opportunities (Smit et al., 2003). Although mitigation and adaptation present two fundamentally dissimilar approaches, however, both ultimately intend to lessen the influences of changing climate (Tol, 2005; Locatelli et al., 2011). The IPCC Fourth and Fifth Assessment Reports make the connection between mitigation and adaptation (IPCC, 2007, 2014). The interest in the interconnections between the two strategies has grown over the years, particularly in the land-use sector (Denton et al., 2014; Duguma et al., 2014). Community-based adaptation approaches are essential to allow local people to take charge and plan for location-specific mitigation measures (Shaw et al., 2015; Pattison et al., 2021).

Climate change mitigation measures

The existing and future global prosperity has relied heavily on the overexploitation of natural resources for energy needs (Sarkodie et al., 2020). Planet has fixed fossil fuel reserves and fossil fuel comprises approximately 80% of the world's primary energy demand (Chu, 2009). It is unlikely that countries will stop coal consumption within the foreseeable future, as such extensive application of capture and long-term storage of CO_2 emissions from fossil fuel power plants combined with increased energy efficiency is a sustainable mitigation measure to reduce GHG emissions (Schrag, 2007; Chu, 2009; Wennersten et al., 2015). A variety of mitigation strategies have been developed to minimize atmospheric CO_2 emissions. Specifically, technological alternatives for reducing CO_2 emissions from fossil fuel power plants into the atmosphere include: (1) using fuels that emit less CO_2 than fuels currently being used, such as biofuels, hydrogen, and natural gas in place of coal; (2) increased use of renewable energy sources or use of low carbon energy to produce electricity; and (3) capturing and sequestrating CO_2 from the atmosphere (Mach and Mastrandrea, 2014; Wennersten et al., 2015; Altawell, 2021). There are several CO_2 removal and storage options available, including soil carbon sequestration, afforestation and reforestation, biochar, accelerated weathering and ocean alkalinity enhancement, ocean storage, ocean fertilization, direct air capture, bioenergy with carbon capture and sequestration, and CO_2-enhanced oil recovery (Jeffery et al., 2020; Emerson, 2019).

Carbon capture and sequestration

The terms carbon capture and sequestration refer to both natural and artificial processes of capturing CO_2 from Earth's atmosphere and storing it in oceans, soils, vegetation, and geologic formations. Carbon sequestration is one method of reducing the quantities of atmospheric CO_2 to reduce global climate change (USGS, 2017). Carbon sequestration is an emergent research topic, which addresses a key part of an overall strategy for carbon management to help mitigate the rising CO_2 emission into Earth's atmosphere (Sheikh and Pandit, 2014; Fuss et al., 2018). The global forest net C sink is approximately 10.7 Gt CO_2-eq/year, and grasslands account for about 26% of the ice-free land area, which holds roughly 34% of the global terrestrial carbon (FAO, 2010; Wang et al., 2021). Planting new trees is an effective way to capture substantial quantities of CO_2 (Lewis et al., 2019). Sustainable forest management practices and techniques, such as the selection of appropriate tree species, rotation length, and fertilization regimes, can increase forest carbon sequestration and soil carbon storage (EUSTAFOR, 2010; Waring et al., 2020). Natural forests sequester and store more carbon owing to complex forest stand structures, the distribution of trees by species and size within a stand, and the accumulation of carbon below ground and on the forest floor (Jandl et al., 2007; Waring et al., 2020).

Climate change adaptation strategies

In addressing climate change adaptation issues, international funding agencies and national governments need to create an enabling environment and provide resources and technical support to deal with the influences of changing climate (Eriksen et al., 2020). However, real and effective adaptation occurs through active engagement in civil society organizations, local government, and private enterprises (Eriksen et al., 2020). To develop appropriate adaptation strategies, a major focus should be on communities with greater susceptibility due to the characteristics of impacted regions and countries (Smit et al., 2003). According to OECD (2009), (1) developing countries are especially susceptible to the influences of changing climate because of their heavy reliance on natural resources and their limited capability to deal with the effects; (2) least developed countries, small island developing states and African states require special attention because they are at higher risk due to increased frequency of floods and flash floods, droughts, hurricanes, tornadoes, and desertification; (3) certain regions of the world require special attention as they are more at risk to the effects of changing climate; and (4) vulnerable population (i.e., children, women and the elderly) warrants more attention as they are particularly at risk.

Climate change adaptation to sea level rise

Many researchers report that under higher GHG emission scenarios, the 21st century's rise in sea level may approach or in the extremes exceed 2 m in the case of early-onset Antarctic ice sheet instability (Wong et al., 2017). Rising sea levels threaten several small island nations, such as the Pacific Nation of Tuvalu, Kiribati, and Maldives, which are roughly 2 m above the sea level (Khan et al., 2002; Aung et al., 2009; Kench et al., 2018). To deal with the rising sea level along with the intense extreme events, many low-lying coastal communities are developing responses to adapt to local circumstances (Tol et al., 2008; Lebbe et al., 2021; Nicholls et al., 2007; Kulp and Strauss, 2019). Numerous studies suggest that climate change adaptation strategies should be mainstreamed into economic growth, and development policies and measures for governments to promote economic development while simultaneously reducing susceptibility to the effects of changing climate (Ayers et al., 2014). A report on the principles of adaptation and resilience by The World Bank presents various methods that can be utilized by nations to develop effective adaptation strategies (WB, 2020). Specifically, the report states that (1) the success of adaptive strategies is dependent on ensuring that highly susceptible communities are provided financial, technical and institutional resources to adapt; (2) it is necessary that governments provide financial assistance to boost the adaptive capacity of households and firms; (3) there is a need for review and update of land-use plans to adapt to long-term climate risks, and to not lock people and assets in *hazard-prone* areas; (4) governments need to develop adapting strategies to assist people and firms recover faster from impacts by strengthening early warning systems; (5) communities should be able to deal with effects of changing climate at the macro level; and (6) governments need to identify and prioritize actions to make them compatible with appropriate resources and climate-specific capacity, create a strong legal and efficient institutional framework, and a progress monitoring system.

How are vulnerable regions adapting to climate change?

The following sections present reviews of climate change mitigation and adaptation along with environmental sustainability in Bangladesh, Maldives, and the US State of Florida, which are highly susceptible to the influences of changing climate.

Case study: Bangladesh

Bangladesh is a low-lying deltaic country in South Asia, located within 20°34′ and 26°38′ N and 88°01′ and 92°41′ E. Bangladesh is a densely populated country, which is scheduled to officially become a developing country in 2026. According to the 2015 Climate Change Vulnerability Index, the country's economy is the most at risk due to the effects of changing climate (WB, 2016). Specifically, 2,500, 8,000, and 14,000 km^2 of land are predicted to be lost as a result of sea level rise of 0.1, 0.3, and 1.0 m, respectively (Hossain, 2010; Fig. 7.1). It is important to note that coastal

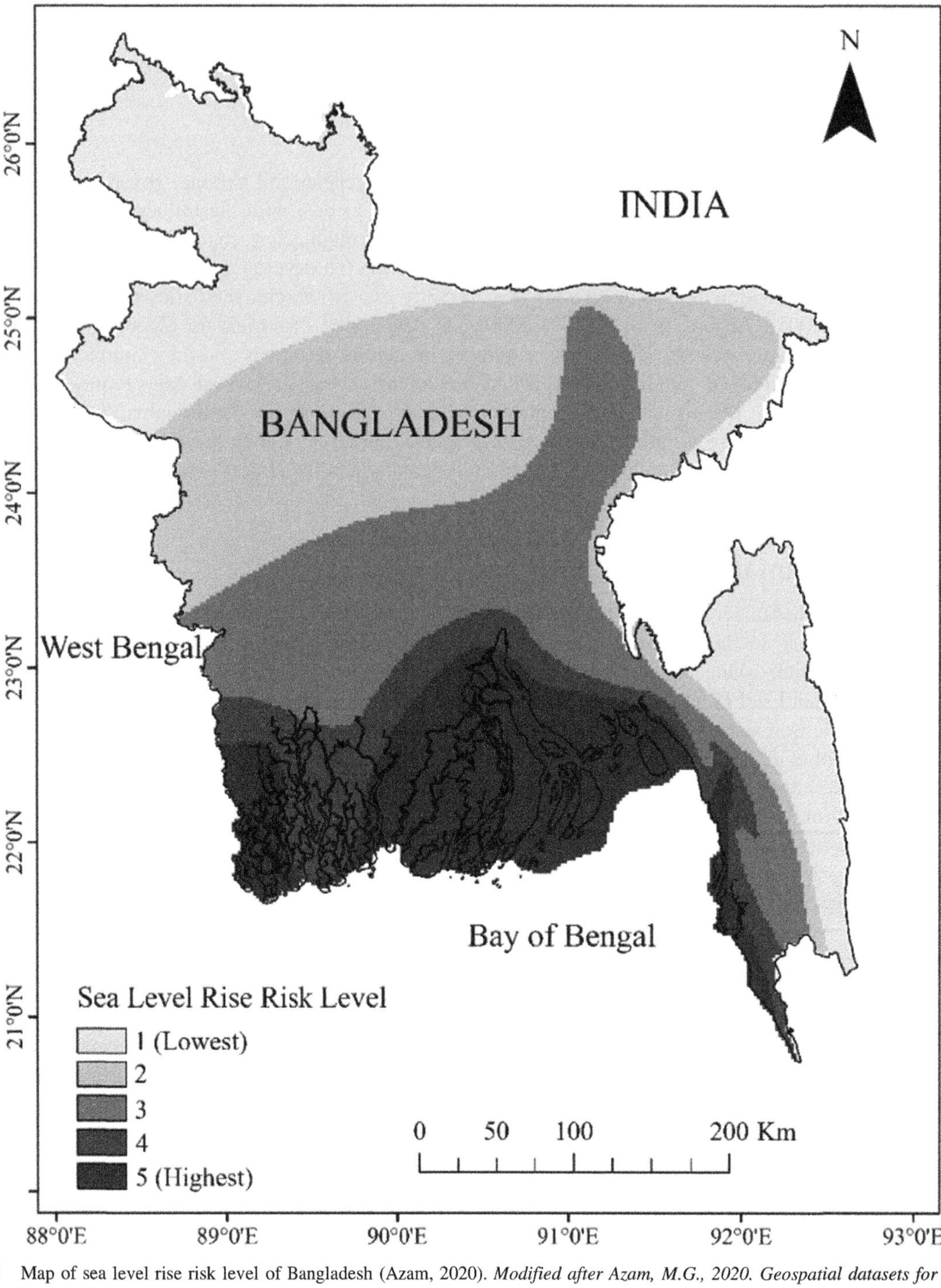

FIGURE 7.1 Map of sea level rise risk level of Bangladesh (Azam, 2020). *Modified after Azam, M.G., 2020. Geospatial datasets for assessing the vulnerability of Bangladesh to climate change and extremes. Mendeley Data, V2. doi: 10.17632/cv6cyfgmcd.2.*

Bangladesh covers approximately 32% of the landmass, and its 35 million residents are anticipated to suffer from the impacts of rising sea levels (Pethick and Orford, 2013; Ahmad, 2019). In the southwestern region of Bangladesh, water resources have been severely impacted by sea level rise and saltwater intrusion, and local communities have adapted to the changing environment in numerous ways (Sanchez et al., 2015). Successful adaptation practices include rainwater-fed ponds, and pond-sand filters along with rainwater harvesting systems (Haque et al., 2020). Past researchers found that in recent years, Bangladesh's cyclone early warning services and evacuation programs have vastly improved and significantly reduced the number of cyclone-related fatalities over the last two decades (Ahsan et al., 2020; Huq, 2022).

On multicentennial timescales, mitigation, and controlled sedimentation strategies will be required to sustain the relative delta height that is crucial to the Ganges–Brahmaputra–Meghna Delta's survival (Brown et al., 2018). Additionally, the Bangladesh Climate Change and Gender Action Plan 2013, formulated by the Bangladesh Government, is a guiding document that incorporates gender considerations into policies, strategies, and interventions associated with changing climate (Haq et al., 2015). During the past three decades, the Bangladesh Government has implemented numerous initiatives to address various influences of changing climate (Fatemi et al., 2020). Bangladesh has already invested over 10 billion USD in climate change actions, and numerous international organizations (the World Bank and British Council) are investing or have invested in actions to mitigate climate change in the southwestern portion of the country (WB, 2016). Additionally, since 2013, The World Bank's Coastal Embankment Improvement Project has assisted the country in mitigating the effects of tidal inundation and storm surges. Due to the rehabilitation and upgrades of older systems, 333,000 Bangladeshis have increased protection from tidal inundation and storm surges, and 424,000 Bangladeshis have access to storm shelters (WB, 2021a). Chowdhury et al. (2021) reviewed the policies and programs for climate change mitigation and environmental sustainability in Bangladesh. These researchers found that there is a national energy policy and several initiatives to curb GHG emissions, such as renewable development, rational and effective use of energy, intended nationally determined contributions, clean development mechanisms, and international and private sector partnerships.

Case study: Maldives

The Maldives (1°55′ N and 73°3′ E), the lowest terrain of any country across the globe, is situated in the Indian Ocean in Asia. Over 80% of the land area of the country is less than 1 m above the mean sea level (UNDP, 2022). The small size and low-lying nature of its 25 low-lying coral atolls and 1190 small tropical islands (averaging roughly 1.5 m above mean sea level), narrow width, and dispersed nature of coral reefs and atolls make Maldives particularly vulnerable to inundation, saltwater intrusion into aquifers, extreme winds and sea level rise (Duvat and Magnan, 2019; Khan et al., 2002; Manzo et al., 2021; WB, 2010) (Fig. 7.2). Environmental problems already present on the Maldivian islands are predicted to get worse even with a small rise in sea level (WB, 2021b). Duvat and Magnan (2019) used insights from the Maldives to study the adaptation pathways for atoll islands and found that the Maldivian islands have been vastly modified by anthropogenic activities. According to the UNDP (2022), adapting to changing climate is a multidimensional goal for Maldives as it prioritizes climate and disaster resilience, and aims to promote sustainable development outcomes. The island nation has recognized the necessity to safeguard island harbors from coastal natural disasters and rising sea levels while simultaneously reducing effects on the natural environment specifically corals that provide natural surge protection (WB, 2021b). Past researchers have found that many Maldivians are not completely aware of or concerned about the timescale of climate change, the influences of changing climate and sea level rise, environmental impact, and the relevant concepts of urgency and crisis (Arnall and Kothari, 2015; Stojanov et al., 2016; Sakamoto et al., 2022). However, international agencies, such as the USAID, The World Bank, and the Global Climate Change Alliance Plus Initiative (funded by the European Union) are contributing, and have contributed in the past in activities to preserve and restore the Maldivian ecosystems (WB, 2018; GCCA +, 2018; USAID, 2022). Malatesta and Friedberg (2017) studied Maldivian environmental policies and climate change susceptibility. These researchers report that the environmental policies indicate the presence of powerful spatial dialectical methods and antitheses.

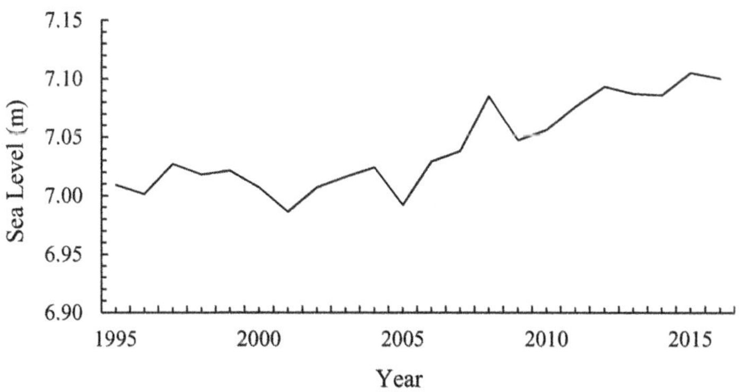

FIGURE 7.2 Tidal gauge data on sea level in Male, Maldives 1995–2016 (Permanent Service for Mean Sea Level, 2020). *Data taken from Permanent Service for Mean Sea Level, 2020. Data and station information for MALE-B, HULULE. [online]. <https://www.psmsl.org/data/obtaining/stations/1753.php> (accessed 24.06.22).*

Case study: the US State of Florida

The US State of Florida (27° 59′ N and 81° 45′ W), is a flat, low-lying peninsula bordered on three sides by the Gulf of Mexico, the Atlantic Ocean, and the Florida Strait. Florida has more than 1700 streams and rivers, over 7700 lakes, 700 springs, 11 million acres of various types of wetlands, including the Everglades in South Florida and Green Swamp in Central Florida, and several aquifers (Purdum et al., 1998; Marella and Dixon, 2018). In Florida, the effects of changing climate have been noticeable for decades (Tompkins and DeConcini, 2013). During the next century, if Earth continues to experience a warming climate, then the sea level along coastal Florida is predicted to rise from 0.30 to 1.2 m (USEPA, 2016; Fig. 7.3). Rising sea levels are expected to gradually inundate wetlands and dry land, erode beaches,

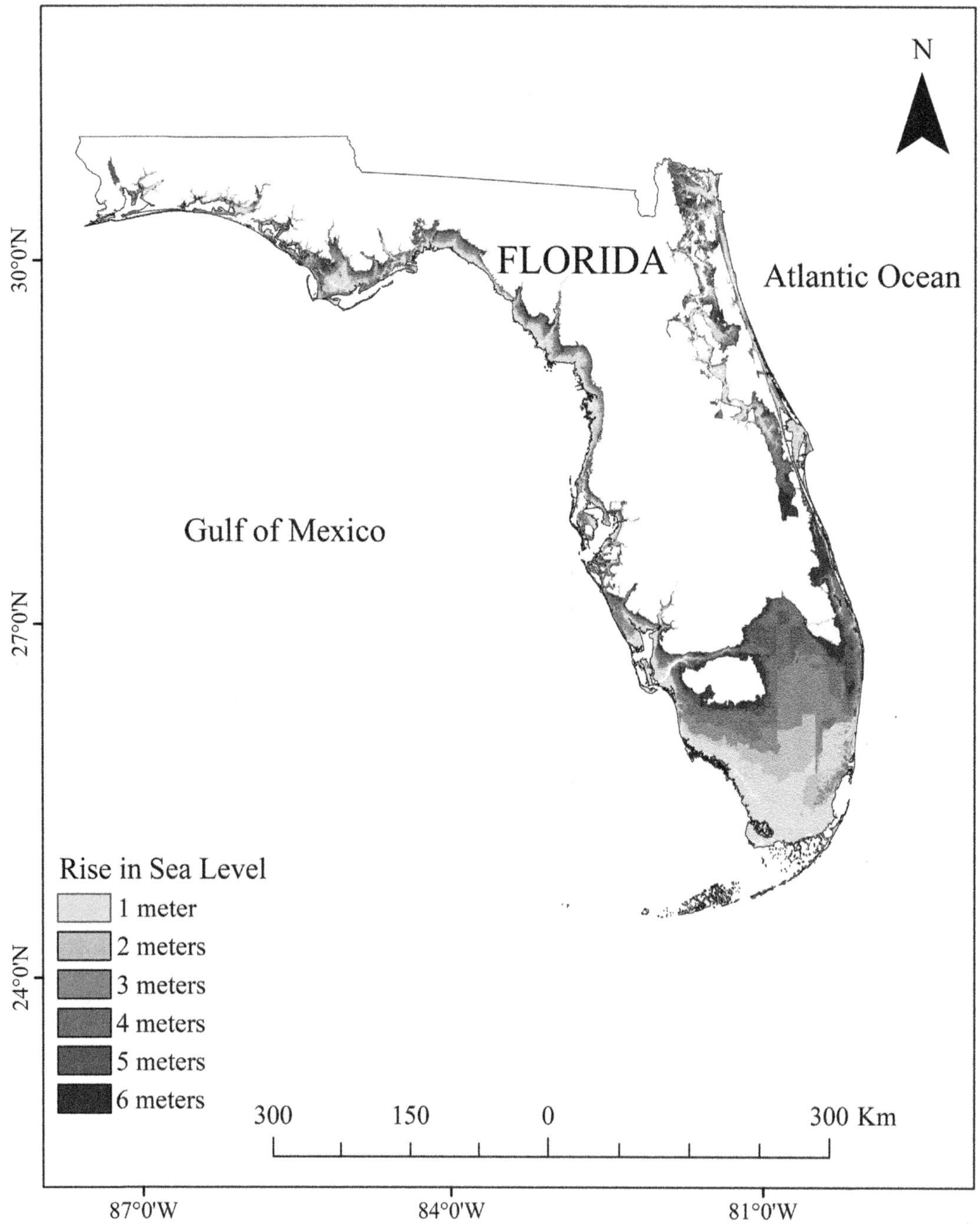

FIGURE 7.3 Map of sea level rise in Florida. *Data taken from Data Basin. Available at: https://databasin.org/datasets/dd54d301894f4322a7a30832572c4a7e/.*

aggravate inundation and exacerbate saltwater intrusion in the state. In Florida, saltwater has already encroached into coastal aquifers because fresh groundwater levels have dropped relative to sea level, which has allowed higher gradient water to move toward the freshwater zones of the aquifers (USGS, 2019). Florida is the third-most populous state in the United States with a population of more than 22 million (Population U.S.A., 2022), and each year the population is increasing by 300,000 persons (O'Brien, 2019). By 2030, Florida's freshwater need is expected to rise by 28% in comparison to levels in 2005 (USEPA, 2013). Local governments struggle to sustainably manage water demand and create water conservation strategies as the aquifers Florida's residents presently rely on for their water usage become vulnerable to the effects of changing climate (Obeysekera et al., 2017). A thorough approach to water planning, which uses alternate water sources and future water use planning based on long-term water demand trends, can make Florida's future water resources more reliable and sustainable (Bloetscher et al., 2010; Asefa et al., 2014). The Florida Everglades is susceptible to climate change, sea level rise and saltwater intrusion as the highest parts of the region are only a few meters above sea level (FOCC 2010; USGS, 2018). In the absence of appropriate infrastructure, the intrusion of saline water is threatening both the natural ecosystem along the urban water supply in the neighboring area (Prinos et al., 2014; Gonzalez, 2018). As temperatures continue to increase, restoration of this largest ecosystem in the world will become more difficult because of the increasing water demand. The state's coral reefs are experiencing the brunt of ocean acidification. To maintain coral reefs, which are also a major player in the state's tourism industry, in a net growth or net-neutral condition, limiting the effects of ocean acidification is crucial. Several mitigation and adaptation strategies have been proposed to control ocean acidification influences on Florida's coral reefs. Albright and Cooley (2019) reviewed interventions intended to address different ecosystem service changes in the Florida Reef Tract and found that the majority of the proposed interventions are expensive and do not address the main cause of ocean acidification (i.e., GHG emissions). Currently, the USEPA is actively pursuing research on actions and best practices to adapt to and mitigate the impacts of the ocean and coastal acidification, such as buffering sediments, restoring seagrasses, and conserving refugia (USEPA, 2022). In terms of adaptation strategies to slow the rate of changing climate, FOCC (2010) suggests (1) encouraging water reuse and more rainwater harvesting; (2) relying on natural buffers against rising sea levels; (3) developing heat health action plans; (4) developing renewable energy sources; and (5) curbing CO_2 emissions. Suman (2017) investigated climate change and security in Florida and revealed that even though the state is susceptible to the effects of changing climate in several sectors, mitigation and adaptation planning at the State level has been underwhelming. Specifically, in 2012, the Natural Resources Defense Council ranked Florida's preparation planning for changing climate in category three out of four categories; which meant that the state is largely unprepared and lagging (NRDC, 2012). The annual Southeast Florida Climate Leadership Summit aims to expedite climate-related alliance and information sharing (SEFRCCC, 2021). In 2010 four counties in Florida (Broward, Miami-Dade, Monroe, and Palm Beach) formed the Southeast Florida Regional Climate Change Compact to organize mitigation and adaptation activities across county lines (FDEP, 2018). Florida's local governments also encourage intergovernmental collaborations and stakeholders to promote environmental sustainability (MCCAP, 2013; Molinaroli et al., 2019). In 2011 the state legislature passed the Community Planning Act, in which, local governments are given the option of developing an Adaptation Action Area within their jurisdiction to address the influences of rising sea levels. In 2013 the National Oceanic and Atmospheric Administration approved a 5-year initiative titled: "Community Resiliency: Planning for Sea Level Rise," which scrutinized the statewide framework and best practices for incorporating adaptation into the existing planning processes (FDEP, 2018). Further, the country's environmental sustainability is weak because corporate lobbying for fossil fuels and large industries has negatively influenced the country's shift to low-carbon technologies (Sachs, 2018).

COVID-19 pandemic's impacts on climate change adaptation

Since the first officially reported case of COVID-19 disease in 2019, the rapidity and intensity of the disease threatened almost all aspects of socioeconomic life across the globe. A literature search indicates that during the pandemic, numerous scientists focused their work on issues concerning the pandemic and climate change issues. Klenert et al. (2020) took a comprehensive approach to draw comparisons between the pandemic and climate change mitigation. During the early stages of the pandemic, these researchers found that delaying mitigation is expensive; policy design should not be influenced by bias; impacts of changing climate are expected to worsen existing inequalities, which can be further aggravated in the absence of prompt action; global problems need various forms of collective action; and transparency in scientific communications are necessary to counter the effects of climate science politicization. Manzanedo and Manning (2020) reported no vaccines or possible treatment options available to solve the climate crisis. Additionally, important lessons learned from the pandemic suggest that urgent actions must be taken to combat climate change and prevent the worst-case consequences. Shulla et al. (2021) studied the effects of the pandemic on reaching the SDGs and reported that climate change does not know national borders, therefore, strong global cooperation is an essential part of realizing the SDGs. In 2020 global CO_2 emissions dropped by −7.4% or −3.0 $MtCO_2$/day in the power industry and by −19% or −4.3 $MtCO_2$/day in the industrial sector (Le Quéré et al., 2020). However, in 2021, global CO_2 emissions were merely 1% less than the record levels of 2019, driven by changes in power and industrial emissions from India and China, and a rebound of CO_2 emissions above pre-pandemic levels (Davis et al., 2022). Despite the rise of usage of cleaner energy sources in many nations, the carbon intensity of electricity has increased noticeably in late 2021, which reflects a revival of fossil fuel-based generation due to the recent increase in the cost of natural gas (Plumer, 2022). Fuentes et al. (2020) compared the similarities and differences between COVID-19 with the climate crisis. According to the Harvard T.H. Chan School of Public Health changing climate modifies how people relate to other species on our planet, which impacts human health and infection risks. Although the changing climate did not cause the pandemic, however, there is increasing evidence that the degradation of environmental reservoirs and wildlife habitats through the impacts of climate change and biodiversity loss has made it easier for pathogens to thrive in new environments (El-Sayed and Kamel, 2020; WHO, 2022). Phillips et al. (2020) reported the Gulf Coast of the United States was experiencing a severe hurricane season and the US State of Louisiana became a hot spot of COVID-19 mortality, to some extent because of racial disparities in medical care. Governments must recognize the intersection between the climate crisis and the pandemic, and develop long-term strategies for both climate adaptation and pandemic preparedness.

Conclusions

Climate change is a global phenomenon, but the effects are not felt equally across the globe. In particular, low-lying coastal settlements are at the forefront of climate change impacts. Several adaptation and mitigation options are currently available to reduce our susceptibility to sea level rise, saltwater intrusion, and other extreme weather events. However, to address the impacts of changing global climate, a single mitigation or adaptation measure by itself is insufficient; an effective response requires thorough planning and integration of appropriate mitigation measures as well as adaptation strategies. Currently, climate change is being factored into many sustainable development plans, nonetheless, further action is still required to avoid or limit many of its deleterious impacts.

References

Ahmad, H., 2019. Bangladesh coastal zone management status and future trends. Journal of Coastal Zone Management 22 (1), ISSN: 2473-3350.

Ahsan, M.N., Khatun, A., Islam, A.S., Vink, K., Ohara, M., 2020. Preferences for improved early warning services among coastal communities at risk in cyclone prone south-west region of Bangladesh. Progress in Disaster Science . Available from: https://doi.org/10.1016/j.pdisas.2020.100065.

Albright, R., Cooley, S., 2019. A review of interventions proposed to abate impacts of ocean acidification on coral reefs. Regional Studies in Marine Science 29, 100612.

Altawell, N., 2021. Energy's general introduction (fossil fuels & renewable energy). Rural Electrification. Academic Press. Available from: https://doi.org/10.1016/C2019-0-04251-3.

Arnall, A., Kothari, U., 2015. Challenging climate change and migration discourse: Different understandings of timescale and temporality in the Maldives. Global Environmental Change 31, 199–206.

Asefa, T., Adams, A., Kajtezovic-Blankenship, I., 2014. A tale of integrated regional water supply planning: Meshing socio-economic, policy, governance, and sustainability desires together. Journal of Hydrology 519, 2632–2641.

Aung, T., Singh, A., Prasad, U., 2009. A study of sea-level changes in the Kiribati area for the last 16 years. Weather 64 (8). Available from: https://doi.org/10.1002/wea.453.

Ayers, J.M., Huq, S., Faisal, A.M., Hussain, S.T., 2014. Mainstreaming climate change adaptation into development: a case study of Bangladesh. WIREs Climate Change 5, 37–51.

Azam, M.G., 2020. Geospatial datasets for assessing vulnerability of Bangladesh to climate change and extremes. Mendeley Data V2, doi: 10.17632/cv6cyfgmcd.2.

Bloetscher, F., Meeroff, D., Heimlich, B., Brown, A., Bayler, D., Loucraft, M., 2010. Improving resilience against the effects of climate change. Journal – American Water Works Association 102 (11), 36–46.

Brown, S., Nicholls, R., Lázár, A., Hornby, D., Hill, C., Hazra, S., et al., 2018. What are the implications of sea-level rise for a 1.5, 2 and 3 °C rise in global mean temperatures in the Ganges-Brahmaputra-Meghna and other vulnerable deltas? Regional Environmental Change 18 (6), 1829–1842. Available from: https://doi.org/10.1007/s10113-018-1311-0.

Carretero, S., Rapaglia, J., Bokuniewicz, H., Kruse, E., 2013. Impact of sea-level rise on saltwater intrusion length into the coastal aquifer, Partido de La Costa, Argentina. Continental Shelf Research 61–62. Available from: https://doi.org/10.1016/j.csr.2013.04.029.

Chowdhury, M., Rahman, S., Abubakar, I., Aina, Y., Hasan, M., Khondaker, A., 2021. A review of policies and initiatives for climate change mitigation and environmental sustainability in Bangladesh. Environment, Development and Sustainability 23 (2), 1133–1161. Available from: https://doi.org/10.1007/s10668-020-00627-y.

Chu, S., 2009. Carbon capture and sequestration. Science 325 (5948), 1599. Available from: https://doi.org/10.1126/science.1181637 (2021). Adaptation: What is it and why do we need it? [online]. <https://climatescience.org/advanced-what-is-adaptation> (accessed 20.05.22).

Davis, S., Liu, Z., Deng, Z., Zhu, B., Ke, P., Sun, T., et al., 2022. Emissions rebound from the COVID-19 pandemic. Nature Climate Change 12 (5), 412–414. Available from: https://doi.org/10.1038/s41558-022-01332-6.

Delgado, C. (2021). How Developing Countries Can Reduce Emissions Without Compromising Growth [online]. Available at: <https://earth.org/how-developing-countries-can-reduce-emissions-without-compromising-growth/> (accessed 03.05.22).

Denton, F., Wilbanks, T.J., Abeysinghe, A.C., Burton, I., Gao, Q., Lemos, M.C., et al., 2014. Climate-resilient pathways: adaptation, mitigation, and sustainable development. In: Field, C.B., Barros, V.R., Dokken, D.J., Mach, K.J., Mastrandrea, M.D., Bilir, T.E., Chatterjee, M., Ebi, K.L., Estrada, Y.O., Genova, R.C., Girma, B., Kissel, E.S., Levy, A.N., MacCracken, S., Mastrandrea, P.R., White, L.L. (Eds.), Climate change 2014: impacts, adaptation, and vulnerability. Part A: Global and sectoral aspects. Contribution of working group II to the fifth assessment report of the Intergovernmental Panel on Climate Change. Cambridge University Press, Cambridge and New York, pp. 1101–1131.

Duguma, L.A., Minang, P.A., van Noordwijk, M., 2014. Climate change mitigation and adaptation in the land use sector: from complementarity to synergy. Environmental Management 54, 420–432. Available from: https://doi.org/10.1007/s00267-014-0331-x.

Duvat, V.K.E., Magnan, A.K., 2019. Contrasting potential for nature-based solutions to enhance coastal protection services in atoll islands. In: Klöck, C., Fink, M. (Eds.), Dealing With Climate Change on Small Islands: Towards Effective and Sustainable Adaptation? Göttingen: Göttingen University Press, pp. 45–75. Available from: https://doi.org/10.17875/gup2019-1211.

El-Sayed, A., Kamel, M., 2020. Climatic changes and their role in emergence and re-emergence of diseases. Environmental Science and Pollution Research International 27 (18), 22336–22352. Available from: https://doi.org/10.1007/s11356-020-08896-w.

Emerson, D., 2019. Biogenic iron dust: a novel approach to ocean iron fertilization as a means of large scale removal of carbon dioxide from the atmosphere. Frontiers in Marine Science 6. Available from: https://doi.org/10.3389/fmars.2019.00022.

Eriksen, S., Klein, R.J.T., Ulsrud, K., Naess, L.O., O'Brien, K., 2007. Climate Change Adaptation and Poverty Reduction: Key Interactions and Critical Measures. Report prepared for the Norwegian Agency for Development Cooperation (Norad), Global Environmental Change and Human Security, University of Oslo.

Eriksen, S., Schipper, E., Scoville-Simonds, M., Vincent, K., Adam, H., Brooks, N., et al., 2020. Adaptation interventions and their effect on vulnerability in developing countries: Help, hindrance or irrelevance? World Development 141, 105383.

EUSTAFOR (European State Forest Association), 2010. European Forestry in the Face of Climate Change. European Forestry House. [online]. Available at: <https://eustafor.eu/uploads/EustaforClimateReport2010_2-1.pdf> (accessed 25.06.22).

FAO, 2010. Global Forest Resources Assessment 2010. FAO Forestry Paper. [online]. Available at: <https://www.fao.org/3/i1757e/i17a57e.pdf> (accessed 25.06.22).

Fatemi, M.N., Okyerem, S.A., Diko, K.S., Kita, M., 2020. Level climate governance in Bangladesh via climate change mainstreaming: lessons for local climate action in Dhaka City. Urban Science 4 (2), 24. Available from: https://doi.org/10.3390/urbansci4020024.

FDEP, 2018. Florida adaptation planning guidebook. [online]. <https://floridadep.gov/sites/default/files/AdaptationPlanningGuidebook.pdf> (accessed 26.06.22).

FOCC (Florida Oceans and Coastal Council), 2010. The effects of climate change on Florida's ocean and coastal resources. A special report to the Florida Energy and Climate Commission and the people of Florida. [online]. <https://floridadep.gov/sites/default/files/The%20Effects%20of%20Climate%20Change%20on%20Florida%27s%20Ocean%20and%20Coastal%20Resources_0.pdf> (accessed 21.06.22).

Fuentes, R., Galeotti, M., Lanza, A., Manzano, B., 2020. COVID-19 and climate change: a tale of two global problems. Sustainability 12 (8560). Available from: https://doi.org/10.3390/su12208560.

Fuss, S., Lamb, W., Callaghan, M., Hilaire, J., Creutzig, F., Amann, T., et al., 2018. Negative emissions—Part 2: costs, potentials and side effects. Environmental Research Letters 13 (6), 063002. Available from: https://doi.org/10.1088/1748-9326/aabf9f.

GCCA + (Global Climate Change Alliance +), 2018. Support to climate change adaptation and mitigation in Maldives [online]. <https://www.gcca.eu/programmes/support-climate-change-adaptation-and-mitigation-maldives> (accessed 18.05.22).

Gonzalez, E.S., 2018. Everglades needs more fresh water to fight salt water intrusion. Florida International University News . Available from: https://news.fiu.edu/2018/everglades-needs-more-fresh-water-to-fight-salt-water-intrusion.

Haque, S., Rahman, M.F., Huq, S., 2020. Adapting to Climate Change in Southwestern International Affairs [online]. Available at: <https://gjia.georgetown.edu/2020/07/28/adapting-to-climate-change-in-southwestern-bangladesh-need-for-transformational-measures/> (accessed 18.05.22).

Haq, M.I., Omar, M.A.T., Zahra, Q.A., Shashi, I.J., Rahman, M.R., 2015. Evaluation of Adaptation Policies in GBM Delta of Bangladesh (Deltas, Vulnerability and Climate Change: Migration and Adaptation [DECCMA] Working Paper). DECCMA Consortium, Southampton.

Harvard T.H. Chan School of Public Health, n.d. Coronavirus and climate change. [online]. <https://www.hsph.harvard.edu/c-change/subtopics/coronavirus-and-climate-change/> (accessed 26.06.22).

Hossain, M.A., 2010. Global warming induced sea level rise on soil, land and crop production loss in Bangladesh. Proceedings of the 19th World Congreve of Soil Science, Solutions for a Changing World 77–80. Available from: https://doi.org/10.1016/j.worlddev.2020.105383. Brisbane, Australia, 2010, August 1–6.

Huq, S. (2022). Adapting to Climate Change: Lessons from Bangladesh. [online] Available at: <https://www.scientificamerican.com/article/adapting-to-climate-change-lessons-from-bangladesh/> (accessed 18.05.22).

IPCC, 2007. Climate change 2007: impacts, adaptation and vulnerability. <https://www.ipcc.ch/site/assets/uploads/2018/03/ar4_wg2_full_report.pdf> (accessed 24.06.22).

IPCC, 2014. Climate change 2014: mitigation of climate change. In: Edenhofer, O., Pichs-Madruga, R., Sokona, Y., Farahani, E., Kadner, S., Seyboth, K., et al.,Contribution of Working Group III to the Fifth Assessment Report of the Intergovernmental Panel on Climate Change. Cambridge University Press, Cambridge and New York, NY.

IPCC, 2021. Climate change 2021: the physical science basis. <https://www.ipcc.ch/report/sixth-assessment-report-working-group-i/> (accessed 24.06.22).

IPCC, 2022. The evidence is clear: the time for action is now we can halve emissions by 2030. [online]. <https://www.ipcc.ch/2022/04/04/ipcc-ar6-wgiii-pressrelease/> (accessed 26.06.22).

Jandl, R., Lindner, M., Vesterdal, L., Bauwens, B., Baritz, R., Hagedorn, F., et al., 2007. How strongly can forest management influence soil carbon sequestration? Geoderma 137 (3–4), 253–268. Available from: https://doi.org/10.1016/j.geoderma.2006.09.003.

Jeffery, L., Höhne, N., Moisio, M., Day, T., Lawless, B., 2020. Options for Supporting Carbon Dioxide Removal. New Climate Institute. [online]. Available at: <https://newclimate.org/wp-content/uploads/2020/07/Options-for-supporting-Carbon-Dioxide-Removal_July_2020.pdf> (accessed 25.06.22).

Kench, P.S., Ford, M.R., Owen, S.D., 2018. Patterns of island change and persistence offer alternate adaptation pathways for atoll nations. Nature Communications 9, 605. Available from: https://doi.org/10.1038/s41467-018-02954-1.

Khan, T., Quadir, D., Murty, T., Kabir, A., Aktar, F., Sarker, M., 2002. Relative sea level changes in Maldives and vulnerability of land due to abnormal coastal inundation. Marine Geodesy 25 (1–2), 133–143. Available from: https://doi.org/10.1080/014904102753516787.

Klenert, D., Funke, F., Mattauch, L., O'Callaghan, B., 2020. Five lessons from COVID-19 for advancing climate change mitigation. Environmental and Resource Economics 76 (4), 751–778. Available from: https://doi.org/10.1007/s10640-020-00453-w.

Kulp, S.A., Strauss, B.H., 2019. New elevation data triple estimates of global vulnerability to sea-level rise and coastal flooding. Nature Communications 10, 4844. Available from: https://doi.org/10.1038/s41467-019-12808-z.

Kundzewicz, Z., Kanae, S., Seneviratne, S., Handmer, J., Nicholls, N., Peduzzi, P., et al., 2013. Flood risk and climate change: global and regional perspectives. Hydrological Sciences Journal 59 (1), 1–28.

Lawler, J.J., 2009. Climate change adaptation strategies for resource management and conservation planning. Annals of the New York Academy of Sciences 1162 (1), 79–98.

Lebbe, T.B., Rey-Valette, H., Chaumillion, É., et al., 2021. Designing coastal adaptation strategies to tackle sea level rise. Frontiers in Marine Science . Available from: https://doi.org/10.3389/fmars.2021.74060203 Sec. Ocean Solutions.

Lewis, S.L., Wheeler, C.E., Mitchard, E.T.A., Koch, A., 2019. Restoring natural forests is the best way to remove atmospheric carbon. Nature 568, 25–28.

Le Quéré, C., Jackson, R., Jones, M., Smith, A., Abernethy, S., Andrew, R., et al., 2020. Temporary reduction in daily global CO2 emissions during the COVID-19 forced confinement. Nature Climate Change 10 (7), 647–653. Available from: https://doi.org/10.1038/s41558-020-0797-x.

Lindsey, R., 2020. Climate Change: Global Sea Level. [online]. Available at: <https://www.climate.gov/news-features/understanding-climate/climate-change-global-sea-level> (accessed 25.06.22).

Lindsey, R., Dahlman, L., 2021. Climate Change: Global Temperature. Available at: <http://www.climate.gov/news-features/understanding-climate/climate-change-global-temperature> (accessed 14.05.22).

Locatelli, B., Evans, V., Wardell, A., Andrade, A., Vignola, R., 2011. Forests and climate change in Latin America: linking adaptation and mitigation. Forests 2, 431–450.

Loubere, P. (2012). The Global Climate System. The Nature Education Knowledge Project. [online] Available at: <https://www.nature.com/scitable/knowledge/library/the-global-climate-system-74649049//> (accessed 24.06.22).

Mach, K., Mastrandrea, M., 2014. Climate Change 2014: Impacts, Adaptation, and Vulnerability, 1. Cambridge University Press, New York.

Malatesta, S., Friedberg, M.S., 2017. Environmental policy and climate change vulnerability in the Maldives: from the 'lexicon of risk' to social response to change. Island Studies Journal 12 (1), 53–70. Available at: https://pdfs.semanticscholar.org/2320/54522b81bff0094cc7a3f6eef5b74-de2311e.pdf.

Manzanedo, R.D., Manning, P., 2020. COVID-19: Lessons for the climate change emergency. Science of the Total Environment 742, 140563. Available from: https://doi.org/10.1016/j.scitotenv.2020.140563.

Manzo, D., Zee, G., Uddin, S., Jovanovic, D., 2021, November 3. Facing dire sea level rise threat, Maldives turns to climate change solutions to survive. ABC News. Available at: <https://abcnews.go.com/International/facing-dire-sea-level-rise-threat-maldives-turns/story?id = 80929487> (accessed 13.05.22).

Marella, R.L., Dixon, J.F., 2018. Data tables summarizing the source-specific estimated water withdrawals in Florida by water source, category, county, and water management district, 2015. U.S. Geological Survey data release . Available from: https://doi.org/10.5066/F7N29W5M.

MCCAP (2013). Monroe Country Climate Action Plan. https://www.monroecounty-fl.gov/DocumentCenter/View/13786/Final-Climate-Action-Plan-adopted-2013?bidId.

Molinaroli, E., Guerzoni, S., Suman, D., 2019. Do the adaptations of Venice and Miami to sea level rise offer lessons for other vulnerable coastal cities? Environmental Management 64, 391–415. Available from: https://doi.org/10.1007/s00267-019-01198-z.

Mondal, M.S.H., 2019. The implications of population growth and climate change on sustainable development in Bangladesh. Sabinet African Journals. Available from: https://hdl.handle.net/10520/EJC-13b8cad1c4 (accessed 26.06.22).

Nicholls, R.J., Wong, P.P., Burkett, V.R., Codignotto, J.O., Hay, J.E., McLean, R.F., et al., 2007. Coastal Systems and low-lying areas. In: Parry, M. L., Canziani, O.F., Palutikof, J.P., van der Linden, P.J., Hanson, C.E. (Eds.), Climate Change 2007: Impacts, Adaptation and Vulnerability.

Contribution of Working Group II to the Fourth Assessment Report of the Intergovernmental Panel on Climate Change. Cambridge University Press, Cambridge, pp. 315–356.

NRDC, 2012. Ready or not: an evaluation of state climate and water preparedness planning. NRDC Issue Brief 12-03-A .

Obeysekera, J., Graham, W., Sukop, M., Asefa, T., Wang, D., Ghebremichael, K., et al., 2017. Implications of climate change on Florida's water resources. Florida's Climate: Changes, Variations, & Impacts . Available from: https://doi.org/10.17125/fci2017.ch03.

O'Brien, E., 2019. Florida's population growing by 900 people a day, report says. WUSF Public Media – WUSF 89.7. [online]. Available at: <https://wusfnews.wusf.usf.edu/news/2019-07-26/floridas-population-growing-by-900-people-a-day-report-says> (accessed 25.09.22).

OECD, 2009. Integrating climate change adaptation into development co-operation. Policy agenda. [online]. <https://www.oecd.org/env/cc/44887764.pdf> (accessed 26.06.22).

Pattison, A., Henke, C.R., Pumilio, J., 2021. Community-based climate action planning as an act of advocacy: a case study of liberal arts education in a rural community. Journal of Environmental Studies and Sciences 11, 183–193. Available from: https://doi.org/10.1007/s13412-020-00655-0.

Permanent Service for Mean Sea Level, 2020. Data and station information for MALE-B, HULULE. [online]. <https://www.psmsl.org/data/obtaining/stations/1753.php> (accessed 24.06.22).

Pethick, J., Orford, J.D., 2013. Rapid rise in effective sea-level in Southwest Bangladesh: Its causes and contemporary rates. Global and Planetary Change 111, 237–245.

Phillips, C., Caldas, A., Cleetus, R., Dahl, K., Declet-Barreto, J., Licker, R., et al., 2020. Compound climate risks in the COVID-19 pandemic. Nature Climate Change 10 (7), 586–588. Available from: https://doi.org/10.1038/s41558-020-0804-2.

Plumer, B., 2022, February 2. How spiking energy prices complicate the fight against global warming. The New York Times. Available at: https://www.nytimes.com/2022/02/02/climate/oil-gas-prices-climate-change.html#:~:text=Across%20the%20globe%2C%20rising%20natural,driving%20up%20planet%2Dwarming%20emissions.

Population U.S.A., 2022. Florida population 2022 [online]. <https://www.usapopulation.org/florida-population/#:~:text=Based%20on%20our%20research%2C%20Florida,last%2010%20years%20since%202011> (accessed 26.06.22).

Prinos, S.T., Wacker, M.A., Cunningham, K.J., Fitterman, D.V., 2014. Origins and Delineation of Saltwater Intrusion in the Biscayne Aquifer and Changes in the Distribution of Saltwater in Miami-Dade County, Florida. US Department of the Interior, US Geological Survey. Available from: https://pubs.usgs.gov/sir/2014/5025/pdf/sir2014-5025.pdf.

Purdum, E.D., Krafft, P.A., Fernald, E.A., Anderson, J.R., 1998. Water Resource Atlas of Florida. Institute of Science and Public Affairs, Florida State University, Tallahassee, FL.

Ramanatham, V., Feng, Y., 2008. On avoiding dangerous anthropogenic interference with the climate system: formidable challenges ahead. Proceeding of the National Academy of Sciences 105 (38), 14245–14250.

Rosen, R.A., Guenther, E., 2015. The economics of mitigating climate change: what can we know? Technological Forecasting and Social Change 91, 93–106. Available from: https://doi.org/10.1016/j.techfore.2014.01.013.

Sachs, J., 2018. 6 Ways America Can Get Back Track With the UN Sustainable Development Goals. News from the Columbia Climate School. [online]. Available at: <https://news.climate.columbia.edu/2018/09/21/america-sustainable-development-goals/> (accessed 28.06.22).

Sakamoto, A., Nishiya, K., Guo, X., Sugimoto, A., Nagasaki, W., Doi, K., 2022. Mitigating impacts of climate change induced sea level rise by infrastructure development: case of the Maldives. Journal of Disaster Research 17 (3). Available from: https://doi.org/10.20965/jdr.2022.p0327.

Sanchez, M., Bashar, F., Janssen, K., Vogels, G.M., Snel, M., Zhou, J., et al., 2015. SWIBANGLA: Managing salt water intrusion impacts in coastal groundwater systems in Bangladesh. Deltares Report Number: 1207671-000-BGS-0016. doi: 10.13140/2.1.1793.1042.

Sarkodie, S., Owusu, P., Leirvik, T., 2020. Global effect of urban sprawl, industrialization, trade and economic development on carbon dioxide emissions. Environmental Research Letters 15 (3), 034049.

Schrag, D.P., 2007. Preparing to capture carbon. Science 315 (5813), 812–813. Available from: https://doi.org/10.1126/science.1137632. PMID: 17289991.

SEFRCCC, 2021. Proceedings of the 13th annual southeast Florida climate compact. [online]. <https://southeastfloridaclimatecompact.org/event/13th-annual-sefl-summit/> (accessed 26.06.22).

Shaw, R., Pulhin, J.M., Pereira, J.J., 2015. Climate change adaptation and disaster risk reduction: overview of issues and challenges. Climate Change Adaptation and Disaster Risk Reduction: Issues and Challenges 1–19. Available from: https://doi.org/10.1108/S2040.

Sheikh, A.Q., Pandit, A.K., 2014. Terrestrial carbon sequestration as a climate change mitigation activity. Journal of Pollution Effects & Control 02 (01). Available from: https://doi.org/10.4172/2375-4397.1000110.

Shulla, K., Voigt, B., Cibian, S., Scandone, G., Martinez, E., Nelkovski, F., et al., 2021. Effects of COVID-19 on the sustainable development goals (SDGs). Discover Sustainability 2 (1). Available from: https://doi.org/10.1007/s43621-021-00026-x.

Smit, B., Pilifosova, O., Burton, I., Challenger, B., Huq, S., Klein, R.J.T., et al., 2003. Adaptation to climate change in the context of sustainable development and equity: climate change 2001: impacts, adaptation, and vulnerability. Semantic Scholar 877–912. Available from: https://www.semanticscholar.org/paper/Adaptation-to-climate-change-in-the-context-of-and-Smit-Pilifosova/63b5c9371ac8dc511e109a02947ffdff6438c179.

Stager, C., 2012. What happens after global warming? Nature Education Knowledge 3 (10), 7.

Stojanov, R., Duží, B., Kelman, Nemec, D., Procháka, D., 2016. Local perceptions of climate change impacts and migration patterns in Male, Maldives. The Geographical Journal 2016. Available from: https://doi.org/10.1111/geoj.12177.

Suman, D., 2017. Climate change and security: the case of Florida. Global Security Review 1. Available from: https://doi.org/10.25148/GSR.1.009611.

Tol, R.S., Klein, R.J., Nicholls, R.J., 2008. Towards successful adaptation to sea-level rise along Europe's coasts. Journal of Coastal Research 24 (2), 432–442.

Tol, R.S.J., 2005. Adaptation and mitigation: trade-offs in substance and methods. Environmental Science & Policy 8, 572–578.

Tompkins, C.F., DeConcini, C., 2013. Frontlines of Climate Change: Florida Leaders Take Action on Sea Level Rise. World Resources Institute. [online]. Available at: <https://www.wri.org/insights/frontlines-climate-change-florida-leaders-take-action-sea-level-rise> (accessed 26.06.22).

UNDP, 2022. Maldives – UNDP climate change adaptation. [online]. <https://www.adaptation-undp.org/explore/maldives> (accessed 11.05.22).

USAID, 2022. Environment and Global Climate Change. USAID. Available at: <https://www.usaid.gov/maldives/environment-and-global-climate-change> (accessed 18.05.22).

USEPA, 2013. Saving water in Florida. WaterSense. [online]. <https://www.epa.gov/sites/default/files/2017-02/documents/ws-ourwater-florida-state-fact-sheet.pdf> (accessed 26.06.22).

USEPA, 2016. What Climate Change Means for Florida. EPA 430-F-16-011. Available from: https://www.epa.gov/sites/default/files/2016-08/documents/climate-change-fl.pdf.

USEPA, (2022). What EPA is Doing to Address Ocean and Coastal Acidification. [online]. Available at: https://www.epa.gov/ocean-acidification/what-epa-doing-address-ocean-and-coastal-acidification. (accessed 26.06.22).

USGS, 2017. What is carbon sequestration? [online]. <https://www.usgs.gov/faqs/what-carbon-sequestration> (accessed 25.06.22).

USGS, 2018. Sea Level Rise and Climate: Impacts on the Greater Everglades Ecosystem and Restoration. Available from: https://www.usgs.gov/centers/florence-bascom-geoscience-center/science/sea-level-rise-and-climate-impacts-greater#overview

USGS, 2019. Saltwater intrusion. [online]. <https://www.usgs.gov/mission-areas/water-resources/science/saltwater-intrusion#overview> (accessed 27.06.22).

Wang, F., Harindintwali, J., Yuan, Z., Wang, M., Wang, F., Li, S., et al., 2021. Technologies and perspectives for achieving carbon neutrality. The Innovation 2 (4), 100180.

Waring, B., Neumann, M., Prentice, I.C., Adams, M., Smith, P., Siegert, M., 2020. Forests and Decarbonization – Roles of Natural and Planted Forests. Frontiers in Forest and Global Change . Available from: https://doi.org/10.3389/ffgc.2020.00058.

WB, 2010. Climate change in the Maldives. [online]. <https://www.worldbank.org/en/news/feature/2010/04/06/climate-change-in-the-maldives#:~:text = THE%20THREAT%20OF%20CLIMATE%20CHANGE&text = With%20the%20melting%20of%20polar,in%20the%20worst%2Dcase%20scenario> (accessed 09.06.22).

WB, 2016. Climate change in the Maldives [online]. <https://www.worldbank.org/en/news/feature/2010/04/06/climate-change-in-the-maldives> (accessed 1105.22).

WB, 2018. Maldives' wetlands help fight climate change [online]. <https://www.worldbank.org/en/news/feature/2018/02/01/first-terrestrial-park-maldives-climate-change-adaptation> (accessed 18.05.22).

WB, 2020. The adaptation principles: 6 ways to build resilience to climate change. [online] <https://www.worldbank.org/en/news/feature/2020/11/17/the-adaptation-principles-6-ways-to-build-resilience-to-climate-change> (accessed 14.06.22).

WB, 2021a. Building costal resilience to protect lives and livelihoods in Bangladesh. [online]. <https://www.worldbank.org/en/results/2021/04/26/building-coastal-resilience-to-protect-lives-and-livelihoods-in-bangladesh> (accessed 29.06.22).

WB, 2021b. Climate risk country profile Maldives. [online]. <https://www.adb.org/sites/default/files/publication/672361/climate-risk-country-profile-maldives.pdf> (accessed 16.06.22).

Wennersten, R., Sun, Q., Li, H., 2015. The future potential for carbon capture and storage in climate change mitigation – an overview from perspectives of technology, economy and risk. Journal of Cleaner Production 103, 724–736. Available from: https://doi.org/10.1016/j.jclepro.2014.09.023.

WHO, 2022. New Report Highlights the Impact of Changes in Environment on One Health. World Health Organization Europe. Available from: https://www.who.int/europe/news/item/01-07-2022-new-report-highlights-the-impact-of-changes-in-environment-on-one-health.

Wong, T.E., Bakker, A.M., Keller, K., 2017. Impacts of Antarctic fast dynamics on sea-level projections and coastal flood defense. Climatic Change 144, 347–364. Available from: https://doi.org/10.1007/s10584-017-2039-4.

Section B

Water sustainability

Chapter 8

Importance of indigenous knowledge in achieving environmental sustainability

Anjali Sharma[1], Devkant Sharma[2], Ajmer Singh Grewal[1], Himani Bajaj[3], Mayank Yadav[3], Ashwani K. Dhingra[1] and Bhawna Chopra[1]

[1]Guru Gobind Singh College of Pharmacy, Yamunanagar, Haryana, India, [2]Ch. Devi Lal College of Pharmacy, Jagadhri, Haryana, India, [3]Adarsh Vijendra Institute of Pharmaceutical Sciences, Shobhit University, Gangoh, Uttar Pradesh, India

Introduction

Indigenous peoples all across the world are continually fighting to protect their integrity, heritage, and insight in a society that is still conquered by a Western viewpoint. They are faced with the dilemma of existing in two forms: indigenous and nonindigenous, which are always at odds with one another, with the latter wielding more authority over the former. Indigenous peoples have managed to survive for generations by adjusting to severe climate circumstances in a variety of ways and developing sustainable livelihood systems (Magni, 2017).

Overview of green chemistry

The Pollution Prevention Act of 1990, a US environmental law, claimed that creating industrial processes that don't produce waste is the best way to reduce pollution. The green chemical process was consequently created. According to the Environmental Protection Agency, "green chemistry" refers to the design of chemical processes and products that significantly reduce the usage or manufacture of potentially hazardous substances (Fig. 8.1).

FIGURE 8.1 Green Chemistry an overview.

Role of Green Chemistry in Ecosystem Restoration to Achieve Environmental Sustainability. DOI: https://doi.org/10.1016/B978-0-443-15291-7.00015-8

This entails fewer waste materials and nontoxic materials, and more effectively, green chemistry, on the other hand, is particularly prone to reducing and preventing pollution at source. Environmental chemistry is the study of how pollutants and the natural environment interact chemically (Anastas and Kirchhoff, 2002).

How green chemistry differs from cleaning up pollution?

By limiting or eliminating the risks associated with chemical feedstocks, reagents, solvents, and end products, green chemistry decreases pollution at its source. Contrary to remediation, which is cleaning up environmental spills and other releases, end-of-the-pipe treatment includes treating waste streams. Hazardous compounds are removed from the environment through remediation, whereas green chemistry prevents the presence of these materials in the first place. The use of hazardous chemicals to remove environmental pollutants must be reduced or eliminated for a method to be considered "green chemistry."

Indigenous peoples' definition

Finding an appropriate definition of indigenous people that would cover all the aspects is the toughest challenge. The two most commonly used definitions are those provided by José Martnez Cobo, in the "Study of the Problem of Discrimination against Indigenous Populations and with reference to the International Labour Organization Convention no. 169 from 1989" Indigenous communities, peoples, and nations, in accordance with Martnez Cobo, are those that "reflect themselves differently from other segments of the societies now predominating in those territories, or parts of them, because they have an ancient continuity with pre-aggression and pre-colonial civilizations that developed on their territories." They are constantly focused on maintaining, improving, and passing down to future generations, thus, they are not currently dominant (Samarawickrama, 2018).

Indigenous and modern knowledge

Model-making regarding the natural world's functioning results in knowledge. All civilizations, try to understand the workings of the world and use that intelligence to perform environmental manipulation activities. Prescientific societies gathered knowledge at a sluggish pace through earlier development of the modern hypothetic approach of scientifically gathering and understanding the natural world's workings. References to environment likings, life cycles, and behavior forms of species such as birds are extremely precise within these constraints (Diamond, 1989; Gadgil et al., 1993). Natural world working models are inextricably related to the worldview of any culture. Current scientific insight has been extremely fruitful in progressing human understanding and management of simpler systems, in addition to its associated worldview of people disregarding the natural environment. The overview that positive science has come up with is of limited value in providing real endorsements for sustainable resource usage since these complex systems vary widely on three-dimensional and chronological scales (Ehrlich, 1987; Slobodkin, 1988). Overuse and simplification of such complex ecological systems have occurred in science-based cultures, resulting in a slew of resource depletion and environmental degradation issues. The wisdom of indigenous communities is significant in this context. People's place and belief system in the world, and showing dignity to the rest of the natural world, are vital for developing sustainable relationships with innate resources (Oldfield and Alcorn, 1991). All societies do not live in peace with the world and not all indigenous peoples do so now, outside of industrial societies (Diamond, 1992). A recognition and collection of information regarding the vital role that species play in creating ecological services and natural resources are one of these methods (Lasserre and Ruddle, 1983; Ruddle and Johannes, 1989).

Sustainable development

There has been a discussion over how to conceptualize sustainable development within the context of sustainability. According to some academics, there was no distinction between the two previously mentioned. Sustainability is the capacity to withstand or adjust to change in an existing environment. It has little concern for the future and is mostly focused on serving the wants of the now. According to (Ayres, 2008), sustainability also entails the existence of laws and other regulations that control how people live their lives. This concept leads to the conclusion that economic development cannot be stopped, which is crucial for efforts to advance sustainable development (Barker, 2006; Haase et al., 2013; Stoddart, 2011).

Importance of indigenous knowledge

Indigenous wisdom has been infused into several parts of local rural people's lives. Soil fertility is maintained by mulching, watermills are used to grind and mill food with water, pests are prevented by using ash in agriculture, and nutrition is enhanced by selective forest fire are some of the examples of indigenous peoples' survival methods. Farmer-friendly, cost-effective, environmentally friendly, and socially acceptable, IK goods are available. (Sharma et al., 2009). It includes, among other things, seed treatments, storage techniques, planting, and harvesting instruments, as well as other activities and technology.

Knowledge transmission systems

The community and the family

Indigenous insight is a biotic phenomenon that passes from generation to generation (Battiste, 2002). The transfer of knowledge depends on many aspects such as age, income, sex, background, family values, occupation, etc. "Depending upon the quality and quantity of knowledge of the family, this may lead to various systems in the society" (Briggs, 2005; Grenier, 1998). Education forms the major basis of knowledge in society. Learning is a continuous process, everyone learns for a lifetime through proper education. Learning by observations through practicals, demonstrations, hands-on training, live sessions, and metaphors are also typically used. Women and elders are the most essential members of the family (Tebtebba Foundation, 2009). These people serve as the main caregivers and guardians of conventional wisdom and culture (Cunningham, 2010b). One of the most crucial elements of transmission, older people were the guardians of indigenous knowledge (Dweba and Mearns, 2011). The implementation of educational programs, curriculum creation, and instructional planning were all significantly aided by the Alaska Rural Systemic Education Initiative (Barnhardt, 2008).

Transmission system for the human being

Understanding cultural evolution requires an understanding of the dynamics of knowledge transmission and acquisition, and how various facets of culture are acquired and embodied by humans (Guglielmino et al., 1995; Hewlett and Cavalli-Sforza, 1986). Vertical transmission, for instance, promotes cultural continuity while limiting innovation (Reyes-García et al., 2009), whereas oblique and horizontal transmission, particularly if contact with transmitters is frequent, might expedite the spread of new cultural features (Hewlett and Cavalli-Sforza, 1986; McElreath and Strimling, 2008).

Formal schooling

Since the start of the 20th century, the historical and current idea of acceptable education systems based on Western ideals that encourage mingling has turned into an unthinkable amount of indigenous knowledge loss, according to studies (Mato, 2015; Stavenhagen, 2015). In many parts of the world, official schooling has been used to include indigenous knowledge and customs. Universities that are multicultural in Latin America (Mato, 2018) and the Alaska Rural Systemic Initiative are two effective examples. Both exhibit the beneficial effects of educational initiatives. They constitute what indigenous peoples want in terms of conventional education, namely, eugenically relevant education with inclusive school syllabi that reflect cultural variety, recognition of indigenous languages, and the use of optional pedagogical approaches (Aikman, 2011; Stavenhagen, 2015).

Sustainable practices for indigenous knowledge

Land, ecotourism, and resource management

The indigenous groups had different perceptions of development than modern societies, which differ from the modern view. The key ideas of indigenous knowledge of the well-being of humanity and the environment are responsibility towards nature, its preservation, and proper management of the available resources. Marketing is not required for natural resources. The community serves as the center of gravity for the lives of indigenous people (Cunningham, 2010a). Similar to other resource management practices, this farming practice integrates fundamental aspects of native cultures, community-based management solutions, and traditional institutions and rituals.

Role of women in land and resource management

Due to differences in specialties between men and women, both of them perform different tasks that are related to their respective domains. Men take responsibility for forest-related activities like hunting, farming, and logging. On the other hand, women are more involved in production and management activities in home gardens and nontimber forests due to their proximity to the domestic and private realm (Schmink and Gómez-García, 2015). Similarly, Shiva (1992) highlights that indigenous women are the guardians of the environment, due to this they play a vital role in food security, environmental sustainability, and sustainable resource management. Furthermore, women have their share of 70% of the work in ever-changing agricultural methods, from seed selection through weeding the fields, harvesting, processing, and selling surplus products.

Strategies of new livelihood: an alternative to sustainable development is ecotourism

Typically, indigenous peoples use their creations to sustain themselves, with surpluses being shared fairly within the group or traded with other civilizations (Cunningham, 2010b). However, access to markets for buying and selling goods has recently transformed many indigenous people's way of life, providing them with more options to generate money. This has resulted in a slew of economic issues, including rising inequalities and the risk of debt, as well as a decline in indigenous populations' traditional livelihoods, as a result of a shift in people's attitudes from community-centered to individualized, and a disregard for traditional sustainable practices (Ford et al., 2010). Some of these tribes have begun to add supplementary occupations to better their living conditions within the context of noncommercial manufacturing and ecosystem management, two essential elements of indigenous social and cultural systems.

Indigenous knowledge and disaster risk reduction

In comparison to other fields, the incorporation of indigenous knowledge into catastrophe risk reduction has been delayed (McAdoo et al., 2009). Indeed, disaster risk reduction research has proved the importance of native traditional knowledge including early warning, awareness, reaction, and postdisaster recovery in addition to preventing and reducing the consequences of natural disasters (Rautela and Karki, 2015). Depending on the natural hazards indigenous groups around the world use a variety of measures. Inhibition techniques based on weather forecasting and changes in agricultural practices to restrict crop indemnities, as well as researchers, have discovered several additional measures to safeguard the populace and cattle.

Traditional practices in animal husbandry

Due to the domestication of various livestock species, the traditional knowledge of animal husbandry can be considered centuries. However, because these techniques are popular in rural India and are rarely documented, there is a risk that these knowledge systems will erode. Crop wastes such as stalks, straw, stovers, tops, and crop thrushes such as wheat, paddy straw, and other crop thrushes, as well as crop by-products, such as straw of wheat, bajra, bran, maize, and husk, are used in traditional feeding practices.

Indigenous knowledge systems in forestry development

Planting exotic trees has been the focus of national and international provisions to re-green deforested slopes for more than 30 years. Plantings of eucalyptus, acacia, and tropical pines were made, in part because of their drought endurance, but also because scientists understood nothing about native plants. The natives have overcome the challenges of firewood and water shortages in only three years after the afforestation initiative began. Other more valuable trees such as Thitka, Chukrasia, and Margose are also intercropped by local people.

Knowledge systems integration and the implication for the Agenda 2030

The fundamental focus of Agenda 2030 is how to ensure that indigenous knowledge is effectively integrated. According to research, engaged approaches are essential for accurately portraying various knowledge systems, notably traditional and scientific information. Additionally, these techniques are essential for fostering cross-cultural comprehension of these knowledge systems, whose efficient complementarity is well known in response to approaches to combating climate change (Kronik and Verner, 2010), the prevention and mitigation of natural hazard risks (Mercer et al., 2010), and biodiversity conservation (Prakash, 2004). In this approach, researchers created interactive situations for the fusion of traditional and scientific knowledge systems (Mercer et al., 2010). For example, four particular procedures are required: (1) Community involvement; (2) Documentation of susceptibility causes; (3) Identification of indigenous and scientific approaches to address the issues impacting indigenous people's deficiencies; and (4) Development of integrated laws and regulations. Another example is working in Indonesia, the Philippines, and Timor-Leste on the integration of indigenous and scientific knowledge for hydrometeorological risks. Developing dialog and creating partnerships with indigenous peoples, civil society, administrations, development partners, management systems, and academics from diverse sectors to encourage the maintenance of indigenous knowledge and the merger of the two knowledge systems (Kronik and Verner, 2010; Krupnik and Ray, 2007; Sillitoe, 1998; Gorjestani, 2004). As a result, efforts should be made to ensure that the information of all stakeholders is regarded as equally legitimate, important, and advantageous (Table 8.1).

Based on numerous variables, including density and atomic weight, heavy metals can be characterized in a variety of ways. In small amounts, several heavy metals like iron, cobalt, and zinc are necessary for human nutrition but are poisonous in larger amounts. However, only a few elements, like lead, cadmium, and mercury are toxic even in trace amounts. Humans are exposed to heavy metals through cutaneous contact, ingestion through the skin, and inhalation from ambient or drinking water. Table 8.2 describes the effects and sources of essential and nonessential metals.

Applications of green chemistry in environmental sustainability

Chemists are searching for more sustainable and environmentally friendly ways to employ in daily life. Below is a discussion of a few cases.

- Biodegradable plastics: Nonbiodegradable plastics are responsible for a vast amount of pollution in our world. Produced in Minnesota are Ingeo food containers composed of polylactic acid. Scientists at Nature Works developed a resin from corn starch that is as tough as rigid petroleum-based plastic and is currently being used to produce vessels such as water bottles and yogurt pots. BASF developed a recyclable polyester film called Eco flow that they blend with calcium carbonate and cassava starch to generate completely biodegradable bags (Vroman et al., 2009).
- Eco-friendly paint: Volatile organic chemicals are released in large quantities as oil-based "alkyd" paints dry and cure (VOCs). These VOCs have significant environmental effects. Sherwin William produced water-based acrylic alkyd paints using soya bean oil, acrylic paint, and PET plastic from recycled soda bottles. These paints were created in sufficient quantities in 2010 to eliminate 362,874 kg of VOCs.

TABLE 8.1 Classification of the documented indigenous technical knowledge (ITK) on coconut in Kerala.

Sr. no.	ITK categories	Practices under each category	Frequency (%)
1.	Soil and season	06	4.7
2.	Selection of mother palm	07	5.4
3.	Collection and storage of seed nuts	26	20.2
4.	Nursery	17	13.2
5.	Seedlings	08	6.2
6.	Preparatory cultivation	04	3.1
7.	Planting in main field	07	5.4
8.	Manuring	18	13.9
9.	Cultural operations	19	14.7
10.	Water management	06	4.7
11.	Intercrops	04	3.1
12.	Yield and harvest	04	3.1
13.	Others	03	2.3
	Total	129	100

TABLE 8.2 Sources and effects of essential and nonessential heavy metals.

Metal	Source	Metal type	Clinical effects
Arsenic (As)	Deposition of atmosphere, use of pesticide, and mining activities	Nonessential	Nasal septum perforation, respiratory cancer, neuropathy, skin cancer
Zinc (Zn)	Plumbing activities, manufacture of brass, and refineries of oil	Essential	Liver and kidney dysfunction and gastrointestinal disease
Cadmium (Cd)	Use of fertilizers and pesticides	Nonessential	Proteinuria, aminoaciduria, osteomalacia, glucosuria
Copper (Cu)	Printing and plating activities copper polishing	Essential	Metabolic syndrome
Iron (Fe)	Intake of iron supplements and oral consumption	Essential	Vomiting, diarrhea, abdominal pain, dehydration, and lethargy
Chromium (Cr)	Electroplating, manufacture of textiles, fabrication of steel	Nonessential	Ulcer, nasal septum perforation, cancer of the respiratory tract, and Alzheimer's.
Cobalt (Co)	Hip alloy replacement case	Essential	Hematological, cardiovascular, hepatic, and endocrine
Lead (Pb)	Batteries, coal combustion, and paint industry	Nonessential	Encephalopathy, peripheral neuropathy, central nervous disorders, anemia

- Green bleaching agents: By soaking tiny pieces of wood in a salt solution and sodium sulfide, accompanied by an interaction with chlorine, the lignin from the wood required to manufacture high-quality white paper is traditionally removed. Because they are carcinogens, these substances are unhealthy. Because they are carcinogens, these substances are unhealthy.
- Putting out flames sustainably: The widely used conventional chemical firefighting foams release hazardous materials into the environment, polluting water and weakening the ozone layer. To successfully put out fires without creating harmful compounds like conventional firefighting supplies, a novel foam named Pyro cool has been developed.
- Green dry cleaning: Perchloroethylene is the solvent most usually used for dry cleaning clothing (perc) it pollutes groundwater and is suspected to be carcinogenic. The new technique known as micelles technology was developed by Timothy Remark, Joseph De Simons, and James Mcclain by employing liquid carbon dioxide as a safer solvent and surfactant to dry clean clothing. This method is currently being used commercially by several dry cleaners. Laundry machines have been modified to use green solvents instead of the carcinogen PERC to adopt this procedure.
- Making rainwater clear in an environmentally friendly way: Traditionally, alum is used to make municipal and industrial sewage water clear. Alzheimer's disease is brought on by harmful ions, which alum has been proven to enhance in treated water. It has been discovered that the powder from the tamarind seed kernels works as well as an alum at making wastewater clear.
- Lighter automobiles prepared, lower prices, and less carbon dioxide emission: It has been discovered that contemporary synthetic polyesters can minimize the amount of foam used in car seats, sufficiently reducing their weight, which lowers the consumption of fuel along with the reduction in emission of atmospheric carbon dioxide.
- Organic light-emitting diodes, which provide more light with less energy usage, are an example of revolutionary lighting technology.
- Use of unleaded fuel: The greater the octane rating, the higher the quality of the fuel. These days, the octane number of gasoline is raised without the inclusion of lead substances like tetra ethyl lead. Methyl tertiary butyl ether, which provides oxygen to gasoline and prevents the production of per-oxy compounds, is added to gasoline to produce unleaded fuel.
- Benzol: This is produced as a by-product of the carbonization of coal. It can also be made by fractionally distilling light oil. It contains several additional hydrocarbons along with a combination of 70% of benzene, 18% toluene, and 16% of xylenes. Due to its excellent anti-knocking value, it can be utilized as a major component of motor fuels, which lowers both the consumption of gasoline and the production of harmful pollutants.

Conclusion

Many concerns that directly impact indigenous peoples are addressed in the new sustainable development agenda. Indigenous people have faced and continue to face several challenges, including poverty, access to justice, climate change, and education. Their wisdom and know-how, which are deeply founded on indigenous peoples' relationships with nature and community, have managed to overcome these challenges; nonetheless, it is insufficient. These populations' survival as well as their belief systems would be in danger if some remedies are not put into place right away because they are caught between environmental dangers and development initiatives. Because of knowledge loss, indigenous peoples' vulnerability and risks have already increased. Therefore, the national and worldwide communities must start seeing indigenous peoples and their knowledge as vital allies in the fight against climate change, sustainable development issues, and the preservation of biodiversity around the world. In light of the post-2015 sustainability strategy, it is urgently necessary to work together to develop and put into practice the necessary innovations that will enable indigenous peoples to uphold and realize their rights while also engaging in policymaking and becoming active agents of transformation.

References

Aikman, S., 2011. Educational and indigenous justice in Africa. International Journal of Educational Development 31 (1), 15–22.

Anastas, P.T., Kirchhoff, M.M., 2002. Origins, current status, and future challenges of green chemistry. Accounts of Chemical Research 35 (9), 686–694.

Ayres, L., 2008. Thematic coding and analysis, The SAGE Encyclopedia of Qualitative Research Methods, 1. Sage, pp. 868–876.

Barker, K., 2006. Barker Review of Land Use Planning: Final Report, Recommendations. The Stationery Office.

Barnhardt, R., 2008. Creating a place for indigenous knowledge in education. In: Gruenewald, D.A., Smith, G.A. (Eds.), Place-Based Education in the Global Age: Local Diversity New York. Psychology Press, pp. 113–133.

Battiste, M., 2002. Indigenous Knowledge and Pedagogy in First Nations Education Aliterature Review With Recommendations. Report prepared for the National Working Group on Education, Indian and Northern Affairs Canada, Ottawa, ON.

Briggs, J., 2005. The use of indigenous knowledge in development: problems and challenges. Progress in Development Studies 5 (2), 99–114.

Cunningham, M., 2010a. Acerca de la visión del "buen vivir" de los pueblos indígenas en Latinoamérica. Asuntos Indigenas 10 (1–2), 52–59.

Cunningham, M., 2010b. Laman laka: Our indigenous path to self-determined development. In: Towards an Alternative Development Paradigm: Tebtebba. Indigenous People's Self-Determined Development, pp. 89–116.

Diamond, J., 1992. The Rise and Fall of the Third Chimpanzee. Vintage.

Diamond, J.M., 1989. This-fellow frog, name belong-him Dawko. Natural History 98 (16), 18–20.

Dweba, T.P., Mearns, M.A., 2011. Conserving indigenous knowledge as the key to the current and future use of traditional vegetables. International Journal of Information Management 31 (6), 564–571.

Ehrlich, P.R., 1987. AIBS News. Bioscience 757–763.

Ford, J.D., Pearce, T., Duerden, F., Furgal, C., Smit, B., 2010. Climate change policy responses for Canada's Inuit population: the importance of and opportunities for adaptation. Global Environmental Change 20 (1), 177–191.

Gadgil, M., Berkes, F., Folke, C., 1993. Indigenous knowledge for biodiversity conservation. Ambio 151–156.

Gorjestani, N., 2004. Indigenous knowledge: the way forward. In: World Bank (Ed.), Indigenous knowledge. Local pathways to Global Development. Marking Five Years of the World Bank Indigenous Knowledge for Development Program. World Bank.

Grenier, L., 1998. Working With Indigenous Knowledge: A Guide for Researchers. International Development Research Centre (IDRC), Ottawa. Available from: http://network.idrc.ca/ev.php?IN = 28703_201&ID2 = DO_TOPIC.

Guglielmino, C.R., Viganotti, C., Hewlett, B., Cavalli-Sforza, L.L., 1995. Cultural variation in Africa: role of mechanisms of transmission and adaptation. Proceedings of the National Academy of Sciences 92 (16), 7585–7589.

Haase, D., Kabisch, N., Haase, A., 2013. Endless urban growth? On the mismatch of population, household and urban land area growth and its effects on the urban debate. PLOS ONE 8 (6), e66531.

Hewlett, B.S., Cavalli-Sforza, L.L., 1986. Cultural transmission among Aka pygmies. American Anthropologist 88 (4), 922–934.

Kronik, J., Verner, D., 2010. Indigenous Peoples and Climate Change in Latin America and the Caribbean. World Bank Publications.

Krupnik, I., Ray, G.C., 2007. Pacific walruses, indigenous hunters, and climate change: bridging scientific and indigenous knowledge. Deep sea Research Part II: Topical Studies in Oceanography 54 (23–26), 2946–2957.

Lasserre, P., Ruddle, K., 1983. Traditional Knowledge and Management of Marine Coastal Systems. IUBS.

Magni, G., 2017. Indigenous knowledge and implications for the sustainable development agenda. European Journal of Education 52 (4), 437–447.

Mato, D., 2015. Pueblos indígenas, Estados y educación superior: Aprendizajes de experiencias en varios países de América Latina potencialmente útiles a los procesos en marcha en Argentina. Cuadernos de Antropología Social 41, 5–23.

Mato, D., 2018. Diversidad cultural e interculturalidad en educación superior. Experiencias en América Latina .

McAdoo, B.G., Moore, A., Baumwoll, J., 2009. Indigenous knowledge and the near field population response during the 2007 Solomon Islands tsunami. Natural Hazards 48 (1), 73–82.

McElreath, R., Strimling, P., 2008. When natural selection favors imitation of parents. Current Anthropology 49 (2), 307–316.

Mercer, J., Kelman, I., Taranis, L., Suchet-Pearson, S., 2010. Framework for integrating indigenous and scientific knowledge for disaster risk reduction. Disasters 34 (1), 214–239.

Oldfield, M.L., Alcorn, J.B., 1991. Biodiversity: Culture, Conservation, and Ecodevelopment. University of Toronto.

Prakash, S., 2004. Using indigenous knowledge to raise agricultural productivity. In: The World Bank (Ed.), Indigenous Knowledge: Local Pathways to Global Development. The World Bank, Washington, DC, pp. 215–218.

Rautela, P., Karki, B., 2015. Weather forecasting: traditional knowledge of the people of Uttarakhand Himalaya. Journal of Geography, Environment and Earth Science International 3 (3), 1–14.

Reyes-García, V., Broesch, J., Calvet-Mir, L., Fuentes-Peláez, N., McDade, T.W., Parsa, S., et al.,and TAPS Bolivian Study Team 2009. Cultural transmission of ethnobotanical knowledge and skills: an empirical analysis from an Amerindian society. Evolution and Human Behavior 30 (4), 274–285.

Ruddle, K., Johannes, R.E. (Eds.), 1989. Traditional Marine Resource Management in the Pacific Basin: An Anthology (No. 2). UNESCO/ROSTSEA.

Samarawickrama, S., 2018. Sustainable Applications Behind Geographically Indigenous Product Cultures: A Study of Village Communities in Central High Valley, Sri Lanka. Proceedings of the 11th International Conference of Faculty of Architecture Research Unit (FARU), University of Moratuwa, Sri Lanka, December 08, 2018.

Schmink, M., Gómez-García, M.A., 2015. Under the Canopy: Gender and Forests in Amazonia. Centre for International Forestry Research.

Sharma, S., Bajracharya, R., Sitaula, B., 2009. Indigenous Technology Knowledge in Nepal—A Review. CSIR.

Shiva, V., 1992. Women's indigenous knowledge and biodiversity conservation. India International Centre Quarterly 19 (1/2), 205–214.

Sillitoe, P., 1998. The development of indigenous knowledge: a new applied anthropology. Current Anthropology 39 (2), 223–252.

Slobodkin, L.B., 1988. Intellectual problems of applied ecology. Bioscience 38 (5), 337–342.

Stavenhagen, R., 2015. Indigenous peoples' rights to education. European Journal of education 50 (3), 254–257.

Stoddart, H. (Ed.), 2011. A Pocket guide to sustainable development governance. In: Stakeholder Forum.

Tebtebba Foundation, 2009, February 24–27. Asia Summit on Climate Change and Indigenous Peoples. Bali, Indonesia.

Chapter 9

Recent advances in sustainability science for environmental conservation

Damanpreet Kaur Lang[1], Varinder Singh[1], Md Altamash Ahmad[1], Balraj Saini[1], Balakumar Chandrasekaran[2] and Rajwinder Kaur[1]

[1]Chitkara College of Pharmacy, Chitkara University, Rajpura, Punjab, India, [2]Faculty of Pharmacy, Philadelphia University, Amman, Jordan

Introduction

In today's world, industrialization is a major contributor to the growth and urbanization of society. This industrialization has very detrimental effects on the environment and is a major cause of land, air, water, and noise pollution (Iyinbor et al., 2018). Globally, and especially in developing nations, there is a considerable environmental deterioration that threatens to be endangering long-term growth prospects. This is true because of their developmental process depends on the utilization of natural resources (Adenuga et al., 2006). Sustainability science focuses on the vulnerabilities, sturdiness, endurance, and stability of the joint human-environment system. The notion "sustainable development" emphasizes meeting the needs of humans, especially those of the world's poor, while also preventing detrimental effects on the environment. In this chapter, we have discussed various methods and recent advances in sustainability science that have aided in conserving the environment.

What is sustainability science?

As described by the International Union for Conservation of Nature (IUCN), sustainable development refers to the ability to maintain a particular process or state for enhancing the quality of human existence while residing inside the maximum limit of ecosystems (Spangenberg, 2011) and the integrated human-environment system's vulnerabilities, sturdiness, endurance, and stability are the main topics of sustainability science. The study of sustainability assesses the dynamic interaction between the society and the environment. The natural sciences are integrated with the social sciences and humanities in this multidisciplinary science.

Soil pollution

Soil pollution can be defined as the accumulation of persistent chemicals, radioactive substances, toxins, or agents that cause diseases and may produce ill effects on humans, animals, and plants (Okrent, 1999). The chief reasons for soil pollution are directly linked to human activities leading to a build-up of contaminants in the soil (Cachada et al., 2018). Chemical substances are exposed to the environment, including soil, at all stages of their life cycles. In early times, people used to manage their home and work waste on their own using small-scale practices like decomposition, vermicompost, use of designated pits, etc. Therefore, sustainable science needs to be incorporated with traditional methods to save and protect soil from all types of pollutants and contaminants.

Novel techniques to reduce soil pollution

Soil can be treated to remove contaminants in two ways, in-situ and ex-situ. The former involves techniques where the soil is not excavated but remains in the ground throughout the treatment. It includes the introduction of constituents or substances or manipulation of the soil. The latter includes the techniques involving excavation of the soil from the ground for its treatment. Soil excavated can either be transported somewhere else for treatment or treated on-site.

Role of Green Chemistry in Ecosystem Restoration to Achieve Environmental Sustainability. DOI: https://doi.org/10.1016/B978-0-443-15291-7.00018-3

Ball milling

Ball milling aids detoxification of insistent pollutants and heavy metals in the soil (Wei et al., 2022). This technology has been proven effective in the treatment of chlorinated, brominated, and fluorinated organics present in the soil (Cagnetta et al., 2016). Generally, during the ball-milling process materials collide with grinding balls with high-velocity rotation, and the nonmetallic particles come into contact with the cogrinding reagents completely (Cagnetta et al., 2018). The ball-milling process causes insufficient debromination if the ball-milling time is too short whereas excessive milling destroys structures of nonmetallic particles and resources.

Phytoremediation

Phytoremediation is a method of removing or reducing heavy metals from the soil. This is a very eco-friendly and promising approach for preserving the soil against pollution (Bhat et al., 2022). The type of plant, environmental conditions, the composition of the root, region of the root, and biological and physicochemical properties of the soil majorly determine the phytoremediation process. Traditional phytoremediation methods have limited application on a large scale. Further, phytoremediation is divided into five types: phytoextraction, phytostabilization, hemofiltration, phytovalorization, and phytostimulation (Fig. 9.1).

Bioventing and biosparging

Bioventing and biosparging are analogous practices that activate the biodegradation of substances that are aerobically degradable using direct air injection into the soil. This method can be divided into two categories: active and passive technology. In passive technology, gas exchange through vent wells was accomplished by the effect of air pressure, whereas in active technology, it was aided by a blower or pump. Bioventing could be used to clean up hydrocarbon-polluted locations, such as spills of crude oil or crude oil products.

Biopile

Biopile is a process that comprises stacking or piling contaminated soils to enhance microbial action through cost-efficient and precise nutrient alteration, exposure to air, and irrigation practices (Fig. 9.2; Whelan et al., 2015). It entails the exclusion of a wide spectrum of petrochemical and other organic chemical pollutants from spoiled soils (Germaine et al., 2015; Pavel and Gavrilescu, 2008). The benefit of this technology is that in a confined space, a large quantity of soil can be processed (Koul and Taak, 2018).

Pyrolysis

Pyrolysis is a technique used in the treatment of contaminated soil, which involves the chemical decomposition of toxins using high pressure, high temperature ($>430°C$), and anoxic conditions (Venderbosch and Prins, 2010). This process converts toxins or pollutants into char, fuel gas, water, or other organic liquids (Koul and Taak, 2018). This is a high-efficiency technology with great pollution-cleaning capacity, reduction in the volume of toxins, no production of carbon dioxide, low cost, and can be easily implemented on a large scale as well as small scale.

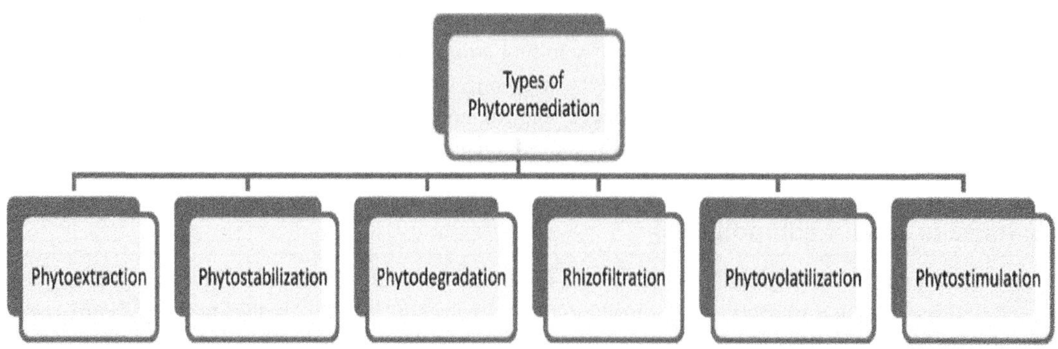

FIGURE 9.1 Types of phytoremediation.

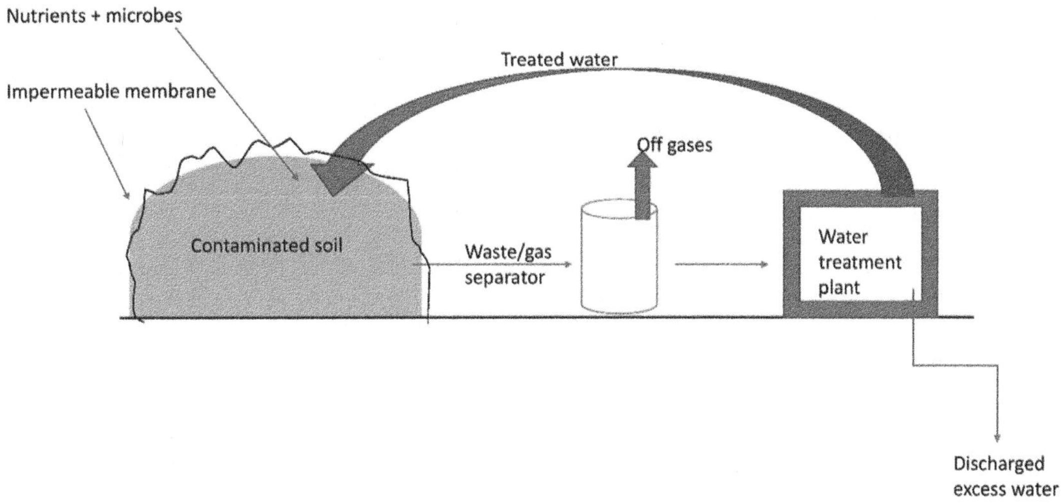

FIGURE 9.2 Soil is decontaminated using biopile.

Thermal remediation

Thermal remediation is a widely used method for the decontamination of volatile toxins from the soil with the aid of heat. The electrical resistance heating technique of thermal remediation involves the passing of an electric current through moisture present in the soil, in between the electrodes and hence producing heat. Conductive heating comprises the process in which the soil surface is supplied with heat and vacuum concurrently to degrade the toxins or contaminants present in the soil. Radio-frequency heating implies in situ heating of contaminated soil using a high-frequency alternating electric field. The vitrification technique helps in the removal, destruction, or immobilization of the contaminants in the sludge or soil. It can be performed with two processes—it comprises vitrification of materials utilizing electrical resistance and electrodes directly placed in the soil. It employs plasma-arc technology, in which electrodes are embedded in the soil and an electric current of very high voltage is utilized to heat it to temperatures between 1600°C and 2000°C.

Electrokinetic remediation

In this type of remediation, an electric field gradient is initiated on both sides of the soil to be decontaminated, using the electric current. Using processes like electrophoresis, electromigration, and electroosmotic flow, the toxins or contaminants are carried out to the poles in the existence of electric current. A solar system can act as a good alternative instead of DC power or electrical energy.

Nano-bioremediation

Particles that have a size of less than 100 nm are called ultrafine or nanoparticles. Types of nanoparticles that can be used in nano-bioremediation are: TiO_2 (titanium dioxide) based nanoparticles serve as photocatalysts for treating polluted water. TiO_2 absorbs numerous dyes from the solution. Immobilized TiO_2 nanoparticles have been proven to remove N-butoxymethyl2-chloro-2, 6-diethylacetanilide contamination found in soil. Arginine-modified TiO_2 aids in the removal of nitrobenzene from the soil (Makarova et al., 2000). In a study conducted by (Xie et al., 2014), bimetallic Ni/Fe was used to reduce decabromodiphenyl ether from the soil. Pd/Fe nanoparticle aids in the removal of trichloroethene (Lien and Zhang, 2001). Nanotubes are small materials cylindrical in shape and hollow from the inside. They contain fullerenes and carbon nanotubes. They have very special properties like intensified adsorption capacity, high strength, thermal conduction, and electrical conduction. Molecules get adsorbed at the pores' surface present in the nanotubes and interact with them. Carbon nanotubes are effective in the removal of organic contaminants present in the soil such as 1,2-dichlorobenzenes and trihalomethanes (Lu et al., 2005). Magnetic chitosan nanoparticle enhances the removal of Cobalt (II) (Chang et al., 2006).

Water pollution

Water pollution can be defined as the presence of any chemical, biological, or physical pollutants in water that cause a damaging effect on the quality of water, marine life, and the population consuming water (Goel, 2006). Sources of pollutants include- chemicals released from industries such as petrochemicals and solvents, biocidal such as pesticides, industrial products such as lubricants naturally existing chemicals such as heavy metals and cyanotoxins, and consumer products such as soaps and detergents. However, some global warming remedial actions for water pollution can be implemented which include capturing and storing carbon dioxide, plantations of bioenergy crops that will produce bioenergy, and afforestation among other actions (Iyinbor et al., 2018).

Novel techniques to reduce water pollution

Treatment of wastewater is a process involving various steps including preliminary, primary, biological, or secondary, and advanced treatment. Primary treatment is a method involving the separation of solid and liquid pollutants in which the solid pollutants tend to settle down as they are heavier than liquid pollutants and then are removed from the further process. Secondary treatment is also called biological treatment as it involves the breakdown of organic wastes by bacteria and other microorganisms and the advanced process involves the removal of trace elements, hormones, steroids, or pharmaceuticals from the wastewater. Advanced treatment is nowadays used more commonly in developed countries for the treatment of water which removes dissolved chemicals like phosphorous or nitrogen. Ludzack et al. (1962) introduced a method that helps in the denitrification of wastewater with the biological treatment process involving an anoxic environment in the activated sludge process. A modified Ludzack and Ettinger process along with a 5-stage process was introduced by Bardenpho which was very effective in removing phosphorous (<0.1 mg/L) and nitrogen (4 mg/L) from wastewater and is widely used nowadays. Various breakthrough techniques for wastewater treatment are described below (Table 9.1).

Physical methods

Physical methods of wastewater treatment involve various methods which aid in the removal of pollutants from effluent. It involves the use of forces that exist in nature like gravity, electrical forces, and van der Waals forces and the methods include adsorption, sedimentation, and floatation. The biosorption method of wastewater treatment involves the utilization of biosorbents of biological origin such as fungi, bacteria, algae, yeast, etc. This microbial biomass possesses the innate ability to requisition metals from wastewater contaminated with heavy metals and decrease its concentration from parts per million to parts per billion level. Biosorption, when compared to other conventional treatment mechanisms, has various advantages such as high efficiency, high selectivity for dissolved heavy metals, and ease of separating the heavy metals from the rest of the effluent.

Physicochemical and chemical methods

Physicochemical and chemical method involves the use of various chemical reactions that are utilized for the treatment of wastewater or polluted water to restore its quality and make it usable for consumption. Advanced oxidation processes involve techniques that are favorable for removing nonbiodegradable pollutants from wastewater. They involve the production of hydroxyl radicals which are highly reactive and powerful oxidants. They are formed on the spot usually by various combinations of UV (ultraviolet radiations), hydrogen peroxide, ozone, and photocatalysts produced by salts of transition metals. AOPs are further subdivided into H_2O_2/pyridine/copper (II) system, ozonation, Fenton's process, ultrasonic radiation, and photocatalysis. H_2O_2/pyridine/copper (II) system is one type of advanced oxidation process that involves the utilization of a transition metal and a chelating ligand which forms a transition metal complex and is used to produce superoxide anions. Hydroxyl radicals have high oxidizing capabilities and reduce various toxic substances from the wastewater provided low pH conditions. This process has advantages like the high capability to

TABLE 9.1 Novel techniques to reduce water pollution.

Wastewater novel treatment methods			
Physical methods	**Physicochemical/chemical methods**	**Biological methods**	**Thermal methods**
Adsorption	Advance-oxidation processes (AOP)	Aerobic treatment	Wet-air oxidation
Biosorption	Liquid–liquid extraction	Anaerobic treatment	Thermolysis
Ion exchange	Coagulation–flocculation and chemical precipitation method	–	–

decrease chemical oxygen demand (COD) and dyes and toxicants from the wastewater. Ultrasonic radiation or rays are used to treat wastewater using highly powerful ultrasound waves that generate strong cavitation in an aqueous solution and produce reactive free radicals and shock waves. Liquid–liquid extraction involves the movement of solute particles to the organic phase from an aqueous phase. Coagulation is a process that involves the addition of a chemical substance that induces neutralization of the charge of the particles present which leads to their aggregate formation.

Biological methods

The biological method of wastewater treatment involves the utilization of microbes such as bacteria which helps in the degradation of the pollutants present in the wastewater. This method of treatment is nontoxic, requires fewer chemical agents and energy, has a lower cost of running, and yields effluent that is not toxic. It is broadly divided into two categories- aerobic and anaerobic methods. Bacteria such as *Aeromonas hydrophilia*, *Bacillus subtilis*, and *Bacillus cereus* are majorly involved in this process. The fungal treatment method involves the use of fungal microbiota which have the function of cleaning the contaminants of wastewater. The anaerobic method involves the use of microorganisms such as bacteria which work in an anaerobic environment for the degradation of contaminants present in the wastewater. It involves the use of a large bioreactor receptacle for the conversion of organic contaminants into biogas such as hydrogen sulfide, carbon dioxide, and methane. It is cheaper than the aerobic process and the production of sludge in bulk is avoided.

Thermal methods

The thermal method of wastewater treatment involves the use of thermal energy to reduce or degrade the pollutants in the wastewater. Two main thermal methods are wet-air oxidation and thermolysis. They work by converting all the organic pollutants to water and carbon dioxide and help reduce the COD of the effluent. This method is also capable of reducing inorganic pollutants such as ammonia and cyanide from wastewater. This treatment method is expensive when compared to other biological or chemical methods of wastewater treatment. Hence, it is not recommended that wastewater be treated at a COD concentration <5 g/L.

Air pollution

As per the World Health Organization (WHO), air pollution may be defined as a contamination of the outdoor or indoor environment, caused by any physical, biological, or chemical agents that alters the original or natural characteristics present in the air or atmosphere. The process of air pollution includes a series of steps: the genesis of pollutants and their lubrication from a source; their transformation and transport in the air: and harmful actions on ecosystems, materials, and human beings. Vegetation is harmed in various ways namely necrosis, and chlorosis followed by changes in growth. In humans, air pollutants mainly target the respiratory system.

Novel techniques

Fabric filter

Fabric filters are also called baghouses, these are abundantly used to isolate particles from steams of dry gas. They also help in dry scrubbing to prepare a substrate for support of adsorbents used to eliminate a specific gas from gas steam. Through a cloth medium, a dirty gas stream is dragged, which only allows the passage of clean gas and arrests the particulate. The construction of a fabric filter is such that the gas-bearing dust is passed unidirectionally via fabric so that the particles are reserved on the dirty gas side or the upstream, while the gas cleaned moves via fabric to the clean gas side or downstream.

Electrostatics and electrostatic precipitation

Electrostatics may be defined as the utilization of electrostatic forces for the removal of liquid droplets (particles carried in suspension) or charged solid particles. Electrostatics are widely utilized for electrostatic photocopying, dry coating applications, mineral extraction, and paint spraying. Advantages are (1) low operational cost, (2) super elimination efficiency of droplets/fine particles, (3) a large volume of gases can be handled with a low-pressure drop, (4) high-temperature susceptibility (700°C), (5) collection of wet mists/fumes or powder materials. Disadvantages are (1) wide space requirement, (2) high capital investments, (3) failure to collect gaseous pollutants, (4) variation in operating conditions.

Wet and dry scrubbing

The process of scrubbing can be referred to as a unit operation in which, absorbent absorbs one or more than one component of a gas stream. In the process of wet scrubbing, water serves as a majorly used absorbent liquid, whereas in dry scrubbing, a semidry slurry or a dry powder is used as an absorbent. The physical criteria for creating a wet scrubber are straightforward: Absorb the pollutant in the gas stream with a liquid that has a high solubility of the pollutant, and increase the number of gas-liquid interaction surfaces.

Thermal oxidation

Thermal oxidation is also called thermal incineration, or afterburn in which oxidation of organic vapors is carried out at high temperatures. In this process, organic contaminants are incinerated with air in the presence of O_2 by heating contaminant-laden waste gas at a high temperature (1000°F or more). Thermal incinerators are widely used for reducing the emissions of nearly all volatile organic compound (VOC) sources, dryers, kilns, solvent operations, and distillation vents. Incinerators have the disadvantage of high operating costs due to expensive supplemental fuel.

Gas-phase activated carbon adsorption

Adsorption is the process in which fluid molecules adhere or stick to the surface of a solid. By adsorption, pollutants or solids can be easily eradicated or captured from gases, airstream, liquids, or solids. Adsorbate is the material being adsorbed, and adsorber is the adsorption system. This process can be used to isolate solids from liquids, gases from gases, dissolved gases from liquids, and ions from the liquid. The interior of the surface includes by far the majority of free surface area due to its very porous nature. As a result, the adsorbed molecule must spread inside the pore.

Ultraviolet photolysis

UV light is been used for various purposes; it has been found that UV light can be beneficial in the demolition of toxic organics found in aqueous solutions. Organics can be UV photolyzed using a broad range of high-intensity UV light. Direct photolysis has demonstrated that it can be effectively used to convert organic toxins into inert or nontoxic by-products. The main advantage of this technology is that it destroys harmful organic chemicals instead of transferring them to another medium, such as carbon. It may not be completely effective for the removal of heavy metals and particulate matter present in the air.

Biological control systems

Steam containing VOCs of low or moderate concentration can be decontaminated with biological systems. The concentrations of VOC must be high enough to provide food for microorganisms, but not so high that they become harmful. It has three types: biofilters, bio-trickling filters, and bio-scrubbers. Biofilms of fungi and bacteria are cultured on a porous medium. The gas is made to pass via a biologically active medium, and there the micro-organism breaks the more complex compounds into their simplified form, eventually into carbon dioxide and methane, if aerobic and anaerobic, respectively.

Noise pollution

Noise pollution is becoming a growing source of concern for environmentalists all over the world, as excessive noise levels can cause animal death as well as human disabilities. The increase of engines, acoustic devices, vehicles, machinery, and various other factors that cause noise pollution is on the verge in today's time making its control and deterrence a critical objective. Machinery, construction activities, music concerts, and various measures of transportation, such as automobiles, planes, and trains, are the primary sources of outdoor noise pollution around the world. Mitigation of noise pollution involves various measures which can be taken to reduce the intensity and harm caused due to noise. Fig. 9.3 describes various noises produced and their loudness in decibels.

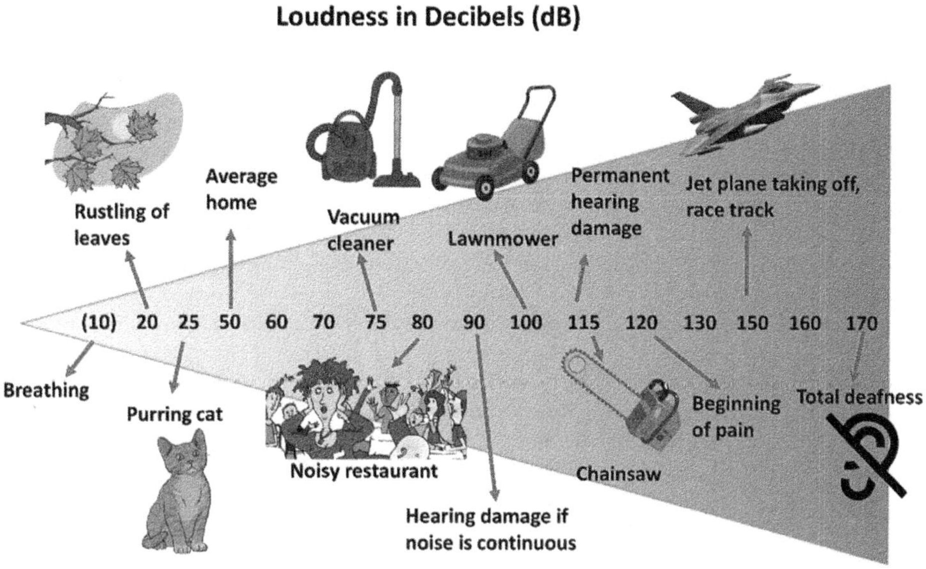

FIGURE 9.3 Various activities with their noise levels in decibels.

Novel methods to control noise pollution

Noise mitigation or control is a set of strategies that helps to reduce noise pollution and its adverse impact on human health and wildlife. Transportation noise management, occupational noise control, urban planning, and architectural design are the main areas of noise moderation. There are a variety of approaches for dampening reverberation from special-purpose spaces such as concert halls, social and entertainment venues, conference rooms, auditoriums, and audio recording rooms as well as exterior walls, ceiling, and floor assemblies. Basic technologies involved in the reduction of noise pollution are sound insulation, sound absorption, vibration damping, and vibration isolation. Sound insulation is done by deterring the transference of noise by blocking with a mass barrier. Vibration isolation is a method of deterring the flow of vibration energy from the source to a receiver by using a flexible element such as springs, or strings or by using a physical break such as rubber mounts and corks.

Role of green chemistry in environmental conservation

Green chemistry refers to the development of chemical procedures and products that minimize or do away with the use or production of hazardous materials. Green chemistry aims to minimize and reduce pollution at the molecular level, it promotes scientific elucidation of actual environmental problems, minimizes the harmful effects of chemical substances and chemical reactions processes on the environment and human health. The two-letter term "green chemistry" and its 12 guiding principles were created by Paul Anastas and John Warner. The use of supercritical carbon dioxide as a green solvent, aqueous hydrogen peroxide for clean oxidations, and the use of hydrogen in asymmetric synthesis were the three major advancements in green chemistry that Ryoji Noyori recognized in 2005. The Green Chemistry idea includes resolving chemical hazards that are dangerous to both human and environmental health, including toxicity, physical risks, climatic change, and the degradation of natural resources. The green chemistry principles can serve as a guide for scientists and engineers as they work to decrease or completely stop the usage and production of hazardous-type chemicals and develop the proper technology to boost social, environmental, and economic advancements.

Conclusion

Sustainability science is undeniably heading towards its gradual omnipresence in our everyday lives through air purifiers, fuel alternatives, organic farming, and much more. Therefore, the role of sustainability science in sustainable development translates into the immense potential it has to contribute to the growth of the latter. Sustainability science is a natural healer for all types of earth diseases. Although sustainability science is in its primary stages, it is this deep whirlpool of opportunities to create and help improve the standard of living for future generations. More and more philosophers and scientists must show interest in sustainability science. Integrating it with technology, experimentation, and research will promote its scope.

References

Adenuga, A.O., Ogujiuba, K., Ohuche, F.K., 2006. Sustainability of the environment and water pollution in Nigeria: Problems, management, and policy options. Global Journal of environmental sciences 5 (1), 49–59.

Bhat, S.A., Bashir, O., Haq, S.A.U., Amin, T., Rafiq, A., Ali, M., et al., 2022. Phytoremediation of heavy metals in soil and water: An eco-friendly, sustainable and multidisciplinary approach. Chemosphere 303, 134788.

Cachada, A., Rocha-Santos, T., Duarte, A.C., 2018. Soil and pollution: an introduction to the main issues. Soil pollution. Academic Press, pp. 1–28.

Cagnetta, G., Robertson, J., Huang, J., Zhang, K., Yu, G., 2016. Mechanochemical destruction of halogenated organic pollutants: A critical review. Journal of Hazardous Materials 313, 85–102.

Cagnetta, G., Zhang, K., Zhang, Q., Huang, J., Yu, G., 2018. Mechanochemical pre-treatment for viable recycling of plastic waste containing haloorganics. Waste Management 75, 181–186.

Chang, Y.C., Chang, S.W., Chen, D.H., 2006. Magnetic chitosan nanoparticles: studies on chitosan binding and adsorption of Co(II) ions. Reactive and Functional Polymers 66 (3), 335–341.

Germaine, K.J., Byrne, J., Liu, X., Keohane, J., Culhane, J., Lally, R.D., et al., 2015. Ecopiling: a combined phytoremediation and passive biopiling system for remediating hydrocarbon impacted soils at field scale. Frontiers in Plant Science 5, 756.

Goel, P.K., 2006. Water Pollution: Causes, Effects and Control. New Age International.

Iyinbor, A.A., Adebesin, B.O., Abimbola, O., Adelani-Akande, A., 2018. Water pollution: effects, prevention, and climatic impact, Water Challenges of an Urbanizing World, 33. IntechOpen.

Koul, B., Taak, P., 2018. Biotechnological Strategies for Effective Remediation of Polluted Soils. Springer Nature.

Lien, H.L., Zhang, W.X., 2001. Nanoscale iron particles for complete reduction of chlorinated ethenes. Colloids and Surfaces A: Physicochemical and Engineering Aspects 191 (1), 97–105.

Lu, C., Chung, Y.L., Chang, K.F., 2005. Adsorption of trihalomethanes from water with carbon nanotubes. Water Research 39 (6), 1183–1189.

Ludzack, F.J., Ettinger, M.B., 1962. Controlling operation to minimize activated sludge effluent nitrogen. Journal (Water Pollution Control Federation) 34 (9), 920–931.

Makarova, O.V., Rajh, T., Thurnauer, M.C., Martin, A., Kemme, P.A., Cropek, D., 2000. Surface modification of TiO2 nanoparticles for photochemical reduction of nitrobenzene. Environmental Science and Technology 34 (22), 4797–4803.

Okrent, D., 1999. On intergenerational equity and its clash with intragenerational equity and on the need for policies to guide the regulation of disposal of wastes and other activities posing very long-term risks. Risk Analysis 19 (5), 877–901.

Pavel, L.V., Gavrilescu, M., 2008. Overview of ex situ decontamination techniques for soil cleanup. Environmental Engineering & Management Journal (EEMJ) 7 (6).

Spangenberg, J.H., 2011. Sustainability science: a review, an analysis, and some empirical lessons. Environmental Conservation 38 (3), 275–287.

Venderbosch, R.H., Prins, W., 2010. Fast pyrolysis technology development. Biofuels, bioproducts and biorefining 4 (2), 178–208.

Wei, M., Wang, B., Chen, M., Lyu, H., Lee, X., Wang, S., et al., 2022. Recent advances in the treatment of contaminated soils by ball milling technology: Classification, mechanisms, and applications. Journal of Cleaner Production 130821.

Whelan, M.J., Coulon, F., Hince, G., Rayner, J., McWatters, R., Spedding, T., et al., 2015. Fate and transport of petroleum hydrocarbons in engineered biopiles in polar regions. Chemosphere 131, 232–240.

Xie, Y., Fang, Z., Cheng, W., Tsang, P.E., Zhao, D., 2014. Remediation of polybrominated diphenyl ethers in soil using Ni/Fe bimetallic nanoparticles: influencing factors, kinetics and mechanism. Science of the Total Environment 485, 363–370.

Chapter 10

Microplastic in the environment: sources, workflow, identification techniques, and impacts on human health

Tharmaraj Vairaperumal[1,2], Devi Selvaraj[3] and Paromita Chakraborty[2,4]

[1]Institute of Clinical Medicine, College of Medicine, National Cheng Kung University, Tainan, Taiwan ROC, [2]Center for Research in Environment, Sustainability Advocacy and Climate Change (REACH), SRM Institute of Science and Technology, Kattankulathur, Tamil Nadu, India, [3]PG and Research Department of Chemistry, Cauvery College for Women (Autonomous), Tiruchirappalli, Tamil Nadu, India, [4]Faculty of Biology and Environmental Protection, University of Lodz, Lodz, Poland

Introduction

Microplastics persistently remain one of the significant residual toxins in our environments (Munyaneza et al., 2022; Rezania et al., 2018; Sarker et al., 2020). Microplastics, ranging from 1 μM to 5 mm in size (Hale et al., 2020; Horton et al., 2017), are derived from both commercial products (primary origin) and mechanical fragmentation processes (secondary origin) (An et al., 2020; Lehtiniemi et al., 2018). These tiny plastic particles, known as primary microplastics, enter diverse environments including land and water bodies through sources such as sewage systems, domestic waste, and industrial effluent (Coyle et al., 2020). Secondary microplastics are generated through the natural decomposition of larger plastic fragments under the influence of UV light, waves, and wind, subsequently entering the surrounding environments (Conti et al., 2021; Wicaksono et al., 2021). The presence of microplastics in the environment poses a challenge for their removal, leading to potential impacts on ecological systems. This is particularly concerning as aquatic organisms, which can directly or indirectly enter the food chain, may pose a risk to human health (Browne et al., 2007). The shape, size, density, surface functionality, and hydrophobicity of microplastics significantly influence the interaction, transformation, and bioavailability of these particles for aquatic organisms (Masura et al., 2015). In the marine ecosystem, water contamination and biological uptake of microplastics exhibit variability, primarily driven by the shape and size of these particles, which in turn affect their surface area density (Bakir et al., 2014; Morét-Ferguson et al., 2010). Microplastic fragments aggregate or interact with environmental matrices and living organisms, including bacteria, resulting in the pollution of aquatic environments (Harrison et al., 2011). Therefore it is crucial to understand the physical properties (such as shape and size) and chemical properties (including microplastic compositions and functional groups) of microplastics when examining environmental samples. Polyvinyl chloride, polystyrene, polyethylene, polypropylene, polyester, polyethylene terephthalate, polyurethane, polyamide, and nylon are among the frequently used plastic materials (Jones et al., 2020). Polystyrene, polyethylene, polypropylene, and polyester are the primary plastic constituents of microplastics, while polyamide is a common secondary microplastic found in beach environments (Efimova et al., 2018). The identification and chemical analysis of microplastics in complex ecological samples involves three sequential steps. Step one is the separation or extraction of microplastics from various heterogeneous samples, including seawater (Miller et al., 2021), freshwater (Razeghi et al., 2021), aquatic organisms (Claessens et al., 2013), tissues (Avio et al., 2015), sediments (Nuelle et al., 2014), and beach sand (Besley et al., 2017). Step two is the visual identification of microplastics through the utilization of physical properties and quantification in aquaculture environments using microscopic techniques such as optical microscopy, scanning electron microscopy (SEM), and TEM analysis (Girão, 2020; Primpke et al., 2020; Xiang et al., 2022). Step three is the chemical composition of microplastics is confirmed through the utilization of FT-IR, Raman spectroscopy techniques, and pyrolysis gas chromatography−mass spectrometry (Pyro-GCMS) techniques (Bouzid et al., 2022; Funck et al., 2020; Schymanski et al., 2021). Microplastic pollution stems from

Role of Green Chemistry in Ecosystem Restoration to Achieve Environmental Sustainability. DOI: https://doi.org/10.1016/B978-0-443-15291-7.00002-X

a wide range of sources, including residential areas, traffic runoff, agricultural practices, wastewater treatment, landfills, improper disposal, manufacturing processes, boat yards, tourism activities, fishing and offshore industries, rivers, ocean currents, and transportation through sea and air routes. This chapter provides on the sources of microplastics, different types of microplastics, and the pathways through which microplastics enter various environmental matrices. The detailed discussion of the sampling techniques, sample preparation methodologies, characterization methods, and chemical identification techniques for analyzing microplastics in environmental samples. Furthermore, the chapter explores the potential exposure and toxicity of microplastics to human health. Finally, strategies for reducing microplastic pollution are also presented.

Sources of microplastics

Microplastic sources can be categorized as primary and secondary sources. Primary microplastics consist of polymeric materials in smaller sizes (≤5 mm) that directly enter the ecosystem (Cole et al., 2011). These microplastics are present in various consumer hygiene products like mouth spray, face wash, body wash, and personal care products. Furthermore, they can also originate from the breakdown of paints, textiles, and plastic objects found in households and offices, including toys, floor coverings, furniture, and dryers, resulting in microplastic dust and fibers. Secondary microplastics, on the other hand, are larger plastic materials such as plastic bags and packaging that undergo fragmentation processes, leading to smaller plastic fragments due to UV photodegradation and weathering (Cooper and Corcoran, 2010; Weinstein et al., 2016). Fig. 10.1 illustrates the schematic diagram depicting the entry of different types of microplastics into the aquatic ecosystem.

Personal care products

Cosmetic and personal care products, including face masks, shower gels, and toothpaste, release plastic microbeads and their associated additives, which act as sorbents (Leslie, 2014). The pollution from microbeads has become a growing concern as they contain toxic chemicals that can accumulate and transfer into the food webs of marine organisms (Rochman, 2015).

Synthetic textile

Natural and synthetic fibers are predominantly derived from textile garments (Ahmed and Mondal, 2021), medical filtration materials, and protective clothing (Zhong, 2011). During the washing process, microfiber fragments, typically measuring around 30–40 μm in size, are primarily released.

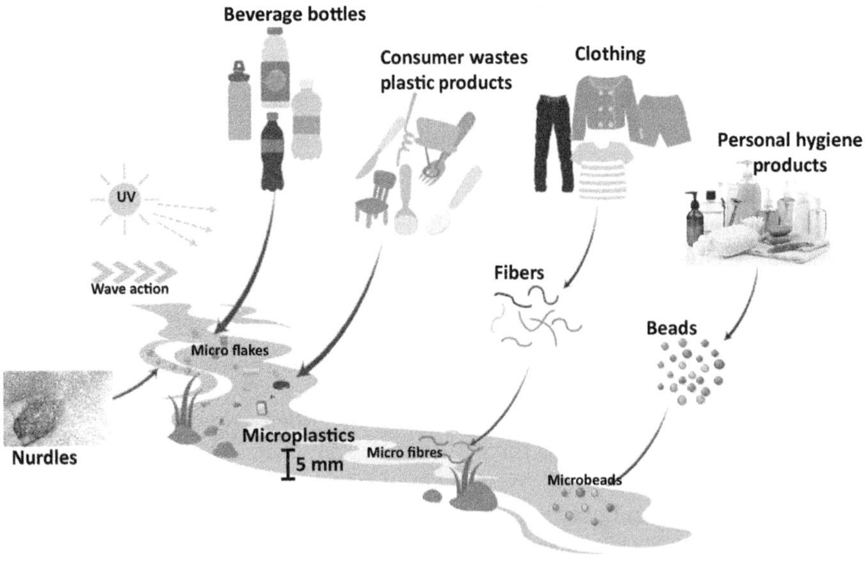

FIGURE 10.1 A schematic diagram illustrating the entry of diverse microplastics into the aquatic ecosystem.

Plastic pellet manufacturers

Plastic pellets, ranging in size from 1 to 6 mm, are produced from granules of various plastic products and come in different shapes. The commonly utilized plastic polymers for these pellets include polyvinyl chloride, polyethylene, and polypropylene, often found in different colors. During the manufacturing process and transportation, plastic pellets have the potential to enter the environment (Andrady, 2011; Endo et al., 2005; Hidalgo-Ruz et al., 2012).

Packaging material and litter

Packaging materials are the primary contributors to microplastic pollution, from littering, including items such as toys, cutlery, shoes, and clothing generating a larger amount of pollutants (Raynaud, 2014). Additionally, both biodegradable and nonbiodegradable agricultural plastic packaging contribute to microplastic pollution as they degrade into smaller fragments that eventually enter surface water.

Foodstuffs packaging materials

Food packaging materials are widely used due to their cost-effectiveness, lightweight nature, durability, and excellent performance. However, the everyday utilization of plastic packaging has led to the release of microplastic particles into food items. Furthermore, Table 10.1 provides comprehensive information on the classification of microplastic sources. They are categorized by source types, main sources, subgroups and product sources.

Types of microplastics

Fragments, fibers, microbeads, film, and foam are among the various types of microplastics that contribute to environmental pollution and pose potential risks to ecosystems (Guo et al., 2019; Pignattelli et al., 2020).

Fragments

Larger plastic materials like single-use plastic products, lids, and cutlery break down into smaller fragments due to UV radiation and wave actions. Specifically, water bottles and plastic bags are major sources of microplastic fragments, which further degrade into micro-sized plastic flakes.

Fibers

Microfibers primarily originate from sources such as washing machines, fleece clothing, diapers, and cigarette butts. Synthetic fabrics, notably polyester, commonly found in athletic wear, tend to release a higher amount of microfibers.

Microbeads

Microbeads, which are nonbiodegradable polymers with a diameter of ≤ 1 mm, are sourced from facial cleansers, exfoliating soaps, and toothpaste. These exceptionally small and nondigestible microbeads can easily enter the food chain, posing a potential risk.

Film

Plastic films are primarily derived from plasticizers such as phthalic acid esters, commonly utilized in agriculture, which can contaminate soil organisms and negatively impact soil quality.

Foam

Styrofoam finds applications in packaging materials, coffee saucers, and as a stuffing material. Plasticizer compounds from Styrofoam can leach into food and beverages, posing concerns for human health. Table 10.2 provides information on the major sources and corresponding polymers, including polyethylene, polystyrene, polypropylene, polyolefins, and polyester.

TABLE 10.1 Classification of microplastics sources by source type, main sources, subgroup, and product sources.

Source type	Main sources	Subgroup	Product sources
Primary sources of microplastics	Consumer care products	PCP	Microbeads: Face wash, body wash, and toothpaste
		Other PCP products	Microspheres: Shave foam, lipstick, sunscreen, and mascara
		Other buyer products	Glitter powder, antiskid powder, and printer toner
	Commercial or industrial use products	Medical applications	Dentist tooth polish and medical applications of plastic beads
		Processing industry	Microbeads: Oil or gas exploration and textile printing
	Microplastic dust emissions from industrial production	Commercial handling or maintenance of plastic items	Dust: Cutting, polishing, and molding plastic items
	Dust and fibers	Household dust emissions	Paint removal by scrubbing and heavy-duty wash
			Textile fibers ripped loose in laundry machines and driers
		Agriculture	Debasing agricultural film/ covering film
		The maritime sector, aquaculture, and fisheries	Scrape and enduring of plastic wires and exteriors in waterfronts
			Sewage from aquaculture and harpooning net scrubbing amenities
	Waste treatment and reprocessing	A plastic constituent part from the grating and breaking of plastic waste	Ship breaking/ decommissioning of ships and offshore installations
			Food waste shredders on ships and institutions
			Composting plastic-contaminated organic waste
Secondary sources	Macro plastic waste handling	Littering in public spaces	Parks, recreational spots, and roadsides were renovated by municipalities
		Industrial and construction waste	Dumping of construction and process waste
		Municipal landfills	Plastic particles in water effluent and not well-covered air drift plastics
	Municipal effluent (sewage system and hurricane water)	Municipal dirt-handling plants	Macro plastics transient motorized screens and city surface water drains
	Runoff from terrestrial sources	Regular riverine involvement	Macro plastic leftover is released straight into the water
	Maritime waste handling	Fishery	Damage of trawls and nets
		Aquaculture	Storm loss of floaters
		Shipping	Waste was thrown overboard

PCP, Personal care products.

TABLE 10.2 Various types of microplastics along with their major sources and corresponding polymer types.

Types of microplastic	Major sources	Polymer types
Fragment/flakes	Beverage bottles, plastic lids, plastic cups, plastic straws, plastic stirrers, plastic forks, plastic knives, and plastic spoons	High-density or Low density of polyethylene
Fiber	Fishing nets and clothing or textiles	Polystyrene
Beads	Toothpaste, shower gel, and facial cleansers	Polyethylene and/or polypropylene
Film	Plastic bags and food wrappers	Polyolefins and polyester
Foam	Foam floats, styrofoam, and cushioning	Polystyrene

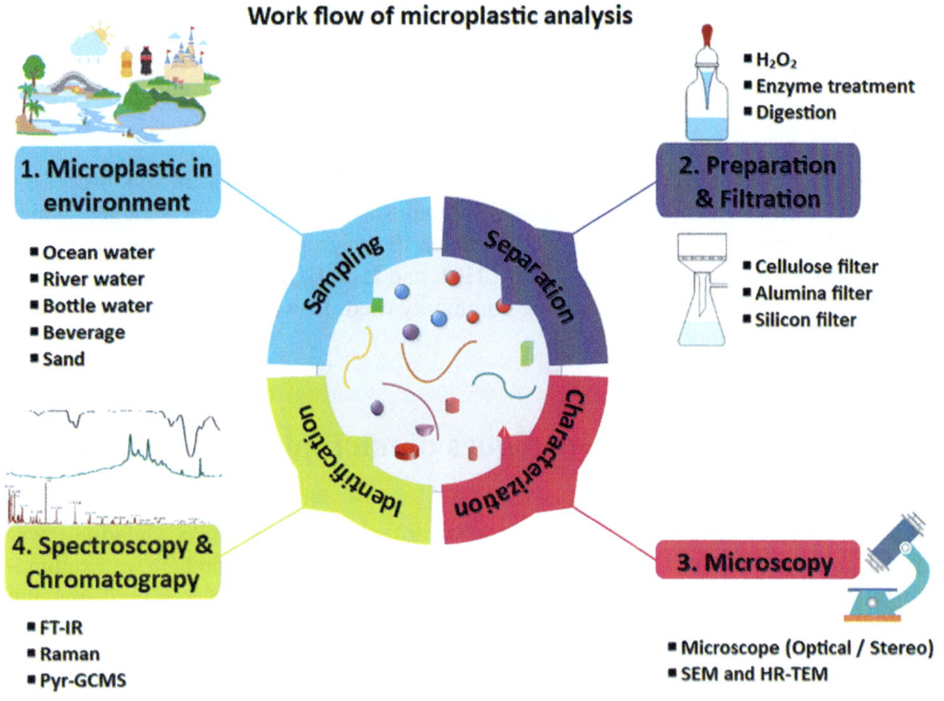

FIGURE 10.2 Illustration of the five-step methodology commonly employed for the investigation of microplastics in ecosystems.

Microplastic identification workflow in environmental samples

The most commonly employed methodology for investigating microplastics in ecosystems involves five steps, as illustrated in Fig. 10.2.

Methods for sampling and preparation

Microplastic sampling has focused on environmental matrices including water and sediments.

Water samples

Water samples are typically collected from the surface, middle, and bottom layers. Various sampling methods, such as nets, filters, and pumps, are commonly employed. Clean water samples with fewer sediment particles allow for direct filtration of larger volumes, facilitating concentration during sampling. Table 10.3 provides an overview of the advantages and disadvantages of different water sample collection methods, including nets, filters, and pumps.

TABLE 10.3 Advantages and disadvantages of sample collectors.

Sample collector	Advantages	Disadvantages
Nets	Easy to use and collected larger sample volume	Expensive and time-consuming process
Filters	Does not require special equipment	Required storage equipment
Pumps	Collecting more samples and saving time	Required more sample containers

Sediment samples

Microplastics exhibit uneven distribution in sediment samples due to various environmental factors, including UV radiation and wind waves. Sediment samples are typically collected from the surface at a depth of 1–5 cm, ranging from open burning areas and textile regions to riversides and tidal zones, extending to the ocean floor. Dry sediment samples are collected directly, while water-based samples are obtained using containers or barrels. Wet sediment samples can be collected using grab samplers, box corers, or gravity cores.

Sample preparation methods

Density separation is employed to separate microplastics from sediments and other solid materials. In this process, a mixture of saturated inorganic metal salts such as $ZnCl_2$, NaI, or sodium polytungstate is added to the samples, which are then shaken and allowed to settle. This initial separation helps concentrate the microplastics. To remove organic matter from the samples, chemical digestions using acids (such as HNO_3, HCl, and H_2SO_4), alkaline solutions (such as NaOH or KOH), or oxidizing agents (such as H_2O_2 or sulfuric acid) are commonly used. However, enzymatic digestion, though effective, can be costly for larger sample quantities.

Characterization and chemical identification techniques of microplastics in water samples

The collection of microplastic particles on filter paper was carried out from different water samples including river water, surface water, groundwater, and packaged water at various locations. Fig. 10.3 shows a comprehensive examination of multiple facets concerning microplastics, accomplished by employing a combination of microscopic imaging and spectroscopic methods. Below, we discuss the five techniques currently utilized for the analysis of microplastics.

Optical microscopy

Optical microscopy characterization can provide valuable information about the color, size ($>100\ \mu m$), and shapes of microplastics. Fig. 10.3A demonstrates the microscopic identification of different microplastics and detailed insights into the visual characteristics and features of these microplastic particles (Song et al., 2015). The identification process includes the examination of both polymeric and nonpolymeric (organic) microplastics. Among the nonpolymeric materials, the images showcase microplastic particles derived from cotton, rayon, and artificial strings.

Scanning electron microscopy coupled with energy-dispersive X-ray spectroscopy

SEM offers a comprehensive structural characterization of microplastic constituents in the nanometer range, which is a significant advantage. However, SEM alone does not provide chemical evidence. On the other hand, when coupled with EDX, SEM can provide statistical information about the elemental compositions. Fig. 10.3B presents a collection of SEM images that show different types of microplastic particles including fibers, spheres, and fragments (Pivokonsky et al., 2018).

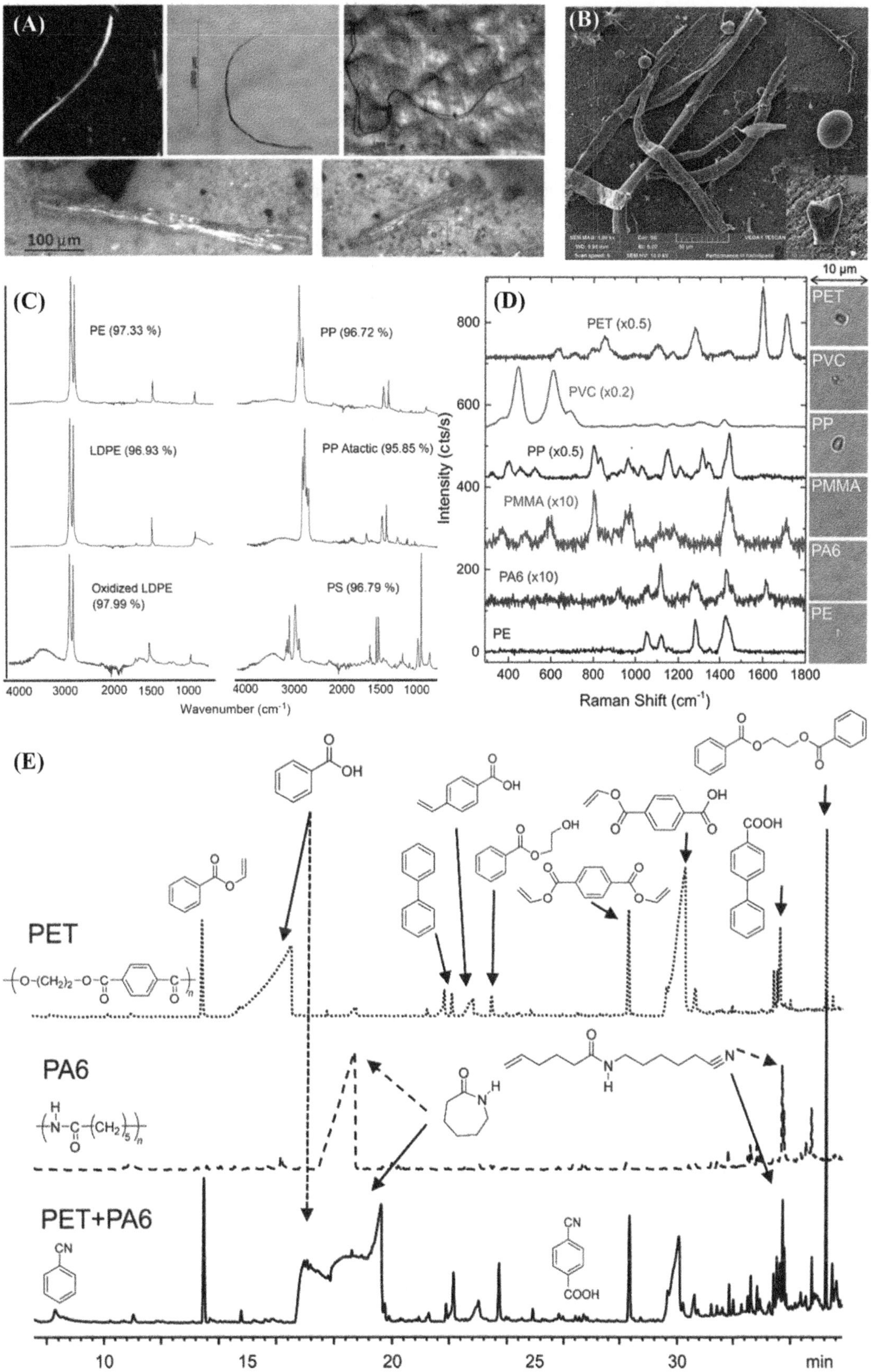

FIGURE 10.3 (A) Microscopic images of natural plastic and nonplastic (Song et al., 2015), (B) SEM image of different shaped microparticles (Pivokonsky et al., 2018), (C and D) FT-IR and Raman spectra of the different types of microplastics (Gillibert et al., 2019; Syakti et al., 2018), and (E) Pyro-GCMS spectra of PET and PA6 (Coralli et al., 2022). *Pyro-GCMS*, Pyrolysis gas chromatography–mass spectrometry; *SEM*, scanning electron microscopy.

Infrared spectroscopy

Infrared spectroscopy is a commonly employed technique for analyzing microplastics, offering numerical analysis and chemical identification to confirm the identity of the microplastic. Fig. 10.3C displays the FT-IR spectra of polyethylene, low-density polyethylene, polypropylene, and polystyrene (Syakti et al., 2018). The spectral features are assigned to carbonyl groups (C = O, C—O, and C = C) at 1713, 1178, and 1631 cm^{-1}, respectively. Asymmetrical methyl C-H bending and C-H rocking bands are observed at approximately 1467 cm^{-1} (δ-CH_2) and 720 cm^{-1} (γ-CH_2), respectively. The stretching of —OH groups is observed in the region of 3050—3570 cm^{-1}.

Micro Raman spectroscopy

Micro Raman spectroscopy is a widely employed analytical technique for microplastic analysis, ranking second in popularity. Raman instruments utilize visible objectives, providing high-quality images along with valuable structural information, numerical analysis, and chemical identification of microplastics through comparison with commercial libraries. The Raman spectra of different polymers, namely polyethylene terephthalate, polyvinyl chloride, polypropylene, polymethyl methacrylate, polyamide-6, and polyethylene, dispersed in seawater were recorded and shown in Fig. 10.3D (Gillibert et al., 2019).

Pyrolysis gas chromatography—mass spectrometry

Pyro-GCMS is a powerful analytical technique used to assess the chemical composition of microplastics by analyzing their programs. This method enables the identification and quantification of microplastic polymers based on their weight per polymer and polymer type, surpassing the limitations of conventional quantification methods. Pyro-GCMS provides a numerical analysis of microplastics, allowing for a comprehensive understanding of their composition and potential exposure in various environments. The pyrolysis product analysis using Pyro-GCMS spectra (Coralli et al., 2022) revealed that the predominant pyrolysis product of poly(ethylene terephthalate), polyamide-6, and the copolymerization of PET with PA6 as shown in Fig. 10.3E. This provides insight into the composition of the analyzed sample, indicating the presence of both PET and PA6 in the microplastic mixture.

Table 10.4 provides a summary of the different analytical methods used for the analysis of microplastics, along with information on sampling sources and types of polymers. It offers an overview of the techniques and resources utilized in studying microplastic contamination in various environmental samples.

Impact of microplastics on human health

Microplastics can adsorb and absorb hazardous chemical pollutants, including heavy metals and toxic ions, from the environment. These contaminants can then be transferred to food products, posing a direct risk to human health (Van Cauwenberghe et al., 2015). The impact of chemical exposure to microplastics has been investigated using human cell lines, examining parameters such as cytotoxicity, immune response, histamine release, oxidative stress, genotoxicity, and cell barrier function. Cytotoxicity studies of plastic chemicals have been conducted on environmental samples using 11 different cell models, with the primary in vitro test employing a colony-forming assay with V79 cells (Furukuma and Fujii, 2016). In the study, nine different polymers ranging from 0.1 to 282 μM in size, with two different shapes, were utilized at concentrations ranging from 0.01 to 100,000 μg/mL. The exposure experiments focused on five specific polymers, namely, low-density polyethylene, polyethylene, polypropylene, polystyrene, and PVC, with circular and asymmetrical shapes ranging from 0.5 μM to 0.1 nm. The exposure duration varied from 2 to 4 days. Another study by Mondal et al. examined synthesized microplastic PET particles at concentrations of 50, 250, 500, 750, and 1000 mg/L using Allium cepa root tips, with an exposure duration of 72 hours (Mondal et al., 2022). Lu et al. conducted a study on the in vitro cytotoxicity effects of polystyrene particles measuring approximately 1 μM in size, using human umbilical vein endothelial cells (Lu et al., 2023). Gautam et al. conducted a study to investigate the toxicity effects of PE particles with sizes of 30.5 ± 10.5 and 6.2 ± 2.0 μM on intestinal epithelial Caco-2 and lung epithelial A549 cells (Gautam et al., 2022). The immune responses to microplastic exposure were examined using polystyrene microplastic (PS-MP) particles of sizes 0.2, 2, and 10 μM respectively. Human microglial HMC-3 cells were employed for the investigation, and the results revealed that the immune response was characterized by microglial apoptosis induced by phagocytosis, particularly for the 0.2 and 2 μM PS-MPs (Kwon et al., 2022). Dong et al. have studied the 1.72-sized PS-MP with 10 and 1000 $\mu g/cm^2$ at 24 hours exposure to Interleukin 6 and 8 proteins (IL-6 and IL-8) (Dong et al., 2020). Cytotoxicity

TABLE 10.4 Overview of analytical methods for microplastic analysis, sampling sources, and polymer types.

Analytical method	Sampling source	Pretreatment	Size/ quantification	Polymer types	Reference
Laser direct infrared spectroscopy	Drinking water	Dune infiltration and sedimentation	20 to 2 μm	PA, PET rubbers, PE, and CPE	Bäuerlein et al. (2022)
μFT-IR	Wastewater treatment plant effluents	Cascade filtration	100 and 50 μm	PE	Al-Azzawi et al. (2022)
μFT-IR	Organic-rich freshwater samples	Centrifugation and digestion	–	PE, PP, PS, PVC, N-6, PET	Monteiro et al. (2022)
FT-IR and Raman	Drinking water	Filtering/sieving	1–500 μM	PE, PS, PVC, and PET	De Frond et al. (2022)
FT-IR	Freshwaters	Digestion	20–300 μm	–	Rowenczyk et al. (2022)
FT-IR	Drinking water treatment plants	Sedimentation/filtration	20 μM	PE, PP, PET, PMMA, PA and PS	Jung et al. (2022)
FT-IR	River water	Filtration	10 μM	Polychloroprene	Devereux et al. (2022)
ATR-FT-IR	River water	Digestion	700 μM–5 mm	PE	de Carvalho et al. (2022)
FT-IR	Surface water	Density separation and digestion	20 μm	PP, EVA, and PET	Arcadio et al. (2023)
FT-IR	Industrial wastewater	Density separation	<0.1–≥2 mm	N-6, cellulose acetate, and PS	Haque et al. (2022)
μ-Raman	wastewater treatment plant	Filtration	38–1000 μM	PP, PE, and PS	Becucci et al. (2022)
μ-Raman	Drinking water	Filtration	10–45 μM	–	Cherniak et al. (2022)
μ-Raman	Freshwater	Filtration	–	PP and PE	Kalogerakis et al. (2022)
Pyro-GCMS	Wastewater treatment plants	Digestion, filtration, and liquid extraction	1 μM	PE, PVC, PET, PP, and PMMA	Okoffo et al. (2023)
Pyro-GCMS	textile laundry wastewater	Digestion, filtration, and liquid extraction	0.2 and 0.6 μg	PET, N-6, and PAN	Lim et al. (2022)

Ethylene-vinyl acetate copolymer; *EVA*, Ethylene-vinyl acetate; *N-6*, Nylon-6; *PA*, polyamide; *PAN*, polyacrylonitrile *PE*, polyethylene; *PET*, polyethylene terephthalate; *PMMA*, polymethyl methacrylate; *PP*, polypropylene; *PVC*, polyvinyl chloride; *Pyro-GCMS*, pyrolysis gas chromatography–mass spectrometry.

of PS-MPs was assessed using adenomatous cells of the small intestine (Magrì et al., 2018; Stock et al., 2019; Wu et al., 2019). The results demonstrated that the cytotoxic effects and oxidative damages were specifically attributed to the individual PS-MP particles.

Strategies to reduce microplastic pollution

The following strategies are contributing to reducing microplastic pollution:

1. Implementing effective filtration systems in wastewater treatment plants can significantly minimize the release of microplastics into aquatic environments.

2. Promoting the use of biodegradable and eco-friendly materials in various industries, such as textiles and packaging, can help reduce the generation of microplastics at the source. Educating the public about proper waste disposal practices, emphasizing the importance of recycling, and avoiding single-use plastics, can prevent microplastics from entering ecosystems.
3. Encouraging eco-friendly manufacturing processes, and opting for natural fibers can contribute to reducing microplastic pollution in the long run.

By adopting these strategies collectively, we can strive to mitigate the harmful impact of microplastics on our environment.

Conclusions

This chapter provides a comprehensive overview of the sampling methods and sample preparation procedures employed for analyzing microplastics in water and sediment samples. The analytical workflow includes the characterization techniques of microscopy and SEM, which enable a detailed examination of microplastic properties. Furthermore, spectroscopic methods such as FT-IR, Raman, and Pyro-GCMS chromatography are utilized to identify the chemical composition of microplastics. The presence of microplastic particles in contaminated food and drinking water has been associated with the occurrence of chronic illnesses in humans. Therefore it is crucial to prioritize the reduction of microplastic contamination in the human ecosystem. This can be achieved by improving waste management practices and raising awareness to minimize littering and plastic waste, ultimately contributing to the restoration of aquatic ecosystems.

Acknowledgments

Dr. Paromita gratefully acknowledges the India-Norway cooperation project on capacity building for reducing plastic and chemical pollution in India (INOPOL) Project Number IND-19/0005. The Research Council of Norway is also acknowledged for financing Dr. Steindal and Dr. Nizzetto's participation in the research through the project ASAP-Asian Scientific Alliance for Plastic Pollution and Value Network Management (Contract no. 302575).

References

Ahmed, F., Mondal, M.I.H., 2021. Introduction to natural fibres and textiles. Fundamentals of Natural Fibres and Textiles 1–32. Available from: https://doi.org/10.1016/B978-0-12-821483-1.00001-2.

Al-Azzawi, M.S.M., Funck, M., Kunaschk, M., der Esch, E., Von, Jacob, O., et al., 2022. Microplastic sampling from wastewater treatment plant effluents: best-practices and synergies between thermoanalytical and spectroscopic analysis. Water Research 219, 118549. Available from: https://doi.org/10.1016/J.WATRES.2022.118549.

Andrady, A.L., 2011. Microplastics in the marine environment. Marine Pollution Bulletin 62, 1596–1605. Available from: https://doi.org/10.1016/J.MARPOLBUL.2011.05.030.

An, L., Liu, Q., Deng, Y., Wu, W., Gao, Y., Ling, W., 2020. Sources of microplastic in the environment. Handbook of Environmental Chemistry 95, 143–159. Available from: https://doi.org/10.1007/698_2020_449/TABLES/2.

Arcadio, C.G.L.A., Navarro, C.K.P., Similatan, K.M., Inocente, S.A.T., Ancla, S.M.B., Banda, M.H.T., et al., 2023. Microplastics in surface water of Laguna de Bay: first documented evidence on the largest lake in the Philippines. Environmental Science and Pollution Research 30, 29824–29833. Available from: https://doi.org/10.1007/S11356-022-24261-5/FIGURES/5.

Avio, C.G., Gorbi, S., Regoli, F., 2015. Experimental development of a new protocol for extraction and characterization of microplastics in fish tissues: first observations in commercial species from Adriatic Sea. Marine Environmental Research 111, 18–26. Available from: https://doi.org/10.1016/J.MARENVRES.2015.06.014.

Bakir, A., Rowland, S.J., Thompson, R.C., 2014. Enhanced desorption of persistent organic pollutants from microplastics under simulated physiological conditions. Environmental Pollution 185, 16–23. Available from: https://doi.org/10.1016/J.ENVPOL.2013.10.007.

Becucci, M., Mancini, M., Campo, R., Paris, E., 2022. Microplastics in the Florence wastewater treatment plant studied by a continuous sampling method and Raman spectroscopy: a preliminary investigation. Science of the Total Environment 808, 152025. Available from: https://doi.org/10.1016/J.SCITOTENV.2021.152025.

Besley, A., Vijver, M.G., Behrens, P., Bosker, T., 2017. A standardized method for sampling and extraction methods for quantifying microplastics in beach sand. Marine Pollution Bulletin 114, 77–83. Available from: https://doi.org/10.1016/J.MARPOLBUL.2016.08.055.

Bouzid, N., Anquetil, C., Dris, R., Gasperi, J., Tassin, B., Derenne, S., 2022. Quantification of microplastics by pyrolysis coupled with gas chromatography and mass spectrometry in sediments: challenges and implications. Microplastics 1, 229–239. Available from: https://doi.org/10.3390/MICROPLASTICS1020016.

Browne, M.A., Galloway, T., Thompson, R., 2007. Microplastic—an emerging contaminant of potential concern? Integrated Environmental Assessment and Management 3, 559–561. Available from: https://doi.org/10.1002/IEAM.5630030412.

Bäuerlein, P.S., Hofman-Caris, R.C.H.M., Pieke, E.N., ter Laak, T.L., 2022. Fate of microplastics in the drinking water production. Water Research 221, 118790. Available from: https://doi.org/10.1016/J.WATRES.2022.118790.

Cherniak, S.L., Almuhtaram, H., McKie, M.J., Hermabessiere, L., Yuan, C., Rochman, C.M., et al., 2022. Conventional and biological treatment for the removal of microplastics from drinking water. Chemosphere 288, 132587. Available from: https://doi.org/10.1016/J.CHEMOSPHERE.2021.132587.

Claessens, M., Van Cauwenberghe, L., Vandegehuchte, M.B., Janssen, C.R., 2013. New techniques for the detection of microplastics in sediments and field collected organisms. Marine Pollution Bulletin 70, 227–233. Available from: https://doi.org/10.1016/J.MARPOLBUL.2013.03.009.

Cole, M., Lindeque, P., Halsband, C., Galloway, T.S., 2011. Microplastics as contaminants in the marine environment: a review. Marine Pollution Bulletin 62, 2588–2597. Available from: https://doi.org/10.1016/J.MARPOLBUL.2011.09.025.

Conti, I., Simioni, C., Varano, G., Brenna, C., Costanzi, E., Neri, L.M., 2021. Legislation to limit the environmental plastic and microplastic pollution and their influence on human exposure. Environmental Pollution 288, 117708. Available from: https://doi.org/10.1016/J.ENVPOL.2021.117708.

Cooper, D.A., Corcoran, P.L., 2010. Effects of mechanical and chemical processes on the degradation of plastic beach debris on the island of Kauai, Hawaii. Marine Pollution Bulletin 60, 650–654. Available from: https://doi.org/10.1016/J.MARPOLBUL.2009.12.026.

Coralli, I., Giorgi, V., Vassura, I., Rombolà, A.G., Fabbri, D., 2022. Secondary reactions in the analysis of microplastics by analytical pyrolysis. Journal of Analytical and Applied Pyrolysis 161, 105377. Available from: https://doi.org/10.1016/J.JAAP.2021.105377.

Coyle, R., Hardiman, G., Driscoll, K.O., 2020. Microplastics in the marine environment: a review of their sources, distribution processes, uptake and exchange in ecosystems. Case Studies in Chemical and Environmental Engineering 2, 100010. Available from: https://doi.org/10.1016/J.CSCEE.2020.100010.

de Carvalho, A.R., Riem-Galliano, L., ter Halle, A., Cucherousset, J., 2022. Interactive effect of urbanization and flood in modulating microplastic pollution in rivers. Environmental Pollution 309, 119760. Available from: https://doi.org/10.1016/J.ENVPOL.2022.119760.

De Frond, H., Thornton Hampton, L., Kotar, S., Gesulga, K., Matuch, C., Lao, W., et al., 2022. Monitoring microplastics in drinking water: an interlaboratory study to inform effective methods for quantifying and characterizing microplastics. Chemosphere 298, 134282. Available from: https://doi.org/10.1016/J.CHEMOSPHERE.2022.134282.

Devereux, R., Westhead, E.K., Jayaratne, R., Newport, D., 2022. Microplastic abundance in the Thames River during the New Year period. Marine Pollution Bulletin 177, 113534. Available from: https://doi.org/10.1016/J.MARPOLBUL.2022.113534.

Dong, C.Di, Chen, C.W., Chen, Y.C., Chen, H.H., Lee, J.S., Lin, C.H., 2020. Polystyrene microplastic particles: in vitro pulmonary toxicity assessment. Journal of Hazardous Materials 385, 121575. Available from: https://doi.org/10.1016/J.JHAZMAT.2019.121575.

Efimova, I., Bagaeva, M., Bagaev, A., Kileso, A., Chubarenko, I.P., 2018. Secondary microplastics generation in the sea swash zone with coarse bottom sediments: laboratory experiments. Frontiers in Marine Science 5, 402880. Available from: https://doi.org/10.3389/FMARS.2018.00313/BIBTEX.

Endo, S., Takizawa, R., Okuda, K., Takada, H., Chiba, K., Kanehiro, H., et al., 2005. Concentration of polychlorinated biphenyls (PCBs) in beached resin pellets: variability among individual particles and regional differences. Marine Pollution Bulletin 50, 1103–1114. Available from: https://doi.org/10.1016/J.MARPOLBUL.2005.04.030.

Funck, M., Yildirim, A., Nickel, C., Schram, J., Schmidt, T.C., Tuerk, J., 2020. Identification of microplastics in wastewater after cascade filtration using pyrolysis-GC–MS. MethodsX 7, 100778. Available from: https://doi.org/10.1016/J.MEX.2019.100778.

Furukuma, S., Fujii, N., 2016. In vitro cytotoxicity evaluation of plastic marine debris by colony-forming assay. Japanese Journal of Environmental Toxicology 19, 71–81. Available from: https://doi.org/10.11403/JSET.19.71.

Gautam, R., Jo, J.H., Acharya, M., Maharjan, A., Lee, D.E., Pramod, P.B., et al., 2022. Evaluation of potential toxicity of polyethylene microplastics on human derived cell lines. Science of the Total Environment 838, 156089. Available from: https://doi.org/10.1016/J.SCITOTENV.2022.156089.

Gillibert, R., Balakrishnan, G., Deshoules, Q., Tardivel, M., Magazzù, A., Donato, M.G., et al., 2019. Raman tweezers for small microplastics and nanoplastics identification in seawater. Environmental Science & Technology 53, 9003–9013. Available from: https://doi.org/10.1021/ACS.EST.9B03105/ASSET/IMAGES/LARGE/ES-2019-031059_0011.JPEG.

Girão, A.V., 2020. SEM/EDS and optical microscopy analysis of microplastics. Handbook of Microplastics in the Environment. Springer, pp. 1–22. Available from: https://doi.org/10.1007/978-3-030-10618-8_7-1.

Guo, X., Chen, C., Wang, J., 2019. Sorption of sulfamethoxazole onto six types of microplastics. Chemosphere 228, 300–308. Available from: https://doi.org/10.1016/J.CHEMOSPHERE.2019.04.155.

Hale, R.C., Seeley, M.E., La Guardia, M.J., Mai, L., Zeng, E.Y., 2020. A global perspective on microplastics. Journal of Geophysical Research: Oceans 125, . Available from: https://doi.org/10.1029/2018JC014719e2018JC014719.

Haque, Md.M., Nupur, F.Y., Parvin, F., Tareq, S.M., 2022. Occurrence and characteristics of microplastic in different types of industrial wastewater and sludge: a potential threat of emerging pollutants to the freshwater of Bangladesh. Journal of Hazardous Materials Advances 8, 100166. Available from: https://doi.org/10.1016/J.HAZADV.2022.100166.

Harrison, J.P., Sapp, M., Schratzberger, M., Mark Osborn, A., 2011. Interactions between Microorganisms and marine micro plastics: a call for research. Marine Technology Society Journal 45, 12–20. Available from: https://doi.org/10.4031/MTSJ.45.2.2.

Hidalgo-Ruz, V., Gutow, L., Thompson, R.C., Thiel, M., 2012. Microplastics in the marine environment: a review of the methods used for identification and quantification. Environmental Science & Technology 46, 3060–3075. Available from: https://doi.org/10.1021/ES2031505/ASSET/IMAGES/LARGE/ES-2011-031505_0005.JPEG.

Horton, A.A., Walton, A., Spurgeon, D.J., Lahive, E., Svendsen, C., 2017. Microplastics in freshwater and terrestrial environments: evaluating the current understanding to identify the knowledge gaps and future research priorities. Science of the Total Environment 586, 127–141. Available from: https://doi.org/10.1016/J.SCITOTENV.2017.01.190.

Jones, J.I., Vdovchenko, A., Cooling, D., Murphy, J.F., Arnold, A., Pretty, J.L., et al., 2020. Systematic analysis of the relative abundance of polymers occurring as microplastics in freshwaters and estuaries. International Journal of Environmental Research and Public Health 2020 17, 9304. Available from: https://doi.org/10.3390/IJERPH17249304.

Jung, J.W., Kim, S., Kim, Y.S., Jeong, S., Lee, J., 2022. Tracing microplastics from raw water to drinking water treatment plants in Busan, South Korea. Science of The Total Environment 825, 154015. Available from: https://doi.org/10.1016/J.SCITOTENV.2022.154015.

Kalogerakis, N., Rytelewska, S., Abrowska, A.D., 2022. The Raman spectroscopy approach to different freshwater microplastics and quantitative characterization of polyethylene aged in the environment. Microplastics 1, 263–281. Available from: https://doi.org/10.3390/MICROPLASTICS1020019.

Kwon, W., Kim, D., Kim, H.Y., Jeong, S.W., Lee, S.G., Kim, H.C., et al., 2022. Microglial phagocytosis of polystyrene microplastics results in immune alteration and apoptosis in vitro and in vivo. Science of the Total Environment 807, 150817. Available from: https://doi.org/10.1016/J.SCITOTENV.2021.150817.

Lehtiniemi, M., Hartikainen, S., Näkki, P., Engström-Öst, J., Koistinen, A., Setälä, O., 2018. Size matters more than shape: ingestion of primary and secondary microplastics by small predators. Food Webs 17, e00097. Available from: https://doi.org/10.1016/J.FOOWEB.2018.E00097.

Leslie, H.A., 2014. Review of Microplastics in Cosmetics: Scientific Background on a Potential Source of Plastic Particulate Marine Litter to Support Decision-Making. VU . Available from: https://hdl.handle.net/1871.1/3266b09c-95e6-499b-aa40-0607152580ce.

Lim, S.J., Park, Y.K., Kim, H., Kwon, J., Moon, H.M., Lee, Y., et al., 2022. Selective solvent extraction and quantification of synthetic microfibers in textile laundry wastewater using pyrolysis-gas chromatography/mass spectrometry. Chemical Engineering Journal 434, 134653. Available from: https://doi.org/10.1016/J.CEJ.2022.134653.

Lu, Yy, Cao, M., Tian, M., Huang, Q., 2023. Internalization and cytotoxicity of polystyrene microplastics in human umbilical vein endothelial cells. Journal of Applied Toxicology 43, 262–271. Available from: https://doi.org/10.1002/JAT.4378.

Magrì, D., Sánchez-Moreno, P., Caputo, G., Gatto, F., Veronesi, M., Bardi, G., et al., 2018. Laser ablation as a versatile tool to mimic polyethylene terephthalate nanoplastic pollutants: characterization and toxicology assessment. ACS Nano 12, 7690–7700. Available from: https://doi.org/10.1021/ACSNANO.8B01331/ASSET/IMAGES/MEDIUM/NN-2018-01331Z_M001.GIF.

Masura, J., Baker, J., Foster, G., Arthur, C., 2015. Laboratory Methods for the Analysis of Microplastics in the Marine Environment: Recommendations for Quantifying Synthetic Particles in Waters and Sediments. https://marinedebris.noaa.gov/sites/default/files/publications-files/noaa_microplastics_methods_manual.pdf.

Miller, M.E., Motti, C.A., Menendez, P., Kroon, F.J., 2021. Efficacy of microplastic separation techniques on seawater samples: testing accuracy using high-density polyethylene. Biological Bulletin 240, 52–66. Available from: https://doi.org/10.1086/710755/ASSET/IMAGES/LARGE/FG4.JPEG.

Mondal, N.K., Kundu, S., Debnath, P., Mondal, A., Sen, K., 2022. Effects of polyethylene terephthalate microplastic on germination, biochemistry and phytotoxicity of *Cicer arietinum* L. and cytotoxicity study on *Allium cepa* L. Environmental Toxicology and Pharmacology 94, 103908. Available from: https://doi.org/10.1016/J.ETAP.2022.103908.

Monteiro, S.S., Rocha-Santos, T., Prata, J.C., Duarte, A.C., Girão, A.V., Lopes, P., et al., 2022. A straightforward method for microplastic extraction from organic-rich freshwater samples. Science of the Total Environment 815, 152941. Available from: https://doi.org/10.1016/J.SCITOTENV.2022.152941.

Morét-Ferguson, S., Law, K.L., Proskurowski, G., Murphy, E.K., Peacock, E.E., Reddy, C.M., 2010. The size, mass, and composition of plastic debris in the western North Atlantic Ocean. Marine Pollution Bulletin 60, 1873–1878. Available from: https://doi.org/10.1016/J.MARPOLBUL.2010.07.020.

Munyaneza, J., Jia, Q., Qaraah, F.A., Hossain, M.F., Wu, C., Zhen, H., et al., 2022. A review of atmospheric microplastics pollution: in-depth sighting of sources, analytical methods, physiognomies, transport and risks. Science of the Total Environment 822, 153339. Available from: https://doi.org/10.1016/J.SCITOTENV.2022.153339.

Nuelle, M.T., Dekiff, J.H., Remy, D., Fries, E., 2014. A new analytical approach for monitoring microplastics in marine sediments. Environmental Pollution 184, 161–169. Available from: https://doi.org/10.1016/J.ENVPOL.2013.07.027.

Okoffo, E.D., Rauert, C., Thomas, K.V., 2023. Mass quantification of microplastic at wastewater treatment plants by pyrolysis-gas chromatography–mass spectrometry. Science of the Total Environment 856, 159251. Available from: https://doi.org/10.1016/J.SCITOTENV.2022.159251.

Pignattelli, S., Broccoli, A., Renzi, M., 2020. Physiological responses of garden cress (L. sativum) to different types of microplastics. Science of the Total Environment 727, 138609. Available from: https://doi.org/10.1016/J.SCITOTENV.2020.138609.

Pivokonsky, M., Cermakova, L., Novotna, K., Peer, P., Cajthaml, T., Janda, V., 2018. Occurrence of microplastics in raw and treated drinking water. Science of the Total Environment 643, 1644–1651. Available from: https://doi.org/10.1016/J.SCITOTENV.2018.08.102.

Primpke, S., Christiansen, S.H., Christiansen, S.H., Christiansen, S.H., Cowger, W., Frond, H., et al., 2020. Critical assessment of analytical methods for the harmonized and cost-efficient analysis of microplastics. Applied Spectroscopy 74 (9), 1012–1047.

Raynaud, J., 2014. Valuing plastic: The business case for measuring, managing and disclosing plastic use in the consumer goods industry. UN. https://wedocs.unep.org/20.500.11822/9238.

Razeghi, N., Hamidian, A.H., Wu, C., Zhang, Y., Yang, M., 2021. Microplastic sampling techniques in freshwaters and sediments: a review. Environmental Chemistry Letters 19, 4225–4252. Available from: https://doi.org/10.1007/S10311-021-01227-6.

Rezania, S., Park, J., Md Din, M.F., Mat Taib, S., Talaiekhozani, A., Kumar Yadav, K., et al., 2018. Microplastics pollution in different aquatic environments and biota: a review of recent studies. Marine Pollution Bulletin 133, 191–208. Available from: https://doi.org/10.1016/J.MARPOLBUL.2018.05.022.

Rochman, C.M., 2015. The complex mixture, fate and toxicity of chemicals associated with plastic debris in the marine environment. Marine Anthropogenic Litter 117–140. Available from: https://doi.org/10.1007/978-3-319-16510-3_5/FIGURES/3.

Rowenczyk, L., Cai, H., Nguyen, B., Sirois, M., Côté-Laurin, M.C., Toupoint, N., et al., 2022. From freshwaters to bivalves: microplastic distribution along the Saint-Lawrence river-to-sea continuum. Journal of Hazardous Materials 435, 128977. Available from: https://doi.org/10.1016/J.JHAZMAT.2022.128977.

Sarker, A., Deepo, D.M., Nandi, R., Rana, J., Islam, S., Rahman, S., et al., 2020. A review of microplastics pollution in the soil and terrestrial ecosystems: a global and Bangladesh perspective. Science of the Total Environment 733, 139296. Available from: https://doi.org/10.1016/J.SCITOTENV.2020.139296.

Schymanski, D., Oßmann, B.E., Benismail, N., Boukerma, K., Dallmann, G., von der Esch, E., et al., 2021. Analysis of microplastics in drinking water and other clean water samples with micro-Raman and micro-infrared spectroscopy: minimum requirements and best practice guidelines. Analytical and Bioanalytical Chemistry 24 (413), 5969–5994. Available from: https://doi.org/10.1007/S00216-021-03498-Y.

Song, Y.K., Hong, S.H., Jang, M., Han, G.M., Rani, M., Lee, J., et al., 2015. A comparison of microscopic and spectroscopic identification methods for analysis of microplastics in environmental samples. Marine Pollution Bulletin 93, 202–209. Available from: https://doi.org/10.1016/J.MARPOLBUL.2015.01.015.

Stock, V., Böhmert, L., Lisicki, E., Block, R., Cara-Carmona, J., Pack, L.K., et al., 2019. Uptake and effects of orally ingested polystyrene microplastic particles in vitro and in vivo. Archives of Toxicology 93, 1817–1833. Available from: https://doi.org/10.1007/S00204-019-02478-7/FIGURES/7.

Syakti, A.D., Hidayati, N.V., Jaya, Y.V., Siregar, S.H., Yude, R., Suhendy, et al., 2018. Simultaneous grading of microplastic size sampling in the Small Islands of Bintan water, Indonesia. Marine Pollution Bulletin 137, 593–600. Available from: https://doi.org/10.1016/J.MARPOLBUL.2018.11.005.

Van Cauwenberghe, L., Claessens, M., Vandegehuchte, M.B., Janssen, C.R., 2015. Microplastics are taken up by mussels (Mytilus edulis) and lugworms (Arenicola marina) living in natural habitats. Environmental Pollution 199, 10–17. Available from: https://doi.org/10.1016/J.ENVPOL.2015.01.008.

Weinstein, J.E., Crocker, B.K., Gray, A.D., 2016. From macroplastic to microplastic: degradation of high-density polyethylene, polypropylene, and polystyrene in a salt marsh habitat. Environmental Toxicology and Chemistry 35, 1632–1640. Available from: https://doi.org/10.1002/etc.3432.

Wicaksono, E.A., Werorilangi, S., Tahir, A., 2021. The influence of weirs on microplastic fate in the riverine environment (case study: Jeneberang River, Makassar City, Indonesia. IOP Conference Series: Earth and Environmental Science 763, 012054. Available from: https://doi.org/10.1088/1755-1315/763/1/012054.

Wu, B., Wu, X., Liu, S., Wang, Z., Chen, L., 2019. Size-dependent effects of polystyrene microplastics on cytotoxicity and efflux pump inhibition in human Caco-2 cells. Chemosphere 221, 333–341. Available from: https://doi.org/10.1016/J.CHEMOSPHERE.2019.01.056.

Xiang, S., Xie, Y., Sun, X., Du, H., Wang, J., 2022. Identification and quantification of microplastics in aquaculture environment. Frontiers in Marine Science 8, 804208. Available from: https://doi.org/10.3389/FMARS.2021.804208/BIBTEX.

Zhong, W., 2011. Textiles for medical filters. Handbook of Medical Textiles. Elsevier, pp. 419–433. Available from: https://doi.org/10.1533/9780857093691.4.419.

Chapter 11

Green synthesis of nanomaterials for the removal of emerging water pollutants

Vinod K. Garg, Nitin Kumar and Aryan Anand

Department of Environmental Sciences and Technology, Central University of Punjab, Bathinda, Punjab, India

Introduction

Water contamination by organic pollutants is a major concern for municipalities, policymakers, scientists, and the general public. Toxic, bio-recalcitrant, and hazardous chemicals are released into the environment due to the rise of polluting industries, such as tanneries, food industries, metallurgies, and pharmaceuticals (Yadav et al., 2021). Due to the improper disposal of industrial wastes and agricultural activities, a variety of emerging pollutants are released into the water bodies and nearby open fields gradually declining the quality of water and soil (Syafrudin et al., 2021; Arif et al., 2021). Emerging adulterants, including pharmaceuticals, pesticides, poly-fluoroalkyl substances (PFAs), and microplastics, are toxic and ruin the quality of water, aquatic living beings, and human health. These pollutants, when enter the living system through water, may cause morbidity and mortality (Chaturvedi et al., 2021). Nanotechnology is a multidisciplinary field that entails the synthesis of nanoscale materials/particles, that is, 1–100 nm (Yadav et al., 2021), which can be applied for the identification, quantification, and remediation of emerging pollutants. Nanoparticles (NPs) can be grafted with functional groups as they have peculiar surface chemistry and so can be used to remove the specific pollutants of interest for efficient clean-up in comparison to conventional techniques (Guerra et al., 2018). NPs are commonly categorized into organic (liposomes, dendrimers, ferritin, micelles), inorganic (metals and metals oxide), and carbon-based (graphene, carbon nanofibers, fullerenes, carbon nanotubes) NPs (Ealias & Saravanakumar, 2017). Because of their excellent biocompatibility and sensitivity, these NPs have a broad range of significance in biomedicine, agriculture, electronics, and environmental remediation (Annamalai et al., 2021; Jain et al., 2022). Metal and semiconductor NPs have distinct optoelectronic properties for photoinduced catalytic processes depending on their size and structure (Sharwani et al., 2022). Nanomaterials are significantly influenced by environmental conditions such as type of water, salinity, pH, and organic chemical constituents that may cause material conglomeration (Saleh, 2020). Scientists are continuously attempting to fabricate new materials by altering their physical and chemical properties for superior efficiency in the remediation of wastewater (Chen et al., 2020; Tian et al., 2020) such as drugs, dyes, personal hygiene products, agrochemicals, endocrine-disrupting chemicals, artificial sweeteners, and other industrial additives (Khan et al., 2022a,b,c,d). Different technologies have been used so far for water treatment including but not limited to electrodialysis, coagulation, adsorption, flotation, ultrafiltration, and electrochemical (Chua et al., 2021), but "Nanotechnology" came out as the most effective and efficient technique with significant advantages (Islam et al., 2022). Biosynthesis of NPs has been attracted in yesteryears to have controlled configuration of nanomaterials about their size, shape, structure, etc. Besides, it is cost-effective, readily available, and environment friendly (Ahmed et al., 2022). The focus of this chapter is on the various mechanisms by which nanomaterials can remove developing/emerging pollutants from water using the green chemistry of plants.

Scope of nanoparticles and their various applications

Nanotechnology is concerned with the study and operation of materials at the atomic and molecular level, and their applications in engineering, medicine, robotics, agriculture, biotechnology, and the environment, to achieve desired outcomes (Bhattacharyya et al., 2009; Prasad et al., 2021). NPs can be used to clean up a variety of contaminated locations by functioning as a "super adsorbent" for a variety of contaminants (Table 11.1). Additionally, they play a

Role of Green Chemistry in Ecosystem Restoration to Achieve Environmental Sustainability. DOI: https://doi.org/10.1016/B978-0-443-15291-7.00017-1

TABLE 11.1 Applications of green synthesized nanomaterials in wastewater treatment.

Plant/plant part used	Nanomaterial used	Targeted pollutant	Removal efficiency (%)	References
Pomegranate peel	Nano-structured activated Biochar	Ciprofloxacin	89.94 (batch reactor), 84.74 (packed bed reactor)	Hamadeen and Elkhatib (2022)
Waste tires cellulose	γ-Fe_2O_3/MWCNTs/Cellulose	Malachite green	87.35	Khalatbary et al. (2022)
Eriobotrya japonica	g-C_3N_4 nanosheets	Bisphenol-A	88	Khan et al. (2022a, b,c,d)
Pomegranate peel	AC/nZVI (activated carbon and nano zerovalent iron)	Furfural	81.87	Rashtbari et al. (2022)
Walnut peel extract	AC/ZnO	Eo-Y and Er-B	95.11/98.31	Rashtbari et al. (2022)
Chitosan	Fe_2O_3/$NiFe_2O_4$	Methyl green dye	96.51	Ansari et al. (2022)
Pumice granatum leaf extract	CuO	Safranin-O dye	90.45	Vidovix et al. (2022)
Ruellia tuberosa	ZnO	Methylene blue and Malachite green	94/92	Vasantharaj et al. (2021)
Lemon leaf extract	ZnO–$CdWO_4$	Congo red dye	98.5	Fatima et al. (2021)
Rice husk	CeO_2@SiO_2	Amoxicillin	100	Pham et al. (2021)
Green tea extract	Mn	Mitoxantrone (MTX)	97.4 (lab trials), 77.3 (real wastewater)	He et al. (2021)
Monsonia burkiana extract	CuO	Methylene blue(MB), Sulfamethoxazole (SMX)	83 (MB), 58 (SMX)	Rambiyana et al. (2021)
Moringa olifera leaves	Fe_3O_4	levofloxacin	86.15	Altaf et al. (2021)
Coffea Arabica seed (CAS) extract	2D platinum (Pt) superstructures	4-Nitrophenol + Methylene Blue	100	Bogireddy et al. (2021)
Green tea leaves	Co_3O_4/MCM-41	Methyl parathion	100	Salam et al. (2020)
Capparis decidua leaf extract	CuO	Lambda-Cyhalothrin (L-CHT)	99	Iqbal et al. (2021)
	NiO		89%	
Epilobium parviflorum extract	$MnFe_2O_4$@SiO_2-Ag	Rhodamine B (RhB), Methylene Blue (MB)	100	Ertürk et al. (2021)
Daucus carota leaves extract	D-Ag	Rhodamine-6G dye (Rh-6G)	89	Joseph et al. (2020)
	D-Pd		98	
Euphorbia maculata	CuO	Methylene Blue (MB) Rhodamine (RhB) Congo Red (CR)	MB (10mg/mL) = 96 RhB (10mg/mL) = 89 CR (10mg/mL) = 85	Pakzad et al. (2019)
	Ni@Fe_3O_4		MB (10mg/mL) = 89 RhB (10mg/mL) = 81 CR (10mg/mL) = 79	
Acacia nilotica (pods extract)	α-Fe_2O_3	Methyl orange (MO)	94	Da'na et al. (2018)

crucial and decisive part in the development of rapid and precise environmental sensors that may be used to detect contaminants at atomic scales as well as to control harmful bacteria (Chakraborty et al., 2021). NPs have a unique ability to have new surface alterations, and to infiltrate through very tiny pores in the subsurface and settle down in groundwater; therefore NPs can travel a long distance and have a wide distribution (Marcon et al., 2021). Plant extracts, as well as microorganisms like bacteria and fungi, can be used to produce the green synthesized NPs, but plant-based synthesis is very easy and fast as compared to the microbes, which are difficult to maintain and preserve due to contamination (Nayantara and Kaur, 2018). On the other hand, biosynthesis is restricted to metal oxide NPs, that is, they are the only ones that can be used, and this method uses intracellular and extracellular enzymes derived from plants and microorganisms to synthesize NPs (Qamar and Ahmad, 2021). These nanoscale materials can be utilized in biosensors, bio-probes, and other biosystems in combination with methods for cleaning up environmental resources (Dhasmana et al., 2019).

Recent advancements in sustainable approaches for pollutant abatement using nanoparticles

Water pollution has increased considerably, which is now a critical global issue and the endurance and existence of hazardous organic pollutants such as dyes, fertilizers, pesticides, heavy metals, and other food processing wastes pose significant environmental and health risks (Rathi et al., 2021). The utilization of plant materials for the building of NPs has rapidly grown in different scientific fields, and this progress has provided several environmentally acceptable approaches in which natural products substitute hazardous chemicals in complex and time-consuming synthesis processes (Punetha et al., 2022). Using green scientific concepts in both the development and execution of nanomaterials meant that this innovation is brought closer in a reliable manner, reducing risks to welfare and the environment while yielding overall profit without affecting quality and execution (Rauwel et al., 2015; Yazdanian et al., 2022). The advancement of nanotechnology/nanoformulations and their potential uses make the green synthesis of "metal" and "metal oxide" NPs an enthusiastic and ideal field for various studies (Singh et al., 2018). The whole process of making NPs through green technology does not require any type of reducing and capping chemicals such as $NaBH_4$ and hydrazine as the plants contain such nature-friendly agents, which finally enables to keep away the employment of unsafe synthetic materials (Bibi et al., 2017).

Approaches for pollutant separation from water through green synthesized nanoparticles

Every day, various industrial sectors (e.g., textiles, fertilizers, agrochemicals, and tanning) generate millions of tons of wastewater, and a considerable percentage of it is released into water bodies without any treatment, which consists of a variety of carcinogenic, poisonous, nondegradable and bioaccumulative inorganic (e.g., heavy metals) and organic pollutants (e.g., dyes and surfactants). Various studies have been carried out to extract these pollutants from water through different processes using numerous materials, and some of them are given below (Table 11.2).

Adsorption

Adsorption is a physical treatment method that has proven to be a highly effective method for eliminating organic pollutants from wastewater. Vander Waals forces, hydrogen bonds, electrostatic contact, and hydrophobic interaction are a few examples of the physical mechanisms that take place primarily because of the polarity of the adsorbate molecule and the adsorbent surface, contrarily, chemical adsorption is selective and entails stronger forces because it results in the formation of chemical bonds through the exchange of electrons between the adsorbent and the adsorbate (Barquilha and Braga, 2021). Also, nanostructures constructed using eco-friendly green techniques have been investigated as adsorbents by inducing complex formation, redox, and ion-exchange affinities with respect to pollutants for their efficacious elimination from wastewater (Hassan et al., 2020). It has been proven that noxious waste with hydroxyl (−OH) groups have higher adsorption capacity than those without hydroxyl (−OH) groups because toxins with OH groups can easily form electrostatic bonding with the surface hydroxyl group of adsorbent, and promote the adsorption process (Aragaw et al., 2021). Conversely, organic pollutants find inner-sphere complexes to exchange ions (OH) with iron oxide or hydroxide (Nayantara and Kaur, 2018). Various researches revealed that green-manufactured nanomaterials such as "Green Iron Oxide Nanomaterials" are also enriched in carbon-oxygen functional groups (Su et al., 2020), providing

TABLE 11.2 Various green synthesized nanomaterials used for "adsorption" and "photocatalysis."

Plant used	Nanomaterial used	Target pollutant	Removal mechanism	References
Zingiber officinale	$TiO_2/g\text{-}C_3N_4$	Sulfadiazine and Bisphenol-A	Photocatalysis	Khan et al. (2022a, b,c,d)
Areca catechu	ZnO	Rhodamine B, Methyline blue and Nigrosine dye	Photocatalysis	Raghavendra et al. (2022)
Acmella uliginosa	$CuO\text{-}Gd_2Ti_2O_7$	Malachite green	Photocatalysis	Halomoan et al. (2022)
Sunflower seed waste	Carbon NPs	Ciprofloxacin and malachite green	Adsorption	Almufarij et al. (2022)
Ramalina sinensis	Fe_3O_4	Lead and cadmium	Adsorption	Arjaghi et al. (2021)
Barberry leaf	nZVI	Cr(VI)	Photocatalysis	Samadi et al. (2021)
Moringa oleifera seed	Ag	Methylene blue, orange, red, and nitrophenol	Photocatalysis	Mehwish et al. (2021)
Phenylalanine	$MnWO_4$	Acid Yellow 23, eosin Y, and crystal violet	Photocatalysis	Zinatloo-Ajabshir et al. (2021)
Sutherlandia frutescence	CdS	Malachite green and sulfisoxazole	Photocatalysis	Munyai et al. (2021)
Carica papaya peel	CuO	Palm oil mill effluent	Photocatalysis	Munyai et al. (2021)
Green tea extract	Fe	AS(V)	Adsorption	Wu et al. (2021)
Ricinus communis seeds	nZVI	Methylene blue	Adsorption	Abdelfatah et al. (2021)
Green tea extract	Fe_3O_4/AC (powdered and granular activated carbon)	Copper	Adsorption	Shahrashoub and Bakhtiari (2021)
Camellia sinensis	SnO_2	Methylene blue, Methyl orange, Rhodamine B	Photocatalysis	Luque et al. (2021)
Spirulina platensis	IONP (recyclable iron oxide NPs)	Cationic crystal violet and anionic methyl orange	Adsorption	Shalaby et al. (2021)
M. oleifera	Fe_3O_4	Pb(II)	Adsorption	Gautam et al. (2020)
Cynodon dactylon and *Muraya koenigii*	Fe_3O_4	Cr(VI)	Adsorption	Vishnu and Dhandapani (2020)
Camellia sinensis	Biochar	Caffeine (CFN)	Adsorption	Keerthanan et al. (2020)
Eichhornia crassipes (Pontederia crassipes)	Co doped TiO_2	Reactive Blue 19 Reactive Red 76	Photo catalysis	El Nemr et al. (2019)
Phaseolus vulgaris	AgNPs	Reactive Red 141	Photocatalysis	Rani et al. (2020)

much more active adsorption sites for both anions and cations than traditionally manufactured NPs, biosynthesized ZnO NPs unveiled excellent adsorption for the Pb(II) ions (Pal et al., 2019), nano biochar for carbamazepine (Naghdi et al., 2019), nZVI for ciprofloxacin (Shao et al., 2018).

Photocatalysis

Photocatalysis is an advanced oxidation process that can produce a variety of reactive species including reactive oxygen species, reactive nitrogen species, oxygen radicals, and hydroxyl radicals. It has recently extended a lot of relevance for its promising application in the break-up of noxious wastes (Podporska-Carroll et al., 2017; Kumar et al., 2020). By using a basic photocatalytic redox mechanism, the engaged semiconductor degrades organic and microbial cells. Regmi developed a Fe-doped $BiVO_4$, an active, visible light semiconductor that effectively removed ibuprofen and *Escherichia coli* from water (Regmi et al., 2017). Semiconductors can easily create electron–hole pairs within the UV–Vis light spectrum due to their structure and the fact that they have an energy gap of around (2.3 ≥ eV) (Karunamoorthy and Velluchamy, 2018). The mechanism of photocatalyst NP has been shown in Fig. 11.1. As a result of the electron excitation, a positively charged hole (h^{+-}) is created in the VB, and an electron is created in the conduction band. The electron in the conduction band is assumed to be a potent reducer; meanwhile, the VB hole is thought to enhance photo-redox chemistry, and as a result, these properties can be recognized once their dimensions are determined on the nanoscale. NiO-NPs have emerged as one of the finest choices for photocatalyst degradation in the current decade, owing to their nontoxicity, less expensive, chemical stability, and ease of access. Photocatalysts have been identified as compounds capable of destroying sewage pollutants and converting them to innocuous molecules such as water and carbon dioxide. Owing to their low pricing, scarcity of high energy, and lack of effluence, some photocatalyst materials, such as TiO_2, ZnO, Fe_2O_3, and NiO, with unique electronic features, incorporate full VB and a hollow conducting bond, are more favored and popular than others (Akbari et al., 2020). Soto-Robles et al. (2019) biosynthesized ZnO NPs using *Hibiscus sabdariffa* extract with different volumes of 1% (ZnO-J1), 4% (ZnO-J4), and 8% (ZnO-J8) to check their photocatalytic activity on the methylene blue and it is revealed that ZnO-J1 has the highest photocatalytic efficiency of 95% as compared to ZnO-J4 and ZnO-J8 which is 91% and 87% respectively (Soto-Robles et al., 2019). Also, Ijaz et al. (2017) produced CuO NPs utilizing the *Abutilon indicum* extract, revealing the efficient photo-degradation of Acid Black 210 (AB) dye under sunlight (Ijaz et al., 2017).

Nanofiltration

To produce safe and clean drinking water from wastewater, the notable advancement in membrane science is the nanofiltration (NF) membrane, which is one of the most substantial and frequently used wastewater treatment techniques. NF is capable of eliminating however almost all organic contaminants from wastewater, along with a sizable number of dangerous

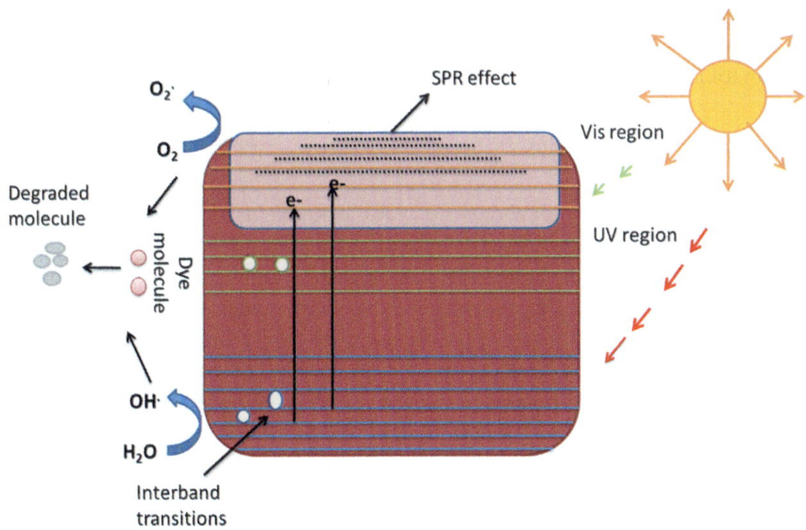

FIGURE 11.1 Photocatalysis for dye degradation is shown schematically.
A photocatalytic reaction starts when a photocatalyst absorbs a photon with an energy greater than its band gap. An electron–hole pair is thus produced, and an oxidation-reduction reaction begins at the metal surface as soon as an electron in the valence bond is driven to the conduction band. After that, oxygen molecules scavenge conduction band electrons to create superoxide ions (O_2^-). Hydroxyl radicals (OH^-) are produced when superoxide ions combine with protons or when water molecules engage with an electron–hole pair. The recurrent reactions of hydroxyl radicals and superoxide ions on organic compounds lead to dye degradation. *Source: From Rani, P. et al., 2020. Highly stable AgNPs prepared via a novel green approach for catalytic and photocatalytic removal of biological and non-biological pollutants. Environment International, 143. https://doi.org/10.1016/j.envint.2020.105924.*

bacteria. Interfacial polymerization (IP) is frequently utilized to create thin film membranes in the NF range. Solvents used in this process are volatile and their emission can cause serious health problems (Falca et al., 2021). Therefore these solvents such as benzene, n-hexane, and cyclohexane must be replaced by greener ones that are eco-friendly. Using α-pinene as an environmentally friendly alternative solvent in IP reaction between ter-phthaloyl chloride and quercetin (Abdellah et al., 2018) developed polyester films. Similarly, ionic liquid, [C4mim][Tf2N], was used as the organic phase to prepare the polyamide layer for the first time, although based on the composition chosen, not all ionic liquids are regarded as environmentally friendly, and a high-cost has to be bear (Zheng et al., 2021). Recently cellulose-based NF membranes known as cellulose nanofibers have been used extensively to remove pollutants (Barhoum et al., 2020) and due to their excellent selectivity, mechanical strength high aspect ratio, low energy consumption, surface area, surface charge, and, nano cellulose-based membranes are very effective and environmentally friendly technology for desalinating and/or purifying water (Hassan et al., 2017; Salama et al., 2021). Variations in the aspect ratio, choice of membrane filtration material, and processing procedures can change the pore size (Tan et al., 2020). The removal of organic contaminants can be accomplished using NF membranes with nanopore sizes ranging from 2 to 1 nm (Choudhury et al., 2020). Inorganic pollutants, heavy metals, and biological species are only a few examples of the various types of water contaminants that have been successfully targeted by surface alterations to alter membrane reactivity (Khumalo et al., 2019).

Future scope and challenges

Due to the intrinsic greenness and viability of the manufacturing processes, as well as their effective performance in reducing environmental toxins, biologically prepared NPs are intriguing substances in light of the usage of dangerous chemicals and substances to produce nano-objects that can catalyze the breakdown of organic pollutants into nontoxic classes or adsorb pollutants from aqueous solution and are environmentally safe, relatively affordable, and energy-efficient to produce (Nasrollahzadeh et al., 2021). Synthesis of nanomaterials from metals and their oxides using a greener method including extracts of bacteria, yeast, fungus, and plants has become an appealing research over the past ten years and the researchers found that the plant extract has great efficiency with regards to its capping and reducing agents for the formation of stable materials (Singh et al., 2018). Although NP production is uncomplicated and safer for environs, more optimization studies are needed for the green approach to synthesize a precise size and shape (Salem and Fouda, 2021). Mechanism and metabolite used for the greener method are required to be investigated to a greater extent that contributes to the transformation of metal salts into nanoscale materials and to scale up the yield of these materials (Ahmed et al., 2022). Although these materials are produced for the removal of contaminants from the environment, the assessment of green NP's potential for toxicity and risk management related to their manufacture, handling, maintenance, and dispersion represent the biggest challenges (Rana et al., 2020). By successfully overcoming the inherent limitations, this nontoxic engineering has a proven ability to benefit all spheres of life greatly.

Conclusion

This chapter provides an overview of the green synthesis of various kinds of NPs to overcome the conventional physical and chemical methods. The use of a different kind of plant extract for the fabrication of distinct nanomaterials has been discussed, which are more sustainable, environmentally safe, easily available, and easy to handle when compared to other natural extracts of microorganisms like fungi, bacteria, and yeast, which needs a well-established lab and are not easy to handle. This chapter was structured to cover the research on the environmentally sustainable synthesis of NPs and their uses in environmental rehabilitation. Further research in the area of green NP synthesis should focus on transferring lab research to an industrial scale while keeping current concerns like the environment and health in mind.

References

Abdelfatah, A.M., et al., 2021. Green synthesis of nano-zero-valent iron using Ricinus communis seeds extract: characterization and application in the treatment of methylene blue-polluted water. ACS Omega 6 (39), 25397–25411. Available from: https://doi.org/10.1021/acsomega.1c03355.

Abdellah, M.H., et al., 2018. Effective interfacially polymerized polyester solvent resistant nanofiltration membrane from bioderived materials. Advanced Sustainable Systems 2 (7), 1–7. Available from: https://doi.org/10.1002/adsu.201800043.

Ahmed, S.F., Mofijur, M., et al., 2022. Green approaches in synthesising nanomaterials for environmental nanobioremediation: Technological advancements, applications, benefits and challenges. Environmental Research 204. Available from: https://doi.org/10.1016/j.envres.2021.111967.

Akbari, A., et al., 2020. Effect of nickel oxide nanoparticles as a photocatalyst in dyes degradation and evaluation of effective parameters in their removal from aqueous environments. Inorganic Chemistry Communications 115. Available from: https://doi.org/10.1016/j.inoche.2020.107867.

Almufarij, R.S., et al., 2022. Optimization, nature, and mechanism investigations for the adsorption of ciprofloxacin and malachite green onto carbon nanoparticles derived from low-cost precursor via a green route. Molecules (Basel, Switzerland) Available from: https://doi.org/10.3390/molecules27144577.

Altaf, S., et al., 2021. Removal of levofloxacin from aqueous solution by green synthesized magnetite (Fe3O4) nanoparticles using *Moringa olifera*: kinetics and reaction mechanism analysis. Ecotoxicology and Environmental Safety 226, 112826. Available from: https://doi.org/10.1016/j.ecoenv.2021.112826.

Annamalai, J., et al., 2021. Recent trends in microbial nanoparticle synthesis and potential application in environmental technology: a comprehensive review. Environmental Science and Pollution Research 28 (36), 49362–49382. Available from: https://doi.org/10.1007/s11356-021-15680-x.

Ansari, M.J., et al., 2022. Preparation of new bio-based chitosan/Fe2O3/NiFe2O4 as an efficient removal of methyl green from aqueous solution. International Journal of Biological Macromolecules 198, 128–134. Available from: https://doi.org/10.1016/j.ijbiomac.2021.12.082.

Arif, Z. et al., 2021 Impact on groundwater quality resources due to industrial effluent. Wiley Online Library. Available at: https://onlinelibrary.wiley.com/doi/abs/10.1002/9781119709732.ch11 (accessed 25.03.2022).

Aragaw, T.A., Bogale, F.M., Aragaw, B.A., 2021. Iron-based nanoparticles in wastewater treatment: A review on synthesis methods, applications, and removal mechanisms. Journal of Saudi Chemical Society 25 (8). Available from: https://doi.org/10.1016/j.jscs.2021.101280.

Arjaghi, S.K., et al., 2021. Retracted article: green synthesis of iron oxide nanoparticles by RS lichen extract and its application in removing heavy metals of lead and cadmium. Biological Trace Element Research 199 (2), 763–768. Available from: https://doi.org/10.1007/s12011-020-02170-3.

Barhoum, A., et al., 2020. Plant celluloses, hemicelluloses, lignins, and volatile oils for the synthesis of nanoparticles and nanostructured materials. Nanoscale 12 (45), 22845–22890. Available from: https://doi.org/10.1039/d0nr04795c.

Barquilha, C.E.R., Braga, M.C.B., 2021. Adsorption of organic and inorganic pollutants onto biochars: challenges, operating conditions, and mechanisms. Bioresource Technology Reports 15. Available from: 10.1016/j.biteb.2021.100728.

Bhattacharyya, D., Singh, S., Satnalika, N., 2009. Nanotechnology, big things from a tiny world: a review. Science and Technology 2 (3), 29–38.

Bibi, I., et al., 2017. Nickel nanoparticle synthesis using Camellia sinensis as reducing and capping agent: growth mechanism and photo-catalytic activity evaluation. International Journal of Biological Macromolecules 103, 783–790. Available from: https://doi.org/10.1016/j.ijbiomac.2017.05.023.

Bogireddy, N.K.R., et al., 2021. Green fabrication of 2D platinum superstructures and their high catalytic activity for mitigation of organic pollutants. Catalysis Today 360, 185–193. Available from: https://doi.org/10.1016/j.cattod.2019.06.044.

Chakraborty, U., Kaur, G., Chaudhary, G.R., 2021. In: Kumar, Rajeev, Kumar, Raman, Kaur, G. (Eds.), Development of Environmental Nanosensors for Detection Monitoring and Assessment BT – New Frontiers of Nanomaterials in Environmental Science. Springer Singapore, Singapore, pp. 91–143. Available from: https://doi.org/10.1007/978-981-15-9239-3_5.

Chaturvedi, M., et al., 2021. Emerging contaminants in wastewater: sources of contamination, toxicity, and removal approaches. Emerging Treatment Technologies for Waste Management 103–132. Available from: https://doi.org/10.1007/978-981-16-2015-7_5.

Chen, J., et al., 2020. Advances in nanomaterials for photodynamic therapy applications: status and challenges. Biomaterials 237, 119827. Available from: https://doi.org/10.1016/j.biomaterials.2020.119827.

Choudhury, R.R., Sahoo, S.K., Gohil, J.M., 2020. Potential of bioinspired cellulose nanomaterials and nanocomposite membranes thereof for water treatment and fuel cell applications. Cellulose 27 (12), 6719–6746. Available from: https://doi.org/10.1007/s10570-020-03253-z.

Chua, S.F., et al., 2021. The emergence of multifunctional adsorbents and their role in environmental remediation. Journal of Environmental Chemical Engineering 9 (1), 104793. Available from: https://doi.org/10.1016/j.jece.2020.104793.

Da'na, E., Taha, A., Afkar, E., 2018. Green synthesis of iron nanoparticles by *Acacia nilotica* pods extract and its catalytic, adsorption, and antibacterial activities. Applied Sciences (Switzerland) 8 (10). Available from: https://doi.org/10.3390/app8101922.

Dhasmana, A., et al., 2019. Scope of nanoparticles in environmental toxicant remediation. Environmental Biotechnology: For Sustainable Future. Springer, pp. 31–44. Available from: http://doi.org/10.1007/978-981-10-7284-0_2.

Ealias, A.M., Saravanakumar, M.P., 2017. A review on the classification, characterisation, synthesis of nanoparticles and their application. IOP Conference Series: Materials Science and Engineering 263 (3). Available from: https://doi.org/10.1088/1757-899X/263/3/032019.

El Nemr, A., et al., 2019. Photocatalytic and biological activities of undoped and doped TiO2 prepared by Green method for water treatment. Journal of Environmental Chemical Engineering 7 (5), 103385. Available from: https://doi.org/10.1016/j.jece.2019.103385.

Ertürk, A.S., Elmaci, G., Gürbüz, M.U., 2021. Reductant free green synthesis of magnetically recyclable MnFe2O4@SiO2-Ag core-shell nanocatalyst for the direct reduction of organic dye pollutants. Turkish Journal of Chemistry 45 (6), 1968–1979. Available from: https://doi.org/10.3906/kim-2108-2.

Falca, G., et al., 2021. Naturally extracted hydrophobic solvent and self-assembly in interfacial polymerization. ACS Applied Materials & Interfaces 13 (37), 44824–44832. Available from: https://doi.org/10.1021/acsami.1c07584.

Fatima, B., et al., 2021. Facile green synthesis of ZnO–CdWO4 nanoparticles and their potential as adsorbents to remove organic dye. Environmental Pollution 271, 116401. Available from: https://doi.org/10.1016/j.envpol.2020.116401.

Gautam, P.K., Shivalkar, S., Banerjee, S., 2020. Synthesis of M. oleifera leaf extract capped magnetic nanoparticles for effective lead [Pb (II)] removal from solution: Kinetics, isotherm and reusability study. Journal of Molecular Liquids 305, 112811. Available from: https://doi.org/10.1016/j.molliq.2020.112811.

Guerra, F.D., et al., 2018. Nanotechnology for environmental remediation: materials and applications. Molecules (Basel, Switzerland) 23 (7), 1–23. Available from: https://doi.org/10.3390/molecules23071760.

Halomoan, I., et al., 2022. Facile preparation of CuO-Gd2Ti2O7 using Acmella uliginosa leaf extract for photocatalytic degradation of malachite green. Materials Research Bulletin 150, 111726. Available from: https://doi.org/10.1016/j.materresbull.2021.111726.

Hamadeen, H.M., Elkhatib, E.A., 2022. New nanostructured activated biochar for effective removal of antibiotic ciprofloxacin from wastewater: adsorption dynamics and mechanisms. Environmental Research 210, 112929. Available from: https://doi.org/10.1016/j.envres.2022.112929.

Hassan, M., et al., 2017. Membranes based on cellulose nanofibers and activated carbon for removal of Escherichia coli bacteria from water. Polymers 9 (8), 1–14. Available from: https://doi.org/10.3390/polym9080335.

Hassan, M., et al., 2020. Critical review of magnetic biosorbents: their preparation, application, and regeneration for wastewater treatment. Science of the Total Environment 702, 134893. Available from: https://doi.org/10.1016/j.scitotenv.2019.134893.

He, F., et al., 2021. Reducing the impact of antibiotics in wastewaters: increased removal of mitoxantrone from wastewater by biosynthesized manganese nanoparticles. Journal of Cleaner Production 293. Available from: https://doi.org/10.1016/j.jclepro.2021.126207.

Ijaz, F., et al., 2017. Green synthesis of copper oxide nanoparticles using abutilon indicum leaf extract: antimicrobial, antioxidant and photocatalytic dye degradation activities. Tropical Journal of Pharmaceutical Research 16 (4), 743–753. Available from: https://doi.org/10.4314/tjpr.v16i4.2.

Iqbal, A., et al., 2021. Green synthesis of flower-shaped copper oxide and nickel oxide nanoparticles via capparis decidua leaf extract for synergic adsorption-photocatalytic degradation of pesticides. Catalysts 11 (7). Available from: https://doi.org/10.3390/catal11070806.

Islam, T., et al., 2022. Chapter 1 – Appraisal of nanotechnology for sustainable environmental remediation. In: Koduru, J.R., et al., (Eds.), Micro and Nano Technologies. Elsevier, pp. 3–31. Available from: https://doi.org/10.1016/B978-0-12-824547-7.00010-2.

Jain, N., Gupta, E., Kanu, N.J., 2022. Plethora of carbon nanotubes applications in various fields – a state-of-the-art-review. Smart Science 10 (1), 1–24. Available from: https://doi.org/10.1080/23080477.2021.1940752.

Joseph Kirubaharan, C., et al., 2020. Green synthesis of Ag and Pd nanoparticles for water pollutants treatment. Water Science and Technology 82 (11), 2344–2352. Available from: https://doi.org/10.2166/wst.2020.498.

Karunamoorthy, S., Velluchamy, M., 2018. Design and synthesis of bandgap tailored porous Ag/NiO nanocomposite: an effective visible light active photocatalyst for degradation of organic pollutants. Journal of Materials Science: Materials in Electronics 29 (23), 20367–20382. Available from: https://doi.org/10.1007/s10854-018-0172-0.

Keerthanan, S., et al., 2020. Engineered tea-waste biochar for the removal of caffeine, a model compound in pharmaceuticals and personal care products (PPCPs), from aqueous media. Environmental Technology and Innovation 19, 100847. Available from: https://doi.org/10.1016/j.eti.2020.100847.

Khalatbary, M., et al., 2022. Adsorption studies on the removal of malachite green by γ-Fe2O3/MWCNTs/Cellulose as an eco-friendly nanoadsorbent. Biomass Conversion and Biorefinery . Available from: https://doi.org/10.1007/s13399-022-02475-4.

Khan, S., et al., 2022a. A review on nanotechnology: properties, applications, and mechanistic insights of cellular uptake mechanisms. Journal of Molecular Liquids 348, 118008. Available from: https://doi.org/10.1016/j.molliq.2021.118008.

Khan, S. et al., 2022b. no date. Emerging contaminants of high concern for the environment: Current trends and future research. Elsevier. Available at: https://www.sciencedirect.com/science/article/pii/S0013935121019101 (accessed 28.03.2022b).

Khan, I., Khan, S., et al., 2022c. Biological inspired green synthesis of TiO2 coupled g-C3N4 nanocomposites and its improved activities for sulfadiazine and bisphenol A degradation. Journal of Cluster Science. Available from: https://doi.org/10.1007/s10876-022-02317-3.

Khan, I., Luo, M., et al., 2022d. Green synthesis of SrO bridged LaFeO3/g-C3N4 nanocomposites for CO2 conversion and bisphenol A degradation with new insights into mechanism. Environmental Research 207, 112650. Available from: https://doi.org/10.1016/j.envres.2021.112650.

Khumalo, N.P., et al., 2019. Dual-functional ultrafiltration nano-enabled PSf/PVA membrane for the removal of Congo red dye. Journal of Water Process Engineering 31, 100878. Available from: https://doi.org/10.1016/j.jwpe.2019.100878.

Kumar, A., et al., 2020. Bio-inspired and biomaterials-based hybrid photocatalysts for environmental detoxification: a review. Chemical Engineering Journal 382, 122937. Available from: https://doi.org/10.1016/j.cej.2019.122937.

Luque, P.A., et al., 2021. Green synthesis of tin dioxide nanoparticles using Camellia sinensis and its application in photocatalytic degradation of textile dyes. Optik 229, 166259. Available from: https://doi.org/10.1016/j.ijleo.2021.166259.

Marcon, L., Oliveras, J., Puntes, V.F., 2021. In situ nanoremediation of soils and groundwaters from the nanoparticle's standpoint: a review. Science of the Total Environment 791. Available from: https://doi.org/10.1016/j.scitotenv.2021.148324.

Mehwish, H.M., et al., 2021. Green synthesis of a silver nanoparticle using Moringa oleifera seed and its applications for antimicrobial and sun-light mediated photocatalytic water detoxification. Journal of Environmental Chemical Engineering 9 (4), 105290. Available from: https://doi.org/10.1016/j.jece.2021.105290.

Munyai, S., et al., 2021. Green synthesis of cadmium sulphide nanoparticles for the photodegradation of Malachite green dye, sulfisoxazole and removal of bacteria. Optik 247, 167851. Available from: https://doi.org/10.1016/j.ijleo.2021.167851.

Naghdi, M., et al., 2019. Pine-wood derived nanobiochar for removal of carbamazepine from aqueous media: adsorption behavior and influential parameters. Arabian Journal of Chemistry 12 (8), 5292–5301. Available from: https://doi.org/10.1016/j.arabjc.2016.12.025.

Nasrollahzadeh, M., et al., 2021. Green-synthesized nanocatalysts and nanomaterials for water treatment: current challenges and future perspectives. Journal of Hazardous Materials 401, 123401. Available from: https://doi.org/10.1016/j.jhazmat.2020.123401.

Nayantara, Kaur, P., 2018. Biosynthesis of nanoparticles using eco-friendly factories and their role in plant pathogenicity: a review. Biotechnology Research and Innovation 2 (1), 63–73. Available from: https://doi.org/10.1016/j.biori.2018.09.003.

Pakzad, K., Alinezhad, H., Nasrollahzadeh, M., 2019. Green synthesis of Ni@Fe3O4 and CuO nanoparticles using Euphorbia maculata extract as photocatalysts for the degradation of organic pollutants under UV-irradiation. Ceramics International 45 (14), 17173–17182. Available from: https://doi.org/10.1016/j.ceramint.2019.05.272.

Pal, G., Rai, P., Pandey, A., 2019. Chapter 1 – Green synthesis of nanoparticles: a greener approach for a cleaner future. In: Shukla, A.K., Iravani, S. (Eds.), Green Synthesis, Characterization and Applications of Nanoparticles: Micro and Nano Technologies. Elsevier, pp. 1–26. Available from: https://doi.org/10.1016/B978-0-08-102579-6.00001-0.

Pham, T.-D., et al., 2021. Synthesis and characterization of novel hybridized CeO_2@SiO_2 nanoparticles based on rice husk and their application in antibiotic removal. Langmuir 37 (9), 2963–2973. Available from: https://doi.org/10.1021/acs.langmuir.0c03632.

Podporska-Carroll, J., et al., 2017. Antibacterial properties of F-doped ZnO visible light photocatalyst. Journal of Hazardous Materials 324, 39–47. Available from: https://doi.org/10.1016/j.jhazmat.2015.12.038.

Prasad, R.D., et al., 2021. A review on concept of nanotechnology in veterinary medicine. ES Food & Agroforestry 28–60. Available from: https://doi.org/10.30919/esfaf481.

Punetha, D.V., et al., 2022. Recent advancements in green synthesis of nanoparticles for improvement of bioactivities: a review. Current Pharmaceutical Biotechnology. Bentham Science Publishers, pp. 904–919. Available from: http://doi.org/10.2174/1389201022666210812115233.

Qamar, S.U.R., Ahmad, J.N., 2021. Nanoparticles: mechanism of biosynthesis using plant extracts, bacteria, fungi, and their applications. Journal of Molecular Liquids 334, 116040. Available from: https://doi.org/10.1016/j.molliq.2021.116040.

Raghavendra, V.B., et al., 2022. Green synthesis of zinc oxide nanoparticles (ZnO NPs) for effective degradation of dye, polyethylene and antibacterial performance in waste water treatment. Journal of Inorganic and Organometallic Polymers and Materials 32 (2), 614–630. Available from: https://doi.org/10.1007/s10904-021-02142-7.

Rambiyana, T., et al., 2021. Photocatalytic degradation of dyes, pharmaceutical and the removal bacterial pollutants using rod shaped green derived CuO nanoparticles. International Journal of Environmental Analytical Chemistry. Available from: https://doi.org/10.1080/03067319.2021.1981303.

Rana, A., Yadav, K., Jagadevan, S., 2020. A comprehensive review on green synthesis of nature-inspired metal nanoparticles: mechanism, application and toxicity. Journal of Cleaner Production 272. Available from: https://doi.org/10.1016/j.jclepro.2020.122880.

Rani, P., et al., 2020. Highly stable AgNPs prepared via a novel green approach for catalytic and photocatalytic removal of biological and non-biological pollutants. Environment International 143. Available from: https://doi.org/10.1016/j.envint.2020.105924.

Rashtbari, Y., et al., 2022. Green synthesis of zero-valent iron nanoparticles and loading effect on activated carbon for furfural adsorption. Chemosphere 287, 132114. Available from: https://doi.org/10.1016/j.chemosphere.2021.132114.

Rathi, B.S., Kumar, P.S., Vo, D.-V.N., 2021. Critical review on hazardous pollutants in water environment: occurrence, monitoring, fate, removal technologies and risk assessment. Science of the Total Environment 797, 149134. Available from: https://doi.org/10.1016/j.scitotenv.2021.149134.

Rauwel, P., et al., 2015. A review on the green synthesis of silver nanoparticles and their morphologies studied via TEM. Advances in Materials Science and Engineering 2015, 682749. Available from: https://doi.org/10.1155/2015/682749.

Regmi, C., et al., 2017. Visible-light-induced Fe-doped BiVO4 photocatalyst for contaminated water treatment. Molecular Catalysis 432, 220–231. Available from: https://doi.org/10.1016/j.mcat.2017.02.004.

Salam, M.A., AbuKhadra, M.R., Mohamed, A.S., 2020. Effective oxidation of methyl parathion pesticide in water over recycled glass based-MCM-41 decorated by green Co3O4 nanoparticles. Environmental Pollution 259. Available from: https://doi.org/10.1016/j.envpol.2019.113874.

Salama, A., et al., 2021. Nanocellulose-based materials for water treatment: adsorption, photocatalytic degradation, disinfection, antifouling, and nanofiltration. Nanomaterials 11 (11). Available from: https://doi.org/10.3390/nano11113008.

Saleh, T.A., 2020. Nanomaterials: classification, properties, and environmental toxicities. Environmental Technology & Innovation 20, 101067. Available from: https://doi.org/10.1016/J.ETI.2020.101067.

Salem, S.S., Fouda, A., 2021. Green synthesis of metallic nanoparticles and their prospective biotechnological applications: an overview. Biological Trace Element Research 199 (1), 344–370. Available from: https://doi.org/10.1007/s12011-020-02138-3.

Samadi, Z., et al., 2021. Facile green synthesis of zero-valent iron nanoparticles using barberry leaf extract (GnZVI@BLE) for photocatalytic reduction of hexavalent chromium. Bioorganic Chemistry 114, 105051. Available from: https://doi.org/10.1016/j.bioorg.2021.105051.

Shahrashoub, M., Bakhtiari, S., 2021. The efficiency of activated carbon/magnetite nanoparticles composites in copper removal: industrial waste recovery, green synthesis, characterization, and adsorption-desorption studies. Microporous and Mesoporous Materials 311, 110692. Available from: https://doi.org/10.1016/j.micromeso.2020.110692.

Shalaby, S.M., et al., 2021. Green synthesis of recyclable iron oxide nanoparticles using Spirulina platensis microalgae for adsorptive removal of cationic and anionic dyes. Environmental Science and Pollution Research 28 (46), 65549–65572. Available from: https://doi.org/10.1007/s11356-021-15544-4.

Shao, Y., et al., 2018. Ecotoxicology and environmental safety preparation of wheat straw-supported nanoscale zero-valent iron and its removal performance on ciprofloxacin. Ecotoxicology and Environmental Safety 158, 100–107. Available from: https://doi.org/10.1016/j.ecoenv.2018.04.020.

Sharwani, A.A., et al., 2022. Photocatalytic degradation activity of goji berry extract synthesized silver-loaded mesoporous zinc oxide (Ag@ZnO) nanocomposites under simulated solar light irradiation. Scientific Reports 12 (1), 10017. Available from: https://doi.org/10.1038/s41598-022-14117-w.

Singh, J., et al., 2018. "Green" synthesis of metals and their oxide nanoparticles: applications for environmental remediation. Journal of Nanobiotechnology 16 (1), 84. Available from: https://doi.org/10.1186/s12951-018-0408-4.

Soto-Robles, C.A., et al., 2019. Study on the effect of the concentration of Hibiscus sabdariffa extract on the green synthesis of ZnO nanoparticles. Results in Physics 15, 102807. Available from: https://doi.org/10.1016/j.rinp.2019.102807.

Su, B., et al., 2020. Impact of green synthesized iron oxide nanoparticles on the distribution and transformation of As species in contaminated soil. Environmental Pollution 258, 113668. Available from: https://doi.org/10.1016/j.envpol.2019.113668.

Syafrudin, M., et al., 2021. Pesticides in drinking water—a review. International Journal of Environmental Research and Public Health 18 (2), 468. Available from: https://doi.org/10.3390/IJERPH18020468.

Tan, H.-F., Ooi, B.S., Leo, C.P., 2020. Future perspectives of nanocellulose-based membrane for water treatment. Journal of Water Process Engineering 37, 101502. Available from: https://doi.org/10.1016/j.jwpe.2020.101502.

Tian, X., et al., 2020. A comprehensive review on toxic petrochemical wastewater pretreatment and advanced treatment'. Journal of Cleaner Production 245, 118692. Available from: https://doi.org/10.1016/j.jclepro.2019.118692.

Vasantharaj, S., et al., 2021. Enhanced photocatalytic degradation of water pollutants using bio-green synthesis of zinc oxide nanoparticles (ZnO NPs). Journal of Environmental Chemical Engineering 9 (4), 105772. Available from: https://doi.org/10.1016/j.jece.2021.105772.

Vidovix, T.B., et al., 2022. Adsorption of Safranin-O dye by copper oxide nanoparticles synthesized from Punica granatum leaf extract. Environmental Technology (United Kingdom) 43 (20), 3047–3063. Available from: https://doi.org/10.1080/09593330.2021.1914180.

Vishnu, D., Dhandapani, B., 2020. Integration of *Cynodon dactylon* and *Muraya koenigii* plant extracts in amino-functionalised silica-coated magnetic nanoparticle as an effective sorbent for the removal of chromium(VI) metal pollutants. IET Nanobiotechnology 449–456. Available from: https://doi.org/10.1049/iet-nbt.2019.0313.

Wu, Z., et al., 2021. Removal of As(V) by iron-based nanoparticles synthesized via the complexation of biomolecules in green tea extracts and an iron salt. Science of the Total Environment 764, 142883. Available from: https://doi.org/10.1016/j.scitotenv.2020.142883.

Yadav, N., et al., 2021. Detection and remediation of pollutants to maintain ecosustainability employing nanotechnology: a review. Chemosphere 280. Available from: https://doi.org/10.1016/j.chemosphere.2021.130792.

Yazdanian, M., et al., 2022. The potential application of green-synthesized metal nanoparticles in dentistry: a comprehensive review'. Bioinorganic Chemistry and Applications 2022, 2311910. Available from: https://doi.org/10.1155/2022/2311910.

Zheng, D., et al., 2021. Fabrication of thin-film composite membranes for organic solvent nanofiltration by mixed monomeric polymerization on ionic liquid/water interfaces. Journal of Membrane Science 636, 119551. Available from: https://doi.org/10.1016/j.memsci.2021.119551.

Zinatloo-Ajabshir, S., Baladi, M., Salavati-Niasari, M., 2021. Sono-synthesis of MnWO4 ceramic nanomaterials as highly efficient photocatalysts for the decomposition of toxic pollutants. Ceramics International 47 (21), 30178–30187. Available from: https://doi.org/10.1016/j.ceramint.2021.07.197.

Further reading

Cheng, H.H., et al., 2020. Recent advancement on biological technologies and strategies for resource recovery from swine wastewater. Bioresource Technology 303, 122861. Available from: https://doi.org/10.1016/j.biortech.2020.122861.

Chowdhary, P., et al., 2020. Role of industries in water scarcity and its adverse effects on environment and human health. Environmental Concerns and Sustainable Development. Springer, Singapore, pp. 235–256. Available from: https://doi.org/10.1007/978-981-13-5889-0_12.

Han, J., Qiu, W., Gao, W., 2010. Potential dissolution and photo-dissolution of ZnO thin films. Journal of Hazardous Materials 178 (1), 115–122. Available from: https://doi.org/10.1016/j.jhazmat.2010.01.050.

https://www.who.int/newsroom/factsheets/detail/drinkingwater?msclkid = 4429378cab4311ecace871659c2535fe (accessed 24.03.2022).

Kanu, I., Achi, O.K., 2011. Industrial effluents and their impact on water quality of receiving rivers in Nigeria. Journal of Applied Technology in Environmental Sanitation 1 (1), 75–86. Available at:. Available from: https://www.researchgate.net/profile/Ome-Achi/publication/287104597_Industrial_effluents_and_their_impact_on_water_quality_of_receiving_rivers_in_Nigeria/links/611ac3e31ca20f6f86287362/Industrial-effluents-and-their-impact-on-water-quality-of-receiving-rivers-in-Nigeria.pdf.

Karimi, H., Rahimpour, A., Shirzad Kebria, M.R., 2016. Pesticides removal from water using modified piperazine-based nanofiltration (NF) membranes. Desalination and Water Treatment 57 (52), 24844–24854. Available from: https://doi.org/10.1080/19443994.2016.1156580.

Khan, F.S.A., et al., 2020. Magnetic nanoparticles incorporation into different substrates for dyes and heavy metals removal—a review. Environmental Science and Pollution Research 27 (35), 43526–43541. Available from: https://doi.org/10.1007/s11356-020-10482-z.

Li, H., et al., 2020. Environment-enhancing process for algal wastewater treatment, heavy metal control and hydrothermal biofuel production: a critical review. Bioresource Technology 298, 122421. Available from: https://doi.org/10.1016/j.biortech.2019.122421.

Nageeb, M., 2013. Adsorption technique for the removal of organic pollutants from water and wastewater. Organic Pollutants – Monitoring, Risk and Treatment. IntechOpen. Available from: https://doi.org/10.5772/54048.

Nosaka, Y., Nosaka, A.Y., 2017. Generation and detection of reactive oxygen species in photocatalysis. Chemical Reviews 117 (17), 11302–11336. Available from: https://doi.org/10.1021/acs.chemrev.7b00161.

Ong, C., et al., 2020. Green synthesis of thin-film composite membranes for organic solvent nanofiltration. ACS Sustainable Chemistry and Engineering 8 (31), 11541–11548. Available from: https://doi.org/10.1021/acssuschemeng.0c02320.

Pi, Y. et al., 2017. 'Abstract : Keywords'. Available from: https://doi.org/10.1016/j.cej.2017.12.092.

Ramakrishnan, B., et al., 2021. Organic farming: does it contribute to contaminant-free produce and ensure food safety. Science of the Total Environment 769, 145079. Available from: https://doi.org/10.1016/J.SCITOTENV.2021.145079.

Sun, W., Xiao, L., Wu, X., 2019. 'Facile synthesis of NiO nanocubes for photocatalysts and supercapacitor electrodes. Journal of Alloys and Compounds 772, 465–471. Available from: https://doi.org/10.1016/j.jallcom.2018.09.185.

Zaimee, M.Z., Sarjadi, M.S., Rahman, M.L., 2021. Heavy metals removal from water by efficient adsorbents. Water . Available from: https://doi.org/10.3390/w13192659.

Zhu, M., et al., 2017. Au/La2Ti2O7 nanostructures sensitized with black phosphorus for plasmon-enhanced photocatalytic hydrogen production in visible and near-infrared light. Angewandte Chemie – International Edition 56 (8), 2064–2068. Available from: https://doi.org/10.1002/anie.201612315.

Chapter 12

Optimized conditions for removal of Ni(II) using sugarcane bagasse biosorbent as an adsorbent

Indu Singh[1], Bushra Yaseen[1], Chinky Gangwar[1], Rashmi Nayak[2], Indresh Kumar[1], Pradeep Kumar Pandey[1], Joy Sarkar[1] and Radhey Mohan Naik[1]

[1]*Department of Chemistry, Lucknow University, Lucknow, Uttar Pradesh, India,* [2]*CSIR-National Botanical Research Institute, Lucknow, Uttar Pradesh, India*

Introduction

Due to the rise in industrial as well as natural phenomena, the release of bulky metals into the environment has roused many folds (Ramachandran et al., 2009; Basaiah et al., 2014). These bulky metals cause acute toxicity to the biological system, that is, fauna and flora (Gautam et al., 2014; Huseen and Mohammed, 2019; Singh Sankhla et al., 2017). The toxicity of mercury causes Minimata disease and Itai-Itai disease is caused due to cadmium toxicity (Wolf and Baynes, 2007). The various organizations of different countries around the globe have specified the limit of these bulky metal concentrations to be released from industrial effluents in the environment with a density greater than 5 mg/cm^2 (Farombi et al., 2007). Particularly exposure to Nickel which is involved in manufacturing stainless steel (Reck and Rotter, 2012), electronic components (Flowers et al., 2017), jewelry, and coins (Zhai et al., 2003), may lead to high blood pressure (Spes et al., 2020), developmental deficiency, a neurological disorder (Nickel and Gu, 2018), cardiovascular disease (Micheu et al., 2020), and lung cancer (Ohiagu et al., 2022). Oxidative damage by exposure to Nickel due to the formation of free radicals is also reported (Mlinaric et al., 2019). Not only this breast cancer (Bai et al., 2018), infertility (Rizvi et al., 2019), and miscarriage are also found to be caused by exposure to the Nickel(II) (Ni(II)). As stated by WHO policies, the extreme permissible concentration of Ni(II) ions in water solution is exclusively 0.02 mg/L for drinking water (BIS, 1994) (R-1). Ni(II) ions bioavailability in hydrated form is attributed to its toxicological effects therefore, the removal of Ni(II) ions from aqueous solution is of significant consideration to minimize the content of Nickel in the wastewater either by removal or remediation. Nowadays diverse methods are known for the removal and recovery of Nickel namely adsorption (O'connell et al., 2006), precipitation (Tsai et al., 2020), reverse osmosis (Al-Absi et al., 2021), electrodialysis (Wang et al., 2020), ion exchange (Ma et al., 2018), and so on. However, these methods involve high costs and technical constraints. Thus there is a necessity to design an economical way for the deduction of Nickel from wastewater. The utilization of the ion-exchange method by the involvement of biomass obtained as industrial /agricultural waste or from microbial biomass has gained momentum as this biomass possesses functional groups such as carboxyl (−COOH), thiol (−SH), amide ($-NH_2$), hydroxyl (−OH), and phosphate ($-PO_4$). These functional groups were encountered to be involved in the biosorption of hefty metals. The use of bacterial strains *Bacillus subtilis* SM 29 (Abdel-Monem et al., 2010), *B. subtilis* WD 90 (Aslam et al., 2020), *Nostoc calcicola* (Ghanbarzadeh et al., 2022), *Rhizobium leguminosarum* (Albareda et al., 2015) fungal derived biosorbent is briefed in the literature for deduction of Nickel from wastewater. The current work is based on using sugarcane bagasse (SB) a solid waste generated by sugar industries as a biosorbent for the expulsion of bulky metal Ni(II) ions from synthetic wastewater.

Materials and methods

Chemicals employed were purchased from Sigma Aldrich company and employed directly as acquired without additional refinement.

Role of Green Chemistry in Ecosystem Restoration to Achieve Environmental Sustainability. DOI: https://doi.org/10.1016/B978-0-443-15291-7.00003-1

Preparation of sugarcane bagasse adsorbent

Raw material SB was collected from the local fruit retailer in Lucknow, India. After collecting, it was bathed numerous times with distilled and then Milli Q water for the removal of particles adhered to the surfaces. After complete washing, it was then sun-dried for about 4–5 days in the open air. For complete dryness, it was further dried in an oven for about 2 hours between 80°C and 100°C. After that, powdered material was sieved into 100 μm mesh size to increase its surface area. The resulting powdered material that was SB was again washed and parched in a heated oven for 24 hours at 70°C and stored for further use.

Adsorbent acid activation

The strained SB powder (100 g) was sponged in 1.0 L of 40% H_2SO_4 in an automated shaker at 200 rpm speed continuously for 12 hours. After complete sponging, SB was filtered and rinsed with DD water to maintain a neutral pH. Moreover, it was further parched in a hot oven for 24 hours at an elevated temperature of 150°C. The acid-activated strained SB powder was then finally reserved in a tight container for additional utilization.

Nickel solution preparation

By adding the required quantity of $NiSO_4 \cdot 0.6H_2O$ in DD water Ni(II) stock solution was prepared. The other standard solutions of Ni(II) of different concentrations were prepared by adding water to the stock solution.

Batch mode adsorption studies

100 mL volumetric flasks were used to carry out the batch adsorption experiments. This method was employed to examine the effectiveness of time of contact (30–180 minutes), pH (2.0–10.0), temperature (20°C–50°C), inception M^+ ion concentration (10–200 mg/L), and adsorbent portion (0.25–5.0 g/L). The adsorption of N(II) metal ions on SB adsorbent was obtained by shaking the 50 mL solutions of Ni(II) of known concentrations with the prescribed adsorbent dose in an automated shaker for a specific period. After equilibrium attainment, the adsorbent was segregated from the solution through filtration. Further, the concentration of Ni(II) ions concentration in the given solution was measured spectrophotometrically. For maintaining the pH of the adsorptive solution NaOH and HCl were used. The most appropriate conditions involved in segregating hefty metals from solution are demonstrated through these studies. The amount of SB biosorbent sponged per unit mass and percentage removal of heavy metals were calculated using Eqs. (12.1) and (12.2), respectively.

$$\%\ \text{Removal of metal ions} = \frac{(Co - Ce)V}{C0} \times 100 \tag{12.1}$$

$$qe = \frac{(Co - Ce)V}{m} \tag{12.2}$$

where qe stands for soaking capability. C_o and C_e stand for inception and terminal equilibrium concentration of the M^+ ions test solution in mg/L, V stands for total M^+ ions present in solution in L.

Results and discussion

Adsorbent characterizations

The prepared SB biosorbent was characterized via different characterization techniques including FT-IR, field emission scanning electron microscopy (FE-SEM), and Brunauer–Emmett–Teller (BET).

Fourier transform infrared spectroscopy analysis of sugarcane bagasse biosorbent

The Fourier transform infrared spectroscopy (FT-IR) spectra of the prepared SB adsorbent were recorded using PerkinElmer Version 10.03.06 FT-IR. The spectrum is represented in Fig. 12.1 from region 4000 to 500 wavenumber. The vibration stretching peak at 3484 cm^{-1} corresponds to chemisorbed water and hydroxyl groups present at the surface.

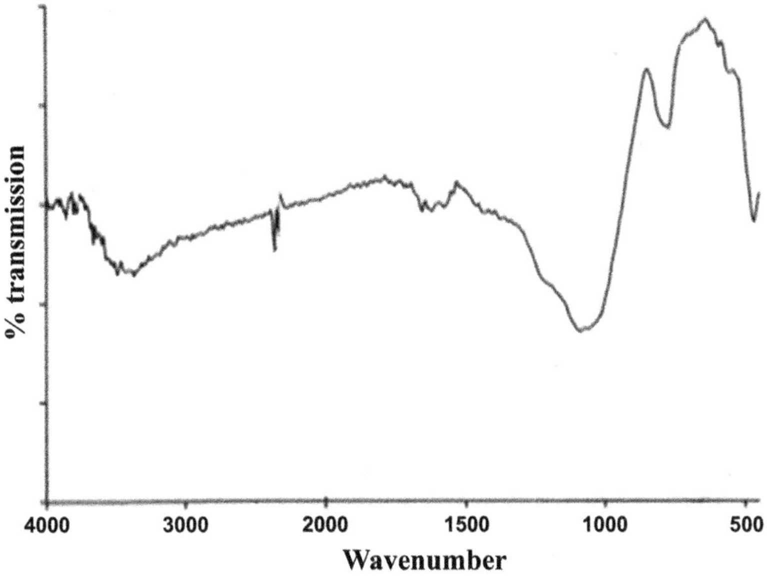

FIGURE 12.1 Fourier transform infrared spectroscopy spectrum of sugarcane bagasse adsorbent.

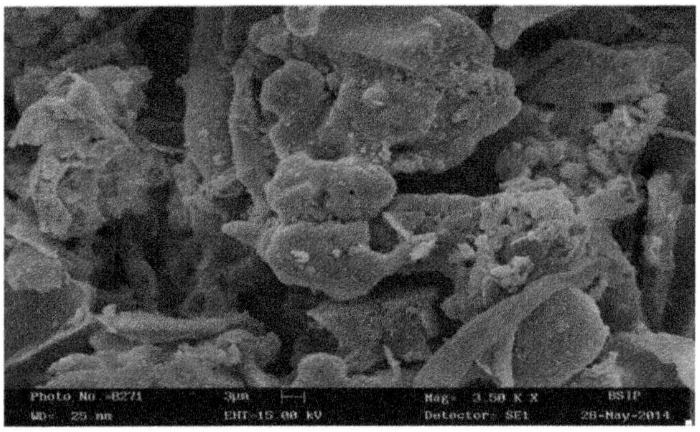

FIGURE 12.2 Field emission scanning electron microscopy image of sugarcane bagasse adsorbent.

Morphological analysis

The surface morphological studies of the prepared adsorbent were examined using FE-SEM. The FE-SEM image shown in Fig. 12.2 demonstrates the exterior structure and porosity of the prepared SB adsorbent.

Surface area (Brunauer–Emmett–Teller) analysis

Superficial area and pore size analysis of the SB powder were evaluated utilizing the BET technique. BET spectra of the prepared biosorbent were obtained from Babasaheb Bhimrao Ambedkar University (BBAU), Lucknow (Uttar Pradesh, India). Fig. 12.3 shows the N_2 isotherm of the prepared adsorbent at 77.0K. The isotherm clearly shows that the biosorbent is mesoporous.

Ni(II) metal ion analysis

Samples of different concentrations were purified through filter paper after purification all the samples were taken and analyzed for M^{++} ion concentration involving a UV-visible spectrophotometer (Systronic 2203). The quantity of metal soaked during the connection time with triggered SB was assessed using the calibration curve (Abs vs. Concentration).

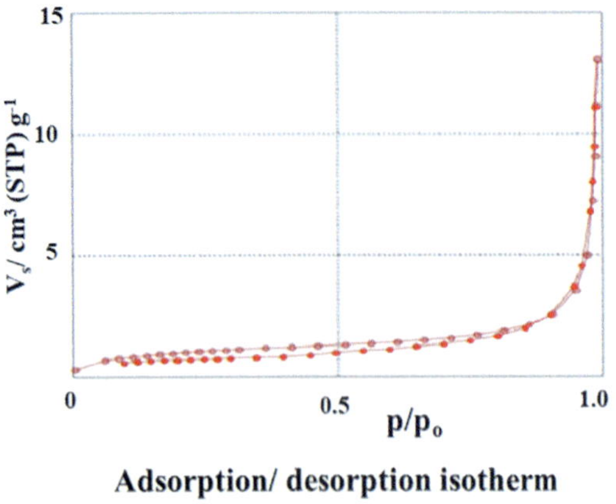

FIGURE 12.3 N_2 isotherm of sugarcane bagasse adsorbent.

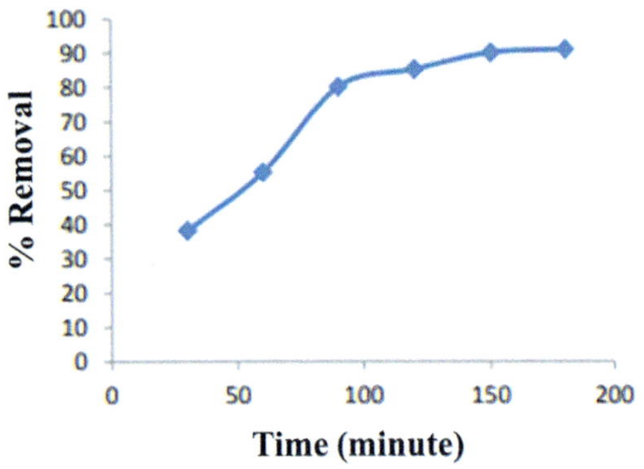

FIGURE 12.4 Contact time effect for percentage removal of Ni(II) by sugarcane bagasse adsorbent.

Contact time effect

For the determination of the contact time effect between 30–180 minutes for the percentage withdrawal of Ni(II), the biosorbent dose used was 2 g/L. The initial metal ion concentration was 50 mg/L at pH 6.0 at 30°C. Fig. 12.4 demonstrates the metal removal efficiency of the biosorbent with respect to time. From the figure, it is evident that on increasing time removal efficiency increases up to 100 minutes and then becomes constant. The optimum time to study metal removal efficiency using SB biosorbent is 120 minutes. Therefore 120 minutes was taken as a contact time for further experiments.

Temperature effect

To evaluate the effect of the temperature, four variable temperatures 20°C, 30°C, 40°C, and 50°C were chosen for the percentage age removal of the Ni(II) ions by SB biosorbent, using 2 g/L biosorbent amount. Inceptive M^+ ion concentration 50 mg/L and contact time of 120 minutes were kept constant for all four temperatures. Fig. 12.5 represents that on escalating the temperature from 20°C to 50°C the metal uptake increases by the biosorbent. This increase in the metal uptake on temperature increment exhibits more interaction between M^+ ions and the biosorbent. As a result, 30°C was found to be the optimum temperature for the uptake of the Ni(II) ions by SB biosorbent.

pH effect

Out turn of pH on the percentage age discard of Ni(II) ions by SB biosorbent is represented in Fig. 12.6. The effect of pH was examined for different pH values starting from acidic to basic (2, 4, 6, 8, and 10) keeping the M^+ ion concentration 50 mg/L, contact time 120 minutes, and operating temperature 30°C constant for all pH. From the result, it was observed that on increasing the pH in the acidic medium the percentage age removal of metal was high and found to be maximum at pH 6.0. After this certain pH, the efficiency for the removal of M ions decreases. The reason behind this result might be ascribed to the protonation of adsorbent at low pH and metal hydroxide precipitation at higher pH.

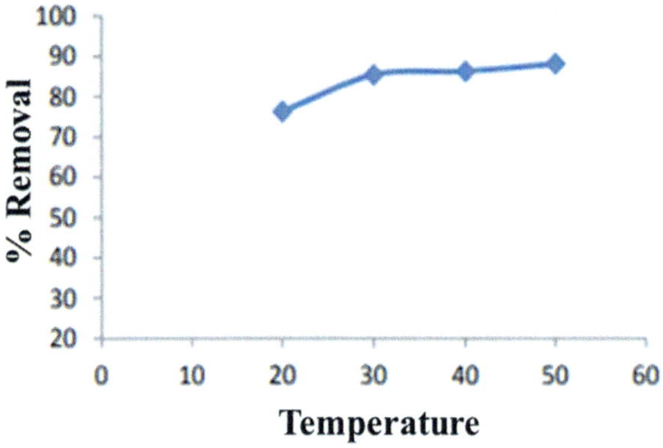

FIGURE 12.5 Temperature effect on percentage age removal of Ni(II) ions via sugarcane bagasse adsorbent.

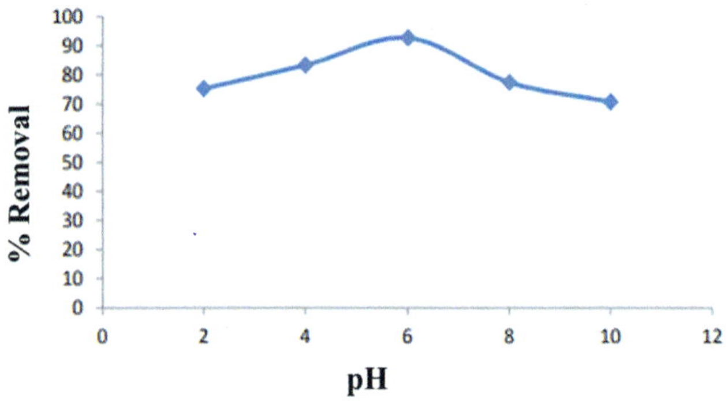

FIGURE 12.6 pH effect on the percentage removal of Nickel(II) ions by sugarcane bagasse adsorbent.

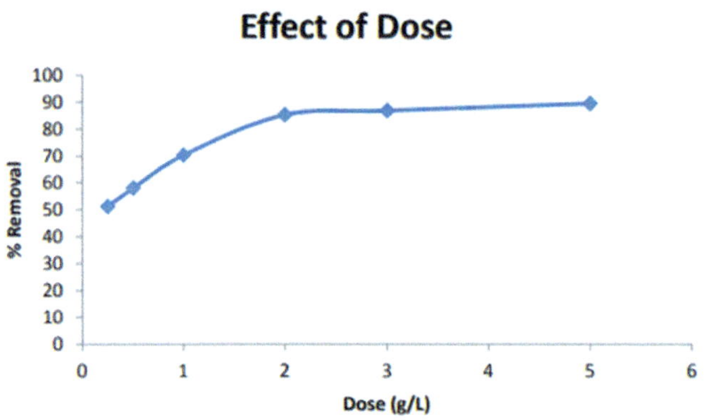

FIGURE 12.7 Sugarcane bagasse adsorbent doses effect on percentage age removal of Nickel(II).

Sugarcane bagasse biosorbent dose effect

To examine the effect of SB biosorbent doses, experiments were performed with varying biosorbent concentrations at six different concentrations from 0.25 to 5.0 g/L. Keeping M ion concentration (50 mg/L), contact time (120 minutes), pH (6.0), and temperature (30°C) constant. A sharp increment in the percentage age removal of the metal was observed with an increase in the concentration of SB biosorbent and reached optimum. It was evident from Fig. 12.7 the optimum value using biosorbent concentration was found to be 2 g/L. On increasing biosorbent concentration the availability of the binding sites to the metal increases.

Inception M^+ ion concentration effect

Inception M^+ ion concentration effect represented in Fig. 12.8, between 10 and 200 mg/L on Nickel(II) ion removal from synthetic wastewater. The effect was evaluated using an adsorbent dose 2 g/L, pH 6.0, and time of contact

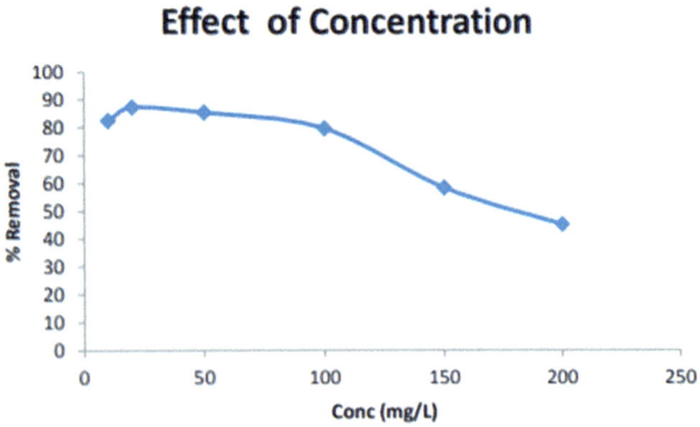

FIGURE 12.8 Inception M ion concentration effect on percentage age discard of Ni(II) ion by sugarcane bagasse biosorbent.

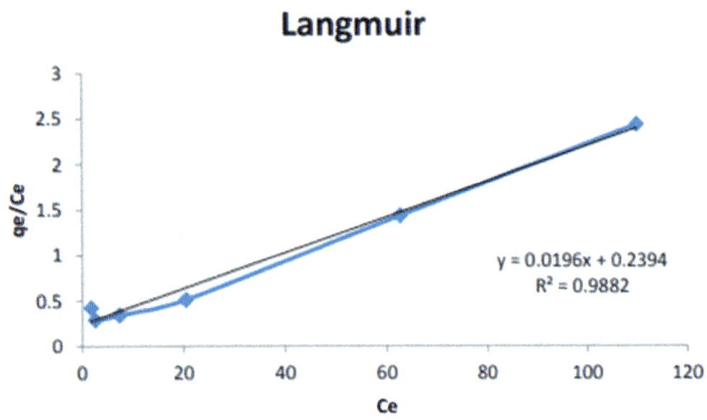

FIGURE 12.9 Langmuir adsorption isotherm.

120 minutes. The percentage age removal of Ni(II) was found to be maximum at 10 mg/L concentration of the initial metal ion. Further increasing the increasing concentration of the metal ion above 10 mg/L the percentage age removal of Ni(II) ions decreases. It was observed that at lesser concentrations, the available sites are higher. While increasing the metal ion concentration, available binding decreases. Thus, the removal of metal ions directly depends on the initial metal ion concentration.

Adsorption isotherm models

Adsorption isotherms including Langmuir and Freundlich mainly represent the equilibrium relationship between the mass of the solute adsorb on adsorbate. In the present investigation, these two adsorption isotherms are utilized for further experimental analysis. Langmuir adsorption was employed for the elucidation of the adsorption occurring at the homogenous site whereas, the Freundlich isotherm was employed for the determination of the adsorption occurring at the heterogeneous sites. Eq. (12.3) is the representation for the Langmuir adsorption isotherm (LAI), whereas Eq. (12.4) is used for the determination of Freundlich adsorption isotherm (FAI).

$$\frac{Ce}{qe} = \frac{Ce}{qmax} + 1/KL \tag{12.3}$$

$$lnqe = \ln KF + \frac{1}{n}\ln Ce \tag{12.4}$$

where q_{max} = maximum uptake capability, K_L = is the Langmuir constant, K_F = Freundlich constant.

The experimental data obtained were directly applied to the above-mentioned equations. To verify the applicability of the LAI, a plot of q_e/C_e versus C_e was obtained, represented in Fig. 12.9. A linear plot was obtained for LAI which verifies the suitability of LAI. Similarly, to verify the FAI a plot of q_e versus C_e was obtained as shown in Fig. 12.10, which also confirms the suitability of the FAI. From the experimental data, the constants for both the isotherms were also evaluated, as reported in Table 12.1. The results from the table suggest that the LAI represents better results than FAI which suggests that for an endothermic process, for an increase in temperature the values of q_{max} and K_L increase.

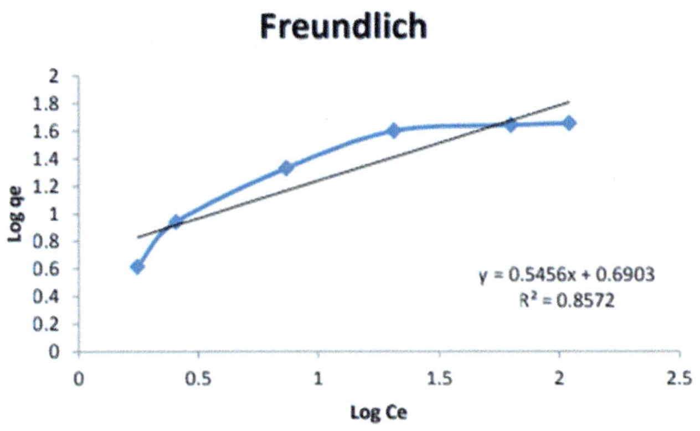

FIGURE 12.10 Freundlich adsorption isotherm.

TABLE 12.1 Representation of adsorption isotherm studies.

LAI			FAI		
q_{max}	LAI constant	Adj. R^2 for LAI	n	FAI constant	Adj. R^2 for FAI
52.0204	0.2394	0.98	1.8328	1.9943	0.86

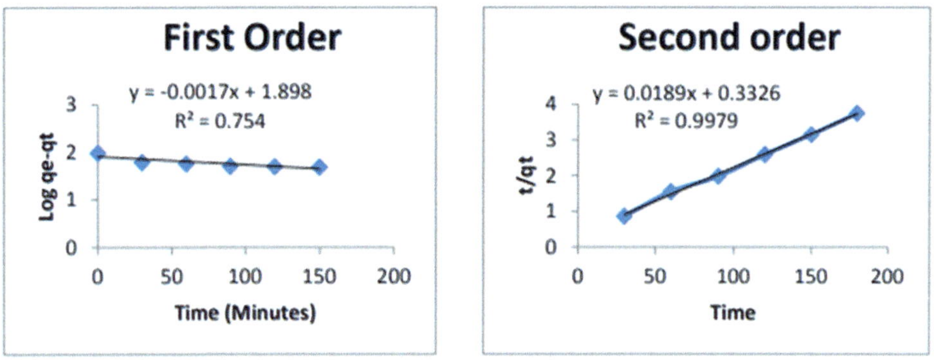

FIGURE 12.11 Kinetic studies of the Ni(II) on the sugarcane biosorbent.

Kinetic studies

The kinetic study for the rate discardation of Ni(II) ions from wastewater was employed using pseudo-second-order kinetics. To analyze the surface assimilation kinetics property, two different equations were employed: Eqs. (12.5) and (12.6).

$$\frac{t}{qt} = \frac{1}{k_2 q_1^2} + \frac{1}{q_1} t \tag{12.5}$$

where q_1 is the maximum uptake capacity, q_t is the quantity of Ni uptake at a time t, k_2 is the equilibrium rate constant for pseudo-second-order uptake.

The intraparticle diffusion model can be represented in Eq. 12.6

$$q_t = k_t t^{1/2} + I \tag{12.6}$$

where k_t stands for intraparticle distribution rate constant, I is the constant that is connected to margin layer thickness.

The kinetic data results suggest the pseudo-second-order model given in Fig. 12.11 describes the adsorption of Ni (II) ions superior to beside models. The pseudo-second-order model works on the assumption that the duplet surface sites might be engaged by a single adsorbate ion. The pseudo-second-order is ineffectual to recognize the diffusion

TABLE 12.2 Representation of kinetic studies of nickel(II) adsorption by sugarcane bagasse biosorbent.

Pseudo first order		Pseudo second order	
Rate constant k_1 (min^{-1})	**Adj. R^2**	**Rate constant k_2 (g/mg.min)**	**Adj. R^2**
0.0131	0.75	2.42×10^{-6}	0.99

mechanism. Because of this, the intraparticle diffusion model comes into the picture. If on extrapolation line passes through the origin for plot t/q_t versus time intraparticle diffusion is found to be the rate-controlling step. Contradictory for the present result intercept was examined which suggests that the edges effect is more prominent. The results obtained for both models suggest that the Adj. R^2 of the pseudo-second-order model is better than compared with the intraparticle diffusion model. Representation of kinetic studies of Nickel(II) adsorption by SB biosorbent is represented in Table 12.2.

Conclusion

The present investigation demonstrates the application of SB as an efficacious adsorbent for the segregation of Ni(II) ions from wastewater. Optimum conditions including pH, temperature, M ion concentration, temperature, and biosorbent doge are also represented in detail. From the results, it was found the ideal pH for the removal of Ni(II) ions was 6.0. Furthermore, it was evident from the result that the removal effectiveness of the metal ion increases with the SB adsorbent dose. Along with this adsorption isotherms for two adsorption isotherms LAI and FAI were employed and the best results were obtained for the LAI. In series to regulate the rate for the segregation of Ni(II) ions from wastewater kinetic study was also performed. Pseudo II and order kinetics well fitted with Adj. $R^2 > 0.99$. From this study, we may conclude that SB biosorbent might be directly applied to remove heavy metals from the effluents discharged from the treatment plants.

Acknowledgment

The authors are grateful to the HOD, Chemistry, Lucknow University, Lucknow (U.P.) India, for providing basic infrastructure facilities, FT-IR spectroscopy, and UV-visible spectrophotometry for performing experimental work. The authors are also grateful to Birbal Sahni Institute of Palaeobotany Lucknow, for FESEM facility. The authors are grateful to BBAU, Lucknow, Uttar Pradesh for BET analysis.

References

Abdel-Monem, M., Al-Zubeiry, A., Al-Gheethi, A., 2010. Biosorption of nickel by Pseudomonas cepacia 120S and Bacillus subtilis 117S. Water Science and Technology: A Journal of the International Association on Water Pollution Research 61, 2994–3007.

Al-Absi, R., Abu-Dieyeh, M., Ben-Hamadou, R., Nasser, M., Al-Ghouti, M., 2021. Novel composite materials of modified roasted date pits using ferrocyanides for the recovery of lithium ions from seawater reverse osmosis brine. Scientific Reports 11, 18896.

Albareda, M., Rodrigue, A., Brito, B., Ruiz-Argüeso, T., Imperial, J., Mandrand-Berthelot, M.-A., et al., 2015. Rhizobium leguminosarum HupE is a highly-specific diffusion facilitator for nickel uptake. Metallomics Integrated Biometal Science 7.

Aslam, F., Yasmin, A., Sohail, S., 2020. Bioaccumulation of lead, chromium, and nickel by bacteria from three different genera isolated from industrial effluent. International Microbiology 23.

Bai, X., Ni, J., Beretov, J., Graham, P., Li, Y., 2018. Cancer stem cell in breast cancer therapeutic resistance. Cancer Treatment Reviews 69, 152–163.

Basaiah, T., C.K., S., Krishnappa, M., 2014. Study on heavy metals biosorption ability of Saccharomyces cerevisiae. International Journal of Biological Research 2.

Farombi, O., Adelowo, O.A., Ajimoko, Y.R., 2007. Biomarkers of oxidative stress and heavy metal levels as indicators of environmental pollution in African cat fish (Clarias gariepinus) from Nigeria Ogun River. International Journal of Environmental Research and Public Health 4, 158–165.

Flowers, P., Reyes, C., Ye, S., Kim, M.J., Wiley, B., 2017. 3D printing electronic components and circuits with conductive thermoplastic filament. Additive Manufacturing 18.

Gautam, R., Sharma, S., Mahiya, S., Chattopadhyaya, M., 2014. Contamination of heavy metals in aquatic media: transport, toxicity and technologies for remediation. Heavy Metals in Water: Presence, Removal and Safety 1–24.

Ghanbarzadeh, M., Niknam, V., Soltani, N., Ebrahimzadeh, H., Shahavi, M.H., 2022. Removal of Phenanthrene by some microalga species and study of antioxidative compounds in Nostoc calcicola ISC 89. Journal of Soils and Sediments 22.

Huseen, H., Mohammed, A., 2019. Heavy metals causing toxicity in fishes. Journal of Physics: Conference Series 1294, 062028.

Ma, A., Abushaikha, A., Allen, S., Mckay, G., 2018. Ion exchange homogeneous surface diffusion modelling by binary site resin for the removal of nickel ions from wastewater in fixed beds. Chemical Engineering Journal 358.

Micheu, M., Birsan, M.-V., Robert, S., Keresztesi, Á., Nita, I.-A., 2020. From air pollution to cardiovascular diseases: the emerging role of epigenetics. Molecular Biology Reports 47.

Mlinaric, M., Durgo, K., Katić, V., Spalj, S., 2019. Cytotoxicity and oxidative stress induced by nickel and titanium ions from dental alloys on cells of gastrointestinal tract. Toxicology and Applied Pharmacology 383, 114784.

Nickel, M., Gu, C., 2018. Regulation of central nervous system myelination in higher brain functions. Neural Plasticity 2018, 1–12.

O'connell, D., Birkinshaw, C., O'dwyer, T., 2006. A modified cellulose adsorbent for the removal of nickel(II) from aqueous solutions. Journal of Chemical Technology and Biotechnology 81, 1820–1828.

Ohiagu, F., Chikezie, P., Chinazaekpere, A., Chikezie, C., 2022. Human exposure to heavy metals: toxicity mechanisms and health implications. Material Science & Engineering International Journal 6, 78–87. Available from: https://doi.org/10.15406/mseij.2022.06.00183.

Ramachandran, S., Coradin, T., Jain, P., Verma, S., 2009. Nostoc calcicola immobilized in silica-coated calcium alginate and silica gel for applications in heavy metal. Silicon 1, 215–223.

Reck, B., Rotter, V., 2012. Comparing growth rates of nickel and stainless steel use in the early 2000s. Journal of Industrial Ecology 16, 518–528.

Rizvi, A., Parveen, S., Khan, S., Naseem, I., 2019. Nickel toxicology with reference to male molecular reproductive physiology. Reproductive Biology 20.

Singh Sankhla, M., Sharma, K., Kumar, R., 2017. Heavy metal causing neurotoxicity in human health. International Journal of Innovative Research in Science, Engineering and Technology 6, 7721–7726.

Spes, K., Chen, W., Krebes, L., Völker, K., Abeßer, M., Negrin, P., et al., 2020. Heart-microcirculation connection: effects of ANP (atrial natriuretic peptide) on pericytes participate in the acute and chronic regulation of arterial blood pressure. Hypertension 76, 12015772. Available from: https://doi.org/10.1161/HYPERTENSIONAHA.120.15772.

Tsai, T.-H., Chou, H.-W., Wu, Y.-F., 2020. Removal of nickel from chemical plating waste solution through precipitation and production of microsized nickel hydroxide particles. Separation and Purification Technology 251, 117315.

Wang, C., Li, T., Yu, G., Deng, S., 2020. Removal of low concentrations of nickel ions in electroplating wastewater by combination of electrodialysis and electrodeposition. Chemosphere 263, 128208.

Wolf, M., Baynes, J., 2007. Cadmium and mercury cause an oxidative stress-induced endothelial dysfunction. Biometals: An International Journal on the Role of Metal Ions in Biology, Biochemistry, and Medicine 20, 73–81.

Zhai, H., Chew, A.L., Bashir, S., Reagan, K., Hostynek, J., Maibach, H., 2003. Provocative use of nickel coins in nickel-sensitized subjects and controls. The British Journal of Dermatology 149, 311–317.

Chapter 13

Application of pyrolysis techniques to produce bio-sorbents for water treatment

Thi Thao Truong[1], Truong Xuan Vuong[1] and Thuong Tuan Nguyen[2]

[1]TNU-University of Sciences, Thai Nguyen University, Tan Thinh Ward, Thai Nguyen City, Thai Nguyen, Vietnam, [2]Institute of Life Sciences, Thai Nguyen University, Quyet Thang Commune, Thai Nguyen City, Thai Nguyen, Vietnam

Introduction

Although about 70% of the Earth's surface is covered by water, 97% of it belongs to the ocean, and only about 3% is freshwater which is valuable for human existence and development. Pollution is increasing and gradually showing apparent effects on humans and the environment. Common pollutants, including heavy metals, anions, dyes, organic substances, and residues of antibiotics, are usually removed by physical and chemical methods, such as adsorption, membrane filtration, solvent extraction, coagulation/flocculation, adsorption, oxidation process (by ozone, hydrogen peroxide, Fenton's reagent), photodegradation, photocatalytic degradation, electrochemical methods, chemical precipitation, or/and biological methods for organic pollutants as an anaerobic process, aerobic process, constructed wetlands (CWs), biocomposting, enzymes (Chowdhary et al., 2018). Among them, adsorption is a method that is both straightforward and effective, and is of great interest in research and application. The most extensively recognized ideal adsorbent is activated carbon, but both their manufacturing and post-adsorption reuse cost are high, or severe pollution if not recovered (AC is persistent black carbonaceous materials) (Lin et al., 2020; Imran-Shaukat et al., 2022). Meanwhile, biochar is typically produced from the thermochemical conversion of biomass (Bamdad et al., 2018), generally produced from plants, aquatic (algae, microalgal), fungi, bacterial, agricultural wastes, and industrial wastes (Macedo et al., 2020). The biochar surface is less carbonized but rich in oxygen-containing groups than AC and usually includes −C, C = C, OH, metal-O/OH, CHO−, COOH, aromatic carbon skeletons, mineral crystal phases, and other oxygen-containing functional groups (Hassan et al., 2020). Biochar is applied as an effective biosorbent for many contaminated objects, such as heavy metal (Cutillas-Barreiro et al., 2014; Lin et al., 2020; Roha et al., 2021; Saeed et al., 2022; Xu et al., 2022), organic (Mohammed et al., 2018; Abatal et al., 2020; Anastopoulos et al., 2020; Oginni and Singh, 2021), dye (Aguayo-Villarreal et al., 2016; Ahmad et al., 2019, 2021; Al-Zawahreh et al., 2022), pesticides (Mandal et al., 2017; Mokrzycki et al., 2020), herbicide (Tulun et al., 2021; Binh et al., 2022), pharmaceutical compounds (Antón-Herrero et al., 2018; Conde-Cid et al., 2019a; Velusamy et al., 2021; Puga et al., 2022), nutrients (NH_4^+ and PO_4^{3-}) (An et al., 2021), anion (Keränen et al., 2013). Pyrolysis is the process in which biomass is heated to 300−900°C for hours without oxygen. This chapter focuses on applying pyrolysis techniques to produce bio-sorbents for water treatment, the characteristics and applications of pyrolysis products.

Pyrolysis process and characteristics of biochar

Feedstocks collected from various resources were washed with distilled water (or dilute alkali, dilute acid, or organic solvents depending on the source of the material, for example, agricultural residues, industrial wastes, food wastes, then washed again by distilled water). Afterward, they were dried in the sun and/or dried in an oven to dehydrated powders; then, the dry biomass was ground, sieved, and stored before pyrolysis. Depending on heating rate, pyrolyzed temperature and time, and biomass sources, the properties and applications of biochar are different.

Role of Green Chemistry in Ecosystem Restoration to Achieve Environmental Sustainability. DOI: https://doi.org/10.1016/B978-0-443-15291-7.00033-X

Sources of biochar

Biochar is the "charred organic matter" obtained from other lignocellulosic biomass. It mainly contains carbon (more than 50% by weight), hydrogen, oxygen, nitrogen (Qiu et al., 2021), and may be S, P, K, Ca, Mg, or mineral elements. In the pyrolysis process, they will be decomposed gradually with temperature, releasing gas and forming the porous carbon skeleton. Various plant's parts have been studied using biochar synthesis such as: rice straw (Huang et al., 2015; Lin et al., 2020; Sakhiya et al., 2021; Wei and Lu, 2021; Xu et al., 2022); stem of eucalyptus (Antón-Herrero et al., 2018; Foong et al., 2020), corn stover (Li et al., 2019a,b), *Eichhornia crassipes* and *Phragmites australis* (Doan et al., 2021), *Solidago canadensis* L. (Nie et al., 2021), tobaco (Zhou et al., 2018), corn stalk (Huang et al., 2019; Yu et al., 2019; Zhuang et al., 2020); leaves of African almond (*Terminalia catappa* L) (Jabar et al., 2022), *Azadirachta indica* (Thangagiri et al., 2022), *Cynara scolymus* (Mahmoud et al., 2021a,b), mango (Vyavahare et al., 2018), waste tea (Wasewar et al., 2009; Uzun et al., 2010; Li et al., 2018a,b; Çelebi, 2020; Altaf et al., 2021; Shirvanimoghaddam et al., 2021), fargesia (Ma et al., 2018), palm (Saremi et al., 2020), camphor (Hu et al., 2019); Roots (Antón-Herrero et al., 2018; Li et al., 2019a,b); bark of tree (*Acacia mangium*) (Zhang et al., 2020a,b), palm (Sun et al., 2013), pine (Conde-Cid et al., 2019b; Cuong Nguyen et al., 2021), cotton (Rumjit et al., 2021); rice husk (Cuong et al., 2019; Roha et al., 2021; Saeed et al., 2022); seed (fruit): soap nut seeds (Velusamy et al., 2021), rape seed (Zhao et al., 2017), tobaco seed (Onorevoli et al., 2018); coffee husks (Cuong Nguyen et al., 2021), grape pomace (Yoon et al., 2021), watermelon (Ahmed et al., 2021); shell of fruit or seed: walnut shell (Georgieva et al., 2020; Geng et al., 2021; Tulun et al., 2021), matamba fruit shell (Obey et al., 2022), peanut shell (Wilson et al., 2006; AL-Othman et al., 2012; Garg et al., 2019; Shan et al., 2020; Yang et al., 2020; Ahmad et al., 2021; An et al., 2021; Aryee et al., 2021), coconut shell, groundnut shell (Praveen et al., 2021), ackee apple pods (Bello et al., 2021), pecan nutshells (Zazycki et al., 2018), badam shell (Jia et al., 2021); grass: *Miscanthus giganteus* grass (Cibati et al., 2017), lemongrass (Putri et al., 2020), elephant grass (Adesemuyi et al., 2020), kans grass (Baig et al., 2014), barley grass (Zhang et al., 2019a,b); aquatic (algal, microalgal), *Chlorella* sp. and *Spirulina* sp. (Shen et al., 2017; Zheng et al., 2017; Amin and Chetpattananondh, 2019; Choi et al., 2020; Law et al., 2022), microalgal (Gan et al., 2018; Yang et al., 2021a,b,c). Some biochars were made up of industrial or agricultural waste, but the essence is still from plants like pine wood sawdust (Zhuang et al., 2020; Manyuchi et al., 2021; Wijeyawardana et al., 2022), poultry litter and wood generated in the construction sector (Rodriguez et al., 2020), paper mill sludge (Yoon et al., 2017; Xu et al., 2021; Wang et al., 2021a,b), industrial waste cork (Wang et al., 2020). Beside, biochar is also derived from animals: swine manure (Rodriguez et al., 2020), cow manure (Kończyk et al., 2022), egg shells (Liu et al., 2020a,b), shells of some crustaceans: shrimp shells (Yu et al., 2021), mussel shell (Conde-Cid et al., 2019b), crab shell (Dai et al., 2018), crayfish shell (Xiao et al., 2017; Zhang et al., 2019a,b; Wang et al., 2021a,b; Sun et al., 2021a,b,c; Zhang et al., 2021), lobster (Ma et al., 2021), oyster (Lian et al., 2021). In addition, the biochar sources are also from industrially manufactured waste of both animal and plant origins, such as waste from food processing plants (Al Ketife et al., 2021) or food waste (Xing et al., 2021); or unrelated to plant or animal origin, such as tire and PVC plastic residues wastes (Rodriguez et al., 2020), waste resultant from wastewater treatment plants (Liu et al., 2020a,b; Zeng et al., 2021).

Pyrolysis process

Pyrolysis is a process for converting biomass into AC in the absence of oxygen and at a specific temperature and time. The sources of raw materials will determine the chemical composition and the structure of biochar. The heating rate and pyrolysis time are crucial factors that strongly influence biochar's characteristics and applicability. The fast pyrolysis process takes place at a high temperature (to 800°C or 900°C) and heating rate, with the residence time being shorter (<2 seconds), and the yield of biochar (about 12%) being small. Fast pyrolysis is the method that is used to produce bio-oil (Sun et al., 2021a,b,c). The vapor does not escape rapidly but is incubated in the solid system for about 5–30 minutes before escaping (Mohan et al., 2014), limiting the liquid formation, increasing the solid formation, and increasing the yield of biochar with large particle sizes (Sun et al., 2021a,b,c). For some other biochar sources from agricultural and industrial wastes: the chemical composition of tire rubber consists of 15.5% acetone extracts, 6.0%, carbon black 29,5%, and rubber hydrocarbon 49% (Cao, 2007); PVC plastic molecule is $(C_2H_3Cl)_n$ and a small number of additives, it means that carbon is about 38% weight; food waste contains protein-rich organic substances (Al Ketife et al., 2021), of which cereal starch accounts for about 50% (Xing et al., 2021). The sludge collected and recycled from a piggery (Guangdong Province, China) contains organic substances rich in C and N elements (Liu et al., 2020a,b). So the pyrolysis process was usually carried out at higher temperatures (up to 700°C–850°C) and longer time (2–8 hours) (Rodriguez et al., 2020; Yang et al., 2021a,b,c; Sun et al., 2021a,b,c; Zhang et al., 2021).

Properties of biochar

The chemical composition of biochar

Biochars' chemical component depends on biomass and pyrolysis temperature. Their main component is carbon, 26.4 wt% (from sawdust) (Manyuchi et al., 2021), 38.4 wt% (from rice straw) (Xu et al., 2022), 46.4% (from spent mushroom substrates) 56.3% (from cow manue) (Kończyk et al., 2022), 51−57 wt% (from corn cob) (Foong et al., 2020), 54%−55.4% (from the root of rose), 62.73% (from energetic tobacco seed waste) (Onorevoli et al., 2018), 63.5%−70.5% (from the stem of eucalyptus) (Khare et al., 2017), 70.4% (from wet distiller grains) (Kończyk et al., 2022), 72.68 wt% (from matamba fruit shell) (Obey et al., 2022), 68.82%−81.23% (from coffee grounds) (Zhang et al., 2020a,b), 64.08%−84.25% (from pine tree residure) (Park et al., 2019), 71.87%−83.27% (from pine fruit shell) (Mohammed et al., 2018), 69.7%−86.4% (from shrimp shell) (Salawu et al., 2022), commonly is O, H and N. As the pyrolysis temperature increase, the C content increased while the H, O, and N contents declined, and the H/C, O/C, and N/C ratios decreased. Besides, plant-based biochar also contains a small number of minerals, such as K, Mg, and Ca (from pine trees) (Park et al., 2019); 0.97 wt% K, 0.46 wt% Mg, 0.37 wt% Ca, 0.31 wt% P and some trace compounds (Si, S) (from matamba fruit shell) (Obey et al., 2022), or elements as P, K, Ca, Mg, S, Fe, Zn, Mn, Na, Cu, Cr Ni, As, Cd, Pb, and Se (Onorevoli et al., 2018), Ca, K, Si, Mg, Al, Fe, Na, S, P, N, Cl (Kończyk et al., 2022), K, Na, Ca, Mg, P, S, N, Si, Cl (Yang et al., 2021a,b,c). N, P, Ca, and Mg content increased with temperature in the form of SiO_2 and $CaNaPO_4$ (Yoon et al., 2021). The biochar surface usually has some functional groups: −OH aliphatic, $-CH_3$ and $-CH_2-$, C−Halkyl groups, γ-CH in aromatic rings, Aromatic C = C, C = C in alkenes, Ketone C = O, Carbonyl C = O, C-O phenolic, C-O alcohol, ester C-O-C (Uzun et al., 2010; Khare et al., 2017; Zazycki et al., 2018; Onorevoli et al., 2018; Zhang et al., 2019a,b; Li et al., 2019a,b; Park et al., 2019; Georgieva et al., 2020; Zhang et al., 2020a,b; Wang et al., 2020; Xing et al., 2021; Liu et al., 2021a,b; Manyuchi et al., 2021; Binh et al., 2022; Xu et al., 2022). As the burning temperature increases, the content of these functional groups decreases (Binh et al., 2022). It is also possible to appear bond/inorganic mineral components as C = O/C-O in carbonates, O-P-O in phosphates (Jia et al., 2021); −NH; −C-N; P-C-O aromatic; P = OOH (Jabar et al., 2022); Si-O-Si, Si-H bond in phenols; benzene ring; sylvite (KCl) and calcite ($CaCO_3$) (Cuong et al., 2019; Dai et al., 2021; Xu et al., 2022); $CaCO_3$ in calcite and aragonite forms (Xiao et al., 2017), O = S = O in sulfate, C = S and S = O in sulfoxide (Huang et al., 2019); the υ(P-O) of the tetrahedral PO_4^{3-}, group υ(C-O) of CO_3^{2-} (755 cm^{-1}) and Si-O-Si in $MgSO_4$, $Ca_3(PO_4)_2$, SiO_2, $K_2MgSi_5O_{12}$, and $Ca_3Mg_2(PO_4)_4$ (Yang et al., 2021a,b,c). The signature of these bonds usually increases with the pyrolysis temperature. These inorganic forms are favorable for ion exchange adsorption with heavy metals or forming bonds with organic functional groups and inorganic anion (as PO_4^{3-}), enhancing the adsorption capacity of organic substances in water, making the adsorption capacity of biochar becomes more abundant and efficient.

The surface morphology

The biochar synthesized from biomass of plant, animal, or industrial origin consists of many pores on the surface. The average pore size changes depending on the biomass's nature, pyrolysis pretreatment process, post-pyrolysis treatment, especially pyrolysis time and temperature, resulting in different pore volumes and specific surface area. The raw biomass surface sample usually does not have pores or low porosity. Most biochars had porous structure: very few biochars had microporous (pore diameter <2 nm) (Wang et al., 2020; Jabar et al., 2022); most biochars had mesoporous structure (pore diameter: 2−50 nm) (Sahoo et al., 2021; Zhang et al., 2019a,b; Li et al., 2018a,b; Manyuchi et al., 2021; Lian et al., 2021; Dai et al., 2021; Ma et al., 2018; Xu et al., 2022; Khairiah et al., 2021; Yang et al., 2021a,b,c). Few biochars had macroporous, the pore diameter >50 nm (Puga et al., 2022); some biochar had both micropores (13.69 ÷ 28.57% of pore in the pore sizes <2 nm) and narrow mesopores (47.25 ÷ 68.18% of pore in 2−20 nm) (Binh et al., 2022); or both mesoporous and macroporous (Liu et al., 2021a,b) (Table 13.1).

The pore diameter and density determine the material's specific surface areas. There are also some biochars with a small surface area is small (< 8 m^2/g) (Xu et al., 2022; Cuong Nguyen et al., 2021; Dai et al., 2021; Lian et al., 2021; Li et al., 2018a,b; Choi et al., 2020; Amin and Chetpattananondh, 2019; Yang et al., 2021a,b,c). They are microalgal biochar and some biochar which pyrolysis at 250°C−400°C less than 1 hour, while most biochar which pyrolysis at 450°C−700°C had a large surface area, from several ten to several hundred m^2/g (Park et al., 2019; Binh et al., 2022; Tulun et al., 2021; Wang et al., 2020; Sahoo et al., 2021; Zhang et al., 2019a,b; Amin and Chetpattananondh, 2019; Puga et al., 2022; Dai et al., 2021; Khairiah et al., 2021; Puga et al., 2022; Amin and Chetpattananondh, 2019; Mandal et al., 2017; Manyuchi et al., 2021; Cuong Nguyen et al., 2021; Puga et al., 2022; Dawood et al., 2017; Georgieva et al., 2020; Ma et al., 2018), even is 816.00 m^2/g of African almond leaves biochar (Jabar et al., 2022). The surface

TABLE 13.1 The pore diameter and the specific surface areas of other biochars.

I.	Pore diameter (nm)	Biochar	References	II.	Specific surface areas (m^2/g)	Biochar	References
1	<2 nm	African almond leaves	Wang et al. (2020), Jabar et al. (2022)	1	0.31 ÷ 2.63	*Spirulina* sp.	Choi et al. (2020)
2	2.54	Vetiver grass	Dai et al. (2021)	2	1.15 ÷ 1.73	Rice straw	Xu et al. (2022)
3	2.37 ÷ 5.66	Pigeon pea stalk and bamboo	Sahoo et al. (2021)	3	1.3	*S. japonica*	Amin and Chetpattananondh (2019)
4	3,98	Crayfish shell	Zhang et al. (2019b)	4	2.62	*Coffea arabica*	Cuong Nguyen et al. (2021)
5	4.44	Used tea leaves	Li et al. (2018a)	5	4.60 ÷ 5.19	Eucalyptus leaves and rice straw	Dai et al. (2021)
6	4.60	Eucalyptus leaves	Dai et al. (2021)	6	2.72 ÷ 6.03	Oyster shells	Lian et al. (2021)
7	9.3	Sawdust	Manyuchi et al. (2021)	7	6.46 ÷ 7.89	*Spirulina* sp.	Yang et al. (2021a,b,c)
8	9.10 ÷ 11.06	Oyster shells	Lian et al. (2021)	8	31	Crayfish shell	Zhang et al. (2019a, b)
9	15.56	Rice straw	Dai et al. (2021)	9	30.038 ÷ 34.2	*Eucheuma* sp.	Amin and Chetpattananondh (2019)
10	2.6–19.2	Fargesis leaves	Ma et al. (2018)	10	34.99 and 38.99	The root of rose and stem of eucalyptus	Khare et al. (2017)
11	32.63÷40.61	Rice straw	Xu et al. (2022)	11	8.4 ÷ 255.8	Pine cones, pine needles, pine bark	Park et al. (2019)
12	36.023	Banana peel	Khairiah et al. (2021)	12	16.9 ÷ 261.78	Pigeon pea stalk	Sahoo et al. (2021)
13	37.48 ÷ 46.41	*Chlorella* sp.	Yang et al. (2021a,b,c)	13	6.10 ÷ 350.22	Corn cob	
14	34.73 ÷ 53.68	*Spirulina* sp.	Yang et al. (2021a,b,c)	14	12.9 ÷ 353.3	Walnut shell	Tulun et al. (2021)
15	53.96	Eucalyptus	Puga et al. (2022)	15	31.8 ÷ 392.5	Waste cork	Wang et al. (2020)
16	59.53	Grapevine cane	Puga et al. (2022)	16	63.52 ÷ 307.10	Bamboo	Sahoo et al. (2021)
17	72.63	Holm tree	Puga et al. (2022)	17	62	Grapevine cane	Puga et al. (2022)
18	9.74 ÷ 92.69	Shrimp shell	Liu et al. (2021a,b)	18	92.88	Vetiver grass	Ma and Li (2020a)
19	17.47 ÷114.34	Corn straw	Liu et al. (2021a,b)	19	179.668	Banana peel	Khairiah et al. (2021)
20	6.89 ÷ 80.68	75% shrimp shell + 25% corn straw	Liu et al. (2021a,b)	20	207	Holm tree	Puga et al. (2022)
				21	266	*Chlorella* sp.	Amin and Chetpattananondh (2019)
				22	300	Sawdust	Manyuchi et al. (2021)
				23	335	Eucalyptus, pine cone	Puga et al. (2022), Ma et al. (2020a)
				24	425	Walnut shell	Georgieva et al. (2020)
				25	461.83	Fargesis leaves	Ma et al. (2018)

area decreases slightly when the temperature and time increase beyond a limit. The pore diameter also depends on the nature of the source material, which increases with pyrolysis temperature and time (Lian et al., 2021; Sahoo et al., 2021; Yang et al., 2021a,b,c; Xu et al., 2022), or vice versa (Liu et al., 2021a,b). Some biochar has a thin layer structure: wet distiller grains biochar (Kończyk et al., 2022); crab shell biochar (Dai et al., 2018). Some biochar remained in the original cell structure of feedstock. Some small pores also collapsed to form large pores that increased with pyrolysis temperature and time (Wang et al., 2020). Some biochars had honeycomb, porous and smoother surface structures with irregular hollow-rounded-inward concaves, but the smaller pores were not open (Ma et al., 2018; Sahoo et al., 2021; Obey et al., 2022).

The pH and the zeta potential of biochar

Almost no biochar was strongly acid; only a few biochar were weakly acid near neutral (6.66 for *Mimosa pigra* biochar) (Cuong Nguyen et al., 2021). Most biochars were alkaline: 7.59–10.57 of rice straw biochar (Xu et al., 2022), 7.25–10.07 of bamboo biochar, 7.90–10.14 of pigeon pea stalk biochar (Sahoo et al., 2021), 7.15–7.84 of pine cones biochar, 8.10–9.84 of pine needles biochar, and 7.21–8.92 of pine bark biochar (Park et al., 2019), 8.92 for *Acacia auriculiformis* biochar, 10.09 for *Coffea arabica* biochar (Cuong Nguyen et al., 2021), 8.59 Rice straw biochar, 8.68 eucalyptus leaves biochar, and 8.42 vetiver grass biochar (Dai et al., 2021), 8.82–10.1 of bamboo chips, corn cob, eucalyptus bark, rice husk and rice straw biochar (Mandal et al., 2017), 9.4–10.3 of lobster shell biochar (Ma et al., 2021); some biochars even were strong alkaline as biochar of crab shell, crayfish shell biochar, oyster shells biochar: pH = 10–14 (Lian et al., 2021), 10.36–10.82 (elephant grass biochar) (Adesemuyi et al., 2020). Previous studies show that the zeta potential of biochar is in the weak acid to the weak base: 5.48 of *A. auriculiformis* biochar, 6.02 of *M. pigra* biochar, 8.82 of *C. arabica* biochar (Cuong Nguyen et al., 2021), 6.7 of frocab shell biochar (Dai et al., 2018), 8.35 (from walnut shell biochar), 7.6 ÷ 8.1 of sawdust biochar and 8.1 ÷ 8.75 of paddy husk biochar (as pyrolysis temperature increases) (Wijeyawardana et al., 2022), 6.5–8.3 (*Spirulina* sp. Biochar) (Choi et al., 2020), 8.2 of pine tree biochar (Naghipour et al., 2018), 8.5 pine cone biochar (Dawood et al., 2017). Depending on the surface charge, the adsorption capacity of biochar favors different types of pollutants.

Modified pyrolysis

All biochar has a porous structure, and the main component is amorphous carbon, which usually contains O, H, and also N, S, and P as well as a small amount of some metal elements. Some biochar had a pretty high adsorption capacity. Still, many other biochar have limitations: poor hydrophilicity, reusability is not high, limited specific surface area, low adsorption capacity of active sites, some plant-based biochar generally shows low yields, some biochar derived from anthropogenic biomass (such as sewage sludge and poultry manure) is the abundant heavy metals, and low C contents lead to small specific surface area and the risk of metal leaching into the environment. Therefore, the biochar activation is necessary, and have been done by many different methods.

Physical methods

Physical methods are biochar activation by physical forces, usually in two ways: (1) blowing a stream of gas (carbon dioxide, steam, or a combination of them) through the biochar at high temperature and pressure; (2) pulverizing biochar at room temperature with the high-efficiency mill. The physical action only drastically destroys the biochar structure, decreases the size of solid particles increases specific surface area, expands nano-pore distribution, and increases mesopore volume but does not change the chemical properties of biochar. Still, the fragmentation of solid particles often increases the density of functional groups on the surface of biochar. The steam (or CO_2) activation process is usually done following the pyrolysis stage. After pyrolysis of the feedstock at 300°C–700°C for 1–2 hours, the system is raised to 800°C–900°C with the steam pass for 0.5–3 hours. The biochar structure changes, but at the same time, some unwanted reactions may occur, such as:

$$C + H_2O \rightarrow CO + H_2$$
$$CO + H_2O \rightarrow CO_2 + H_2$$
$$CO_2 + C \rightarrow 2CO$$

With pulverizing biochar, the process can be done using various equipment, such as the planetary ball mill, the size of solid particles was down to about 1 nm by mechanical force, the specific surface area was lager, and the adsorption capacity was higher. However, with the small size (and increase in the density of the functional group), the solubility of biochar increases, so it becomes tough to separate them from the water after adsorption; therefore, this method is rarely used.

Chemical methods

The chemical method means that the activation of biochar by chemicals leads to a significant change in the components, structure, and properties of biochar. The modified biochar process is mainly classified into three categories: (1) modifying the feedstock; (2) modifying the prepared biochar; (3) combining modified feedstock and prepared biochar.

Modified the feedstock

The feedstocks are first impregnated with one/some chemical activators at a specific temperature and time. Second, the impregnated materials (were washed or not) were dried and pyrolyzed. The activator acts as a dehydrating agent and/or oxidant to remove moisture, reduce the pyrolysis temperature, enhance the formation of porous structures, or create another chemical component attached to the biochar. The activators are usually metals, salts, oxide/hydroxide (alkaline, alkaline earth), acids, and transition metal oxide (usually magnetic, such as Fe_2O_3, Fe_3O_4, and MFe_2O_4). The impregnation with a chemical one step such as KOH (AL-Othman et al., 2012; Mokrzycki et al., 2020); NaOH (Zhang and Leiviskä, 2020); $Ca(OH)_2$, Fe powder, $FeCl_3$ (Agra et al., 2014; Ignácio et al., 2018); H_3PO_4 (Momčilović et al., 2011; Abatal et al., 2020; Jia et al., 2021); $CaCl_2$ (Aguayo-Villarreal et al., 2016); $ZnCl_2$ (Bello et al., 2021); montmorillonite (Fu et al., 2020); the mixture of CaO and $ZnCl^+$ (Machado et al., 2020), then dried and heated. For example, bamboo wood was boiled in dilute ammonia, washed, dried, and then impregnated with ferric solution (3 times) after pyrolysis (form nano-α-Fe_2O_3/Fe_3O_4 on bamboo biochar) (Zhu et al., 2018). Kans grass biomass materials were suspended in $FeSO_4$ and $FeCl_3$ solutions, and then a certain amount of $NH_3.H_2O$ was added into the system at 30°C for 2 hours, filtered, washed with DI water, dried and copyrolysis (Baig et al., 2014).

Modified the prepared biochar

Activate biochar with an activator to change the structure of the synthesized biochar. The prepared biochar was impregnated with an activator solution and then washed and dried. Activators commonly used as alkali: NaOH or KOH (Laksaci et al., 2017; Cai et al., 2018; Cuong et al., 2019; Ahmad et al., 2020; Birer et al., 2021; Mokrzycki et al., 2020); acid: H_2SO_4 (Cataldo et al., 2018), H_3PO_4 (Ma et al., 2020a); strong oxidizing agent: H_2O_2 (Ahmed et al., 2021), HNO_3 (Song, 2019; Anastopoulos et al., 2020; Lin et al., 2020); methanol (Mokrzycki et al., 2020); or some activators as biochar was mixed with $KMnO_4$ solution, dried and mixed with KOH before to calcined (An et al., 2021). Activated by attaching another component to biochar, prepared biochar was impregnated in a solution (or performed a reaction process), dried, and then heated. The process can be done in one stage, such as impregnating biochar in that chemical solution: $Fe(NO_3)_3$ (Godlewska et al., 2020; Altaf et al., 2021); $CuCl_2$ or $FeCl_3$ (Tomin et al., 2021); $(NH_4Fe)_2(SO_4)_3$ (Shirvanimoghaddam et al., 2021); $ZnCl_2$ (Yang et al., 2021a,b,c); $AlCl_3$ (Cuc et al., 2022); melamine, urea, ammonium chloride, ammonium nitrate (Kasera et al., 2021); natural attapulgite (Li et al., 2016); $ZnCl_2 + FeCl_3$ (Ma et al., 2020a); $KMnO_4 + Fe(NO_3)_3 + LaCl_3$ (Song, 2019); adding biochar into a solution of H_2O_2 and KSCN (1 mol) and was shaken at 25°C for 24 hours (Macedo et al., 2020); tetradecyl-*N*, N-trimethyl ammonium bromide (Mahmoud et al., 2016). For example: attach metal to biochar by putting biochar in the reaction solution to form that metal: biochar was added to a $FeSO_4$ solution, then an excess $NaBH_4$ was added to the mixture, stirred for 30 minutes, finally, the suspension was filtered, washed, and dried, the product is zero-valent iron biochar (Devi and Saroha, 2014; Ji et al., 2022); the cobalt and iron mixture solution (molar ratio 1:2) in deionized water were poured slowly into the solution of Banana pseudostem biochar dispersed in deionized water; next, this mixture was added dropwise by hot 1 M NaOH at 70°C for 45 minutes under magnetic stirrer, finally, centrifuged, washed precipitate to neutralization, dried, obtained $CoFe_2O_4$-biochar hybrid nanocomposites (Chakhtouna et al., 2021); first, synthesize biochar of Pineapple peel; secondly, biochar magnetization step: carried out the synthesis of Fe_2O_3 in the solution in the presence of Pineapple peel biochar; thirdly, carrying out the synthesis of $La(OH)_3$ in the solution in the presence of magnetic biochar (Li et al., 2018a,b). The maize straw biochar and ZnO (50:50 ratio by weight) were mixed, then transferred to an agate jar with agate balls (1:15 ratio by weight) in a planetary ball mill, milled for 72 hours; biochar and ZnO soaked in ethanol under sonication and stirred (Hu et al., 2019).

Combine modified the feedstock and modified biochar

Some studies have used a combination of modified feedstock and modified prepared biochar. For example, Lignin (5 g) was suspended in 50 mL 20 wt/wt% H_3PO_4 solution for 12 hours before drying at 105°C, then was pyrolysized at 400°C for 1 hour, after that was washed by 100 mL HCl (1.0 M) deionized water till to neutral pH, dried, was added to $MnCl_2$ solution (0.002 M) vigorous stirring (600 rpm) for 6 hours and droped add $KMnO_4$ sollution until the suspension turned to dark brown; or prepared biochar was added to $KMnO_4$ or $MnSO_4$ solution; finaly, the suspension was filtered, washed and dried (Liu et al., 2021a,b); corn straw powder was impregnated with $MgCl_2$ at 105°C for 23 hours, the pyrolysized at 400°C for 3 hours, prepared biochar the modified by chitosan (Li et al., 2022); Biochar of wheat straw was magnetized by coprecipitation method, and then forming composite with bimetallic organic framework (Fe and Mg with Terephthalat) MBC@Fe/Mg-MIL-88B (Lu et al., 2022); lotus stem were immersed into 2% NaOH solution stirred for 24 hours then washed with deionized water, dried, soaked in 10% $(NH_4)_2HPO_4$ solution, oxidized, pyrolysized at 500°C for 1 hour, then biochar powder was soaked in $FeCl_3$ solution, the dried mixture was heated to 700°C for 1.5 hours, cooled, washed (Feng et al., 2018); Peanuts shell were mixed with H_3PO_4 (88%) before pyrolysis, and biochar was soaked overnight in 1% solution of $NaHCO_3$, finally washed and dried (Garg et al., 2019).

Application of biochar in adsorption for water treatment

Cation heavy metal removal

Heavy metal is generally defined as metals and metalloids with relatively high density, atomic weights (or atomic numbers) (the atomic number is usually greater than 20 and atomic density above 4 g/cm^3), and are toxic from at ppb levels. They include Cu, Pb, Zn, Cd, Hg, Fe, Co, Cr, Ni, Pd, Ag, Pt, Sn and Mn. The rich functional groups on the surface as hydroxyl, carbonyl, carboxyl, amino, π bond, and aromatic ring play an essential role in its ability to adsorb heavy metals since they are rich in electrons, often negatively charged, strengthening the electrostatic attraction between biochar surface and heavy metal ions, increasing to form stable complexes between them, or providing electron for the reduction of high valence metals as As(V) and Cr (VI). Therefore, the number of studies using biochar as a metal adsorbent is quite a lot. Some results are listed in Table 13.2.

Published studies reported that many types of biochar and modified biochar could individually or simultaneously adsorb many different metals with high efficiency: Lobster shell biochar adsorbed Cu (II) by ion exchange mechanism and Cd (II) followed cation exchange and interaction with functional groups mechanism (Ma et al., 2021); Cu (II), Pb (II), and Zn (II) from stormwater were adsorbed by sawdust and paddy husk biochar, the maximum adsorption capacities were 10.27, 6.48, and 17.57 mg/g respectively (Wijeyawardana et al., 2022); Cd (II), Cr (III), Cu (II), and Pb (II)

TABLE 13.2 Biochar adsorbed heavy metals and mechanism.

Heavy metal	Biochar	Efficiency (capacity)	Mechanism	References
Cu	Tobacco stem	428 mg/g	Surface complexation	Zhou et al. (2018)
	Agro-waste	195.1 mg/g	Precipitation, reduction, and complexation	Yang et al. (2018)
	Paper mill sludge	40.92%		Xu et al. (2021)
	Peanut shells	0.793 mmol/g		Wilson et al. (2006)
	Crayfish shell	715.10 mg/g	Precipitation as carbonates and oxides	Wang et al. (2021a,b)
Pb	Tobacco stem	399 mg/g	Precipitation	Zhou et al. (2018)
	Agro-waste	161.9	Precipitation reduction and complexation	Yang et al. (2018)
	Food waste		Electrostatic adsorption and complexation precipitation	Xing et al. (2021)
	Crayfish shell	190.7 mg/g		Xiao et al. (2017)
	Peanut shells	0.94 mmol/g	Ion exchange complexation	Wilson et al. (2006)
	Crayfish shell	1166.44 mg/g		Sun et al. (2021a,b,c)
	Crayfish shell	1682.87 mg/g	Precipitation and C-π interaction	Wang et al. (2021a,b)
			Precipitation as carbonates and oxides	

(Continued)

TABLE 13.2 (Continued)

Heavy metal	Biochar	Efficiency (capacity)	Mechanism	References
Cd	Tobacco stem	472 mg/g	Cation-π interaction	Zhou et al. (2018)
	Paper mill sludge	41.6 mg/g		Yoon et al. (2017)
	Peanut shells	0.433 mmol/g	Electrostatic adsorption and complexation	Wilson et al. (2006)
	Peanut shells	55.43 mg/g		Maria and Bezerra (2018)
		47.39 mg/g		Tao et al. (2021)
	Corn stalks	217.41 mg/g	Complexation electrostatic attraction, ion exchange and surface complexation	Shen et al. (2017)
	Chlorella sp.			
	Agro-waste	24.76		Saravanan et al. (2021)
Cr (III/VI)	Sludge	291.54 mg/g	Chemical adsorption of monolayer and supplemented by electrostatic attraction	Zeng et al. (2021)
Cr (III/VI)	$FeCl_3$ and $ZnCl_2$-modified corn stalks	138.89 mg/g	Electrostatic, coprecipitation and ion exchange	Yu et al. (2019)
		58.54 mg/g	Electrostatic mode	Thangagiri et al. (2022)
	Azadirachta indica leaves	350.42 mg/g	Electrostatic adsorption (ion-pair), pore flling, reduction and coordination	Mahmoud et al. (2021a, b)
	Modified magnetic shrimp shell			
	Polyaniline modifed date seed	12.58 mg/g	Intra-particle diffusion	Tripathy et al. (2021)
As(V)	Paper mill sludge	22.8 mg/g	Complexation	Yoon et al. (2017)
		94.99 mg/g		Xu et al. (2021)
	Paper mill sludge	766 and 922 μg/g	Hybrid adsorption	Zoroufchi et al. (2021)
As(III)	Electrochemically modified Canola straw	20 and 50 mg/g	Adsorption and oxidation	Ignácio et al. (2018)
	Fe-coated biochar	15.34 mg/g		Song (2019)
Zn	Agro-waste	109.7	Precipitation, reduction and complexation	Yang et al. (2018)
	Paper mill sludge	87.93%		Xu et al. (2021)
	Peanut shells	0.870 mmol/g		Wilson et al. (2006)
	Tea waste	8.9 mg/h		Wasewar et al. (2009)
Ni	Peanut shells	0.449 mmol/g		Wilson et al. (2006)
Hg (II)	Modified bamboo	99%	Chemisorption	Sun et al. (2021a,b,c)
	Sulfurized corn stalk	117.7 and 268.5 mg/g	Chemisorption	Huang et al. (2019)
Sm (III)	*Cynara scolymus*	650 μmol/g	Complexation, ion exchange	Mahmoud et al. (2021a)
La (III)	Modified *Sargassum fusiforme*	362.32 mg/g		Wang et al. (2016)
V (V)	Wet distiller grains	1.61 mg/g		Kończyk et al. (2022)
Sb	Zero-valent iron rice straw biochar	98.25 mg/g	Sb combined with the surface functional groups adsorbent form antimonite/antimonate; Sb(III)/Sb(V) precipitated with Fe(II)/Fe(III)	Ji et al. (2022)

were adsorbed by root of rose and eucalyptus stem biochar $q_{e,max}$ up to 60 mg/g follows second order kinetics, first surface adsorption, then intra-particle diffusion, follows hydrogen exchange mechanism (Khare et al., 2017); Mn, Pb, Zn, Fe were adsorbed by banana peel biochar (Khairiah et al., 2021); peanut shell biochar adsorbed Pb (II) > Cu (II) > Cd (II) > Ni (II) > Zn (II) (Shan et al., 2020); Cr (III), Fe (III), Zn (II), Ni (II), Pb (II), Mn (II), and Cu (II) were adsorbed by sawdust biochar (Manyuchi et al., 2021); multi-element wastewater contained Cd, Pb, As, Cr, Cu, Fe, Ni, Se, Zn, Mn, Li, Be, Co, Mo were adsorbsed by copper-modifed activated bamboo biocharreached equilibrium within 3 hours (Thotagamuge et al., 2021).

Anion inorganic removal

Biochar not only adsorbs metal cation well but also adsorbs many inorganic anions, such as phosphate (Zhu et al., 2018; Yang et al., 2021a,b,c; Kang et al., 2021; Wang et al., 2021a,b; Jiang et al., 2022; Shan et al., 2022); nitrat (Adesemuyi et al., 2020); uranium (VI) ($UO_2(NO_3)_2$) (Ahmed et al., 2021); Cr (VI) in $K_2Cr_2O_7$ (AL-Othman et al., 2012; Feng et al., 2018; Jia et al., 2021); Se (VI) (Godlewska et al., 2020); mixture of chromium Cr (VI) and nitrate (Herath et al., 2021); mixture of nitrate and phosphate (Cuc et al., 2022); mixture of ammonium and phosphate (Li et al., 2022); mixture of cation heavy metals (Ni (II), Co (II) and nutrients (NH_4^+ and PO_4^{3-}) (Shirvanimoghaddam et al., 2021); As (V) in Na_2HAsO_4 in mixture with other cation heavy metals (Yoon et al., 2017; Ignácio et al., 2018; Xu et al., 2021; Zoroufchi et al., 2021); Removal organic/inorganic arsenic from water by porous biochar-supported $MnFe_2O_4$ magnetic (Wen et al., 2021) or adsorbed arsenate and fluoride by Al-and/or Mg-oxides to magnetic biochars (Shen et al., 2021). Most adsorbents are metal-modified biochar: Fe, Al, Ca, etc. provide the adsorbents with effective adsorption sites following chemical coprecipitation.

Organic contaminants

Due to the increasing amount and type of waste organics released into the environment, their toxic levels are becoming highly dangerous. Some studies using biochar as an adsorbent to treat organic pollutants are presented in Table 13.3.

TABLE 13.3 Organic pollutants adsorption and mechanism.

organic pollutants	Biochar	Efficiency	Mechanism	References
Atrazine	Modified corn cob	11.31÷19.58 mg/g	H-bonding interaction	Binh et al. (2022)
Formaldehyde	Acidmine drainage sludge and spent coffee waste	1008–1811 mg/g	Physicochemical properties	Ahn et al. (2021)
Phenol	Pine fruit shells	26.738 mg/g	Physisorption	Oginni and Singh (2021)
Phenol (Phen), 4-chlorophenol (4-ClPhen) and 4-nitrophenol (4-NPhen)	Bark of H. campechianum (HC)	84.8 ÷ 94.09 mg/g		Abatal et al. (2020)
p-nitrophenols	Microalgal biochar	240.8 mg/g	Highly polarized microalgal biochar	Zheng et al. (2017)
Phenolic	H_3PO_4 modified *Acacia mangium* bark	96.92 mg/g	Intra-particle diffusion	Zhang et al. (2020a,b)
2-chlorophenol	Chemically and physically activated biochars of barley malt bagasse	150 mg/g, remain for 5 cycles	Physisorption	Machado et al. (2020)
Bisphenol A	Sewage sludge biochar (SSB) and KOH or KOH/MeOH modified SSB	88.1%	Hydrogen 514 bonding and π-π interaction	Birer et al. (2021)
Aniline	Magnetic modified biochar	75.7 mg/g	Physicochemical properties	Prochazkova et al. (2022)
Caffeine	White pine and Norway spruce needles	9.27 and 11.85 mg/g		Oginni and Singh (2021)
	Pine needles	6.54 mg/g	Electrostatic interactions and the formation of outer-sphere complexes	Anastopoulos et al. (2020)
Benzo[a]pyrene	Sesame straw biochar modified and $ZnCl_2$-modified biochar	92.67%÷96.13%	Porous structure and the π-π conjugation interaction	Yang et al. (2021a,b,c)
Phenanthrene	HNO_3 oxidation Grass and manure biochar		Pore-filling and π-π electron donor-acceptor interactions	Jin et al. (2017)
Crystal violet dye	Mango leaves	99.85% at 2500 mg/L	Interaction mechanism between cationic dye molecules and functional groups on adsorbents	Vyavahare et al. (2018)

(Continued)

TABLE 13.3 (Continued)

organic pollutants	Biochar	Efficiency	Mechanism	References
Methylene blue dye	Anaerobic digestion residue, palm bark, and eucalyptus	99.5, 99.3, and 86.1% at 5 mg/L MB.	Chemical and physical adsorption	Sun et al. (2013)
	African almond	98% from 50 mg/L MB		Jabar et al. (2022)
	Sawdust	27.47 ÷ 78.13 mg/g	Physisorption and chemisorption Process physisorption	Nath et al. (2021)
	Peanut shell	555.6 mg/g		Ahmad et al. (2020)
Crystal violet	Lemongrass leaf	36.10 mg/g	Electrostatic interaction (Hydrogen bonding, ion-dipole force)	Putri et al. (2020)
Malachite green and congo red	Crab shell (rich in calcium)	12,502 and 20,317 mg/g	Electrostatic attraction, hydrogen bonding and π-π interaction	Dai et al. (2018)
Methyl orange dye	Wattle bark, mimosa, and coffee husks	12.3 mg/g	Electrostatic adsorption and pore diffusion	Cuong Nguyen et al. (2021)
Methyl violet dye	Chitin	1000 mg/g	Physical adsorption	Zazycki et al. (2019)
Methyl red dye	Lemongrass leaf	63.87% at 25 mg/L MR	Diffusion rate, electrostatic attraction	Ahmad et al. (2019)
Basic Violet 10 and direct blue	pine bark	112.6 and 127.1 mg/g		Al-Zawahreh et al. (2022)
Congo red	NaOH modified bamboo	899.44 mg/g	Electrostatic interaction and hydrogen bonding	Wu et al. (2021)
Basic blue 41	Co-pyrolysis of Saccharina japonica and goethite	1494 mg/g	Ion exchanges, hydrogen bonding, π-π-interaction and pore-filling	Damertey et al. (2021)
Acid yellow 36	Peanut shells	49.275 mg/g	Chemical forces	Garg et al. (2019)
Acid fuchsin dye	Peanut-shell magnetized biocarbon	99.35%	Physio-chemical adsorption process	Renita et al. (2021)
Methylene blue and reactive blue 19	*Eichhornia crassipes* and *Phragmites australis*	39.36 ÷ 88.89 mg/g	Electrostatic interaction, ion exchange, π−π stacking interaction, *n*−π interaction, hydrogen bonding, and pore-flling	Doan et al. (2021)
Tetracycline hydrochloride	Ball-milled crayfish shell biochar	60.7 mg/g	Physisorption via pore filling, followed by chemisorption including hydroxyl complexation, H bonding, and electrostatic interaction	Zhang et al. (2021)
Tetracycline	Modified Municipal sewage sludge	286.9129 mg/g	Strong π -π stacking interaction and pore-filling	Tang et al. (2018)
	Date palm leaves	76.92 mg/g	Ionic and electrostatic interactions	Saremi et al. (2020)
	Fargesia leaves	79.8÷90.8%		Ma et al. (2018)
Sulfadiazine and sulfamethoxazole	Coffee ground	121.5 μg/g and 130.1 μg/g	π-π electron donoracceptor interactions	Zhang et al. (2020a,b)
Sulfadiazine, sulfamethazine and sulfachloropyridazine	HNO_3 modified Walnut shell	32, 46, and 40 mg/g	π-π EDA, Electrostatic interaction, Lewis acid-base interaction, hydrophobic interaction, and H-bond.	Geng et al. (2021)
Chloramphenicol	Ammonium polyphosphate-modified peanut shells biochar	423.7 mg/g	Micro-pore-filling, Van der Waals force, p-p interaction, and hydrogen-bonding interaction	Yang et al. (2020)
Naproxen	Peanut shells	105÷324 mg/g	Pore filling, π-π interaction, hydrogen bonding formations, n-π interaction, van der Waals force, and electrostatic attraction	Tomul et al. (2020)
Ciprofloxacin	Used tea leaves	238.10 mg/g	π−π interactions, hydrogen bonding, and electrostatic attraction	Li et al. (2018a,b)
	Magnetic ZnO doped biochar	449.40 mg/L	Physical adsorption and chemisorption involving intense π-π stacking interaction, electrostatic interaction, cation exchange interaction	Hu et al. (2019)

(*Continued*)

TABLE 13.3 (Continued)

organic pollutants	Biochar	Efficiency	Mechanism	References
Diclofenac	Pine tree	54.67 mg/g		Naghipour et al. (2018)
Venlafaxine Trazodone and Fluoxetine	Eucalyptus, grapevine cane, and holm tree biochar)		Electron donor/acceptor interaction	Puga et al. (2022)
Acetaminophen	$MgCl_2$, $ZnSO_4$, $ZnCl_2$, and KOH-modified tree bark biochar	91.9%, decreased 50% after 4 cycles	Pore-filling mechanisms mainly, in addition are hydrogen bonding, π-π interactions and the aromatics of biochar, n-π interaction, van der Waals interactions	dos Reis et al. (2022)
Levofloxacin	Ferromanganese biochar	~ 170 mg/g, 65% do'n to 50%	Pore filling, π - π stacking interaction, hydrogen bonding forces	Xiang et al. (2019)
Sulfamethoxazole	Sludge biochar (SBC) and multi-walled carbon nanotube (CNT) composite	2.35×10^4μg/g	Physisorption and chemisorption	Ma et al. (2020b)
Norfloxacin	Clay-biochar composite	5.24 mg/g	Electrostatic interactions	Li et al. (2016)
Pentachlorophenol	Magnetic paper mill sludgebiochar composites	79.7% after 240 min while 100% PCP removal after 480 min	Simultaneous adsorption and dechlorination mechanisms	Devi and Saroha (2014)
Amoxicillin	$CoFe_2O_4$-modifed banana pseudostembiochar	99.99 mg/g	Electrostatic interactions, stacking of π-π, and bonds of hydrogen	Chakhtouna et al. (2021)
Azole fungicides	Bimetallic organic frameworks and magnetic biochar	86.11 and 89.87 mg/g	Electrostatic interaction, hydrogen bonding, covalent bonding, and π-π stacking	Lu et al. (2022)
Cymoxanil pesticide	Grape pomace	161 mg/g	metal and hydrophilic interaction; chemical sorption and multilayer formation	Yoon et al. (2021)
Atrazine and imidacloprid	Eucalyptus bark, corn cob, bamboo chips, rice husk, rice straw, and acid-treated	atrazine (37.5÷70.7%) imidacloprid (39.9÷77.8%)	Aromaticity, polarity, pore diameter, pH, and weak acid fraction	Mandal et al. (2017)
Chlorpyrifos	Walnut shell	3.536 mg/g	Hydrophobic bonding, electrostatic attraction, van der Waals affinity, hydrogen bonding, covalent bonding, entrapment in micropores,	Tulun et al. (2021)

Conclusion and outlook

In the role of green absorbent, biochar has many advantages: safe, environmentally friendly; the source material is abundant, low cost, and easy to find; the pyrolysis-derived biochars are porous, developed specific surface, rich in surface functional groups, and easy to produce. Before the pyrolysis of produced biochar from biomass, many biochars were suitable adsorbents. Still, some other biochars are medium adsorbent due to the limited specific surface (such as microalgae microalgae-derived biochar), and all are difficult to recover and reuse. So far, most pyrolysis has been changed to co-pyrolysis by a feedstock/ biochar with one or more activators or, forming modified biochar with outstanding features, overcoming the limitations of pristine biochar, not only increasing the max adsorption capacity, expanding the type of adsorbates but also enhanced the recovery and reuse. The pyrolysis process consumes energy; therefore, it is necessary to consider research to improve the biochar field and study the use of other products effectively to reduce biochar costs. In addition, there are limited studies on the pyrolysis mechanism related to biochar quality. Most of them are pyrolysis experiments at some defined condition, then select a product with desired properties to be used as an adsorbent. For each feedstock, repeat the same test procedure. The same goes for co-pyrolysis. More systematic studies are needed, thereby building a complete theoretical basis for the pyrolysis method.

References

Abatal, M., et al., 2020. Carbonaceous material obtained from bark biomass as adsorbent of phenolic compounds from aqueous solutions. Journal of Environmental Chemical Engineering 8 (3), 103784. Available from: https://doi.org/10.1016/j.jece.2020.103784.

Adesemuyi, M.F., et al., 2020. Preparation and characterisation of biochars from elephant grass and their utilisation for aqueous nitrate removal: effect of pyrolysis temperature. Journal of Environmental Chemical Engineering 8 (6), 104507. Available from: https://doi.org/10.1016/j.jece.2020.104507.

Agra, E., Kalderis, D., Diamadopoulos, E., 2014. Ca and Fe modified biochars as adsorbents of arsenic and chromium in aqueous solutions. Journal of Environmental Management 146, 444–450. Available from: https://doi.org/10.1016/j.jenvman.2014.07.029.

Aguayo-Villarreal, I.A., et al., 2016. Effect of surface chemistry of carbons from pine sawdust for the adsorption of acid, basic and reactive dyes and their bioregeneration using Pseudomona putida. Ecological Engineering 95, 112–118. Available from: https://doi.org/10.1016/j.ecoleng.2016.06.056.

Ahmad, M.A., et al., 2019. Sorption studies of methyl red dye removal using lemon grass (*Cymbopogon citratus*). Chemical Data Collections 22, 100249. Available from: https://doi.org/10.1016/j.cdc.2019.100249.

Ahmad, M.A., et al., 2020. Adsorption of methylene blue from aqueous solution by peanut shell based activated carbon. Materials Today: Proceedings 47 (xxxx), 1246–1251. Available from: https://doi.org/10.1016/j.matpr.2021.02.789.

Ahmad, M.A., et al., 2021. Adsorptive potentials of lemongrass leaf for methylene blue dye removal. Chemical Data Collections 31, 100578. Available from: https://doi.org/10.1016/j.cdc.2020.100578.

Ahmed, W., Mehmood, S., Núñez-Delgado, A., et al., 2021. Enhanced adsorption of aqueous Pb(II) by modified biochar produced through pyrolysis of watermelon seeds. Science of the Total Environment 784, 147136. Available from: https://doi.org/10.1016/j.scitotenv.2021.147136.

Ahn, Y., et al., 2021. Efficient removal of formaldehyde using metal-biochar derived from acid mine drainage sludge and spent coffee waste. Journal of Environmental Management 298, 113468. Available from: https://doi.org/10.1016/j.jenvman.2021.113468.

Al Ketife, A.M.D., Almomani, F., Znad, H., 2021. Sustainable removal of copper from wastewater using chemically treated bio-sorbent: characterization, mechanism and process kinetics. Environmental Technology and Innovation 23, 101555. Available from: https://doi.org/10.1016/j.eti.2021.101555.

AL-Othman, Z.A., Ali, R., Naushad, M., 2012. Hexavalent chromium removal from aqueous medium by activated carbon prepared from peanut shell: adsorption kinetics, equilibrium and thermodynamic studies. Chemical Engineering Journal 184, 238–247. Available from: https://doi.org/10.1016/j.cej.2012.01.048.

Altaf, A.R., et al., 2021. One-step synthesis of renewable magnetic tea-biochar derived from waste tea leaves for the removal of Hg^0 from coal-syngas. Journal of Environmental Chemical Engineering 9 (4), 105313. Available from: https://doi.org/10.1016/j.jece.2021.105313.

Al-Zawahreh, K., et al., 2022. Competitive removal of textile dyes from solution by pine bark-compost in batch and fixed bed column experiments. Environmental Technology and Innovation 27, 102421. Available from: https://doi.org/10.1016/j.eti.2022.102421.

Amin, M., Chetpattananondh, P., 2019. Biochar from extracted marine Chlorella sp. residue for high efficiency adsorption with ultrasonication to remove Cr(VI), Zn(II) and Ni(II). Bioresource Technology 289 (April), 121578. Available from: https://doi.org/10.1016/j.biortech.2019.121578.

An, Q., et al., 2021. Ammonium removal from groundwater using peanut shell based modified biochar: mechanism analysis and column experiments. Journal of Water Process Engineering 43 (July), 102219. Available from: https://doi.org/10.1016/j.jwpe.2021.102219.

Anastopoulos, I., Katsouromalli, A., Pashalidis, I., 2020. Oxidized biochar obtained from pine needles as a novel adsorbent to remove caffeine from aqueous solutions. Journal of Molecular Liquids 304, 112661. Available from: https://doi.org/10.1016/j.molliq.2020.112661.

Antón-Herrero, R., et al., 2018. Comparative adsorption of tetracyclines on biochars and stevensite: looking for the most effective adsorbent. Applied Clay Science 160 (July), 162–172. Available from: https://doi.org/10.1016/j.clay.2017.12.023.

Aryee, A.A., et al., 2021. A review on functionalized adsorbents based on peanut husk for the sequestration of pollutants in wastewater: modification methods and adsorption study. Journal of Cleaner Production 310 (May), 127502. Available from: https://doi.org/10.1016/j.jclepro.2021.127502.

Baig, S.A., et al., 2014. Effect of synthesis methods on magnetic Kans grass biochar for enhanced As(III, V) adsorption from aqueous solutions. Biomass and Bioenergy 71 (V), 299–310. Available from: https://doi.org/10.1016/j.biombioe.2014.09.027.

Bamdad, H., Hawboldt, K., MacQuarrie, S., 2018. A review on common adsorbents for acid gases removal: focus on biochar. Renewable and Sustainable Energy Reviews 81 (May), 1705–1720. Available from: https://doi.org/10.1016/j.rser.2017.05.261.

Bello, M.O., et al., 2021. Isotherm and kinetic studies of adsorption of methylene blue using activated carbon from ackee apple pods. Chemical Data Collections 31, 100607. Available from: https://doi.org/10.1016/j.cdc.2020.100607.

Binh, Q.A., Nguyen, V.H., Kajitvichyanukul, P., 2022. Influence of pyrolysis conditions of modified corn cob bio-waste sorbents on adsorption mechanism of atrazine in contaminated water. Environmental Technology and Innovation 26, 102381. Available from: https://doi.org/10.1016/j.eti.2022.102381.

Birer, A.M., Gözmen, B., Sönmez, Ö., Kalderis, D., 2021. Evaluation of sewage sludge biochar and modified derivatives as novel SPE adsorbents for monitoring of bisphenol A. Chemosphere 268, 128866. Available from: https://doi.org/10.1016/j.chemosphere.2020.128866.

Cai, W., et al., 2018. Synthesis of peanut shell based magnetic activated carbon with excellent adsorption performance towards electroplating wastewater. Chemical Engineering Research and Design 140, 23–32. Available from: https://doi.org/10.1016/j.cherd.2018.10.008.

Cao, W., 2007. Study on properties of recycled tire rubber modified asphalt mixtures using dry process. Construction and Building Materials 21 (5), 1011–1015. Available from: https://doi.org/10.1016/j.conbuildmat.2006.02.004.

Cataldo, S., et al., 2018. Biochar from byproduct to high value added material – a new adsorbent for toxic metal ions removal from aqueous solutions. Journal of Molecular Liquids 271, 481–489. Available from: https://doi.org/10.1016/j.molliq.2018.09.009.

Çelebi, H., 2020. Recovery of detox tea wastes: usage as a lignocellulosic adsorbent in Cr_6^+ adsorption. Journal of Environmental Chemical Engineering 8 (5), 104310. Available from: https://doi.org/10.1016/j.jece.2020.104310.

Chakhtouna, H., et al., 2021. Functional $CoFe_2O_4$-modified biochar derived from banana pseudostem as an efficient adsorbent for the removal of amoxicillin from water. Separation and Purification Technology 266 (March), 118592. Available from: https://doi.org/10.1016/j.seppur.2021.118592.

Choi, Y.K., et al., 2020. Adsorption behavior of tetracycline onto Spirulina sp. (microalgae)-derived biochars produced at different temperatures. Science of the Total Environment 710, 136282. Available from: https://doi.org/10.1016/j.scitotenv.2019.136282.

Chowdhary, P., Raj, A., Bharagava, R.N., 2018. Environmental pollution and health hazards from distillery wastewater and treatment approaches to combat the environmental threats: a review. Chemosphere 194, 229–246. Available from: https://doi.org/10.1016/j.chemosphere.2017.11.163.

Cibati, A., et al., 2017. Assessment of *Miscanthus × giganteus* derived biochar as copper and zinc adsorbent: study of the effect of pyrolysis temperature, pH and hydrogen peroxide modification. Journal of Cleaner Production 162, 1285–1296. Available from: https://doi.org/10.1016/j.jclepro.2017.06.114.

Conde-Cid, M., et al., 2019a. Experimental data and model prediction of tetracycline adsorption and desorption in agricultural soils. Environmental Research 177 (February), 108607. Available from: https://doi.org/10.1016/j.envres.2019.108607.

Conde-Cid, M., et al., 2019b. Competitive adsorption/desorption of tetracycline, oxytetracycline and chlortetracycline on pine bark, oak ash and mussel shell. Journal of Environmental Management 250 (September), 109509. Available from: https://doi.org/10.1016/j.jenvman.2019.109509.

Cuc, T., et al., 2022. Case studies in chemical and environmental engineering equilibrium single and co-adsorption of nutrients from aqueous solution onto aluminium-modified biochar. Case Studies in Chemical and Environmental Engineering 5, 100181. Available from: https://doi.org/10.1016/j.cscee.2022.100181.

Cuong Nguyen, X., et al., 2021. Sustainable carbonaceous biochar adsorbents derived from agro-wastes and invasive plants for cation dye adsorption from water. Chemosphere 282 (June 2020), 131009. Available from: https://doi.org/10.1016/j.chemosphere.2021.131009.

Cuong, D.V., et al., 2019. Meso/micropore-controlled hierarchical porous carbon derived from activated biochar as a high-performance adsorbent for copper removal. Science of the Total Environment 692 (1), 844–853. Available from: https://doi.org/10.1016/j.scitotenv.2019.07.125.

Cutillas-Barreiro, L., et al., 2014. Pine bark as bio-adsorbent for Cd, Cu, Ni, Pb and Zn: batch-type and stirred flow chamber experiments. Journal of Environmental Management 144, 258–264. Available from: https://doi.org/10.1016/j.jenvman.2014.06.008.

Dai, L., et al., 2018. Calcium-rich biochar from crab shell: an unexpected super adsorbent for dye removal. Bioresource Technology 267, 510–516. Available from: https://doi.org/10.1016/j.biortech.2018.07.090.

Dai, W., et al., 2021. Characteristics and quantification of mechanisms of Cd^{2+} adsorption by biochars derived from three different plant-based biomass. Arabian Journal of Chemistry 14 (5), 103119. Available from: https://doi.org/10.1016/j.arabjc.2021.103119.

Damertey, D., Han, S., Sung, D., 2021. Biochar from the co-pyrolysis of Saccharina japonica and goethite as an adsorbent for basic blue 41 removal from aqueous solution. Science of the Total Environment 797, 149160. Available from: https://doi.org/10.1016/j.scitotenv.2021.149160.

Dawood, S., Sen, T.K., Phan, C., 2017. Synthesis and characterization of slow pyrolysis pine cone bio-char in the removal of organic and inorganic pollutants from aqueous solution by adsorption: kinetic, equilibrium, mechanism and thermodynamic. Bioresource Technology 246, 76–81. Available from: https://doi.org/10.1016/j.biortech.2017.07.019.

Devi, P., Saroha, A.K., 2014. Synthesis of the magnetic biochar composites for use as an adsorbent for the removal of pentachlorophenol from the effluent. Bioresource Technology 169, 525–531. Available from: https://doi.org/10.1016/j.biortech.2014.07.062.

Doan, V.D., et al., 2021. Comparative study on adsorption of cationic and anionic dyes by nanomagnetite supported on biochar derived from *Eichhornia crassipes* and *Phragmites australis* stems. Environmental Nanotechnology, Monitoring and Management 16 (August), 100569. Available from: https://doi.org/10.1016/j.enmm.2021.100569.

dos Reis, G.S., et al., 2022. A comparative study of chemical treatment by $MgCl_2$, $ZnSO_4$, $ZnCl_2$, and KOH on physicochemical properties and acetaminophen adsorption performance of biobased porous materials from tree bark residues. Colloids and Surfaces A: Physicochemical and Engineering Aspects 642 (February), 128626. Available from: https://doi.org/10.1016/j.colsurfa.2022.128626.

Feng, Z., et al., 2018. Mechanisms of Cr(VI) removal by $FeCl_3$-modified lotus stem-based biochar ($FeCl_3$@LS-BC) using mass-balance and functional group expressions. Colloids and Surfaces A: Physicochemical and Engineering Aspects . Available from: https://doi.org/10.1016/j.colsurfa.2018.04.054 [Preprint], (Vi).

Foong, S.Y., et al., 2020. Production of biochar for potential catalytic and energy applications via microwave vacuum pyrolysis conversion of cassava stem. Materials Science for Energy Technologies 3, 728–733. Available from: https://doi.org/10.1016/j.mset.2020.08.002.

Fu, C., et al., 2020. The single / co-adsorption characteristics and microscopic adsorption mechanism of biochar-montmorillonite composite adsorbent for pharmaceutical emerging organic contaminant atenolol and lead ions. Ecotoxicology and Environmental Safety 187 (August 2019), 109763. Available from: https://doi.org/10.1016/j.ecoenv.2019.109763.

Gan, Y.Y., et al., 2018. Torrefaction of microalgal biochar as potential coal fuel and application as bio-adsorbent. Energy Conversion and Management 165 (January), 152–162. Available from: https://doi.org/10.1016/j.enconman.2018.03.046.

Garg, D., et al., 2019. Application of waste peanut shells to form activated carbon and its utilization for the removal of Acid Yellow 36 from wastewater. Groundwater for Sustainable Development 8 (December 2018), 512–519. Available from: https://doi.org/10.1016/j.gsd.2019.01.010.

Geng, X., et al., 2021. Carboxyl-functionalized biochar derived from walnut shells with enhanced aqueous adsorption of sulfonamide antibiotics. Journal of Environmental Management 280 (August), 111749. Available from: https://doi.org/10.1016/j.jenvman.2020.111749.

Georgieva, V.G., Gonsalvesh, L., Tavlieva, M.P., 2020. Thermodynamics and kinetics of the removal of nickel (II) ions from aqueous solutions by biochar adsorbent made from agro-waste walnut shells. Journal of Molecular Liquids 312, 112788. Available from: https://doi.org/10.1016/j.molliq.2020.112788.

Godlewska, P., Bogusz, A., Dobrzyn, J., 2020. Engineered biochar modified with iron as a new adsorbent for treatment of water contaminated by selenium. Journal of Saudi Chemical Society V24 (Issue 11), 824–834. Available from: https://doi.org/10.1016/j.jscs.2020.07.006.

Hassan, M., et al., 2020. Influences of feedstock sources and pyrolysis temperature on the properties of biochar and functionality as adsorbents: a meta-analysis. Science of the Total Environment 744, 140714. Available from: https://doi.org/10.1016/j.scitotenv.2020.140714.

Herath, A., et al., 2021. Biochar-supported polyaniline hybrid for aqueous chromium and nitrate adsorption. Journal of Environmental Management 296 (July), 113186. Available from: https://doi.org/10.1016/j.jenvman.2021.113186.

Hu, Y., et al., 2019. An efficient adsorbent: simultaneous activated and magnetic ZnO doped biochar derived from camphor leaves for ciprofloxacin adsorption. Bioresource Technology 288 (April), 1–8. Available from: https://doi.org/10.1016/j.biortech.2019.121511.

Huang, Y.F., et al., 2015. Microwave pyrolysis of rice straw to produce biochar as an adsorbent for CO_2 capture. Energy 84, 75–82. Available from: https://doi.org/10.1016/j.energy.2015.02.026.

Huang, S., et al., 2019. Sulfurized biochar prepared by simplified technic with superior adsorption property towards aqueous Hg(II) and adsorption mechanisms. Materials Chemistry and Physics 238 (April). Available from: https://doi.org/10.1016/j.matchemphys.2019.121919.

Ignácio, J., et al., 2018. High efficiency removal of As(III) from waters using a new and friendly adsorbent based on sugarcane bagasse and corncob husk Fe-coated biochars. Ecotoxicology and Environmental Safety 162 (March), 616–624. Available from: https://doi.org/10.1016/j.ecoenv.2018.07.042.

Imran-Shaukat, M., Wahi, R., Ngaini, Z., 2022. The application of agricultural wastes for heavy metals adsorption: a meta-analysis of recent studies. Bioresource Technology Reports 17 (November 2021), 100902. Available from: https://doi.org/10.1016/j.biteb.2021.100902.

Jabar, J.M., et al., 2022. African almond (*Terminalia catappa* L) leaves biochar prepared through pyrolysis using H_3PO_4 as chemical activator for sequestration of methylene blue dye. Results in Engineering 14, 100385. Available from: https://doi.org/10.1016/j.rineng.2022.100385.

Ji, J., et al., 2022. Optimisation of preparation conditions and removal mechanism for trivalent antimony by biochar-supported nano zero-valent iron. Environmental Technology & Innovation 26, 102240. Available from: https://doi.org/10.1016/j.eti.2021.102240.

Jia, X., et al., 2021. Mesopore-rich badam-shell biochar for efficient adsorption of Cr(VI) from aqueous solution. Journal of Environmental Chemical Engineering 9 (4), 105634. Available from: https://doi.org/10.1016/j.jece.2021.105634.

Jiang, M., et al., 2022. Removal of phosphate by a novel activated sewage sludge biochar: equilibrium, kinetic and mechanism studies. Applications in Energy and Combustion Science 9, 100056. Available from: https://doi.org/10.1016/j.jaecs.2022.100056.

Jin, J., et al., 2017. Effects of chemical oxidation on phenanthrene sorption by grass- and manure-derived biochars. Science of the Total Environment 598, 789–796. Available from: https://doi.org/10.1016/j.scitotenv.2017.04.160.

Kang, J.K., et al., 2021. Fe-loaded biochar obtained from food waste for enhanced phosphate adsorption and its adsorption mechanism study via spectroscopic and experimental approach. Journal of Environmental Chemical Engineering 9 (4), 105751. Available from: https://doi.org/10.1016/j.jece.2021.105751.

Kasera, N., Hall, S., Kolar, P., 2021. Effect of surface modification by nitrogen-containing chemicals on morphology and surface characteristics of N-doped pine bark biochars. Journal of Environmental Chemical Engineering 9 (2), 105161. Available from: https://doi.org/10.1016/j.jece.2021.105161.

Keränen, A., et al., 2013. Preparation of novel anion exchangers from pine sawdust and bark, spruce bark, birch bark and peat for the removal of nitrate. Chemical Engineering Science 98, 59–68. Available from: https://doi.org/10.1016/j.ces.2013.05.007.

Khairiah, K., et al., 2021. Data on characterization, model, and adsorption rate of banana peel activated carbon (*Musa acuminata*) for adsorbents of various heavy metals (Mn, Pb, Zn, Fe). Data in Brief 39, 107611. Available from: https://doi.org/10.1016/j.dib.2021.107611.

Khare, P., et al., 2017. Plant refuses driven biochar: application as metal adsorbent from acidic solutions. Arabian Journal of Chemistry 10, S3054–S3063. Available from: https://doi.org/10.1016/j.arabjc.2013.11.047.

Kończyk, J., Kluziak, K., Kołodyńska, D., 2022. Adsorption of vanadium (V) ions from the aqueous solutions on different biomass-derived biochars. Journal of Environmental Management 313 (January). Available from: https://doi.org/10.1016/j.jenvman.2022.114958.

Laksaci, H., et al., 2017. Synthesis and characterization of microporous activated carbon from coffee grounds using potassium hydroxides. Journal of Cleaner Production 147, 254–262. Available from: https://doi.org/10.1016/j.jclepro.2017.01.102.

Law, X.N., et al., 2022. Microalgal-based biochar in wastewater remediation: its synthesis, characterization and applications. Environmental Research 204 (PA), 111966. Available from: https://doi.org/10.1016/j.envres.2021.111966.

Li, Y., et al., 2016. Removal of Norfloxacin from aqueous solution by clay-biochar composite prepared from potato stem and natural attapulgite. Colloids and Surfaces A: Physicochemical and Engineering Aspects . Available from: https://doi.org/10.1016/j.colsurfa.2016.11.064 [Preprint].

Li, J., et al., 2018a. Study of ciprofloxacin removal by biochar obtained from used tea leaves. Journal of Environmental Sciences (China) 73, 20–30. Available from: https://doi.org/10.1016/j.jes.2017.12.024.

Li, T., et al., 2018b. $La(OH)_3$-modified magnetic pineapple biochar as novel adsorbents for efficient phosphate removal. Bioresource Technology 263 (February), 207–213. Available from: https://doi.org/10.1016/j.biortech.2018.04.108.

Li, L., et al., 2019a. Oxygen-rich biochar from torrefaction: a versatile adsorbent for water pollution control. Bioresource Technology 294 (September), 122142. Available from: https://doi.org/10.1016/j.biortech.2019.122142.

Li, N., et al., 2019b. Mechanisms of U(VI) removal by biochar derived from *Ficus microcarpa* aerial root: a comparison between raw and modified biochar. Science of the Total Environment 697, 134115. Available from: https://doi.org/10.1016/j.scitotenv.2019.134115.

Li, L., et al., 2022. A novel chitosan modified magnesium impregnated corn straw biochar for ammonium and phosphate removal from simulated livestock wastewater. Environmental Technology and Innovation 26, 102519. Available from: https://doi.org/10.1016/j.eti.2022.102519.

Lian, W., et al., 2021. Influence of pyrolysis temperature on the cadmium and lead removal behavior of biochar derived from oyster shell waste. Bioresource Technology Reports 15 (April), 100709. Available from: https://doi.org/10.1016/j.biteb.2021.100709.

Lin, S., et al., 2020. Recycling application of waste long-root *Eichhornia crassipes* in the heavy metal removal using oxidized biochar derived as adsorbents. Bioresource Technology 314, 123749. Available from: https://doi.org/10.1016/j.biortech.2020.123749.

Liu, H., et al., 2020a. Egg shell biochar-based green catalysts for the removal of organic pollutants by activating persulfate. Science of the Total Environment 745, 141095. Available from: https://doi.org/10.1016/j.scitotenv.2020.141095.

Liu, J., et al., 2020b. Resource utilization of swine sludge to prepare modified biochar adsorbent for the efficient removal of Pb(II) from water. Journal of Cleaner Production 257, 120322. Available from: https://doi.org/10.1016/j.jclepro.2020.120322.

Liu, J., et al., 2021a. Mixed biochar obtained by the co-pyrolysis of shrimp shell with corn straw: co-pyrolysis characteristics and its adsorption capability. Chemosphere 282 (May), 131116. Available from: https://doi.org/10.1016/j.chemosphere.2021.131116.

Liu, X., Li, M., Singh, S.K., 2021b. Manganese-modified lignin biochar as adsorbent for removal of methylene blue. Journal of Materials Research and Technology 12, 1434–1445. Available from: https://doi.org/10.1016/j.jmrt.2021.03.076.

Lu, Z., et al., 2022. Integration of bimetallic organic frameworks and magnetic biochar for azole fungicides removal. Environmental Advances 7, 100152. Available from: https://doi.org/10.1016/j.envadv.2021.100152.

Ma, C., et al., 2018. Honeycomb tubular biochar from fargesia leaves as an effective adsorbent for tetracyclines pollutants. Journal of the Taiwan Institute of Chemical Engineers 91, 299–308. Available from: https://doi.org/10.1016/j.jtice.2018.05.032.

Ma, J., et al., 2021. The utilization of lobster shell to prepare low-cost biochar for high-efficient removal of copper and cadmium from aqueous: sorption properties and mechanisms. Journal of Environmental Chemical Engineering 9 (1), 104703. Available from: https://doi.org/10.1016/j.jece.2020.104703.

Ma, Y., Li, P., et al., 2020a. Iron/zinc and phosphoric acid modified sludge biochar as an efficient adsorbent for fluoroquinolones antibiotics removal. Ecotoxicology and Environmental Safety 196. Available from: https://doi.org/10.1016/j.ecoenv.2020.110550.

Ma, Y., Yang, L., et al., 2020b. Carbon nanotube supported sludge biochar as an efficient adsorbent for low concentrations of sulfamethoxazole removal. Science of the Total Environment 20, 718. Available from: https://doi.org/10.1016/j.scitotenv.2020.137299.

Macedo, C.A., et al., 2020. Organosulphur-modified biochar : an effective green adsorbent for removing metal species in aquatic systems São Paulo State University (UNESP), Institute of Chemistry, Rua Prof. Francisco São Paulo State University (UNESP), Institute of Science and. Surfaces and Interfaces 100822. Available from: https://doi.org/10.1016/j.surfin.2020.100822.

Machado, L.M.M., et al., 2020. J Treatment of effluents containing 2-chlorophenol by adsorption onto chemically and physically activated biochars. Journal of Environmental Chemical Engineering 8 (6), 104473. Available from: https://doi.org/10.1016/j.jece.2020.104473.

Mahmoud, M.E., et al., 2016. Kinetics, isotherm, and thermodynamic studies of the adsorption of reactive red 195 A dye from water by modified Switchgrass Biochar adsorbent. Journal of Industrial and Engineering Chemistry 37, 156–167. Available from: https://doi.org/10.1016/j.jiec.2016.03.020.

Mahmoud, M.E., Abou-Ali, S.A.A., Elweshahy, S.M.T., 2021a. Efficient and ultrafast removal of Cd(II) and Sm(III) from water by leaves of *Cynara scolymus* derived biochar. Materials Research Bulletin 141 (November 2020), 111334. Available from: https://doi.org/10.1016/j.materresbull.2021.111334.

Mahmoud, M.E., Mohamed, A.K., Salam, M.A., 2021b. Self-decoration of N-doped graphene oxide 3-D hydrogel onto magnetic shrimp shell biochar for enhanced removal of hexavalent chromium. Journal of Hazardous Materials 408 (August 2020), 124951. Available from: https://doi.org/10.1016/j.jhazmat.2020.124951.

Mandal, A., Singh, N., Purakayastha, T.J., 2017. Characterization of pesticide sorption behaviour of slow pyrolysis biochars as low cost adsorbent for atrazine and imidacloprid removal. Science of the Total Environment 577, 376–385. Available from: https://doi.org/10.1016/j.scitotenv.2016.10.204.

Manyuchi, M.M., et al., 2021. Influence of sawdust based biochar on gold tailings wastewater heavy metal contaminants removal. South African Journal of Chemical Engineering 37 (May), 81–91. Available from: https://doi.org/10.1016/j.sajce.2021.05.003.

Maria, M., Bezerra, M., 2018. Mono and binary component adsorption of phenol and cadmium using adsorbent derived from peanut shells. Journal of Cleaner Production 201, 219–228. Available from: https://doi.org/10.1016/j.jclepro.2018.07.291.

Mohammed, N.A.S., et al., 2018. Phenol adsorption on biochar prepared from the pine fruit shells: equilibrium, kinetic and thermodynamics studies. Journal of Environmental Management 226 (May), 377–385. Available from: https://doi.org/10.1016/j.jenvman.2018.08.033.

Mohan, D., et al., 2014. Organic and inorganic contaminants removal from water with biochar, a renewable, low cost and sustainable adsorbent - a critical review. Bioresource Technology 160, 191–202. Available from: https://doi.org/10.1016/j.biortech.2014.01.120.

Mokrzycki, J., et al., 2020. Hydrothermal and pyrolytic biochars from waste milk thistle (*Silybum marianum*) extrudates as precursors for production of effective isoproturon adsorbents. Journal of Water Process Engineering 37 (June), 101459. Available from: https://doi.org/10.1016/j.jwpe.2020.101459.

Momčilović, M., et al., 2011. Removal of lead(II) ions from aqueous solutions by adsorption onto pine cone activated carbon. Desalination 276 (1–3), 53–59. Available from: https://doi.org/10.1016/j.desal.2011.03.013.

Naghipour, D., et al., 2018. Characterization, kinetic, thermodynamic and isotherm data for diclofenac removal from aqueous solution by activated carbon derived from pine tree. Data in Brief 18, 1082–1087. Available from: https://doi.org/10.1016/j.dib.2018.03.068.

Nath, H., et al., 2021. Bioresource technology reports removal of methylene blue from water using okra (*Abelmoschus esculentus* L.) mucilage modified biochar. Bioresource Technology Reports 14 (March), 100689. Available from: https://doi.org/10.1016/j.biteb.2021.100689.

Nie, Z., et al., 2021. Characterization of pyridine biodegradation by two Enterobacter sp. strains immobilized on *Solidago canadensis* L. stem derived biochar. Journal of Hazardous Materials 414 (December 2020), 125577. Available from: https://doi.org/10.1016/j.jhazmat.2021.125577.

Obey, G., Adelaide, M., Ramaraj, R., 2022. Biochar derived from non-customized matamba fruit shell as an adsorbent for wastewater treatment. Journal of Bioresources and Bioproducts 7 (2), 109–115. Available from: https://doi.org/10.1016/j.jobab.2021.12.001.

Oginni, O., Singh, K., 2021. Effect of carbonization temperature on fuel and caffeine adsorption characteristics of white pine and Norway spruce needle derived biochars. Industrial Crops and Products 162 (February 2020), 113261. Available from: https://doi.org/10.1016/j.indcrop.2021.113261.

Onorevoli, B., et al., 2018. Characterization of feedstock and biochar from energetic tobacco seed waste pyrolysis and potential application of biochar as an adsorbent. Journal of Environmental Chemical Engineering 6 (1), 1279–1287. Available from: https://doi.org/10.1016/j.jece.2018.01.039.

Park, J.H., et al., 2019. Cadmium adsorption characteristics of biochars derived using various pine tree residues and pyrolysis temperatures. Journal of Colloid and Interface Science 553, 298–307. Available from: https://doi.org/10.1016/j.jcis.2019.06.032.

Praveen, S., et al., 2021. Techno-economic feasibility of biochar as biosorbent for basic dye sequestration. Journal of the Indian Chemical Society 98 (8), 100107. Available from: https://doi.org/10.1016/j.jics.2021.100107.

Prochazkova, J., et al., 2022. Magnetically responsive low-cost adsorbents for aniline removal. Cleaner Engineering and Technology 6, 100394. Available from: https://doi.org/10.1016/j.clet.2021.100394.

Puga, A., et al., 2022. Continuous adsorption studies of pharmaceuticals in multicomponent mixtures by agroforestry biochar. Journal of Environmental Chemical Engineering 10 (1). Available from: https://doi.org/10.1016/j.jece.2021.106977.

Putri, K.N.A., Keereerak, A., Chinpa, W., 2020. Novel cellulose-based biosorbent from lemongrass leaf combined with cellulose acetate for adsorption of crystal violet. International Journal of Biological Macromolecules 156, 762–772. Available from: https://doi.org/10.1016/j.ijbiomac.2020.04.100.

Qiu, B., et al., 2021. Biochar as a low-cost adsorbent for aqueous heavy metal removal: a review. Journal of Analytical and Applied Pyrolysis 155 (December 2020), 105081. Available from: https://doi.org/10.1016/j.jaap.2021.105081.

Renita, A.A., Amarnath, D.J., Lakshmi, S., 2021. Synthesis of peanut-shell magnetized biocarbon for acid fuchsin dye removal. Materials Today: Proceedings 43, 3075–3078. Available from: https://doi.org/10.1016/j.matpr.2021.01.407.

Rodriguez, J.A., et al., 2020. Influence of pyrolysis temperature and feedstock on the properties of biochars produced from agricultural and industrial wastes. Journal of Analytical and Applied Pyrolysis 149 (March), 104839. Available from: https://doi.org/10.1016/j.jaap.2020.104839.

Roha, B., et al., 2021. Model sorption of industrial wastewater containing Cu^{2+}, Cd^{2+}, and Pb^{2+} using individual and mixed rice husk biochar. Environmental Technology and Innovation 24. Available from: https://doi.org/10.1016/j.eti.2021.101900.

Rumjit, N.P., et al., 2021. Kinetic and isotherm studies on adsorptive removal of sulfates by cotton shell derived biochar: recovery of sulfates from marcasite soil. Sustainable Chemistry and Pharmacy 20 (December 2020), 100361. Available from: https://doi.org/10.1016/j.scp.2020.100361.

Saeed, A.A.H., et al., 2022. Removal of cadmium from aqueous solution by optimized rice husk biochar using response surface methodology. Ain Shams Engineering Journal 13 (1), 101516. Available from: https://doi.org/10.1016/j.asej.2021.06.002.

Sahoo, S.S., et al., 2021. Production and characterization of biochar produced from slow pyrolysis of pigeon pea stalk and bamboo. Cleaner Engineering and Technology 3, 100101. Available from: https://doi.org/10.1016/j.clet.2021.100101.

Sakhiya, A.K., et al., 2021. A comparative study of physical and chemical activation of rice straw derived biochar to enhance Zn^{+2} adsorption. Bioresource Technology Reports 15 (July), 100774. Available from: https://doi.org/10.1016/j.biteb.2021.100774.

Salawu, O.A., Han, Z., Adeleye, A.S., 2022. Shrimp waste-derived porous carbon adsorbent: performance, mechanism, and application of machine learning. Journal of Hazardous Materials 129266. Available from: https://doi.org/10.1016/j.jhazmat.2022.129266.

Saravanan, A., et al., 2021. Ultrasonic assisted agro waste biomass for rapid removal of Cd(II) ions from aquatic environment: mechanism and modelling analysis. Chemosphere 271, 129484. Available from: https://doi.org/10.1016/j.chemosphere.2020.129484.

Saremi, F., et al., 2020. Adsorption of tetracycline antibiotic from aqueous solutions onto vitamin B6-upgraded biochar derived from date palm leaves. Journal of Molecular Liquids 318, 114126. Available from: https://doi.org/10.1016/j.molliq.2020.114126.

Shan, R., et al., 2020. Single and competitive adsorption affinity of heavy metals toward peanut shell-derived biochar and its mechanisms in aqueous systems. Chinese Journal of Chemical Engineering 28 (5), 1375–1383. Available from: https://doi.org/10.1016/j.cjche.2020.02.012.

Shan, X., et al., 2022. Biochar / Mg-Al spinel carboxymethyl cellulose-La hydrogels with cationic polymeric layers for selective phosphate capture. Journal of Colloid and Interface Science 606, 736–747. Available from: https://doi.org/10.1016/j.jcis.2021.08.078.

Shen, Y., et al., 2017. Microalgal-biochar immobilized complex: a novel efficient biosorbent for cadmium removal from aqueous solution. Bioresource Technology 244, 1031–1038. Available from: https://doi.org/10.1016/j.biortech.2017.08.085.

Shen, Z., et al., 2021. Anchoring Al- and / or Mg-oxides to magnetic biochars for Co-uptake of arsenate and fluoride from water. Journal of Environmental Management 293 (May), 112898. Available from: https://doi.org/10.1016/j.jenvman.2021.112898.

Shirvanimoghaddam, K., Czech, B., Tyszczuk-rotko, K., 2021. Sustainable synthesis of rose flower-like magnetic biochar from tea waste for environmental applications. Journal of Advanced Research 34, 13–27. Available from: https://doi.org/10.1016/j.jare.2021.08.001.

Song, Z., 2019. Synthesis and adsorption of FeMn La-impregnated biochar composite as an adsorbent for As (III) removal from aqueous. Environmental Pollution 247, 128–135. Available from: https://doi.org/10.1016/j.envpol.2019.01.044.

Sun, L., Wan, S., Luo, W., 2013. Biochars prepared from anaerobic digestion residue, palm bark, and eucalyptus for adsorption of cationic methylene blue dye: characterization, equilibrium, and kinetic studies. Bioresource Technology 140, 406–413. Available from: https://doi.org/10.1016/j.biortech.2013.04.116.

Sun, L., et al., 2021a. Preparation of sorbents derived from bamboo and bromine flame retardant for elemental mercury removal. Journal of Hazardous Materials 410 (October 2020), 124583. Available from: https://doi.org/10.1016/j.jhazmat.2020.124583.

Sun, T., et al., 2021b. Crayfish shell biochar for the mitigation of Pb contaminated water and soil: characteristics, mechanisms, and applications. Environmental Pollution 271, 116308. Available from: https://doi.org/10.1016/j.envpol.2020.116308.

Sun, Y., et al., 2021c. The potential of biochar and lignin-based adsorbents for wastewater treatment: comparison, mechanism, and application—a review. Industrial Crops and Products 166 (April), 113473. Available from: https://doi.org/10.1016/j.indcrop.2021.113473.

Tang, L., et al., 2018. Sustainable efficient adsorbent: alkali-acid modified magnetic biochar derived from sewage sludge for aqueous organic contaminant removal. Chemical Engineering Journal . Available from: https://doi.org/10.1016/j.cej.2017.11.048.

Tao, Q., et al., 2021. An integrated method to produce fermented liquid feed and biologically modified biochar as cadmium adsorbents using corn stalks. Waste Management 127, 112–120. Available from: https://doi.org/10.1016/j.wasman.2021.04.027.

Thangagiri, B., et al., 2022. Removal of hexavalent chromium by biochar derived from *Azadirachta indica* leaves: batch and column studies. Chemosphere 286 (P1), 131598. Available from: https://doi.org/10.1016/j.chemosphere.2021.131598.

Thotagamuge, R., et al., 2021. Copper modified activated bamboo charcoal to enhance adsorption of heavy metals from industrial wastewater. Environmental Nanotechnology, Monitoring & Management 16 (August), 100562. Available from: https://doi.org/10.1016/j.enmm.2021.100562.

Tomin, O., Vahala, R., Roza, M., 2021. Tailoring metal-impregnated biochars for selective removal of natural organic matter and dissolved phosphorus from the aqueous phase. Microporous and Mesoporous Materials 328 (October), 111499. Available from: https://doi.org/10.1016/j.micromeso.2021.111499.

Tomul, F., et al., 2020. Peanut shells-derived biochars prepared from different carbonization processes: comparison of characterization and mechanism of naproxen adsorption in water. Science of the Total Environment 726. Available from: https://doi.org/10.1016/j.scitotenv.2020.137828.

Tripathy, S., et al., 2021. Efficient removal of Cr (VI) by polyaniline modified biochar from date (*Phoenix dactylifera*) seed. Groundwater for Sustainable Development 15 (April), 100653. Available from: https://doi.org/10.1016/j.gsd.2021.100653.

Tulun, Ş., et al., 2021. Adaptive neuro-fuzzy interference system modelling for chlorpyrifos removal with walnut shell biochar. Arabian Journal of Chemistry 14 (12). Available from: https://doi.org/10.1016/j.arabjc.2021.103443.

Uzun, B.B., et al., 2010. Synthetic fuel production from tea waste: characterisation of bio-oil and bio-char. Fuel 89 (1), 176–184. Available from: https://doi.org/10.1016/j.fuel.2009.08.040.

Velusamy, K., et al., 2021. Analysis on the removal of emerging contaminant from aqueous solution using biochar derived from soap nut seeds. Environmental Pollution 287 (May), 117632. Available from: https://doi.org/10.1016/j.envpol.2021.117632.

Vyavahare, G., et al., 2018. Strategies for crystal violet dye sorption on biochar derived from mango leaves and evaluation of residual dye toxicity. Journal of Cleaner Production 207, 296–305. Available from: https://doi.org/10.1016/j.jclepro.2018.09.193.

Wang, Y., et al., 2016. Ammonium citrate-modified biochar : an adsorbent for La (III) ions from aqueous solution. Colloids and Surfaces A: Physicochemical and Engineering Aspects 509, 550–563. Available from: https://doi.org/10.1016/j.colsurfa.2016.09.060.

Wang, Q., et al., 2020. Converting industrial waste cork to biochar as Cu (II) adsorbent via slow pyrolysis. Waste Management 105, 102–109. Available from: https://doi.org/10.1016/j.wasman.2020.01.041.

Wang, S., et al., 2021a. Highly adsorptive pristine and magnetic biochars prepared from crayfish shell for removal of Cu(II) and Pb(II). Journal of the Taiwan Institute of Chemical Engineers 127, 175–185. Available from: https://doi.org/10.1016/j.jtice.2021.08.004.

Wang, Z., et al., 2021b. From wastes to functions: a paper mill sludge-based calcium-containing porous biochar adsorbent for phosphorus removal. Journal of Colloid and Interface Science 593, 434–446. Available from: https://doi.org/10.1016/j.jcis.2021.02.118.

Wasewar, K.L., et al., 2009. Batch adsorption of zinc on tea factory waste. DES 244 (1–3), 66–71. Available from: https://doi.org/10.1016/j.desal.2008.04.036.

Wei, L., Lu, J., 2021. Adsorption of microcystin-LR by rice straw biochars with different pyrolysis temperatures. Environmental Technology & Innovation 23, 101609. Available from: https://doi.org/10.1016/j.eti.2021.101609.

Wen, Z., et al., 2021. Porous biochar-supported $MnFe_2O_4$ magnetic nanocomposite as an excellent adsorbent for simultaneous and effective removal of organic / inorganic arsenic from water. Journal of Hazardous Materials 411 (September 2020), 124909. Available from: https://doi.org/10.1016/j.jhazmat.2020.124909.

Wijeyawardana, P., et al., 2022. Removal of Cu, Pb and Zn from stormwater using an industrially manufactured sawdust and paddy husk derived biochar. Environmental Technology & Innovation 28, 102640. Available from: https://doi.org/10.1016/j.eti.2022.102640.

Wilson, K., et al., 2006. Select metal adsorption by activated carbon made from peanut shells. Bioresource Technology 97, 2266–2270. Available from: https://doi.org/10.1016/j.biortech.2005.10.043.

Wu, Z., et al., 2021. Synthesis of polyethyleneimine modified $CoFe_2O_4$-loaded porous biochar for selective adsorption properties towards dyes and exploration of interaction mechanisms. Separation and Purification Technology 277 (July), 119474. Available from: https://doi.org/10.1016/j.seppur.2021.119474.

Xiang, Y., et al., 2019. A sustainable ferromanganese biochar adsorbent for effective levo fl oxacin removal from aqueous medium. Chemosphere 237, 124464. Available from: https://doi.org/10.1016/j.chemosphere.2019.124464.

Xiao, Y., et al., 2017. Sorption of heavy metal ions onto crayfish shell biochar: effect of pyrolysis temperature, pH and ionic strength. Journal of the Taiwan Institute of Chemical Engineers 0, 1–8. Available from: https://doi.org/10.1016/j.jtice.2017.08.035.

Xing, Y., et al., 2021. A novel eco-friendly recycling of food waste for preparing biofilm-attached biochar to remove Cd and Pb in wastewater. Journal of Cleaner Production 311 (May), 127514. Available from: https://doi.org/10.1016/j.jclepro.2021.127514.

Xu, Z., et al., 2021. Adsorption behaviors of paper mill sludge biochar to remove Cu, Zn and As in wastewater. Environmental Technology & Innovation 23, 101616. Available from: https://doi.org/10.1016/j.eti.2021.101616.

Xu, Z., et al., 2022. Optimizing pyrolysis temperature of contaminated rice straw biochar: heavy metal(loid) deportment, properties evolution, and Pb adsorption/immobilization. Journal of Saudi Chemical Society 26 (2), 101439. Available from: https://doi.org/10.1016/j.jscs.2022.101439.

Yang, F., et al., 2018. Fabrication and characterization of hydrophilic corn stalk biochar-supported nanoscale zero-valent iron composites for efficient metal removal. Bioresource Technology . Available from: https://doi.org/10.1016/j.biortech.2018.06.029 [Preprint].

Yang, J., et al., 2020. High-yield and high-performance porous biochar produced from pyrolysis of peanut shell with low-dose ammonium polyphosphate for chloramphenicol adsorption. Journal of Cleaner Production 264, 121516. Available from: https://doi.org/10.1016/j.jclepro.2020.121516.

Yang, H., et al., 2021a. Effect of pyrolysis conditions on food waste conversion to biochar as a coagulant aid for wastewater treatment. Journal of Water Process Engineering 41 (April), 102081. Available from: https://doi.org/10.1016/j.jwpe.2021.102081.

Yang, Q., et al., 2021b. Cellulose derived biochar: preparation, characterization and Benzo [a] pyrene adsorption capacity. Grain & Oil Science and Technology 4 (4), 182–190. Available from: https://doi.org/10.1016/j.gaost.2021.10.001.

Yang, Z., et al., 2021c. The effect of carbonization temperature on the capacity and mechanisms of Pb(II) adsorption by microalgae residue-derived biochar. Ecotoxicology and Environmental Safety 225, 112750. Available from: https://doi.org/10.1016/j.ecoenv.2021.112750.

Yoon, K., et al., 2017. Fabrication of engineered biochar from paper mill sludge and its application into removal of arsenic and cadmium in acidic water. Bioresource Technology 1–7. Available from: https://doi.org/10.1016/j.biortech.2017.07.020.

Yoon, J., et al., 2021. Assessment of adsorptive behaviors and properties of grape pomace-derived biochar as adsorbent for removal of cymoxanil pesticide. Environmental Technology & Innovation 21, 101242. Available from: https://doi.org/10.1016/j.eti.2020.101242.

Yu, Y., et al., 2019. Unraveling sorption of Cr (VI) from aqueous solution by $FeCl_3$ and $ZnCl_2$ - modi fi ed corn stalks biochar: implicit mechanism and application. Bioresource Technology (174), . Available from: https://doi.org/10.1016/j.biortech.2019.122466.

Yu, J., et al., 2021. Insight into the key factors in fast adsorption of organic pollutants by hierarchical porous biochar. Journal of Hazardous Materials 403 (August 2020), 123610. Available from: https://doi.org/10.1016/j.jhazmat.2020.123610.

Zazycki, M.A., et al., 2018. New biochar from pecan nutshells as an alternative adsorbent for removing reactive red 141 from aqueous solutions. Journal of Cleaner Production 171, 57–65. Available from: https://doi.org/10.1016/j.jclepro.2017.10.007.

Zazycki, M.A., et al., 2019. Chitin derived biochar as an alternative adsorbent to treat colored effluents containing methyl violet dye. Advanced Powder Technology 30 (8), 1494–1503. Available from: https://doi.org/10.1016/j.apt.2019.04.026.

Zeng, B., et al., 2021. Preparation of sludge biochar rich in carboxyl / hydroxyl groups by quenching process and its excellent adsorption performance for Cr(VI). Chemosphere 285 (July), 131439. Available from: https://doi.org/10.1016/j.chemosphere.2021.131439.

Zhang, R., Leiviskä, T., 2020. Surface modifi cation of pine bark with quaternary ammonium groups and its use for vanadium removal. Chemical Engineering Journal 385 (November 2019), 123967. Available from: https://doi.org/10.1016/j.cej.2019.123967.

Zhang, Y., et al., 2019a. Pyrolysis-temperature depended quinone and carbonyl groups as the electron accepting sites in barley grass derived biochar. Chemosphere 232, 273–280. Available from: https://doi.org/10.1016/j.chemosphere.2019.05.225.

Zhang, J., et al., 2019b. Desorption of calcium-rich crayfish shell biochar for the removal of lead from aqueous solutions. Journal of Colloid and Interface Science 554, 417–423. Available from: https://doi.org/10.1016/j.jcis.2019.06.096.

Zhang, F., et al., 2020a. Utilization of bark waste of *Acacia mangium*: the preparation of activated carbon and adsorption of phenolic wastewater. Industrial Crops & Products (April), 113157. Available from: https://doi.org/10.1016/j.indcrop.2020.113157.

Zhang, X., et al., 2020b. Characterization and sulfonamide antibiotics adsorption capacity of spent coffee grounds based biochar and hydrochar. Science of the Total Environment 716. Available from: https://doi.org/10.1016/j.scitotenv.2020.137015.

Zhang, D., et al., 2021. Enhanced adsorption for the removal of tetracycline hydrochloride (TC) using ball-milled biochar derived from crayfish shell. Colloids and Surfaces A: Physicochemical and Engineering Aspects 615 (December 2020), 126254. Available from: https://doi.org/10.1016/j.colsurfa.2021.126254.

Zhao, B., et al., 2017. Effect of pyrolysis temperature, heating rate, and residence time on rapeseed stem derived biochar. Journal of Cleaner Production 174, 977–987. Available from: https://doi.org/10.1016/j.jclepro.2017.11.013.

Zheng, H., et al., 2017. Adsorption of p-nitrophenols (PNP) on microalgal biochar: analysis of high adsorption capacity and mechanism State Key Laboratory of Urban Water Resource and Environment, Harbin Institute of Department of Chemical Engineering, National Cheng Kung Uni. Bioresource Technology . Available from: https://doi.org/10.1016/j.biortech.2017.05.025 [Preprint].

Zhou, Z., et al., 2018. Effect of pyrolysis condition on the adsorption mechanism of lead, cadmium and copper on tobacco stem biochar. Journal of Cleaner Production 187, 995–1005. Available from: https://doi.org/10.1016/j.jclepro.2018.03.268.

Zhu, Z., et al., 2018. Journal of Water Process Engineering A hierarchical porous adsorbent of nano- α -Fe_2O_3/Fe_3O_4 on bamboo biochar (HPA-Fe / C-B) for the removal of phosphate from water. Journal of Water Process Engineering 25 (January), 96–104. Available from: https://doi.org/10.1016/j.jwpe.2018.05.010.

Zhuang, Z., Wang, L., Tang, J., 2020. Efficient removal of volatile organic compound by ball-milled biochars from different preparing conditions. Journal of Hazardous Materials 124676. Available from: https://doi.org/10.1016/j.jhazmat.2020.124676.

Zoroufchi, K., Soltan, J., Mcphedran, K.N., 2021. Electrochemically modified adsorbents for treatment of aqueous arsenic: pore diffusion in modified biomass vs. biochar. Chemical Engineering Journal 423 (April), 130061. Available from: https://doi.org/10.1016/j.cej.2021.130061.

Section C

Waste management and environment protection

Chapter 14

Mycogenic synthesis of nanoparticles and their application in dye degradation

Aakansha Singh and Brijesh Kumar Mishra
Department of Environmental Science and Engineering, Indian Institute of Technology (Indian School of Mines), Dhanbad, Jharkhand, India

Introduction

According to Composite Water Management Index report, it is stated that by 2030, the demand for water in India will double the available supply, which implies acute water scarcity for hundreds of millions of people. It is estimated that nearly 40% of the population in India by 2030 will not have access to freshwater. Moreover, excessive exploitation and imprudent use of freshwater sources due to rapid industrialization and inappropriate discharge of industrial wastewater is a cause for major concern. Among various industries, the textile industry is considered to have a massive impact on polluting the water bodies. It is one of the largest producers of industrial liquid waste and consumes a surplus amount of water during its unit operations (Noman et al., 2020; Buscio et al., 2019). Along with the textile industry, other industries like paper, tannery, pharmaceuticals, printing ink, and, food are also major users of dyes in India. The average worldwide production of dyes is estimated to be around 7×10^8 kg annually (Adhikari et al., 2017). The basic structure possesses a chromophore (i.e., color inducing groups), and an auxochrome (i.e., color helpers) like amino, hydroxyl, and carboxylic acid groups, which attributes to the solubility and fixation of the dyes. Chromophores must be involved in the conjugated system of the structure to impart colors (dos Santos et al., 2007; Sonal et al., 2018). Nitro and Nitroso dyes, Azo dyes, triarylmethane dyes, Anthraquinone dyes, and indigo dyes are classified based on different chromophores present in their structure. Among them commonly used commercial dyes are azo dyes (having nitrogen as the azo group $-N=N-$) which contribute the most (nearly 70%) followed by anthraquinone dye (Benkhaya et al., 2020). In addition to their wide-scale applications, dyes play a substantial role in polluting the environment by generating an enormous amount of wastewater. The wastewater generated from these industries contains highly suspended solids, organic chemicals, color, and other soluble substances depending on the fiber and the process used (dos Santos et al., 2007). In the aquatic ecosystem, the presence of dyes hampers the photosynthetic activities of the algae and flora leading to the depletion of oxygen levels, eventually disturbing the ecological balance (Nandhini et al., 2019). Generally toxicity level increases with the increase of benzene rings in the dye structure (Islam et al., 2019; Karagoz, 2019; Lellis et al., 2019). Degradation of dyes is quite challenging owing to their synthetic origin, complex structure, and, high solubility (Gupta and Suhas, 2009; Varjani et al., 2020). Hence, a prerequisite treatment before discharge is fundamentally required to be implemented. Till now various physical, chemical, and biological methods have been developed to combat the toxicity produced by dyes. However, due to various limitations like high maintenance, sludge generation, gradual loss of efficiency, faced during the physio-chemical process along with varying growth conditions during biological handling, these methods are in urge for replacement with more ecologically friendly and sustainable methods (Varjani et al., 2020). Nanotechnology can provide better and novel applications for dye removal. Norio Taniguchi, in 1974, first used the term nanotechnology illuminating its dimensional accuracy and its benefits in different applications (Ali et al., 2017). High adsorption, better catalytic reactions, and large surface area are a few characteristics of nanoparticles that have lately attracted the attention of researchers in wastewater treatment. High-energy ball milling, laser ablation, electrospraying, and physical vapor depositions are some of the physical methods of synthesis. Chemical reduction by organic and inorganic reducing agents, coprecipitation, pyrolysis, solvothermal are few chemical

Role of Green Chemistry in Ecosystem Restoration to Achieve Environmental Sustainability. DOI: https://doi.org/10.1016/B978-0-443-15291-7.00028-6

methods of synthesis (Küünal et al., 2018; Raina et al., 2020). It is a need of the era for the development of a biocompatible, clean, nontoxic methodology that can curb all the disadvantages of the above-mentioned methods (Malik et al., 2014). The reason for increased attention toward biogenic synthesis is that it can be conducted at ambient temperature and pressure, is environmentally benign, has better control over size and shape characteristics, does not use chemicals and almost minimum sludge is produced (Das and Chatterjee, 2019). Fungi provide some additional attributes during synthesis also termed mycosynthesis over bacteria and plants exhibiting easy handling, a faster rate of growth, production of various extra and intracellular enzymes, and greater wall binding capacity (Balaji et al., 2009; Du et al., 2015; Guilger-Casagrande and Lima, 2019). Hence, this book chapter encompasses a brief description of mycogenic approach for nanoparticles and its various metal and metal oxides synthesis, along with factors affecting the synthesis process, characterization, and their application in dye degradation.

Biogenic approach: mycosynthesis

The kingdom fungi usually consist of a multicellular eukaryotic organism that lacks chlorophyll. They show symbiotic associations with plants and bacteria and can reproduce both sexually and asexually (Reid and Webster, 1973). Fungi are recognized as suitable nanofactories that can synthesize various metals and metal oxides of a particular size and morphology, and the process is entitled as mycosynthesis. A variety of fungal strains have been explored for the synthesis of metal nanoparticles, among them the use of *Fusarium*, *Penicillium*, and *Aspergillus* species are widely explored (Chhipa, 2019). The two approaches used for the synthesis of nanoparticles are the top-down approach and the bottom-up approach as illustrated in Fig. 14.1 (Jamkhande et al., 2019; Zielonka and Klimek-Ochab, 2017).

Polysaccharides, proteins, mycelia, and fruiting bodies of fungi are the parts generally used for the synthesis of metal nanoparticles (Owaid, 2019). Synthesis of nanoparticles may take place intracellularly or extracellularly (Ahmad et al., 2003; Shankar Naik, 2020). The plausible reason is the electrostatic interaction between the metal ions and the negatively charged cell wall (Sneha et al., 2010). During extracellular synthesis, the metal ion salts are added to the biomass filtrate containing fungal biomolecules, the formed nanoparticles are further capped and stabilized followed in the dispersion media containing organic biomolecules as illustrated in Fig. 14.2.

Nearly 30 elements including Ag, Au, Ti, Zn, Ce, Fe, Mg, Pt, P, Cu, Cd, Ti metals, and their metal oxides have been fabricated till now by exploring different strains of fungi (Chhipa, 2019; Shankar Naik, 2020; Vágó et al., 2016; Yosri et al., 2018; Li et al., 2022). Among them, silver and gold nanoparticles have contributed the most as they are antibacterial, and antifungal and also exhibit a wide-scale application.

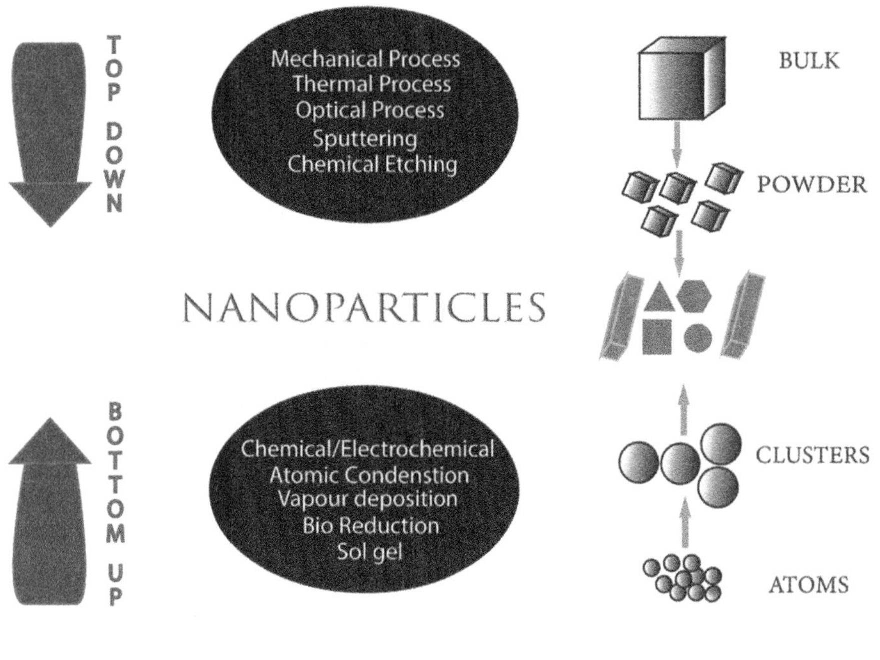

FIGURE 14.1 Approach for nanoparticle synthesis.

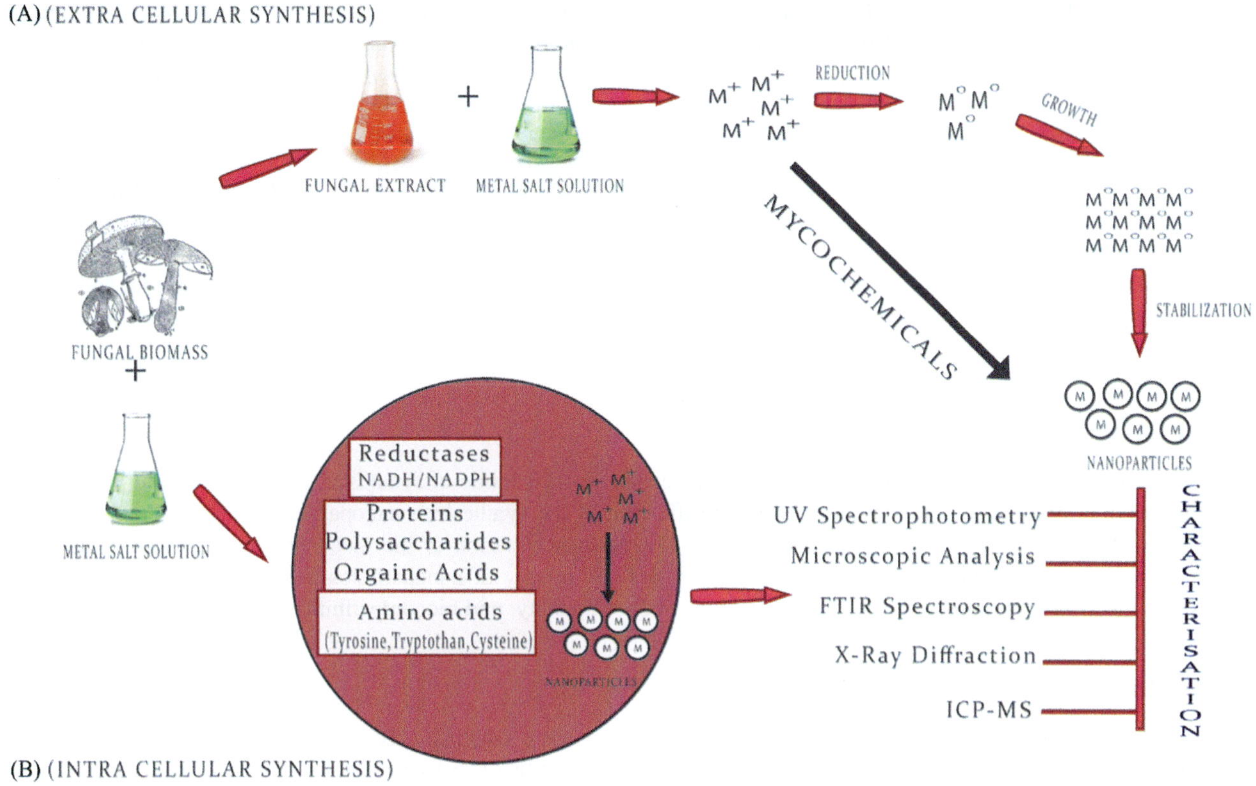

FIGURE 14.2 Mycogenic synthesis of nanoparticles.

Types of green nanostructures

Metal and metal oxide nanoparticles

Engender of various metal and metal oxide nanoparticles by mycogenesis is quite promising. Biosorption and bioreduction are two methods by which fungi are capable of producing nanoparticles when exposed to metal salts. The surface of the cell wall is the site of synthesis in the former case while bioreduction takes place by the release of certain enzymes and biomolecules (Siddiqi and Husen, 2016).

Silver nanoparticles

Biosynthesis and characterization of silver nanoparticles employing *Aspergillus* sp. (Fatima et al., 2016), *Fusarium oxysporum* (Srivastava et al., 2019), *Trichoderma reesei* (Vahabi *et al.*, 2011), *Coriolus versicolor* (Sanghi and Verma 2009b), and *Trichoderma viride* (Fayaz et al., 2010) have been reported by many researchers. Owaid (2019) synthesized silver nanoparticles exhibiting spherical shape and crystalline nature using macrofungi *Pleurotus* (oyster mushroom). Vigneshwaran et al. (2007), synthesized silver nanoparticles on the surface of fungal mycelium using *Aspergillus flavus*. UV–visible spectroscopy is used to observe surface plasmon resonance bands which reflect the optical properties of the material (Ahmad et al., 2003; Vahabi et al., 2011). In addition to its extraordinary performance as an antimicrobial agent, it is also widely used in wastewater treatments, and exhibits catalytic activities, and antifungal medicine. Short-term exposures to silver nanoparticles are considered to be quite safe compared to its long term exposure, which needs to be accessed at a larger scale (Firdhouse and Lalitha, 2015).

Gold nanoparticles

Gold nanoparticles act as nanochromophores at an absorption band at near 500–550 nm wavelengths (Abdelhalim et al., 2012; Chen et al., 2018). Clarance et al. (2020) have demonstrated the formation of gold nanoparticles from *Fusarium solani* ATLOY-8. In another study, 21 types of filamentous microscopic fungi were selected to study the one-step synthesis of gold nanoparticles at room temperature (Vágó et al., 2016). Other fungi which have been explored for the synthesis of gold nanoparticles include *Cystoseira baccata* (González-Ballesteros et al., 2017), *Bipolaris tetramera* (Fatima et al., 2015), *Penicillium citrinum* (Goswami and Ghosh 2013), *F. solani* (Gopinath and Arumugam 2014), *F. oxysporum* (Ahmad et al., 2003). Nanotriangular gold structures were bio-fabricated using *Aspergillus clavatus* (Verma et al., 2012).

Other metal nanoparticles

Vanti et al. (2019) fabricated chitosan copper nanoparticles using the fungus *Pythium aphanidermatum* and *Rhizoctonia solani*. *F. oxysporum* and *Verticillium* sp. were used to synthesize Fe_3O_4 nanoparticles (Bharde et al., 2006). Sanghi and Verma (2009a) analyzed the detoxification potential of *C. versicolor* fungi and implemented it to synthesize CdS nanoparticles with the help of the thiol group present in the cysteine amino acid through sequestration, reduction, and intoxication of Cd^{2+} ions. *Phanerochaete chrysosporium* was used to generate Pd nanoparticles by adsorption into fungal mycelium (Tarver et al., 2019).

Factors affecting nanoparticle synthesis

The characteristics of synthesized nanoparticles majorly depend on their mode of synthesis, the substrate used, the purpose of the application, and finally, the method used for purification of the synthesized nanoparticle (Kuchibhatla et al., 2012; Patra and Baek 2014). Parameters such as temperature, pH, the concentration of the metal precursor, culture media, and amount of biomass predominantly regulate the characteristics and quality of nanoparticles and must be optimized for the synthesis of metal nanoparticles to achieve good stability, monodispersity, and biocompatibility (Balakumaran et al., 2015).

Temperature

Temperature is an indispensable factor as it greatly affects the reaction kinetics. It is reported in studies that higher temperature increases the rate of synthesis due to an increase in the collision between the reacting species (Phanjom and Ahmed 2017). In most cases, it is observed that the synthesis of nanoparticles by fungi requires a temperature between 10°C–80°C. An increase in temperature caused an increase in size due to the agglomeration of the formed nanoparticles (AbdelRahim et al., 2017).

pH

The literature survey presents that the pH of the reaction medium influences the size and texture of the synthesized nanoparticles. Usually, absorption spectra are measured from pH 2 to 12 and variations in particle size are subsequently recorded. pH change alters the properties of the electrical double layer thus influencing zeta potential and stability of the nanoparticles. Anigol et al. (2017) reported uniformity in the size of the synthesized particles at basic pH when compared to acidic pH. The reason could be attributed to the denaturation of protein at lower pH. The above was inferred from the X-ray diffraction pattern which indicated a narrow peak at basic pH.

Concentration of metal precursor

To achieve nanoparticles with desired characteristics, the concentration of metal precursors needs to be optimized. Commonly the concentrations of metal ions vary from 10^{-2} to 10^{-3} M while keeping other parameters constant. An irregular morphology with large-sized nanostructures was reported as a result of an excess amount of metal ions used during the synthesis. Using different concentrations of silver nitrate (1–10 mM), *Aspergillus oryzae* was used to synthesize silver nanoparticles of size 7.22–17.06 nm at 8 Mm metal ion concentration. It was also reported that the intensity of the color of dispersion is directly proportional to the concentration of metal ion precursor (Phanjom and Ahmed, 2017).

Effect of culture media

Substrate-specific culture medium is required for nanoparticle synthesis by fungi as it induces the production of enzymes required for the reduction of metal ions for the generation of nanoparticles (Husseiny et al., 2015). *Duddingyonia flagans* biomass was cultured in pure water and water containing insect carapaces. It was observed that filtrate having insect carapaces synthesized a greater number of nanoparticles. Because insect carapaces act as a natural source of chitin which is an important part of the fungal cell wall (Costa Silva et al., 2017).

Amount of biomass

Some research shows that nanoparticle synthesis is directly proportional to the amount of biomass, as a greater amount of enzymes are secreted, while some studies also demonstrate an indirect relationship between the two. Shahzad et al. (2019) studied the silver nanoparticle synthesis by using wet biomass of different weights 1, 4, 7, and 10 g. He observed that among all weights 7 g biomass weight gave rise to more mono-dispersed and small-size nanoparticles (94 nm).

Other factors

In nonbiogenic methods, chemicals or surfactants like sodium borohydride, hydrazine, and polyvinyl alcohol are used as an additional step after the synthesis of nanoparticles. In contrast to biogenic synthesis, in which the capping of nanoparticles occurs along with its synthesis by the release of biomolecules from the organisms used during the process (Chhipa 2019). Elgorban et al. (2016) with the aid of microscopic analysis revealed the presence of a thin layer of organic material around the synthesized particle which indicated the presence of a capping agent. In one of the studies, a peak at 280 nm was observed in the filtrate indicating the presence of amino acid residues corresponding to tryptophan and tyrosine during silver nanoparticle synthesis by *Aspergillus* species (Devi and Joshi, 2015).

Characterization of nanoparticles

Surface area, pore size, porosity, particle agglomeration and aggregation, adsorption potential, shape, size, crystallinity, fractal dimensions, orientation, intercalation, and dispersion of nanoparticles are some of the properties which are needed to be accessed during synthesis. A few important techniques for the characterization of synthesized nanoparticles are addressed below.

UV–visible spectroscopy

Nanoparticles have unique optical properties that are proportional to their shape and size and also help in determining their application. A color change is observed in the solution due to the excitation of surface Plasmon vibrations and the absorption peak is directly proportional to particle size. For example, silver shows an absorption peak specific around 400–450 nm while gold nanoparticles show a specific peak around 500–550 nm (Patra and Baek, 2014; Sachdeva et al., 2020).

Microscopic techniques

Scanning electron and transmission electron microscopy provides direct visualization of the particle synthesized while atomic force microscopy helps in quantitative information of the nanoparticles like their size (length, width, and height), and also helps in determining their physical properties like surface texture. In one of the researches, the size and shape of synthesized gold nanoparticles varied accordingly by 54.8, 110.13, and 261.73 nm showing spherical, triangular, and hexagonal shapes, respectively (Fatima et al., 2015).

Fourier transform infrared spectroscopy

It determines the nature of the functional group present on nanoparticles which helped in stabilization and capping of the same. The peak size of the spectrum represents the different types of functional groups present in the nanomaterials. It gives information based on the chemical composition and the physical state of the sample. The plausible reason is the bioreduction of Au(III) ions into Au(0) by the biomolecules released during the formation of nanoparticles which also provided capping to prevent agglomeration (González-Ballesteros et al., 2017; Patra and Baek 2014).

Energy-dispersive X-ray spectroscopy

It is used to determine the abundance of specific elements by analyzing the energy spectrum. It can be attached to a scanning electron microscope to analyze the energy of X-rays emitted which is a characteristic feature of each element. For example, the copper and oxygen weight compositions in the synthesized CuO nanoparticles were reported 80.65% and 16.96%, respectively (Patra and Baek, 2014; Ren et al., 2009).

Dynamic light scattering and zeta potential analysis

Dynamic light scattering measures the hydrodynamic diameter of nanoparticles, aggregation state, and/or its dispersion in solution. Zeta potential is used to determine the stability of nanoparticles in a colloidal solution and for surface charge determination of the particle. The higher the zeta potential, the greater the repulsion between particles which ultimately leads to lower agglomeration of the nanoparticles. Ganesan et al. (2020) observed sterling stability of ZnO nanoparticles because of a high zeta potential value of −88.6 mV.

Other techniques

Photon correlation spectroscopy is used for the study of the size distribution of nanoparticles and their motion in the medium by measuring the diffusion coefficient of the nanoparticle (Patra and Baek, 2014).

Application of mycosynthesis-based nanoparticles for dye degradation

An array of fungal strains and their various processes have been explored in recent years for treating various organic and inorganic contaminants. Along with white-rot fungi like *P. chrysosporium* other, species like *Aspergillus, Rhizopus, Pichia* have been found effective in the decolorization of various aromatic dyes. fungal bioreactors, biosorption, and myconanoparticles are the most promising mechanisms implemented for the degradation of dyes.

Fungal bioreactors

Bioreactors refer to a device or a system where organisms are cultivated either by submerging in liquid media or are attached as a biofilm over the supporting surface which is integrated into a reactor. Among the different supporting materials available for use in the system, polyurethane foam is considered the best, as it provides greater adhesion of biomass, shows mechanical resistance, and is economical (Mielgo et al., 2002; Sen et al., 2016). In another work, immobilization of *P. chrysosporium* for degradation of orange II dye in a continuous packed bed bioreactor was monitored. MnP activity showed a breakdown of chromophoric and aromatic groups exhibiting nearly complete decolorization with a hydraulic retention time of 24 hours (Mielgo et al., 2001). Hai et al. (2013) demonstrated the degradation of acid orange 7 operated in a batch mode membrane bioreactor by the formation of pellets of *C. versicolor* that exhibited excellent enzymatic activity and achieved greater than 95% decolorization. HPLC was used to detect the same which revealed a gradual reduction in the peak height found associated with the dye in the visible and UV region of wavelength.

Biosorption

Chitin being an important constituent of the fungal cell wall is considered one of the potent biosorbents of metals. In one of the researches, dead fungal biomass of *Aspergillus fumigates* was studied for the biosorption of methylene blue. Ambient room temperature, alkaline pH, and, 120 minutes of contact time were found suitable for achieving nearly 93.43% biosorption efficiency, but the biomass particle size showed an inverse relation to dye absorption capacity. The reason is greater biosorption on the wider surface of smaller-sized sorbent particles (Kabbout and Taha, 2014). A comparative biosorption study of 13 different dyes over dead biomass of three fungal strains namely *Aspergillus niger*, *A. oryzae*, and *Rhizopus arrhizus* were studied where pH was found to be an important parameter affecting the biosorption process of various dyes (Saraf and Vaidya, 2015).

Myconanoparticles

The mushroom *Flammulina velutipes* extract was used for the synthesis of crystalline structured gold nanoparticles arranged on the inner surface of the plasma membrane. It was tested for its efficiency toward the degradation of methylene blue as a function of time, the solution gradually became colorless due to the formation of leucomethylene blue denoted by a reduction in absorption peak (Narayanan et al., 2015). Literature surveys indicate that fungi exhibit higher metal tolerance, this ability of *Cladosporium oxysporum* AJPO3 was explored for the synthesis of gold nanoparticles and, catalysis of rhodamine B mediated by sodium borohydride (Bhargava et al., 2016). Protein-capped zinc oxide nanoparticles were manufactured using extracellular proteins synthesized by the fungal species *Aspergillus* and demonstrated around twofold higher photocatalytic activity. This higher catalytic efficiency was observed due to the protein synthesized by fungal mycelia and the absorption of dye with a lower rate of recombination of e^-/h^+ pairs (Jain et al., 2014). Mycosorption and mycoremediation are processes that act as potential alternatives and can be applied for the decolorization and degradation of dyes. Laccase (Lac), Manganese Peroxidase (MnP), and Lignin peroxidase (LiP) are a few of the oxidative enzymes that can be considered as important factors for dye degradation. In one of the works *Funalia trogii*, a white-rot fungus was harvested for pellets and was evaluated for effective decolorization of astrazon red dye. It revealed that a change in pH (6–11) does not affect the efficiency of degradation. Laccase, an enzyme that participated in catalytic reaction could be employed for the degradation of dyes (Yesilada et al., 2002). Ranimol et al. (2018) explored the fungus *Trichoderma hazianum* for the production of partly processed laccase enzyme using guaiacol as a substrate. They concluded that the constant flow of H_2O_2 in the former case led to greater decolorization (Jaouani et al., 2003). As discussed earlier, the synthesized nanoparticles possess some cytotoxicity. Hence to combat that, (Gao et al., 2022) recently developed eco-friendly vault nanoparticles where enzyme laccase was encapsulated. By using an extracellular extract from *Dictyota dichotoma*, ZnO nanoparticles were synthesized which resulted in 90% decolorization of fast green dye within 180 minutes (Kumar et al., 2022). The exploration of fungal strains for various nanoparticle formations and their subsequent roles in particular dye degradation is discussed in Table 14.1.

TABLE 14.1 Exploration of fungal strains for nanoparticle formation and dye degradation.

Fungi	Nanoparticle	Synthesis mechanism	Size (nm) and shape	Dye degradation	References
Aspergillus niger	ZnO	Extracellular	84–91, Spherical	Bismark brown	Kalpana et al. (2018)
Agaricus and *Phlebia* genera	–	Extracellular enzyme	ND	Rhodamine B	Hermann et al. (2020)
Penecillium oxalicum	Ag	Extracellular	4, Spherical	Methylene blue	Du et al. (2015)
Agaricus bisporus and *Ganoderma lucidum*	Ag	Extracellular extract	10–80	Direct blue 71	Sriramulu and Sumathi (2017)
Cladosporium oxysporum	Au	Extracellular synthesis	72.32 ± 21.80, Quasi-spherical fcc packing	Rhodamine B	Bhargava et al. (2016)
Flammulina velutipes	Au	Extracellular	74.32, triangular, spherical, and irregular	Methylene blue	Narayanan et al. (2015)
Agaricus bisporous	Au	Extracellular	53, oval, spherical, drum-like, hexagonal, and triangular shapes	Methylene blue	Dheyab et al. (2020)
Aspergillus species	Protein-capped ZnO	Extracellular	80–120, quasi-spherical, symmetrical, polydisperse	Methylene blue	Jain et al. (2014)
Trichoderma harzianum	ND	Extracellular	ND	Congo red, methylene blue, crystal violet and malachite green	Ranimol et al. (2018)
Funalia trogii	ND	Extracellular	ND	Astrazon red dye	Yesilada et al. (2002)
Pestalotiopsis versicolor	Ag	Extracellular	5–50, spherical	Congo red, Rhodamine B and Orange G	Rajput et al. (2017)
Rhodotorula glutinis	Ag	Extracellular	15.45, spherical	Methylene Blue	Cunha et al. (2018)
Penicillium oxalicum	Fe	Extracellular	140	Methylene Blue	Mathur et al. (2021)
Dictyota dichotoma	Zno	Extracellular	80	Fast green	Kumar et al. (2022)

ND, Not detected.

Conclusion

In today's world scenario, various metal and metal oxide nanoparticles find their applicability in the fields of medicine, industries, pharmaceuticals, and cosmetics. It is a one-pot synthesis method and is devoid of any harmful chemicals. This chapter demonstrates the effectiveness of myconanoparticles in the treatment of organic pollutants, and its antibiotic and antioxidant capability. Some factors like pH, temperature, amount of biomass, culture media, and metal ion concentration which play a crucial role in the synthesis of nanoparticles are also discussed. With the help of these factors, we can unravel the synthesis of green nanoparticles for better stability. However, a multidisciplinary approach and a deep understanding of nanomaterials, and their synthesis processes are required to determine their fate and impact on the environment.

Acknowledgment

The authors are grateful to the Department of Environmental Science and Engineering, Indian Institute of Technology (ISM) Dhanbad, India, for providing technical and financial support.

References

Abdelhalim, M.A.K., Mady, M.M., Ghannam, M.M., 2012. Physical properties of different gold nanoparticles: ultraviolet-visible and fluorescence measurements. Journal of Nanomedicine and Nanotechnology 3 (3). Available from: https://doi.org/10.4172/2157-7439.1000133.

AbdelRahim, K., et al., 2017. Extracellular biosynthesis of silver nanoparticles using Rhizopus stolonifer. Saudi Journal of Biological Sciences 24 (1), 208–216. Available from: https://doi.org/10.1016/j.sjbs.2016.02.025.

Adhikari, S., et al., 2017. Kinetics and mechanism of dye adsorption on WO3 nanoparticles. Applied Surface Science 420, 472–482. Available from: https://doi.org/10.1016/j.apsusc.2017.05.191.

Ahmad, A., et al., 2003. Extracellular biosynthesis of silver nanoparticles using the fungus Fusarium oxysporum. Colloids and Surfaces B: Biointerfaces 28 (4), 313–318. Available from: https://doi.org/10.1016/S0927-7765(02)00174-1.

Ali, J., et al., 2017. Insight into eco-friendly fabrication of silver nanoparticles by Pseudomonas aeruginosa and its potential impacts. Journal of Environmental Chemical Engineering 5 (4), 3266–3272. Available from: https://doi.org/10.1016/j.jece.2017.06.038.

Anigol, L.B., Charantimath, J.S., Gurubasavaraj, P.M., 2017. Effect of concentration and pH on the size of silver nanoparticles synthesized by green chemistry. Organic and Medicinal Chemistry International Journal Biosynthesis 3 (5), 1–5. Available from: https://doi.org/10.19080/OMCIJ.2017.03.555622.

Balaji, D.S., et al., 2009. Extracellular biosynthesis of functionalized silver nanoparticles by strains of Cladosporium cladosporioides fungus. Colloids and Surfaces B: Biointerfaces 68 (1), 88–92. Available from: https://doi.org/10.1016/j.colsurfb.2008.09.022.

Balakumaran, M.D., Ramachandran, R., Kalaichelvan, P.T., 2015. Exploitation of endophytic fungus, Guignardia mangiferae for extracellular synthesis of silver nanoparticles and their in vitro biological activities. Microbiological Research 178, 9–17. Available from: https://doi.org/10.1016/j.micres.2015.05.009.

Benkhaya, S., M'rabet, S., Harfi, A., 2020. A review on classifications, recent synthesis and applications of textile dyes. Inorganic Chemistry Communications 115, 107891.

Bharde, A., et al., 2006. Extracellular biosynthesis of magnetite using fungi. Small (Weinheim an der Bergstrasse, Germany) 2 (1), 135–141. Available from: https://doi.org/10.1002/smll.200500180.

Bhargava, A., et al., 2016. Utilizing metal tolerance potential of soil fungus for efficient synthesis of gold nanoparticles with superior catalytic activity for degradation of rhodamine B. Journal of Environmental Management 183, 22–32. Available from: https://doi.org/10.1016/j.jenvman.2016.08.021.

Buscio, V., Lopez-Grimau, V., Alvarez, M.D., Gutierrez-Bouzan, C., et al., 2019. Reducing the environmental impact of textile industry by reusing residual salts and water: ECUVal system. Chemical Engineering Journal 373. Available from: https://doi.org/10.1016/j.cej.2019.04.146.

Chen, H., Zhou, K., Zhao, G., 2018. Gold nanoparticles: from synthesis, properties to their potential application as colorimetric sensors in food safety screening. Trends in Food Science and Technology 78, 83–94. Available from: https://doi.org/10.1016/j.tifs.2018.05.027.

Chhipa, H., 2019. Mycosynthesis of nanoparticles for smart agricultural practice: a green and eco-friendly approach. Green Synthesis, Characterization and Applications of Nanoparticles. Elsevier Inc. Available from: https://doi.org/10.1016/b978-0-08-102579-6.00005-8.

Clarance, P., et al., 2020. Green synthesis and characterization of gold nanoparticles using endophytic fungi Fusarium solani and its in-vitro anticancer and biomedical applications. Saudi Journal of Biological Sciences 27 (2), 706–712. Available from: https://doi.org/10.1016/j.sjbs.2019.12.026.

Costa Silva, L.P., et al., 2017. Extracellular biosynthesis of silver nanoparticles using the cell-free filtrate of nematophagous fungus Duddingtonia flagrans. International Journal of Nanomedicine 12, 6373–6381. Available from: https://doi.org/10.2147/IJN.S137703.

Cunha, F.A., et al., 2018. Biogenic synthesis of multifunctional silver nanoparticles from Rhodotorula glutinis and Rhodotorula mucilaginosa: antifungal, catalytic and cytotoxicity activities. World Journal of Microbiology and Biotechnology 34 (9), 127. Available from: https://doi.org/10.1007/s11274-018-2514-8.

Das, M., Chatterjee, S., et al., 2019. Green synthesis of metal/metal oxide nanoparticles toward biomedical applications: Boon or bane. Green Synthesis, Characterization and Applications of Nanoparticles. Elsevier, pp. 265–301.

Devi, L., Joshi, S., 2015. Ultrastructures of silver nanoparticles biosynthesized using endophytic fungi. Journal of Microscopy and Ultrastructure 3 (1), 29. Available from: https://doi.org/10.1016/j.jmau.2014.10.004.

Dheyab, M.A., et al., 2020. Mycosynthesis of gold nanoparticles by the Portabello mushroom extract, Agaricaceae, and their efficacy for decolorization of Azo dye. Environmental Nanotechnology, Monitoring and Management 14, 100312. Available from: https://doi.org/10.1016/j.enmm.2020.100312.

dos Santos, A.B., Cervantes, F.J., van Lier, J.B., 2007. Review paper on current technologies for decolourisation of textile wastewaters: Perspectives for anaerobic biotechnology. Bioresource Technology 98 (12), 2369–2385. Available from: https://doi.org/10.1016/j.biortech.2006.11.013.

Du, L., et al., 2015. Synthesis of small silver nanoparticles under light radiation by fungus Penicillium oxalicum and its application for the catalytic reduction of methylene blue. Materials Chemistry and Physics 160, 40–47. Available from: https://doi.org/10.1016/j.matchemphys.2015.04.003.

Elgorban, A.M., et al., 2016. Extracellular synthesis of silver nanoparticles using Aspergillus versicolor and evaluation of their activity on plant pathogenic fungi. Mycosphere 7 (6), 844–852. Available from: https://doi.org/10.5943/mycosphere/7/6/15.

Fatima, F., et al., 2015. Antimicrobial and immunomodulatory efficacy of extracellularly synthesized silver and gold nanoparticles by a novel phosphate solubilizing fungus Bipolaris tetramera. BMC Microbiology 15 (1), 1–10. Available from: https://doi.org/10.1186/s12866-015-0391-y.

Fatima, F., et al., 2016. Extracellular mycosynthesis of silver nanoparticles and their microbicidal activity. Journal of Global Antimicrobial Resistance 7, 88–92. Available from: https://doi.org/10.1016/j.jgar.2016.07.013.

Fayaz, M., et al., 2010. Blue orange light emission from biogenic synthesized silver nanoparticles using Trichoderma viride. Colloids and Surfaces B: Biointerfaces 75 (1), 175–178. Available from: https://doi.org/10.1016/j.colsurfb.2009.08.028.

Firdhouse, M.J., Lalitha, P., 2015. Biosynthesis of silver nanoparticles and its applications. Journal of Nanotechnology 2015. Available from: https://doi.org/10.1155/2015/829526.

Ganesan, V., et al., 2020. Periconium sp. (endophytic fungi) extract mediated sol-gel synthesis of ZnO nanoparticles for antimicrobial and antioxidant applications. Materials Science in Semiconductor Processing 105, 104739. Available from: https://doi.org/10.1016/j.mssp.2019.104739.

Gao, Y., et al., 2022. Decolorization and detoxification of synthetic dye compounds by laccase immobilized in vault nanoparticles. Bioresource Technology 351, 127040. Available from: https://doi.org/10.1016/j.biortech.2022.127040.

González-Ballesteros, N., et al., 2017. Green synthesis of gold nanoparticles using brown algae Cystoseira baccata: its activity in colon cancer cells. Colloids and Surfaces B: Biointerfaces 153, 190–198. Available from: https://doi.org/10.1016/j.colsurfb.2017.02.020.

Gopinath, K., Arumugam, A., 2014. Extracellular mycosynthesis of gold nanoparticles using Fusarium solani. Applied Nanoscience (Switzerland) 4 (6), 657–662. Available from: https://doi.org/10.1007/s13204-013-0247-4.

Goswami, A.M., Ghosh, S., 2013. Biological synthesis of colloidal gold nanoprisms using *Penicillium citrinum* MTCC9999. Journal of Biomaterials and Nanobiotechnology 4 (2), 20–27. Available from: https://doi.org/10.4236/jbnb.2013.42a003.

Guilger-Casagrande, M., Lima, R.D., 2019. Synthesis of silver nanoparticles mediated by fungi: a review. Frontiers in Bioengineering and Biotechnology 7, 1–16. Available from: https://doi.org/10.3389/fbioe.2019.00287.

Gupta, V.K., Suhas, 2009. Application of low-cost adsorbents for dye removal – A review. Journal of Environmental Management 90 (8), 2313–2342.

Hai, F.I., et al., 2013. Degradation of azo dye acid orange 7 in a membrane bioreactor by pellets and attached growth of Coriolus versicolour. Bioresource Technology 141, 29–34. Available from: https://doi.org/10.1016/j.biortech.2013.02.020.

Hermann, K.L., et al., 2020. Discoloration of rhodamine b dye by white-rot fungi in solid bleached sulfate paperboard coated with polyethylene terephthalate: scale-up into non-sterile packed-bed bioreactor. Journal of Environmental Chemical Engineering 8 (3), 103685. Available from: https://doi.org/10.1016/j.jece.2020.103685.

Husseiny, S.M., Salah, T.A., Anter, H.A., 2015. Biosynthesis of size controlled silver nanoparticles by Fusarium oxysporum, their antibacterial and antitumor activities. Beni-Suef University Journal of Basic and Applied Sciences 4 (3), 225–231. Available from: https://doi.org/10.1016/j.bjbas.2015.07.004.

Islam, M.A., et al., 2019. Removal of dye from polluted water using novel nano manganese oxide-based materials. Journal of Water Process Engineering 32, 100911. Available from: https://doi.org/10.1016/j.jwpe.2019.100911.

Jain, N., Bhargava, A., Panwar, J., 2014. Enhanced photocatalytic degradation of methylene blue using biologically synthesized "protein-capped" ZnO nanoparticles. Chemical Engineering Journal 243, 549–555. Available from: https://doi.org/10.1016/j.cej.2013.11.085.

Jamkhande, P.G., et al., 2019. Metal nanoparticles synthesis: an overview on methods of preparation, advantages and disadvantages, and applications. Journal of Drug Delivery Science and Technology 53, 101174. Available from: https://doi.org/10.1016/j.jddst.2019.101174.

Jaouani, A., Sayadi, S., Vanthournhout, M., Penninckx, M.J., 2003. Potent fungi for decolourisation of olive oil mill wastewaters. Enzyme and Microbial Technology 33 (6), 802–809.

Kabbout, R., Taha, S., 2014. Biodecolorization of textile dye effluent by biosorption on fungal biomass materials. Physics Procedia 55, 437–444. Available from: https://doi.org/10.1016/j.phpro.2014.07.063.

Kalpana, V.N., et al., 2018. Biosynthesis of zinc oxide nanoparticles using culture filtrates of Aspergillus niger: antimicrobial textiles and dye degradation studies. OpenNano 3, 48–55. Available from: https://doi.org/10.1016/j.onano.2018.06.001.

Karagoz, I., 2019. Cmbebih 2019. IFMBE Proceedings – CMBEBIH 73, 159–163. Available from: https://doi.org/10.1007/978-3-030-17971-7.

Kuchibhatla, S.V.N.T., et al., 2012. Influence of aging and environment on nanoparticle chemistry: Implication to confinement effects in nanoceria. Journal of Physical Chemistry C 116 (26), 14108–14114. Available from: https://doi.org/10.1021/jp300725s.

Kumar, R.V., et al., 2022. Synthesis of zinc oxide nanoparticles mediated by Dictyota dichotoma endophytic fungi and its photocatalytic degradation of fast green dye and antibacterial applications. South African Journal of Botany 151, 1–8. Available from: https://doi.org/10.1016/j.sajb.2022.03.016.

Küünal, S., Rauwel, P., Rauwel, E., 2018. Plant extract mediated synthesis of nanoparticles. Emerging Applications of Nanoparticles and Architectural Nanostructures: Current Prospects and Future Trends. Elsevier Inc. Available from: https://doi.org/10.1016/B978-0-323-51254-1.00014-2.

Lellis, B., et al., 2019. Effects of textile dyes on health and the environment and bioremediation potential of living organisms. Biotechnology Research and Innovation 3 (2), 275–290. Available from: https://doi.org/10.1016/j.biori.2019.09.001.

Li, Q., et al., 2022. Nanoparticle and nanomineral production by fungi. Fungal Biology Reviews 41, 31–44. Available from: https://doi.org/10.1016/j.fbr.2021.07.003.

Malik, P., et al., 2014. Green chemistry based benign routes for nanoparticle synthesis. Journal of Nanoparticles 2014, 1–14. Available from: https://doi.org/10.1155/2014/302429.

Mathur, P., et al., 2021. Endophytic fungi mediated synthesis of iron nanoparticles: Characterization and application in methylene blue decolorization. Current Research in Green and Sustainable Chemistry 4, 100053. Available from: https://doi.org/10.1016/j.crgsc.2020.100053.

Mielgo, I., et al., 2001. A packed-bed fungal bioreactor for the continuous decolourisation of azo-dyes (Orange II. Journal of Biotechnology 89 (2–3), 99–106. Available from: https://doi.org/10.1016/S0168-1656(01)00319-4.

Mielgo, I., et al., 2002. Biodegradation of a polymeric dye in a pulsed bed bioreactor by immobilised Phanerochaete chrysosporium. Water Research 36 (7), 1896–1901. Available from: https://doi.org/10.1016/S0043-1354(01)00384-0.

Nandhini, N.T., Rajeshkumar, S., Mythili, S., 2019. The possible mechanism of eco-friendly synthesized nanoparticles on hazardous dyes degradation. Biocatalysis and Agricultural Biotechnology 19, 101138. Available from: https://doi.org/10.1016/j.bcab.2019.101138.

Narayanan, K.B., Park, H.H., Han, S.S., 2015. Synthesis and characterization of biomatrixed-gold nanoparticles by the mushroom Flammulina velutipes and its heterogeneous catalytic potential. Chemosphere 141, 169–175. Available from: https://doi.org/10.1016/j.chemosphere.2015.06.101.

Noman, M., Shahid, M., Taimoor, A., Niazi, M., Hussain, S., Song, F., Manzoor, I., 2020. Use of biogenic copper nanoparticles synthesized from a native sp. as Escherichia photocatalysts for azo dye degradation and treatment of textile effluents. Environmental Pollution 257, 113514.

Owaid, M.N., 2019. Green synthesis of silver nanoparticles by Pleurotus (oyster mushroom) and their bioactivity: review. Environmental Nanotechnology, Monitoring and Management 12, 100256. Available from: https://doi.org/10.1016/j.enmm.2019.100256.

Patra, J.K., Baek, K.H., 2014. Green nanobiotechnology: factors affecting synthesis and characterization techniques. Journal of Nanomaterials 2014. Available from: https://doi.org/10.1155/2014/417305.

Phanjom, P., Ahmed, G., 2017. Effect of different physicochemical conditions on the synthesis of silver nanoparticles using fungal cell filtrate of Aspergillus oryzae (MTCC No. 1846) and their antibacterial effect. Advances in Natural Sciences: Nanoscience and Nanotechnology 8 (4). Available from: https://doi.org/10.1088/2043-6254/aa92bc.

Raina, S., Roy, A., Bharadvaja, N., 2020. Degradation of dyes using biologically synthesized silver and copper nanoparticles. Environmental Nanotechnology, Monitoring and Management 13, 100278. Available from: https://doi.org/10.1016/j.enmm.2019.100278.

Rajput, K., Sharma, J., Agrawal, P.K., 2017. Mycosynthesis of silver nanoparticles using endophytic fungus Pestalotiopsis versicolor and investigation of its antibacterial and azo dye degradation efficacy. Kavaka 71 (49), 65–71.

Ranimol, G., Venugopal, T., Gopalakrishnan, S., Sunkar, S., 2018. Production of laccase from Trichoderma harzianum and its application in dye decolourisation. Biocatalysis and Agricultural Biotechnology 16, 400–404.

Reid, D.A., Webster, J., 1973. Introduction to fungi. Kew Bulletin 28 (1), 166. Available from: https://doi.org/10.2307/4117096.

Ren, G., et al., 2009. Characterisation of copper oxide nanoparticles for antimicrobial applications. International Journal of Antimicrobial Agents 33 (6), 587–590. Available from: https://doi.org/10.1016/j.ijantimicag.2008.12.004.

Sachdeva, A., Singh, S., Singh, P.K., 2020. Synthesis, characterisation and synergistic effect of ZnO nanoparticles to antimicrobial activity of silver nanoparticle. Materials Today: Proceedings 34, 649–653. Available from: https://doi.org/10.1016/j.matpr.2020.03.150.

Sanghi, R., Verma, P., 2009a. A facile green extracellular biosynthesis of CdS nanoparticles by immobilized fungus. Chemical Engineering Journal 155 (3), 886–891. Available from: https://doi.org/10.1016/j.cej.2009.08.006.

Sanghi, R., Verma, P., 2009b. Biomimetic synthesis and characterisation of protein capped silver nanoparticles. Bioresource Technology 100 (1), 501–504. Available from: https://doi.org/10.1016/j.biortech.2008.05.048.

Saraf, S., Vaidya, V.K., 2015. Comparative study of biosorption of textile dyes using fungal biosorbents. International Journal of Current Microbiology and Applied Sciences 2 (2), 357–365. Available from: https://www.ijcmas.com/special/2/Sukhada%20Saraf%20and%20Varsha%20K.%20Vaidya.pdf.

Sen, S.K., et al., 2016. Fungal decolouration and degradation of azo dyes: a review. Fungal Biology Reviews 30 (3), 112–133. Available from: https://doi.org/10.1016/j.fbr.2016.06.003.

Shahzad, A., et al., 2019. Size-controlled production of silver nanoparticles by Aspergillus fumigatus BTCB10: likely antibacterial and cytotoxic effects. Journal of Nanomaterials 2019. Available from: https://doi.org/10.1155/2019/5168698.

Shankar Naik, B., 2020. Biosynthesis of silver nanoparticles from endophytic fungi and their role in plant disease management. Microbial Endophytes. Elsevier Inc. Available from: https://doi.org/10.1016/b978-0-12-819654-0.00012-0.

Siddiqi, K.S., Husen, A., 2016. Fabrication of metal nanoparticles from fungi and metal salts: scope and application. Nanoscale Research Letters 11 (1), 1–15. Available from: https://doi.org/10.1186/s11671-016-1311-2.

Sneha, K., et al., 2010. Corynebacterium glutamicum-mediated crystallization for silver ions through sorption and reduction process. Chemical Engineering Journal 162, 989–996.

Sonal, S., Singh, A., Mishra, B.K., 2018. Decolorization of reactive dye Remazol Brilliant Blue R by zirconium oxychloride as a novel coagulant: optimization through response surface methodology. Water Science and Technology 78 (2), 379–389. Available from: https://doi.org/10.2166/wst.2018.307.

Sriramulu, M., Sumathi, S., 2017. Photocatalytic, antioxidant, antibacterial and anti-inflammatory activity of silver nanoparticles synthesised using forest and edible mushroom. Advances in Natural Sciences: Nanoscience and Nanotechnology 8 (4). Available from: https://doi.org/10.1088/2043-6254/aa92b5.

Srivastava, S., et al., 2019. Production, characterization and antibacterial activity of silver nanoparticles produced by Fusarium oxysporum and monitoring of protein-ligand interaction through in-silico approaches. Microbial Pathogenesis 129, 136–145. Available from: https://doi.org/10.1016/j.micpath.2019.02.013.

Tarver, S., et al., 2019. Biomineralization of Pd nanoparticles using Phanerochaete chrysosporium as a sustainable approach to turn platinum group metals (PGMs) wastes into catalysts. International Biodeterioration and Biodegradation 143, 104724. Available from: https://doi.org/10.1016/j.ibiod.2019.104724.

Vágó, A., et al., 2016. One-step green synthesis of gold nanoparticles by mesophilic filamentous fungi. Chemical Physics Letters 645, 1–4. Available from: https://doi.org/10.1016/j.cplett.2015.12.019.

Vahabi, K., Mansoori, G.A., Karimi, S., 2011. Biosynthesis of silver nanoparticles by fungus Trichoderma reesei (a route for large-scale production of AgNPs). Insciences Journal 1 (1), 65–79. Available from: https://doi.org/10.5640/insc.010165.

Vanti, G.L., et al., 2019. Synthesis and application of chitosan-copper nanoparticles on damping off causing plant pathogenic fungi. International Journal of Biological Macromolecules 156, 1387–1395. Available from: https://doi.org/10.1016/j.ijbiomac.2019.11.179.

Varjani, S., et al., 2020. Microbial degradation of dyes: an overview. Bioresource Technology 123728. Available from: https://doi.org/10.1016/j.biortech.2020.123728.

Verma, V.C., et al., 2012. Biofabrication of anisotropic gold nanotriangles using extract of endophytic Aspergillus clavatus as a dual functional reductant and stabilizer. Nanoscale Research Letters 7, 1–7.

Vigneshwaran, N., et al., 2007. Biological synthesis of silver nanoparticles using the fungus Aspergillus flavus. Materials Letters 61 (6), 1413–1418. Available from: https://doi.org/10.1016/j.matlet.2006.07.042.

Yesilada, O., Cing, S., Asma, D., 2002. Decolourisation of the textile dye Astrazon Red FBL by Funalia trogii pellets. Bioresource Technology 81 (2), 155–157. Available from: https://doi.org/10.1016/S0960-8524(01)00117-1.

Yosri, M., Abdel-Aziz, M.M., Sayed, R.M., 2018. Larvicidal potential of irradiated myco-insecticide from Metarhizium anisopliae and larvicidal synergistic effect with its mycosynthesized titanium nanoparticles (TiNPs). Journal of Radiation Research and Applied Sciences 11 (4), 328–334.

Zielonka, A., Klimek-Ochab, M., 2017. Fungal synthesis of size-defined nanoparticles. Advances in Natural Sciences: Nanoscience and Nanotechnology 8 (4). Available from: https://doi.org/10.1088/2043-6254/aa84d4.

Chapter 15

Heavy metals in water: challenges and remediation

S. Senthilkumar[1], V. Siva[2], Arumugam Murugan[3], C.R. Ravikumar[4], Ponnusamy Thillai Arasu[5], Arumugam Manohar[6] and Segni Asafa Ashana[5]

[1]*Department of Chemistry, Ayya Nadar Janaki Ammal College, Sivakasi, Tamil Nadu, India,* [2]*Department of Physics, Karpagam Academy of Higher Education, Coimbatore, Tamil Nadu, India,* [3]*Department of Chemistry, North Eastern Regional Institute of Science and Technology, Nirjuli, Itanagar, Arunachal Pradesh, India,* [4]*Research Centre, Department of Chemistry, East West Institute of Technology, Bangalore, Karnataka, India,* [5]*College of Natural and Computational Sciences, Wollega University, Nekemte, Oromiya, Ethiopia,* [6]*Department of Chemistry, Periyar Maniammai Institute of Science and Technology, Thanjavur, Tamil Nadu, India*

Introduction

The availability of clean fresh water is a fundamental requirement for drinking purposes for wildlife and humans. However, water demand increases global level year by year, and the different types of pollutants have compromised the favorable water resources. Furthermore, scientists have established that a major impression of environmental changes and water cycle changes may lead to flood, drought, and toxicity of chemical pollutants in the atmosphere (Hua et al., 2012; Vargas-Solano et al., 2022). The pollution of water with metals with higher atomic weight is one of the unhealthiest pollutants in the world due to its indestructible characteristics (Fei and Hu, 2022). Due to the modernization and fast metropolitan of metal plating, tanneries, mining, batteries, painting, printing, paper industries, car radiator production, photography industries, and fertilizer industries, the existence of heavy metal ions in wastewater has increased and those directly or indirectly affecting living things and the environment (Pohl, 2020). Specifically, in developing countries, a major problem is the heavy metal-contaminated sewage in the environment such as lakes, rivers, streams, and oceans, and it generates a hazard to the ecosystem and wellness (Arora, 2019). Based on the guidance of the WHO, health problems started by unacceptable stage and permissible degrees in beverages, the regular metals with heavy atomic weight present in wastewater and their origin should be treated well to minimize the hazardous effect (Peng et al., 2014).

Sources of metal pollution

The main sources of water quality contaminants include mining, processing of petroleum, steel, smelting facilities, pulp and paper mills, textile and agricultural businesses, etc. The wastewater gets into natural resources and may cause changes in the physicochemical composition of water and ultimately becomes unsuitable for use (Förstner and Wittmann, 1979; Clarkson, 1997). The main natural sources of mercury are vulcanism on land and under the sea, as well as weathering processes. Mercury concentrations of 0.05–0.31 ppb were found in thermal water from Yellowstone National Park (White et al., 1970). The physical consequence of mercury toxin has been registered in a study proposed by the Swedish government (Nordberg et al., 2014). In 1969, Berglund and Berlin highlighted a noticeable indication of mercury toxin present in the range of 0.2–0.6 ppm (Tsuchiya, 1969). Because of the increasing use of nickel in industries such as alloy preparation and electroplating, nickel finds its way into natural bodies of water discharged in sewers (Malik et al., 2014; Patterson, 1978). Due to the accumulation of cadmium in crop plants and eventual transmission to humans, fertilizers, which carry up to 8 ppm cadmium, are another origin of pollution (Genchi et al., 2020). Natural quantities of the manganese complex in soils, plants, and animals are frequently too large that significant increases by human commercial behavior would be difficult (Pinsino et al., 2012).

Role of Green Chemistry in Ecosystem Restoration to Achieve Environmental Sustainability. DOI: https://doi.org/10.1016/B978-0-443-15291-7.00014-6

Existing water treatment processes

Exposure to heavy metals is dangerous to humans. Therefore, it is crucial to remove those undesired heavy metals thoroughly and completely from water systems. To eliminate massive metals from the polluted water by several methods presented in Fig. 15.1.

Ion exchange

Ionic exchange can be interpreted as the reciprocating of ions between the substrate state and the neighboring space and is mostly reversible, the ion exchanger can be utilized by several times. Adhesives are commonly made in the shape spherical and have compression and strain-free forms to withstand corrosion (Bashir et al., 2019). These inventions resist decomposition by antioxidant factors namely chlorine and can endure strain and stress that break down weak long-lasting substances (Víctor-Ortega et al., 2016). For more rigid partitions, little particle dimension or smaller interconnected materials are required. Ion exchangers are extensively employed in various industries and used in pollution control (Dąbrowski et al., 2004). The list of industrially recovered and refined metals by ion exchange includes uranium, thorium, gold, platinum, silver, copper, cobalt, nickel, zinc, chromium, and tungsten.

Photocatalytic process

A lot of attention to wastewater treatment such as photocatalytic technology because of its distant reaction level, conservation of energy, high efficiency, and vigorous redox efficiency (Deng et al., 2019). This novel method has been used in numerous research areas, including selective oxidation, and disinfection of microorganisms, including the decomposition of organic compounds, the production of hydrogen from water decomposition, the production of liquid chemical fuels, and photosynthesis (Deliyanni et al., 2009). Nanosized metal oxides, nano synthetic manganese oxides, titanium oxides, ferric oxides, magnesium oxides, cerium oxides, and aluminum oxides are among the adsorbents that are readily used because of their large surfaces and increased performance due to the calibration effect (Pradeep and Anshup, 2009; El-Sayed, 2001; Zhao et al., 2011).

Adsorption process

The adsorption technique is commonly utilized to eradicate HMs from wastewater (WW) since it is affordable, accessible, and ecologically beneficial. Both economic chemisorption and bio-chemisorption are utilized to separate HM from

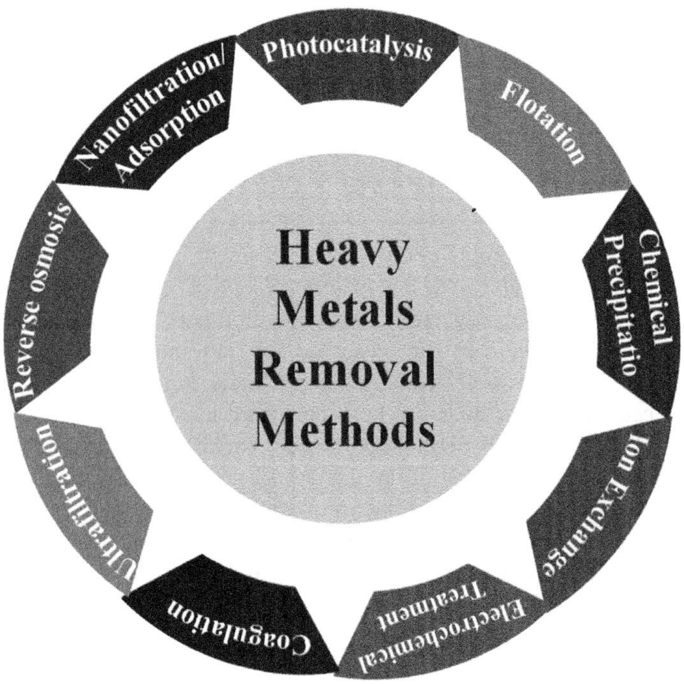

FIGURE 15.1 Various methods of the heavy metal removal process.

WW with a huge elimination capacity (Gopalakrishnan et al., 2015). Graphene has a high specific surface area is a carbon-based nanomaterial and holds higher stability of the structure (Thangavel and Venugopal, 2014).

Membrane filtration

In an aqueous solution, heavy metal ions can be captured by chemical/physical adsorption. Because it has strong contact with heavy metals and high absorption capacity (Algieri et al., 2021). To remove heavy metals from wastewater, many types of membrane filters can be used, depending on the size of the retained particles (Zhao et al., 2021). Due to its unusual behavior, UF can pass through water and less-molecular mass solutions while holding onto molecules that are heavier than the pore dimension of the membrane. The ultrafiltration was built of cellulose YM10 produced in Amigon. At pHs of 8.5 and 9.5, the Cu (II) and Zn (II) ions, correspondingly, had rejection rates of about 100% and 95%. The outcomes indicate that the removal of cytosine metals was significantly improved 6–10 times, rather than membrane use alone (Fujioka et al., 2015). Nanofiltration (NF) and reverse osmosis methods are also utilized to extract heavy metals from sewage. RO membranes remove all ions, including monovalent ions, which are highly removable; For example, commercial RO membranes used to desalinate seawater exhibit 99.5%–99.8% rejection values for sodium chloride (Petrinic et al., 2015). Recently, NF and RO thin film (TFC) membranes have become increasingly popular for better salt rejection, higher water flow, and interesting mechanical resistance.

Advantages and disadvantages of existing treatments

The advantages and disadvantages of various heavy metal removal techniques are given in Table 15.1 (Yang et al., 2019; Pakdel and Peighambardoust, 2018).

Removal of heavy metals by nanomaterials methods

Pollution is described as the harmful reaction to a limited environmental form by the inclusion of the waste products of a community (Bockris, 1977; Hänsch and Mendel, 2009). Electrochemical methods use metal ion removal, which keeps up to improve to connect the dispute between lower consent ranges and more complex sewage formation (Mishra et al., 2022; Zehra et al., 2020). Disposal of the heavy metal ions from a water medium includes physical techniques, such as reverse osmosis, electrodialysis, adsorption by the activated carbon, *etc.*, chemical techniques such as sulfide precipitation and ion-exchange, solvent extraction, and biological methods such as adsorption by aquatic plants like water *Hyacinth*, duckweed, *etc.* Disposal of the low concentration of heavy metal ions is more effective by ion exchange or adsorption on solid adsorbents (Sis and Uysal, 2014; Dąbrowski, 2001). Of the various methods attainable for the elimination of HM ions, the adsorption technique is the most commonly used method (Rouquerol et al., 2013). The adsorption process is rapidly gaining prominence as a method of treating industrial effluents and for the separation of the HM ions from the wastewater since it appears to be the most effective, quick, sludge-free, and quick operation (Vidali et al., 1991). There are two types of adsorption methods, which are physisorption or physical adsorption or van der Waals adsorption and chemisorption (Muscat and Newns, 1978). The adsorbed molecules are held to the surface via chemisorption (Yang et al., 2019) by covalent forces of the same common class as those that exist between bound atoms in molecules.

TABLE 15.1 Benefits and drawbacks of water treatment approaches.

Method	Advantages	Disadvantages
Adsorption	Flexibility, ease of use, simplicity of design, and immunity to hazardous contaminants	Adsorbents need to be renewed
Ion exchange	Decent removal suitable for a variety of heavy metals	Adsorbents need to be recycled or thrown away
Membrane filtration technologies	Respectable heavy metal removal	Costly and concentrated sludge manufacturing
Electrochemical treatment	A process that is quick and efficient for some metal ions	Large particle formation and high energy costs
Biological treatment	Effective towards some HM	High production of sludge

Materials available for the elimination of the HM ions

The nanomaterials that have been widely used in the elimination of HM ions include metal oxide-related nanomaterials (oxides of zinc, iron, aluminum, manganese, and titanium materials), carbon-related nanomaterials (graphene and carbon nanotubes), silica-related nanomaterials, zero-valent metal-based nanomaterials (Ag nanoparticles, and Au nanoparticles), and nanoparticles with zero charges (inorganic-based nanocomposites, organic polymer-based nanocomposites, and magnetic nanocomposites) (Keng et al., 2014). To use these nanomaterials more effectively in water treatment, however, significant obstacles still need to be removed. The majority of nanomaterials are unstable and have a propensity to agglomerate, which lowers their capability for removal. The nanoscale size of the nanomaterials makes it often difficult to quickly and effectively remove them from the aqueous solution. To lessen the harmful outcome of nanomaterials on both the people and the environment, regulations on their use are urgently needed (Ince and Ince, 2017). When creating nanomaterials, it is significant to examine the hazards and consequences (Lim et al., 2018).

Graphene oxide

Additionally, graphene can be produced using both bottom-up and top-down approaches, including chemical exfoliation, chemical vapor deposition, and other processes, allowing it to be further functionalized to create nanomaterials based on GO (Liu et al., 2019; Karabörk et al., 2022). The material's recovery efficiency was found to be 98.65%−101.75% for zinc(II) ions and 98.55%−104% for cobalt(II) ions, respectively (Lee et al., 2022). GO has more reactive hydroxyl, epoxy, and carboxyl groups on its surface, yet it is unable to eliminate negatively charged species like anion metals. Gadolinium oxide added to GO makes it easier to remove heavy metals that are both cations and anions. However, due to the light flakes, friable, and thin, the result is almost unusable. To fix these issues, GO-Gd_2O_3 is bound into granules using a polymer combination of carboxymethyl cellulose and sodium alginate, whose capacity to adsorb Cr(III), As(V), and Pb(II) has been confirmed. In agreement with the thermodynamic data, Pb(II) adsorption is exothermic, but Cr(III) and As(V) adsorptions are endothermic (Verma et al., 2022). The utilization of GO and 13X zeolite *via* a method of chemical reduction, a composite material called 13X zeolite/reduced 13X-rGO with a hierarchically porous structure has been effectively developed. Under conditions of calcium substance of 20 ppm and earliest lead concentration of 150 ppb, the waste lead concentration would be reduced to the WHO-allowed level (10 ppb) in 30 seconds. The 13X-rGO simultaneously demonstrated remarkable adsorption performance, exceeding the WHO requirement (6.07 mg/g), by roughly 2.23 times the amount of one 13X zeolite (2.72 mg/g). Due to its quick adsorption, cheap cost, and the possibility of recycling, 13X-rGO can therefore be established as a viable adsorption material for the refinement of consumable water polluted with lead (Liu et al., 2022). The amine and carboxyl groups of CS/GO substances may coordinate U(VI) and produce an electrical double layer (EDL) as part of the electrosorption mechanism. Further evidence for the effective use of CS/GO electrode substances for the electrosorption elimination of U(VI) from radioactive wastewater came from the electrosorption-desorption cycles (Singh et al., 2022). In a study, solvothermal synthesis is used to create enhanced ZnO on GO nanoparticles. The nanocomposite or nanohybrid's Cr(VI) adsorption kinetics and elimination potential are improved by the deposition of ZnO nanoparticles on GO for the nanohybrid (ZnO-GO). The efficacy of elimination decreases under alkaline conditions while increasing from a high acidic to a neutral pH. According to the thermodynamics analysis, Cr(VI) adsorption on ZnO-GO is entropy-driven, endothermic, and spontaneous. Gibb's Free Energy's negative value illustrates the thermodynamic automation and viability of the sorption activities. The created nanocomposites have shown to be great choices for removing Cr(VI) from natural wastewater systems and water bodies (Guan et al., 2022).

Layered double hydroxides

Layered double hydroxides (LDHs) are ionic laminar composites built using layers that resemble brucite and are positively charged, with charged reimbursed anions and solvation fragments in the interlayer area. The problems and views would be put forth based on a detailed analysis of the most recent advancements, providing encouraging insights on environmental purification using LDHs-based materials (Dong et al., 2022; Roy et al., 2022; Huang et al., 2022). When herbicides glyphosate (GLY)/2-methyl-4-chlorophenoxyacetic acid (MCP) and heavy metal Cu^{2+}/Pb^{2+} elimination accomplishment of LDH@Bt for soil is tested, the elimination rates of LDH@Bt composite for 4 pollutants are greater than those of LDHs and Bt. Additionally, various adsorption processes for the precipitation-based removal of Cu^{2+}/Pb^{2+} and the surface-based removal of GLY/MCP are proposed. By conducting column leaching studies and magnetic collectability tests, the application effectiveness and retrieval capacity of the LDH@Bt composite in soil is assessed. A favorable technique for correction and eradication of HMs and pesticide pollutants from the soil is made possible by good elimination accomplishment and magnetic collectability (Cheng et al., 2022).

Electrocoagulation

Due to its simplicity of use, adaptability, eco-friendliness, and small environmental impact, electrocoagulation (EC) is a well-known WW refinement technology that has been extensively investigated for a wide span of WW contaminants (Mollah et al., 2004; Song et al., 2022). For the refinement of Cu-EDTA WW, a combination electrooxidation and electrocoagulation (EO-EC) reactor utilizing RuO_2-IrO_2/Ti and Al electrodes has been constructed (Jain et al., 2022). On EO-EC performance, the outcome of current density, electrolyte, NaCl concentration, pH, and starting concentration is examined. A larger current density resulted in greater removal efficiency of Cu-EDTA. The electrolyte type played a key influence in the EO-EC process. The performance of the EO-EC reactor is better in neutral or alkaline solution than in acidic solution. Cu and COD removal efficiencies achieved 99.85% and 85.01%, correspondingly, within 60 minutes at the conditions of current density 10.29 mA cm^2, C_0(NaCl) 1 g L^{-1}, C_0(Cu) 50 mg L^{-1}, and pH 7. The removal effectiveness and energy usage of the electrocoagulation process are investigated to determine the best conditions for removing lead (Pb), cobalt (Co), and manganese (Mn) ions from simulated wastewater (Dong et al., 2022). In a study, iron Fe (II) removal from an aqueous medium is explored using electrocoagulation with electrodes made of aluminum (Das and Nandi, 2022). The electrocoagulation-based simultaneous elimination of co-occurring Fe(II), As(V), and fluoride ions from groundwater (EC) is investigated in a recent study (Golder et al., 2007). Other water quality measures throughout the EC process, including residual TDS, sulfate, nitrate, chloride, and phosphate ions, as well as hardness, turbidity, and alkalinity, are discovered to change depending on the operating circumstances. The coexistence of Fe(II), As(V), and fluoride ions in the water increases the effectiveness of contaminant ion removal. The capture of F, Fe, and as ions as well as other ions like Ca, Na, Cl, and Al by the metal hydroxide flocs is supported by EDAX analysis of the sludge.

Metal-organic frameworks

In the last three decades, metal-organic frameworks (MOFs) have been used as an absorbent material and it is constructed of organic linkers and secondary building units of metal ions or metal clusters linked into three-dimensional lattices (Howarth et al., 2016; Reddy et al., 2020). MOFs have applications in drug delivery, separation, catalysis, and gas storage (Siva et al., 2020; Shi et al., 2019) and the advantages of MOFs are high surface area and porosity. Commercial heavy metal removal methods using MOF, such as chemical precipitation, sorbents, and membranes, have many disadvantages, such as large commercial and energy prices, difficult regeneration, low elimination efficiency, and/or fouling, and production of a large quantity of chemical sediment (Kobielska et al., 2018). Fig. 15.2 illustrates MOF-related micro/nanoscaled substances originated for HM ions elimination. An inexpensive, water-stable metal-organic polymer composite, Fe-BTC/PDA, shows fast, particular elimination of numerous amounts of HMs such as Pb^{2+} and Hg^{2+} from natural global water specimens (Sun et al., 2018). PDA is attached to the inner MOF surface and acquires external permeability, associated with a composite that is attached to 1634 mg Hg^{2+} and 394 mg Pb^{2+} and eliminates more than 99.8% of these ions from a 1 ppm solution. Delivers a consumable dose in a short time.

Titanium dioxide (TiO_2)

Many metal oxides and metal oxides-based composites can act as adsorbents in surface chemistry when properly prepared and activated. Titanium dioxide is the most widely studied metal oxide due to its high stability, high photocatalytic activity, non-toxicity, and excellent dielectric properties (Gebru and Das, 2017). Magnesium oxide is non-toxic, non-corrosive, thermally stable, and environmentally friendly (Taghipour et al., 2019; Engates and Shipley, 2011). Among the nanomaterials used for environmental remediation, TiO_2 NPs were extensively studied. Because, these NPs show good capabilities for high reactivity, photocatalysis, and chemical stability, and have been successfully used to modulate the mobility and toxicity of heavy metals in soil, water, and sediment. Bulk and nanoparticulate TiO_2 anatase show various chemical properties, catalytic reactivity, and surface acidity related to various surface planes. A specific examination of Pb adsorption on TiO_2 nanoparticles reveals aggregates showing higher adsorption and huge adsorption potential when normalized to surface area but these variables need to be more comprehensively cataloged to determine trends (Dong et al., 2015).

Nano zero-valent iron (nZVI)

To remediate large amounts of organic pollutants in the environment by using ZVI as an effective material. The advantages of ZVI are magnetic and catalytic properties, low cost, and relatively low toxicity (Mironyuk et al., 2019). nZVI has more activity and high reducing capacity, strong absorption properties, and attractive mobility than microscale iron

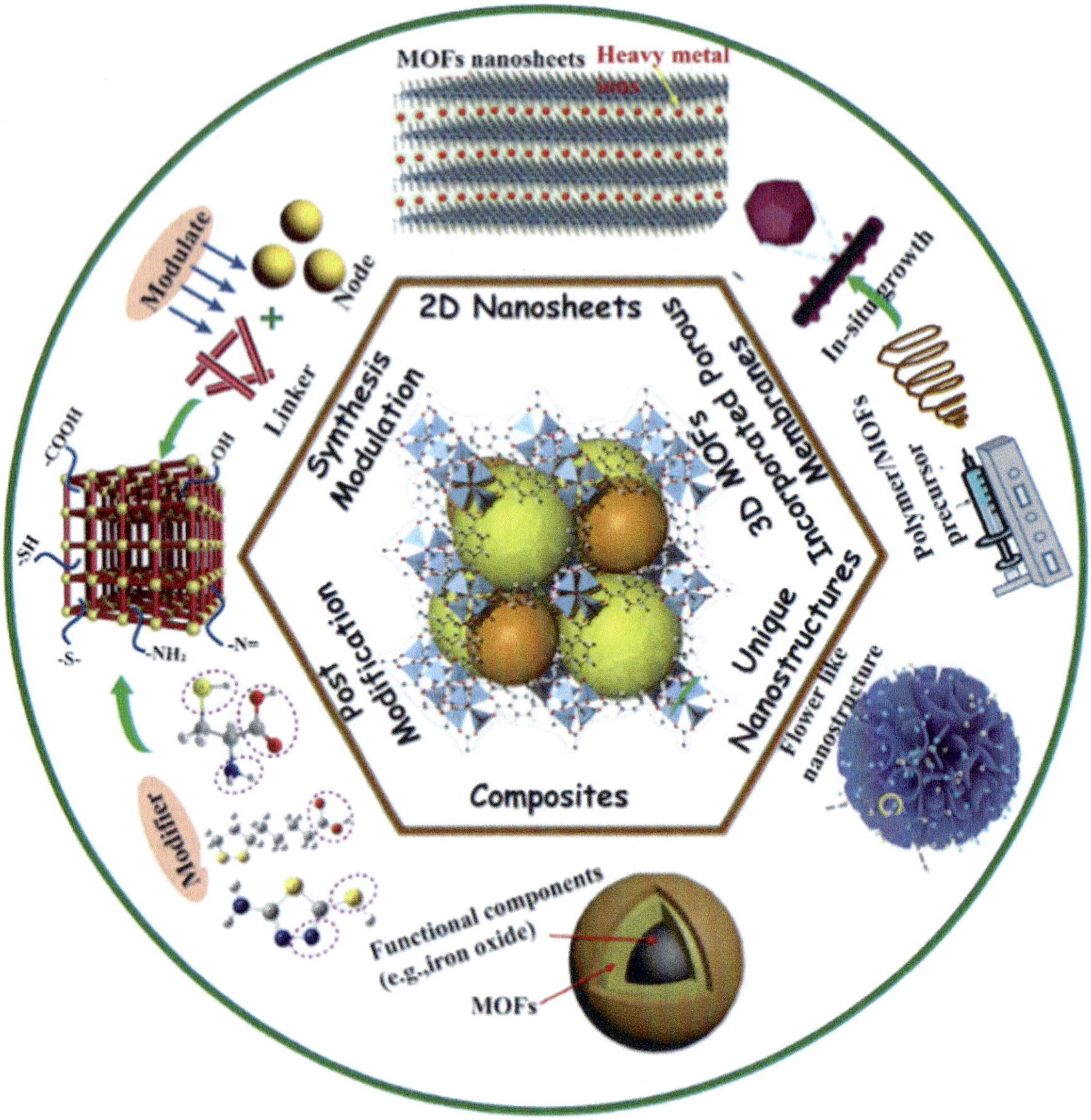

FIGURE 15.2 Heavy metal ion removal materials built on MOF at the micro- and nanoscale (Xu et al., 2021). *Reproduced from with permission of Elsevier, Xu, G.R., An, Z.H., Xu, K., Das, R., Zhao, H.L., 2021. Metal organic framework(MOF)-based micro/nanoscaled materials for heavy metal ions removal: the cutting-edge study on designs, synthesis and applications. Coordination Chemistry Reviews 427, 213554, https://doi.org/10.1016/j.ccr.2020.213554.*

particles. These properties allow nZVI to act as a constructive solution representative for a broad area of pollutants (Tang et al., 2019) (Fig. 15.3).

Role of green chemistry in heavy metal removal from water

A biological material's ability to absorb heavy metals from water or wastewater through metabolic or physicochemical processes is known as biosorption. The type of biomass and its characteristics, the chemical composition of the water or wastewater, and the environmental factors all affect the biosorption mechanism. In a biosorption mechanism that is not dependent on metabolism, metals are absorbed through physical adsorption, ion exchange, and chemical sorption (Joshi, 2018). When removing HM ions from WW, the use of non-living microbes appears to be preferable to the use of living cells. Utilizing the drained biomass powder of the cyanobacteria *Oscillatoria laete-virens* that was gathered from a WW canal, biosorption studies were conducted. Both Cr(VI) and Ni(II) adsorption by the biosorbent is controlled by ionic attraction at pH 5 (Suman, 2012). Values for maximum biosorption potential revealed that some biosorbents are

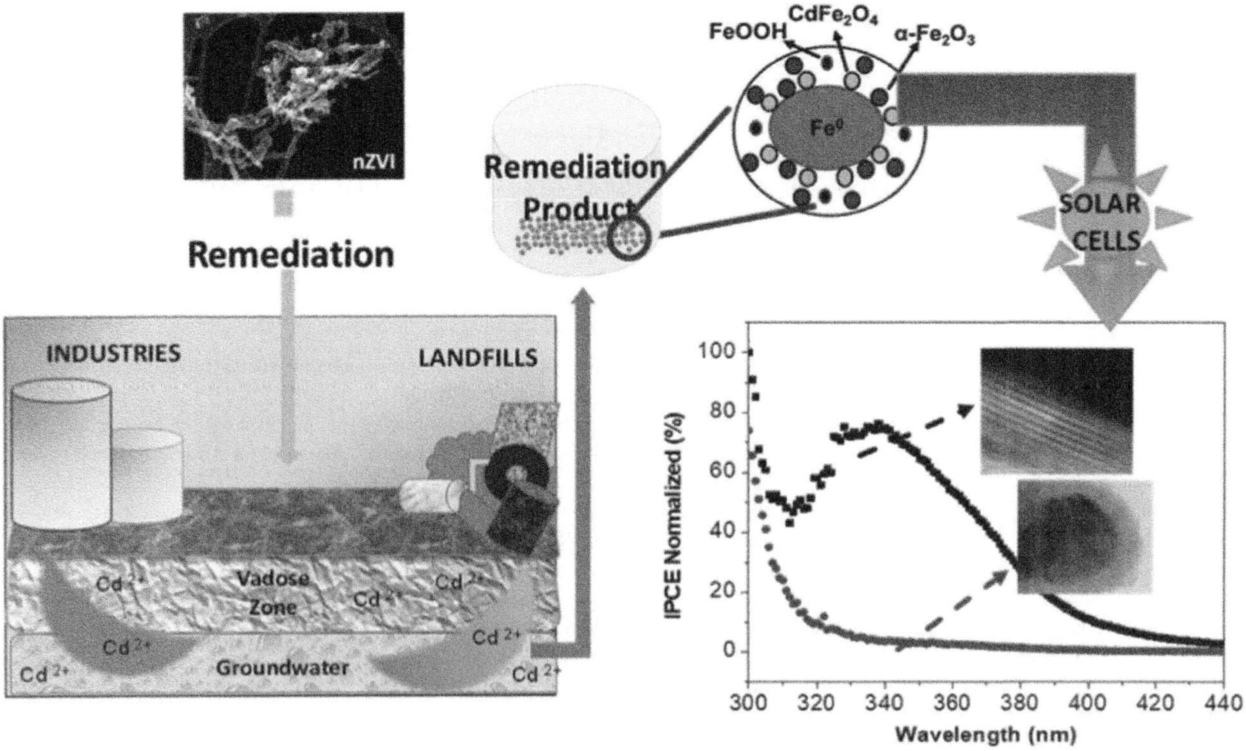

FIGURE 15.3 Utilizing zero-valent iron at the nanoscale, cadmium water remediation for photoelectrochemical solar cells (Soto-Hidalgo and Cabrera, 2018). *Reproduced from Open Access Journal under the term of CCAC Copyright ©2018, Soto-Hidalgo, K. T., Cabrera, C. R., 2018. Nanoscale Zero Valent Iron for Environmental Cadmium Metal Treatment. In Saleh, H.E.M., Koller, M. (eds.), Green Chemistry, IntechOpen, London. https://doi.org/10.5772/intechopen.72737.*

particularly successful at recovering or removing mercury ions from water. It was found that *Mucor rouxii* IM-80, *M. rouxii* mutant, *Mucor* sp. 1, and *Mucor* sp. 2 are extremely favorable biomaterials for the elimination of the metal ion investigated in terms of production efficiency and economic factors (Martinez-Juarez et al., 2012).

Conclusions and future prospective

In developing countries, increased water scarcity and contamination take part in an important lack of access to pure consumable water. Although there are many different types of water pollution, a heavy metal pollutant in a consumable water reservoir is a developing consideration. The HM occurs naturally all-round the Earth's crust, anthropogenic activities such as mining and smelting, factory manufacturing and utilization, and domestic and agricultural use of metals and metal-comprising elements create atmospheric pollution and human exposure. Physical properties such as temperature, phase association, adsorption, and sequestration affect their bioavailability. Metals like Co, Cu, Cr, Fe, Mg, Mn, and Zn have been recognized as important elements required for various biochemical and physiological functions. To conclude, brief research should be carried out to identify the possible factors affecting heavy metal removal efficiency. Establishing low-cost materials and methods for HM elimination from WW should also be considered in further studies. Further pilot-scale research is also needed to test and implement the best heavy metal recovery process. Improved approaches for the elimination of HM in WW with low environmental impact and low cost should be considered.

References

Algieri, C., Chakraborty, S., Candamano, S., 2021. A way to membrane-based environmental remediation for heavy metal removal. Environment 8 (6), 52. Available from: https://doi.org/10.3390/environments8060052.

Arora, R., 2019. Adsorption of heavy metals—a review. Materials Today: Proceedings 18, 4745—4750. Available from: https://doi.org/10.1016/j.matpr.2019.07.462.

Bashir, A., Malik, L.A., Ahad, S., Manzoor, T., Bhat, M.A., Dar, G.N., et al., 2019. Removal of heavy metal ions from aqueous system by ion-exchange and biosorption methods. Environmental Chemistry Letters 17 (2), 729—754. Available from: https://doi.org/10.1007/s10311-018-00828-y.

Bockris, J.M. (Ed.), 1977. Environmental Chemistry. Springer, Boston, MA.

Cheng, X., Deng, J., Li, X., Wei, X., Shao, Y., Zhao, Y., 2022. Layered double hydroxides loaded sludge biochar composite for adsorptive removal of benzotriazole and Pb(II) from aqueous solution. Chemosphere 287, 131966. Available from: https://doi.org/10.1016/j.chemosphere.2021.131966.

Clarkson, T.W., 1997. The toxicology of mercury. Critical Reviews in Clinical Laboratory Sciences 34 (4), 369–403. Available from: https://doi.org/10.3109/10408369708998098.

Dąbrowski, A., 2001. Adsorption—from theory to practice. Advances in Colloid and Interface Science 93 (1–3), 135–224. Available from: https://doi.org/10.1016/S0001-8686(00)00082-8.

Dąbrowski, A., Hubicki, Z., Podkościelny, P., Robens, E., 2004. Selective removal of the heavy metal ions from waters and industrial wastewaters by ion-exchange method. Chemosphere 56, 91–106. Available from: https://doi.org/10.1016/j.chemosphere.2004.03.006.

Das, D., Nandi, B.K., 2022. Removal of co-existing Fe(II), As(V) and fluoride ions from groundwater by electrocoagulation. Groundwater for Sustainable Development 17, 00752. Available from: https://doi.org/10.1016/j.gsd.2022.100752.

Deliyanni, E.A., Peleka, E.N., Matis, K.A., 2009. Modelling the sorption of metal ions from aqueous solution by iron-based adsorbents. Journal of Hazardous Materials 172, 550–558. Available from: https://doi.org/10.1016/j.jhazmat.2009.07.130.

Deng, Y., Feng, C., Tang, L., Zeng, G., Chen, Z., Zhang, M., 2019. Nanohybrid photocatalysts for heavy metal pollutant control. In: Tang, L., Deng, Y., Wang, J., Wang, J., Zeng, G. (Eds.), Micro and Nano Technologies, Nanohybrid and Nanoporous Materials for Aquatic Pollution Control. Elsevier, Amsterdam, pp. 125–153. Available from: https://doi.org/10.1016/B978-0-12-814154-0.00005-0.

Dong, H., Zeng, G., Tang, L., Fan, C., Zhang, C., He, X., et al., 2015. An overview on limitations of TiO_2-based particles for photocatalytic degradation of organic pollutants and the corresponding countermeasures. Water Research 79, 128–146. Available from: https://doi.org/10.1016/j.watres.2015.04.038.

Dong, W., Gu, X., Shu, Y., Cao, D., Yu, J., Abdel-Fatah, M.A., et al., 2022. Pulse electrocoagulation combined with a coagulant to remove antimony in wastewater. Journal of Water Process Engineering 47, 102749. Available from: https://doi.org/10.1016/j.jwpe.2022.102749.

El-Sayed, M.A., 2001. Some interesting properties of metals confined in time and nanometer space of different shapes. Accounts of Chemical Research 34, 257–264. Available from: https://doi.org/10.1021/ar960016n.

Engates, K.E., Shipley, H.J., 2011. Adsorption of Pb, Cd, Cu, Zn, and Ni to titanium dioxide nanoparticles: effect of particle size, solid concentration, and exhaustion. Environmental Science and Pollution Research International 18 (3), 386–395. Available from: https://doi.org/10.1007/s11356-010-0382-3.

Fei, Y., Hu, Y.H., 2022. Design, synthesis and performance of adsorbents for heavy metal removal from wastewater: a review. Journal of Materials Chemistry A 10, 1047–1085. Available from: https://doi.org/10.1039/D1TA06612A.

Förstner, U., Wittmann, G.T., 1979. Metal Pollution in the Aquatic Environment. Springer Berlin, Heidelberg. Available from: https://doi.org/10.1007/978-3-642-96511-1.

Fujioka, T., Oshima, N., Suzuki, R., Price, W.E., Nghiem, L.D., 2015. Probing the internal structure of reverse osmosis membranes by positron annihilation spectroscopy: gaining more insight into the transport of water and small solute. Journal of Membrane Science 486, 106–118. Available from: https://doi.org/10.1016/j.memsci.2015.02.007.

Gebru, K.A., Das, C., 2017. Removal of Pb (II) and Cu (II) ions from wastewater using composite electrospun cellulose acetate/titanium oxide (TiO_2) adsorbent. Journal of Water Process Engineering 16, 1–13. Available from: https://doi.org/10.1016/j.jwpe.2016.11.008.

Genchi, G., Sinicropi, M.S., Lauria, G., Carocci, A., Catalano, A., 2020. The effects of cadmium toxicity. International Journal of Environmental Research and Public Health 17 (11), 3782. Available from: https://doi.org/10.3390/ijerph17113782.

Golder, A.K., Samanta, A.N., Ray, S., 2007. Removal of trivalent chromium by electrocoagulation. Separation and Purification Technology 53 (1), 33–41. Available from: https://doi.org/10.1016/j.seppur.2006.06.010.

Gopalakrishnan, A., Krishnan, R., Thangavel, S., Venugopal, G., Kim, S.J., 2015. Removal of heavy metal ions from pharma-effluents using graphene-oxide nanosorbents and study of their adsorption kinetics. Journal of Industrial and Engineering Chemistry 30, 14–19. Available from: https://doi.org/10.1016/j.jiec.2015.06.005.

Guan, X., Yuan, X., Zhao, Y., Wang, H., Wang, H., Bai, J., et al., 2022. Application of functionalized layered double hydroxides for heavy metal removal: a review. The Science of the Total Environment 838, 155693. Available from: https://doi.org/10.1016/j.scitotenv.2022.155693.

Hänsch, R., Mendel, R.R., 2009. Physiological functions of mineral micronutrients (Cu, Zn, Mn, Fe, Ni, Mo, B, Cl). Current Opinion in Plant Biology 12 (3), 259–266. Available from: https://doi.org/10.1016/j.pbi.2009.05.006.

Howarth, A.J., Liu, Y., Li, P., Li, Z., Wang, T.C., Hupp, J.T., et al., 2016. Chemical, thermal and mechanical stabilities of metalorganic frameworks. Nature Reviews Materials 1, 15018. Available from: https://doi.org/10.1038/natrevmats.2015.18.

Hua, M., Zhang, S., Pan, B., Zhang, W., Lv, L., Zhang, L.Q., 2012. Heavy metal removal from water/wastewater by nanosized metal oxides: a review. Journal of Hazardous Materials 211, 317–331. Available from: https://doi.org/10.1016/j.jhazmat.2011.10.016.

Huang, H., Xia, C., Liang, D., Xie, Y., Kong, F., Fu, J., et al., 2022. Removal and magnetic recovery of heavy metals and pesticides from soil by layered double hydroxides modified biotite. Chemical Engineering Journal 431, 134113. Available from: https://doi.org/10.1016/j.cej.2021.134113.

Ince, M., Ince, O.K., 2017. An overview of adsorption technique for heavy metal removal from water/wastewater: a critical review. International Journal of Pure and Applied Sciences 3 (2), 10–19.

Jain, A., Rai, S., Srinivas, R., Al-Raoush, R.I., 2022. Bioinspired modelling and biogeography-based optimization of electrocoagulation parameters for enhanced heavy metal removal. Journal of Cleaner Production 338, 130622. Available from: https://doi.org/10.1016/j.jclepro.2022.130622.

Joshi, N.C., 2018. Biosorption: a green approach for heavy metal removal from water and wastewaters. Research Journal of Life Sciences, Bioinformatics, Pharmaceutical and Chemical Sciences 4, 59–69. Available from: https://doi.org/10.26479/2018.0401.05.

Karabörk, M., Muhammed, B.A., Tümer, M., Uruş, S., 2022. Organosilane- functionalized graphene oxide hybrid material: efficient adsorbent for heavy metal ions in drinking water. Phosphorus, Sulfur, and Silicon and the Related Elements 197 (2), 133–143. Available from: https://doi.org/10.1080/10426507.2021.2012676.

Keng, P.S., Lee, S.L., Ha, S.T., Hung, Y.T., Ong, S.T., 2014. Removal of hazardous heavy metals from aqueous environment by low-cost adsorption materials. Environmental Chemistry Letters 12 (1), 15–25. Available from: https://doi.org/10.1007/s10311-013-0427-1.

Kobielska, P.A., Howarth, A.J., Farha, O.K., Nayak, S., Sanjit, 2018. Metal–organic frameworks for heavy metal removal from water. Coordination Chemistry Reviews 358, 92–107. Available from: https://doi.org/10.1016/j.ccr.2017.12.010.

Lee, S., Lingamdinne, L.P., Yang, J.K., Koduru, J.R., Chang, Y.Y., Naushad, M., 2022. Biopolymer mixture-entrapped modified graphene oxide for sustainable treatment of heavy metal contaminated real surface water. Journal of Water Process Engineering 46, 102631. Available from: https://doi.org/10.1016/j.jwpe.2022.102631.

Lim, J.Y., Mubarak, N.M., Abdullah, E.C., Nizamuddin, S., Khalid, M., Inamuddin, 2018. Recent trends in the synthesis of graphene and graphene oxide based nanomaterials for removal of heavy metals—a review. Journal of Industrial and Engineering Chemistry 66, 29–44. Available from: https://doi.org/10.1016/j.jiec.2018.05.028.

Liu, X., Ma, R., Wang, X., Ma, Y., Yang, Y., Zhuang, L., et al., 2019. Graphene oxide-based materials for efficient removal of heavy metal ions from aqueous solution: a review. Environmental Pollution (Barking, Essex: 1987) 252, 62–73. Available from: https://doi.org/10.1016/j.envpol.2019.05.050.

Liu, Y., Tang, X., Zhou, L., Liu, Z., Ouyang, J., Dai, Y., et al., 2022. Nanofabricated chitosan/graphene oxide electrodes for enhancing electrosorptive removal of U(VI) from aqueous solution. Separation and Purification Technology 290, 120827. Available from: https://doi.org/10.1016/j.seppur.2022.120827.

Malik, D., Singh, S., Thakur, J., Singh, R.K., Kaur, A., Nijhawan, S., 2014. Heavy metal pollution of the Yamuna River: an introspection. International Journal of Current Microbiology and Applied Science 3 (10), 856–863.

Martinez-Juarez, V.M., Cardenas-Gonzalez, J.F., Torre-Bouscoulet, M.E., Acosta- Rodriguez, I., 2012. Biosorption of Mercury (II) from aqueous solutions onto fungal biomass. Bioinorganic Chemistry and Applications 2012, 156190. Available from: https://doi.org/10.1155/2012/156190.

Mironyuk, I., Tatarchuk, T., Naushad, M., Vasylyeva, H., Mykytyn, I., 2019. Highly efficient adsorption of strontium ions by carbonated mesoporous TiO_2. Journal of Molecular Liquids 285, 742–753. Available from: https://doi.org/10.1016/j.molliq.2019.04.111.

Mishra, R., Kumar, A., Singh, E., Kumar, S., Tripathi, V.K., Jha, S.K., et al., 2022. Current status of available techniques for removal of heavy metal contamination in the river ecosystem. In: Madhav, S., Kanhaiya, S., Srivastav, A., Singh, V., Singh, P. (Eds.), Ecological Significance of River Ecosystems. Elsevier, Amsterdam, pp. 217–234. Available from: https://doi.org/10.1016/B978-0-323-85045-2.00007-8.

Mollah, M.Y.A., Morkovsky, P., Gomes, J.A.G., Kesmez, M., Parga, J., Cocke, D.L., 2004. Fundamentals, present and future perspectives of electrocoagulation. Journal of Hazardous Materials 114 (1–3), 199–210. Available from: https://doi.org/10.1016/j.jhazmat.2004.08.009.

Muscat, J., Newns, D.M., 1978. Chemisorption on metals. Progress in Surface Science 9 (1), 1–43. Available from: https://doi.org/10.1016/0079-6816(78)90005-9.

Nordberg, G.F., Fowler, B.A., Nordberg, M. (Eds.), 2014. Handbook on the Toxicology of Metals. Academic Press, Amsterdam, Netherlands. Available from: https://doi.org/10.1016/C2011-0-07884-5.

Pakdel, P.M., Peighambardoust, S.J., 2018. Review on recent progress in chitosan- based hydrogels for wastewater treatment application. Carbohydrate Polymers 201, 264–279. Available from: https://doi.org/10.1016/j.carbpol.2018.08.070.

Patterson, J.W., 1978. Industrial Wastewater Treatment Technology. Ann Arber Publishers, USA.

Peng, C., Jin, R., Li, G., Li, F., Gu, Q., 2014. Recovery of nickel and water from wastewater with electrochemical combination process. Separation and Purification Technology 136, 42–49. Available from: https://doi.org/10.1016/j.seppur.2014.08.025.

Petrinic, I.P., Korenak, J., Povodnik, D., Hélix-Nielsen, C., 2015. A feasibility study of ultrafiltration/reverse osmosis (UF/RO)-based wastewater treatment and reuse in the metal finishing industry. Journal of Cleaner Production 101, 292–300. Available from: https://doi.org/10.1016/j.jclepro.2015.04.022.

Pinsino, A., Matranga, V., Roccheri, M.C., 2012. Manganese: a new emerging contaminant in the environment. In: Srivastava, J. (Ed.), Environmental Contamination. IntechOpen, London, pp. 17–36. Available from: 10.5772/31438.

Pohl, A., 2020. Removal of heavy metal ions from water and wastewaters by sulfur-containing precipitation agents. Water, Air, and Soil Pollution 231, 1–17. Available from: https://doi.org/10.1007/s11270-020-04863-w.

Pradeep, T., Anshup, 2009. Noble metal nanoparticles for water purification: a critical review. Thin Solid Films 517, 6441–6478. Available from: https://doi.org/10.1016/j.tsf.2009.03.195.

Reddy, C.V., Reddy, K.R., Harish, V.V.N., Shim, J., Shankar, M.V., Shetti, N.P., et al., 2020. Metal-organic frameworks (MOFs)-based efficient heterogeneous photocatalysts: synthesis, properties and its applications in photocatalytic hydrogen generation, CO reduction and photodegradation of organic dyes. International Journal of Hydrogen Energy 45, 7656–7679. Available from: https://doi.org/10.1016/j.ijhydene.2019.02.144.

Rouquerol, J., Rouquerol, F., Llewellyn, P., Maurin, G., Sing, K., 2013. Adsorption by Powders and Porous Solids: Principles, Methodology and Applications. Academic Press, Cambridge.

Roy, A.S., Pillai, S.K., Ray, S.S., 2022. Layered double hydroxides for sustainable agriculture and environment: an overview. ACS Omega 7 (24), 20428–20440. Available from: https://doi.org/10.1021/acsomega.2c01405.

Shi, Y., Yang, A.F., Cao, C.S., Zhao, B., 2019. Applications of MOFs: recent advances in photocatalytic hydrogen production from water. Coordination Chemistry Reviews 390, 50–75. Available from: https://doi.org/10.1016/j.ccr.2019.03.012.

Singh, S., Anil, A.G., Khasnabis, S., Kumar, V., Nath, B., Adiga, V., et al., 2022. Sustainable removal of Cr(VI) using graphene oxide-zinc oxide nanohybrid: adsorption kinetics, isotherms and thermodynamics. Environmental Research 203, 111891. Available from: https://doi.org/10.1016/j.envres.2021.111891.

Sis, H., Uysal, T., 2014. Removal of heavy metal ions from aqueous medium using Kuluncak (Malatya) vermiculites and effect of precipitation on removal. Applied Clay Science 95, 1–8. Available from: https://doi.org/10.1016/j.clay.2014.03.018.

Siva, V., Shameem, V.,A., Murugan, A., Athimoolam, S., Vinitha, G., Bahadur, S.A., 2020. Structural, thermal and electro-optical properties of guanidine based Metal-Organic Framework (MOF). Chinese Journal of Physics 68, 764–777. Available from: https://doi.org/10.1016/j.cjph.2020.09.036.

Song, P., Sun, C., Wang, J., Ai, S., Dong, S., Sun, J., et al., 2022. Efficient removal of Cu-EDTA complexes from wastewater by combined electrooxidation and electrocoagulation process: performance and mechanism study. Chemosphere 287, 131971. Available from: https://doi.org/10.1016/j.chemosphere.2021.131971.

Soto-Hidalgo, K.T., Cabrera, C.R., 2018. Nanoscale zero valent iron for environmental cadmium metal treatment. In: Saleh, H.E.M., Koller, M. (Eds.), Green Chemistry. IntechOpen, London. Available from: 10.5772/intechopen.72737.

Suman, D., 2012. Biosorption of chromium and nickel by dried biomass of cyanobacterium *Oscillatoria laete-virens*. International Journal of Environmental Science and Technology 3, 341–352. Available from: https://doi.org/10.6088/ijes.2012030131032.

Sun, D.T., Peng, L., Reeder, W.S., Moosavi, S.M., Tiana, D., Britt, D.K., et al., 2018. Rapid, selective heavy metal removal from water by a metal-organic framework/polydopamine composite. ACS Central Science 4 (3), 349–356. Available from: https://doi.org/10.1021/acscentsci.7b00605.

Taghipour, S., Hosseini, S.M., Ataie-Ashtiani, B., 2019. Engineering nanomaterials for water and wastewater treatment: review of classifications, properties and applications. New Journal of Chemistry 43, 7902–7927. Available from: https://doi.org/10.1039/C9NJ00157C.

Tang, J., Zeng, G., Tang, L., Dong, H., Yang, G., Feng, H., et al., 2019. Iron-based nanohybrids for aquatic recalcitrant pollutant treatment. In: Tang, L., Deng, Y., Wang, J., Wang, J., Zeng, G. (Eds.), In Micro and Nano Technologies, Nanohybrid and Nanoporous Materials for Aquatic Pollution Control. Elsevier, Amsterdam, pp. 201–232. Available from: https://doi.org/10.1016/B978-0-12-814154-0.00007-4.

Thangavel, S., Venugopal, G., 2014. Understanding the adsorption property of graphene-oxide with different degrees of oxidation levels. Powder Tech 257, 141–148. Available from: https://doi.org/10.1016/j.powtec.2014.02.046.

Tsuchiya, K., 1969. Epidemic of poisoning in the Agano river area—an introductory review. The Keio Journal of Medicine 18, 213–227.

Vargas-Solano, S.V., Rodríguez-González, F., Martínez-Velarde, R., Morales- García, S.S., Jonathan, M.P., 2022. Removal of heavy metals present in water from the Yautepec River Morelos México, using Opuntiaficus-indica mucilage. Environmental Advance 7, 100160. Available from: https://doi.org/10.1016/j.envadv.2021.100160.

Verma, M., Lee, I., Oh, J., Kumar, V., Kim, H., 2022. Synthesis of EDTA-functionalized graphene oxide-chitosan nanocomposite for simultaneous removal of inorganic and organic pollutants from complex wastewater. Chemosphere 287, 132385. Available from: https://doi.org/10.1016/j.chemosphere.2021.132385.

Víctor-Ortega, M.D., Ochando-Pulido, J.M., Martínez-Ferez, A., 2016. Iron removal and reuse from Fenton-like pretreated olive mill wastewater with novel strong- acid cation exchange resin fixed-bed column. Journal of Industrial and Engineering Chemistry 36, 298–305. Available from: https://doi.org/10.1016/j.jiec.2016.02.019.

Vidali, G., Ihm, G., Kim, H.Y., Cole, M.W., 1991. Potentials of physical adsorption. Surface Science Reports 12 (4), 135–181. Available from: https://doi.org/10.1016/0167-5729(91)90012-M.

White, D.E., Hinkle, M.E., Barnes, I., 1970. Mercury contents of natural thermal and mineral fluids. Mercury in the Environment. Geological Survey Professional Paper 713, 25–28.

Xu, G.R., An, Z.H., Xu, K., Das, R., Zhao, H.L., 2021. Metal organic framework(MOF)-based micro/nanoscaled materials for heavy metal ions removal: the cutting-edge study on designs, synthesis and applications. Coordination Chemistry Reviews 427, 213554. Available from: https://doi.org/10.1016/j.ccr.2020.213554.

Yang, J., Hou, B., Wang, J., Tian, B., Bi, J., Wang, N., et al., 2019. Nanomaterials for the removal of heavy metals from wastewater. Nanomaterials 9 (3), 424. Available from: https://doi.org/10.3390/nano9030424.

Zehra, A., Meena, M., Swapnil, P., Raytekar, N.A., Upadhyay, R.S., 2020. Sustainable approaches to remove heavy metals from water. Environmental Microbiology and Biotechnology 127–146. Available from: https://doi.org/10.1007/978-981-15-2817-0_6.

Zhao, S., Liao, Z., Fane, A., Li, J., Tang, C., Zheng, C., et al., 2021. Engineering antifouling reverse osmosis membranes: a review. Desalin 499, 114857. Available from: https://doi.org/10.1016/j.desal.2020.114857.

Zhao, X., Lv, L., Pan, B., Zhang, W., Zhang, S., Zhang, Q., 2011. Polymer-supported nanocomposites for environmental application: a review. Chemical Engineering Journal 170, 381–394. Available from: https://doi.org/10.1016/j.cej.2011.02.071.

Section D

Agricultural sustainability

Chapter 16

Hazardous consequences of pharmaceutical wastes to groundwater

Alka Sahu[1], Markandeya[2], Neeraj Kumar Singh[3], Manish Yadav[3], Sheo Prasad Shukla[4] and Devendra Mohan[5]

[1]Department of Environmental Science and Engineering, Indian Institute of Technology (Indian School of Mine), Dhanbad, Jharkhand, India, [2]Ex-Department of Civil Engineering, Indian Institute of Technology (Banaras Hindu University), Varanasi, Uttar Pradesh, India, [3]Central Mine Planning and Design Institute Limited (CMPDIL), Regional Institute-7, Bhubaneswar, Odisha, India, [4]Rajkiya Engineering College, Banda, Uttar Pradesh, India, [5]Department of Civil Engineering, Indian Institute of Technology (Banaras Hindu University), Varanasi, Uttar Pradesh, India

Introduction

Groundwater is one of the most precious resources available on this earth. It is being used for different purposes like irrigation, water supply, industrial use, etc. The rocks which keep the groundwater are called aquifers. They are generally made up of sand, sandstone, or limestone. These formations generally have large connected spaces, which makes them permeable. A water table can be defined as the depth from the surface where we get the water (Suhag, 2022). As per water and related statistics 2015, for the whole nation, the entire natural runoff accumulating in the river is about 1869 billion cubic meter/year. Out of this total available water resource, only 1123 BCM/year is usable (Fig. 16.1). Rest of the water could not be used due to the odd distribution of river basins and uneven topography. Out of this usable water resource, 61.44% share is surface water, and the remaining 38.56% is groundwater (Srinivasamoorthy et al., 2014).

From the above Fig. 16.1 and data, it can be witnessed that groundwater is a fragile and precious resource. Rainfall is the major contributor to annual groundwater resources; it contributes around 68% and the remaining 32% is from other sources, for example, return flow of irrigation, canal seepage, etc. In India, the major consumer of groundwater is the irrigation sector, followed by domestic usage and industrial use (Fig. 16.2).

During the years 2000 to 2010, a rapid increase (about 36%) has been observed in global consumption of antibiotics. Developing countries are amongst the leading consumers of antibiotics with India being the biggest consumer with 12.9 × 109 units (10 · 7 units per capita), trailed by China with 10 · 0 × 109 units (7 · 5 units per capita) and the USA with 6 · 8 × 109 units (22 units per capita) in 2010. In India, during the period from 2000 to 2015, the consumption of antibiotics increased from 3.2 to 6.5 billion Daily Defined Doses (DDDs) (Aktan et al., 2018; Van Boeckel et al., 2014).

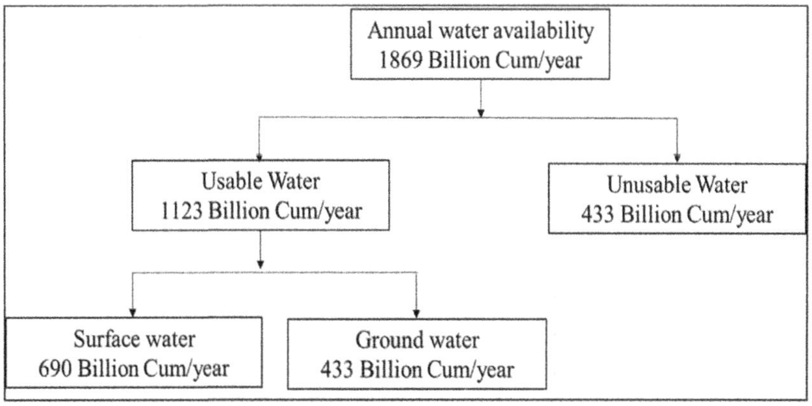

FIGURE 16.1 Annual water availability of India.

Role of Green Chemistry in Ecosystem Restoration to Achieve Environmental Sustainability. DOI: https://doi.org/10.1016/B978-0-443-15291-7.00011-0

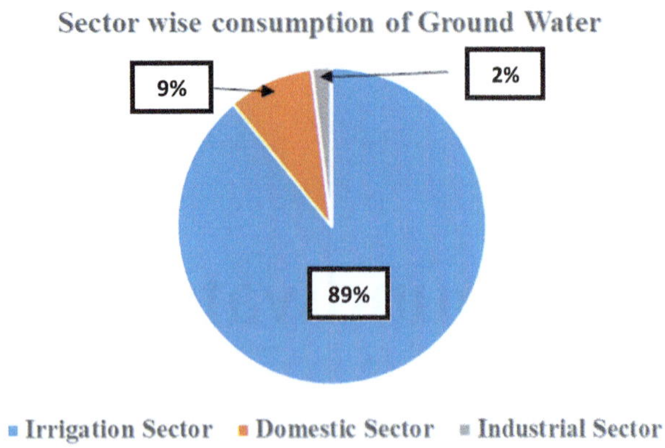

FIGURE 16.2 Sector-wise consumption of groundwater.

Since the consumption of antibiotics, antibacterial, antimalarial, and antifungal drugs is steadily increasing; it becomes necessary to study the future of these medicines in the environment. Due to this huge consumption of drugs, its traces are also observed in human and animal excreta and urine, ultimately polluting different environmental components through various activities. The pharma sector is amongst the fastest-growing sectors in the current world scenario. Indian healthcare sector is one of the fastest-rising sectors, and it is projected that it will cross US$ 372 billion by 2022 (Louise et al., 2021). From this, we can see that the pharma industry has a great future which brings not only prosperity but also some great environmental challenges. Therefore, proper disposal and treatment of pharmaceutical waste is the need of the hour; otherwise, it will become one of the most challenging emerging pollutants with dire consequences. The presence of traces of pharmaceutical constituents in groundwater, surface water bodies, potable water, and soil have raised concerns, as they pose potential hazards to human health as well as the environment. As a consequence of the deleterious impact on the structure and function of the ecosystem, these pharmaceuticals have turned out to be a group of emerging contaminants that the scientific community is increasingly concerned about. A portion of these products are flushed or dumped continuously as domestic waste and finally go into sewage treatment plants, which has not been planned for the removal of these contaminants.

Pharmaceutical waste composition

Pharmaceutical waste includes expired drugs, syringes, and waste materials containing drug residues. Due to inappropriate disposal of these wastes over landfill sites, it has become a potential pollution threat to nearby water bodies around these landfill sites (Hu et al., 2003). Apart from human and veterinary medical treatments, its use in aquaculture and agricultural products has also increased the additional accumulation of pharmaceutical chemicals in our environment. It is projected that the yearly global consumption of antibiotics (antimicrobials per kg of animal generated) in livestock, chicken, and pigs will increase by 67% from the year 2010 to 2030 (Van Boeckel et al., 2014).

As demonstrated in Fig. 16.3, pharmaceutical waste can go into the groundwater through various routes which include human or animal excreta, dumping of expired drugs to landfill sites, leaching from soil to water excreta, etc. (WHO, 2018).

Green chemistry for pharmaceutical wastes

Pharmaceutical manufacturing is one of the fastest-growing industries both in India as well as the global level (Veleva et al., 2018). Consequently, it puts a burden on specific types of pollution, which has a severe impact on both, environmental and human health. New chemical entities and active pharmaceutical ingredients (APIs) are two main categories of pharmaceutical products. APIs amount to higher environmental pollution due to characteristic complexities involved in their synthesis (Sharma et al., 2020). Limitations of various conventional techniques employed for the treatment of pharmaceutical wastes indicate that only the "end of the pipe" method may not be adequate to meet the challenges posed by the wide variety of wastes coming out of the industry and therefore, efficient management of the wastes may be one way of handling the problem. In this connection, green chemistry can help to provide a more eco-friendly and cost-effective approach because it has the potential to become an alternative to conventional techniques for partial or complete treatment of pharmaceutical wastes. Green chemistry is defined as "the design of chemical products and processes that reduce or eliminate the generation of hazardous substances" (USEPA, 2022) and it is based on 12 principles

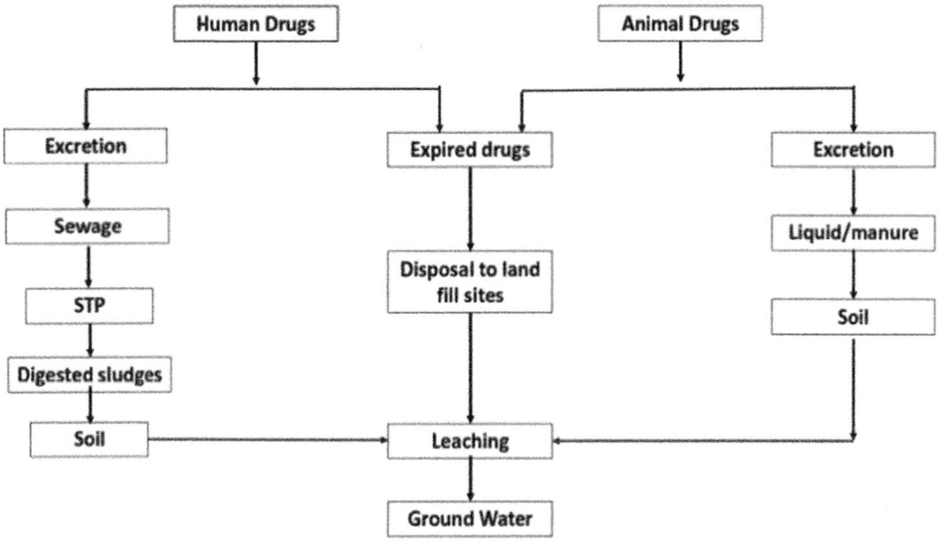

FIGURE 16.3 Transport of pharmaceutical waste in groundwater.

(Anastas and Eghbali, 2010; ACS, 2022). The principle of atom economy, that is, incorporating all materials used in the synthesis into the final products results in fewer byproducts and thus, in relatively lower costs involved in storage and disposal. As per a report by the US Environmental Protection Agency (USEPA), there has been a decrease in the release of various industrial wastes by 9% between 2007 and 2018, and 3% less total wastes were disposed of in 2018 compared to those in 2017 (Sharma et al., 2020) after the adoption of green chemistry in the manufacturing process. Despite numerous advantages, the application of green chemistry also has some disadvantages like challenges in research to develop new products and processes, achieving the required societal and economic benefits (Clark, 1999). Moreover, the mere development of more environment-friendly technology or products in laboratories does not guarantee that these can essentially be adopted at an industrial scale (Anastas and Kirchhoff, 2002) at some later stage.

Environmental impacts

The world population is suffering from water scarcity since the available water is not fit for consumption and other utilities without treatment. Various sectors are continuously degrading the water quality as water is available in the form of surfaces such as rivers, ponds, and canals; underground water in the form of springs and water available on ice poles in the form of ice. During antibiotic production, about 95% of antibiotic composites get discharged and remain unchanged into the sewerage (Fig. 16.4). Further, greater concentrations of antibiotics can alter the composition of the microbial group which eventually will impact the food chains. Clofibric acid is the most frequently occurring drug in the water bodies followed by propyphenazone, ibuprofen, and diclofenac. Moreover, diclofenac is very toxic for animals and birds such as cattle and vultures (Louise et al., 2021).

The third is from the manufacturing facilities of pharmaceutical products. In the above cases, this waste ends up in STPs that are not designed to eliminate such types of pollutants (Majumder et al., 2020). Nowadays the lakes and rivers that receive treated wastewater consist of pharmaceutical concentration, which varies from 0.1 to 1.0 mg/L. At such low concentrations, it is difficult for researchers to determine the impact of drugs on the ecosystem. It is difficult to establish a direct link between the extinction of vultures and pharmaceutical waste but reports are there which establish the highest correlation between extinction and drug among other environmental impacts (Richards et al., 2018). The other ecotoxicological example is the use of the anti-parasite drug (ivermectin) prescribed by veterinary consultants for cows, which has been found to reduce the number of insects and microbes in cow dung which lowers the decay process of cow dung and also reduce the significant amount of food available to birds and bats (Richards et al., 2018). However, the outcomes are generally insignificant, and a less clear cause-and-effect relationship is observed. In the laboratory, it has been observed that even extremely low concentrations of this ingredient can upset the sexual growth of male fish. Therefore, several intersex fish have been observed in rivers everywhere in the world that fall downstream from sewage plants. In the nearby zones of pharma industries, it has been speculated that health as well as the environment is getting affected directly or indirectly by pharmacological wastewater (Halling-Sørensen et al., 1998). Drinking

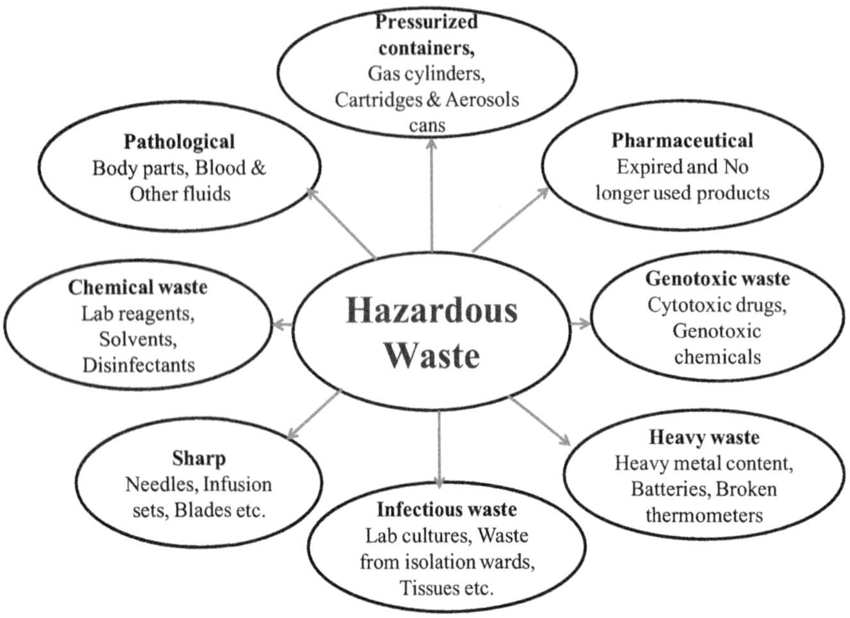

FIGURE 16.4 Different categories of pharmaceutical waste.

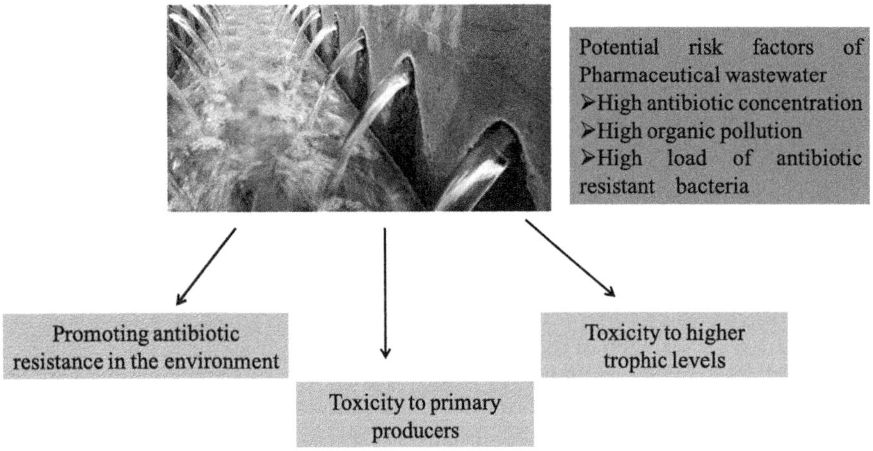

FIGURE 16.5 Environmental Impacts of pharmaceutical wastewater.

water sources are getting polluted due to the discharge of unprocessed or partially processed effluents by pharma industries (Fig. 16.5).

It is generally assumed that the concentration of pharmaceutical substances in the environment is lower than the actual levels which results in a short-term impact on fauna, including aquatic life. Migliore et al. (2003) assessed the lethal impacts of a few veterinary drugs on flora.

Health impacts

To date, no study suggests any ill effects on human health due to pharmaceutical waste in water (Majumder et al., 2020). Further, studies are required if there are any significant population-level health impacts arising due to long-term and low-level exposure to pharmaceutical waste in drinking waters. The following are the major issues related to pharma waste.

Carcinogenicity

When any new drugs get introduced in the market, it is one of the most anxiety-making side effects. Since it generally has a large incubation period before manifesting, it needs long-term, generally lifespan, testing over faunae to get some idea regarding the carcinogenic possibility of a new pharmaceutical (Richards et al., 2018).

Mutagenicity

A drug can also alter the DNA and it can be considered a toxic effect. If this change affects the somatic cells (somatic mutation); it can also lead to cancer. Aging and atherosclerosis are other effects that are associated with DNA alteration.

Teratogenicity

Teratogenic drugs can be classified as drugs in which teratogen agents are present, these agents can disrupt the growth of the embryo or fetus. Teratogens can also stop the pregnancy or generate a congenital malformation (a birth defect). Different classes of teratogens contain maternal infections, radiation, chemicals, and drugs (Kumar et al., 2018).

Allergic effects

Rashes, itchy eyes, swollen lips, tongue, or faces are general allergic reactions to penicillin. However, in a few cases, it may be life-threatening due to an anaphylactic reaction. This type of reaction usually occurs within an hour after taking penicillin.

Remediation

The existing chemical and physical treatment techniques are filtration, coagulation/flocculation, and advanced oxidation processes like the use of hydrogen peroxide, ozone, and UV light, but due to their costly nature, they are not always applicable and may result in secondary pollution. Conventional methods such as chemical treatment are not effective in treatment of the complex pharmaceutical compounds. Chemical and physicochemical purification methods rely on expensive chemicals and remediation units that are very hard to manage on an industrial scale (Amin et al., 2013). Several bioremediation strategies are now being explored and proposed for the elimination of pharmaceutical pollutants from the environment, owing to the limitations of physical and chemical treatment strategies in the management of xenobiotics.

Conventional remediation approaches

To further improve the water quality of potable water and industrial-grade water different remediation methods have taken a forward step to mitigate and deploy the contamination in water resources (Karigar and Rao, 2011). Conventional technologies provide direct treatment to different types of toxic wastes, organic and inorganic wastes, through biological oxidation, adsorption, chemical oxidation, and incineration (Kumar et al., 2018).

Bioremediation

Bioremediation involves the degradation of contaminants by enzymatic activity, which leads to the transformation of toxic contaminants into nontoxic or less toxic substances. In this process, a wide range of organisms such as algae, fungi, bacteria, and plants are employed to control the contamination (Neina, 2019). However, variations in environmental parameters like temperature, pH, nutrition levels, oxygen, and unexpected changes in the toxicity level of influent might make biological treatment approaches vulnerable. Microorganisms break down the waste or transform it into another form. The product's biodegradation is influenced by several parameters, including toxicity, stereochemistry, concentration, microbial strain efficiency, degradation conditions, retention duration, and the presence and concentration of other chemicals. Different biological methods are effective for the elimination of different pharmaceuticals. Many pharmaceuticals can be removed with the optimum performance using novel microalgae-based technologies. Raj and Anjaneyulu (2005) used the biodegradation technique to get the maximum COD depletion (86.6%) in pharmaceutical wastewater. The bioremediation procedure significantly reduced COD, TSS, TDS, and sulfate in pharmaceutical wastewater, according to Das et al. (2012). In addition, a microbial consortium consisting of bacterial and fungal cultures was found to help diminish toxicity from pharmaceutical effluent (Rosen et al., 1998). Aerobic and anaerobic treatment can be applied for the biological treatment of pharmaceutical wastewater. For full-scale biological treatment, a variety of high-rate reactors have been devised. Wei et al. (2012) used the membrane aerated biofilm reactor (MABR) technique to treat pharmaceutical wastewater, achieving COD treatment efficiencies of 90% and ammonical nitrogen treatment efficiencies of 98%.

Anaerobic treatment

Anaerobic reactors have been demonstrated to treat pharmaceutical effluent more efficiently in studies. Rao et al. (2004) used an Upflow Anaerobic Sludge Reactor to remove 90% COD from Penicillin G-containing pharmaceutical effluent. Another study used anaerobic baffled reactors to treat ampicillin and Aureomycin wastewater, resulting in partial degradation (16%–30%). In anaerobic treatment, the high concentration of organic waste present in industrial wastewater converts into CO_2 and methane with the help of microorganisms. Numerous reactor designs have been developed to treat wastewater from various industries, including the upflow anaerobic sludge blanket reactor (UASB), anaerobic contact reactor (ACR), anaerobic fixed film reactor (AFFR), and fluidized bed reactor (FBR). Because of the slow growth rate and low pH, the hydraulic retention time (HRT) is prolonged. Rao et al. (2004) investigated the anaerobic biological treatment of pharmacological wastewater and found that it removed 60%–70% of COD.

Upflow anaerobic sludge blanket reactors

For the management of industrial wastewater, UASB reactors are generally used. For the treatment of chemical synthesis-based pharmaceutical wastewater, several researchers have used a hybrid UASB reactor. In 2009, Sreekanth et al. (2009) reported using a hybrid UASB reactor to treat bulk drug industry wastewater using thermophilic strain. The effectiveness of a laboratory-scale UASB reactor for treating chemical synthesis-based pharmacological wastewater under a range of operating circumstances has also been investigated.

Anaerobic fixed film reactor

Rao et al. (2004) investigated the use of an AFFR to treat wastewater from a bulk drug manufacturing facility with highly suspended solids, concluding that the AFFR could be used effectively to treat wastewater from bulk drug manufacturing facilities with high BOD (80%–90%), COD (60%–70%), TDS and TSS removal.

Aerobic treatment

In a biological reactor, Khan and Mostafa (2011) investigated the aerobic treatment of pharmacological effluent. Vanerkar et al. (2013) used an activated sludge system to treat wastewater from the herbal pharmaceutical sector, attaining COD diminutions of 68.98%–91.02%, BOD diminutions of 76.90%–97.26% and suspended solids diminutions of 74.61%–95.54%. For the treatment of high-strength mixed pharmaceutical wastewater, Elmolla et al. (2012) optimized SBR operating settings by achieving more than 60% BOD removal and 83% COD removal after 24 hours of HRT. Ng et al. (1989) used the sequencing batch reactor (SBR) technology for the treatment of pharmacological wastewater and attained a substantial decline in COD and BOD levels. Altaf and Ali (2010) used a SBR for the treatment of wastewater and developed a biological methodology for the determination of relative toxicity using dissolved oxygen 2.0 mg/L at varying pH levels.

Fungal treatment

Various fungal strains are useful in the treatment of industrial wastewater, but they have limits due to their extended growth cycles and spore formation. For the treatment of pharma industry wastewater, Spina et al. (2012) equated activated sludge treatment to fungal treatment. They utilized a fungal strain called Bjerk and eraadusta MUT 2295 in this investigation, which decreased COD by 91% when compared to activated sludge, which only decreased COD by 78%. Laccase secreted by *Trametes hirsuta* has been explored for the removal of emerging contaminants from wastewater. The immobilized laccase was able to degrade pharmaceuticals at neutral pH (Hachi et al., 2017). *Mucor hiemalis* was explored for pharmaceutical degradation following the success of myco remediation and the adaptability of fungi (Esterhuizen-Londt et al., 2016).

Bacterial treatment

Some of the strains of bacteria including Pseudomonas, Streptomonas, Enterobactor, Acinetobactor, Aeromonas, and Klebsiella demonstrate up to a 44% reduction in COD. Chaturvedi et al. (2006) isolated 15 rhizosphere bacteria in less than 30 days which lowered color by 76% and BOD and COD by 85%–86%. The bacterial community must deliver all metabolic abilities for the complete mineralization of hazardous organic pollutants. Das et al. (2012) treated industrial effluent by use of bacterial consortia and obtained a reduction in the level of sulfates, TDS, total suspended solids, and COD. Using a mixed bacterial culture, some researchers were able to remove a large amount of COD from pharmaceutical wastewater (62% at 30°C and 38% at 60°C).

Phytoremediation

Phytoremediation is a bioremediation process whereby green plants are used to remove, transfer, stabilize, and demolish the pollutants from the environment and render them non-available (Neina, 2019). Floating treatment wetlands are a sustainable approach to remediating various contaminates from wastewater (Afzal et al., 2019). Since natural or transgenic plants can bioaccumulate toxins, phytoremediation is shown to be a better treatment option for bio-treatment (Amin et al., 2013). Aquatic plants have a high capability for reducing harmful metals, BOD, and total solids in wastewater. Billore et al. (2001) used plants Typha latipholia and Phragmitiskarka to clean industrial wastewater. Under laboratory-scale conditions, the removal capacity of two aquatic plants (Lemna sp. and Spirogyra sp.) was investigated. After 20 days of incubation, the maximum removal efficiency was achieved in uncovered planted systems utilizing secondary-treated effluent, with removal efficiencies ranging from 31% to 100% (Garcia et al., 1995). In membrane bioreactors (MBR), the removal of pharmaceutical chemicals by microbial sludge with infinite and limited solid retention times (SRT) was studied. During MBR operation, most pharmaceutical compounds had moderate to high removal efficiencies. Hydrophobic compounds were mostly detected in coarse particles, where the majority of adsorption sites were available, whereas hydrophilic and moderately hydrophobic compounds were kept in tiny particles and gel-like substances, and were thereafter rejected by the membrane. When the MBR was operated at a greater biomass concentration under lengthy sludge age conditions, the removal rates of medicinal chemicals were also improved (Prasertkulsak et al., 2019). Antibiotic pollution has received a lot of attention, and removing antibiotics from water is critical. The biochars were employed to extract tetracycline (TC) from aqueous solution, and the impact of various experimental conditions on TC removal was studied. Rice straw biochar at a high temperature can operate as an adsorbent to extract tetracycline from an aqueous solution (Fan et al., 2018).

Membrane bioreactor

New oxidation methods like ozonation, oxidation, perozonation, TiO_2 photocatalysis, direct photolysis, solar photocatalysis, Fenton reaction, and ultrasonic irradiation are among the promising new treatments. These methods considerably improve the rate of pharmaceutical removal and biodegradability from wastewater. These recalcitrant organic compounds are mineralized via biological processes and meet the wastewater discharge standards. Such a combination of processes will be efficient and have the potential to be used on a large scale. Anaerobic sludge reactors, anaerobic film reactors, and anaerobic filters are different examples of anaerobic processes (Cheng et al., 2018). A combined treatment process for pharmaceutical wastewater was adopted by Sirtori et al. (2009) that resulted in a 95% reduction of DOC. The solar photo-Fenton treatment achieved a 33% reduction, while the immobilized biomass reactor achieved a 62% reduction. Hybrid methods like MBR followed by reverse osmosis, nanofiltration, or ultrafiltration are preferred for the elimination of pharmaceuticals. The removal of beta blockers and pharmaceuticals has been reported to be effective when ozonation is followed by a biologically activated carbon hybrid system. Other hybrid methods like activated sludge followed by ultrafiltration or activated sludge and gamma radiations are very cost-effective for the elimination of some analgesics.

Oxidation

Chlorine, chlorine dioxide (ClO_2), and ozone (O_3) are strongly oxidizing, disinfection, and bleaching agent, and act as a microbiocide against waterborne pathogens. For over 70 years, chlorination has been the primary disinfection product in the United States. Although O_3 is a strong oxidizing gas with a short half-life, in practical treatment applications, it tends to attack the double bonds of fatty acids in bacterial cell walls and the protein capsid of viruses (Al Rmalli et al., 2005). O_3 and UV radiation can be used for decontamination separately. Moreover, combinatorial treatments O_3/UV/H_2O_2 commonly involve two oxidation/photolysis courses to reduce unwanted elements by an oxidation process through hydroxyl free radicals generated in situ.

Electrochemical

The electrochemical remediation method can be extremely beneficial for the remediation of wastewater. Principally, the electrical potential difference consists of electrodes, or a network of electrodes, which are inserted in diverse formations in the polluted water. And at the outlet section treated water flowed out (Khansorthong and Hunsom, 2009).

Chemical extraction

Chelating agents like ethylene diamine tetra acetic acid (EDTA), nitrilo triacetic acid (NTA), and chelating resin are the various organic composites that possess metal ions binding places at their structure (Gagnon et al., 2005; Xie, 2004). So, these agents are frequently used in the remediation process. A study was done using 3% hydrogen peroxide (H_2O_2) and 0.1 M EDTA to extract 60% of the Cu and Pb and 70% of the petroleum in the soils, respectively (Guerra et al., 2018). Nowadays, a biodegradable agent such as diacetic glutamic acid (GLDA) is prepared via green chemistry to remove the Cd, Ni, and Cu-contaminated sludge at a low pH of 4. When sample to chelater used in the ratio of 1:3 the removal efficiency was witnessed as 89% for Cd, 82% for Ni, and 84% for Cu. Adsorption, Absorption, Ion exchange, and supercritical water oxidation (SCWO), Sonochemical degradation, Flocculation process (Morris, 2009; Streche et al., 2018).

Application in green chemistry

The "Green Chemistry" or "Green Pharmacy" approach can be an efficient tool or mechanism to manage pharmaceutical waste. In this methodology the entire life cycle of a product is monitored along with its production routes, to make it as sustainable as possible by consuming less energy and material, generating fewer unwanted byproducts, or by producing easily biodegradable byproducts. Chemical oxidation procedures including ozonation/H_2O_2, UV photolysis/H_2O_2, and photo-Fenton can effectively remove up to 100% of beta blockers and medicines. Some pharmaceuticals can be removed completely using a combination ozonation-ultrasound device (Ahmed et al., 2017). Adishkumar et al. (2012) investigated the combined solar photo-Fenton method with an aerobic SBR for the treatment of pharmacological wastewater. Although the UV treatment effectively reduced antibiotic concentrations, it also increased the effluent's toxicity. However, the UV-algae combined treatment resulted in relatively high removal efficiency and lowered effluent toxicity at the same time (Du et al., 2015).

Conclusions

From the above discussions, it can be visualized that the pharmaceutical industries are going to be one of the most growing sectors in the coming future. It will also put additional pollutants burden on different components of the environment. Pharmaceutical waste has become an emerging pollutant for groundwater quality and has severe impacts over the environment and human health. Proper redressal of pharmaceutical waste is requisite of the time. Existing sewage treatment plants are not fully equipped to mitigate pharmaceutical traces present in sewage. Adaptation of new emerging technologies and mitigative measures needs to be implemented. Further investigations and studies also need to be continued to understand the faith of pharmaceutical waste on the environment.

References

ACS (American Chemical Society). (2022). 12 principles of green chemistry. Available at: https://www.acs.org/content/acs/en/greenchemistry/principles/12-principles-of-green-chemistry.html.

Adishkumar, S., Sivajothi, S., Banu, J.R., 2012. Coupled solar photofenton process with aerobic sequential batch reactor for treatment of pharmaceutical wastewater. Desalination and Water Treatment 48, 89–95.

Afzal, M., Rehman, K., Shabir, G., Tahseen, R., Hashmat, A.J., 2019. Large-scale remediation of oil-contaminated water using floating treatment wetlands. NPJ Clean Water 2 (1), 1–10.

Ahmed, M.B., Zhou, J.L., Ngo, H.H., Guo, W., Thomaidis, N.S., Xu, J., 2017. Progress in the biological and chemical treatment technologies for emerging contaminant removal from wastewater: A critical review. Journal of Hazardous Materials 323, 274–298.

Aktan, C.K., Uzunhasanoglu, A.E., Yapsakli, K., 2018. Speciation of nickel and zinc, its short-term inhibitory effect on anammox, and the associated microbial community composition. Bioresource Technology 268, 558–567.

Al Rmalli, S.W., Harrington, C.F., Ayub, M., Haris, 2005. A biomaterial based approach for arsenic removal from water. Journal of Environmental Monitoring 7 (4), 279–282.

Altaf, M.S., Ali, T.A., 2010. Wastewater treatment using sequential batch reactor and development of microbiological method for the analysis of relative toxicity. Pakistan Journal of Nutrition 9 (6), 574–576.

Amin, A., Ramchandra, A.T.R., Azhar, M., Nayak, H., 2013. Bioremediation of different waste waters—a review. Journal of Fisheries and Aquatic Sciences 7 (2), 7–17.

Anastas, P.T., Eghbali, N., 2010. Green chemistry: Principles and practice. Chemical Society Reviews 39, 301–312. Available from: https://doi.org/10.1039/B918763B.

Anastas, P.T., Kirchhoff, M.M., 2002. Origins currents status and future challenges of green chemistry. Accounts of Chemical Research 35 (9), 686–694.

Billore, S.K., Singh, N., Ram, H.K., Sharma, J.K., Singh, V.P., Nelson, R.M., et al., 2001. Treatment of a molasses based distillery effluent in a constructed wetland in central India. Water Science and Technology 44 (11–12), 441–448.

Chaturvedi, S., Chandra, R., Rai, V., 2006. Isolation and characterization of *Phragmitis australis* (L) rhizosphere bacteria from contaminated site for bioremediation of colored distillery effluent. Ecological Engineering 27, 202–207.

Cheng, Y., Chen, Y., Lu, J., Nie, J., Liu, Y., 2018. Fenton treatment of bio-treated fermentation-based pharmaceutical wastewater: Removal and conversion of organic pollutants as well as estimation of operational costs. Environmental Science and Pollution Research 25 (12), 12083–12095.

Clark, J.H., 1999. Green chemistry: Challenges and opportunities. Green Chemistry 1 (1), 1–8. Available from: https://doi.org/10.1039/A807961G.

Das, M.P., Bashwant, M., Kumar, K., Das, J., 2012. Control of pharmaceutical effluent parameters through bioremediation. Journal of Chemical and Pharmaceutical Research 4 (2), 1061–1065.

Du, Y., Zhang, S., Guo, R., Chen, J., 2015. Understanding the algal contribution in combined UV-algae treatment to remove antibiotic cefradine. RSC Advances 5 (74), 59953–59959.

Elmolla, E.S., Ramdas, N., Chaudhuri, M., 2012. Optimization of Sequencing Batch Reactor operating conditions for treatment of high strength pharmaceutical wastewater. Environmental Science & Technology 5 (6), 452–459.

Esterhuizen-Londt, M., Schwartz, K., Pflugmacher, S., 2016. Using aquatic fungi for pharmaceutical bioremediation: Uptake of acetaminophen by *Mucor hiemalis* does not result in an enzymatic oxidative stress response. Fungal Biology 120 (10), 1249–1257.

Fan, S., Wang, Y., Li, Y., Wang, Z., Xie, Z., Tang, J., 2018. Removal of tetracycline from aqueous solution by biochar derived from rice straw. Environmental Science and Pollution Research 25 (29), 29529–29540.

Gagnon, G., Rand, J.L., O'Leary, K.C., Rygel, A.C., Chauret, C., Andrews, R.C., 2005. Disinfectant efficacy of chlorite and chlorine dioxide in drinking water biofilms. Water Research 39 (9), 1809–1817.

Garcia, A., Rivas, H.M., Figueroa, J.L., Monroe, A.L., 1995. Case history: Pharmaceutical wastewater treatment plant upgrade, Smith Kline Beecham Pharmaceuticals Company. Desalination 102 (1–3), 255–263.

Guerra, F.D., Attia, M.F., Whitehead, D.C., Alexis, F., 2018. Nanotechnology for environmental remediation: Materials and applications. Molecules (Basel, Switzerland) 23 (7), 1760.

Hachi, M., Chergui, A., Yeddou, A.R., Selatnia, A., Cabana, H., 2017. Removal of acetaminophen and carbamazepine in single and binary systems with immobilized laccase from *Trametes hirsuta*. Biocatalysis and Biotransformation 35 (1), 51–62.

Halling-Sørensen, B., Nielsen, S.N., Lanzky, P.F., Ingerslev, F., Lützhøft, H.C.F., Jørgensen, S.E., 1998. Occurrence, fate and effects of pharmaceutical substances in the environment- a review. Chemosphere 36 (2), 357–393.

Hu, Z., Chandran, K., Grasso, D., Smets, B.F., 2003. Impact of metal sorption and internalization on nitrification inhibition. Environmental Science &Technology 37 (4), 728–734.

Karigar, C.S., Rao, S.S., 2011. Role of microbial enzymes in the bioremediation of pollutants: A review. Enzyme Research 2011.

Khan, M.Z.H., Mostafa, M.G., 2011. Aerobic treatment of pharmaceutical wastewater in a biological reactor. International Journal of Environmental Science 1 (7), 1797–1805.

Khansorthong, S., Hunsom, M., 2009. Remediation of wastewater from pulp and paper mill industry by the electrochemical technique. Chemical Engineering Journal 151 (1–3), 228–234.

Kumar, V., Shahi, S., Singh, S., 2018. Bioremediation: An eco-sustainable approach for restoration of contaminated sites. Microbial bioprospecting for sustainable development. Springer, pp. 115–136.

Louise, K.M., Ghaly, T.M., Gillings, M.R., 2021. A survey of sub-inhibitory concentrations of antibiotics in the environment. Journal of Environmental Sciences 99, 21–27.

Majumder, S., Chatterjee, Basnet, P., Mukherjee, J., 2020. ZnO based nanomaterials for photocatalytic degradation of aqueous pharmaceutical waste solutions – a contemporary review. Environmental Nanotechnology, Monitoring & Management 100386.

Migliore, L., Cozzolino, S., Fiori, M., 2003. Phytotoxicity to and uptake of enrofloxacin in crop plants. Chemosphere 52 (7), 1233–1244.

Morris, J.C., 2009. Formation of halogenated organics by chlorination of water supplies: A review, Office of Research and Development, Vol. 1. US Environmental Protection Agency, 1975.

Neina, D., 2019. The role of soil pH in plant nutrition and soil remediation. Applied and Environmental Soil Science 2019.

Ng, W.J., Yap, M.G.S., Sivadas, M., 1989. Biological treatment of a pharmaceutical wastewater. Biological Wastes 29 (4), 299–311.

Prasertkulsak, S., Chiemchaisri, C., Chiemchaisri, W., Yamamoto, K., 2019. Removals of pharmaceutical compounds at different sludge particle size fractions in membrane bioreactors operated under different solid retention times. Journal of Hazardous Materials 368, 124–132.

Raj, D.S.S., Anjaneyulu, Y., 2005. Evaluation of biokinetic parameters for pharmaceutical wastewaters using aerobic oxidation integrated with chemical treatment. Process Biochemistry 40 (1), 165–175.

Rao, A.G., Naidu, G.V., Prasad, K.K., Rao, N.C., Mohan, S.V., Jetty, A., et al., 2004. Anaerobic treatment of wastewater with high suspended solids from a bulk drug industry using fixed film reactor (AFFR). Bioresource Technology 93, 241–247.

Richards, N.L., Gilbert, M., Taggart, M., Naidoo, V., 2018. A cautionary tale: Diclofenac and its profound impact on vultures. In: Dellasala, D.A., Goldstein, M.I. (Eds.), Encyclopedia of the anthropocene. Elsevier, pp. 247–255.

Rosen, M., Welander, T., Löfqvist, A., Holmgren, J., 1998. Development of a new process for treatment of a pharmaceutical wastewater. Water Science & Technology 37 (9), 251–258.

Sharma, S., Das, J., Braje, W.M., Dash, A.K., Handa, S., 2020. A glimpse into green chemistry practices in the pharmaceutical industry. ChemSusChem . Available from: https://doi.org/10.1002/cssc.202000317.

Sirtori, C., Zapata, A., Oller, I., Gernjak, W., Agüera, A., Malato, S., 2009. Decontamination industrial pharmaceutical wastewater by combining solar photo-Fenton and biological treatment. Water Research 43, 661–668.

Spina, F., Anastasi, A.E., Prigione, V., Tigini, V., Varese, G.C., 2012. Biological treatment of industrial wastewaters: A fungal approach. Chemical Engineering Transactions 27, 175–180.

Sreekanth, D., Sivaramakrishna, D., Himabindu, Y., Anjaneyulu, Y., 2009. Thermophilic treatment of bulk drug pharmaceutical industrial wastewaters by using hybrid up flow anaerobic sludge blanket reactor. Bioresource Technology 100 (9), 2534–2539.

Srinivasamoorthy, K., Gopinath, M., Chidambaram, S., Vasanthavigar, M., Sarma, V.S., 2014. Hydrochemical characterization and quality appraisal of groundwater from Pungar sub basin, Tamilnadu. India, Journal of King Saud University – Science 26 (1), 37–52.

Streche, C., Cocârţă, D.M., Istrate, I., Badea, A.A., 2018. Decontamination of petroleum-contaminated soils using the electrochemical technique: Remediation degree and energy consumption. Scientific Reports 8 (1), 3272.

Suhag, R. (2022) Overview of ground water in India. Working Papers id: 9504, 2016. eSocialSciences.

USEPA. (2022) Basics of green chemistry. Available at: https://www.epa.gov/greenchemistry/basics-green-chemistry.

Van Boeckel, T.P., Gandra, S., Ashok, A., Caudron, Q., Grenfell, B.T., Levin, S.A., et al., 2014. Global antibiotic consumption 2000 to 2010: An analysis of national pharmaceutical sales data. The Lancet Infectious Diseases 14 (8), 742–750.

Vanerkar, A.P., Satyanarayan, S., Dharmadhikari, D.M., 2013. Full scale treatment of herbal pharmaceutical industry wastewater. International Journal of Chemical and Physical Sciences 2, 52–62.

Veleva, V.R., Cue, B.W., Todorova, S., Thakor, H., Mehta, N.H., Padia, K.B., 2018. Benchmarking green chemistry adoption by the Indian pharmaceutical supply chain. Green Chemistry Letters and Reviews 11 (4), 439–456. Available from: https://doi.org/10.1080/17518253.2018.1530802.

Wei, X., Li, B., Zhao, S., Wang, L., Zhang, H., Li, C., et al., 2012. Mixed pharmaceutical wastewater treatment by integrated membraneaerated biofilm reactor (MABR) system—a pilot-scale study. Bioresource Technology 122, 189–195.

WHO, WHO Pharmaceuticals Newsletter No. 1, 2018. https://www.who.int/medicines/publications/PharmaNewsletter1_18/en/.

Xie, Y., 2004. Disinfection byproducts in drinking water: Formation, analysis, and control. 2003. CRC Press.

Chapter 17

Sustainable approaches to heavy metal removal from water

Richa Soni[1], Dericks Praise Shukla[2] and Nadjib Drouiche[3]

[1]Physics Department, Lancaster University, Lancaster, United Kingdom, [2]School of Engineering, Indian Institute of Technology, Mandi, Himachal Pradesh, India, [3]CRTSE-Division CCPM- N°2, Alger, Algeria

Introduction

Recent estimates have indicated that more than 1.5 billion people across the globe are devoid of clean water access-one of the most vital elements of life (Barlow and Clarke, 2017). The discharge of waste coming out from the industries is increasing water contamination, especially heavy metals. Heavy metals are elements having densities greater than 5 g/cm^3 and are considered common pollutants (Saif and Khan, 2017; Prieto et al., 2018). Heavy metals possess a half-life of more than 20 years and therefore a threat to the environment (Kapoor and Singh, 2021). Continuing exposure to heavy metals in the environment is a real hazard. Heavy metal toxicity not only causes ill effects to the health but also disturbs the soil balance reducing the fertility (Wieczorek-Dąbrowska et al., 2013). Cadmium, lead and mercury are toxic and hence nonessential while manganese, iron, copper and zinc in trace quantities are required and therefore considered to be essential (Ali et al., 2019). The erratic release of heavy metals into the water is a notable concern around the globe, as the toxic nature of the heavy metals have lasting consequences on the biological systems (Mishra et al., 2018). The decrease in the growth of plants is due to the heavy metals which reduces the enzymatic activity governing photosynthesis (Nematian and Kazemeini, 2013). Heavy metals above the acceptable limits will have detrimental effects on humans, other organisms and environment (Mansourri and Madani, 2016). The toxicity level of metals for humans follows the sequence $Co < Al < Cr < Pb < Ni < Zn < Cu < Cd < Hg$. Mercury, cadmium and lead are the ones which are perceived more as a matter of concern (Mansourri and Madani, 2016). Lead is very toxic and causes irreversible brain damage and encephalopathic symptoms (Ashizawa et al., 2012). Initial exposure of lead may cause headaches, loss of memory, irritation and dullness, with prolonged exposure leading to disturbed hemoglobin synthesis and anemia (Järup, 2003). The WHO recommended limits of lead in wastewater is 0.01 ppm (Ayeni, 2014; Chiroma et al., 2014). Cadmium contacts causes dysfunction of kidney, degeneration of bone, liver and blood damage and shows satisfactory evidence of being carcinogenic (Abadin et al., 2007). The WHO recommended limits of cadmium in wastewater is 0.003 ppm (Chiroma et al., 2014; Aneyo et al., 2016). Toxic nature of mercury is well known, it may affect nervous system and prolonged exposure results into neurological damage (Bernhoft, 2012). The WHO recommended limits of mercury in wastewater is 0.001 ppm (Onuegbu et al., 2013). Copper at high levels may cause damage to the kidneys and even death. The minor symptoms also include irritation to nose, eyes and mouth (Gautam et al., 2014). Excess of zinc may cause nausea, vomiting, anemia, and cholesterol issues in humans (Gautam et al., 2014).

Conventional methods for heavy metal removal from water

Several traditional approaches to remove heavy metals from water are present: microfiltration, chemical precipitation method, reverse osmosis, sedimentation, nanofiltration, adsorption, electrodialysis and ion exchange (Sayqal and Ahmed, 2021; Kaur and Roy, 2021). Recent innovative adsorbents, photocatalysts and ion exchange resins are not utilized yet for large water treatment facilities and cost will also be an aspect to be considered. The conventional approaches for the removal of heavy metals are promising but usually require a lot of manpower and are expensive (Gardea-Torresdey et al., 2005; Malaviya and Singh, 2011). Consecutively, low-cost, sustainable options for heavy

Role of Green Chemistry in Ecosystem Restoration to Achieve Environmental Sustainability. DOI: https://doi.org/10.1016/B978-0-443-15291-7.00020-1

TABLE 17.1 Advantages and limitations of conventional treatment methods used for heavy metal removal (Monteiro et al., 2012; Cheng et al., 2019; Goswami et al., 2021)

Methods	Advantages	Limitations
Adsorption	• May be low cost (depending on the type) • Process is fast. • May be regenerated and reused. • Wide usage, ease of operation.	• Highly efficient adsorbents may be costly. • Regeneration might not be easy and may require use of chemicals. • Disposal of adsorbed and exhausted adsorbent.
Ion exchange	• Effective method with high regeneration capacity. • Metal selectivity is possible.	• Operation cost is high.
Membrane filtration	• Possess high efficiency. • Easy to use and does not require a lot of infrastructure.	• Membranes are costly. • Initial capital cost and maintenance cost is high. • Energy may be required for filtration.
Reverse osmosis	• Very efficient, able to remove most of the heavy metals	• Expensive and energy intensive
Chemical precipitation	• Facile technique	• Huge quantities of sludge is generated which need to be disposed of thereby affecting cost. • Requires chemical.
Electrochemical method	• Requires minimum chemicals. • Pure metal can be retrieved.	• Energy intensive. • Requires disposal of sludge. • Expensive process.

metal treatment are required. Biological methods like biosorption/bioaccumulation of heavy metals are much sought-after techniques these days and much research is being carried out in this area (Razzak et al., 2022). The advantages and limitations of the conventional treatment methods used for heavy metal removal are presented in Table 17.1.

Bioremediation

This chapter discusses the bioremediation process which is a sustainable method of heavy metal removal and uses microorganisms or plants to degrade the toxic contaminants. Bioremediation is cost-effective when compared to chemical treatment methods, it is also non-obtrusive.

Bioremediation using microbes (microbial bioremediation)

Microbial Bioremediation is a treatment method involving microbial application to depollute contaminants (Ayangbenro and Babalola, 2017). The microbes clean up the contaminants from the water, soil, and other sources in turn using the contaminants as a nutritional and energy source (Ostrem Loss and Yu, 2018). It is low cost when compared to conventional methods as the biosystem used in this process is made from waste biomass or naturally available biomass of bacteria, fungi, or algae (Ayangbenro and Babalola, 2017). Microorganisms do not generate any by-products and are effective even at low concentrations of metal (Das and Dash, 2014). A pictorial illustration of Microbial bioremediation is shown in Fig. 17.1. A variety of microorganisms are available to be used for the process of bioremediation. Microbes contain resistant genes in their DNA to forms of heavy metals and undergo several mechanisms to uptake heavy metals and survive (Sayqal and Ahmed, 2021). Recent studies have also highlighted the use of microbial consortia as a replacement for pure cultures owing to their synergistic metabolism, which increases the efficiency and speed of the degradation process (Das and Dash, 2014).

Bioattenuation

The bioattenuation process is the natural way to degrade heavy metals through the presence of a metabolic variety of fundamental microorganisms present at the site of contamination (Abatenh et al., 2017). As the site of contamination is varied, the time taken for natural attenuation to occur is dependent on the condition of the site, concerned contaminants present, and existing microbial flora on the site (Azubuike et al., 2016). Bioattenuation is efficient and cost-effective on

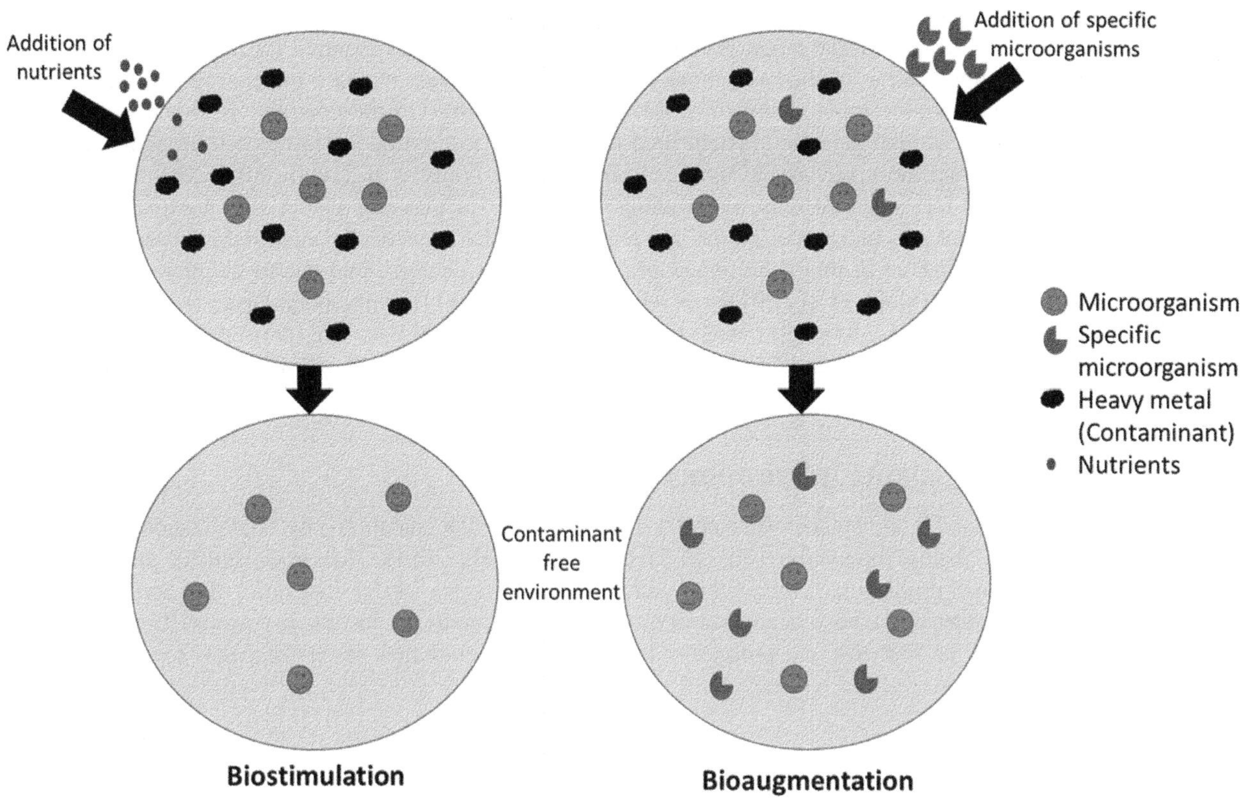

FIGURE 17.1 Representation of microbial bioremediation.

sites that do not require any rigorous treatment and as no anthropogenic involvement is required costs involving post clean-up are substantially reduced. The limitation of this method is that alone this method becomes insufficient due to the lack of appropriate microorganisms present at the contaminated site (Zehra et al., 2020)

Biostimulation

Biostimulation, as the name suggests is the stimulation to increase the microbes in the site of contamination by the addition of relevant nutrients. Typically, the addition of nutrients like carbon, phosphorus, and nitrogen in a suitable proportion is required to enhance the degrading ability of microorganisms (Ying, 2018). Some common nutrients used for biostimulation are water-soluble compounds (e.g., KNO_3, $NaNO_3$, NH_3NO_3, K_2HPO_4, and $MgNH_4PO_4$) (Nikolopoulou et al., 2013).

Bioaugmentation

Bioaugmentation is the method of adding relevant microorganisms in the water with the ability to degrade heavy metals. The microorganisms added in this process could either be a single strain or a group of multiple strains (Niu et al., 2009). The process of bioaugmentation is dependent on biotic and abiotic factors. Biotic factors include a choice of specific microorganisms to degrade the heavy metal and compete with the intrinsic flora at the site (Abatenh et al., 2017). Abiotic factors include the chemical and physical properties of the site, the nature of the contaminant, and its concentration (Goswami et al., 2021). Varied species are being used for this purpose like *Bacillus, Pseudomonas, Sphingobium*, etc. The fungi belonging to the genera *Aspergillus*, *Penicillium*, and *Verticillium* have been used to treat heavy metals from wastewater (Bahobil et al., 2017).

Engineered microorganisms

The metabolic potential of microbes offers an effective mechanism for removing pollutants. Microbes have not previously been exposed to contaminants; when they are exposed the microbes gradually change their metabolic pathway and learn to survive in a stressed environment. The recombinant DNA and molecular biology methods have enabled: modification/blocking/amplification of the target gene which encodes the enzymes required for the metabolic functioning of microorganisms; minimization of the restrictions of metabolic pathways, increase the energy generation, introduction heterologous gene for new traits (Shimizu, 2002; Liu et al., 2006). On the contrary, genetically engineered microorganisms are superior as their genome is engineered to develop a strong import-storage system allowing uptake of heavy metals and hence are more versatile (Gupta and Singh, 2017). Valls et al. (2000) informed a fusion protein (β-domain of IgA protease) for enhanced biosorption along the transfer of plasmids comprising the essential genetic material between recombinant and indigenous strains for improved results. Genetically modified microorganisms also have their own set of limitations as at times the fitness of these microbes decrease and due to the presence of foreign gene the energy demands rise (Singh et al., 2011), so introducing such modified microorganisms to the environment may be treated as a threat until all information about them has been secured. Researchers have also pointed to a safer route of introducing suicidal genetically modified microorganisms which will have programmed death genes and after remediation, they will die (Pandey et al., 2005).

Bioremediation using plants (phytoremediation)

Several processes are involved in the uptake of metals through plants: phytodegradation, phytostabilization, phytovolatilization, phytoextraction, phytoaccumulation (Fig. 17.2) (Tangahu et al., 2011). Numerous aquatic plants such as *Ludwigia stolonifera* (water primrose), *Lemna gibba* (duckweed), *crassipes* (water hyacinth), *Eichhornia Veronica anagallis-aquatica* (blue water well), have been used to proficiently remove heavy metals from water. The major benefits of this method include the low cost and processing ease; the only requirement is for the plants to grow naturally (Saleh et al., 2020).

Phytoextraction

Also referred to as phytoaccumulation, phytosorption, or phytosequestration; is the uptake of toxins by roots to the shoots. The shoots and roots are then collected to remove the toxins (Ali et al., 2013). This method proves to be more sustainable in comparison to conventional cleaning techniques. (Chandra and Kumar, 2017), proved that twelve suitable native plants demonstrated their capacity to uptake iron when growing on sludge. *Amaranthus spinosus* L. was found to accumulate Zinc and Manganese in the shoot. All other plants except for *A. spinosus* L. and *Ricinus communis*, were found to be leaf accumulators for zinc, manganese, and iron. Asgari Lajayer et al. (2019) have verified the possibility of ornamental plants for phytoextraction of heavy metals from contaminated sites.

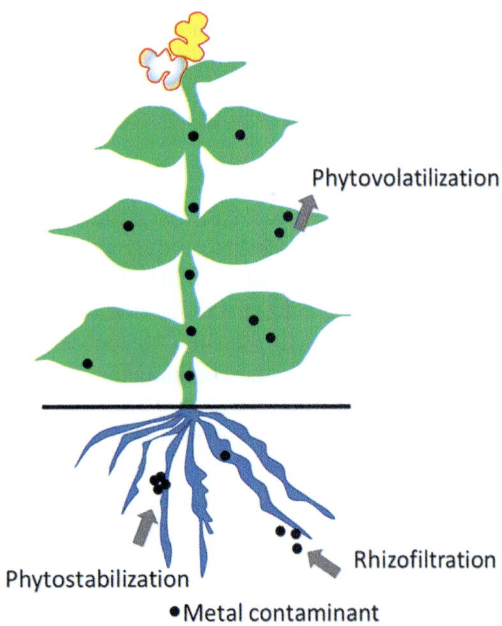

FIGURE 17.2 Representation of phytoremediation.

Phytostabilization

Phytostabilization uses plants that can tolerate metals to decrease their bioavailability by immobilizing the heavy metals and preventing their movement into the ecosystem and decreasing the chances of their entrance into the food chain (Wang et al., 2020). Luo et al. (2019) examined the suitability of using four woody plants in combination with organic modifications to phytostabilize zinc smelting slag.

Phytofiltration

This method utilizes various parts of plants for heavy metal removal from water: roots (rhizofiltration), seedlings (blastofiltration), or shoots (caulofiltration). The most prevalent is rhizofiltration, and plants used for this method are first grown in clean water to form an effective root system and then they are planted in the polluted waters for removal of heavy metals. Posttreatment the roots are gathered and disposed-off. For the treatment of wetland water, aquatic species such as cattail, hyacinth, duckweed, and azolla, are generally used owing to their high accumulation of heavy metals, rapid growth, high tolerance, and high production of biomass (Hooda, 2007).

Phytovolatilization

Phytovolatilization includes the uptake of contaminants from the roots of the plants and then volatilize and releasing them into the atmosphere. *Astragalus racemosus* may convert selenium into dimethyl diselenide whereas *Arabidopsis thaliana*, boosts the volatility of mercury and converts Hg^{2+} to Hg^0 (Awa and Hadibarata, 2020).

Engineered approaches for phytoremediation

The process of phytoremediation is at times very slow; it is dependent on plant growth which limits its fast and large-scale application of this method (Dhankher et al., 2012). Post recombination of DNA, the traits could be observed in the plant of concern. Traditional breeding has limitations, but genetic engineering allows to transfer of genes to sexually incompatible plants (Marques et al., 2009). Heavy metals tolerance via genetic engineering is increased by overexpression of genes whose function is in the uptake, sequestration, and translocation of heavy metals (Das et al., 2016). Plants, such as *Silene cucubalus*, *A. thaliana*, *Nicotiana tabacum*, *Brassica juncea*, *Populus angustifolia* have been effectively genetically engineered with diverse bacterial genes, resulting in improved accumulation of heavy metal and transformation as related with corresponding native plants (Gupta and Singh, 2017; Mosa et al., 2016). It was reviewed that modifying genes to enhance the synthesis of metal chelation agents increases the tolerance of heavy metals (Mosa et al., 2016). Selenium is taken up by the plants and converted into dimethyl selenide which is volatile. Arabidopsis was grown in sodium arsenate (125 M) medium, coexpressing two bacterial genes, the transgenic plants accumulated three times more arsenic in aboveground biomass (Dhankher et al., 2002). The limitations of the application of phytoremediation are the environment of the site. Temperature affects the growth of plants, metabolism, and transpiration rate and therefore affects heavy metal absorption and removal (Bhargava et al., 2012). Another problem is that it is difficult to seek approval for testing the engineered plants owing to the risk involved with ecosystem and food safety. Table 17.2 lists the advantages and limitations of microbial remediation and phytoremediation.

Biosorption

Biosorption is a kind of adsorption where the adsorbent is a biological matrix. Biosorption is a physico-chemical method, that allows biomass to accumulate passively heavy metals into their structure (Fomina and Gadd, 2014). It uses various microorganisms, plant and animal biomass, and their derived materials, wastes, and biopolymers. Owing to the presence of a plethora of functional groups, biomaterials have a great affinity for metal ions (Volesky, 2007). Usually, biosorbents are prepared from dead biomass (Baysal et al., 2009; Selatnia et al., 2004; Arıca et al., 2004; Michalak et al., 2013). Bioadsorption is a promising technique as it is eco-friendly and attracted a lot of researchers to find cost-effective options for biosorbents that are locally available to remove the contaminant. Also, reusability and low volume of sludge are a benefit of the process (Maximous et al., 2010). Several low-cost adsorbents have a low adsorption capacity and thus further investigations are required to alter their structural properties to enhance their efficiency (Pourrahim et al., 2020). Various green wastes have been investigated like corn cobs, tree leaves, wheat bran, rice husks, tree barks, and aquatic weeds (Malakahmad et al., 2016). However, some issues are present to regenerate it, and that increases the cost of treatment.

TABLE 17.2 Advantages and limitations of microbial remediation and phytoremediation (Liu et al., 2020; Cheng & Li, 2007).

Method	Advantages	Limitations
Microbial bioremediation Kulshreshtha et al. (2014)	• Natural process, microbes thrive in presence of contaminants and declines when contaminant degrades. • Less energy required when compared to other conventional technologies. • Cost-effective • Complete destruction of contaminant is possible without transferring from one medium to another	• Microbes might have a lot of specificity to degrade. • Degradation also depends on factors like: environmental conditions, nutrient level, type of contaminant • Time take by remediation process is longer than conventional treatments. • Hard to predict the prolonged effectiveness of the process.
Phytoremediation Dubchak and Bondar (2019)	• Cost effective • Monitoring of plants is easy. • Possibility of recovery of reusable metals (Phytomining) • Uses naturally present microorganisms, least harmful process.	• Remediation reach is limited to the depth occupied by the roots and surface area. • Plant growth affects the remediation. • Contaminants from plants can leach out. • Toxicity of contaminants may hamper plant growth. • Safe disposal of plants containing contaminants.

Nanotechnology

Nanomaterials can be used for in situ remediation of groundwater and can substitute the costlier technologies (Kaufmann et al., 2003). Nanomaterials have a high specific surface area and reactivity to interact with heavy metal contaminants via adsorption. Ion-exchange, redox reactions, and electrostatic interaction (Ekrami et al., 2022). The use of nanomaterials should not interfere with the ecological concerns of the environment, as the toxicity of nanomaterials to the environment is still unclear (Ekrami et al., 2022). The development of biodegradable nanoparticles is attracting the attention of researchers to make this process of treatment more sustainable in the near future. Activated carbon has a large surface area, porous structure, and high affinity due to which it is used for organic and inorganic removal from water (Dias et al., 2007; Karnib et al., 2014). Carbon nanotubes became popular as they could be functionalized and composites could be made (Park et al., 2007; Choi & Jang, 2008). Carbon nanotubes when compared to activated carbon have more available adsorption sites per unit mass and shorter interparticle diffusion distance because of which the efficiency of heavy metal removal is greater than activated carbon (Rao et al., 2007). Graphene is another very popular nanosorbent and is used in conjugation with oxides of metals to adsorb heavy metals from water (Sreeprasad et al., 2011; Sun et al., 2015; Cui et al., 2015; Dong et al., 2015; Luo et al., 2014). Zeolite-based nanosorbents: Zeolites are crystalline hydrated aluminosilicate having uniform pores. Zeolites are usually bigger but their pore sizes lie in the range of 0.4 and 1 nm and therefore they are considered nanomaterials (Hu & Apblett, 2014). One of the most widely used materials is granular ferric hydroxide which is used for arsenic removal commercially (Bolisetty et al., 2019).

Emerging technologies

Graphene-based microbots

These devices are self-propelled robots that can devour heavy metals from water. Vilela et al. demonstrated an effective microbot that was able to reduce lead from water from 1000 ppb to 50 ppb in 2 hours (Vilela et al., 2016). The bots were self-propelled and had magnetic behavior which made it easy to collect after the treatment and also could be regenerated if pH adjustments were made.

Metal–organic frameworks (MOFs)

MOFs also known as porous coordination polymers (PCPs), are multifunctional materials evolving as promising candidates in a multitude of different applications (Zhu and Xu, 2014) These synthetic materials have high porosity, owing to self-assembled metal moieties and organic linkers (Efome et al., 2018). Farha et al. (2012) synthesized MOF with 7000

m^2/g and computer simulations increased the surface area to 14,600 m^2/g (Kobielska et al., 2018). Sun et al. (2018) developed an eco-friendly MOF polymer composite with a high efficiency of 99.8% in removing mercury and lead.

Membrane-based removal

Membrane-based removals are very reliable. Membrane treatment does not require the addition of chemicals and is a relatively simpler technique in terms of operation, control, and scale-up (Zhu et al., 2014; Qu et al., 2013). Several membrane technologies are popular like ultrafiltration, reverse osmosis, and nanofiltration. Another advancement in membrane-based technologies is nanohybrid technology. These membranes are developed by integrating inorganic particles in the membrane precursors (Peydayesh et al., 2019). Mixed matrix membranes are made by incorporating porous inorganic materials like carbon-based sieves, and zeolites in the main membranes so that the advantages of both materials can be incorporated into the removal system (Pendergast and Hoek, 2011). Nanocomposite membranes are made by integrating membranes with nanomaterials like silver oxide, enhancing the surface properties of the membranes (Yin & Deng, 2015).

Electrodialysis

Another membrane-based technique of heavy metal removal is electrodialysis in which the membrane process is driven by electricity. Chehayeb & Nayar (2018), experimented to reduce the cost of the electrodialysis process aiming to achieve a uniform ion distribution across the membrane. Researchers are now developing models to estimate the energy consumption of electrodialysis treatment systems and then tailor the treatment by computer simulations before the system is built (Shah et al., 2018).

Conclusions

This chapter presented the sustainable practices available for heavy metal remediation. The existing physical and chemical methods that are available have some limitations and are not cost-effective and eco-friendly. Several bioadsorbents, microbial populations, and plants are utilized to remove/degrade heavy metals and are treated as sustainable options for the removal of heavy metals. Sustainability, itself is a very inclusive term and includes a lot of conditions to be kept in mind. A sustainable water treatment technology should be less expensive (in terms of both recurring as well as non-recurring costs, less energy requirements, along with a maximum volume of treated water and zero requirements of posttreatment of treated water). It would not be correct that the use of only biomaterials in treatment would make a process sustainable instead a process that could in totality solve the purpose of providing treated water to masses and fulfill most of the criteria of being a sustainable technology: technical aspects, efficiency performance, scalability, and sustainability would be called as an ideal candidate of treatment.

Future prospects

The deterioration of water quality and the introduction of contaminants especially heavy metals has increased. In the current times, when all are trying to move towards sustainable options; even treatment /remediation methods are being looked upon from the perspective of being sustainable. The involvement of naturally available materials and microbes for remediation purposes is the new focus of research. These alternatives are both energy as well as cost-saving. The current limitations are the inability of microorganisms to degrade lethal contaminants to harmless ones. Research progress is required in the areas of genetically modified microorganisms which can provide in situ remediation and can degrade a variety of heavy metals under stressed conditions. Also, genetically modified organisms should not pose a secondary threat to the environment and should be harmless to the environment. The unwillingness among the public to accept genetically modified organisms for bioremediation also needs to be considered in future studies, and their non-toxic nature must be evidenced. In addition, the use of microbes in conjugation with other physical methods can prove to be a better remediation technique (Muthusaravanan et al., 2018). Advancements in nanotechnology and its application in remediation also show a lot of future scope.

References

Abadin, H., Ashizawa, A., Llados, F., Stevens, Y.W., 2007. Toxicological profile for lead.

Abatenh, E., Gizaw, B., Tsegaye, Z., Wassie, M., 2017. The role of microorganisms in bioremediation-a review. Open Journal of Environmental Biology 2 (1), 038–046.

Ali, H., Khan, E., Ilahi, I., 2019. Environmental chemistry and ecotoxicology of hazardous heavy metals: environmental persistence, toxicity, and bioaccumulation. Journal of Chemistry 2019.

Ali, H., Khan, E., Sajad, M.A., 2013. Phytoremediation of heavy metals—concepts and applications. Chemosphere 91 (7), 869–881.

Aneyo, I.A., Doherty, F.V., Adebesin, O.A., Hammed, M.O., 2016. Biodegradation of pollutants in waste water from pharmaceutical, textile and local dye effluent in Lagos, Nigeria. Journal of Health and Pollution 6 (12), 34–42.

Arıca, M.Y., Bayramoğlu, G., Yılmaz, M., Bektaş, S., Genç, Ö., 2004. Biosorption of Hg2 + , Cd2 + , and Zn2 + by Ca-alginate and immobilized wood-rotting fungus *Funalia trogii*. Journal of Hazardous Materials 109 (1-3), 191–199.

Asgari Lajayer, B., Khadem Moghadam, N., Maghsoodi, M.R., Ghorbanpour, M., Kariman, K., 2019. Phytoextraction of heavy metals from contaminated soil, water and atmosphere using ornamental plants: mechanisms and efficiency improvement strategies. Environmental Science and Pollution Research 26 (9), 8468–8484.

Ashizawa, A., Faroon, O., Ingerman, L., Jenkins, K., Tucker, P., Wright, S., 2012. Toxicological profile for cadmium.

Awa, S.H., Hadibarata, T., 2020. Removal of heavy metals in contaminated soil by phytoremediation mechanism: a review. Water, Air, & Soil Pollution 231 (2), 1–15.

Ayangbenro, A.S., Babalola, O.O., 2017. A new strategy for heavy metal polluted environments: a review of microbial biosorbents. International Journal of Environmental Research and Public Health 14 (1), 94.

Ayeni, O., 2014. Assessment of heavy metals in wastewater obtained from an industrial area in Ibadan, Nigeria. RMZ–M&G [Internet] 61, 19–24.

Azubuike, C.C., Chikere, C.B., Okpokwasili, G.C., 2016. Bioremediation techniques–classification based on site of application: principles, advantages, limitations and prospects. World Journal of Microbiology and Biotechnology 32 (11), 1–18.

Bahobil, A., Bayoumi, R.A., Atta, H.M., El-Sehrawey, M.M., 2017. Fungal biosorption for cadmium and mercury heavy metal ions isolated from some polluted localities in KSA. International Journal of Current Microbiology and Applied Science 6 (6), 2138–2154.

Barlow, M., Clarke, T., 2017. Blue Gold: The Battle Against Corporate Theft of the World's Water. Routledge.

Baysal, Z., Cinar, E., Bulut, Y., Alkan, H., Dogru, M., 2009. Equilibrium and thermodynamic studies on biosorption of Pb (II) onto *Candida albicans* biomass. Journal of Hazardous Materials 161 (1), 62–67.

Bernhoft, R.A., 2012. Mercury toxicity and treatment: a review of the literature. Journal of Environmental and Public Health 2012.

Bhargava, A., Carmona, F.F., Bhargava, M., Srivastava, S., 2012. Approaches for enhanced phytoextraction of heavy metals. Journal of Environmental Management 105, 103–120.

Bolisetty, S., Peydayesh, M., Mezzenga, R., 2019. Sustainable technologies for water purification from heavy metals: review and analysis. Chemical Society Reviews 48 (2), 463–487.

Chandra, R., Kumar, V., 2017. Phytoextraction of heavy metals by potential native plants and their microscopic observation of root growing on stabilised distillery sludge as a prospective tool for in situ phytoremediation of industrial waste. Environmental Science and Pollution Research 24 (3), 2605–2619.

Chehayeb, K.M., Nayar, K.G., 2018. On the merits of using multi-stage and counterflow electrodialysis for reduced energy consumption. Desalination 439, 1–16.

Cheng, G., Li, P., 2007. Phytoremediation and microbial remediation of petroleum contaminated soil. Chinese Journal of Environmental Engineering 1 (6), 91–96.

Cheng, S.Y., Show, P.L., Lau, B.F., Chang, J.S., Ling, T.C., 2019. New prospects for modified algae in heavy metal adsorption. Trends in Biotechnology 37 (11), 1255–1268.

Chiroma, T.M., Ebewele, R.O., Hymore, F.K., 2014. Comparative assessment of heavy metal levels in soil, vegetables and urban grey waste water used for irrigation in Yola and Kano. International Refereed Journal of Engineering and Science 3 (2), 01–09.

Choi, M., Jang, J., 2008. Heavy metal ion adsorption onto polypyrrole-impregnated porous carbon. Journal of Colloid and Interface Science 325 (1), 287–289.

Cui, L., Wang, Y., Gao, L., Hu, L., Yan, L., Wei, Q., et al., 2015. EDTA functionalized magnetic graphene oxide for removal of Pb (II), Hg (II) and Cu (II) in water treatment: adsorption mechanism and separation property. Chemical Engineering Journal 281, 1–10.

Das, N., Bhattacharya, S., Maiti, M.K., 2016. Enhanced cadmium accumulation and tolerance in transgenic tobacco overexpressing rice metal tolerance protein gene OsMTP1 is promising for phytoremediation. Plant Physiology and Biochemistry 105, 297–309.

Das, S., Dash, H.R., 2014. Microbial bioremediation: a potential tool for restoration of contaminated areas. Microbial Biodegradation and Bioremediation. Elsevier, pp. 1–21.

Dhankher, O.P., Li, Y., Rosen, B.P., Shi, J., Salt, D., Senecoff, J.F., et al., 2002. Engineering tolerance and hyperaccumulation of arsenic in plants by combining arsenate reductase and γ-glutamylcysteine synthetase expression. Nature Biotechnology 20 (11), 1140–1145.

Dhankher, O.P., Pilon-Smits, E.A., Meagher, R.B., Doty, S., 2012. Biotechnological approaches for phytoremediation. Plant Biotechnology and Agriculture. Academic Press, pp. 309–328.

Dias, J.M., Alvim-Ferraz, M.C., Almeida, M.F., Rivera-Utrilla, J., Sánchez-Polo, M., 2007. Waste materials for activated carbon preparation and its use in aqueous-phase treatment: a review. Journal of Environmental Management 85 (4), 833–846.

Dong, Z., Zhang, F., Wang, D., Liu, X., Jin, J., 2015. Polydopamine-mediated surface-functionalization of graphene oxide for heavy metal ions removal. Journal of Solid State Chemistry 224, 88–93.

Dubchak, S., Bondar, O., 2019. Bioremediation and phytoremediation: best approach for rehabilitation of soils for future use. Remediation Measures for Radioactively Contaminated Areas. Springer, Cham, pp. 201–221.

Efome, J.E., Rana, D., Matsuura, T., Lan, C.Q., 2018. Metal–organic frameworks supported on nanofibers to remove heavy metals. Journal of Materials Chemistry A 6 (10), 4550–4555.

Ekrami, E., Pouresmaieli, M., sadat Hashemiyoon, E., Noorbakhsh, N., Mahmoudifard, M., 2022. Nanotechnology: a sustainable solution for heavy metals remediation. Environmental Nanotechnology, Monitoring & Management 100718.

Farha, O.K., Eryazici, I., Jeong, N.C., Hauser, B.G., Wilmer, C.E., Sarjeant, A.A., et al., 2012. Metal−organic framework materials with ultrahigh surface areas: is the sky the limit? Journal of the American Chemical Society 134 (36), 15016−15021.

Fomina, M., Gadd, G.M., 2014. Biosorption: current perspectives on concept, definition and application. Bioresource Technology 160, 3−14.

Gardea-Torresdey, J.L., Peralta-Videa, J.R., De La Rosa, G., Parsons, J.G., 2005. Phytoremediation of heavy metals and study of the metal coordination by X-ray absorption spectroscopy. Coordination Chemistry Reviews 249 (17-18), 1797−1810.

Gautam, R.K., Sharma, S.K., Mahiya, S., Chattopadhyaya, M.C., 2014. Contamination of heavy metals in aquatic media: transport, toxicity and technologies for remediation.

Goswami, R.K., Agrawal, K., Shah, M.P., Verma, P., 2021. Bioremediation of heavy metals from wastewater: a current perspective on microalgae-based future. Letters in Applied Microbiology .

Gupta, S., Singh, D., 2017. Role of genetically modified microorganisms in heavy metal bioremediation. Advances in Environmental Biotechnology. Springer, Singapore, pp. 197−214.

Hooda, V., 2007. Phytoremediation of toxic metals from soil and waste water. Journal of Environmental Biology 28 (2), 367.

Hu, A., Apblett, A. (Eds.), 2014. Nanotechnology for Water Treatment and Purification. Springer International Publishing, Switzerland.

Järup, L., 2003. Hazards of heavy metal contamination. British Medical Bulletin 68 (1), 167−182.

Kapoor, D., Singh, M.P., 2021. Heavy metal contamination in water and its possible sources. Heavy Metals in the Environment. Elsevier, pp. 179−189.

Karnib, M., Kabbani, A., Holail, H., Olama, Z., 2014. Heavy metals removal using activated carbon, silica and silica activated carbon composite. Energy Procedia 50, 113−120.

Kaufmann, R.B., Staes, C.J., Matte, T.D., 2003. Deaths related to lead poisoning in the United States, 1979−1998. Environmental Research 91 (2), 78−84.

Kaur, S., Roy, A., 2021. Bioremediation of heavy metals from wastewater using nanomaterials. Environment, Development and Sustainability 23 (7), 9617−9640.

Kobielska, P.A., Howarth, A.J., Farha, O.K., Nayak, S., 2018. Metal−organic frameworks for heavy metal removal from water. Coordination Chemistry Reviews 358, 92−107.

Kulshreshtha, A., Agrawal, R., Barar, M., Saxena, S., 2014. A review on bioremediation of heavy metals in contaminated water. *IOSR Journal of Environmental Science*. Toxicology and Food Technology 8 (7), 44−50.

Liu, S., Yang, B., Liang, Y., Xiao, Y., Fang, J., 2020. Prospect of phytoremediation combined with other approaches for remediation of heavy metal-polluted soils. Environmental Science and Pollution Research 27 (14), 16069−16085.

Liu, Z., Hong, Q., Xu, J.H., Jun, W., Li, S.P., 2006. Construction of a genetically engineered microorganism for degrading organophosphate and carbamate pesticides. International Biodeterioration & Biodegradation 58 (2), 65−69.

Luo, S., Xu, X., Zhou, G., Liu, C., Tang, Y., Liu, Y., 2014. Amino siloxane oligomer-linked graphene oxide as an efficient adsorbent for removal of Pb (II) from wastewater. Journal of Hazardous Materials 274, 145−155.

Luo, Y., Wu, Y., Qiu, J., Wang, H., Yang, L., 2019. Suitability of four woody plant species for the phytostabilization of a zinc smelting slag site after 5 years of assisted revegetation. Journal of Soils and Sediments 19 (2), 702−715.

Malakahmad, A., Tan, S., Yavari, S., 2016. Valorization of wasted black tea as a low-cost adsorbent for nickel and zinc removal from aqueous solution. Journal of Chemistry 2016.

Malaviya, P., Singh, A., 2011. Physicochemical technologies for remediation of chromium-containing waters and wastewaters. Critical Reviews in Environmental Science and Technology 41 (12), 1111−1172.

Mansourri, G., Madani, M., 2016. Examination of the level of heavy metals in wastewater of Bandar Abbas Wastewater Treatment Plant. Open Journal of Ecology 6 (2), 55−61.

Marques, A.P., Rangel, A.O., Castro, P.M., 2009. Remediation of heavy metal contaminated soils: phytoremediation as a potentially promising clean-up technology. Critical Reviews in Environmental Science and Technology 39 (8), 622−654.

Maximous, N.N., Nakhla, G.F., Wan, W.K., 2010. Removal of heavy metals from wastewater by adsorption and membrane processes: a comparative study. International Journal of Environmental and Ecological Engineering 4 (4), 125−130.

Michalak, I., Chojnacka, K., Witek-Krowiak, A., 2013. State of the art for the biosorption process—a review. Applied Biochemistry and Biotechnology 170 (6), 1389−1416.

Mishra, S., Kumar, A., Yadav, S., Singhal, M.K., 2018. Assessment of heavy metal contamination in water of Kali River using principle component and cluster analysis, India. Sustainable Water Resources Management 4 (3), 573−581.

Monteiro, C.M., Castro, P.M., Malcata, F.X., 2012. Metal uptake by microalgae: underlying mechanisms and practical applications. Biotechnology Progress 28 (2), 299−311.

Mosa, K.A., Saadoun, I., Kumar, K., Helmy, M., Dhankher, O.P., 2016. Potential biotechnological strategies for the cleanup of heavy metals and metalloids. Frontiers in Plant Science 7, 303.

Muthusaravanan, S., Sivarajasekar, N., Vivek, J.S., Paramasivan, T., Naushad, M., Prakashmaran, J., et al., 2018. Phytoremediation of heavy metals: mechanisms, methods and enhancements. Environmental Chemistry Letters 16 (4), 1339−1359.

Nematian, M.A., Kazemeini, F., 2013. Accumulation of Pb, Zn, Cu and Fe in plants and hyperaccumulator choice in Galali iron mine area, Iran. International Journal of Agriculture and Crop Sciences (IJACS) 5 (4), 426−432.

Nikolopoulou, M., Pasadakis, N., Kalogerakis, N., 2013. Evaluation of autochthonous bioaugmentation and biostimulation during microcosm-simulated oil spills. Marine Pollution Bulletin 72 (1), 165−173.

Niu, G.L., Zhang, J.J., Zhao, S., Liu, H., Boon, N., Zhou, N.Y., 2009. Bioaugmentation of a 4-chloronitrobenzene contaminated soil with *Pseudomonas putida* ZWL73. Environmental Pollution 157 (3), 763–771.

Onuegbu, T.U., Umoh, E.T., Onwuekwe, I.T., 2013. Physico—chemical analysis of effluents from Jacbon chemical industries limited, makers of bona-lux emulsion and gloss paints. International Journal of Science and Technology 2 (2), 169–173.

Ostrem Loss, E.M., Yu, J.H., 2018. Bioremediation and microbial metabolism of benzo (a) pyrene. Molecular Microbiology 109 (4), 433–444.

Park, H.G., Kim, T.W., Chae, M.Y., Yoo, I.K., 2007. Activated carbon-containing alginate adsorbent for the simultaneous removal of heavy metals and toxic organics. Process Biochemistry 42 (10), 1371–1377.

Pandey, G., Paul, D., Jain, R.K., 2005. Conceptualizing "suicidal genetically engineered microorganisms" for bioremediation applications. Biochemical and Biophysical Research Communications 327 (3), 637–639.

Pendergast, M.M., Hoek, E.M., 2011. A review of water treatment membrane nanotechnologies. Energy & Environmental Science 4 (6), 1946–1971.

Peydayesh, M., Mohammadi, T., Bakhtiari, O., 2019. Water desalination via novel positively charged hybrid nanofiltration membranes filled with hyperbranched polyethyleneimine modified MWCNT. Journal of Industrial and Engineering Chemistry 69, 127–140.

Pourrahim, S., Salem, A., Salem, S., Tavangar, R., 2020. Application of solid waste of ductile cast iron industry for treatment of wastewater contaminated by reactive blue dye via appropriate nano-porous magnesium oxide. Environmental Pollution 256, 113454.

Prieto, M.J., Acevedo, S.O.A., Prieto, G.F., González, N.T., 2018. Phytoremediation of soils contaminated with heavy metals. Biodiversity International Journal 2, 362–376.

Qu, X., Alvarez, P.J., Li, Q., 2013. Applications of nanotechnology in water and wastewater treatment. Water Research 47 (12), 3931–3946.

Rao, G.P., Lu, C., Su, F., 2007. Sorption of divalent metal ions from aqueous solution by carbon nanotubes: a review. Separation and Purification Technology 58 (1), 224–231.

Razzak, S.A., Farooque, M.O., Alsheikh, Z., Alsheikhmohamad, L., Alkuroud, D., Alfayez, A., et al., 2022. A comprehensive review on conventional and biological-driven heavy metals removal from industrial wastewater. Environmental Advances 100168.

Saif, S., Khan, M.S., 2017. Assessment of heavy metals toxicity on plant growth promoting rhizobacteria and seedling characteristics of *Pseudomonas putida* SFB3 inoculated greengram. Acta Scientific Agriculture 1, 47–56.

Saleh, H.M., Moussa, H.R., Mahmoud, H.H., El-Saied, F.A., Dawoud, M., Wahed, R.S.A., 2020. Potential of the submerged plant *Myriophyllum spicatum* for treatment of aquatic environments contaminated with stable or radioactive cobalt and cesium. Progress in Nuclear Energy 118, 103147.

Sayqal, A., Ahmed, O.B., 2021. Advances in heavy metal bioremediation: an overview. Applied Bionics and Biomechanics 2021.

Selatnia, A., Madani, A., Bakhti, M.Z., Kertous, L., Mansouri, Y., Yous, R., 2004. Biosorption of Ni2 + from aqueous solution by a NaOH-treated bacterial dead Streptomyces rimosus biomass. Minerals Engineering 17 (7-8), 903–911.

Shah, S.R., Wright, N.C., Nepsky, P.A., 2018. Cost-optimal design of a batch electrodialysis system for domestic desalination of brackish groundwater. Desalination 443, 198–211.

Shimizu, H., 2002. Metabolic engineering—integrating methodologies of molecular breeding and bioprocess systems engineering. Journal of Bioscience and Bioengineering 94 (6), 563–573.

Singh, J.S., Abhilash, P.C., Singh, H.B., Singh, R.P., Singh, D.P., 2011. Genetically engineered bacteria: an emerging tool for environmental remediation and future research perspectives. Gene 480 (1-2), 1–9.

Sreeprasad, T.S., Maliyekkal, S.M., Lisha, K.P., Pradeep, T., 2011. Reduced graphene oxide–metal/metal oxide composites: facile synthesis and application in water purification. Journal of Hazardous Materials 186 (1), 921–931.

Sun, D.T., Peng, L., Reeder, W.S., Moosavi, S.M., Tiana, D., Britt, D.K., et al., 2018. Rapid, selective heavy metal removal from water by a metal–organic framework/polydopamine composite. ACS Central Science 4 (3), 349–356.

Sun, J., Liang, Q., Han, Q., Zhang, X., Ding, M., 2015. One-step synthesis of magnetic graphene oxide nanocomposite and its application in magnetic solid phase extraction of heavy metal ions from biological samples. Talanta 132, 557–563.

Tangahu, B.V., Sheikh Abdullah, S.R., Basri, H., Idris, M., Anuar, N., Mukhlisin, M., 2011. A review on heavy metals (As, Pb, and Hg) uptake by plants through phytoremediation. International Journal of Chemical Engineering 2011.

Valls, M., Atrian, S., de Lorenzo, V., Fernández, L.A., 2000. Engineering a mouse metallothionein on the cell surface of *Ralstonia eutropha* CH34 for immobilization of heavy metals in soil. Nature Biotechnology 18 (6), 661–665.

Vilela, D., Parmar, J., Zeng, Y., Zhao, Y., Sánchez, S., 2016. Graphene-based microbots for toxic heavy metal removal and recovery from water. Nano Letters 16 (4), 2860–2866.

Volesky, B., 2007. Biosorption and me. Water Research 41 (18), 4017–4029.

Wang, Y., Luo, Y., Zeng, G., Wu, X., Wu, B., Li, X., et al., 2020. Characteristics and in situ remediation effects of heavy metal immobilizing bacteria on cadmium and nickel co-contaminated soil. Ecotoxicology and Environmental Safety 192, 110294.

Wieczorek-Dąbrowska, M., Tomza-Marciniak, A., Pilarczyk, B., Balicka-Ramisz, A., 2013. Roe and red deer as bioindicators of heavy metals contamination in north-western Poland. Chemistry and Ecology 29 (2), 100–110.

Yin, J., Deng, B., 2015. Polymer-matrix nanocomposite membranes for water treatment. Journal of Membrane Science 479, 256–275.

Ying, G.G., 2018. Remediation and mitigation strategies. Integrated Analytical Approaches for Pesticide Management. Academic Press, pp. 207–217.

Zehra, A., Meena, M., Swapnil, P., Raytekar, N.A., Upadhyay, R.S., 2020. Sustainable approaches to remove heavy metals from water. Microbial Biotechnology: Basic Research and Applications 127–146.

Zhu, Q.L., Xu, Q., 2014. Metal–organic framework composites. Chemical Society Reviews 43 (16), 5468–5512.

Zhu, W.P., Sun, S.P., Gao, J., Fu, F.J., Chung, T.S., 2014. Dual-layer polybenzimidazole/polyethersulfone (PBI/PES) nanofiltration (NF) hollow fiber membranes for heavy metals removal from wastewater. Journal of Membrane Science 456, 117–127.

Further reading

Amrose, S., Burt, Z., Ray, I., 2015. Safe drinking water for low-income regions. Annual Review of Environment and Resources 40 (1), 203–231.

Gebauer, H., Saul, C.J., 2014. Business model innovation in the water sector in developing countries. Science of the Total Environment 488, 512–520.

Intelligence, G.W., Yearbook, I.D., Summit, G.W., Card, R., 2011. Global water intelligence. Global Water Intelligence 12 (10), 1–72.

Luo, H., Law, W.W., Wu, Y., Zhu, W., Yang, E.H., 2018. Hydrothermal synthesis of needle-like nanocrystalline zeolites from metakaolin and their applications for efficient removal of organic pollutants and heavy metals. Microporous and Mesoporous Materials 272, 8–15.

Luo, J., Ye, Y., Gao, Z., Wang, W., 2014. Essential and nonessential elements in the red-crowned crane *Grus japonensis* of Zhalong Wetland, northeastern China. Toxicological & Environmental Chemistry 96 (7), 1096–1105.

US Department of Health and Human Services, 2007. *Agency for Toxic Substances and Disease Registry: Toxicological profile for Lead.* Atlanta, GA: US Department of Health and Human Services.

Chapter 18

Remediation strategies for the removal of microplastics from the water

Lata Rani[1,2], Gagandeep Kaur[2], Parul Sood[2], Jyotsna Kaushal[1] and Arun Lal Srivastav[3]

[1]*Centre for Water Sciences, Chitkara University Institute of Engineering & Technology, Chitkara University, Rajpura, Punjab, India,* [2]*Chitkara School of Pharmacy, Chitkara University, Baddi, Himachal Pradesh, India,* [3]*Chitkara University School of Engineering and Technology, Chitkara University, Baddi, Himachal Pradesh, India*

Introduction

After 50 years of growth, plastics manufacturing around the world reached 38.1 crore tonnes in 2015 (Ritchie and Roser, 2018). It is alarming that this rise in consumption has led to an increase in global plastic pollution (MacArthur et al., 2016). Many plastic-based items are exfoliated and broken down, releasing microplastics into the environment in nanometer-sized fragments (Zhang et al., 2018). Ocean sediment (Van Cauwenberghe et al., 2013), developed and underdeveloped areas (Hirai et al., 2011), freshwaters (Faure et al., 2015), and seawaters (Van Cauwenberghe et al., 2013) have been found to contain microplastics (Law et al., 2014). Aquatic habitats are becoming more contaminated with microplastics and their breakdown products due to a buildup of debris in the water (Andrady, 2011; Sun et al., 2019). Secondary microplastics, formed when raw plastic particles are degraded by physicochemical and biological processes in the environment, are distinct from primary ones, which are commonly found in households for personal care (Galgani et al., 2013). Microplastics have become a common problem because of poor plastic waste management (Gilani et al., 2019; Thompson, 2015). The toxicity of microplastics can be mitigated in a variety of ways depending on their chemical composition and the polymerization additives used (Meeker et al., 2009; Sussarellu et al., 2016). It is estimated that 13 million metric tonnes of plastic waste were dumped into the oceans in 2018, an increase from the year before (Plastics Europe, E.P.R.O., 2019; Shen et al., 2019b, 2020a,b). By 2025, an estimated 250 million tonnes of plastics are expected to be released (Jambeck et al., 2015). According to how they're used, plastics can last anywhere from 1 to 15 years before they become plastic garbage (Plastics Europe, E.P.R.O., 2017). Microplastics are formed when plastics break down in a variety of environments (Auta et al., 2017; Shen et al., 2019a). The term "microplastics" refers to particles made up of plastics having a diameter of between 0.1 and 5 mm (Anderson et al., 2017; Thompson et al., 2004; Wen et al., 2018), while "nano plastics" refers to plastic particles with a diameter of less than 0.1 mm (da Costa et al., 2016; Enfrin et al., 2019; Shen et al., 2019b). It's not yet clear if this classification of micro and nano plastics is official or not (Frias and Nash, 2019; Hartmann et al., 2019a,b; Stark, 2019). Various rivers and lakes including oceans also showed evidence for the presence of microplastics (Horton et al., 2017b; Mendoza et al., 2017). Surface and groundwater are the principal sources of consumption by society (Yi et al., 2018). Microplastic pipelines and sink tanks can be made from closed and open freshwater systems. As far-flung areas around the globe have been found to contain the highest levels of microplastics (Free et al., 2014). Microplastics entering drinking water have not been regulated by the government, nor have there been any direct methods to remove them. Some recent methods for eradicating microplastics from drinking water include coagulation and membrane partition (Ma et al., 2018, 2019b).

Transportation of microplastic in the aquatic environment

Among the principal causes of pollution are cosmetics and cleaning goods, plastic pellets or powders used primarily for air blowing, and feedstocks used to make plastic things (Eriksen et al., 2013). As a result, a wider variety of animals will be able to consume plastic rubbish, raising the possibility of new environmental hazards (Thompson et al., 2004).

Role of Green Chemistry in Ecosystem Restoration to Achieve Environmental Sustainability. DOI: https://doi.org/10.1016/B978-0-443-15291-7.00004-3

Microplastics can also be found in wastewater treatment systems (Long et al., 2019). While large plastic particles are effectively removed during wastewater treatment, microplastics regularly enter and collect in the aquatic environment due to their small size (Murphy et al., 2016). Aside from being close to the ocean, several water treatment plants contribute significantly to the discharge of microplastics. The effluents from 3340 wastewater treatment plants (WWTPs), with a combined daily capacity of 78,106 million cubic meters, are positioned along the coast, where they may be dumped into aquatic environments either directly or indirectly (Jin et al., 2014). Experts are investigating how microplastics enter water treatment plants, as well as how to identify and remove them from the water supply (Sun et al., 2019). Microplastic debris transported by rivers to the oceans comprises soil particles, according to early research (Horton et al., 2017a; Shao et al., 2019). Kosuth et al. (2018) discovered microplastics in 81% of 159 samples of drinking water from around the world (156 tap water samples from 14 countries and 3 bottled water samples from the USA). Microplastic fibers detected at a depth of 25 cm (Zubris and Richards, 2005) are relatively long-lasting, allowing for gradual migration into solid media and a subsequent transfer into liquid media (Fig. 18.1).

Health effect of microplastic

The accumulation of MPs and nanoplastics in tissues results in swelling and obstruction (Wang et al., 2016; Wright et al., 2013). According to in vitro testing, MPs concentrate on the crabs' gills, stomachs, and metabolic systems and change the way fish cells function (Karbalaei et al., 2018; Brennecke et al., 2015). Additionally, it has been demonstrated that they can transfer contaminants and microorganisms (Manzoor et al., 2020; Wang et al., 2016). The degree of exposure and sensitivity of the individual largely dictated such adverse effects (Prata et al., 2020). Oxidative damage, cytotoxicity, and translocation to other tissues have all been associated with MP exposure (Galloway, 2015; Schirinzi et al., 2017; Wright and Kelly, 2017; Anbumani and Kakkar, 2018). In the ecosystem and among living things, MPs persist for a very long time. As a result, the animals are exposed to MPs for a long time, which could result in immune cell damage, chronic pain, swelling, and cell proliferation and death. Patients with MPs had significantly greater rates of inflammatory bowel illness than healthy individuals (Yan et al., 2021). Over time, PS MPs slowed the development of Caco-2 cells (Wu et al., 2020). PS MPs, according to Wu et al., (2019) perturbed the mitochondrial membrane potential. MPs could also act as vectors for other bacteria (Kirstein et al., 2016). They can absorb things from their environment or eject substances from their matrixes (Crawford and Quinn, 2017).

Remediation and transformation strategies

Microbial biodegradation

Biodegradation offers a low carbon footprint, cost-effectiveness, and low energy required compared to other abiotic purifying processes. Microorganisms can utilize plastics (Montazer et al., 2020; Zheng et al., 2005).

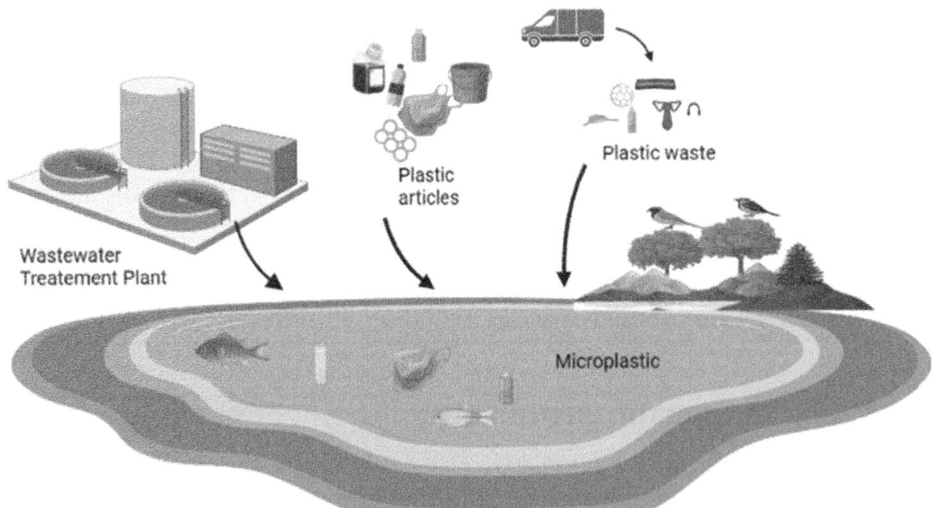

FIGURE 18.1 Microplastic contamination in the aqueous phase.

Bacteria

Many Bacillus species, including *Rhodococcus*, *Kocuria palustris* M16 from the class of Actinobacterias, can decompose MPs. Degradation rates varied from 1% to 10% over 30 days and a year for the bulk of the samples. (Roager and Sonnenschein, 2019). After 6 weeks at 30°C, the bacterium *Ideonella sakaiensis* 201-F6 that Yoshida et al. (2016) obtained from consortia of microbes was able to break down PET almost completely. A PETase enzyme and a mono-(2-hydroxy-ethyl) terephthalic acid (MHET) for cell differentiation are used to accomplish this. Over 10 hours, the system depolymerized at least 90% of PET into monomers, outperforming all current PET hydrolases, including *I. sakaiensis* and its improved versions. With the depolymerized monomers, PET with the same properties as petrochemical PET can be recycled to close the twist of the circular economy (Tournier et al., 2020). PET's low crystallinity and quick chain mobility further contribute to its rapid degradation rate (Yang et al., 2016; Marten et al., 2005). The deterioration of PET has progressed, making it increasingly susceptible to biodegradation. Non-hydrolyzable polymers like PE and PP, which are found in a wide range of environments and are resistant to biological degradation, necessitate rapid action (Jacquin et al., 2019; Krueger et al., 2015; Pathak, 2023). The degradation of polyethylene terephthalate (PE) and polyethylene terephthalate (PET) has been the topic of various studies. As a result of these studies, researchers have found that bacteria can degrade plastic waste in a range of settings, including cold-water marine environments (Urbanek et al., 2018) and soils (Helmberger et al., 2020; Kumar et al., 2020).

Fungi

Fungi, like bacteria, can release enzymes to utilize MPs as their source of carbon and energy. (Sánchez, 2020). Intercellular detoxification processes, fungal adaptability, and exogenous enzyme processes that create hydrolases are all possible reasons for fungi's drug-degrading abilities. Several other solutions, such as the addition of pro-oxidant species, the decrease of biocides, and the use of antioxidant stabilizers during the production of plastics (Sánchez, 2020). Fungi were shown to be more effective at breaking down PE than bacteria (Muhonja et al., 2018). However, only a small amount of study has been conducted on enzymes that might be involved in the breakdown of MPs (Sánchez, 2020). ASTM Standards (the American Society for Testing and Materials) say that measuring CO_2 evolution is preferable to weight loss measurement for evaluating the biodegradability of bacteria and fungi (ASTM International, 2012; ASTM, 2017). A combination of CO_2 evaluation and isotope labeling is used to determine the level of assimilation (Amaral-Zettler et al., 2020). Because polyolefin chains lack functional groups and have a large molecular weight, biodegradation studies are severely limited therefore abiotic treatment becomes essential (Gewert et al., 2015). A "plastisphere" microbial community can be formed when natural bacteria mix with MPs. Access to the deep sea may be a problem for biochemists who are interested in using deep-sea bacteria as a source of potential enzymes (Wei et al., 2020).

Advanced oxidation processes

Photodegradation/photocatalytic degradation

Understanding photolysis in the natural original atmospheric conditions is required before understanding photocatalytic activities. During light rays-initiated degradation, the carbon-carbon backbone of plastics passes through three phases: initiation, propagation, and termination (Gewert et al., 2015). Natural photodegradation activities generate peroxyl radicals, which are essential for the formation of radicals on the backbone. Additionally, chemical chain breakage, addition reaction with O_2 or H_2O, and EPFR synthesis in MPs occur postquarter of photoirradiation (Zhu et al., 2019). A reduction in molecular weight and the formation of multiple small molecular products occur as a result of the EPFRs producing ROS (Zhu et al., 2019). Chemical constituents of MPs influence the production of EPFRs to a significant extent. When exposed to light, EPFRs are more likely to develop in PS and phenol formaldehyde resin (PF) with numerous benzene rings, but aliphatic polymers including polyvinylchloride and PE are photo-inert (Wang et al., 2020a).

Solid-phase photocatalytic breakdown

Solid-phase photocatalytic degradation occurs when discarded plastic is exposed to the sun's rays in the open air. Research into the photodegradation of solid-state plastics dates around the 1970s (Allen and Mckellar, 1975). Before this, the utmost popular way for performing photodegradation tests was to make composite films by suspending photocatalysts in plastic's stock solution. The common approach for evaluating deterioration efficiency is to monitor the loss in weight of composites, which indicates the efficiency of transforming polymeric carbons to carbon dioxide by mineralization.

Water-based photocatalysis

Degrading polyvinyl chloride (95%) and polyvinylidene chloride (5%), respectively, using TiO_2 and ZnO acting as photocatalysts was reported (Hidaka et al., 1996). In the beginning, the dechlorination rate of ZnO outpaced that of TiO_2, but it soon leveled off. Surprisingly the addition of oxidizing agents ($K_2S_2O_8$ or H_2O_2) boosted the dechlorination effectiveness of TiO_2 but hampered ZnO-dechlorination because of considerable catalyst degradation during photo-oxidation. When ZnO nanorods were used to improve the surface-to-volume ratio of LDPE for 175 hours, the carbonyl index of LDPE rose by 30%. Pt NPs were added to ZnO nanorods (ZnO-Pt) to limit electron-hole pair recombination and speed up the absorption of visible light (Tofa et al., 2019b).

Fenton or Fenton-like systems

Recovering polymers from trash using Fenton systems is particularly common (Wang et al., 2017a,b). Advanced oxidation processes (AOP) technologies have recently been employed in MP remediation because of the redox potentials of the produced hydroxyl and sulfate radicals ($^\bullet OH$, $E^0 = 2.7$ V and $SO_4^\bullet$, $E^0 = 3.1$ V vs NHE, respectively). Fenton and heat-activated PDS procedures improved the aging behavior of PS and HDPE MPs (Liu et al., 2019b). There was a wide range of PS and PE sizes before treatment with AOPs. In the first 30 days of PDS therapy, 80.1% of PS and 97.4% of PE were less than 20 m in diameter. Due to their smaller dimensions and raised O/C ratio, MPs with a higher degree of alteration will have greater adsorption capacities toward ciprofloxacin, showing that AOPs can degrade MPs, but the increased adsorption capacities of MPs may increase their toxicity (Liu et al., 2019b). Reactive oxygen species (ROS) and MP oxidation kinetics are enhanced by increasing the reaction temperature and employing a catalyst, which in turn increases the performance of AOPs systems. Manganese nanoparticles embedded in a carbon nanotube (Mn@NCNT) were created by Kang et al. (2019).

Thermal treatment

Earlier municipal solid trash was upgraded via heat treatment since it conserves landfill area and reduces sanitary problems caused by waste putridness (Lombardi et al., 2015). Thermal treatment now has the added benefit of reclaiming some of the heat it generates (Lombardi et al., 2015). Owing to its high calorific value and huge output, plastics represent a potential availability of high hydrogen and carbon elements for fuel generation. Several recent reviews detail several ways of turning plastics into energy, including gasification, pyrolysis, cracking in refineries, liquefaction, and reforming (Lopez et al., 2018; Nanda and Berruti, 2021; Panda et al., 2010; Wong et al., 2015). The carbons in plastic trash can also be turned into extra value-enhanced nano-sized particles (e.g., carbon nanotubes), which have remarkable physical and chemical characteristics and can be employed in a wide range of applications (Bazargan and McKay, 2012). MPs have been converted using thermal techniques in the recent past. At temperatures between 400 and 1200°C, Tavares et al. (2018) used several gasification agents to co-gasify PET-based MPs and biomass (air, O_2, and steam). Optimum hydrogen yields (63%–66%) can be achieved with a mixture of 90% PET and 10% biomass, whereas the lower heating value (LHV) of 9.2 MJ/Nm^3 may be achieved with a mixture of 50% PET and 50% biomass, according to the study. A high-pressure LDPE pyrolysis was carried out to explore the impact of stress on product distribution and transformation pathways to evaluate the effect on fuel quality (Cheng et al., 2020).

MPs removal in wastewater treatment plants

MPs can be removed at each point of WWTP treatment even though they were not initially meant to do so. According to Sun et al. (2019), WWTPs in various nations have been obtained and the efficacy of MPS removal has been assessed at each treatment step. Preliminary, secondary, and tertiary treatment are all independent phases in the WWTP process. MPs are removed in amounts ranging from 35% to 99% throughout the first two stages of treatment. Low-density MPs float to the surface due to the air flotation process, which makes mechanical skimming much more efficient. To collect highly dense MPs in solid colloidal particles for further treatment, sedimentation techniques are used (Ngo et al., 2019). Lighter MPs (such as PA, PP, and PE) are less likely to be obtained in sediment than highly dense MPs (such as PET and PES). Biological treatment and clearing mechanisms can reduce the elimination percentage of MPs by 0.2%–14.2%

in the next phase. Extracellular polymers and flocs from bacterial strains contribute to MP buildup and sinking in the clarification tanks, where bigger MPs are eliminated (Sun et al., 2019). In the subsequent tertiary therapy, a variety of methods are used to lower the MPs to 0.25%. Comparing various tertiary treatment processes, Talvitie and colleagues (2017) found that the membrane bioreactor (MBR) had the best removal effectiveness of any of the other methods tested (99.9%) namely reverse osmosis (RO), ultrafiltration (UF), and dynamic membrane technology (DM) (Poerio et al., 2019). Anaerobic and aerobic processes, such as hydrolysis-acidification, hydrogen, and methane gas generation, and the digestion of waste-activated sludge, have all been found to be negatively affected by MPs (Wei et al., 2019; Zhang et al., 2020a,b,c). While MPs have been researched for their effect on the ability of nitrite and ammonium-oxidizing bacteria, denitrifiers, and polyphosphate-accumulating bacterial organisms (PAOs), but still the results were insignificant (Liu et al., 2019a). PVC MPs, for example, caused membrane fouling in an MBR system, making it difficult to remove organic materials and ammonia (Li et al., 2020). In both UF and RO processes, membrane breakdown can be caused by irregular MPs (Zhang et al., 2020c). The removal of MPs from large WWTPs has not yet been figured out, which could be the focus of future research.

Separation processes

Fast granular filtration

The method of entrapment between the pores of gravel or anthracite grains could be used to get rid of MPs in water in addition to adsorbing them onto the silica particles themselves (Rajala et al., 2020). Fast granular filters can be rejuvenated by backwashing frequently, although the hydroxyl groups present on the surface of the MPs make it difficult to remove (Enfrin et al., 2019). Three WWTPs of South Korea were researched by Hidayaturrahman and Lee (2019) and it was found that 74% of the MPs in sewage could be eradicated using large-scale High granular filtration. For the elimination of MPs, Talvitie et al. (2017) studied an adsorbent filtering material with 1 m sand of 3–5 mm particle size and quartz with 0.1–0.5 mm grain size. The efficiency of MPs eradication was 97%, and the MP's frequency dropped from 0.7 to 0.02 MPs/L. The granular filtration technique efficiently eliminated 60%–80% of the MPs in a laboratory study, whereas biochar and sand filtration eradicated 95%–100% (Wang et al., 2020d). Quick granular filtration's efficacy in removing contaminants is dependent on particle size (Zhang et al., 2020b,c).

Membrane filtration

To remove MPs from the membrane, pores could be made smaller than the diameter of the particles. Murray and Örmeci (2020) used nylon syringe filters with 0.22 and 0.7 m pore dimensions to isolate MPs from each other. By using lab-scale ultrafiltration and microfiltration, polycarbonate and nylon filter sheets with pore sizes of 0.22 and 25 mm successfully removed 99.3% of the MPs (mostly 10–1000 m) discovered in synthetic saltwater (Yaranal et al., 2021). It is possible to remove MPs via membrane filtration, however, the economic viability of nanofiltration is difficult due to its high pressure and low treatment rate requirements (Enfrin et al., 2020). Membrane fouling by MPs is modest at the start of treatment and can eventually progress to complete blockage. Internal pore blockage is the primary mechanism of membrane clogging over time, while cake formation is a possible side effect (Hou et al., 2021).

Membrane bioreactor

Membrane filtration with an MBR has been one of the subjects of most studies and shows the most promising way to remove MPs. Combining membrane filtration techniques with either aerobic or anaerobic bioreactors is one of the possibilities for constructing MBR. Because aeration is needed to activate microorganisms in AeMBR, it frequently has a higher cost of operation than Anaerobic MBR (Zaouri et al., 2021). AnMBR is often more efficient than AeMBR in eliminating MPs and NPs. For example, for AnMBR, 94% of MPs were eliminated within six hours, whereas for AeMBR, 98.3% of MPs were eliminated after 3 months (Pittura et al., 2021). Membrane fouling is the underlying issue in all of these methods. Membrane fouling occurs during the separation of MPs and NPs when operating circumstances, membrane characteristics, and wastewater factors interact (Maliwan et al., 2021). Overall, AeMBR versus AnMBR procedures differ in terms of fouling and permeability. Compared to other wastewater treatment methods, for MPs, MBR normally has one of the best removal rates (Shen et al., 2020a,b). Only 25% of MPs were eradicated using an MBR (Leslie et al., 2017). There are long fibers that can cross membranes longitudinally and strong pressure is exerted in MBRs, which could explain this phenomenon. Abrasion, on the other hand, can cause the membrane's pores to grow, reducing its effectiveness over time. However, the MBR was a realistic mechanism for removing MPs (Hou et al., 2021; Lares et al., 2018; Lv et al., 2019; Michielssen et al., 2016; Ngo et al., 2019; Padervand et al., 2020). It is MBRs' long hydraulic retention period (24–100 hours), which encourages microbial growth and the capture of MPs (Talvitie et al., 2017).

Flocculation, electrocoagulation, and coagulation

The in-situ creation of metal and hydroxide ions via electrodes distinguishes electrocoagulation from traditional coagulation. Examples of cations that neutralize the charges of MPs and NPs include Fe and Al, as well as H_2O. For these contaminants to be separated from the water, they form larger, more stable aggregates. MPs and NPs can be efficiently removed using coagulation or electrocoagulation in this case (Ma et al., 2019a,b; Perren et al., 2018; Rajala et al., 2020; Shahi et al., 2020; Wang et al., 2020c; Lapointe et al., 2020). To top it all off, it is also suggested that mixing in some sand to the coagulation process could boost MP clearance by around 27% (Shahi et al., 2020). More than 99% of poly-(methyl methacrylate) and poly-(vinyl acetate) NPs were removed from the water after 30 minutes of lab-scale coprecipitation by calcium carbonate (Batool and Valiyaveettil, 2020). After just three minutes, the clearance rate was more than 96%, which was a success of this technique. Coagulation, electrocoagulation, and membrane filtering can be used to remove MPs and other contaminants from secondary effluents (Enfrin et al., 2019).

Air flotation

When using air flotation for basic and secondary water treatment, flocs and suspended compounds like MPs are removed. Air bubbles 20–70 m in diameter are generated by infusing pressurized air into the water and before being deposited on the surface of the water, they get themselves attached to MPs (Talvitie et al., 2017). Conventional air flotation methods rely heavily on the hydrophilic/hydrophobic interaction to remove MPs. It was discovered that the adherence of PET MPs and PE MPs to the microbubbles was stronger than the attachment of PA, which possesses hydrophilic groups on its surface (Wang et al., 2020b). Surfactants like cetyltrimethylammonium bromide induce charge attraction to take precedence over other considerations. Air flotation, for example, could remove 20–25 m-sized MPs at 100% efficiency by raising the saturation pressure by 0.50 MPa (Wang et al., 2020b). 95% of MPs (20–300 m) were eradicated using a flocculant of 40 mg/L poly aluminum chloride and air flotation at a unit atmospheric pressure (Talvitie et al., 2017).

Conclusion

Microplastics are pervasive pollutants around the globe. Microplastics are plastics smaller than 5 mm. Many plastic-based items are exfoliated and broken down, releasing microplastics into the environment as nanometer fragments. Ocean sediment developed and underdeveloped areas of freshwaters and seawaters have been found to contain microplastics. Aquatic habitats are becoming more contaminated with microplastics and their breakdown products due to a buildup of debris in the water. Secondary microplastics, formed when raw plastic particles are degraded by physicochemical and biological processes in the environment, are distinct from primary ones, which are commonly found in households for personal care. The polymeric structure of microplastics and the ease with which they can move between different ecosystems make them a major source of concern for biologists and environmentalists.

References

Allen, N.S., McKellar, J.F., 1975. Photodegradation and stabilization of commercial polyolefins. Chemical Society Reviews 4 (4), 533–547.

Amaral-Zettler, L.A., Zettler, E.R., Mincer, T.J., 2020. Ecology of the plastisphere. Nature Reviews. Microbiology 18 (3), 139–151.

Anbumani, S., Kakkar, P., 2018. Ecotoxicological effects of microplastics on biota: a review. Environmental Science and Pollution Research 25, 14373–14396.

Anderson, P.J., Warrack, S., Langen, V., Challis, J.K., Hanson, M.L., Rennie, M.D., 2017. Microplastic contamination in lake Winnipeg, Canada. Environmental Pollution 225, 223–231.

Andrady, A.L., 2011. Microplastics in the marine environment. Marine Pollution Bulletin 62 (8), 1596–1605. Available from: https://doi.org/10.1016/j.marpolbul.2011.05.030.

ASTM International, 2012. ASTM D7473-12, standard test method for weight attrition of plastic materials in the marine environment by open system aquarium incubations. ASTM.

ASTM, 2017. Standard Test Method for Determining Aerobic Biodegradation of Plastic Materials in the Marine Environment by a Defined Microbial Consortium or Natural Sea Water Inoculum. ASTM, West Conshohocken, PA.

Auta, H.S., Emenike, C.U., Fauziah, S.H., 2017. Distribution and importance of microplastics in the marine environment: a review of the sources, fate, effects, and potential solutions. Environment international 102, 165–176.

Batool, A., Valiyaveettil, S., 2020. Coprecipitation—an efficient method for removal of polymer nanoparticles from water. ACS Sustainable Chemistry & Engineering 8 (35), 13481–13487.

Bazargan, A., McKay, G., 2012. A review–synthesis of carbon nanotubes from plastic wastes. Chemical Engineering Journal 195, 377–391.

Brennecke, D., Ferreira, E.C., Costa, T.M., Appel, D., da Gama, B.A., Lenz, M., 2015. Ingested microplastics (> 100 μm) are translocated to organs of the tropical fiddler crab Uca rapax. Marine Pollution Bulletin 96 (1–2), 491–495.

Cheng, L., Gu, J., Wang, Y., Zhang, J., Yuan, H., Chen, Y., 2020. Polyethylene high-pressure pyrolysis: better product distribution and process mechanism analysis. Chemical Engineering Journal 385, 123866.

Crawford, C.B., Quinn, B., 2017. Plastic production, waste and legislation. Microplastic Pollutants 30, 39–56.

da Costa, J.P., Santos, P.S., Duarte, A.C., Rocha-Santos, T., 2016. (Nano) plastics in the environment–sources, fates and effects. Science of the Total Environment 566, 15–26.

Enfrin, M., Dumée, L.F., Lee, J., 2019. Nano/microplastics in water and wastewater treatment processes–origin, impact and potential solutions. Water Research 161, 621–638.

Enfrin, M., Lee, J., Le-Clech, P., Dumée, L.F., 2020. Kinetic and mechanistic aspects of ultrafiltration membrane fouling by nano-and microplastics. Journal of Membrane Science 601117890.

Eriksen, M., Mason, S., Wilson, S., Box, C., Zellers, A., Edwards, W., et al., 2013. Microplastic pollution in the surface waters of the Laurentian Great Lakes. Marine Pollution Bulletin 77 (1–2), 177–182.

Faure, F., Demars, C., Wieser, O., Kunz, M., De Alencastro, L.F., 2015. Plastic pollution in Swiss surface waters: nature and concentrations, interaction with pollutants. Environmental Chemistry 12 (5), 582–591. Available from: https://doi.org/10.1071/EN14218.

Free, C.M., Jensen, O.P., Mason, S.A., Eriksen, M., Williamson, N.J., Boldgiv, B., 2014. High-levels of microplastic pollution in a large, remote, mountain lake. Marine Pollution Bulletin 85 (1), 156–163.

Frias, J.P., Nash, R., 2019. Microplastics: finding a consensus on the definition. Marine Pollution Bulletin 138, 145–147.

Galgani, F., Hanke, G., Werner, S., De Vrees, L., 2013. Marine litter within the European marine strategy framework directive. ICES Journal of Marine Science 70 (6), 1055–1064. Available from: https://doi.org/10.1093/icesjms/fst122.

Galloway, T.S., 2015. Micro-and nano-plastics and human health. Marine Anthropogenic Litter 343–366.

Gewert, B., Plassmann, M.M., MacLeod, M., 2015. Pathways for degradation of plastic polymers floating in the marine environment. Environmental Science: Processes & Impacts 17 (9), 1513–1521.

Gilani, A.G., Taghvaei, V., Rufchahi, E.M., Mirzaei, M., 2019. Tautomerism, solvatochromism, preferential solvation, and density functional study of some heteroarylazo dyes. Journal of Molecular Liquids 273, 392–407. Available from: https://doi.org/10.1016/j.molliq.2018.10.054.

Hartmann, N.B., Huffer, T., Thompson, R.C., Hassellöv, M., Verschoor, A., Daugaard, A.E., et al., 2019a. Are we speaking the same language? Recommendations for a definition and categorization framework for plastic debris.

Hartmann, N.B., Hüffer, T., Thompson, R.C., Hassellöv, M., Verschoor, A., Daugaard, A.E., et al., 2019b. Response to the letter to the editor regarding our feature "are we speaking the same language? Recommendations for a definition and categorization framework for plastic debris.". Environmental Science & Technology 53 (9), 4678–4679.

Helmberger, M.S., Tiemann, L.K., Grieshop, M.J., 2020. Towards an ecology of soil microplastics. Functional Ecology 34 (3), 550–560.

Hidaka, H., Suzuki, Y., Nohara, K., Horikoshi, S., Hisamatsu, Y., Pelizzetti, E., et al., 1996. Photocatalyzed degradation of polymers in aqueous semiconductor suspensions. I. Photooxidation of solid particles of polyvinylchloride. Journal of Polymer Science Part A: Polymer Chemistry 34 (7), 1311–1316.

Hidayaturrahman, H., Lee, T.G., 2019. A study on characteristics of microplastic in wastewater of South Korea: identification, quantification, and fate of microplastics during treatment process. Marine Pollution Bulletin 146, 696–702.

Hirai, H., Takada, H., Ogata, Y., Yamashita, R., Mizukawa, K., Saha, M., et al., 2011. Organic micropollutants in marine plastics debris from the open ocean and remote and urban beaches. Marine Pollution Bulletin 62 (8), 1683–1692. Available from: https://doi.org/10.1016/j.marpolbul.2011.06.004.

Horton, A.A., Svendsen, C., Williams, R.J., Spurgeon, D.J., Lahive, E., 2017a. Large microplastic particles in sediments of tributaries of the River Thames, UK–Abundance, sources and methods for effective quantification. Marine Pollution Bulletin 114 (1), 218–226.

Horton, A.A., Walton, A., Spurgeon, D.J., Lahive, E., Svendsen, C., 2017b. Microplastics in freshwater and terrestrial environments: evaluating the current understanding to identify the knowledge gaps and future research priorities. Science of the Total Environment 586, 127–141.

Hou, L., Kumar, D., Yoo, C.G., Gitsov, I., Majumder, E.L.W., 2021. Conversion and removal strategies for microplastics in wastewater treatment plants and landfills. Chemical Engineering Journal 406126715.

Jacquin, J., Cheng, J., Odobel, C., Pandin, C., Conan, P., Pujo-Pay, M., et al., 2019. Microbial ecotoxicology of marine plastic debris: a review on colonization and biodegradation by the "Plastisphere.". Frontiers in Microbiology 10865.

Jambeck, J.R., Geyer, R., Wilcox, C., Siegler, T.R., Perryman, M., Andrady, A., et al., 2015. Plastic waste inputs from land into the ocean. Science (New York, N.Y.) 347 (6223), 768–771.

Jin, L., Zhang, G., Tian, H., 2014. Current state of sewage treatment in China. Water Research 66, 85–98.

Kang, J., Zhou, L., Duan, X., Sun, H., Ao, Z., Wang, S., 2019. Degradation of cosmetic microplastics via functionalized carbon nanosprings. Matter 1 (3), 745–758.

Karbalaei, S., Hanachi, P., Walker, T.R., Cole, M., 2018. Occurrence, sources, human health impacts and mitigation of microplastic pollution. Environmental Science and Pollution Research 25, 36046–36063.

Kirstein, I.V., Kirmizi, S., Wichels, A., Garin-Fernandez, A., Erler, R., Löder, M., et al., 2016. Dangerous hitchhikers? Evidence for potentially pathogenic Vibrio spp. on microplastic particles. Marine Environmental Research 120, 1–8.

Kosuth, M., Mason, S.A., Wattenberg, E.V., 2018. Anthropogenic contamination of tap water, beer, and sea salt. PLoS One 13 (4), e0194970.

Krueger, M.C., Harms, H., Schlosser, D., 2015. Prospects for microbiological solutions to environmental pollution with plastics. Applied Microbiology and Biotechnology 99 (21), 8857–8874.

Kumar, A., von Wolff, N., Rauch, M., Zou, Y.Q., Shmul, G., Ben-David, Y., et al., 2020. Hydrogenative depolymerization of nylons. Journal of the American Chemical Society 142 (33), 14267–14275.

Lapointe, M., Farner, J.M., Hernandez, L.M., Tufenkji, N., 2020. Understanding and improving microplastic removal during water treatment: impact of coagulation and flocculation. Environmental Science & Technology 54 (14), 8719–8727.

Lares, M., Ncibi, M.C., Sillanpää, M., Sillanpää, M., 2018. Occurrence, identification and removal of microplastic particles and fibers in conventional activated sludge process and advanced MBR technology. Water Research 133, 236–246.

Law, K.L., Morét-Ferguson, S.E., Goodwin, D.S., Zettler, E.R., DeForce, E., Kukulka, T., et al., 2014. Distribution of surface plastic debris in the eastern Pacifc Ocean from an 11-year data set. Environmental Science & Technology 48 (9), 4732–4738. Available from: https://doi.org/10.1021/es4053076.

Leslie, H.A., Brandsma, S.H., Van Velzen, M.J.M., Vethaak, A.D., 2017. Microplastics en route: field measurements in the Dutch river delta and Amsterdam canals, wastewater treatment plants, North Sea sediments and biota. Environment International 101, 133–142.

Li, L., Liu, D., Song, K., Zhou, Y., 2020. Performance evaluation of MBR in treating microplastics polyvinylchloride contaminated polluted surface water. Marine Pollution Bulletin 150110724.

Liu, H., Zhou, X., Ding, W., Zhang, Z., Nghiem, L.D., Sun, J., et al., 2019a. Do microplastics affect biological wastewater treatment performance? Implications from bacterial activity experiments. ACS Sustainable Chemistry & Engineering 7 (24), 20097–20101.

Liu, P., Qian, L., Wang, H., Zhan, X., Lu, K., Gu, C., et al., 2019b. New insights into the aging behavior of microplastics accelerated by advanced oxidation processes. Environmental Science & Technology 53 (7), 3579–3588.

Lombardi, L., Carnevale, E., Corti, A., 2015. A review of technologies and performances of thermal treatment systems for energy recovery from waste. Waste Management 37, 26–44.

Long, Z., Pan, Z., Wang, W., Ren, J., Yu, X., Lin, L., et al., 2019. Microplastic abundance, characteristics, and removal in wastewater treatment plants in a coastal city of China. Water Research 155, 255–265.

Lopez, G., Artetxe, M., Amutio, M., Alvarez, J., Bilbao, J., Olazar, M., 2018. Recent advances in the gasification of waste plastics. A critical overview. Renewable and Sustainable Energy Reviews 82, 576–596.

Lv, X., Dong, Q., Zuo, Z., Liu, Y., Huang, X., Wu, W.M., 2019. Microplastics in a municipal wastewater treatment plant: fate, dynamic distribution, removal efficiencies, and control strategies. Journal of Cleaner Production 225, 579–586.

Ma, B., Xue, W., Ding, Y., Hu, C., Liu, H., Qu, J., 2018. Removal characteristics of microplastics by Fe-based coagulants during drinking water treatment. Journal of Environmental Sciences 78, 267–275.

Ma, B., Xue, W., Ding, Y., Hu, C., Liu, H., Qu, J., 2019a. Removal characteristics of microplastics by Fe-based coagulants during drinking water treatment. Journal of Environmental Sciences 78, 267–275.

Ma, B., Xue, W., Hu, C., Liu, H., Qu, J., Li, L., 2019b. Characteristics of microplastic removal via coagulation and ultrafiltration during drinking water treatment. Chemical Engineering Journal 359, 159–167.

MacArthur, D.E., Waughray, D., Stuchtey, M., 2016. The new plastics economy, rethinking the future of plastics. World Economic Forum.

Maliwan, T., Pungrasmi, W., Lohwacharin, J., 2021. Effects of microplastic accumulation on floc characteristics and fouling behavior in a membrane bioreactor. Journal of Hazardous Materials 411124991.

Manzoor, J., Sharma, M., Sofi, I.R., Dar, A.A., 2020. Plastic waste environmental and human health impacts. In: Handbook of research on environmental and human health impacts of plastic pollution, IGI global, pp. 29–37.

Marten, E., Müller, R.J., Deckwer, W.D., 2005. Studies on the enzymatic hydrolysis of polyesters. II. Aliphatic–aromatic copolyesters. Polymer Degradation and Stability 88 (3), 371–381.

Meeker, J.D., Sathyanarayana, S., Swan, S.H., 2009. Phthalates and other additives in plastics: human exposure and associated health outcomes. Philosophical Transactions of the Royal Society B 364 (1526), 2097–2113. Available from: https://doi.org/10.1098/rstb.2008.0268.

Mendoza, D.J.C., Nieves, C.S., Castañeda, S.S., 2017. Late-onset anterolateral thigh free flap failure in buccal carcinoma reconstruction. Philippine Journal of Otolaryngology Head and Neck Surgery 32 (2), 47–50.

Montazer, Z., Habibi Najafi, M.B., Levin, D.B., 2020. Challenges with verifying microbial degradation of polyethylene. Polymers 12 (1), 123.

Muhonja, C.N., Makonde, H., Magoma, G., Imbuga, M., 2018. Biodegradability of polyethylene by bacteria and fungi from Dandora dumpsite Nairobi-Kenya. PLoS One 13 (7), e0198446.

Murphy, F., Ewins, C., Carbonnier, F., Quinn, B., 2016. Wastewater treatment works (WwTW) as a source of microplastics in the aquatic environment. Environmental Science & Technology 50 (11), 5800–5808.

Murray, A., Örmeci, B., 2020. Removal effectiveness of nanoplastics (< 400 nm) with separation processes used for water and wastewater treatment. Water 12 (3), 635.

Nanda, S., Berruti, F., 2021. Thermochemical conversion of plastic waste to fuels: a review. Environmental Chemistry Letters 19 (1), 123–148.

Ngo, P.L., Pramanik, B.K., Shah, K., Roychand, R., 2019. Pathway, classification and removal efficiency of microplastics in wastewater treatment plants. Environmental Pollution 255113326.

Padervand, M., Lichtfouse, E., Robert, D., Wang, C., 2020. Removal of microplastics from the environment. A review. Environmental Chemistry Letters 18 (3), 807–828.

Panda, A.K., Singh, R.K., Mishra, D.K., 2010. Thermolysis of waste plastics to liquid fuel: a suitable method for plastic waste management and manufacture of value added products—A world prospective. Renewable and Sustainable Energy Reviews 14 (1), 233–248.

Pathak, V.M., 2023. Exploitation of bacterial strains for microplastics (LDPE) biodegradation. Chemosphere 316, 137845.

Perren, W., Wojtasik, A., Cai, Q., 2018. Removal of microbeads from wastewater using electrocoagulation. ACS omega 3 (3), 3357–3364.

Pittura, L., Foglia, A., Akyol, Ç., Cipolletta, G., Benedetti, M., Regoli, F., et al., 2021. Microplastics in real wastewater treatment schemes: comparative assessment and relevant inhibition effects on anaerobic processes. Chemosphere 262128415.

Plastics Europe, E.P.R.O., 2017. Plastics—the facts 2016. An analysis of European plastics production, demand and waste data, pp. 1–15.

Plastics Europe, E.P.R.O., 2019. Plastics—The Facts 2019. An Analysis of European Plastics Production, Demand and Waste Data. PlasticEurope https://www.plasticseurope.org/en/resources/publications/1804-plastics-facts-2019.

Poerio, T., Piacentini, E., Mazzei, R., 2019. Membrane processes for microplastic removal. Molecules (Basel, Switzerland) 24 (22), 4148.

Prata, J.C., da Costa, J.P., Lopes, I., Duarte, A.C., Rocha-Santos, T., 2020. Environmental exposure to microplastics: an overview on possible human health effects. Science of the total environment 702, 134455.

Rajala, K., Grönfors, O., Hesampour, M., Mikola, A., 2020. Removal of microplastics from secondary wastewater treatment plant effluent by coagulation/flocculation with iron, aluminum and polyamine-based chemicals. Water Research 183116045.

Ritchie, H., Roser, M., 2018. Plastic pollution. https://ourworldindata.org/plastic-pollution.

Roager, L., Sonnenschein, E.C., 2019. Bacterial candidates for colonization and degradation of marine plastic debris. Environmental Science & Technology 53 (20), 11636–11643.

Sánchez, C., 2020. Fungal potential for the degradation of petroleum-based polymers: an overview of macro-and microplastics biodegradation. Biotechnology Advances 40107501.

Schirinzi, G.F., Pérez-Pomeda, I., Sanchís, J., Rossini, C., Farré, M., Barceló, D., 2017. Cytotoxic effects of commonly used nanomaterials and microplastics on cerebral and epithelial human cells. Environmental Research 159, 579–587.

Shahi, N.K., Maeng, M., Kim, D., Dockko, S., 2020. Removal behavior of microplastics using alum coagulant and its enhancement using polyamine-coated sand. Process Safety and Environmental Protection 141, 9–17.

Shao, H., Chen, M., Fei, X., Zhang, R., Zhong, Y., Ni, W., et al., 2019. Complete genome sequence and characterization of a polyethylene biodegradation strain, Streptomyces albogriseolus LBX-2. Microorganisms 7 (10), 379.

Shen, M., Zeng, G., Zhang, Y., Wen, X., Song, B., Tang, W., 2019a. Can biotechnology strategies effectively manage environmental (micro) plastics? Science of the Total Environment 697134200.

Shen, M., Zhang, Y., Zhu, Y., Song, B., Zeng, G., Hu, D., et al., 2019b. Recent advances in toxicological research of nanoplastics in the environment: a review. Environmental Pollution 252, 511–521.

Shen, M., Song, B., Zhu, Y., Zeng, G., Zhang, Y., Yang, Y., et al., 2020a. Removal of microplastics via drinking water treatment: current knowledge and future directions. Chemosphere 251126612.

Shen, M., Ye, S., Zeng, G., Zhang, Y., Xing, L., Tang, W., et al., 2020b. Can microplastics pose a threat to ocean carbon sequestration? Marine Pollution Bulletin 150110712.

Stark, M., 2019. Letter to the editor regarding "Are We Speaking the Same Language? recommendations for a definition and categorization framework for plastic debris.". Environmental Science & Technology 53 (9), 4677.

Sun, J., Dai, X., Wang, Q., van Loosdrecht, M.C., Ni, B.-J., 2019. Microplastics in wastewater treatment plants: detection, occurrence and removal. Water Research 152, 21–37. Available from: https://doi.org/10.1016/j.watres.2018.12.050.

Sussarellu, R., Suquet, M., Thomas, Y., Lambert, C., Fabioux, C., Pernet, M.E.J., et al., 2016. Oyster reproduction is affected by exposure to polystyrene microplastics. Proceedings of the National Academy of Sciences of the USA 113 (9), 2430–2435. Available from: https://doi.org/10.1073/pnas.1519019113.

Talvitie, J., Mikola, A., Koistinen, A., Setälä, O., 2017. Solutions to microplastic pollution–Removal of microplastics from wastewater effluent with advanced wastewater treatment technologies. Water Research 123, 401–407.

Tavares, R., Ramos, A., Rouboa, A., 2018. Microplastics thermal treatment by polyethylene terephthalate-biomass gasification. Energy Conversion and Management 162, 118–131.

Thompson, R.C., 2015. Microplastics in the marine environment: sources, consequences and solutions. Marine Anthropogenic Litter. Springer, Cham, pp. 185–200. Available from: https://doi.org/10.1007/978-3-319-16510-3_7.

Thompson, R.C., Olsen, Y., Mitchell, R.P., Davis, A., Rowland, S.J., John, A.W., et al., 2004. Lost at sea: where is all the plastic? Science 304 (5672), 838. -838.

Tofa, T.S., Ye, F., Kunjali, K.L., Dutta, J., 2019b. Enhanced visible light photodegradation of microplastic fragments with plasmonic platinum/zinc oxide nanorod photocatalysts. Catalysts 9 (10), 819.

Tournier, V., Topham, C.M., Gilles, A., David, B., Folgoas, C., Moya-Leclair, E., et al., 2020. An engineered PET depolymerase to break down and recycle plastic bottles. Nature 580 (7802), 216–219.

Urbanek, A.K., Rymowicz, W., Mirończuk, A.M., 2018. Degradation of plastics and plastic-degrading bacteria in cold marine habitats. Applied Microbiology and Biotechnology 102 (18), 7669–7678.

Van Cauwenberghe, L., Vanreusel, A., Mees, J., Janssen, C.R., 2013. Microplastic pollution in deep-sea sediments. Environmental Pollution (Barking, Essex: 1987) 182, 495–499. Available from: https://doi.org/10.1016/j.envpol.2013.08.013.

Wang, J., Tan, Z., Peng, J., Qiu, Q., Li, M., 2016. The behaviors of microplastics in the marine environment. Marine Environmental Research 113, 7–17.

Wang, J.C., Wang, H., Huang, L.L., Wang, C.Q., 2017a. Surface treatment with Fenton for separation of acrylonitrile-butadiene-styrene and polyvinylchloride waste plastics by flotation. Waste Management 67, 20–26.

Wang, W., Ndungu, A.W., Li, Z., Wang, J., 2017b. Microplastics pollution in inland freshwaters of China: a case study in urban surface waters of Wuhan, China. Science of the Total Environment 575, 1369–1374.

Wang, R., Zhang, L., Chen, B., Zhu, X., 2020a. Low-pressure driven electrospun membrane with tuned surface charge for efficient removal of polystyrene nanoplastics from water. Journal of Membrane Science 614118470.

Wang, X., Zheng, H., Zhao, J., Luo, X., Wang, Z., Xing, B., 2020b. Photodegradation elevated the toxicity of polystyrene microplastics to grouper (*Epinephelus moara*) through disrupting hepatic lipid homeostasis. Environmental Science & Technology 54 (10), 6202–6212.

Wang, Z., Lin, T., Chen, W., 2020c. Occurrence and removal of microplastics in an advanced drinking water treatment plant (ADWTP). Science of the Total Environment 700134520.

Wang, Z., Sedighi, M., Lea-Langton, A., 2020d. Filtration of microplastic spheres by biochar: removal efficiency and immobilisation mechanisms. Water Research 184116165.

Wei, W., Huang, Q.S., Sun, J., Dai, X., Ni, B.J., 2019. Revealing the mechanisms of polyethylene microplastics affecting anaerobic digestion of waste activated sludge. Environmental Science & Technology 53 (16), 9604–9613.

Wei, R., Tiso, T., Bertling, J., O'Connor, K., Blank, L.M., Bornscheuer, U.T., 2020. Possibilities and limitations of biotechnological plastic degradation and recycling. Nature Catalysis 3 (11), 867–871.

Wen, X., Du, C., Xu, P., Zeng, G., Huang, D., Yin, L., et al., 2018. Microplastic pollution in surface sediments of urban water areas in Changsha, China: abundance, composition, surface textures. Marine Pollution Bulletin 136, 414–423.

Wong, S.L., Ngadi, N., Abdullah, T.A.T., Inuwa, I.M., 2015. Current state and future prospects of plastic waste as source of fuel: a review. Renewable and Sustainable Energy Reviews 50, 1167–1180.

Wright, S.L., Kelly, F.J., 2017. Plastic and human health: a micro issue? Environmental Science & Technology 51 (12), 6634–6647.

Wright, S.L., Thompson, R.C., Galloway, T.S., 2013. The physical impacts of microplastics on marine organisms: a review. Environmental Pollution 178, 483–492.

Wu, S., Wu, M., Tian, D., Qiu, L., Li, T., 2020. Effects of polystyrene microbeads on cytotoxicity and transcriptomic profiles in human Caco-2 cells. Environmental Toxicology 35 (4), 495–506.

Yan, Z., Liu, Y., Zhang, T., Zhang, F., Ren, H., Zhang, Y., 2021. Analysis of microplastics in human feces reveals a correlation between fecal microplastics and inflammatory bowel disease status. Environmental Science Technology 56 (1), 414–421.

Yang, Y., Yang, J., Jiang, L., 2016. Comment on "A bacterium that degrades and assimilates poly (ethylene terephthalate).". Science (New York, N.Y.) 353 (6301), 759. -759.

Yaranal, N.A., Subbiah, S., Mohanty, K., 2021. Identification, extraction of microplastics from edible salts and its removal from contaminated seawater. Environmental Technology & Innovation 21, 101253.

Yi, H., Jiang, M., Huang, D., Zeng, G., Lai, C., Qin, L., et al., 2018. Advanced photocatalytic Fenton-like process over biomimetic hemin-Bi2WO6 with enhanced pH. Journal of the Taiwan Institute of Chemical Engineers 93, 184–192.

Yoshida, S., Hiraga, K., Takehana, T., Taniguchi, I., Yamaji, H., Maeda, Y., et al., 2016. A bacterium that degrades and assimilates poly (ethylene terephthalate). Science (New York, N.Y.) 351 (6278), 1196–1199.

Zaouri, N., Cheng, H., Khairunnisa, F., Alahmed, A., Blilou, I., Hong, P.Y., 2021. A type dependent effect of treated wastewater matrix on seed germination and food production. Science of the Total Environment 769144573.

Zhang, K., Shi, H., Peng, J., Wang, Y., Xiong, X., Wu, C., et al., 2018. Microplastic pollution in China's inland water systems: a review of fndings, methods, characteristics, efects, and management. The Science of the Total Environment 630, 1641–1653. Available from: https://doi.org/10.1016/j.scitotenv.2018.02.300.

Zhang, Y.T., Wei, W., Sun, J., Xu, Q., Ni, B.J., 2020a. Long-term effects of polyvinyl chloride microplastics on anaerobic granular sludge for recovering methane from wastewater. Environmental Science & Technology 54 (15), 9662–9671.

Zhang, Y.T., Wei, W., Huang, Q.S., Wang, C., Wang, Y., Ni, B.J., 2020b. Insights into the microbial response of anaerobic granular sludge during long-term exposure to polyethylene terephthalate microplastics. Water Research 179115898.

Zhang, X., Chen, J., Li, J., 2020c. The removal of microplastics in the wastewater treatment process and their potential impact on anaerobic digestion due to pollutants association. Chemosphere 251126360.

Zheng, Y., Yanful, E.K., Bassi, A.S., 2005. A review of plastic waste biodegradation. Critical Reviews in Biotechnology 25 (4), 243–250.

Zhu, K., Jia, H., Zhao, S., Xia, T., Guo, X., Wang, T., et al., 2019. Formation of environmentally persistent free radicals on microplastics under light irradiation. Environmental Science & Technology 53 (14), 8177–8186.

Zubris, K.A.V., Richards, B.K., 2005. Synthetic fibers as an indicator of land application of sludge. Environmental Pollution 138 (2), 201–211.

Chapter 19

Application of graphene oxide for wastewater treatment

Subhakanta Dash[1], Rudra Prasanna Nayak[1], Piyush Gupta[2], Itishree Mohanty[1], Laxmidhar Panda[3], Biswabandita Kar[3] and Ratnakar Dash[4]

[1]*Department of Chemistry, Synergy Institute of Engineering and Technology, Dhenkanal, Odisha, India,* [2]*Department of Chemistry, Faculty of Engineering and Technology, SRM Institute of Science and Technology, NCR Campus, Delhi-NCR Campus, Ghaziabad, Uttar Pradesh, India,* [3]*Department of Chemistry, School of Applied Sciences, KIIT Deemed to be University, Bhubaneswar, Odisha, India,* [4]*Department of Chemistry, Mahima Mahavidyalaya Joranda, Atinda, Odisha, India*

Introduction

The contamination of water is a major concern because it disrupts the hormonal and genetic systems of mankind (Basheer, 2018). Graphene is a honeycomb-like material with a (2D) hexagonal thin sheet structure made of sp^2 hybridized carbon atoms. There are many fascinating properties of graphene viz., immense mechanical strength, electrical conductivity, molecular barrier abilities, and other important properties. The treatment of wastewater is an emerging topic concerning the sustainability of clean water resources and human health. Graphene oxide (GO) is a very good adsorbent material for the removal of inorganic and organic water pollutants (Ng et al., 2012; Chang and Wu, 2013; Cao and Zhang, 2015; Li et al., 2019). GO can be modified to act as a catalyst, adsorbent, and membrane for the treatment of wastewater by combining metal-metal oxide with nanoparticles and polymer (Kemp et al., 2013; Perreault et al., 2015; Smith and Rodrigues, 2015; Zhu et al., 2017). The modification of GO can be achieved by the combination of metal-metal oxide with nanoparticles and polymer, and it is effectively used as an adsorbent, catalyst, and membrane for wastewater treatment (Wang et al., 2021; Sanchez et al., 2018). Joshi et al. (2022) reported that polyvinyl alcohol (PVA) assisted GO and Kaolinite clay (C) sheet porous composite aerogel material (CGP) has been used in wastewater treatment. The CGP aerogel removed 98% of methylene blue (MB) and other organic pollutants from industrial wastewater (Joshi et al., 2022). Yadav et al. (2022) studied the commercial UA-60 nanofiltration coated with kappa-carrageenan and GO solution. The newly developed composite membrane material UA60-κ-CGN-GO showed higher selectivity and antifouling properties (Yadav et al., 2022). To treat textile wastewater including the cationic dyes Nile Blue (NB), MB, Malachite green (MG), and Basic Fuchsin, hydrogel-based biocomposite materials based on GO and agar-biopolymer were developed. The adsorption capacities of these dyes were found as 226.46 mg/g (NB), 79.51 mg/g (MB), 58.25 mg/g (MG), and 38.11 mg/g (BF) (Araujo et al., 2022). Similarly, Zeolitic-Imidazolate-Framework-8 @ reduced GO (ZIF-8@rgo) aerogel was developed to remove tetracycline (TC) from wastewater. The (ZIF-8@rgo) aerogel decontaminated TC and showed an excellent adsorption capacity of 1776.26 mg/g (Liu et al., 2022). To remove tartrazine from an aqueous media, sodium alginate (SA) and GO hybrid aerogel beads were created using the gelation process. The highest tartrazine adsorption capacity of 420.36 mg/g was attained using SA-GO aerogel (Cha et al., 2022; Domyati, 2022; Lemos et al., 2021; Lentz et al., 2022). This chapter highlights the various routes adopted for synthesis, properties, and applications of GO.

Oxidation of graphite to GO

In this method, various oxidizing agents have been used to convert Graphite (Gr) to GO. Various oxygen-containing functional groups, such as hydroxyl and epoxy, are added to the graphene layer's basal plane, while carboxyl groups are added at the layer's edge (Yu et al., 2016). When oxidation occurs, the GO layer loses its conjugation and the

Role of Green Chemistry in Ecosystem Restoration to Achieve Environmental Sustainability. DOI: https://doi.org/10.1016/B978-0-443-15291-7.00026-2

carbon atom, which is sp^2 hybridized (planar shape) becomes converted into sp^3 hybridized (tetrahedral shape). The interlayer interaction decreases due to the presence of oxygen functionalities resulting in d-space increases. The interlayer spacing increases in GO, that is more than 0.625 and 0.335 nm in Gr by oxidation method (He et al., 1998; Uhl and Wilkie, 2004; Hontoria-Lucas et al., 1995).

Brodie's method

GO can be synthesized from the slurry of graphene in the presence of fuming nitric acid with potassium chlorate at a temperature of 60°C for 4 days (Brodie, 1860). The product obtained in this process is named graphic acid, which was later renamed GO. There is a risk of explosion during the chemical reaction as potassium chlorate is used in this process. However, this process leads to a basis of determination of the conversion of Graphite into Graphene sheet by the oxidation process.

Staudenmaier's method

Brodies' work was later modified by Stauden Maiers to enhance the yield and safety of the method. For the oxidation of graphene, a mixture of H_2SO_4/HNO_3 is used in the process (Staudenmaier, 1898). In this process, the slurry of graphene was examined with a mixture of H_2SO_4/HNO_3 in the presence of $KClO_3$ at a temperature of 90°C for 4 days. This process is termed the Staudenmair-Hoffman-Hamdi method (Ojha et al., 2014). This process is carried out at a high temperature with a prolonged period (around 4 days) with the possibility of explosion (as $KClO_3$ is used) (Berthelot, 1870; Luzi, 1891; Charpy, 1909; Weinschenk, 1897; Kohlschitter and Haenni, 1919).

Hummers' method

After 40 years, Hummers' proposed a modified method for the preparation of GO. In this process, oxidation of graphene has been carried out in the presence of Mn_2O_7 at 66°C and then allowed for cooling at 0°C. The mixture of H_2SO_4 and $KMnO_4$ is mainly responsible for the oxidative exfoliation procedure of graphene (Fig. 19.1). At last, H_2O is slowly added to the reaction mixture, and the fine suspension of the GO is collected and dried. This method is widely accepted because it's very fast and operates at low temperatures with zero risk of explosion (Hummers and Offeman, 1958). The major drawbacks of these methods are the low-yield of GO and the release of various toxic gases like NO_2 and N_2O_4 generated during the oxidation process. Also, it is very difficult to separate the residual part of Na^+ and NO_3^- ions in the water along with wastewater (Ren and Cheng, 2014; Ye et al., 2019; Ikram et al., 2020).

Tour method

Later on, Hummers' method was partially modified by Dimiev and Tour. They used high concentrations of $KMnO_4$ and employed an excess of H_2SO_4 (6:1 volume) for the oxidation of graphite without adopting $NaNO_3$. This method is very

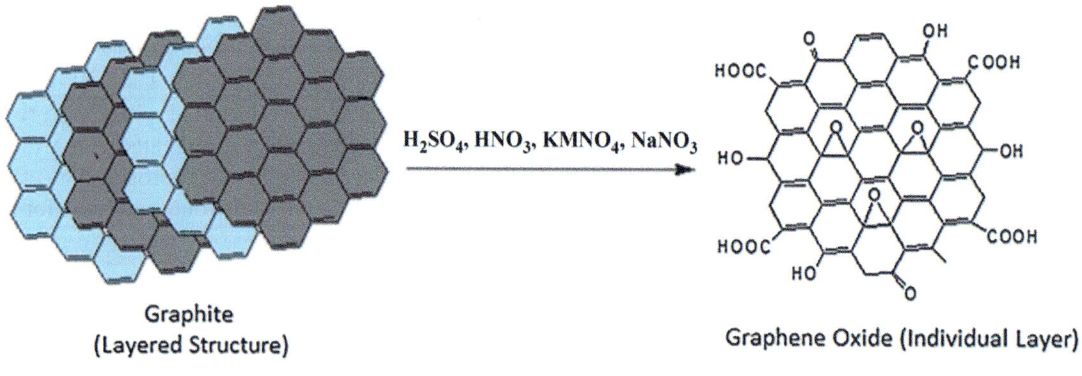

FIGURE 19.1 Graphite oxidation to oxidized graphene. *Permission from Ikram, R., Jan, B.M. and Ahmad, W., 2020. An overview of industrial scalable production of graphene oxide and analytical approaches for synthesis and characterization. Journal of Materials Research and Technology, 9(5), 11587–11610.*

useful for reducing the release of harmful NO_2 and N_2O_4 gases and increasing the product yield also recommends a new reagent H_3PO_3 (Dimiev and Tour, 2014).

Advanced Hummer's method

Chen et al. (2013) studied that oxidation of graphene is carried out at a low temperature in the presence of $KMnO_4$ and concentration H_2SO_4 followed by the inclusion of water and H_2O_2, with the elimination of $NaNO_3$. Using the small size of graphite precursors (1200 mesh size) with numerous centrifugations and washing steps resulted in high yielding of GO (Chen et al., 2013). According to Sun et al. (2015), peroxidation of graphite using impure MnO_2 in a mix of Conc. H_2SO_4 and P_2O_5 can effectively improve GO synthesis when $KMnO_4$ is used as the oxidant. The massive exfoliation and oxidation of graphite was attributed to Mn ions (primarily Mn^{3+}) produced by the reaction of MnO_2 and a mixture of H_2SO_4 and P_2O_5 (Sun et al., 2015). Kovtyukhova et al. (1999) reported that GO can be synthesized from graphite by a two-step oxidation process. The first step includes the oxidation of graphite in the presence of $H_2SO_4/K_2S_2O_8/P_2O_5$. The peroxidized graphite is collected and again oxidized by the Hummers' method practising $KMnO_4$ and Conc. H_2SO_4 at 0°C, which led to the entire oxidation of graphite and yielded balanced GO (Kovtyukhova et al., 1999). In another method, graphite was oxidized using enough $KMnO_4$ in a 9:1 mix of H_2SO_4 and H_3PO_4. GO is efficiently oxidized in the Improved Hummers' method rather than the Hummers' method (Marcano et al., 2010; Emiru and Ayele, 2017). Liou et al. (2015) synthesized GO from graphite in a combination of H_2SO_4 and H_3PO_4 in the ratio of 9:1 with the oxidizing agent $KMnO_4$ (Liou et al., 2015).

Properties of GO

Electrical properties

GO has mostly poor thermal stability and low electrical conductivity because of the disturbance of sp^2 hybridized bonds, broad deflection and oxygen functionalities. According to Loh et al. (2010) GO performed attractive photoluminescence properties which fall in the wavelength range of near UV to near IR. The presence of deflection and functionalities has enhanced the chemical activity of GO, which supports the competence of reduction (Loh et al., 2010). They performed an analysis of the electrical properties of reduced GO (rGO) and GO thin films. He concluded that the electrical properties of GO is reduced, that is, GO (6.86×10^{-7}A) which was less than the electrical properties of GO (6.32×10^{-4}A) is due to the presence of a small functional group of oxygen in the reduced GO.

Mechanical properties

The mechanical characteristics of GO are hardness, plasticity, ductility, rigidity, brittleness, elasticity, toughness, intrinsic strength, and stress. The decreasing value of Young's modulus and intrinsic strength and the further narrowing of its band gap under uniaxial tensile strain is due to the splitting of the sp^2 carbon network in GO. GO has a good stiffness value of 40 GPa with a strength of 120 MPa, and also a homogeneous strength distribution within GO (Gupta, 2018). The strength and modulus of the substance can be determined using conventional mechanical measurement by tension (Xu et al., 2013).

Thermal properties

Graphene oxide's thermal characteristics are due to an increase in temperature, the initiation of a reaction, a change of physical or chemical properties, and phase transition. Taking the thermal conductivity of GO into consideration and for its synthesis, Graphite is being used as a reactant. Fang et al. (2016) investigated GO's thermal conductivity and they found that the value was in the range of 0.5–1 $WM^{-1}K^{-1}$ which was low. The thermal conductivity of synthesized GO can be improved by the inclusion of a polymer. The thermal conductivity of rGO film can be enhanced from 3 to 61 $WM^{-1}K^{-1}$ by annealing GO at a high temperature of 1000°C (Fang et al., 2016). Siva et al. (2019) found that the synthesized nanocomposites had a thermal conductivity value of 19.5 $WM^{-1}K^{-1}$ due to the incorporation of polyvinylidene fluoride-hexafluoro propylene into rGO which is a profound enhancement over the thermal conductivity of GO. The thermal conductivity of GO can also be increased by coupling with the oxide of metal nanoparticles such as ZnO, TiO_2, and WO_3. The high orientation of the broadsheet size, which limits thermal transfer along the boundaries, can also be used to improve the thermal properties of reduced GO (Fig. 19.2) (Siva et al., 2019).

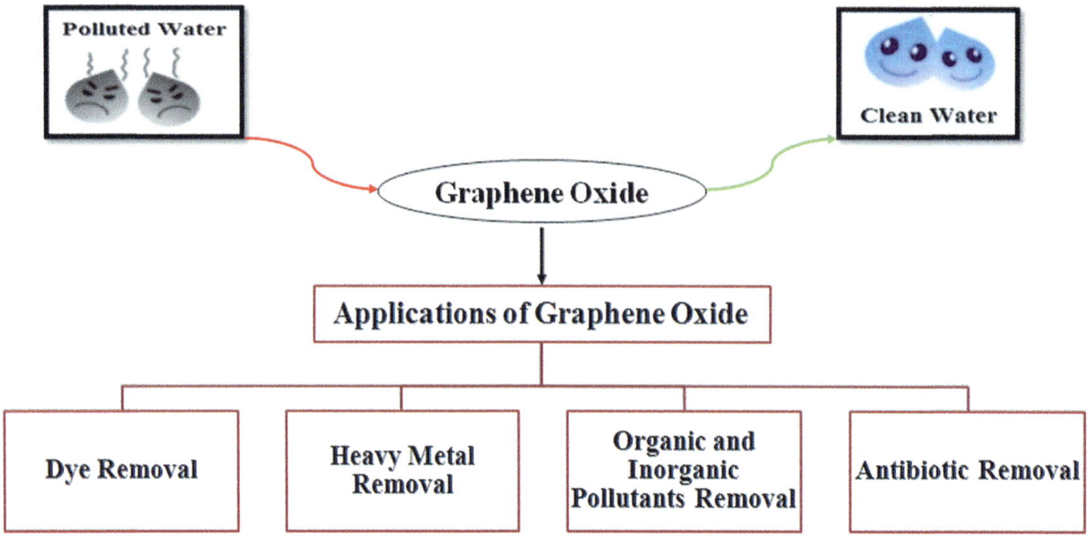

FIGURE 19.2 GO used for the removal of Various Pollutants.

Applications of GO

Dye removal

The GO-based polymeric material is a suitable material for dye removal from wastewater (Wang et al., 2021; Srivastava and Sofi, 2020). The use of GO-based activated carbon (GO-AC) composites was reported as an excellent material for the removal of Methylene Blue (MB) and Crystal Violet (CV) from an aqueous solution. The GO-AC composite exhibited a maximum adsorption value of 70 and 147 mg/g for CV and MB (Abd-Elhamid et al., 2019). Sun et al. (2020) developed a UiO-66-$(OH)_2$/GO composite material for the eradication of MB using a hydrothermal process (Sun et al., 2020). A carboxymethyl cellulose (CMC) composite material was created by Eitaweli et al. (2020) to remove cationic MB at a rate of 180.30 mg/g according to the Langmuir isotherm (Eltaweil et al., 2020). Badhai et al. (2020) introduced a hybrid of magnetite nanoparticles (Fe_3O_4-GO) with 68% elimination of phenol red harmful dyes (Badhai et al., 2020). By layer-by-layer self-assembly of polyvinyl alcohol and GO, Sunset yellow dye removal was achieved through the development of microfiltration polyethersulfone membranes (Januário et al., 2022). For the removal of antacid- orange RL dye, GO-polyaniline, GO-polypyrrole, and GO-polystyrene were used as promising adsorbent materials. When an optimum adsorbent dose of 0.01 g was used at a temperature of 35°C, the maximum removal efficiency of 39.54, 44.76, and 21.43 mg/g was obtained (Noreen et al., 2021). A new variety of GO composite membrane Perylene-3,4,9,10-tetracarboxylic dianhydride (PTCDA) and UiO-66-NH_2 (PTCDA-UiO-66NH_2) were used for the removal of methyl blue, congo red, crystal violet and disperse black by close to 100% (Xiao et al., 2022). A GO-chitosan composite material was synthesized and its performance was observed well in removing MB (Bryan et al., 2022). According to a report, MB was removed from an aqueous solution using GO composite material. With an adsorption capacity of 116.69 mg/g, the adsorption's kinetics were of pseudo-second-order (de Farias et al., 2022).

Removal of heavy metals

GO-based materials play an important role in the efficient removal of heavy metal ions. Cd (II), Cu (II), and Ni (II) can be removed sequentially from wastewater using Fe_3O_4-GO-(o-MWCNTs) nanocomposite material (Wang and Chen, 2015; Rajeshkumar et al., 2018). The maximum removal capacities were 625.00, 574.71, and 384.62 mg/g respectively (Wang et al., 2018). Yu et al. (2020) developed GO/carboxy methyl cellulose nanofibre (CMCNF) composite fiber (CF) for Pb(II) removal. During the wet-spinning process, Fe^{3+} ion (as a coagulant) was continuously added to the cross-linked GO/CMCNF composite fiber and it was found that the maximum removal of Pb(II) is 101.4 mg/g (Yu et al., 2020). Various filler materials viz. rice husk ash, red brick waste, and fumed silica were loaded with GO-based nanoparticles for the removal of various industrial wastes. The maximum removal of color (94.36%), TOC (97.02%), COD (98.66%), turbidity (99.24%), nitrate (98.35%), TP (96.43%), and BOD (98.46%) was achieved in the presence of red brick waste. Adsorption results for SBA-15/GO in the presence of RB-waste showed impressive efficacy in the removal of color (94.80%), TSS (94.36%), TOC (97.02%), COD (98.66%), turbidity (99.24%), nitrate (98.35%), TP (96.43%), and BOD_5 (98.46%) (Aboelfetoh et al., 2021). A Fe_2O_3 fixed mrGO nanocomposite is used for the removal of Pb (II) (Sahu et al., 2022). A mesoporous composite adsorbent material PCN-222/GO-COOH could effectively recover U(VI) and have an excellent adsorption capacity of 426 mg/g (Bi et al., 2021).

Antibiotic removal

The removal of tetracycline hydrochloride was studied by Wang et al. (2020a) using Uio-66$(COOH)_2$/GO composite material at a concentration of 164.91 mg/g and discovered that the composite material has potential application for wastewater treatment (Wang et al., 2020). 99.6% (MB) and 94.88% (tetracycline hydrochloride) were effectively absorbed by (Uio-66-$(OH)_2$/GO) (Sun et al., 2020). Tabrizian et al. (2019) investigated the effect of GO-promoted nZVI/copper bimetallic nanoparticles (BNPs) on tetracyclines (TC) replacement. The results show that the inclusion of cu in the nZVI shows better performance for the replacement of tetracycline by 13% as compared to pure nZVI (Tabrizian et al., 2019). Sahoo et al. (2020) fabricated a nanocomposite adsorbent material (GO/g-C_3N_4-Fe_3O_4) for the removal of tetracycline antibiotics and MB. The maximum absorption capacity of 120 mg/g was achieved at $pH = 3$ for TC and 220 mg/g for MB at $pH = 9$ (Sahoo et al., 2020; Ma et al., 2022). Mesoporous Zeolite-A reduced GO was also used for the removal of MB and lead ions from toxic wastewater. The results show that the removal efficiency of MB and Pb^{2+} was achieved during the first 20 minutes at 98.5% and 93.9% respectively (Farghalia et al., 2021). Similarly, 1% of GO was used to develop an aminated lignin aerogel for the removal of MG. The results indicated that a maximum adsorption capacity of 113.5 mg/g and 91.72% was obtained under optimal conditions (Chen et al., 2020).

Removal of inorganic and organic pollutants

GO and manganese oxide are good adsorbent materials for the removal of para-cresol and para-tertiary-butyl-phenol. Graphene oxide/manganese oxide composite materials showed maximum adsorption capacities of 107.68 mg/g for para-cresol and 135.41 mg/g for para-tertiary-butyl-phenol. The removal efficiency of 90% for para-tertiary-butyl-phenol can be achieved even after five recycles (Zhang et al., 2019). For the treatment of phenolic contaminants in wastewater, the RGO-PEG-ZnO composite material can be regarded as an effective adsorbent (Rout and Jena, 2022). In the sol-gel method, a Montmorillonite-reduced graphene aerogel was developed to remove cadmium and methylene blue (MB) from wastewater. The adsorption method followed Langmuir and pseudo-second-order (Zhou et al., 2022). Novel CaO-based GO nanocomposite materials were developed for the elimination of Cr(VI). The adsorption capacity of nanocomposite material was found to be 38.04 mg/g (Singh et al., 2022). The GO-PEI-NCH (graphene Oxide-Polyethylenimine-Nanocomposite Hydrogel) has the potential application to treat Cr(VI) from polluted water. The GO-PEI-NCH had a maximum Cr(VI) adsorption capacity of 443.8 mg/g (Mittal et al., 2021).

Conclusion and future perspectives

In this chapter, various synthesis routes, properties, and applications of GO-based materials have been discussed. GO is a potential material used in various fields viz. dye removal, removal of heavy metals, antibiotic removal, removal of inorganic + organic pollutants, and wastewater treatment. The continuous studies on GO will help the researchers to find out the new synthesis route of GO. This chapter also summarizes the use of GO-based nanocomposite materials for the removal of different pollutants viz. heavy metals, and dyes. Aside from bioremediation, continuous research on graphene oxide material has demonstrated a new scope in the field of biomedical biosensors. With ongoing investigation, the scientific community will be able to foster advanced integrated techniques for making materials based on graphene and using them widely in a range of applications.

References

Abd-Elhamid, A.I., Kamoun, E.A., El-Shanshory, A.A., Soliman, H.M., Aly, H.F., 2019. Evaluation of graphene oxide-activated carbon as effective composite adsorbent toward the removal of cationic dyes: composite preparation, characterization and adsorption parameters. Journal of Molecular Liquids 279, 530–539.

Aboelfetoh, E.F., Zain Elabedien, M.E., Ebeid, E.-Z.M., 2021. Effective treatment of industrial wastewater applying SBA-15 mesoporous silica modified with graphene oxide and hematite nanoparticles. Journal of Environmental Chemical Engineering 9 (1), 104817.

Araujo, C.M.B.de, Ghislandi, M.G., Rios, A.G., Bezerra da Costa, G.R., Nascimento, B.F.do, Ferreira, A.F.P., et al., 2022. Wastewater treatment using recyclable agar-graphene oxide biocomposite hydrogel in batch and fixed-bed adsorption column: bench experiments and modeling for the selective removal of organics. Colloids and Surfaces A: Physicochemical and Engineering Aspects 639 (20), 128357.

Badhai, P., Kashyap, S., Behera, S.K., 2020. Adsorption of phenol red onto GO-Fe_3O_4 hybrids in aqueous media. Environmental Nanotechnology, Monitoring & Management 13, 100282.

Basheer, A.A., 2018. New generation nano-adsorbents for the removal of emerging contaminants in water. Journal of Molecular Liquids 261, 583–593.

Berthelot, M., 1870. Recherches sur les états du carbone. Annales de Chimie Et de Physique 19, 392–426.

Bi, C., Zhang, C., Ma, F., Zhang, X., Yang, M., Nian, J., et al., 2021. Growth of a mesoporous Zr-MOF on functionalized graphene oxide as an efficient adsorbent for recovering uranium (VI) from wastewater. Microporous and Mesoporous Materials 323, 111223.

Brodie, B.C., 1860. II. On the atomic weight of graphite. Proceedings of the Royal Society of London 10, 11–12.

Bryan, M.Y.K., Chai, P.V., Law, J.Y., Mahmoudi, E., 2022. Graphene oxide-chitosan composite material as adsorbent in removing methylene blue dye from synthetic wastewater. Materialtoday Proceeding 64 (Part 4), 1587–1596.

Cao, N., Zhang, Y., 2015. Study of reduced graphene oxide preparation by Hummers' method and related characterization. Journal of Nanomaterials 2, 1687–4110.

Cha, M., Boo, C., Song, I.H., Park, C., 2022. Investigating the potential of ammonium retention by graphene oxide ceramic nanofiltration membranes for the treatment of semiconductor wastewater. Chemosphere 286, 131745.

Chang, H., Wu, H., 2013. Graphene-based nanocomposites: preparation, functionalization, and energy and environmental applications. Energy & Environmental Science 6 (12), 3483–3507.

Charpy, G., 1909. Sur la formation de l'oxyde graphitique et la définition du graphite. Comptes Rendus Hebdomadaires des Seances de l Academie des Sciences 148 (5), 920–923.

Chen, H., Liu, T., Meng, Y., Cheng, Y., Lu, J., Wang, H., 2020. Novel graphene oxide/aminated lignin aerogels for enhanced adsorption of malachite green in wastewater. Colloids and Surfaces A: Physicochemical and Engineering Aspects 603 (20), 125281.

Chen, J., Yao, B., Li, C., Shi, G., 2013. An improved Hummers method for eco-friendly synthesis of graphene oxide. Carbon 64, 225–229.

Dimiev, A.M., Tour, J.M., 2014. Mechanism of graphene oxide formation. ACS Nano 8 (3), 3060–3068.

Domyati, D., 2022. Chemical and thermal study of metal chalcogenides (zinc sulfide), aluminum oxide, and graphene oxide based nanocomposites for wastewater treatment. Ceramics International . Available online 22 September.

Eltaweil, S.A., Elgarhy, S.G., El-Subruiti, G.M., Omar, A.M., 2020. Carboxymethyl cellulose/carboxylated graphene oxide composite microbeads for efficient adsorption of cationic methylene blue dye. International Journal of Biolgy macromolecule 154, 307–318.

Emiru, T.F., Ayele, D.W., 2017. Controlled synthesis, characterization and reduction of graphene oxide: a convenient method for large scale production. Egyptian Journal of Basic and Applied Sciences 4 (1), 74–79.

Fang, H., Bai, S.L., Wong, C.P., 2016. "White graphene"–hexagonal boron nitride based polymeric composites and their application in thermal management. Composites Communications 2, 19–24.

Farghalia, M.A., Abo-Aly, M.M., Salaheldin, T.A., 2021. Modified mesoporous zeolite-A/reduced graphene oxide nanocomposite for dual removal of methylene blue and Pb^{2+} ions from wastewater. Inorganic Chemistry Communications 126, 108487.

de Farias, L.M.S., Ghislandi, M.G., de Aguiar, M.F., Silva, D.B.R.S., Leal, A.N.R., Silva, F.A.O., et al., 2022. Electrospun polystyrene/graphene oxide fibers applied to the remediation of dye wastewater. Materials Chemistry and Physics 276 (15), 125356.

Gupta, T., 2018. Graphene, Carbon. Springer, pp. 197–228.

He, H., Klinowski, J., Forster, M., Lerf, A., 1998. A new structural model for graphite oxide. Chemical Physics Letters 287 (1–2), 53–56.

Hontoria-Lucas, C., López-Peinado, A.J., López-González, J.D.D., Rojas-Cervantes, M.L., Martín-Aranda, R.M., 1995. Study of oxygen-containing groups in a series of graphite oxides: physical and chemical characterization. Carbon 33 (11), 1585–1592.

Hummers Jr, W.S., Offeman, R.E., 1958. Preparation of graphitic oxide. Journal of the American Chemical Society 80 (6), 1339. –1339.

Ikram, R., Jan, B.M., Ahmad, W., 2020. An overview of industrial scalable production of graphene oxide and analytical approaches for synthesis and characterization. Journal of Materials Research and Technology 9 (5), 11587–11610.

Januário, E.F.D., Vidovix, T.B., Calsavara, M.A., Bergamasco, R., Vieira, A.M.S., 2022. Membrane surface functionalization by the deposition of polyvinyl alcohol and graphene oxide for dyes removal and treatment of a simulated wastewater. Chemical Engineering and Processing - Process Intensification 170, 108725.

Joshi, P., Raturi, A., Srivastava, M., Khatri, O.P., 2022. Graphene oxide, kaolinite clay and PVA-derived nanocomposite aerogel as a regenerative adsorbent for wastewater treatment applications. Journal of Environmental Chemical Engineering 10 (6), 108597.

Kemp, K.C., Seema, H., Saleh, M., Le, N.H., Mahesh, K., Chandra, V., et al., 2013. Environmental applications using graphene composites: water remediation and gas adsorption. Nanoscale 5 (8), 3149–3171.

Kohlschitter, V., Haenni, P., 1919. Zur Kenntnis des graphitischen kohlenstoffs und der graphitsaure. Zeitschrift für Anorganische und Allgemeine Chemie 105 (1), 121. -44.

Kovtyukhova, N.I., Ollivier, P.J., Martin, B.R., Mallouk, T.E., Chizhik, S.A., Buzaneva, E.V., et al., 1999. Layer-by-layer assembly of ultrathin composite films from micron-sized graphite oxide sheets and polycations. Chemistry of Materials 11 (3), 771–778.

Lemos, H.G., Ragio, R.A., Conceição, A.C.S., Venancio, E.C., Mierzwa, J.C., Subtil, E.L., 2021. Assessment of mixed matrix membranes (MMMs) incorporated with graphene oxide (GO) for co-treatment of wastewater and landfill leachate (LFL) in a membrane bioreactor (MBR). Chemical Engineering Journal 425 (1), 131772.

Lentz, L., Mayer, D.A., Dogenski, M., Ferreira, S.R.S., 2022. Hybrid aerogels of sodium alginate/graphene oxide as efficient adsorbents for wastewater treatment. Materials Chemistry and Physics 283 (1), 125981.

Liou, Y.J., Tsai, B.D., Huang, W.J., 2015. An economic route to mass production of graphene oxide solution for preparing graphene oxide papers. Materials Science and Engineering: B 193, 37–40.

Liu, Y., Fu, J., He, J., Wang, B., He, Y., Luo, L., et al., 2022. Synthesis of a superhydrophilic coral-like reduced graphene oxide aerogel and its application to pollutant capture in wastewater treatment. Chemical Engineering Science 260 (12), 117860.

Li, Z., Wang, L., Li, Y., Feng, Y., Feng, W., 2019. Carbon-based functional nanomaterials: preparation, properties and applications. Composites Science and Technology 179, 10–40.

Loh, K.P., Bao, Q., Eda, G., Chhowalla, M., 2010. Graphene oxide as a chemically tunable platform for optical applications. Nature Chemistry 2 (12), 1015–1024.

Luzi, W., 1891. Beitrage Zur Kenntniss Des Graphitkohlenstoffes: gebauer-Schwetschke'sche buchdr.

Marcano, D.C., Kosynkin, D.V., Berlin, J.M., Sinitskii, A., Sun, Z., Slesarev, A., et al., 2010. Improved synthesis of graphene oxide. ACS nano 4 (8), 4806–4814.

Ma, Q., Ming, J., Sun, X., Liu, N., Chen, G., Yang, Y., 2022. Visible light active graphene oxide modified $Ag/Ag_2O/BiPO_4/Bi_2WO_6$ for photocatalytic removal of organic pollutants and bacteria in wastewater. Chemosphere 306, 135512.

Mittal, H., Al Alili, A., Morajkar, P.P., Alhassan, S.M., 2021. Crosslinked hydrogels of polyethylenimine and graphene oxide to treat Cr(VI) contaminated wastewater. Colloids and Surfaces A: Physicochemical and Engineering Aspects 630 (5), 127533.

Ng, Y.H., Ikeda, S., Matsumura, M., Amal, R., 2012. A perspective on fabricating carbon-based nanomaterials by photocatalysis and their applications. Energy & Environmental Science 5 (11), 9307–9318.

Noreen, S., Tahira, M., Ghamkhar, M., Hafiz, I., Bhatti, H.N., Nadeem, R., et al., 2021. Treatment of textile wastewater containing acid dye using novel polymeric graphene oxide nanocomposites (GO/PAN, GO/PPy, GO/PSty). Journal of Materials Research and Technology 14, 25–35.

Ojha, K., Anjaneyulu, O., Ganguli, A.K., 2014. Graphene-based hybrid materials: synthetic approaches and properties. Current Science 397–418.

Perreault, F., De Faria, A.F., Elimelech, M., 2015. Environmental applications of graphene-based nanomaterials. Chemical Society reviews 44 (16), 5861–5896.

Rajeshkumar, S., Liu, Y., Zhang, X., Ravikumar, B., Bai, G., Li, X., 2018. Studies on seasonal pollution of heavy metals in water, sediment, fish and oyster from the Meiliang Bay of Taihu Lake in China. Chemosphere 191, 626–638.

Ren, W., Cheng, H.M., 2014. The global growth of graphene. Nature Nanotechnology 9 (10), 726–730.

Rout, D.R., Jena, H.M., 2022. Polyethylene glycol functionalized reduced graphene oxide coupled with zinc oxide composite adsorbent for removal of phenolic wastewater. Environmental Research 214 (Part 3), 114044.

Sahoo, S.K., Padhiari, S., Biswal, S.K., Panda, B.B., Hota, G., 2020. Fe_3O_4 nanoparticles functionalized $GO/g\text{-}C_3N_4$ nanocomposite: an efficient magnetic nanoadsorbent for adsorptive removal of organic pollutants. Materials Chemistry and Physics 244, 122710.

Sahu, P.S., Verma, R.P., Sah, B., 2022. Synthesis of magnetite-graphene nanocomposite for wastewater treatment. Materials Today Proceeding 62 (Part 10), 6042–6048.

Sanchez, L.M., Ollier, R.P., Gonzalez, J.S., Alvarez, V.A., 2018. Nanocomposite materials for dyes removal. Handbook of Nanomaterials for Industrial Applications Micro and Nano Technologies 922–951.

Singh, S., Sunil Kumar Naik, T.S., Anil, A.G., Khasnabis, S., Nath, B., Basavaraju, U., et al., 2022. A novel CaO nanocomposite cross linked graphene oxide for Cr(VI) removal and sensing from wastewater. Chemosphere 301, 134714.

Siva, D., Karu, J., Hemalatha, J., 2019. Magnetoelectric P (VDF-HFP)-$CoFe_2O_4$ films and their giant magnetoresistance properties. Journal of Magnetism and Magnetic Materials 492, 165689.

Smith, S.C., Rodrigues, D.F., 2015. Carbon-based nanomaterials for removal of chemical and biological contaminants from water: a review of mechanisms and applications. Carbon 91, 122–143.

Srivastava, R., Sofi, I.R., 2020. Impact of synthetic dyes on human health and environment. Impact of Textile Dyes on Public Health and the Environment. IGI Global, pp. 146–161.

Staudenmaier, L., 1898. Verfahren zur darstellung der graphitsäure. Berichte der Deutschen Chemischen Gesellschaft 31 (2), 1481–1487.

Sun, Y., Chen, M., Liu, H., Zhu, Y., Wang, D., Yan, M., 2020. Adsorptive removal of dye and antibiotic from water with functionalized zirconium-based metal organic framework and graphene oxide composite nanomaterial Uio-66-(OH)2/GO. Applied Surface Science 525, 146614.

Sun, J., Yang, N., Sun, Z., Zeng, M., Fu, L., Hu, C., et al., 2015. Fully converting graphite into graphene oxide hydrogels by preoxidation with impure manganese dioxide. ACS Applied Materials & Interfaces 7 (38), 21356–21363.

Tabrizian, P., Ma, W., Bakr, A., Rahaman, M.S., 2019. pH-sensitive and magnetically separable Fe/Cu bimetallic nanoparticles supported by graphene oxide (GO) for high-efficiency removal of tetracyclines. Journal of Colloid and Interface Science 534, 549–562.

Uhl, F.M., Wilkie, C.A., 2004. Preparation of nanocomposites from styrene and modified graphite oxides. Polymer Degradation and Stability 84 (2), 215–226.

Wang, J., Chen, C., 2015. The current status of heavy metal pollution and treatment technology development in China. Environmental Technology Reviews 4 (1), 39–53.

Wang, L., Hu, D., Kong, X., Liu, J., Li, X., Zhou, K., et al., 2018. Anionic polypeptide poly (γ-glutamic acid)-functionalized magnetic Fe_3O_4-GO-(o-MWCNTs) hybrid nanocomposite for high-efficiency removal of Cd (II), Cu (II) and Ni (II) heavy metal ions. Chemical Engineering Journal 346, 38–49.

Wang, K., Wu, J., Zhu, M., Zheng, Y.Z., Tao, X., 2020. Highly effective pH-universal removal of tetracycline hydrochloride antibiotics by UiO-66-(COOH)2/GO metal–organic framework composites. Journal of Solid State Chemistry 284, 121200.

Wang, J., Zhang, J., Han, L., Wang, J., Zhu, L., Zeng, H., 2021. Graphene-based materials for adsorptive removal of pollutants from water and underlying interaction mechanism. Advances in Colloid and Interface Science 289, 102360.

Weinschenk, E., 1897. XVII. Ueber den Graphitkohlenstoff und die gegenseitigen Beziehungen zwischen Graphit, Graphitit und Graphitoid. Zeitschrift für Kristallographie - Crystalline Materials 28 (1–6), 291–304.

Xiao, F., Cao, M., Chu, R., Hu, X., Shi, W., Chen, Y., 2022. NovelPerylene-3, 4, 9, 10-tetracarboxylic dianhydride modified Zr-MOFs/Graphene oxide membrane for dye wastewater treatment. Journal of Colloid and Interface Science 610 (15), 671–686.

Xu, Z., Liu, Z., Sun, H., Gao, C., 2013. Highly electrically conductive Ag-doped graphene fibers as stretchable conductors. Advanced Materials 25 (23), 3249–3253.

Yadav, S., Ibrar, I., Altaee, A., Samal, A.K., Zhou, J., 2022. Surface modification of nanofiltration membrane with kappa carrageenan/graphene oxide for leachate wastewater treatment. Journal of Membrane Science 659 (5), 120776.

Ye, J., Wu, Y.C., Xu, K., Ni, K., Shu, N., Taberna, P.L., et al., 2019. Charge storage mechanisms of single-layer graphene in ionic liquid. Journal of the American Chemical Society 141 (42), 16559–16563.

Yu, H., Hong, H.J., Kim, S.M., Ko, H.C., Jeong, H.S., 2020. Mechanically enhanced graphene oxide/carboxymethyl cellulose nanofibril composite fiber as a scalable adsorbent for heavy metal removal. Carbohydrate Polymers 240, 116348.

Yu, H., Zhang, B., Bulin, C., Li, R., Xing, R., 2016. High-efficient synthesis of graphene oxide based on improved hummers method. Scientific Reports 6 (1), 1–7.

Zhang, B., Zhao, R., Sun., D., Li, Y., Wu, T., 2019. Sustainable fabrication of graphene oxide/manganese oxide composites for removing phenolic compounds by adsorption-oxidation process. Journal of Cleaner Production 215, 165–174.

Zhou, S., Yin, J., Ma, Q., Baihetiyaer, B., Sun, J., Zhang, Y., et al., 2022. Montmorillonite-reduced graphene oxide composite aerogel (M – rGO): a green adsorbent for the dynamic removal of cadmium and methylene blue from wastewater. Separation and Purification Technology 296 (1), 121416.

Zhu, X., Yang, K., Chen, B., 2017. Membranes prepared from graphene-based nanomaterials for sustainable applications: a review. Environmental Science: Nano 4 (12), 2267–2285.

Chapter 20

Antibiotic waste in water: impact and remediation strategies

Prateek Sharma[1], **Bhawna Chopra**[2], **Anjali Sharma**[2], **Geeta Deswal**[2], **Priyanka Kriplani**[2], **Peeyush Kaushik**[2], **Ajmer Singh Grewal**[2] **and Ashwani K. Dhingra**[2]

[1]*Government Pharmacy College, Nagrota Bagwan, Kangra, Himachal Pradesh, India,* [2]*Guru Gobind Singh College of Pharmacy, Yamunanagar, Haryana, India*

Introduction

Since human habitation has increased, antibiotics have become more used in the treatment of sickness in humans, animals, and poultry, among other applications. Antibiotics, on the other hand, can't be entirely degraded by animals (Wang et al., 2019b). Consequently, releasing antibiotic residues puts human health at risk and metabolites into the surrounding environment (Baquero et al., 2008). Human and veterinary antibiotic overuse and misuse are significant contributors to the development of antibiotic-resistant bacteria (ARBs) and genes (ARGs) in the marine ecosystem, which is a cause for global concern (Manaia et al., 2016). By the year 2050, it is anticipated that antibiotic resistance (AR) will result in 10 million fatalities and cost the US economy almost \$100 trillion. ARB and ARGs have been observed in several environmental contexts such as hospital effluent, aquaculture farms, and wastewater treatment facilities (Manaia et al., 2018). Human health is seriously threatened by antibiotic resistance because it may spread from environmental compartments to sources of drinking water. In untreated drinking water sources like wells, as well as in tap water and bottled water, many analyses have shown residues of ARB and ARGs (Ashbolt, 2013; Zhang et al., 2015). Abiotic factors, such as disinfectants and chemical copollutants (including metals and biocides), as well as biotic factors, like the adaptability of bacteria and the activation of a stress response, can take part in the proliferation of ARBs in drinking water (Wu et al., 2018). Metals often enter drinking water systems via distribution pipes that have undergone corrosion or leaching as well as through source water that has been polluted by human activities. Drinking water has been found to have concentrations ranging from 0.1 to 3000 g/L, according to the reports (Chowdhury et al., 2016). Bacterial mutations caused by disinfectants and their metabolites and byproducts might lead to antibiotic resistance (Zhang et al., 2017). The spread of antibiotic resistance in microbial communities (especially those with high bacterial cell density, such as biofilms) may be facilitated via horizontal gene transfer from ARGs to other bacteria (Wang et al., 2019a). Understanding the sources of AR, the mechanisms of AR dissemination, and the role of anthropogenic and environmental factors in AR emergence in the aquatic environment may aid in the development of new strategies for monitoring and reducing the proliferation and persistence of AR in drinking water systems (Gwenzi et al., 2020). Wastewater treatment plants (WWTPs) are a major source of antibiotics and ARGs (Qiao et al., 2018). However, traditional treatment techniques for wastewater have a negligible impact on antibiotics and the ARGs that correlate to them. In the standard wastewater treatment facilities, the removal efficiency for tetracyclines and sulfonamides was only 14.3% and 19.6%, respectively (Karkman et al., 2018; Hassan et al., 2020).

Main sources of antibiotic waste

Domestic misuse

In areas with constraints, rules for better practices have improved the availability of antibiotics throughout the 1990s decade. Despite certain nations not adhering to the limits, the general public requests antibiotics from doctors (Anderson, 1999). Due to household overuse, antibiotics continue to show up in wastewater. For example, the first reports from the United States, where a baseline was evaluated, demonstrate how it has been preserved throughout the

Role of Green Chemistry in Ecosystem Restoration to Achieve Environmental Sustainability. DOI: https://doi.org/10.1016/B978-0-443-15291-7.00007-9

last 6 years, from 2001 to 2007 (McClellan and Halden, 2010). There is no assurance that a patient will utilize antibiotics appropriately or that they won't throw away any that are left over, even if they have access to them and have been prescribed to them (Anwar et al., 2020). Additionally, even though patients take antibiotics appropriately, these drugs are not well absorbed by the body, and between 30% and 90% of them are expelled unaltered in the feces or urine, where they eventually enter the domestic sewage system (Yuan et al., 2016).

Industrial use

The world's entire output of antibiotics is mostly used by the agriculture business, for example, China treats animals with 52% of the world's antibiotics (Zhang et al., 2015). The delay in use in agriculture is widely documented, and the Food and Drug Administration (FDA) in the USA recently issued a veterinary feed regulation in contradiction to European policy, where the region's agricultural use of antibiotics declined from 65% to 27% (Centner, 2016). Similar to this, aquaculture (fish and other marine creatures farming) is a significant worldwide industry, especially in China, the biggest producer of aquaculture in the world (with 58% of output in 2016, which is also a rising activity for other nations) (Tacon, 2018). Antibiotic residues have been discovered in wastewater discharged from pharmaceutical firms in recent investigations in certain Asian nations. The levels of antibiotics found in effluents were greater than those found in samples of wastewater from the hospital and aquaculture facilities (Thai et al., 2018).

Medical garbage

While the body partially metabolizes some of the antibiotics administered to hospital patients, the remainder is excreted and contributes to the hospital's effluents. Antibiotics that are no longer needed are also disposed of in hospital waste. All of these eventually lead to antibiotic fragments in waste materials related to hospitals. Hospital effluents include "emerging contaminants," or substances lacking regulatory status, whose impacts on both the environment and the health of the general people are only partially understood. Hospitals create certain hazardous materials that are processed as trash and disposed of in accordance with the regulations (Fig. 20.1) (Carraro et al., 2016).

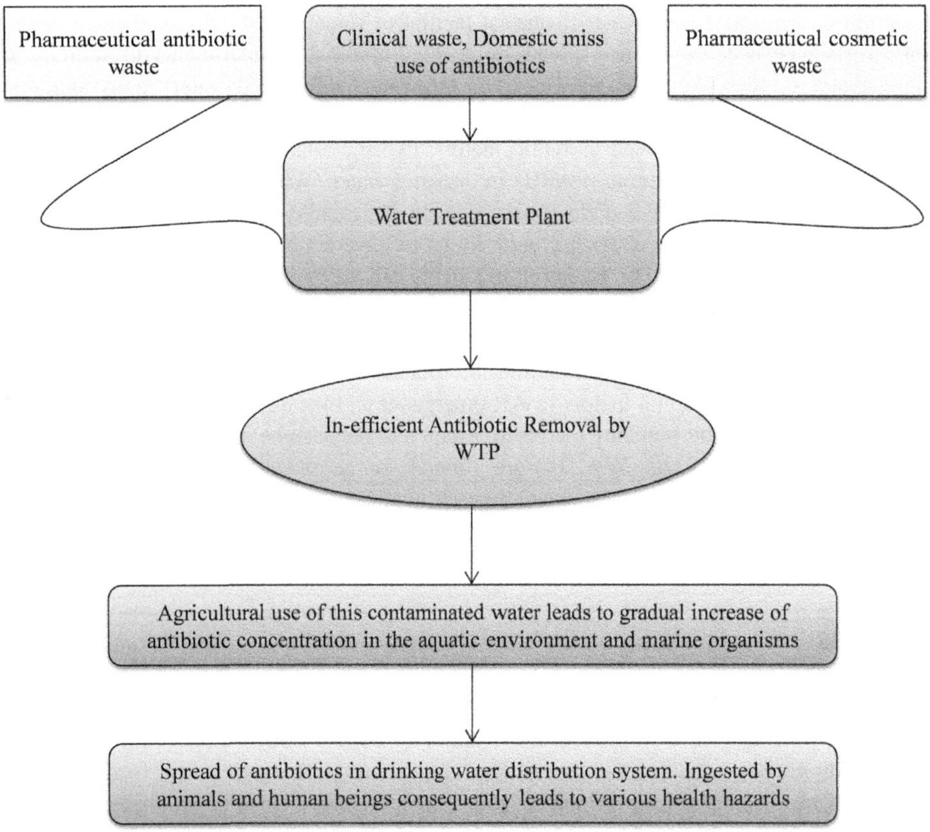

FIGURE 20.1 Schematic representation of transmission of antibiotic waste.

COVID-19 therapy

Due to the requirement to treat 95% of patients with COVID-19 who are suffering from bacterial or fungal illnesses to manage COVID-19, the usage of antibiotics has grown globally. Although, a significant fraction of COVID-19 patients are receiving unnecessary antibiotic treatment (Usman et al., 2020). A recent study that was carried out at a hospital in the United Kingdom suggests that antibiotics should be used strategically, with doxycycline serving as the first line of defense and amoxicillin serving as the second line of defense. Additionally, it forecasts an increase in drug residue releases from WWTPs to rivers and coastal waterways in the UK (Comber et al., 2020; Nguyen et al., 2020). Since many individuals have chosen to self-medicate to combat the infection, the usage of antibiotics has also grown. Additionally, there has been a rise in the use of antibiotics and disinfection products, including biocides that will make their way into wastewater (Malik et al., 2020; Arshad et al., 2020). Health systems sanitation recommends more regular hand washing with soap, such as hand washing for at least 20 seconds each time you enter a building (World Health Organization, 2020).

Impact of antibiotics waste

Increased virulence, pathogenicity, disease outbreaks, and transmission brought on by antibiotic resistance may result in extended hospital stays, increased morbidity, and even death (Berendonk et al., 2015). Controlling human infections and illnesses becomes difficult and expensive as bacteria grow resistant to standard first-line medicines. Because of this, it can be necessary to employ drugs that are the last option or even to develop brand-new treatments. Even though AR is a worldwide issue, impoverished nations are probably more in danger to human health than industrialized ones. This is due to the underdeveloped healthcare systems of developing nations, which are characterized by a high illness load characteristic of tropical conditions and a lack of basic facilities and drugs, necessitating frequent use of antibiotics (Gwenzi and Sanganyado, 2019; Truong et al., 2021). These circumstances lead to a self-triggering or positive feedback mechanism that encourages antibiotic resistance to persist and spread, with a significant adverse impact on people's health. At least three mechanisms exist through which AR in water may be harmful to people: Drinking contaminated water can cause: (1) direct infection with an antibiotic-resistant pathogen, but no sustained human-to-human transmission; (2) sustained human-to-human transmission after direct infection with a pathogenic or opportunistic ARB; and (3) environmental ARGs in drinking water that mutate into human pathogens (Chang et al., 2015). In 2017, the World Health Organization (WHO) compiled a list of medium-to-critical priority pathogenic ARBs to help identify the pathogens that pose the greatest threat to human health. The World Health Organization has urged widespread surveillance of these particularly dangerous ARBs (Pham et al., 2021).

Methods for eliminating harmful substances from drinking water

Traditional strategies

Metals and nanoparticles

The removal of antibiotics using carbon nanotubes, metal or nanoparticles, or sediment adsorption has been the subject of much investigation (Azhar et al., 2017; Chen et al., 2019; Wang et al., 2018). Fe_3O_4-red mud (RM)-nanoparticles (NPs) were tested by Aydin et al. (2019) to see how efficiently they would be able to rid wastewater samples from traditional municipal sewage treatment plants of antibiotics. In that study, intermittent adsorption experiments were utilized to examine how the drug ciprofloxacin was affected by Fe_3O_4-RM-NPs about the quantity of red mud present, pH, the length of the contact, the dosage of the adsorbent, temperature, and other operational factors. Recent studies have shown the potential for using ZnO-MgO nanocomposites and MgO NPs for the elimination of antibiotics (Fakhri and Behrouz, 2015). Zhang et al. (2020) investigated the adsorption of linezolid antibiotics on MgO NPs and ZnO nanocomposites. The combined effect of ZnO-MgO nanocomposites and MgO NPs is a sensitive, beneficial, and convenient adsorption strategy for the removal of linezolid, according to the findings, because a substantial portion of linezolid could be effectively eliminated (90%) by simple procedures in a short period with an environmentally safe adsorbent.

Carbon materials

A carbonaceous substance called biochar may be utilized in agriculture and for carbon sequestration (Chen et al., 2015). Despite being costly and energy-intensive to produce, biochar is ecologically favorable when used for things like carbon sequestration and the cleanup of organic waste (Li et al., 2018). By using biochar made from apple trees,

Kim et al. (2020) tested the efficacy of an activated-sludge system in removing ciprofloxacin (CIP). There were three different amounts of AB applied to the activated sludge bioreactors, each varying from 10% to 20% to 40%. Adsorption by AB was largely responsible for the decrease and removal of CIP (up to 94%). There was a direct correlation between the quantity of AB added and the pace at which the CIP was removed. Physically removing antibiotics from water using activated carbon (AC) has been frequently employed (Yu et al., 2016). Porosity and surface area are two of the most important properties of powdered activated carbon (PAC) (Zhang et al., 2016). Zhang et al. (2019) studied how PAC creates methane and removes antibiotics throughout anaerobic digestion (AD) employing food waste and chicken manure as substrates. Bacterial deactivation and PAC's physical adsorption were the major causes of the decline in antibiotic concentrations.

Advanced oxidation

Pharmaceutical wastewater is an important source of ARGs since it often includes high levels of antibiotic residues (Baker-Austin et al., 2006). However, present pharmaceutical wastewater treatment facilities (PWWTPs) are not very successful at removing antibiotics and ARGs (Al-Jassim et al., 2015; Chang et al., 2010). An anoxic-oxic (A/O) tank, an up-flow anaerobic sludge bed (UASB), and four individual advanced oxidation processes (AOPs) were all part of the laboratory simulation reactor that Hou et al. built (Hou et al., 2019b). Additionally, using a hydrothermal process, Xu et al. successfully constructed expanded graphite immobilized nickel ferrite ($NiCo_2O_4$) and applied it to the degradation of SMX in model wastewater using the PMS activation method (Xu et al., 2020). They accomplished this by using the approach of PMS activation. It was shown to have a high mineralization rate of 32.4%, high catalytic activity (99.9%), and a low rate of ionic leaching. Generally, the science behind advanced oxidation is always being upgraded, and as new types of materials are discovered and developed, the scope of this method's potential applications will continue to expand (He et al., 2020).

Filtration

Due to the distinct lamellar structure of lamellar double hydroxide (LDH), it has been claimed that LDH can absorb organic contaminants that are present in water (Dalla Nora et al., 2020). Despite this, there hasn't been a lot of research done on how LDH can be used to get rid of antibiotics yet. Anionic clay or hydrotalcite are two names that are used interchangeably to refer to LDH, which is a unique multilayered substance that comprises positive-charged metallic ions and negatively charged anions. Sharma et al. (2017) researched the rates of removal of ofloxacin and norfloxacin, and as a result, an innovative approach including integrated adsorption membrane filtration was developed. As a consequence of this, the Ni-Al LDH adsorbent that had been treated with sodium dodecyl sulfate was able to effectively separate ofloxacin and norfloxacin. The two antibiotics had maximum adsorption capacities of 98.7% (norfloxacin) and 94.6% (ofloxacin) at a pH of 7, which was quite close to the adsorbent's point of zero charge value (Causserand et al., 2005). In WWTP systems, where a variety of chemicals may interact at the membrane surface, this is especially true. The removal efficiency of NF and RO membranes has been said to be increased by interactions with naturally occurring organics. Koyuncu and colleagues' research indicates that nanofiltration membranes own a high tetracycline adsorption capacity. When compared to the tetracycline adsorption rate, the hormone rate was much lower (between 22% and 46%) (Koyuncu et al., 2008). On the surface of the membrane, about 80% of chlortetracycline and approximately 50% of polycyclic chemicals were adsorbed, respectively. When compared to hormones and tetracycline, the rejection response caused by sulfonamides was much milder. Reverse osmosis and forward osmosis membranes mostly depend on the size/spatial repulsion to retain pollutants that are a rising problem, but NF and ultrafiltration membranes can still exhibit significant hydrophobic (adsorption) and electrostatic (attraction) interactions (Kim et al., 2018). Free DNA from wastewater treatment facilities contains ARGs, which raises the possibility that they are one of the elements promoting the horizontal gene transfer that increases the prevalence of antibiotic resistance in the environment (Sharma et al., 2016). Slipko and colleagues researched to investigate the possibility that membrane filtering may remove antibiotic resistance genes from free DNA (Slipko et al., 2019).

Flocculation

The process of flocculation has the potential to be useful in environmental engineering because it has several benefits, including low operating costs and straightforward procedures (Martínez-Quiroz et al., 2018). Another area that is intriguing but has not yet been addressed is the impact of the thermo-responsive flocculant architecture (e.g., flocculant graft length). As a result, Hou et al. conducted research on four different types of composite wastewater, which included sulfadiazine (SDZ)-Cu (II), SDZ-Zn (II), TC-Zn (II), and TC-Cu (II). The targeted pollutants used to construct these composite wastewaters were Zn(II), Cu(II), SDZ, and TC, respectively. To remediate the wastewater, CS-g-PNNPAM was used. This approach might work for a small number of typical antibiotics and heavy metal complex pollutants, but it is unlikely to work for a significant number of pollutants. The binding energy of antibiotics and heavy metals is another key factor that influences the performance of flocculants and determines whether or not flocculants are feasible to employ. This is in addition to the binding impact that happens between flocculants and pollutants (Hou et al., 2019a).

High temperature

Composting at high temperatures is receiving a lot of attention both in the United States and in other countries because it is recognized as an efficient way to manage organic solid waste (such as manure from animals, sewage sludge, and garbage from urban households). Recent research on the removal of antibiotics from animal manure composting has mostly concentrated on tetracyclines, whereas other forms of antibiotics have received less attention (Xia et al., 2019). According to studies on the aerobic composting of animal manure, which proved the viability and effectiveness of adding microbial agents, the introduction of high temperature-resistant exogenous bacteria was favorable for the removal of antibiotics in cattle and poultry dung. To remove fluoroquinolone antibiotics from chicken manure, Meng Lei et al. used high heat. The results revealed that high-temperature composting is suitable for this task; when compared to a standard composting system, the removal of fluoroquinolone antibiotics was improved by the exogenous addition of bacteria; and high-temperature composting failed to remove residues (more than 90%) of these drugs effectively (Lei et al., 2015).

Modern technology

MIEX resin

Since the majority of drugs and pharmaceutical products for personal care (PPCPs) are in the form of ions and have pKa values that are close to neutral pH, ion treatment is often recommended for the elimination of PPCPs. For instance, treatment with MIEX may be utilized to successfully get rid of PPCPs in their ion form (Robberson et al., 2006). Using ion treatment, Choi et al. looked at how antibiotic self-decomposition and elimination were affected. SAs and TAs gradually decomposed themselves, but even after 15 days, there were still a lot of antibiotic residues present. Oxytetracycline-HCl and minocycline-HCl degraded quickly, were the exceptions to this norm. In general, when the concentration of dissolved organic matter in raw water is > 0.8 mg/L, the removal efficacy of these antibiotics is reduced due to competition with organic molecules. This is due to the greater abundance of organic molecules (Choi et al., 2007).

Antibiotic-removal beads

Sodium alginate has been used in association with iron as a biocompatible, nontoxic, nonimmunogenic, and degradable polysaccharide in the wet hydrogen peroxide oxidation (CWPO) catalytic process. It has been demonstrated that this mixture is catalytically efficient (Dong et al., 2011). In the beginning, the catalyst that was used was alginate which had been coupled to Fe(II). It is possible to enhance the adherence of metal ions to alginate polymers via the use of an effective method that involves the inclusion of coordinating agents. Cyclohexane dinitrotetraacetic acid (CDTA), also known as cyclohexane dinitrotetraacetic acid, is a chemical that is often employed as a chelating agent for metals, for example, manganese, iron, and copper ions (Norkus et al., 2006). As a consequence of this, CDTA may act as an iron-binding surface agent for alginate and improve how well ofloxacin is extracted from the proclinate gel (Titouhi and Belgaied, 2016). The results indicated that CDTA was successful in destroying ofloxacin.

Biological removal of antibiotics

Pseudomonas sp. H117 was used for the removal of magnetic BMO, biogenic manganese oxides (BMO), TC, Mn(II), and nitrate, in a study that was conducted by Bai et al. (MBMO) (Bai et al., 2020). According to the findings, the removal rates of TC for BMO and MBMO at 12 hours were 91.29% and 96.63%, respectively, whereas this strain was able to remove a large amount of TC (68.86%), nitrate (100%), and Mn(II) (64.64%) in 96 hours. The findings of this study provided details on the application of biological techniques to the removal of numerous toxins from the surroundings of water bodies. The effective management of the transmission of antimicrobial resistance genes is one indirect strategy that has been considered in recent years to limit the antibiotics' scope of action and to lessen their burden (ARGs) (Paulus et al., 2019; He et al., 2018). Nevertheless, it is still unknown if airborne particles from the area around composting facilities will have an impact on the quantity of ARGs present during the composting process. To determine how the air environment of composting facilities affected the abundance of ARG, Cui et al. conducted tests employing thermophilic composting covered by a semipermeable membrane and conventional thermophilic composting of sludge (Cui et al., 2020). After undergoing treatment with semipermeable membrane-covered thermophilic composting, the overall abundance of ARGs was found to be 42.1% lower than the findings obtained with traditional thermophilic composting. When it comes to cleaning water, increased hydrolysis is almost always a method that comes out on top. Yi and associates looked into enhanced hydrolysis as a potential preprocessing strategy. They found that this pretreatment technique can successfully encourage the anaerobic treatment of wastewater from high-concentration oxytetracycline manufacturing. As a result, the amount of ARGs produced was greatly decreased (Yi et al., 2017). Microbiological tests, on the other hand, can determine both the efficacy and the activity of antibiotics at the same time (Dafale et al., 2016). To improve the prediction of the relative potencies of antibiotics, Francisco et al. (2016) suggested using partial least squares regression. This technique improved the linear range of turbidimetric biometrics, enabling us to create a prediction model without overfitting, and evaluating the relative potencies of adriamycin and vancomycin.

Combined strategies

A sol-gel method

Recently, research has been done on the removal of medicines using photocatalysts, molecular filters, activated carbon, and other adsorbent materials (Cao et al., 2013; Prieto-Rodriguez et al., 2012). To overcome the restrictions of the individual components, one study described the use and production of photocatalyst/adsorbent amalgams for the removal of medicines or chemicals from aquatic structures (Zhao et al., 2010). It was claimed that TiO_2-zeolite complex materials were created and employed for the removal of sulfonamide antibiotics from secondary effluent. As a result, sulfamethazine (SMZ) and its analogs were eliminated using TiO_2-zeolite composites (Yap et al., 2012). Nomura et al. (2020) investigated the elimination capabilities of a rotating advanced oxidation contactor (RAOC) outfitted with activated carbon (AC)/TiO_2 complex panes on sulfamonomethoxine.

Ozone and photocatalytic methods

The viability of ozonation and UVC photocatalysis for levofloxacin (LEVO) elimination in aquatic structures was reported by Nasuhoglu et al. (2012). According to the findings, hydroxyl radicals were essential for the photocatalytic elimination of LEVO but had minimal impact on ozonation-based elimination. SMT-containing wastewater was treated by several oxidation techniques, including ozonation, photodissociation, and catalysis, under various circumstances (Beltran et al., 2008). The TOC at each step of the O_3/UVA/TiO_2 process consumes the least ozone and is the least harmful. SMT removal is mostly accomplished by direct ozone reaction, while TOC removal is primarily accomplished through free radical reaction. Hydrogen peroxide is essential for the mineralization of TOC. Similar to O_3/UVA/TiO_2, O_2/UVA/TiO_2 may likewise considerably reduce SMT, but then again it necessitates an extended reaction period (Beltran et al., 2008).

UV exposure and photocatalytic processing

Advanced oxidation approaches including ozonation/hydrogen peroxide, photocatalysis, and membrane technology, were proposed to eliminate antibiotics in wastewater due to the intermittent usage of biological approaches for the degradation of antibiotics. Additionally, there are three conventional sewage handling and sterilization methods, that is, ozone treatment, H_2O_2 (HP) treatment, UV-C radiation, and chloride treatment (Lüddeke et al., 2015; Ben et al., 2017). At concentrations typically utilized for disinfection, traditional methods were demonstrated to be highly poor in removing ARB and their linked genetic materials (Chang et al., 2016). In terms of the mineralization parameters of the iron-dependent AOPs, the elimination frequency of DOC underneath UV-A radiation situations rose from 10%−13% to 19%−21% in the occurrence of (dark) Fenton and Fenton-like procedures. When compared to results achieved with the Fenton/Fenton oxidation method supported by UV-A, the removal of DOC utilizing the UV-C-based approach employing peroxides was greatly enhanced. In the oxidation procedures using peroxide (PS)/UV-C, persulfate (HP)/UV-C, and PMS/UV-C, the removal rates of DOC rose to 31%, 40%, and 59%, respectively. Even for inactivating bacteria, UV-C and oxidant/UV-C radiation procedures seemed to be more successful (Arslan-Alaton et al., 2021).

Conclusion

Modernization and industrialization have also produced new toxins that might be harmful to the environment. Pharmaceuticals, which are widely utilized in veterinary medicine and medical therapies, are regularly found in drinking water, earth, deposits, and superficial water. Because they are resistant to degradation and are not completely digested by living things, antibiotics hurt living things. The antibiotics cannot be removed via orthodox water handling facilities. To decrease the concentration of antibiotics released into the environment and lower the toxicity, WWTPSs and STPs must be improved. Although drugs and resistance coexist, these nations are also contributing to the global development of antibiotic resistance. Therefore, we need to eliminate antibiotic pollution either at the sinking phase (e.g., wastewater) or the source phase (e.g., hospital runoff, poultry manure, and vet runoffs).

References

Al-Jassim, N., Ansari, M.I., Harb, M., Hong, P.Y., 2015. Removal of bacterial contaminants and antibiotic resistance genes by conventional wastewater treatment processes in Saudi Arabia: is the treated wastewater safe to reuse for agricultural irrigation? Water Research 73, 277–290.

Anderson, R.M., 1999. The pandemic of antibiotic resistance. Nature Medicine 5 (2), 147–149.

Anwar, M., Iqbal, Q., Saleem, F., 2020. Improper disposal of unused antibiotics: an often overlooked driver of antimicrobial resistance. Expert Review of Anti-Infective Therapy 18 (8), 697–699.

Arshad, S., Kilgore, P., Chaudhry, Z.S., Jacobsen, G., Wang, D.D., Huitsing, K., et al., 2020. Treatment with hydroxychloroquine, azithromycin, and combination in patients hospitalized with COVID-19. International Journal of Infectious Diseases 97, 396–403.

Arslan-Alaton, I., Karatas, A., Pehlivan, Ö., Ucun, O.K., Ölmez-Hancı, T., 2021. Effect of UV-A-assisted iron-based and UV-C-driven oxidation processes on organic matter and antibiotic resistance removal in tertiary treated urban wastewater. Catalysis Today 361, 152–158.

Ashbolt, N.J., 2013. Human health risk assessment (HHRA) for environmental development and transfer of antibiotic resistance. Environmental Health Perspectives 121, 993–1002.

Aydin, S., Aydin, M.E., Beduk, F., Ulvi, A., 2019. Removal of antibiotics from aqueous solution by using magnetic Fe_3O_4/red mud-nanoparticles. Science of the Total Environment 670, 539–546.

Azhar, M.R., Abid, H.R., Periasamy, V., Sun, H., Tade, M.O., Wang, S., 2017. Adsorptive removal of antibiotic sulfonamide by UiO-66 and ZIF-67 for wastewater treatment. Journal of Colloid and Interface Science 500, 88–95.

Bai, Y., Su, J., Wen, Q., Li, G., Xue, L., Huang, T., 2020. Removal of tetracycline by denitrifying Mn (II)-oxidizing bacterium Pseudomonas sp. H117 and biomaterials (BMO and MBMO): efficiency and mechanisms. Bioresource Technology 312, 123565.

Baker-Austin, C., Wright, M.S., Stepanauskas, R., McArthur, J.V., 2006. Co-selection of antibiotic and metal resistance. Trends in Microbiology 14 (4), 176–182.

Baquero, F., Martínez, J.L., Cantón, R., 2008. Antibiotics and antibiotic resistance in water environments. Current Opinion in Biotechnology 19 (3), 260–265.

Beltran, F.J., Aguinaco, A., García-Araya, J.F., Oropesa, A., 2008. Ozone and photocatalytic processes to remove the antibiotic sulfamethoxazole from water. Water Research 42 (14), 3799–3808.

Ben, W., Wang, J., Cao, R., Yang, M., Zhang, Y., Qiang, Z., 2017. Distribution of antibiotic resistance in the effluents of ten municipal wastewater treatment plants in China and the effect of treatment processes. Chemosphere 172, 392–398.

Berendonk, T.U., Manaia, C.M., Merlin, C., Fatta-Kassinos, D., Cytryn, E., Walsh, F., et al., 2015. Tackling antibiotic resistance: the environmental framework. Nature Reviews. Microbiology 13 (5), 310–317.

Cao, H., Lin, X., Zhan, H., Zhang, H., Lin, J., 2013. Photocatalytic degradation kinetics and mechanism of phenobarbital in TiO_2 aqueous solution. Chemosphere 90 (4), 1514–1519.

Carraro, E., Bonetta, S., Bertino, C., Lorenzi, E., Bonetta, S., Gilli, G., 2016. Hospital effluents management: chemical, physical, microbiological risks and legislation in different countries. Journal of Environmental Management 168, 185–199.

Causserand, C., Aimar, P., Cravedi, J.P., Singlande, E., 2005. Dichloroaniline retention by nanofiltration membranes. Water Research 39 (8), 1594–1600.

Centner, T.J., 2016. Recent government regulations in the United States seek to ensure the effectiveness of antibiotics by limiting their agricultural use. Environment International 94, 1–7.

Chang, X., Meyer, M.T., Liu, X., Zhao, Q., Chen, H., Chen, J.A., et al., 2010. Determination of antibiotics in sewage from hospitals, nursery and slaughter house, wastewater treatment plant and source water in Chongqing region of Three Gorge Reservoir in China. Environmental Pollution 158 (5), 1444–1450.

Chang, Q., Wang, W., Regev-Yochay, G., Lipsitch, M., Hanage, W.P., 2015. Antibiotics in agriculture and the risk to human health: how worried should we be? Evolutionary Applications 8 (3), 240–247.

Chang, Q., Li, K.K., Hu, S.L., Dong, Y.G., Yang, J.L., 2016. Hydroxyapatite supported N-doped carbon quantum dots for visible-light photocatalysis. Materials Letters 175, 44–47.

Chen, C., Zhou, W., Lin, D., 2015. Sorption characteristics of N-nitrosodimethylamine onto biochar from aqueous solution. Bioresource Technology 179, 359–366.

Chen, M., Zhu, M., Zhu, Y., Wang, D., Li, Z., Zeng, G., et al., 2019. Collision of emerging and traditional methods for antibiotics removal: taking constructed wetlands and nanotechnology as an example. NanoImpact 15, 100175.

Choi, K.J., Son, H.J., Kim, S.H., 2007. Ionic treatment for removal of sulfonamide and tetracycline classes of antibiotic. Science of the Total Environment 387 (1–3), 247–256.

Chowdhury, S., Mazumder, M.J., Al-Attas, O., Husain, T., 2016. Heavy metals in drinking water: occurrences, implications, and future needs in developing countries. Science of the Total Environment 569, 476–488.

Comber, S.D., Upton, M., Lewin, S., Powell, N., Hutchinson, T.H., 2020. COVID-19, antibiotics and One Health: a UK environmental risk assessment. Journal of Antimicrobial Chemotherapy 75 (11), 3411–3412.

Cui, P., Bai, Y., Li, X., Peng, Z., Chen, D., Wu, Z., et al., 2020. Enhanced removal of antibiotic resistance genes and mobile genetic elements during sewage sludge composting covered with a semi-permeable membrane. Journal of Hazardous Materials 396, 122738.

Dafale, N.A., Semwal, U.P., Rajput, R.K., Singh, G.N., 2016. Selection of appropriate analytical tools to determine the potency and bioactivity of antibiotics and antibiotic resistance. Journal of Pharmaceutical Analysis 6 (4), 207–213.

Dalla Nora, F.B., Lima, V.V., Oliveira, M.L., Hosseini-Bandegharaei, A., de Lima Burgo, T.A., Meili, L., et al., 2020. Adsorptive potential of Zn–Al and Mg–Fe layered double hydroxides for the removal of 2–nitrophenol from aqueous solutions. Journal of Environmental Chemical Engineering 8 (4), 103913.

Dong, Y., Dong, W., Cao, Y., Han, Z., Ding, Z., 2011. Preparation and catalytic activity of Fe alginate gel beads for oxidative degradation of azo dyes under visible light irradiation. Catalysis Today 175 (1), 346–355.

Fakhri, A., Behrouz, S., 2015. Comparison studies of adsorption properties of MgO nanoparticles and ZnO–MgO nanocomposites for linezolid antibiotic removal from aqueous solution using response surface methodology. Process Safety and Environmental Protection 94, 37–43.

Francisco, F.L., Saviano, A.M., Almeida, T.D.S.B., Lourenço, F.R., 2016. Kinetic microplate bioassays for relative potency of antibiotics improved by partial Least Square (PLS) regression. Journal of Microbiological Methods 124, 28–34.

Gwenzi, W., Sanganyado, E., 2019. Recurrent cholera outbreaks in sub-Saharan Africa: moving beyond epidemiology to understand the environmental reservoirs and drivers. Challenges 10 (1), 1.

Gwenzi, W., Musiyiwa, K., Mangori, L., 2020. Sources, behaviour and health risks of antimicrobial resistance genes in wastewaters: a hotspot reservoir. Journal of Environmental Chemical Engineering 8 (1), 102220.

Hassan, M., Ashraf, G.A., Zhang, B., He, Y., Shen, G., Hu, S., 2020. Energy-efficient degradation of antibiotics in microbial electro-Fenton system catalysed by M-type strontium hexaferrite nanoparticles. Chemical Engineering Journal 380, 122483.

He, Y., Nurul, S., Schmitt, H., Sutton, N.B., Murk, T.A., Blokland, M.H., et al., 2018. Evaluation of attenuation of pharmaceuticals, toxic potency, and antibiotic resistance genes in constructed wetlands treating wastewater effluents. Science of the Total Environment 631, 1572–1581.

He, Y., Zhang, J., Zhou, H., Yao, G., Lai, B., 2020. Synergistic multiple active species for the degradation of sulfamethoxazole by peroxymonosulfate in the presence of CuO@ FeOx@ Fe0. Chemical Engineering Journal 380, 122568.

Hou, T., Du, H., Yang, Z., Tian, Z., Shen, S., Shi, Y., et al., 2019a. Flocculation of different types of combined contaminants of antibiotics and heavy metals by thermo-responsive flocculants with various architectures. Separation and Purification Technology 223, 123–132.

Hou, J., Chen, Z., Gao, J., Xie, Y., Li, L., Qin, S., et al., 2019b. Simultaneous removal of antibiotics and antibiotic resistance genes from pharmaceutical wastewater using the combinations of up-flow anaerobic sludge bed, anoxic-oxic tank, and advanced oxidation technologies. Water Research 159, 511–520.

Karkman, A., Do, T.T., Walsh, F., Virta, M.P., 2018. Antibiotic-resistance genes in waste water. Trends in Microbiology 26 (3), 220–228.

Kim, S., Chu, K.H., Al-Hamadani, Y.A., Park, C.M., Jang, M., Kim, D.H., et al., 2018. Removal of contaminants of emerging concern by membranes in water and wastewater: a review. Chemical Engineering Journal 335, 896–914.

Kim, D.G., Choi, D., Cheon, S., Ko, S.O., Kang, S., Oh, S., 2020. Addition of biochar into activated sludge improves removal of antibiotic ciprofloxacin. Journal of Water Process Engineering 33, 101019.

Koyuncu, I., Arikan, O.A., Wiesner, M.R., Rice, C., 2008. Removal of hormones and antibiotics by nanofiltration membranes. Journal of Membrane Science 309 (1–2), 94–101.

Lei, M., Bing, Y., Nandong, X., 2015. Effect of high temperature composting on removal of fluoroquinolones in chicken manures. Journal of Agro-Environment Science 2, 377–383.

Li, J., Yu, G., Pan, L., Li, C., You, F., Xie, S., et al., 2018. Study of ciprofloxacin removal by biochar obtained from used tea leaves. Journal of Environmental Sciences 73, 20–30.

Lüddeke, F., Heß, S., Gallert, C., Winter, J., Güde, H., Löffler, H., 2015. Removal of total and antibiotic resistant bacteria in advanced wastewater treatment by ozonation in combination with different filtering techniques. Water Research 69, 243–251.

Malik, M., Tahir, M.J., Jabbar, R., Ahmed, A., Hussain, R., 2020. Self-medication during Covid-19 pandemic: challenges and opportunities. Drugs & Therapy Perspectives (Gerontological Nursing Association (Canada)) 36 (12), 565–567.

Manaia, C.M., Macedo, G., Fatta-Kassinos, D., Nunes, O.C., 2016. Antibiotic resistance in urban aquatic environments: can it be controlled? Applied Microbiology and Biotechnology 100, 1543–1557.

Manaia, C.M., Rocha, J., Scaccia, N., Marano, R., Radu, E., Biancullo, F., et al., 2018. Antibiotic resistance in wastewater treatment plants: tackling the black box. Environment International 115, 312–324.

Martínez-Quiroz, M., López-Maldonado, E.A., Ochoa-Terán, A., Pina-Luis, G.E., Oropeza-Guzman, M.T., 2018. Modification of chitosan with carbamoyl benzoic acids for testing its coagulant-flocculant and binding capacities in removal of metallic ions typically contained in plating wastewater. Chemical Engineering Journal 332, 749–756.

McClellan, K., Halden, R.U., 2010. Pharmaceuticals and personal care products in archived US biosolids from the 2001 EPA national sewage sludge survey. Water Research 44 (2), 658–668.

Nasuhoglu, D., Rodayan, A., Berk, D., Yargeau, V., 2012. Removal of the antibiotic levofloxacin (LEVO) in water by ozonation and TiO_2 photocatalysis. Chemical Engineering Journal 189, 41–48.

Nguyen, N.T., Dao, T.H., Truong, T.T., Nguyen, T.M.T., Pham, T.D., 2020. Adsorption characteristic of ciprofloxacin antibiotic onto synthesized alpha alumina nanoparticles with surface modification by polyanion. Journal of Molecular Liquids 309, 113150.

Nomura, Y., Fukahori, S., Fujiwara, T., 2020. Removal of sulfamonomethoxine and its transformation byproducts from fresh aquaculture wastewater by a rotating advanced oxidation contactor equipped with zeolite/TiO_2 composite sheets. Process Safety and Environmental Protection 134, 161–168.

Norkus, E., Vaičiūnienė, J., Vuorinen, T., 2006. Removal of transition metals from alkaline suspensions of cellulose pulp using CDTA as chelating agent. Carbohydrate Polymers 66 (3), 316–320.

Paulus, G.K., Hornstra, L.M., Alygizakis, N., Slobodnik, J., Thomaidis, N., Medema, G., 2019. The impact of on-site hospital wastewater treatment on the downstream communal wastewater system in terms of antibiotics and antibiotic resistance genes. International Journal of Hygiene and Environmental Health 222 (4), 635–644.

Pham, T.D., Le, T.M.A., Pham, T.M.Q., Dang, V.H., Vu, K.L., Tran, T.K., et al., 2021. Synthesis and characterization of novel hybridized CeO_2@ SiO_2 nanoparticles based on rice husk and their application in antibiotic removal. Langmuir: The ACS Journal of Surfaces and Colloids 37 (9), 2963–2973.

Prieto-Rodriguez, L., Miralles-Cuevas, S., Oller, I., Agüera, A., Puma, G.L., Malato, S., 2012. Treatment of emerging contaminants in wastewater treatment plants (WWTP) effluents by solar photocatalysis using low TiO_2 concentrations. Journal of Hazardous Materials 211, 131–137.

Qiao, M., Ying, G.G., Singer, A.C., Zhu, Y.G., 2018. Review of antibiotic resistance in China and its environment. Environment International 110, 160–172.

Robberson, K.A., Waghe, A.B., Sabatini, D.A., Butler, E.C., 2006. Adsorption of the quinolone antibiotic nalidixic acid onto anion-exchange and neutral polymers. Chemosphere 63 (6), 934–941.

Sharma, V.K., Johnson, N., Cizmas, L., McDonald, T.J., Kim, H., 2016. A review of the influence of treatment strategies on antibiotic resistant bacteria and antibiotic resistance genes. Chemosphere 150, 702–714.

Sharma, V., Kumar, R.V., Pakshirajan, K., Pugazhenthi, G., 2017. Integrated adsorption-membrane filtration process for antibiotic removal from aqueous solution. Powder Technology 321, 259–269.

Slipko, K., Reif, D., Wögerbauer, M., Hufnagl, P., Krampe, J., Kreuzinger, N., 2019. Removal of extracellular free DNA and antibiotic resistance genes from water and wastewater by membranes ranging from microfiltration to reverse osmosis. Water Research 164, 114916.

Tacon, A.G., 2018. Global trends in aquaculture and compound aquafeed production. World Aquaculture 49 (2), 33–46.

Thai, P.K., Binh, V.N., Nhung, P.H., Nhan, P.T., Hieu, N.Q., Dang, N.T., et al., 2018. Occurrence of antibiotic residues and antibiotic-resistant bacteria in effluents of pharmaceutical manufacturers and other sources around Hanoi, Vietnam. Science of the Total Environment 645, 393–400.

Titouhi, H., Belgaied, J.E., 2016. Removal of ofloxacin antibiotic using heterogeneous Fenton process over modified alginate beads. Journal of Environmental Sciences 45, 84–93.

Truong, T.T.T., Vu, T.N., Dinh, T.D., Pham, T.T., Nguyen, T.A.H., Nguyen, M.H., et al., 2021. Adsorptive removal of cefixime using a novel adsorbent based on synthesized polycation coated nanosilica rice husk. Progress in Organic Coatings 158, 106361.

Usman, M., Farooq, M., Hanna, K., 2020. Environmental side effects of the injudicious use of antimicrobials in the era of COVID-19. Science of the Total Environment 745, 141053.

Wang, T., Ai, S., Zhou, Y., Luo, Z., Dai, C., Yang, Y., et al., 2018. Adsorption of agricultural wastewater contaminated with antibiotics, pesticides and toxic metals by functionalized magnetic nanoparticles. Journal of Environmental Chemical Engineering 6 (5), 6468–6478.

Wang, X., Yin, R., Zeng, L., Zhu, M., 2019a. A review of graphene-based nanomaterials for removal of antibiotics from aqueous environments. Environmental Pollution 253, 100–110.

Wang, H., Hu, C., Shen, Y., Shi, B., Zhao, D., Xing, X., 2019b. Response of microorganisms in biofilm to sulfadiazine and ciprofloxacin in drinking water distribution systems. Chemosphere 218, 197–204.

World Health Organization, 2020. Water, sanitation, hygiene, and waste management for SARS-CoV-2, the virus that causes COVID-19: interim guidance, 29 July 2020 (No. WHO/2019-nCoV/IPC_WASH/2020.4). World Health Organization.

Wu, Q., Li, S., Zhao, X., Zhao, X., 2018. Interaction between typical sulfonamides and bacterial diversity in drinking water. Journal of Water and Health 16 (6), 914–920.

Xia, H., Chen, J., Chen, X., Huang, K., Wu, Y., 2019. Effects of tetracycline residuals on humification, microbial profile and antibiotic resistance genes during vermicomposting of dewatered sludge. Environmental Pollution 252, 1068–1077.

Xu, M., Zhou, H., Wu, Z., Li, N., Xiong, Z., Yao, G., et al., 2020. Efficient degradation of sulfamethoxazole by $NiCo_2O_4$ modified expanded graphite activated peroxymonosulfate: characterization, mechanism and degradation intermediates. Journal of Hazardous Materials 399, 123103.

Yap, P.S., Cheah, Y.L., Srinivasan, M., Lim, T.T., 2012. Bimodal N-doped P25-TiO_2/AC composite: preparation, characterization, physical stability, and synergistic adsorptive-solar photocatalytic removal of sulfamethazine. Applied Catalysis A: General 427, 125–136.

Yi, Q., Zhang, Y., Gao, Y., Tian, Z., Yang, M., 2017. Anaerobic treatment of antibiotic production wastewater pretreated with enhanced hydrolysis: simultaneous reduction of COD and ARGs. Water Research 110, 211–217.

Yu, F., Li, Y., Han, S., Ma, J., 2016. Adsorptive removal of antibiotics from aqueous solution using carbon materials. Chemosphere 153, 365–385.

Yuan, S.F., Liu, Z.H., Huang, R.P., Yin, H., Dang, Z., 2016. Levels of six antibiotics used in China estimated by means of wastewater-based epidemiology. Water Science and Technology 73 (4), 769–775.

Zhang, Q.Q., Ying, G.G., Pan, C.G., Liu, Y.S., Zhao, J.L., 2015. Comprehensive evaluation of antibiotics emission and fate in the river basins of China: source analysis, multimedia modeling, and linkage to bacterial resistance. Environmental Science & Technology 49 (11), 6772–6782.

Zhang, X., Guo, W., Ngo, H.H., Wen, H., Li, N., Wu, W., 2016. Performance evaluation of powdered activated carbon for removing 28 types of antibiotics from water. Journal of Environmental Management 172, 193–200.

Zhang, Y., Gu, A.Z., He, M., Li, D., Chen, J., 2017. Subinhibitory concentrations of disinfectants promote the horizontal transfer of multidrug resistance genes within and across genera. Environmental Science & Technology 51, 570–580.

Zhang, J., Chua, Q.W., Mao, F., Zhang, L., He, Y., Tong, Y.W., et al., 2019. Effects of activated carbon on anaerobic digestion–methanogenic metabolism, mechanisms of antibiotics and antibiotic resistance genes removal. Bioresource Technology Reports 5, 113–120.

Zhang, A., Li, X., Xing, J., Xu, G., 2020. Adsorption of potentially toxic elements in water by modified biochar: a review. Journal of Environmental Chemical Engineering 8 (4), 104196.

Zhao, C., Deng, H., Li, Y., Liu, Z., 2010. Photodegradation of oxytetracycline in aqueous by 5A and 13X loaded with TiO_2 under UV irradiation. Journal of Hazardous Materials 176 (1–3), 884–892.

Chapter 21

Impact and remediation strategies for emerging organic water pollutants: an overview

Deepak Kumar[1], Manjul Gupta[2], Lata Rani[3], Arun Lal Srivastav[4] and Siyaram[5]

[1]SJVN Limited, Shimla, Himachal Pradesh, India, [2]PARAMARSH Consultancy Services, Ghaziabad, Uttar Pradesh, India, [3]Chitkara School of Pharmacy, Chitkara University, Baddi, Himachal Pradesh, India, [4]Chitkara University School of Engineering and Technology, Chitkara University, Baddi, Himachal Pradesh, India, [5]School of Sciences, Indira Gandhi Open National University, New Delhi, India

Introduction

The rapid growth of the population in developed and developing countries is creating pressure on our existing water resources required for day-to-day needs. That is further aggravated by the known and unknown pollutants getting released in water from various sources without treatment rendering water unfit for consumption (UNWWAP, 2003). The problem getting even worse due to poor monitoring, lack of data on their fate, impact on biota and human health, and treatment methods of emerging organic water pollutants (EOWPs) and their metabolite is still in the initial stage of development. The concentration of these pollutants may vary from nanogram to microgram per liter in wastewater (Field et al., 2006). EOWPs have potential impacts on the environment, biota, and human beings that are not commonly monitored. These water pollutants either do not have standards regulated by national and international laws or have evolving regulatory standards called EOWPs (Du et al., 2015). These emerging pollutants are now being identified because of growing awareness, improvements in detection technologies, and recent research about the risks and impacts of EOWPs (Ahmed et al., 2017). The first emerging pollutant was discovered in the 1800s in the aquatic environment (Du et al., 2015). The number of EOWPs can be increased in the future with advancements in monitoring techniques and the use of the newly invented compound, in various sectors. These emerging pollutants are heavy metals, pesticides, fertilizers (polychlorinated biphenyls, poly aromatic hydrocarbons, personal care products, pharmaceuticals, endocrine disrupters, artificial sweeteners, illicit drugs, plasticizers, dyes, flame retardants, etc.) (Snow et al., 2017; Tran et al., 2018). The existence of these emerging pollutants in our water resources depends on the pattern of their use by the socio-economic community, and high solubility (WHO, 2011; Rodriguez-Narvaez et al., 2017). Many pollutants may be present in areas where their use is negligible or very less because of bioaccumulation and the ability of long-term persistence during transport (Gavrilescu et al., 2015; Sorlini et al., 2019). Worldwide the consumption of organic pesticides has increased from 5 to 6.8 million between 2011 and 2015 (Wang et al., 2015). These pesticides have an impact on insects, rodents, fungi mites, fungi, etc (Velazquez-Fernandez et al., 2012). Organic phosphate, which is an ester form of phosphate is considered safe because of the faster degradation rate. Water resource contamination by pesticides is a serious concern and requires remedial measures to reduce the adverse impacts. A significant portion of pharmaceuticals get discharged unaltered or as an active metabolite and find their way into wastewater in wastewater treatment plants (Chang and Lin, 2001; Dixit and Parmar, 2013; Boxall., 2004). Further medical waste from pharmaceutical industries and illegal drug disposal can add various compounds to our water resources (Brozinski et al., 2013; Jiang and Zhou, 2013). Personal care products (PCP) are cosmetics, UV filters, disinfectants, biocides, etc., used widely to improve quality of life. PCPs have a low potential for long-range transport as PCPs rapidly undergo photolysis and have short half-lives ranging from seconds to hours (Wong et al., 2019). Dyes are widely used in various industries, such as the leather, textile, pharmaceutical, cosmetic, food, and printing industries (Hou et al., 2017; Ooi et al., 2017). The frequent use of azodye in textile industries is because of easiness of use, presence in a wide range of brilliant

Role of Green Chemistry in Ecosystem Restoration to Achieve Environmental Sustainability. DOI: https://doi.org/10.1016/B978-0-443-15291-7.00005-5

shades, photolytic stability, strong binding with textile fibers, and lesser energy requirement (Brüschweiler and Merlot, 2017; Raffi et al., 1997). The most common dyes which are in use are azo, anthraquinone, and deoxidizing dyes. It is estimated that over 10,000 types of dyes are in use by the industries (Kisku et al., 2015).

Sources and transportation of emerging water pollutants

These EOWPs get access to surface and groundwater through point sources such as discharges from various industries, raw or treated wastewater from effluent treatment plants, or diffuse sources such as agricultural runoff, landfill, leachate, sludge, surface, and groundwater runoff, etc. (Tran and Gin, 2017; Ali et al., 2018; Tran et al., 2018). Transportation of EOWPs depends upon their physiochemical properties, which include polarity, adsorption properties, volatile nature, persistence, and interaction with the immediate environment and discharge pattern in the vicinity areas (Debska et al., 2004; Patel et al., 2019; Muller et al., 2020). Wastewater treatment plants, ETPs, and CETPS are the main contributors to various emerging pollutants in the aquatic environment in urban areas (Wang et al., 2016; Tran and Gin, 2017). Treated wastewater still has the presence of EOWPs in such a concentration that can disrupt the aquatic ecosystem (Tran and Gin, 2017; Ali et al., 2018). EOWPs laden water is used for irrigation purposes and EOWPs laden sludge as a fertilizer.

Toxicity and degradation of emerging water pollutants

Research of the last few decades assess short and long-term effects of EOWPs on aquatic life and human health and reveals that EOWP's presence in water may generate adverse impact on aquatic and terrestrial biota on long-term consumption of contaminated food over the permissible limit and by disturbing the normal metabolic activity (Fu et al., 2003). Increased use of pesticides has adverse impacts on the flora and fauna of aquatic systems of nearby areas (Asad et al., 2007). Pesticide adversely impacts the photosynthetic activity of plants by affecting the functions of PSII-associated protein (Jena et al., 2012; Zobiole et al., 2012), nutrition uptake (Zobiole et al., 2010), chlorophyll biosynthesis by an inhibitory effect on δ-aminolevulinic acid (ALA) (Serra et al., 2013; Krenchinski et al., 2017), formation of amino acids, metabolism of nitrogen (Zobiole et al., 2010), photochemical reaction (Vivancos et al., 2011) and by inducing oxidative stress (Vagi et al., 2018), etc. The reactive oxygen species (RoS) generated in the plant is controlled by antioxidants such as ascorbate peroxidase (APX), superoxidase dismutase, and glutathione-S-transferase (GST) (Mishra et al., 2015; Panduranga et al., 2005). Exposure to EOWPS induces behavioral, physiological, and histopathological alternation in the organs of exposed organisms (Singh et al., 2009; Hussain et al., 2016; Sidhu et al., 2019). A decrease in total lipids, total proteins, and glycogen content was also observed in the blood of C. catla (Hussain et al., 2015). The liver was characterized by vacuolation of hepatocytes, narrowing of the lumen, blood sinusoids, severe hemorrhage, dilated lumen, hypertrophied epithelial cells, and necrosis (Subburaj et al., 2018; Sinha et al., 2018). Pesticide exposure in frogs shows sex-linked differences, changes in liver histology, and changes in antioxidant and thyroid hormone levels (Zhang et al., 2020). Studies show that fenvalerate reduces testosterone synthesis, alters the expression of circadian rhythm mRNA, and increases intracellular calcium ion levels in mouse Leydig cells (Guo et al., 2017). Bioaccumulation of these pesticides in sediments and soils may be transmittable to the drinking water supply system (Salter-Blanc et al., 2016). Excessive use of medicines, untreated, partially treated, or allegedly treated pharmaceuticals is polluting our water resources. In India, most of the river stretches are polluted by industries and are major sources of emerging pollutants that are creating a threat to our aquatic and terrestrial life (Lokhande et al., 2011). Once released into the water ecosystem, these compounds have an impact on microbial cells (Lawrence et al., 2007). The presence of compounds like Diclofenac is harmful for birds, but also has impacts on human health and the ecosystem (Green et al., 2004; Hussain et al., 2008). Exposure to gemfibrozil in rodents and eels was found to lead to liver cancer and detrimental effects on nontarget aquatic organisms (Lyssimachou et al., 2014; Hampel et al., 2010). Antidepressants such as fluoxetine have an impact on reproductive function and sex hormones in fish (Mennigen et al., 2008). Antibiotics are the most commonly used drugs (Kümmerer, 2009; Liu et al., 2009; Watanabe et al., 2010). Further, increasing the use of antiseptics and disinfectants may lead to an alteration in endocrine function in humans. It is widely used in household cleaners, as well as in commercial products. The presence of triclosan and some of its metabolites is reported in wastewater, surface waters, sediments, and biological samples (Russell, 2004; Kim et al., 2011). Triclocarban and triclosan tend to accumulate in biosolids, sediments, soil plants, and aquatic species (Prosser et al., 2014; Venkatesan et al., 2012). Triclosan has been found toxic to algae (Meade et al., 2001) and has the potential to disrupt the nitrogen cycle in soil ecosystems (Cha and Cupples, 2010).

Analytical techniques for EOWPS detection

Various analytical techniques, such as Liquid/Ultra high liquid chromatography (Cotton et al., 2016; Rousis et al., 2017), FTIR (Du et al., 2010), UV-Vis (Yuan et al., 2015), mass spectrometry (Lorenzo and Pico, 2017). X-ray diffraction, electrochemical and NMR Spectroscopy (Hiscock et al., 2015; Kumar et al., 2013) are in use for the identification and quantification of various emerging pollutants (Yuan et al., 2015; Kumar et al., 2013). Fig. 21.1 depicts the analytical techniques that can detect the concentrations of EOWPs in water samples.

Treatment options for emerging water pollutants

The increasing number and concentration of various EOWPs in our water resources, now make it necessary to ensure the efficient removal of these pollutants to reduce impacts on the environment and humans. EOWPs released from point sources are easy to regulate, monitor, and treat in comparison to those released from diffuse sources (Malkoske et al., 2016). Conventional methods, such as chlorination, ionization, etc. are not very effective in removing these contaminants from the water resources. Fig. 21.2 shows various types of approaches for water decontamination.

High energy requirements and production of toxic substances are the main disadvantages of treatment by various physical methods, advanced oxidation process, and activated carbon adopted for reduction of organic load in our water resources (Schwarzenbach et al., 2006; Homem and Santos, 2011). However, biodegradation methods such as bioremediation, bio-adsorption, bioaccumulation, and bio-uptake utilizing the microorganisms provide feasible methods of removing contaminants (Guerra et al., 2014). Currently, biological and advanced treatment method in combination or oxidation or reduction method is being utilized for the treatment of EWOPs (Fast et al., 2017). Advanced oxidation and conventional methods are found efficient in the removal of contaminants, but they can generate intermediates, which are difficult to identify and may be more toxic than the initial compounds. EOWPs can be removed from secondary or tertiary effluents by the biosorption treatment (Basheer, 2018). Biosorption is influenced by the nature of biomass, the absorption capacity of sorbents, and the ability to modify the sorbent surface to improve the sorption capacity and regeneration of exhausted bio-sorbents (Adewuyi, 2020). Wastewater treatment with microalgae provides a cost-effective remediation method for EOWPs (Craggs et al., 2012). The microalgal/bacterial biomass can be bio-refined to recover a range of both high-value and low-value products, such as fertilizer, algal plastics, fibers, or protein-rich feed (Sutherland et al., 2018). Hydrophobic, cationic EOWPs have strong, and hydrophilic anionic EOWPs have weak

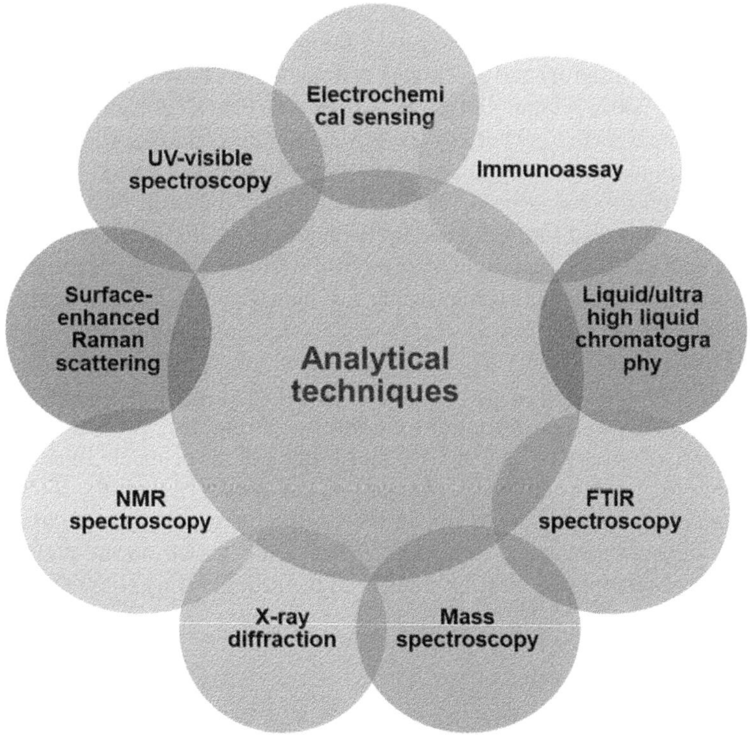

FIGURE 21.1 Various analytical techniques for detection and quantification EOWPs.

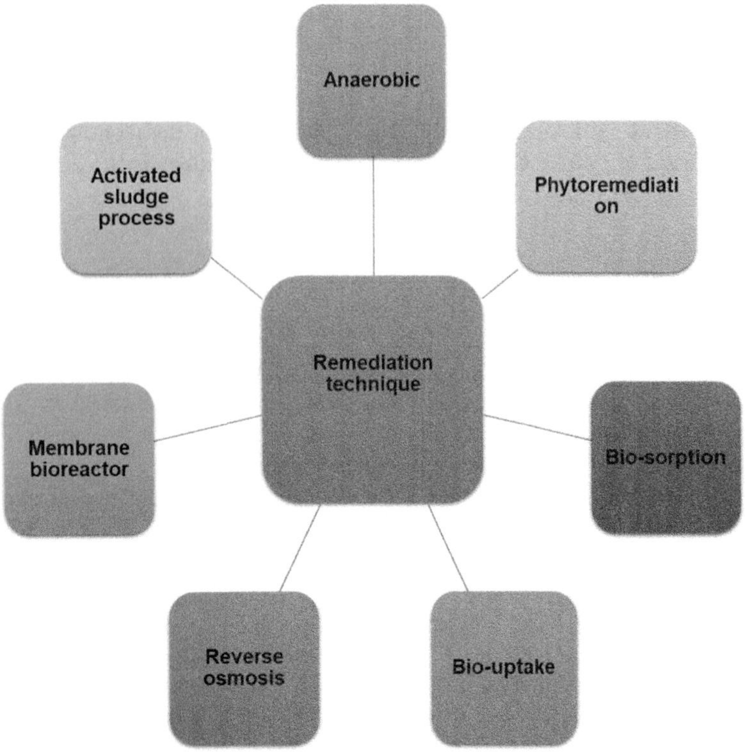

FIGURE 21.2 Various methods used for the treatment of EOWPs.

bioabsorption affinity toward the microalgal cell surface through electrostatic interactions (Xiong et al., 2017). Studies show that lipophilic compounds such as pharmaceuticals, Biperiden, Trihexyphenidyl, etc. have average bioabsorption rates >70%, whereas the hydrophilic pharmaceuticals, fluconazole, trimethoprim, tramadol, etc. had low (<20%) bioabsorption microalgalrates (Gojkovic et al., 2019). Bioabsorption involves various processes such as ion exchange reactions, adsorption reactions, various groups on the microalgal surface, chelation, and microprecipitation (Donmez et al., 1999). Extracted microalgal biomass can be reused with reduced operational cost because of no requirement to maintain viable microalgal cultures (Dixit and Singh, 2014). Various studies reported the removal of 17ά-ethinylestradiol by the green alga Desmodesmus subspicatus and of triclosan by the green alga Nannochloris sp. through the Bio-uptake method (Bai and Acharya, 2019; Coogan et al., 2007; Vernouillet et al., 2010). Aerobic degradation of an organic compound by microalgae-bacteria coupled treatment method occurs by oxygen supply through microalgal photosynthesis, which, in turn, provides CO_2 required for photosynthesis by microalgae (Sutherland et al., 2015). Various combinations of microalgae and bacteria consortiums are also being used for the treatment of industrial wastewater (Safonova et al., 1999). Degradation of EWOPs by bacteria is also supplemented by the microalgae, through the release of dissolved organic matter, which provides requisite substrates for bacterial cometabolism of these emerging pollutants (Wolfaardt et al., 1994). Studies show that the addition of cosubstrates such as starch for cometabolism for remediation of DDT by microbial consortium under nitrogen-fixing conditions plays a very effective role complete mineralization and detoxification of DDT (Khan et al., 2015). Various enzymes such as hydrolases, ligases, transferases, isomerases, esterases, oxidases, and peroxidases have been identified, which play a crucial role in the degradation of pesticides by leading oxidation, hydrolysis, reduction of the nitro group to the amino group by cleavage of a ring, etc. (Verma et al., 2014). Various studies to assessability of biodegradation efficiency of natural river biofilm on pharmaceuticals show no variation in the degrading efficiency of river biofilms for pharmaceuticals during various seasons (Tien et al., 2013). The microbes-based treatment process uses these contaminants as a source of their energy by complete mineralization (Pan et al., 2018). MFC has a cathode and microorganisms that are known as electricians are grown on the anode. These microbes lead to a reduction of the pollutants by promoting the electron transfer to the cathode (Logan, 2009).

Conclusion

Increasing discharges of EWOPs with rapid urbanization and industrialization have severe adverse impacts on aquatic life and human health. Every year thousands of new unknown compounds of unknown toxicity are being added to our environment. The condition is further aggravated by reducing water levels and the unavailability of safe drinking water for day-to-day requirements. At present, these EWOPs are even not being identified and monitored by the regulatory authority, because of limitations of available technologies, standards, and higher costs. Further, variation in various environmental factors, physiochemical properties of emerging contaminants, and removal efficiency of different treatment processes is of great concern in the selection and development of optimal and affordable treatment technologies. The bioremediation process holds great promise for the treatment of EOWPs, as they produce almost no waste and may be cost-effective in combination with other physiochemical and advanced oxidation/anaerobic processes. Bioremediation-based processes hold the seeds for providing new treatment technologies, which will not only help in the protection of our water resources but also sustainable development.

References

Adewuyi, A., 2020. Chemically modified biosorbents and their role in the removal of emerging pharmaceutical waste in the water system. Water 12, 1551.

Ahmed, M.B., Zhou, J.L., Ngo, H.H., Guo, W., Thomaidis, N.S., Xu, J., 2017. Progress in the biological and chemical treatment technologies for emerging contaminant removal from wastewater: a critical review. Journal of Hazardous Materials 323, 274–298.

Ali, M.E., El-Aty, A.M.A., Badawy, M.I., Ali, R.K., 2018. Removal of pharmaceutical pollutants from synthetic wastewater using chemically modified biomass of green alga Scenedesmusobliquus. Ecotoxicology and Environmental Safety 151, 144–152.

Asad, S., Amoozegar, M., Pourbabaee, A.A., Sarbolouki, M., Dastgheib, S., 2007. Decolorization of textile azo dyes by newly isolated halophilic and halotolerant bacteria. Bioresource Technology 98 (11), 2082–2088.

Bai, X., Acharya, K., 2019. Removal of seven endocrine disrupting chemicals (EDCs) from municipal wastewater effluents by a freshwater green alga. Environmental Pollution (Barking, Essex 1987) 247, 534–540.

Basheer, A., 2018. New generation nano-adsorbents for the removal of emerging contaminants in water. Journal of Molecular Liquids 261, 583–593.

Boxall, A.B.A., 2004. The environmental side effects of medication-how are human and veterinary medicines in soils and water bodies affecting human and environmental health? EMBO Reports 5, 1110–1116.

Brozinski, J.M., Lahti, M., Meierjohann, A., et al., 2013. The anti-inflammatory drugs diclofenac, naproxen, and ibuprofen are found in the bile of wild fish caught downstream of a wastewater treatment plant. Environmental Science & Technology 47, 342–348.

Brüschweiler, B.J., Merlot, C., 2017. Azo dyes in clothing textiles can be cleaved into a series of mutagenic aromatic amines which are not regulated yet. Regulatory Toxicology and Pharmacology 88 (Supplement C), 214–226.

Chang, J.S., Lin, C.Y., 2001. Decolorization kinetics of recombinant *Escherichia coli* strain harboring azo dye decolorization determinants for Rhodococcus sp. Biotechnology Letters 23, 63–66.

Cha, J., Cupples, A.M., 2010. Triclocarban and triclosan biodegradation at field concentrations and the resulting leaching potentials in three agricultural soils. Chemosphere 81, 494–499.

Coogan, M.A., Edziyie, R.E., La Point, T.W., Venables, B.J., 2007. Algal bioaccumulation of triclocarban, triclosan, and methyl-triclosan in a North Texas wastewater treatment plant receiving stream. Chemosphere 67 (10), 1911–1918.

Cotton, J., Leroux, F., Broudin, S., Poirel, M., Corman, B., Junot, C., et al., 2016. Development and validation of a multiresidue method for the analysis of more than 500 pesticides and drugs in water based on on-line and liquid chromatography coupled to high resolution mass spectrometry. Water Research 104, 2027.

Craggs, R., Sutherland, D., Campbell, H., 2012. Hectare-scale demonstration of high rate algal ponds for enhanced wastewater treatment and biofuel production. Journal of Applied Phycology 24 (3), 329–337.

Debska, J., Kot-Wasik, A., Namiesnik, J., 2004. Fate and analysis of pharmaceutical residues in the aquatic environment. Critical Reviews in Analytical Chemistry 34 (1), 51–67.

Dixit, D., Parmar, N., 2013. Treatment of pharmaceutical wastewater by electro-coagulation and natural coagulation process: review. International Journal of Scientific & Technology Research 4 (5), 79–88.

Dixit, S., Singh, D.P., 2014. An evaluation of phytoremediation potential of cyanobacterium *Nostoc muscorum*: characterization of heavy metal removal efficiency. Journal of Applied Phycology 26 (3), 1331e1342.

Donmez, G.Ç., Aksu, Z., Oztürk, A., Kutsal, T., 1999. A comparative study on heavy metal biosorption characteristics of some algae. Process Biochemistry 34 (9), 885–892.

Du, B., Haddad, S.P., Scott, W.C., Chambliss, C.K., Brooks, B.W., 2015. Pharmaceutical bioaccumulation by periphyton and snails in an effluent-dependent stream during an extreme drought. Chemosphere 119, 927–934.

Fast, S.A., Gude, V.G., Truax, D.D., Martin, J., Magbanua, B.S., 2017. A critical evaluation of advanced oxidation processes for emerging contaminants removal. Environmental Processes 4, 283–302.

Field, J.A., Johnson, C.A., Rose, J.B., 2006. What is "emerging"? Environmental Science & Technology 40, 7105.

Fu, J., Mai, B., Sheng, G., Zhang, G., Wang, X., Peng, P., et al., 2003. Persistent organic pollutants in environment of the Pearl River Delta, China: an overview. Chemosphere 52, 1411–1422.

Gavrilescu, M., Demnerová, K., Aamand, J., Agathos, S., Fava, F., 2015. Emerging pollutants in the environment: present and future challenges in biomonitoring, ecological risks, and bioremediation. New Biotechnology 32, 147–156.

Gojkovic, Z., Lindberg, R.H., Tysklind, M., Funk, C., 2019. Northern green algae have the capacity to remove active pharmaceutical ingredients. Ecotoxicology and Environmental Safety 170, 644–656.

Green, R.E., Newton, I., Shultz, S., Cunningham, A.A., Gilbert, M., et al., 2004. Diclofenac poisoning as a cause of vulture population declines across the Indian subcontinent. Journal of Applied Ecology 41, 793–800.

Guerra, P., Kim, M., Shah, A., Alaee, M., Smyth, S.A., 2014. Occurrence and fate of antibiotic, analgesic/anti-inflammatory, and antifungal compounds in five wastewater treatment processes. The Science of the Total Environment 473, 235e243.

Guo, Y., Shen, O., Han, J., Duan, H., Yang, S., Zhu, Z., et al., 2017. Circadian rhythm genes mediate fenvalerate-induced inhibition of testosterone synthesis in mouse Leydig cells. Journal of Toxicology and Environmental Health, Part A 80 (23–24), 1314–1320.

Hampel, M., Alonso, E., Aparicio, I., Bron, J.E., Santos, J.L., Taggart, J.B., et al., 2010. Potential physiological effects of pharmaceutical compounds in Atlantic salmon (Salmosalar) implied by transcriptomic analysis. Environmental Science and Pollution Research International 17, 917–933.

Hiscock, J.R., Sambrook, M.R., Wells, N.J., Gale, P.A., 2015. Detection and remediation of organophosphorus compounds by oximate containing organogels. Chemical Science 6, 5680–5684. Available from: https://doi.org/10.1039/C5SC01864.

Homem, V., Santos, L., 2011. Degradation and removal methods of antibiotics from aqueous matrices–a review. Journal of Environmental Management 92 (10), 2304–2347.

Hou, Y., Zhang, R., Yu, Z., Huang, L., Liu, Y., Zhou, Z., 2017. Accelerated azo dye degradation and concurrent hydrogen production in the single-chamber photocatalytic microbial electrolysis cell. Bioresource Technology 224 (Supplement C), 63–68.

Hussain, I., Khan, M.Z., Khan, A., Javed, I., Saleemi, M.K., 2008. Toxicological effects of diclofenac in four avian species. Avian Pathology Journal of the W.V.P.A 37, 315–321.

Hussain, M.I., Kumar, B., Ahmad, M., 2015. Acute toxicity, behavioral response and biochemical composition of blood of common carp, Catlacatla (Hamilton) to an organophosphate insecticide, dimethoate. International Journal of Current Microbiology and Applied Sciences 4, 1189–1199.

Hussain, M.I., Kumar, B., Ahmad, M., 2016. Effect of organophosphate insecticide, Dimethoate on physiology of common carp, Catlacatla (Hamilton) and Labeorohita. International Journal of Current Microbiology and Applied Sciences 5, 322–341. Available from: https://doi.org/10.20546/ijcmas.2016.505.034.

Jena, S., Acharya, S., Mohapatra, P.K., 2012. Variation in effects of four OP insecticides on photosynthetic pigment fluorescence of *Chlorella vulgaris* Beij. Ecotoxicology and Environmental Safety 80, 111–117. Available from: https://doi.org/10.1016/j.ecoenv.2012.02.016.

Jiang, J.Q., Zhou, Z., 2013. Removal of pharmaceutical residues by ferrate (VI). PLoS One 8, 55729.

Khan, M.I., Lee, J., Yoo, K., Kim, S., Park, J., 2015. Improved TNT detoxification by starch addition in a nitrogen-fixing Methylophilus-dominant aerobic microbial consortium. Journal of Hazardous Materials 300, 873–881.

Kim, Y.M., Murugesan, K., Schmidt, S., Bokare, V., Jeon, J.R., Kim, E.J., et al., 2011. Susceptibility and co-metabolism- a comparison for three aerobic pollutant-degrading bacteria. Bioresource Technology 102, 2206–2212.

Kisku, G.C., Markandeya, Shukla, S.P., Singh, D.S., Murthy, R.C., 2015. Characterization and adsorptive capacity of coal fly ash from aqueous solutions of disperse blue and disperse orange dyes. Environmental Earth Sciences 74 (2), 1125–1135. Available from: https://doi.org/10.1007/s12665-015-4098-z.

Krenchinski, F.H., Albrecht, L.P., Albrecht, A.J.P., Cesco, V.J.S., Rodrigues, D.M., Portz, R.L., et al., 2017. Glyphosate affects chlorophyll, photosynthesis, and water use of four Intacta RR2 soybean cultivars. Acta Physiologiae Plantarum 39, 63. Available from: https://doi.org/10.1007/s11738-017-2358-0.

Kumar, V., Upadhay, N., Wasit, A., Singh, S., Kaur, P., 2013. Spectroscopic methods for the detection of organophosphate pesticides—a preview. Current World Environment Journal 8, 313–318. Available from: https://doi.org/10.12944/CWE.8.2.19.

Kümmerer, K., 2009. Antibiotics in the aquatic environment – a review – part I. Chemosphere 75, 417–434.

Lawrence, J.R., Swerhone, G.D.W., Topp, E., Korber, D.R., Neu, T.R., Wassenaar, L.I., 2007. Structural and functional responses of river biofilm communities to the nonsteroidal anti-inflammatory diclofenac. Environmental Toxicology and Chemistry / SETAC 26, 573–582.

Liu, F., Ying, G.G., Tao, R., Zhao, J.L., Yang, J.F., Zhao, L.F., 2009. Effects of six selected antibiotics on plant growth and soil microbial and enzymatic activities. Environmental Pollution (Barking, Essex 1987) 157, 1636–1642.

Logan, B.E., 2009. Exoelectrogenic bacteria that power microbial fuel cells. Nature Reviews. Microbiology 7 (5), 375–381.

Lokhande, R.S., Singare, P.U., Pimple, D.S., 2011. Toxicity study of heavy metals pollutants in wastewater effluents samples collected from Taloja Industrial Estate of Mumbai, India. Resource Environmental 1 (1), 13–19.

Lorenzo, M., Pico, Y., 2017. Gas chromatography and mass spectroscopy. Techniques for the detection of chemical. Contaminants and residues in foods. In Chemical Contaminants and Residues in Food, second ed., pp. 15–50.

Lyssimachou, A., Thibaut, R., Gisbert, E., Porte, C., 2014. Gemfibrozil modulates cytochrome P450 and peroxisome proliferation-inducible enzymes in the liver of the yellow European eel (*Anguilla anguilla*). Environmental Science and Pollution Research International 21, 862–871.

Malkoske, T., Tang, Y., Xu, W., Yu, S., Wang, H., 2016. A review of the environmental distribution, fate, and control of tetrabromobisphenol A released from sources. The Science of the Total Environment 569–570, 1608–1617. Available from: https://doi.org/10.1016/j.scitotenv.2016.06.062.

Meade, M.J., Waddell, R.L., Callahan, T.M., 2001. Soil bacteria Pseudomonas putrid and *Alcaligenes xylosoxidans* subsp. dentrificans inactivate triclosan in liquid and solid substrates. FEMS Microbiology Letters 204, 45–48.

Mennigen, J.A., Martyniuk, C.J., Crump, K., Xiong, H., Zhao, E., Popescu, et al., 2008. Effects of fluoxetine on the reproductive axis of female goldfish (*Carassius auratus*). Physiological Genomics 35, 273–282.

Mishra, I.P., Sabat, G., Mohanty, B.K., 2015. Phytotoxicity of Profenofos 50% EC (curacron 50 EC) to Vignaradiata, L. seedlings: II. Studies on biochemical parameters. International Journal of Applied Sciences and Biotechnology 3 (1), 101–105. Available from: https://doi.org/10.3126/ijasbt.v3i1.12063.

Muller, A., Osterlund, H., Marsalek, J., Viklander, M., 2020. The pollution conveyed by urban runoff: a review of sources. The Science of the Total Environment 709, 136125. Available from: https://doi.org/10.1016/j.scitotenv.2019.136125.

Ooi, J., Lee, L.Y., Hiew, B.Y.Z., Thangalazhy-Gopakumar, S., Lim, S.S., Gan, S., 2017. Assessment of fish scales waste as a low cost and eco-friendly adsorbent for removal of an azo dye: equilibrium, kinetic and thermodynamic studies. Bioresource Technology 245 (Part A), 656–664.

Panduranga, M.G., Mahadeva, P.G., Sudarshana, M.S., 2005. Toxicity of different imbibitions periods of dimethoate on germination, chlorophyll a/b, and dry matter of *Glycine max* (L) Merrill. Cv. 548 KHSB-2, during early seedlings growth. Journal of Physiological Research 18, 199–201. Available from: https://doi.org/10.1007/s10646-015-1591-9.

Pan, L.-j, Li, J., Li, C.-x, Yu, G.-w, Wang, Y., 2018. Study of ciprofloxacin biodegradation by a Thermus sp. isolated from pharmaceutical sludge. Journal of Hazardous Materials 343, 59–67.

Patel, M., Kumar, R., Kishor, K., Mlsna, T., Pittman, C.U., Mohan, D., 2019. Pharmaceuticals of emerging concern in aquatic systems: chemistry, occurrence, effects, and removal methods. Chemical Reviews 119, 3510–3673. Available from: https://doi.org/10.1021/acs.chemrev.8b00299.

Prosser, R.S., Lissemore, L., Topp, E., Sibley, P.K., 2014. Bioaccumulation of triclosan and triclocarban in plants grown in soils amended with municipal dewatered biosolids. Environmental Toxicology and Chemistry / SETAC 33, 975–984.

Raffi, F., Hall, J.D., Cerniglia, C.E., 1997. Mutagenicity of azo dyes used in foods, drugs, and cosmetics before and after reduction by Clostridium species from the human intestinal tract. Food and Chemical Toxicology an International Journal Published for the British Industrial Biological Research Association 35, 897–901.

Rodriguez-Narvaez, O.M., Peralta-Hernandez, J.M., Goonetilleke, A., Bandala, E.R., 2017. Treatment technologies for emerging contaminants in water: a review. Chemical Engineering Journal 323, 361–380.

Rousis, N.I., Bade, R., Bijlsma, L., Zuccato, E., Sancho, J.V., Hernandez, F., et al., 2017. Monitoring a large number of pesticides and transformation products in water samples from Spain and Italy. Environmental Research 156, 3138.

Russell, A.D., 2004. Whither triclosan? The Journal of Antimicrobial Chemotherapy 53, 693–695. Available from: https://doi.org/10.1093/jac/dkh171.

Safonova, E.T., Dmitrieva, I.A., Kvitko, K.V., 1999. The interaction of algae with alcanotrophic bacteria in black oil decomposition. Resources, Conservation & Recycling 27 (12), 193–201.

Salter-Blanc, A.J., Bylaska, E.J., Lyon, M.A., Ness, S.C., Tratnyek, P.G., 2016. Structure-activity relationships for rates of aromatic amine oxidation by manganese dioxide. Environmental Science & Technology 50 (10), 5094–5102.

Schwarzenbach, R.P., Escher, B.I., Fenner, K., Hofstetter, T.B., Johnson, C.A., Von Gunten, U., et al., 2006. The challenge of micropollutants in aquatic systems. Science (New York, N.Y.) 313 (5790), 1072–1077.

Serra, A.A., Nuttens, A., Larvor, V., Renault, D., Couee, I., Sulmon, C., et al., 2013. Low environmentally relevant levels of bioactive xenobiotics and associated degradation products cause cryptic perturbations of metabolism and molecular stress responses in *Arabidopsis thaliana*. Journal of Experimental Botany 64, 2753–2766. Available from: https://doi.org/10.1093/jxb/ert119.

Sidhu, G.K., Singh, S., Kumar, V., Dhanjal, D.S., Datta, S., Singh, J., 2019. Toxicity, monitoring and biodegradation of organophosphate pesticides: a review. Critical Reviews in Environmental Science and Technology 49 (13), 1135–1187.

Singh, R.N., Pandey, R.K., Singh, N.N., Das, V.K., 2009. Acute toxicity and behavioral responses of common carp Cyprinuscarpio (Linn.) to an organophosphate (Dimethoate). World Journal of Zoology 4, 70–75.

Sinha, S., Yadav, G., Kaushik, B.R., Dounde, S., Janghel, D., 2018. Histopathological Impact of Malathion on the testicular cells of freshwater crabs *Barytelhusa cunicularis* (WESTWOOD, 1836). World Journal of Pharmacy and Pharmaceutical Sciences 7, 997–1007. Available from: https://doi.org/10.20959/wjpps20183-10783.

Snow, D.D., Cassada, D.A., Larsen, M.L., Mware, N.A., Li, X., D'Alessio, M., et al., 2017. Detection, occurrence and fate of emerging contaminants in agricultural environments. Water Environment Research a Research Publication of the Water Environment Federation 89, 897–920.

Sorlini, S., Collivignarelli, M.C., Miino, M.C., 2019. Technologies for the control of emerging contaminants in drinking water treatment plants. Environmental Engineering and Management Journal 18, 2203–2213.

Subburaj, A., Jawahar, P., Jayakumar, N., Srinivasan, A., Ahilan, B., 2018. Histopathological investigations in the liver and kidney of the fish, *Oreochromis mossambicus* (Tilapia) exposed to acute Malathion (EC 50%) toxicity. Journal of Experimental Zoology India 21 (1), 77–81.

Sutherland, D.L., Heubeck, S., Park, J., Turnbull, M.H., Craggs, R.J., 2018. Seasonal performance of a full-scale wastewater treatment enhanced pond system. Water Research 136, 150–159.

Sutherland, D.L., Howard-Williams, C., Turnbull, M.H., Broady, P.A., Craggs, R.J., 2015. Enhancing microalgal photosynthesis and productivity in wastewater treatment high rate algal ponds for biofuel production. Bioresource Technology 184, 222–229.

Tien, C.J., Lin, M.C., Chiu, W.H., Chen, C.S., 2013. Biodegradation of carbamate pesticides by natural river biofilms in different seasons and their effects on biofilm community structure. Environmental Pollution (Barking, Essex 1987) 179, 95–104.

Tran, N.H., Gin, K.Y.H., 2017. Occurrence and removal of pharmaceuticals, hormones, personal care products, and endocrine disrupters in a full-scale water reclamation plant. The Science of the Total Environment 599, 1503–1516.

Tran, N.H., Reinhard, M., Gin, K.Y.H., 2018. Occurrence and fate of emerging contaminants in municipal wastewater treatment plants from different geographical regions-a review. Water Research 133, 182–207.

UN WWAP, 2003. Water for people, water for life (2003): 3rd World Water Forum in Kyoto, Japan. http://www.norman-network.net.

Vagi, M.C., Petsas, A.S., Pavlaki, M.D., Smaragdaki, N.M., Kostopoulou, M.N., 2018. Toxic effects of the Organophosphorus insecticide Fenthion on growth and chlorophyll production activity of unicellular marine microalgae *Tetraselmis suecica*: comparison between observed and predicted endpoint toxicity data. doi:10.5772/ intechopen.72321.

Velazquez-Fernandez, J.B., Martínez-Rizo, A.B., Ramírez-Sandoval, M., DomínguezOjeda, D., 2012. Biodegradation and bioremediation of organic pesticides. Pesticides: Recent Trends in Pesticide Residue Assay 253–272.

Venkatesan, A.K., Pycke, B.F., Barber, L.B., Lee, K.E., Halden, R.U., 2012. Occurrence of triclosan, triclocarban, and its lesser chlorinated congeners in Minnesota freshwater sediments collected near wastewater treatment plants. Journal of Hazardous Materials 229–230, 29–35.

Verma, J.P., Jaiswal, D.K., Sagar, R., 2014. Pesticide relevance and their microbial degradation: a-state-of-art. Reviews in Environmental Science and Bio/Technology 13 (4), 429–466.

Vernouillet, G., Eullaffroy, P., Lajeunesse, A., Blaise, C., Gagne, F., Juneau, P., 2010. Toxic effects and bioaccumulation of carbamazepine evaluated by biomarkers measured in organisms of different trophic levels. Chemosphere 80 (9), 1062–1068.

Vivancos, P.D., Driscoll, S.P., Bulman, C.A., Ying, L., Emami, K., Treumann, A., et al., 2011. Perturbations of amino acid metabolism associated with glyphosate dependent inhibition of shikimic acid metabolism affect cellular redox homeostasis and alter the abundance of proteins involved in photosynthesis and photorespiration. Plant Physiology 157 (1), 256–268. Available from: https://doi.org/10.1104/pp.111.18102.

Wang, Y., Ho, S.H., Cheng, C.L., Guo, W.Q., Nagarajan, D., Ren, N.Q., et al., 2016. Perspectives on the feasibility of using microalgae for industrial wastewater treatment. Bioresource Technology 222, 485e497.

Wang, R., Tang, J., Xie, Z., Mi, W., Chen, Y., Wolschke, H., et al., 2015. Occurrence and spatial distribution of organophosphate ester flame retardants and plasticizers in 40 rivers draining into the Bohai Sea, North China. Environmental Pollution 198, 172–178. Available from: https://doi.org/10.1016/j.envpol.2014.12.037.

Watanabe, N., Bergamaschi, B.A., Loftin, K.A., Meyer, M.T., Harter, T., 2010. Use and environmental occurrence of antibiotics in free stall dairy farms with manure forage fields. Environmental Science & Technology 44, 6591–6600.

WHO, 2011.Pharmaceuticals in Drinking Water. World. Health Organization Report HO/HSE/WSH/11.05.

Wolfaardt, G.M., Lawrence, J.R., Robarts, R.D., Caldwell, D.E., 1994. The role of interactions, sessile growth, and nutrient amendments on the degradative efficiency of a microbial consortium. Canadian Journal of Microbiology (Seoul, Korea) 40 (5), 331–340.

Wong, F., Robson, M., Melymul, L., Shunthirasingham, C., et al., 2019. Urban sources of synthetic.

Xiong, J.Q., Kurade, M.B., Kim, J.R., Roh, H.S., Jeon, B.H., 2017. Ciprofloxacin toxicity and its co-metabolic removal by a freshwater microalga *Chlamydomonas mexicana*. Journal of Hazardous Materials 323, 212e219.

Yuan, X., Lacorte, S., Cristale, J., Dantas, R.F., Sans, C., Esplugas, S., et al., 2015. Removal of organophosphate esters from municipal secondary effluent by ozone and UV/H_2O_2 treatments. Separation and Purification Technology 156, 1028–1034. Available from: https://doi.org/10.1016/j.seppur.2015.09.052.

Zhang, W., Deng, Y., Chen, L., Zhang, L., Wang, Z., Liu, R., et al., 2020. Effect of triadimefon and its metabolite on adult amphibians *Xenopus laevis*. Chemosphere 243, 125288.

Zobiole, L.H.S., de Oliveira, R.S., Huber, D.M., Constantin, J., de Castro, C., de Oliveira, F.A., et al., 2010. Glyphosate reduces shoot concentrations of mineral nutrients in glyphosate-resistant soybeans. Plant and Soil 328 (1–2), 57–69. Available from: https://doi.org/10.1007/s11104-009-0081-3.

Zobiole, L.H.S., Kremer, R.J., de Oliveira Jr., R.S., Constantin, J., 2012. Glyphosate effects on photosynthesis, nutrient accumulation, and nodulation in glyphosate-resistant soybean. Journal of Plant Nutrition and Soil Science 175, 319–330. Available from: https://doi.org/10.1002/jpln.201000434.

Chapter 22

Sustainable approaches for heavy metal removal from water

Swayam Siddha[1] and Manoj Kumar[2]

[1]*Department of Environmental Studies, The MS University of Baroda, Vadodara, Gujrat, India,* [2]*Department of Hydro and Renewable Energy, Indian Institute of Technology, Roorkee, Uttrakhand, India*

Introduction

Water covers two-thirds of the earth's surface, which is essential for all life forms. In an era of rapidly growing populations and rapid urbanization, industrialization, and commercial development, a sufficient quantity and quality of water will be needed for human consumption, sanitation, agriculture, irrigation, and manufacturing (Rai, 2012). To meet the water needs of cities, farms, and industries, societies draw vast amounts of water from rivers, lakes, wetlands, and underground aquifers. Since water is currently one of the most critical resources of the 21st century, water managers, policymakers, and researchers anticipate a next world war over water (Rai, 2010). Remediation of contaminated water might be an appropriate resolution that retrieves purified water from industrial wastewater, sewage water, polluted surfaces, and groundwater (Hossain et al., 2020). Wastewater from industrial processes often contains heavy metals, dyes, pesticides, herbicides, pharmaceuticals, and other aromatic composites as contaminants (Price and Heberling, 2018). The risk of consumption of polluted water is increasing day by day in numerous developing countries. Accordingly, the maximum allowable oxygen content in treated wastewater was 40 mg/L for BOD and 120 mg/L for COD (Saravanan et al., 2021). Pb, Hg, Ni, Cr, As, Cu, and Zn are generally considered heavy metals due to their high density and lower atomic number (>20) (Shikha and Singh, 2021). Animal growth and public health are both threatened by heavy metal pollution, as well as aquatic life. A combination of these factors leads to restricted development in addition to anomalous physiological as well as biochemical fluctuations, over-expression of genetic factors behavioral changes, additionally toxicologic paraphernalia, together with scarce metabolism (Sun et al., 2022). Infiltration of urban runoff into still waters can be caused by pesticides, fertilizers, landfill leaks, spills, and industrial runoffs (Raza et al., 2018). Heavy metals can be extracted or remedied from water using several physicochemical methods. Nevertheless, the treatment approach chosen is determined by the quantity of pollution, the polluted locations, the varieties of contamination, as well as other considerations. Numerous ecological advantages are also related to the biological treatment process; for example, plants can provide contamination-free land without disrupting the ecology and can also boost the aesthetic significance of the environment through vegetation cover (Cristaldi et al., 2017). As a result, biological treatment approaches are the key tactical route to take. Currently, cost-effective biological treatment approaches such as plants (constructed wetlands), fungi, and bacteria for hazardous heavy metals removal through soil and water are becoming preferable (Yang et al., 2020).

Sources of heavy metal in aquatic ecosystems

The release of industrial wastes into the environment has an impact on the characteristics of soil, air, and water (Panahi et al., 2018). Due to anthropogenic activities and industrialization, water quality deteriorates, resulting in water pollution (Zeng et al., 2020). The majority of waste disposed of in lakes, rivers, and streams comes from domestic, municipal, and industrial sources. As a result of hazardous chemicals released into water directly from textile, petrochemical, pharmaceutical, refineries and mines, humans, plants, animals and aquatic organisms end up suffering serious health problems (Deletic and Wang, 2019). Microbes pass on most of what they need to survive to humans, but some can be

Role of Green Chemistry in Ecosystem Restoration to Achieve Environmental Sustainability. DOI: https://doi.org/10.1016/B978-0-443-15291-7.00023-7

lethal (Murtaza et al., 2022). Bioaccumulated metals may be found in meat and fish products that are coming from contaminated soil or from contaminated soil that has been sprayed with metals (Madgett et al., 2021). In small concentrations, many heavy metals are toxic to humans. Anthropogenic or natural methods are usually used to enter the environment and deposit heavy metals on water resources. Contaminated water passes through the soil when heavy metals are attracted to humus, an organic matter found in soil (Joseph et al., 2019). Bioaccumulation occurs when heavy metals or pesticides accumulate in living organisms and when aquatic organisms absorb contaminants from the surrounding water faster than their bodies can excrete them (Cao et al., 2020). It is also possible for humans to bioaccumulate contaminants from their food, air, and water, including those found in aquatic organisms and in our drinking water. It can take a long time for heavy metals to degrade in our bodies because they don't biodegrade (Liu et al., 2019). Microorganisms often bind heavy metals to their surfaces and sometimes enter their cells (Hu et al., 2021). It is microorganisms that feed insects and zooplankton, zooplankton that feed fish, and humans that eat tuna (Jiang et al., 2022). By the time heavy metals reach us, we consume them at high concentrations because they bioaccumulate in the bodies of all living organisms (Tessema et al., 2020).

Toxic effects of heavy metals

Physical and chemical aspects of contaminated water pose serious risks to aquatic life and people's health, as they contain a vast number of pathogenic microbes and xenobiotic substances (Jeevanantham et al., 2019). Both consumers and farmers suffer health effects from nutrient-rich wastewater used for food production (Alengebawy et al., 2021). Surface water and freshwater resources become less accessible as a result of the deteriorating quality of water due to numerous complex interactions (Hoang et al., 2021). When contaminated water is consumed by humans, it can cause respiratory, renal, cardiovascular and neurological problems in the lungs, kidneys, liver, and bladder (Proshad et al., 2021). Water pollution can be caused by chemicals, nutrients, microbes, and oil. Growing demand for chemicals drives the development of chemical industries around the world. When it comes to environmental rules, the chemical sector has a substantial impact on the environment and human health (Melnyk et al., 2014). In drinking water, heavy metal particles are absorbed into the sediments, which accumulate over time, increasing the risk of health complications (Yin et al., 2018). It is possible to induce oxidative stress in human beings, as well as affect cellular organelles by ingesting metal ions. Consuming arsenic, for instance, may facilitate abnormal gene expression and DNA hypomethylation. Wilson's disease may be caused by Cu, and Cr damages DNA through oxidative stress (Jacob et al., 2018). Major heavy metal pollutants and their toxic effects on the environment are listed in Table 22.1.

Water treatment technologies

Various water treatment technologies have been developed over the years that purify water to varying degrees, particularly for drinking and domestic purposes (Muga and Mihelcic, 2008; Adeola and Forbes, 2021; Reddy and Lee, 2012). Furthermore, great effort has been placed into the invention of more reliable, lower-cost, robust, energy-efficient, scalable water treatment systems that do not further damage the environment or risk the population's well-being (Sun et al., 2021).

Conventional water treatment technology

As part of water treatment, contaminants are purified and removed, but deficient ingredients are added to enhance the water's natural properties (Nikoladze et al., 1989; Rajasulochana and Preethy, 2016).

Coagulation-flocculation

Coagulation-flocculation can be used in conjunction with other traditional methods to remove metal effluents from water waste. Brewer's yeast was utilized as a flocculant (de Santana et al., 2018) to remove Cd^{2+}, Cr^{3}, Cu^{2+}, Zn^{2+}, and Ni^{2+} from artificial effluents. Further, the effectiveness of yeast as a flocculation agent was investigated, particularly for the removal of Cu^{2+} from water (Soares et al., 2002).

Flotation

Xanthopoulos et al. (2021) used the ion flotation method to extract Cu from ammoniacal leachates of samples of microwave-roasted sulfidic tailings, under ideal circumstances, this technique allowed for the best leaching efficiencies

TABLE 22.1 Major heavy metals and their toxic effects on the environment (Giles and Mann, 2022; Ali et al., 2022; Natasha et al., 2021; Schaefer et al., 2022; Rasafi et al., 2022; Bjerregaard et al., 2022).

Heavy metal	Toxic effects on	
	Human beings	Flora and fauna
As	Affects the skin, liver, lungs, and kidneys, causing skin cancer, and cancers of the bladder and lungs. Mental retardation and developmental disorders can result from prolonged exposure to arsenic.	• May cause developmental toxicity in four species (hamsters, mice, rats, and rabbits). • Death, inhibition of growth, photosynthesis, and reproduction, and behavioral effects.
Pb	Anemia, weakness, high blood pressure, and kidney and brain damage. Lead exposure at very high levels can be fatal. The growing neurological system of a newborn can be harmed by lead.	• Injured animals display symptoms of injury to the central nervous system. • Chlorosis, stunted growth, and root system blackening.
Cd	Kidney disease, osteoporosis, cardiovascular diseases, and cancer.	• Affects the kidney, liver, lung, bone, blood, nervous system, and the immunological system. • Overproduction of reactive oxygen species (ROS) damages plant membranes, destroys cell biomolecules and organelles, and causes chlorosis in the leaves.
Cr	Chromium (VI) is a human carcinogen, resulting in an increased risk of nasal, and sinus cancer.	• Lymphocytosis, anemia, eosinophilia, bronchial and renal lesions (fishes). • Decreases plant growth by inducing ultrastructural modifications of the cell membrane and chloroplast.
Cu	Headache, nausea, dizziness, and diarrhea, irritate your nose, mouth, and eyes.	• Depression, weakness, weakness, recumbency, anorexia, thirst, dyspnea, hemoglobinuria, and jaundice. • Stunting, leaf discoloration, chlorosis, and necrosis.
Manganese	Manganism is a chronic neurological condition characterized by tremors, difficulties walking, and facial muscle spasms.	• Cause developmental problems. • Necrotic patches, chlorosis (pale or yellow color).
Hg	Deadly effects on the nervous, digestive, and immune systems, kidneys, and lungs.	• Anorexia, stomatitis, pharyngitis, vomiting, diarrhea, pain; as well as shock, dyspnea, and dehydration.

of Zn 86% and Cu 75%. Ion flotation's removal of Cu (II) by varying the nano collector's degree of oxidation (using graphene) was also investigated. They found in this work that raising graphene's degree of oxidation might improve Cu (II) removal (Chang et al., 2021).

Precipitation

Precipitation has been the most popular metal technique in industrial applications (Brown and Allen, 2000). Activated carbons made from a variety of materials are used as adsorbents to remove contaminants, particularly heavy metals, from wastewater (Salman, 2014), astragalus (Ince and Kaplan, 2017), carbon nanotubes (Sheibani et al., 2015), and large number of biosorbents (Gutha et al., 2015) have been included into several research in the literature (Fahad et al., 2019; Rajasulochana and Preethy, 2016). Several major cities produce more than 1 million cubic meters per day of treated water such as Moscow, Leningrad, Kyiv, Chicago, Tokyo, London, and Paris (EC, 2017).

Nonconventional water treatment technology

Biological treatment techniques for municipal wastewater that do not follow conventional techniques are low-cost; they are easier to maintain and operate than conventional techniques (Ngo et al., 2002). Contrary to conventional biological methods, these methods concentrate on land. Pathogens are usually more effectively eliminated by these methods when they are correctly planned (Huang and Lee, 2015).

Nanotechnology-based green synthesis approach for water purification

Every year, acute diarrhea brought on by contaminated water kills over 829,000 individuals globally. The elimination of pollutants including viruses, poisonous heavy metals, pesticides, and other persistent and hazardous compounds presents several technical difficulties that can be overcome with the help of nanotechnology. Due to their better processing efficiency and unique properties, such as a high reaction rate and surface-to-mass ratio, nanomaterials offer effective, affordable, and environmentally acceptable alternatives to current treatment materials (Patanjali et al., 2019). Organic and inorganic processes can be used to create nanoparticles (NPs) (Barabadi et al., 2019a). In general, parameters including biological source type, pH, temperature, reaction media, and surface charge affect the properties of the produced nanoparticles during the biosynthetic creation of nanoparticles (Barabadi et al., 2019b).

Sustainable groundwater remediation

Groundwater resources have been experiencing severe stress (Gill et al., 2017). The difficulty of delivering dependable water supplies is expected to worsen, which may result in increased pressure on groundwater resources, resulting in a drop in the groundwater table (Syafiuddin et al., 2020). There is now much worry about the existence of toxins in groundwater since groundwater is sensitive to both organic and inorganic contaminants (Ma and Wu, 2008; Thomann et al., 2020). Groundwater is often treated by digging water wells and transferring dirty water to ground systems using various therapeutic approaches like air stripping, treatment towers, and granular activated carbon (GAC) (Al-Hashimi et al., 2021). The most successful remediation process is determined by the properties of contaminants and pollutants, as well as the reactive medium available (Vallero, 2019). In situ methods especially bioremediation are among the most widely used subsurface treatment procedures in the world because they use microorganisms to break down toxins into less dangerous compounds (Albers et al., 2015). The biostimulation approach creates substrates capable of providing favorable circumstances for microbial activity. In bioaugmentation, microorganisms are put into groundwater contaminations (Němeček et al., 2018).

Green and sustainable approaches for heavy metal treatment

Green chemistry, in contrast to traditional chemistry, is the creation of chemical products and processes that minimize or eliminate the usage and manufacture of harmful compounds.

Constructed wetland

In constructed wetlands (CWs), wetland vegetation is used to treat wastewater more meticulously than in natural wetlands because they're planned and constructed to accomplish this. (Kumar and Singh, 2017). As well as eliminating suspended solids, organic matter, nutrients, heavy metals, pathogens, and toxic pollutants, CWs are suitable for several other purposes (Faulwetter et al., 2009). CWs are relatively close approaches with completely engineered alternatives for which numerous recent studies have been continuously developed throughout the last generations (Vymazal, 2014). CWs are procedures that rely mostly on naturally occurring coupled and generally dependant biochemical advancement that operates by microbiological and photovoltaic advancements and does so in a structured environment (Kumar and Singh, 2017). CWs are therapeutic operations that mimic the processes that occur in aquatic habitats, which comprise substrate, macrophytes, and microbial assemblages (Fig. 22.1). CW has been shown to have various advantages over traditional approaches (Foladori et al., 2015; Chen et al., 2008).

Algae

Considering the present emphasis on zero carbon emissions and the circular economy, algal-based wastewater treatment technologies have an intriguing possibility since algae is a possible source of biosorbent (Rangabhashiyam and Balasubramanian, 2019). Specific algal biomass has better biosorption capabilities for specific ions than available commercial adsorbents such as activated carbon, synthetic ion exchange resins, as well as natural zeolite (Doshi et al., 2006). Seaweed output globally exceeded 3 million tonnes per year (Chen, 2012). The benefits of using algal biomass as removal of metal ions involve greater productivity, simple operation, minimal fertilizer needs, low cost, efficient biomass extraction, biomass affordability in both fresh and saltwater, and a relatively large specific surface area and greater binding affinity toward the heavy metals (Singh et al., 2022). Furthermore, algal cells can be immobilized, algal consortia developed, and nanomaterials designed based on microalgae can enhance the efficiency of heavy metal removal. A more

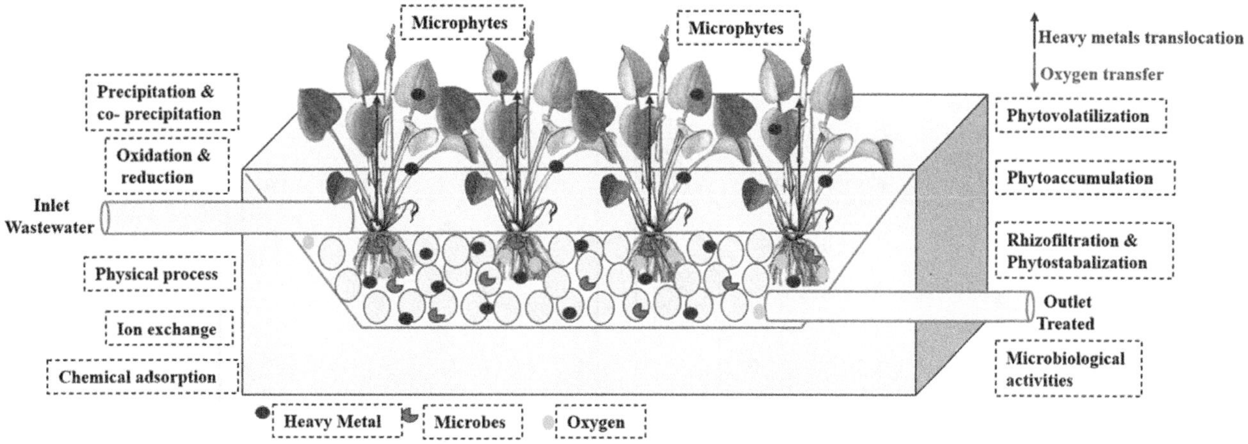

FIGURE 22.1 Role of constructed wetlands for heavy metal removal from water and wastewater.

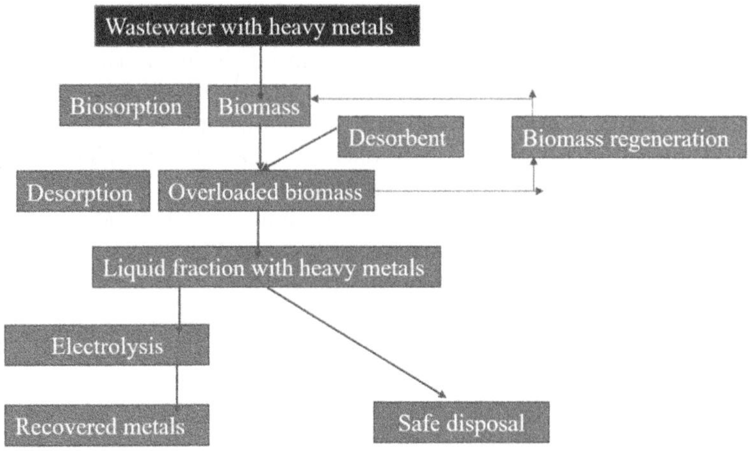

FIGURE 22.2 Role of microbes for heavy metal removal from water and wastewater.

eco-friendly, low-cost, and sustainable method might be achieved by using microalgae-based ligands to design nanocomposite materials. In metabolic engineering, genes are altered to perform a certain function (Mehariya et al., 2021). Microalgae may be able to remove heavy metals from their environment (Cheng et al., 2019; Li et al., 2019).

Microbes

On an industrial level, bacteria perform significant functions in the remediation of hazardous substances (Fig. 22.2). The degradation of organic compounds in wastewater is mostly mediated by microorganisms including their enzymes (Li et al., 2020). Bioremediation is a technologically advanced technology that uses microorganisms to break down contaminants in wastewater and soil settings (Liu et al., 2016). This involves a method of converting biodegradable complicated hazardous chemicals into nontoxic as well as desirable final by-products such as CO_2 and H_2O via biological metabolisms. The nonsettling and dispersed colloids are collected and integrated as biological floc and biofilm (Rani et al., 2019). Vital nutrients, precious materials, and particular organic ingredients can all be extracted and reclaimed. It is a less expensive technique that produces no trash as a consequence. Bioremediation, in addition to becoming an economical and environmentally beneficial technology, however, offers the finest replacement for traditional therapeutic approaches (Aghalari et al., 2020).

Bioelectrochemical system (microbial fuel cell and microbial electrochemical system)

A cutting-edge and effective wastewater treatment method is the utilization of bioelectrochemical systems (BES) in the production of energy and hydrogen (Fig. 22.3) (Kumar and Singh, 2019). Another unique method for generating power from organic waste, such as wastewater and lignocellulosic biomass, is the use of biology with electrochemistry called

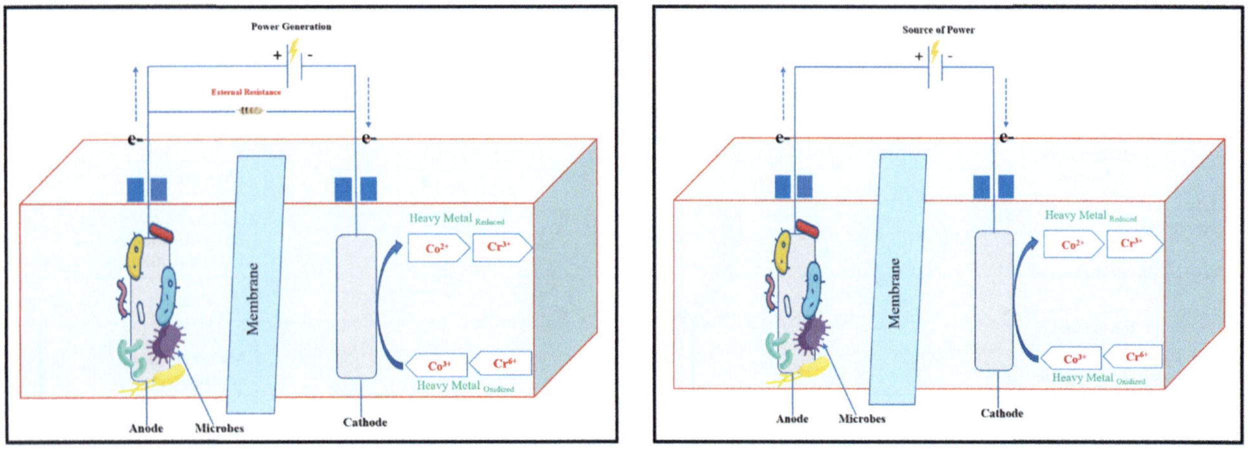

Microbial fuel cell Microbial electrolysis cell

FIGURE 22.3 Role of bioelectrochemical system for heavy metal removal from water and wastewater.

BESs (Kumar and Singh, 2020). Because they are accumulating in the soft tissues, heavy metals including Hg, Pb, Cr, and Cd are generating particular poisoning symptoms even at low concentrations of roughly 1.0–10 mg/L (Nies, 2000; Roberts, 1999; Okereafor et al., 2020; Hasan Abadi et al., 2019). Human health is at risk from hazardous metals found in anthropogenic waste streams, such as Cr (VI) and Cu (II) in electroplating wastewaters (Li et al., 2008), which cause the death of aquatic life and primary producers (Singh and Kalamdhad, 2011), can adversely affect a water body's visual impacts (Hogsden and Harding, 2012). Chemical redox reactions, coagulation, precipitation, adsorption, membrane filtration, and biological accumulation are examples of conventional treatments for metal pollutants (Fu and Wang, 2011). These processes typically result in large quantities of metal sludge or concentrated brines and demand heavy chemical inputs (Djedidi et al., 2009; Pérez-González et al., 2012). BESs have also been researched as an emerging treatment approach for eliminating dangerous metals (Wang and Ren, 2014; Heijne et al., 2010).

Challenges and future prospects

Water contamination threatens the health, environment, and climate of about 80% of the world's population. A vital part of fighting is restoring its water, nutrients, and energy. Additionally, these discharges greatly increase the amount of methane and nitrous oxide greenhouse gas emissions. Besides resulting in food purchases, these discharges also pose a health risk. A conventional medicinal plant produces three times the amount of pollution. There is a need for more action due to population growth. Government and the private sector should work together to empower cities to transform resources. In addition to membrane bioreactors, ballasted reaction processors, and integrated film processes, there are new therapies that address this problem. Bioaerated filters use sinks with immersible media as contact points for biological treatment and filters for separating solids from contaminated water. Aeration using aum-bubbles is used for washing back. A good water decontamination process requires sophisticated technologies, such as complex stabilization and process control, to ensure a reliable and stable operation. Through the use of analytical tools, automation of the treatment process can significantly increase performance and save administrative expenses, human-machine interactions, and clever programmatic controls.

Concluding remarks

Perhaps the most important environmental protection strategy that has to be promoted globally is water purification. Industrial effluent wastewater from factories, refineries, and other sources is often treated on-site. As an alternative, physical waste treatment facilities treat wastewater using synthetic reactions in addition to physical ones. Biological treatment systems are the best for handling sewage from homes and commercial buildings. Factory wastewater is often treated in physical wastewater treatment plants, industries, and manufacturing organizations. Planning, activity, design, storage, and operation are necessary for the attractive implementation of wastewater reuse options in a water resources management system. Techniques for recovering water use a variety of safeguards to reduce the risks to the environment posed by various reuse applications. Green remediation approaches, including the utilization of green materials, resource recovery strategies, nature-based solutions, and energy-efficient strategies, have received much attention with a view to the remediation of water contaminated with heavy metals.

References

Adeola, A.O., Forbes, P.B.C., 2021. Advances in water treatment technologies for removal of polycyclic aromatic hydrocarbons: existing concepts, emerging trends, and future prospects. Water Environment Research 93 (3), 343–359.

Aghalari, Z., et al., 2020. Effectiveness of wastewater treatment systems in removing microbial agents: a systematic review. Global Health 16 (13), 1–11.

Albers, C.N., Feld, L., Ellegaard-Jensen, L., Aamand, J., 2015. Degradation of trace concentrations of the persistent groundwater pollutant 2,6-dichlorobenzamide (BAM) in bioaugmented rapid sand filters. Water Resource 83, 61–70.

Alengebawy, A., Abdelkhalek, S.T., Qureshi, S.R., Wang, M.Q., 2021. Heavy metals and pesticides toxicity in agricultural soil and plants: ecological risks and human health implications. Toxics 9 (3), 42.

Al-Hashimi, O., et al., 2021. A comprehensive review for groundwater contamination and remediation: occurrence, migration and adsorption modelling. Molecules 26, 5913.

Ali, S., et al., 2022. Harnessing plant microbiome for mitigating arsenic toxicity in sustainable agriculture. Environmental Pollution 300, 118940.

Barabadi, H., Damavandi Kamali, K., Jazayeri Shoushtari, F., Tajani, B., Mahjoub, M.A., Alizadeh, A., et al., 2019a. Emerging theranostic silver and gold nanomaterials to combat prostate cancer: a systematic review. Journal of Cluster Science 30 (6), 1375–1382.

Barabadi, H., Tajani, B., Moradi, M., Damavandi Kamali, K., Meena, R., Honary, S., et al., 2019b. Penicillium family as emerging nanofactory for biosynthesis of green nanomaterials: a journey into the world of microorganisms. Journal of Cluster Science 30 (4), 843–856.

Bjerregaard, P., Andersen, C., Andersen, O., 2022. Chapter 26 - Ecotoxicology of metals—sources, transport, and effects on the ecosystem. Handbook on the Toxicology of Metals. Academic Press, US, pp. 593–627.

Brown, P.A.G.S.A., Allen, J., 2000. Metal removal from wastewater using peat. Water Research 34, 3907–3916.

Cao, D.Q., et al., 2020. Removal of heavy metal ions by ultrafiltration with recovery of extracellular polymer substances from excess sludge. Journal of Membrane Science 606 (118103).

Chang, L., et al., 2021. Enhancing the ion flotation removal of Cu (II) via regulating the oxidation degree of nano collector-graphene oxide. Journal of Cleaner Production 295, 126397.

Chen, J.P., 2012. Decontamination of Heavy Metals: Processes, Mechanisms, and Applications. CRC Press, Boston.

Chen, Z.M., et al., 2008. A vertical subsurface-flow constructed wetland in Beijing. Communications in Nonlinear Science and Numerical Simulation 13 (9), 1986–1997.

Cheng, S., et al., 2019. New prospects for modified algae in heavy metal adsorption. Trends Biotechnology 37, 1255–1268.

Cristaldi, A., et al., 2017. Phytoremediation of contaminated soils by heavy metals and PAHs. A brief review. Environmental Technology & Innovation 8, 309–326.

de Santana, M., Zanoelo, E.B.C., Freire, F., 2018. Electrochemical treatment of wastewater from a bakery industry: experimental and modeling study. Process Safety and Environmental Protection 116, 685–692.

Deletic, A., Wang, H., 2019. Water pollution control for sustainable development. Engineering 5 (5), 839–840.

Djedidi, Z., Khaled, J.B., Cheikh, R.B., Blais, J.F., Mercier, G., Tyagi, R.D., 2009. Comparative study of dewatering characteristics of metal precipitates generated during treatment of monometallic solutions. Hydrometallurgy 95 (1–2), 61–69.

Doshi, H., Ray, A., Kothari, I.L., Gami, B., 2006. Spectroscopic and scanning electron microscopy studies of bioaccumulation of pollutants by algae. Current Microbiology 53, 148–157.

EC, 2017. Ninth Report on the implementation status and the programmes for implementation (as required by Article 17) of Council Directive 91/271/EEC concerning urban waste water treatment. Brussels: European Commission.

Fahad, A., Mohamed, R.M.S., Radhi, B., Al-Sahari, M., 2019. Wastewater and its treatment techniques: an ample review. Indian Journal of Science and Technology 12 (25), 1–13.

Faulwetter, J.L., et al., 2009. Microbial processes influencing performance of treatment wetlands: a review. Ecological Engineering 35 (6), 987–1004.

Foladori, P., Bruni, L., Tamburini, S., 2015. Bacteria viability and decay in water and soil of vertical subsurface flow constructed wetlands. Eological Engineering 82, 49–56.

Fu, F., Wang, Q., 2011. Removal of heavy metal ions from wastewaters: a review. Journal of Environmental Management 92 (3), 407–418.

Giles, B., Mann, K., 2022. Arsenic as an immunotoxicant. Toxicology and Applied Pharmacology 454, 116248.

Gill, B., et al., 2017. Economic, social and resource management factors influencing groundwater trade: evidence from Victoria, Australia. Journal of Hydrology 550, 253–267.

Gutha, Y., Munagapati, V., Naushad, M., Abburi, K., 2015. Removal of Ni (II) from aqueous solution by *Lycopersicum esculentum* (Tomato) leaf powder as a low-cost biosorbent. Desalination and Water Treatment 54, 200–208.

Hasan Abadi, N., Danesh Pazhooh, M., Mahdavi Meymand, Z., 2019. Determination of heavy and mineral metals in raw milk produced from Livestock in Khatoon-Abad, Shahr-e-Babak, Kerman. Health and Development Journal 8 (2), 152–162.

Heijne, A.T., Liu, F., Weijden, R.V.D., Weijma, J., Buisman, C.J., Hamelers, H.V., 2010. Copper recovery combined with electricity production in a microbial fuel cell. Environmental Science & Technology 44 (11), 4376–4381.

Hoang, H.G., et al., 2021. Human health risk simulation and assessment of heavy metal contamination in a river affected by industrial activities. Environmental Pollution 285, 117414.

Hogsden, K.L., Harding, J.S., 2012. Consequences of acid mine drainage for the structure and function of benthic stream communities: a review. Freshwater Science 31 (1), 108–120.

Hossain, N., et al., 2020. Waste materials for wastewater treatment and waste adsorbents for biofuel and cement supplement applications: a critical review. Journal of Cleaner Production 255, 120261.

Hu, C., et al., 2021. Trophic transfer of heavy metals through aquatic food web in a seagrass ecosystem of Swan Lagoon, China. Science of the Total Environment 762, 143139.

Huang, L., Lee, D.J., 2015. Membrane bioreactor: a mini review on recent R&D works. Bioresource Technology 194, 383–388.

Ince, M., Kaplan, I.O., 2017. Box–Behnken design approach for optimizing removal of copper from wastewater using a novel and green adsorbent. Atomic Spectroscopy 38, 200–207.

Jacob, J.M., et al., 2018. Biological approaches to tackle heavy metal pollution: a survey of literature. Journal of Environmental Management 217, 56–70.

Jeevanantham, S., et al., 2019. Removal of toxic pollutants from water environment by phytoremediation: a survey on application and future prospects. Environmental Technology & Innovation 13, 264–276.

Jiang, X., et al., 2022. Assessment of heavy metal accumulation in freshwater fish of Dongting Lake, China: effects of feeding habits, habitat preferences and body size. Journal of Environmental Sciences 112, 355–365.

Joseph, L., et al., 2019. Removal of heavy metals from water sources in the developing world using low-cost materials: a review. Chemosphere 229, 142–159.

Kumar, M., Singh, R., 2017. Performance evaluation of semi continuous vertical flow constructed wetlands (SC-VF-CWs) for municipal wastewater treatment. Bioresource Technology 232, 321–330.

Kumar, M., Singh, R., 2019. Sewage water treatment with energy recovery using constructed wetlands integrated with a bioelectrochemical system. Environmental Science: Water Research & Technology 6 (3), 795–808.

Kumar, M., Singh, R., 2020. Installation of a bioelectrochemical system as a pre-cleaner in a constructed wetland with higher pollutant loading under continuous mode. Environmental Science: Water Research & Technology 6 (12), 3326–3340.

Li, H., et al., 2019. Environment-enhancing process for algal wastewater treatment, heavy metal control and hydrothermal biofuel production: a critical review. Bioresource Technology 298, 122421.

Li, Z., et al., 2020. Water treatment effect, microbial community structure, and metabolic characteristics in a field-scale aquaculture wastewater treatment system. Frontiers in Microbiology 11, 1–13.

Li, Z., Zhang, X., Lei, L., 2008. Electricity production during the treatment of real electroplating wastewater containing Cr6 + using microbial fuel cell. Process Biochemistry 43 (12), 1352–1358.

Liu, J., et al., 2016. Impact of plant species on spatial distribution of metabolic potential and functional diversity of microbial communities in a constructed wetland treating aquaculture wastewater. Ecological Engineering 94, 564–573.

Liu, J., Cao, L., Dou, S., 2019. Trophic transfer, biomagnification and risk assessments of four common heavy metals in the food web of Laizhou Bay, the Bohai Sea. Science of The Total Environment 670, 508–522.

Ma, C., Wu, Y., 2008. Dechlorination of perchloroethylene using zero-valent metal and microbial community. Environmental Geology 55 (1), 47–54.

Madgett, A.S., et al., 2021. The concentration and biomagnification of trace metals and metalloids across four trophic levels in a marine food web. Marine Pollution Bulletin 173 (112929).

Mehariya, S., Goswami, R.K., Karthikeysan, O., Verma, P., 2021. Microalgae for high-value products: a way towards green nutraceutical and pharmaceutical compounds. Chemosphere 280, 130553.

Melnyk, A., Kuklinska, K., Wolska, L., Namiesnik, J., 2014. Chemical pollution and toxicity of water samples from stream receiving leachate from controlled municipal solid waste (MSW) landfill. Environmental Research 135, 253–261.

Muga, H.E., Mihelcic, J.R., 2008. Sustainability of wastewater treatment technologies. Journal of Environmental Management 88 (3), 437–447.

Murtaza, G., et al., 2022. Biomagnification of potentially toxic elements in animals consuming fodder irrigated with sewage water. Environmental Geochemistry and Health 44 (2), 1–16.

Natasha, Shahid, M., Khalid, S., Saleem, M., 2021. Unrevealing arsenic and lead toxicity and antioxidant response in spinach: a human health perspective. Environmental Geochemistry and Health 44, 487–496.

Němeček, J., et al., 2018. Thermally enhanced in situ bioremediation of groundwater contaminated with chlorinated solvents – a field test. Science of Total Environment 743–755.

Ngo, H.H., et al., 2002. A comparison of conventional and non-conventional treatment technologies on arsenic removal from water. Water Supply 2 (5–6), 119–125.

Nies, D.H., 2000. Microbial heavy-metal resistance. Applied Microbiology and Biotechnology 51, 451–460.

Nikoladze, G., Mints, D., Kastalsky, A., 1989. Water Treatment for Public and Industrial Supply. MIR Publishers, Moscow.

Okereafor, U., Makhatha, M., Mekuto, L., Uche-Okereafor, N., Sebola, T., Mavumengwana, V., 2020. Toxic metal implications on agricultural soils, plants, animals, aquatic life and human health. International Journal of Environmental Research and Public Health 17 (7), 2204.

Panahi, Y., et al., 2018. Biotechnological applications of nanomaterials for air pollution and water/wastewater treatment. Materials Today 5, 15550–15558.

Patanjali, P., Singh, R., Kumar, A., Chaudhary, P., 2019. Nanotechnology for water treatment: a green approach. Green Synthesis, Characterization and Applications of Nanoparticles. Elsevier, pp. 485–512.

Pérez-González, A., Urtiaga, A.M., Ibáñez, R., Ortiz, I., 2012. State of the art and review on the treatment technologies of water reverse osmosis concentrates. Water Research 46 (2), 267–283.

Price, J.I., Heberling, M.T., 2018. The effects of source water quality on drinking water treatment costs: a review and synthesis of empirical literature. Ecological Economics 151, 195–209.

Proshad, R., et al., 2021. Appraisal of heavy metal toxicity in surface water with human health risk by a novel approach: a study on an urban river in vicinity to industrial areas of Bangladesh. Toxin Reviews 40 (4), 803–819.

Rai, P.K., 2010. Heavy metal pollution in lentic ecosystem of sub-tropical industrial region and its phytoremediation. International Journal of Phytoremediation 12 (3), 226–242.

Rai, P.K., 2012. An eco-sustainable green approach for heavy metals management: two case studies of developing industrial region. Environmental Monitoring Assessement 184, 421–448.

Rajasulochana, P., Preethy, V., 2016. Comparison on efficiency of various techniques in treatment of waste and sewage water - a comprehensive review. Resource-Efficient Technologies 2, 175–184.

Rangabhashiyam, S., Balasubramanian, P., 2019. Characteristics, performances, equilibrium and kinetic modeling aspects of heavy metal removal using algae. Bioresource Technology Reports 5, 261–279.

Rani, N., et al., 2019. Chapter 5 - microbes: a key player in industrial wastewater treatment. Microbial Wastewater Treatment. Elsevier, Spain, pp. 83–102.

Rasafi, T., et al., 2022. Cadmium stress in plants: a critical review of the effects, mechanisms, and tolerance strategies. Critical Reviews in Environmental Science and Technology 52, 675–726.

Raza, W., et al., 2018. Removal of phenolic compounds from industrial waste water based on membrane-based technologies. Journal of Industrial and Engineering Chemistry 71, 1–18.

Reddy, D.H.K., Lee, S.M., 2012. Water pollution and treatment technologies. Environmental & Analytical 2 (4), 1–2.

Roberts, J.R., 1999. Metal Toxicity in Children. Training Manual on Pediatric Environmental Health: Putting It Into Practice. Children's Environmental Health Network, Emeryville, CA, p. 1.

Salman, J., 2014. Optimization of preparation conditions for activated carbon from palm oil fronds using response surface methodology on removal of pesticides from aqueous solution. Arabian Journal of Chemistry 7, 101–108.

Saravanan, A., et al., 2021. Effective water/wastewater treatment methodologies for toxic pollutants removal: processes and applications towards sustainable development. Chemosphere 280.

Schaefer, H.R., et al., 2022. A systematic review of adverse health effects associated with oral cadmium exposure. Regulatory Toxicology and Pharmacology 134, 105243.

Sheibani, M., Ghaedi, M., Marahel, F., Ansari, A., 2015. Congo red removal using oxidized multiwalled carbon nanotubes: kinetic and isotherm study. Desalination and Water Treatment 53, 844–852.

Shikha, D., Singh, P.K., 2021. In situ phytoremediation of heavy metal–contaminated soil and groundwater: a green inventive approach. Environmental Science and Pollution Research 28, 4104–4124.

Singh, A., et al., 2022. Biological remediation technologies for dyes and heavy metals in wastewater treatment: new insight. Bioresource Technology 343, 126154.

Singh, J., Kalamdhad, A.S., 2011. Effects of heavy metals on soil, plants, human health and aquatic life. International Journal of Research in Chemistry and Environment 1 (2), 15–21.

Soares, E., De Coninck, G., Duarte, F., Soares, H., 2002. Use of *Saccharomyces cerevisiae* for Cu^{2+} removal from solution: the advantages of using a flocculent strain. Biotechnology Letters 24, 663–666.

Sun, M., et al., 2021. Electrified membranes for water treatment applications. ACS EST Engineering 1 (4), 725–752.

Sun, Q., et al., 2022. Heavy metals induced mitochondrial dysfunction in animals: molecular mechanism of toxicity. Toxicology 469, 153136.

Syafiuddin, A., Boopathy, R., Hadibarata, T., 2020. Challenges and solutions for sustainable groundwater usage: pollution control and integrated management. Current Pollution Reports 6, 310–327.

Tessema, K., Lemma, B., Fetahi, T., Kebede, E., 2020. Accumulation of heavy metals in the physical and biological systems of Lake Koka, Ethiopia: implications for potential health risks. Lake Reservoirs 25 (3), 314–325.

Thomann, J.A., Werner, A.D., Irvine, D.J., Currell, M.J., 2020. Adaptive management in groundwater planning and development: a review of theory and applications. Journal of Hydrology 586, 124871.

Vallero, D.A., 2019. Hazardous wastes-chapter 31, Waste, second ed. Academic Press, Cambridge, pp. 585–630.

Vymazal, J., 2014. Constructed wetlands for treatment of industrial wastewaters: a review. Ecological Engineering 73, 724–751.

Wang, H., Ren, Z.J., 2014. Bioelectrochemical metal recovery from wastewater: a review. Water Research 66, 219–232.

Xanthopoulos, P.K.D., Kamariah, N., Bussé, J.D.W.S.J.B.K., 2021. Recovery of copper from ammoniacal leachates by ion flotation. Journal of Sustainable Metallurgy 7, 1552–1564.

Yang, C., Ho, Y.N., Inoue, C., Chien, M.F., 2020. Long-term effectiveness of microbe-assisted arsenic phytoremediation by *Pteris vittata* in field trials. Science of Total Environment 740.

Yin, K., Wang, Q., Lv, M., Chen, L., 2018. Microorganism remediation strategies towards heavy metals. Chemical Engineering Journal 360, 1553–1563.

Zeng, Y., Cai, Y., Tan, Q., Dai, C., 2020. An integrated modeling approach for identifying cost-effective strategies in controlling water pollution of urban watersheds. Journal of Hydrology 581 (124373).

Chapter 23

Challenges and remediation of emerging organic pollutants from water

Bidyutlata Patra[1,2], Jiban Kumar Behera[2], Pabitra Mishra[2], Manojit Bhattacharya[2], Ajaya Kumar Rout[3], Jyotirmayee Pradhan[1] and Bhaskar Behera[3]

[1]*Kuntala Kumari Sabat Women's College, Fakir Mohan University, Balasore, Odisha, India,* [2]*P.G. Department of Zoology, Fakir Mohan University, Balasore, Odisha, India,* [3]*P.G. Department of Bioscience and Biotechnology, Fakir Mohan University, Balasore, Odisha, India*

Introduction

Water is an essential part of life since it allows us to exist and sustains our lives. Everyone should drink pure, clean water. However, due to rising urbanization, industry, and associated geogenic disruption, the water supply has degraded. In recent years, there have been several research studies have been performed about the prevalence of emerging contaminants (ECs) or emerging pollutants (EPs) in wastewater, surface water, and groundwater. Recently, emerging contaminants are more commonly recognized, and at the low concentration stage, they generate known or suspected detrimental impacts on human health and the environment. Novel emerging contaminants are produced by synthetic organic compounds or are sometimes exposed to usage and disposal products (Du et al., 2015). All over the world, India has the highest number of groundwater users, with around 80% and 50% of the rural and urban populations using untreated groundwater for residential reasons in India (Chakraborti et al., 2011). Many researchers have performed their studies to survey the availability of chemical contaminants in sludge, soil, air, sediments, and surface water bodies (Gani and Kazmi, 2017; Philip et al., 2018). A number of investigations revealed the presence of abundant pesticides in water bodies from different parts of India (Agarwal et al., 2015). Simultaneously, the presence of diverse micropollutants in India's aquatic environment was also conducted by many researchers (Gani and Kazmi, 2017). Pharmaceuticals and pentachlorophenol (PCP) were found in effluents from rivers, lakes, groundwater, and water treatment plants (Balakrishna et al., 2017). Based on micropollutants and biological samples in aquatic bodies (water, air, soil, sediment, and sludge), India is divided into six different zones (Philip et al., 2018). Many emerging contaminants are still uncontrolled, posing analytical and institutional difficulties faced by the environment (Kavanaugh, 2003). An anthropogenic micro-pollutant monitoring system in river basin areas is urgently necessary for national legislation with the overarching goal of protecting and improving water quality (Stuart et al., 2012). The United States Environmental Protection Agency (EPA) has developed a base guideline for contaminants in drinking water and its micro-organic pollutants (excluding pesticides) (USEPA, 2009). The EPA released a new list of potential contaminants in 2009 called CCL-3, which included three drugs; perfluorooctanoic acid (PFOA), perfluorooctane sulfonic acid (PFOS), and eight hormones (Petrovic and Barceló, 2006).

Resources of emerging organic water pollutants

Emerging contaminants can be defined as chemicals that are new to the ecosystem, such as industrial compounds that have just been created and substances that have previously entered the environment, but have only lately been detected its negative consequences for ecosystems or humanity. The characteristics of emerging contaminants, including their absorption, polarity, volatility, permanence, and interactions with adjacent compartments, determine how they travel from the source sites to the drain (Fig. 23.1) (Barceló and Petrovic, 2006; Farré et al., 2010; Geissen et al., 2010).

Role of Green Chemistry in Ecosystem Restoration to Achieve Environmental Sustainability. DOI: https://doi.org/10.1016/B978-0-443-15291-7.00001-8

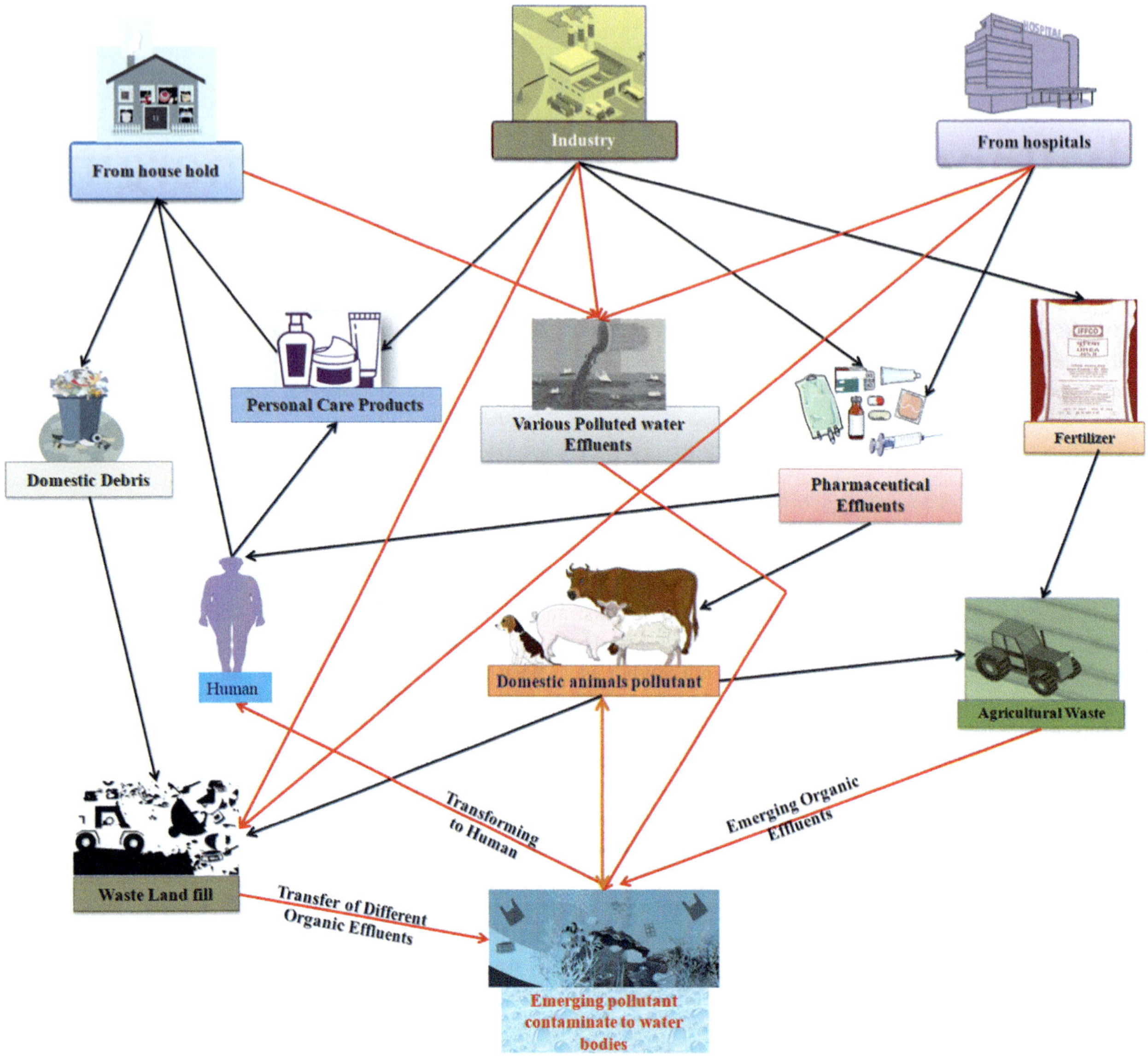

FIGURE 23.1 The surface water bodies polluted by organic contaminants at varied levels.

Endocrine disruptors

Nowadays various artificial, exogenous mixtures of chemicals interlink with the hormonal action of the body and create hazardous diseases (Gore, 2007). This group includes industrial solvents, lubricants and their by-products, fungicides, plasticizers, pesticides, dioxins, detergents, some pharmaceuticals, and flame retardants (Kabir et al., 2015). These man-made compounds after entering the body, either mimic or block hormones, disrupting normal body functioning. According to the EPA, endocrine disruptors are an external factor that interfere with the creation of endogenous hormones that regulates major function like growth, and reproduction with body homeostasis of an organism which is related to biological activity (Crisp et al., 1998; Snyder et al., 2003). Polychlorinated biphenyls (PCBs) are a type of flame retardant found in fabrics, plastics, and furniture foams. These are mostly used to reduce the dangers of fire (Rahman et al., 2001). These compounds are utilized in water, grease, and dirt-resistant plating, leather and textile sprays, and PTFE (Teflon) coatings on non-stick cookware (Houtman, 2010). Industrial chemicals have anti-androgenic properties which are made by acetone and phenol, these estrogenic chemicals are 2,2-bis(4hydroxyphenyl) propane and bisphenol A (BPA) (Sohoni and Sumpter, 1998).

Pharmaceuticals

According to research, many of these molecules can penetrate the environment, distribute, and survive to a higher extent than previously thought. Sometimes micropollutants, for example, pharmaceuticals, lifestyle compounds, and biogenic hormones produced by the industries are directly released into the water bodies (Halling-Sørensen et al., 1998; Liu and Wong, 2013). The concentrations of pharmaceuticals in aquatic systems range from ppt (ng/L) to ppb (lg/L) (Dai et al., 2015). For feeding and treatment purposes (veterinary) pharmaceuticals may be released into the environment alongside animal waste because of excess use or leakage from storage or overflow (Daughton, 2016). Pandemics cause an increase in the usage of a certain medicine in a specific area for a specific amount of time, which causes a significant rise in the drug's load in wastewater (Slater et al., 2011).

Personal care products

Various types of personal care products are available in the market and these are sunscreen filters, insect repellents, bactericides, preservatives, and fragrances (Kosma et al., 2010; Liu and Wong, 2013; Yang et al., 2017). Parabens are para-hydroxybenzoic acids commonly used as preservatives for pharmaceuticals, beverages, and personal products (Guo and Kannan, 2013). DEET is a widely used active ingredient in insect repellents that has been shown to last for lengthy periods in aquatic bodies (Yang et al., 2017). Synthetic perfume products and synthetic musks (SMs) which is used in every household and personal care such as perfumes, fabric softeners, skin care lotions, detergents, and shampoos (Reiner et al., 2007). SMs are divided into four classes such as polycyclic, alicyclic musks, macrocyclic, and nitro types based on cost-effective alternatives to natural counterparts (Homem et al., 2015a,b). The SM polycyclic and nitro musks are available in large amounts in the market. SMs are one of the emerging contaminants pollutants in the category of personal care products due to have possible carcinogenicity, endocrine-disrupting impact, and extended persistence possibility (Johnson and Sumpter, 2001; Homem et al., 2015a,b; Gao et al., 2016).

Artificial sweeteners

Consuming an enormous number of artificial sweeteners (such as saccharin, acesulfame, sucralose, and cyclamate) is a relatively new family of toxins that are emerging in humans as a result of their calorific sugar content in a wide range of foods as well as beverages (Richardson, 2009). Acesulfame is used in baking and soft drinks due to having their long expiration dates and its ability to stay in a liquid medium (Houtman, 2010). Several types of sweeteners cannot be destroyed by the human body, posing a problem because they may wind up in significant concentrations in water bodies (Scheurer et al., 2009).

Surfactants

Surfactants (surface-active agents) are a broad collection of compounds that are structurally divided into water-soluble (polar) and insoluble in water (non-polar hydrocarbon tail) (Ivanković and Hrenović, 2010). Because of their poor bioremediation and general release, biosurfactants have the potential to pollute water. Due to their destruction of cellular cell membranes, biochemical processes, or other cellular constituents, as well as their binding to peptides and proteins, biosurfactants could be especially poisonous to biota, the environment, and human impacts (Doan et al., 2021).

Petrochemicals

The petrochemical sector is a vital industry that contributes to the national economy of any country and support a variety of other industry including agriculture, energy transportation (Jafarinejad, 2016). The wastewater generated by the petrochemical industry comes from a variety of sources, including discharge from raw materials, industrial rains, cooling water, and home sewage (Rojas and Horcajada, 2020). Maleic anhydride petrochemicals are found in agricultural chemicals (Keeley and Townsend, 2011). BTEX contaminants in surface and groundwater are mostly caused by leaks from petroleum storage tanks (Boopathy, 2004; Bonvicini et al., 2015), spills from production wells (Hazrati et al., 2015), refineries (Karadag et al., 2016), pipelines (Lan and Minh, 2013), and storage and distribution terminals, as well as other sources (Louis et al., 2016).

Effects of organic water pollutants in various systems

Effects on aquatic systems

Despite being present in very small concentrations, organic pollutants have harmful effects on aquatic life, domestic animals, and people. At both low and high quantities, diclofenac, carbamazepine, ibuprofen, fluoxetine, and several antibiotics are hazardous to environmental organism (Boxall et al., 2003). The chronic lowest observed effect concentration (LOEC) of medicines is generally around 2 orders of magnitude higher than their average effluent concentrations (Carlsson et al., 2009). Diclofenac's LOEC value for fish toxicity is also within the range of concentrations seen in WWTP effluents (Schwaiger et al., 2004; Kumar et al., 2016). Verapamil and staurosporine disrupt the transport mechanism, exposing vulnerability to other dangerous substances (Smital and Kurelec, 1998). Three steroid estrogens (17a-ethinylestradiol, 17b-estradiol, and estrone) are frequently found in wastewater at low quantities and these have low ecotoxicity thresholds and have an exceedingly small impact on fish fertility (Baronti et al., 2000).

Effects on human body

Pharmaceuticals have been found in lakes, rivers, treated effluents, and groundwater (Thomas and Hilton, 2004). Several antibiotics such as sarafloxacin, oxytetracycline, erythromycin, and thiamphenicol are essentially used for health concerns of both animal and plant (Chen et al., 2020). There is evidence that more toxic chemicals have an impact on the endocrine system, progesterone, and testosterone (Bilal et al., 2020). Various mutagenic chemical compounds such as chloramphenicol, triamterene, sulfamethazine, etc. are used for mutagenesis in various organisms (Chaturvedi et al., 2021). UV filters (sunscreens) are commonly found in personal products like lipstick, fragrances, hair sprays, hair dye, and moisturizers and non-personal products are plastic, detergents, and furniture chemical (Krause et al., 2012). Various biological function like reproduction, metabolism, and development as well as the hypothalamic pituitary thyroid axis are disrupted by cosmetic products, according to numerous in vitro and in vivo studies of researchers (Krause et al., 2012). Gasoline is a carcinogenic substance that is made up of more than 500 different ingredients, out of these major components are MTBE, benzene, and 1,3-butadiene (Kane and Newton, 2010). According to numerous studies, benzene causes serious health problems like cancers and fibrosis (Melo and Barnes, 2007; Bahadar et al., 2014; Edokpolo et al., 2015; Wen et al., 2016). Total trihalomethanes in drinking water and bladder cancer mortality among 65 people in a Taiwanese prefecture were examined using a matched-pair study (Chang et al., 2007). Babies born to mothers exposed to 200 g/L chlorate were more likely to have cleft palates, urinary anomalies, and spinal bifida (Ema et al., 2010).

Remediation strategies

Several studies have looked into novel techniques such as chlorination, biodegradation, activated carbon, photocatalysis, and ionization for eliminating ECs from water in recent years (Table 23.1, Fig. 23.2) (Lapworth et al., 2012; Rojas and

TABLE 23.1 Different wastewater management technologies and methods employed for removing growing organic pollutants.

Emerging organic pollutant	Techniques	Method	Maximum removal efficiency	References
Congo red	Adsorption Membranes Phytoremediation	Ce-UiO-66 MOF TA/GOQDs TFN γ-$Al_2O_3\gamma$-Al_2O_3 – SiCl *Lemna minor*	99.9% 99.8% 94.86% 99.28% 51%	Rego et al. (2021) Zhang et al. (2017) Mahmoud et al. (2018) Al-Zurfi et al. (2020)
Methylene blue	Adsorption Phytoremediation	Ce-UiO-66 MOF AC	90% 64%	Rego et al. (2021) Azam et al. (2020)

(Continued)

TABLE 23.1 (Continued)

Emerging organic pollutant	Techniques	Method	Maximum removal efficiency	References
	Membranes Photocatalysis	MAC MMAC FAC *Scirpus grossus* *Eichhornia crassipes* TA/GOQDs TFN ZOGP-50 HFMs Pd-TiO_2	67% 82% 48.9% 86% 90.8% 97.6% 98.6% 85.9%	Gong and Ye (2013) Almaamary et al. (2017) Sharma et al. (2021) Zhang et al. (2017) Modi and Bellare (2019) Nguyen et al. (2018)
Rhodamine B	Adsorption Phytoremediation Membranes Photocatalysis	Fe_3O_4/HA g-C_3N_4-TiO_2-GA without light irradiation With light irradiation *Eichhornia crassipes* ZOGP-50 HFMs CuI/g-C_3N_4 nanocomposite	98.5% 96.5% 98.4% 84.8% 98.5% 98.5%	Peng et al. (2012) Zhang et al. (2018) Sharma et al. (2021) Modi and Bellare (2019) Ghanbari and Salavati-Niasari (2021)
Methyl orange	Adsorption Phytoremediation Photocatalysis	AC MAC MMAC Chitosan microspheres Ni/Al@PAB BC and Fe_2O_3-BC from PPS *Eichhornia crassipes* CuI/g-C_3N_4 nanocomposite 8% graphene $CaCu_3Ti_4O_{12}$ Pd-TiO_2	66.9% 76% 98.5% 98.5% 75% 100% 62.8% 98% 89% 77.1%	Azam et al. (2020) Zhai et al. (2018) Chen et al. (2018) Chaukura et al. (2017) Sharma et al. (2021) Abdollahi et al. (2018) Nguyen, Fu et al. (2018) Nguyen et al. (2018), Ghanbari and Salavati-Niasari (2021)
Disperse orange 25	Adsorption	CFA Cenospheres nanosyntactic foam Chitosan cenospheres Nanocomposite	75% 90% 97.3%	Kisku et al. (2015) Dhiman et al. (2018) Dhiman et al. (2017)
Cresol red	Phytoremediation	*Eichhornia crassipes*	33.3%	Sharma et al. (2021)
Rose bengal	Phytoremediation	*Eichhornia crassipes*	87.4%	Sharma et al. (2021)
Crystal violet	Phytoremediation	*Eichhornia crassipes*	87.2%	Sharma et al. (2021)
Auramine O	Phytoremediation	*Eichhornia crassipes*	79%	Sharma et al. (2021)
Xylenol orange	Phytoremediation	*Eichhornia crassipes*	46.2%	Sharma et al. (2021)
Phenol red	Phytoremediation	*Eichhornia crassipes*	44.4%	Sharma et al. (2021)
Sun set yellow	Membranes	PVI	>99%	Hu et al. (2018)
Bisphenol A	Adsorption Membranes	Bacterial cellulose/ cyclodextrin Oligomer composites Graphene Fe_3O_4@SiO_2/CTAB-SiO_2 NF and RO	34% 86% 93.2% ≥98%	Liu et al. (2021) Xu et al. (2012) Gong et al. (2021) Yüksel et al. (2013)
Triclosan	Catalysis Adsorption	Ozonation 2C18-BT	100% 100%	Orhon et al. (2017) Phuekphong et al. (2020)

(Continued)

TABLE 23.1 (Continued)

Emerging organic pollutant	Techniques	Method	Maximum removal efficiency	References
Atrazine	Adsorption Membranes	TS-MIL FOTFC	90% 97.3%	Moeini et al. (2019) Rastgar et al. (2020)
Nonylphenol	Oxidation	K_2FeO_4 containing Fe (VI)	98%	Limmun et al. (2019)
Dibupylphthalate	Membranes	Fe,Ag-ZnO/VIS-LED NF	95% 91.5%	Akbari-Adergani et al. (2018) Wei et al. (2016)
Propylparaben	Membranes	Removal by: DOC resin + NF-90 DOC resin + DESAL-HL UVA/PS process ZVI/PS process	100% 100% 94.8% 98.5%	López-Ortiz et al. (2018) Palharim et al. (2020)
Methylparaben	Membranes	Removal by: DOC resin + NF-90 DOC resin + DESAL-HL	91% 92%	López-Ortiz et al. (2018)
Butyl paraben	Membranes	Removal by: DOC resin + NF-90 DOC resin + DESAL-HL	100% 100%	López-Ortiz et al. (2018)
Ethyl paraben	Membranes	Removal by: DOC resin + NF-90 DOC resin + DESAL-HL	96% 97%	López-Ortiz et al. (2018)
Dimethyl phthalate	Membranes	Dimethyl phthalate	82.3%	Wei et al. (2016)
Diethyl phthalate	Membranes	Diethyl phthalate	86.7%	Wei et al. (2016)
Di-*n*-octyl phthalate	Membranes	Di-*n*-octyl phthalate	95.1%	Wei et al. (2016)
Diethylhexyl phthalate	Membranes	Diethylhexyl phthalate	95.4%	Wei et al. (2016)
Dichlorofluorescein	Adsorption	Ce-UiO-66 MOF AC from *Reynoutria japonica* *Moringa oleifera* pods AC from sewage sludge	99.9% 58% 72.4% 96.34%	Rego et al. (2021) Koutník et al. (2020) Viotti et al. (2019) dos Reis et al. (2016)
Amoxicillin	Adsorption	AC-nZVI 3DES	97.9% 98.8%	Ali et al. (2020) Shi et al. (2020)
Ibuprofen	Catalysis Adsorption Biotransformation	Ozonation Iron nanoadsorbent *Desulfovibrio vulgaris*	99% 92% 100%	He et al. (2021) Ali et al. (2016) Martins et al. (2017)
Ciprofloxacin	Catalysis	Removal with Bio-Pt	70%	Martins et al. (2017)
Naproxen	Adsorption	PEC with $ZnO/TiO_2/Ag_2Se$ and visible light GON	100% 65.28%	Changanaqui et al. (2020) Ciğeroğlu et al. (2020)
Paracetamol	Oxidation Electrooxidation Photocatalysis	OMS-2/PS With graphite as anode $ATiO_2$ under UV irradiation	99.5% >90% 99%	Tepe et al. (2020) Periyasamy and Muthuchamy (2018) Jayasree and Remya (2020)

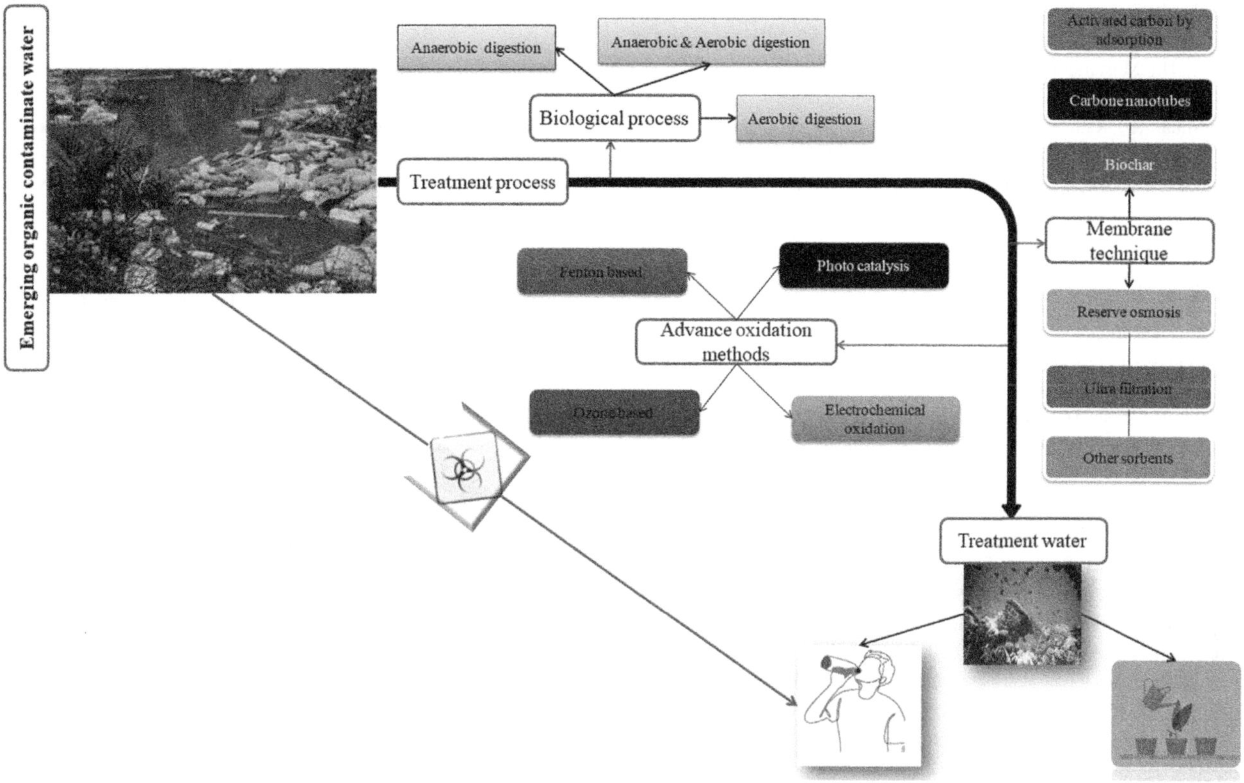

FIGURE 23.2 Wastewater treatment methods to get rid of developing organic pollution in water bodies and reuse it.

Horcajada, 2020). These methods are utilized to get rid of industrial chemicals as well as colors, medications, plasticizers, herbicides, and insecticides (Haq and Raj, 2019).

Conclusion

New poisons are introduced to humans and ecosystems everywhere in a variety of recognized and unknown ways. The use of certain technology for water purification cannot be regarded as commonly applicable to all organic micropollutants due to the wide variety of compounds and the deep variations in their behavior in water. Certain knowledge about the degree of pollution, its exposure sources, the clean-up plan, and specific implications of such pollutants for the aquatic environment is needed to understand. To develop broad remediation theories and practices for developing toxins in the environment and the human body, further approaches and research will be required in the upcoming days.

References

Abdollahi, Y., Sabbaghi, S., et al., 2018. A new achievement in green degradation of aqueous organic pollutants under visible-light irradiation. Water Science and Technology 77 (6), 1493–1504.

Agarwal, A., Prajapati, R., et al., 2015. Pesticide residue in water—a challenging task in India. Environmental Monitoring and Assessment 187 (2), 1–21.

Akbari-Adergani, B., Saghi, M., et al., 2018. Removal of dibutyl phthalate from aqueous environments using a nanophotocatalytic Fe, Ag-ZnO/VIS-LED system: modeling and optimization. Environmental Technology 39 (12), 1566–1576.

Ali, I., Afshinb, S., et al., 2020. Green preparation of activated carbon from pomegranate peel coated with zero-valent iron nanoparticles (nZVI) and isotherm and kinetic studies of amoxicillin removal in water. Environmental Science and Pollution Research 27 (29), 36732–36743.

Ali, I., AL-Othman, Z.A., et al., 2016. Synthesis of composite iron nano adsorbent and removal of ibuprofen drug residue from water. Journal of molecular liquids 219, 858–864.

Almaamary, E., Abdullah, S., et al., 2017. Rawatan metilena biru dalam air sisa menggunakan *Scirpus grossus*. Malaysian Journal of Analytical Science 21 (1), 182–187.

Al-Zurfi, S.K.L., Albdairi, N.A., et al., 2020. Phytoremediation of congo red and brilliant green using lemna minor. Plant Cell Biotechnology and Molecular Biology 1–10.

Azam, K., Raza, R., et al., 2020. Development of recoverable magnetic mesoporous carbon adsorbent for removal of methyl blue and methyl orange from wastewater. Journal of Environmental Chemical Engineering 8 (5), 104220.

Bahadar, H., Mostafalou, S., et al., 2014. Current understandings and perspectives on non-cancer health effects of benzene: a global concern. Toxicology and Applied Pharmacology 276 (2), 83–94.

Balakrishna, K., Rath, A., et al., 2017. A review of the occurrence of pharmaceuticals and personal care products in Indian water bodies. Ecotoxicology and Environmental Safety 137, 113–120.

Barceló, D., Petrovic, M., 2006. Application of liquid chromatography/quadrupole time-of-flight mass spectrometry (LC-QqTOF-MS) in the environmental analysis.

Baronti, C., Curini, R., et al., 2000. Monitoring natural and synthetic estrogens at activated sludge sewage treatment plants and in a receiving river water. Environmental Science & Technology 34 (24), 5059–5066.

Bilal, M., Mehmood, S., et al., 2020. Antibiotics traces in the aquatic environment: persistence and adverse environmental impact. Current Opinion in Environmental Science & Health 13, 68–74.

Bonvicini, S., Antonioni, G., et al., 2015. Quantitative assessment of environmental risk due to accidental spills from onshore pipelines. Process Safety and Environmental Protection 93, 31–49.

Boopathy, R., 2004. Anaerobic biodegradation of no. 2 diesel fuel in soil: a soil column study. Bioresource Technology 94 (2), 143–151.

Boxall, A.B., Kolpin, D.W., et al., 2003. Peer reviewed: are veterinary medicines causing environmental risks? Environmental Science & Technology 37 (15), 286A–294A.

Carlsson, G., Örn, S., et al., 2009. Effluent from bulk drug production is toxic to aquatic vertebrates. Environmental Toxicology and Chemistry 28 (12), 2656–2662.

Chakraborti, D., Das, B., et al., 2011. Examining India's Groundwater Quality Management. ACS Publications.

Chang, C.-C., Ho, S.-C., et al., 2007. Bladder cancer in Taiwan: relationship to trihalomethane concentrations present in drinking-water supplies. Journal of Toxicology and Environmental Health, Part A 70 (20), 1752–1757.

Changanaqui, K., Alarcon, H., et al., 2020. Blue LED light-driven photoelectrocatalytic removal of naproxen from water: kinetics and primary by-products. Journal of Electroanalytical Chemistry 867, 114192.

Chaturvedi, P., Shukla, P., et al., 2021. Prevalence and hazardous impact of pharmaceutical and personal care products and antibiotics in environment: a review on emerging contaminants. Environmental Research 194, 110664.

Chaukura, N., Murimba, E.C., et al., 2017. Synthesis, characterisation and methyl orange adsorption capacity of ferric oxide–biochar nano-composites derived from pulp and paper sludge. Applied Water Science 7 (5), 2175–2186.

Chen, J., Sun, R., et al., 2020. Antibiotics and food safety in aquaculture. Journal of Agricultural and Food Chemistry 68 (43), 11908–11919.

Chen, S., Huang, Y., et al., 2018. Simultaneous and efficient removal of Cr (VI) and methyl orange on LDHs decorated porous carbons. Chemical Engineering Journal 352, 306–315.

Ciğeroğlu, Z., Özdemir, O.K., et al., 2020. Naproxen adsorption onto graphene oxide nanopowders: equilibrium, kinetic, and thermodynamic studies. Water, Air, & Soil Pollution 231 (3), 1–10.

Crisp, T.M., Clegg, E.D., et al., 1998. Environmental endocrine disruption: an effects assessment and analysis. Environmental Health Perspectives 106 (suppl 1), 11–56.

Dai, G., Wang, B., et al., 2015. Occurrence and source apportionment of pharmaceuticals and personal care products in the Beiyun River of Beijing, China. Chemosphere 119, 1033–1039.

Daughton, C.G., 2016. Pharmaceuticals and the Environment (PiE): evolution and impact of the published literature revealed by bibliometric analysis. Science of the Total Environment 562, 391–426.

Dhiman, N., Shukla, S., et al., 2017. Statistical optimization of process parameters for removal of dyes from wastewater on chitosan cenospheres nano-composite using response surface methodology. Journal of Cleaner Production 149, 597–606.

Dhiman, N., Shukla, S.P., et al., 2018. Comprehensive remediation study of disperse dyes in wastewater using cenospheres nanosyntactic foam. Journal of Cleaner Production 182, 206–216.

Doan, T.H.Y., Le, T.T., et al., 2021. Simultaneous adsorption of anionic alkyl sulfate surfactants onto alpha alumina particles: experimental consideration and modeling. Environmental Technology & Innovation 24, 101920.

dos Reis, G.S., Bin Mahbub, M.K., et al., 2016. Activated carbon from sewage sludge for removal of sodium diclofenac and nimesulide from aqueous solutions. Korean Journal of Chemical Engineering 33 (11), 3149–3161.

Du, B., Haddad, S.P., et al., 2015. Pharmaceutical bioaccumulation by periphyton and snails in an effluent-dependent stream during an extreme drought. Chemosphere 119, 927–934.

Edokpolo, B., Yu, Q.J., et al., 2015. Health risk characterization for exposure to benzene in service stations and petroleum refineries environments using human adverse response data. Toxicology Reports 2, 917–927.

Ema, M., Kobayashi, N., et al., 2010. Reproductive and developmental toxicity studies of manufactured nanomaterials. Reproductive Toxicology 30 (3), 343–352.

Farré, M., Pérez, S., et al., 2010. First determination of C60 and C70 fullerenes and N-methylfulleropyrrolidine C60 on the suspended material of wastewater effluents by liquid chromatography hybrid quadrupole linear ion trap tandem mass spectrometry. Journal of Hydrology 383 (1–2), 44–51.

Gani, K.M., Kazmi, A.A., 2017. Contamination of emerging contaminants in Indian aquatic sources: first overview of the situation. Journal of Hazardous, Toxic, and Radioactive Waste 21 (3), 04016026.

Gao, Y., Ji, Y., et al., 2016. Bioaccumulation and ecotoxicity increase during indirect photochemical transformation of polycyclic musk tonalide: a modeling study. Water Research 105, 47–55.

Geissen, V., Ramos, F.Q., et al., 2010. Soil and water pollution in a banana production region in tropical Mexico. Bulletin of Environmental Contamination and Toxicology 85 (4), 407–413.

Ghanbari, M., Salavati-Niasari, M., 2021. Copper iodide decorated graphitic carbon nitride sheets with enhanced visible-light response for photocatalytic organic pollutant removal and antibacterial activities. Ecotoxicology and Environmental Safety 208, 111712.

Gong, Y., Liu, G., et al., 2021. Synthesis of a novel mesoporous Fe_3O_4@ SiO_2/CTAB-SiO_2 composite material and its application in the efficient removal of bisphenol A from water. Colloid and Polymer Science 299 (5), 807–822.

Gong, R., Ye, J., et al., 2013. Adsorptive removal of methyl orange and methylene blue from aqueous solution with finger-citron-residue-based activated carbon. Industrial & Engineering Chemistry Research 52 (39), 14297–14303.

Gore, A.C., 2007. Introduction to endocrine-disrupting chemicals. Endocrine-Disrupting Chemicals. Springer, pp. 3–8.

Guo, Y., Kannan, K., 2013. A survey of phthalates and parabens in personal care products from the United States and its implications for human exposure. Environmental Science & Technology 47 (24), 14442–14449.

Halling-Sørensen, B., Nielsen, S.N., et al., 1998. Occurrence, fate and effects of pharmaceutical substances in the environment-A review. Chemosphere 36 (2), 357–393.

Haq, I., Raj, A., 2019. Endocrine-disrupting pollutants in industrial wastewater and their degradation and detoxification approaches. Emerging and Eco-Friendly Approaches for Waste Management. Springer, pp. 121–142.

Hazrati, S., Rostami, R., et al., 2015. BTEX in indoor air of waterpipe cafés: levels and factors influencing their concentrations. Science of the Total Environment 524, 347–353.

He, Y., Wang, L., et al., 2021. Catalytic ozonation for metoprolol and ibuprofen removal over different MnO_2 nanocrystals: efficiency, transformation and mechanism. Science of the Total Environment 785, 147328.

Homem, V., Silva, E., et al., 2015a. Scented traces–dermal exposure of synthetic musk fragrances in personal care products and environmental input assessment.". Chemosphere 139, 276–287.

Homem, V., Silva, J.A., et al., 2015b. Long lasting perfume–a review of synthetic musks in WWTPs.". Journal of Environmental Management 149, 168–192.

Houtman, C.J., 2010. Emerging contaminants in surface waters and their relevance for the production of drinking water in Europe. Journal of Integrative Environmental Sciences 7 (4), 271–295.

Hu, F., Wang, Z., et al., 2018. Poly (*N*-vinyl imidazole) gel-filled membrane adsorbers for highly efficient removal of dyes from water. Journal of Chromatography A 1563, 198–206.

Ivanković, T., Hrenović, J., 2010. Surfactants in the environment. Arhiv za higijenu rada i toksikologiju 61 (1), 95–109.

Jafarinejad, S., 2016. Petroleum Waste Treatment and Pollution Control. Butterworth-Heinemann.

Jayasree, P., Remya, N., 2020. Photocatalytic degradation of paracetamol using aluminosilicate supported TiO_2. Water Science and Technology 82 (10), 2114–2124.

Johnson, A.C., Sumpter, J.P., 2001. Removal of endocrine-disrupting chemicals in activated sludge treatment works. Environmental Science & Technology 35 (24), 4697–4703.

Kabir, E.R., Rahman, M.S., et al., 2015. A review on endocrine disruptors and their possible impacts on human health. Environmental Toxicology and Pharmacology 40 (1), 241–258.

Kane, E.V., Newton, R., 2010. Occupational exposure to gasoline and the risk of non-Hodgkin lymphoma: a review and meta-analysis of the literature. Cancer Epidemiology 34 (5), 516–522.

Karadag, K., Yati, I., et al., 2016. Effective clean-up of organic liquid contaminants including BTEX, fuels, and organic solvents from the environment by poly (alkoxysilane) sorbents. Journal of Environmental Management 174, 45–54.

Kavanaugh, M.C., 2003. Unregulated and emerging chemical contaminants: technical and institutional challenges. Proceedings of the Water Environment Federation 2003 (12), 1–19.

Keeley, J., Townsend, B., 2011. Petrochemicals.

Kisku, G., Shukla, S., et al., 2015. Characterization and adsorptive capacity of coal fly ash from aqueous solutions of disperse blue and disperse orange dyes. Environmental Earth Sciences 74 (2), 1125–1135.

Kosma, C.I., Lambropoulou, D.A., et al., 2010. Occurrence and removal of PPCPs in municipal and hospital wastewaters in Greece. Journal of Hazardous Materials 179 (1–3), 804–817.

Koutník, I., Vráblová, M., et al., 2020. Reynoutria japonica, an invasive herb as a source of activated carbon for the removal of xenobiotics from water. Bioresource Technology 309, 123315.

Krause, M., Klit, A., et al., 2012. Sunscreens: are they beneficial for health? An overview of endocrine disrupting properties of UV-filters. International Journal of Andrology 35 (3), 424–436.

Kumar, A., Batley, G.E., et al., 2016. Derivation of water quality guidelines for priority pharmaceuticals. Environmental Toxicology and Chemistry 35 (7), 1815–1824.

Lan, T.T.N., Minh, P.A., 2013. BTEX pollution caused by motorcycles in the megacity of HoChiMinh. Journal of Environmental Sciences 25 (2), 348–356.

Lapworth, D., Baran, N., et al., 2012. Emerging organic contaminants in groundwater: a review of sources, fate and occurrence. Environmental Pollution 163, 287–303.

Limmun, W., Ito, A., et al., 2019. Removal of nonylphenol and nonylphenol monoethoxylate from water and anaerobically digested sewage sludge by Ferrate (VI). Chemosphere 236, 124399.

Liu, F., Chen, C., et al., 2021. Film-like bacterial cellulose/cyclodextrin oligomer composites with controllable structure for the removal of various persistent organic pollutants from water. Journal of Hazardous Materials 405, 124122.

Liu, J.-L., Wong, M.-H., 2013. Pharmaceuticals and personal care products (PPCPs): a review on environmental contamination in China. Environment International 59, 208–224.

López-Ortiz, C.M., Sentana-Gadea, I., et al., 2018. The use of combined treatments for reducing parabens in surface waters: ion-exchange resin and nanofiltration. Science of the Total Environment 639, 228–236.

Louis, C., Liu, Y., et al., 2016. PAH, BTEX, carbonyl compound, black-carbon, NO_2 and ultrafine particle dynamometer bench emissions for Euro 4 and Euro 5 diesel and gasoline passenger cars. Atmospheric Environment 141, 80–95.

Mahmoud, M.E., Abdou, A.E., et al., 2018. Sustainable super fast adsorptive removal of Congo red dye from water by a novel technique based on microwave-enforced sorption process. Journal of Industrial and Engineering Chemistry 57, 28–36.

Martins, M., Mourato, C., et al., 2017. Biogenic platinum and palladium nanoparticles as new catalysts for the removal of pharmaceutical compounds. Water Research 108, 160–168.

Melo, J.V., Barnes, D.J., 2007. Chronic myeloid leukaemia as a model of disease evolution in human cancer. Nature Reviews Cancer 7 (6), 441–453.

Modi, A., Bellare, J., 2019. Efficient removal of dyes from water by high flux and superior antifouling polyethersulfone hollow fiber membranes modified with ZnO/cGO nanohybrid. Journal of Water Process Engineering 29, 100783.

Moeini, Z., Azhdarpoor, A., et al., 2019. Removal of atrazine from water using titanium dioxide encapsulated in salicylaldehyde NH_2MIL-101 (Cr): adsorption or oxidation mechanism. Journal of Cleaner Production 224, 238–245.

Nguyen, C.H., Fu, C.-C., et al., 2018. Degradation of methylene blue and methyl orange by palladium-doped TiO_2 photocatalysis for water reuse: efficiency and degradation pathways. Journal of Cleaner Production 202, 413–427.

Orhon, K.B., Orhon, A.K., et al., 2017. Triclosan removal from surface water by ozonation-Kinetics and by-products formation. Journal of Environmental Management 204, 327–336.

Palharim, P.H., Graça, C.A., et al., 2020. Comparison between UVA-and zero-valent iron-activated persulfate processes for degrading propylparaben. Environmental Science and Pollution Research 27 (18), 22214–22224.

Peng, L., Qin, P., et al., 2012. Modifying Fe_3O_4 nanoparticles with humic acid for removal of Rhodamine B in water. Journal of Hazardous Materials 209, 193–198.

Periyasamy, S., Muthuchamy, M., 2018. Electrochemical oxidation of paracetamol in water by graphite anode: effect of pH, electrolyte concentration and current density. Journal of Environmental Chemical Engineering 6 (6), 7358–7367.

Petrovic, M., Barceló, D., 2006. Application of liquid chromatography/quadrupole time-of-flight mass spectrometry (LC-QqTOF-MS) in the environmental analysis. Journal of Mass Spectrometry 41 (10), 1259–1267.

Philip, J.M., Aravind, U.K., et al., 2018. Emerging contaminants in Indian environmental matrices—a review. Chemosphere 190, 307–326.

Phuekphong, A.F., Imwiset, K.J., et al., 2020. Organically modified bentonite as an efficient and reusable adsorbent for triclosan removal from water. Langmuir 36 (31), 9025–9034.

Rahman, F., Langford, K.H., et al., 2001. Polybrominated diphenyl ether (PBDE) flame retardants. Science of the Total Environment 275 (1–3), 1–17.

Rastgar, M., Shakeri, A., et al., 2020. Removal of trace organic contaminants by melamine-tuned highly cross-linked polyamide TFC membranes. Chemosphere 238, 124691.

Rego, R.M., Sriram, G., et al., 2021. Cerium based UiO-66 MOF as a multipollutant adsorbent for universal water purification. Journal of Hazardous Materials 416, 125941.

Reiner, J.L., Wong, C.M., et al., 2007. Synthetic musk fragrances in human milk from the United States. Environmental Science & Technology 41 (11), 3815–3820.

Richardson, S.D., 2009. Water analysis: emerging contaminants and current issues. Analytical Chemistry 81 (12), 4645–4677.

Rojas, S., Horcajada, P., 2020. Metal–organic frameworks for the removal of emerging organic contaminants in water. Chemical Reviews 120 (16), 8378–8415.

Scheurer, M., Brauch, H.-J., et al., 2009. Analysis and occurrence of seven artificial sweeteners in German waste water and surface water and in soil aquifer treatment (SAT). Analytical and Bioanalytical Chemistry 394 (6), 1585–1594.

Schwaiger, J., Ferling, H., et al., 2004. Toxic effects of the non-steroidal anti-inflammatory drug diclofenac: Part I: histopathological alterations and bioaccumulation in rainbow trout. Aquatic Toxicology 68 (2), 141–150.

Sharma, R., Saini, H., et al., 2021. Removal of organic dyes from wastewater using Eichhornia crassipes: a potential phytoremediation option. Environmental Science and Pollution Research 28 (6), 7116–7122.

Shi, H., Wang, Q., et al., 2020. Highly efficient removal of amoxicillin from water by three-dimensional electrode system within granular activated carbon as particle electrode. Journal of Water Process Engineering 38, 101656.

Slater, F.R., Singer, A.C., et al., 2011. Pandemic pharmaceutical dosing effects on wastewater treatment: no adaptation of activated sludge bacteria to degrade the antiviral drug Oseltamivir (Tamiflu®) and loss of nutrient removal performance. FEMS Microbiology Letters 315 (1), 17–22.

Smital, T., Kurelec, B., 1998. The chemosensitizers of multixenobiotic resistance mechanism in aquatic invertebrates: a new class of pollutants. Mutation Research/Fundamental and Molecular Mechanisms of Mutagenesis 399 (1), 43–53.

Snyder, S.A., Westerhoff, P., et al., 2003. Pharmaceuticals, personal care products, and endocrine disruptors in water: implications for the water industry. Environmental Engineering Science 20 (5), 449–469.

Sohoni, P., Sumpter, J., 1998. Several environmental oestrogens are also anti-androgens. Journal of Endocrinology 158 (3), 327–340.

Stuart, M., Lapworth, D., et al., 2012. Review of risk from potential emerging contaminants in UK groundwater. Science of the Total Environment 416, 1–21.

Tepe, O., Tunç, Z., et al., 2020. Efficient removal of paracetamol by manganese oxide octahedral molecular sieves (OMS-2) and persulfate. Water, Air, & Soil Pollution 231 (5), 1–15.

Thomas, K.V., Hilton, M.J., 2004. The occurrence of selected human pharmaceutical compounds in UK estuaries. Marine Pollution Bulletin 49 (5–6), 436–444.

USEPA, C., 2009. Water: Contaminant Candidate List 3. United States Environmental Protection Agency.

Viotti, P.V., Moreira, W.M., et al., 2019. Diclofenac removal from water by adsorption on Moringa oleifera pods and activated carbon: mechanism, kinetic and equilibrium study. Journal of Cleaner Production 219, 809–817.

Wei, X., Shi, Y., et al., 2016. Removal of trace phthalate esters from water by thin-film composite nanofiltration hollow fiber membranes. Chemical Engineering Journal 292, 382–388.

Wen, H., Yuan, L., et al., 2016. Effects of combined exposure to formaldehyde and benzene on immune cells in the blood and spleen in Balb/c mice. Environmental Toxicology and Pharmacology 45, 265–273.

Xu, J., Wang, L., et al., 2012. Decontamination of bisphenol A from aqueous solution by graphene adsorption. Langmuir 28 (22), 8418–8425.

Yang, Y., Ok, Y.S., et al., 2017. Occurrences and removal of pharmaceuticals and personal care products (PPCPs) in drinking water and water/sewage treatment plants: a review. Science of the Total Environment 596, 303–320.

Yüksel, S., Kabay, N., et al., 2013. Removal of bisphenol A (BPA) from water by various nanofiltration (NF) and reverse osmosis (RO) membranes. Journal of Hazardous Materials 263, 307–310.

Zhai, L., Bai, Z., et al., 2018. Fabrication of chitosan microspheres for efficient adsorption of methyl orange. Chinese Journal of Chemical Engineering 26 (3), 657–666.

Zhang, C., Wei, K., et al., 2017. Graphene oxide quantum dots incorporated into a thin film nanocomposite membrane with high flux and antifouling properties for low-pressure nanofiltration. ACS Applied Materials & Interfaces 9 (12), 11082–11094.

Zhang, J.-J., Fang, S.-S., et al., 2018. High-efficiency removal of rhodamine B dye in water using g-C_3N_4 and TiO_2 co-hybridized 3D graphene aerogel composites. Separation and Purification Technology 194, 96–103.

Chapter 24

Utilisation of agricultural waste biopolymer for soil erosion management

Soumya Pandey and Neeta Kumari
Department of Civil and Environmental Engineering, Birla Institute of Technology, Mesra, Ranchi, Jharkhand, India

Introduction

Soil erosion is one of the greatest challenges for sustainable soil management. It directly threatens global soil productivity and food security (FAO, 2015, 2019). Soil erosion is a natural process of land formation but the process has enhanced severely due to anthropogenic activities (contributing about 60%) (Lee and Heo, 2011; Rahmani et al., 2018). Globally, Africa, Asia, Latin America, and North Africa have been reported to have a degrading pattern of soil erosion, while Europe, North America, and the Southwest Pacific had an improving trend of erosion. The most suffering zone in the world has been reported as Southeast Asia (Meshesha et al., 2012; Sathya et al., 2021). As per FAO (2015), 33% of the global soil reserves have already deteriorated and by 2050 about 90% will deteriorate (FAO and ITPS, 2015; IPBES, 2018). Among the targets of Sustainable Development Goals (SDGs) for 2030 by the United Nations, SDG 15: life on land aims at "protection, conservation, sustainable promotion of terrestrial ecosystem combat desertification, halt land degradation and biodiversity loss." Physical land degradation (Addressed in SDG 15.3) is majorly caused by soil erosion by agents like water and wind resulting in massive environmental and economic losses globally. For instance, the global average soil erosion rate is 2.4 t/ha/year (FAO, 2021). Soil erosion is a process of weathering the top soil or surface particles under the influence of mechanical forces such as rain, runoff, and wind (Rostami et al., 2022). The soil particles get weathered mostly because of water and wind (Bhattacharya et al., 2019). Water (rainfall) causes many types of erosion such as rill erosion, sheet erosion, scalding, tunnel, stream bank, landslide, rock avalanches, wind erosion, etc. (Bhattacharyya et al., 2015; Prăvălie et al., 2021; Poesen, 2018; Sran et al., 2012). Other than this some other activities like land leveling, overgrazing, soil mining, crop cultivation, trench and fill are also sources of erosion (Donovan and Monaghan, 2021; Imran and Mehmood, 2020; Sran et al., 2012). Mass movement or wasting in the form of landslides, avalanches, slumping, debris flow, and landslides are more prominent sources in Hilly or mountainous areas (Wood et al., 2020).

Soil erosion

Soil erosion majorly occurs when the shear strength of the rain and wind exceeds the shear strength of the soil particles (Fenta et al., 2016; Filho et al., 2001; Gyssels et al., 2005; Vigiak et al., 2015; Wang et al., 2021). Water erosion among all forms contributes largely to soil erosion. The naturally weathered material stays in its parent location and gets transported through the water (Scholz et al., 2021; Singh et al., 2012). Usually, the detachment of particles happens as:

Splash erosion

The high raindrop intensity acts on the soil surface causing splashes and the particle gets detached (Bojie et al., 2018; Liu et al., 2018).

Surface erosion

When the flowing surface runoff exerts forces on the soil surface the soil particles move along with flowing water (Moreno-Jiménez et al., 2020; Ustaoğlu et al., 2021).

Role of Green Chemistry in Ecosystem Restoration to Achieve Environmental Sustainability. DOI: https://doi.org/10.1016/B978-0-443-15291-7.00019-5

Wind erosion

The detachment of the soil occurs because of the forces exerted by the wind on the soil surface, and because of the effect of detached soil bouncing off the soil surface downwind of the point of initial detachment (saltation) (Zhao et al., 2022).

Tillage erosion

Both the detachment and transport of soil by tillage implements cause the thinning of soils on upper slope areas and can result in over-thickened depositional soils in lower slope positions (Vasu et al., 2019; Wolka et al., 2018).

Factors contributing to soil loss

Properties of soil

Soil strength, structure, nature, Mineralogy. All these properties are interdependent (Almajed et al., 2021). The shear strength of soil determines its ability to protect the soil against shear failure. The mineralogy of soil and its particle size distribution are also crucial factors (Adnan et al., 2021; Zhao et al., 2018). Cohesionless soil (sand and silt soil) is more prone to erosion compared to cohesive soil (clay) (De Baets and Poesen, 2010). Clay mineralogy affects the soil-water interaction (Nabiollahi et al., 2017). Smectite group soil is the most expansive soil as compared to Kaolinite and Illite which are usually non-plastic clays. Other than these properties, the presence of higher organic content prevents soil from getting eroded (Anglade et al., 2022; Deb and Sahu, 2010; Singh et al., 2018).

External forces

Wind and rain are two common external forces that can easily exert forces higher than the shear strength of soil, depending upon their intensity (Bhattacharyya et al., 2015).

Land use land cover

The roots of crops provide both lateral and vertical strengthening to surface soil (Thakur et al., 2020). These covers improve the infiltration rate of water, increase the organic content of the soil, and control mineral and nutrient leaching. The lack of vegetation makes the soil vulnerable to erosion (Gikas et al., 2020).

Topographic features

The slope and slope length play a crucial role in enhancing the erosion process. The areas with higher slopes are more vulnerable to erosion (Thapa, 2020). The eroded particles flow down with runoff (which also flows with higher intensity on slopes) from the highlands and settle in the lowlands.

Why soil erosion is a global threat?

Soil erosion is a massive threat because of multiple reasons. As the soil formation processes occur at a sluggard rate, erosion, and mining occur at a humungous scale (Wu et al., 2020a, b). In general, the topsoil is the richest in nutrients essential for vegetation growth. Erosion usually affects the surface soil thus, depriving us of soil fertility and threatening food security (Vasu et al., 2019). The eroded soil flows to the lowlands and water bodies, clogging the soil surface. Clogging also results in decreased infiltration. In recent years, this problem has started to gain attention globally (Rostami et al., 2022). Soil erosion is estimated using various Geospatial models like USLE (universal soil loss equation) (Patel and Kathwas, 2012), RUSLE (revised universal soil loss equation) (Fig. 24.1), MUSLE (modified universal soil loss equation) (Kumar et al., 2020), SWAT (soil and water assessment tool) (Ricci et al., 2020), WEPP (water erosion prediction project) model (Alewell et al., 2019), etc., are commonly used. These models require processed geospatial data as input and using those the estimation is done. Common input data are soil data (soil quality and strength properties, rainfall data), elevation, slope, flow accumulation, land use, and land cover (Choudhary et al., 2012), and normalized difference vegetation index (Imran and Mehmood, 2020). The use of machine learning has been gaining popularity in soil loss estimation. For instance, artificial neural networks (ANN) combined with different models like

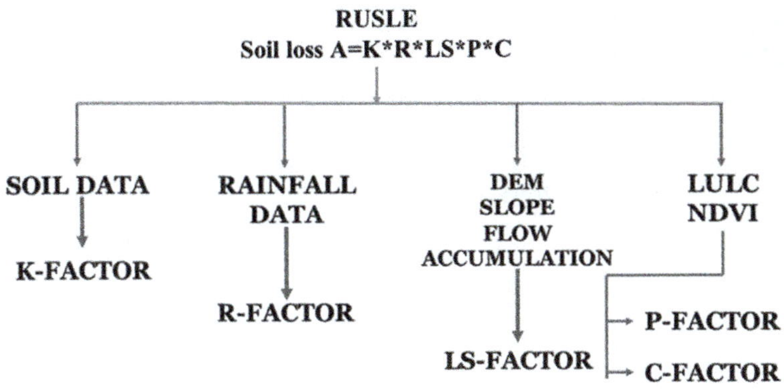

FIGURE 24.1 An example of a soil erosion estimation model: the revised universal soil loss equation (RUSLE model).

WEPP are being studied for their accuracy (Albaradeyia et al., 2011). Some of the advanced techniques used to control erosion are the use of concrete, geosynthetic lining, soil nailing, afforestation, and enhanced vegetation conservation (mulching, agroforestry). Although many traditional and non-traditional practices are being followed globally (Mishra et al., 2017).

Overview of soil conservation techniques

Some of the commonly used techniques for soil conservation are Rubble Machinery (Esiukova et al., 2019), rubbles, or large-size stones placed across the coastal line or shorelines so that the coastal banks do not get eroded by the intensity of water. The rubble controls the water erosive action due to high-intensity flow (Panagos and Katsoyiannis, 2019) and fluctuation in water levels. Check dams are civil construction done on any active rill or gully that may enhance runoff water (Reddy et al., 2021). The water flow does not get obstructed as the safe discharge is allowed through a spillway (Pandey and Singh, 2017). Furthermore, modern materials such as geosynthetics (geotextiles, geomembrane, geocells, geo grids) are made from polymers of polypropylene, polyester, polyethylene, nylon, polyvinylchloride (Esiukova et al., 2019; Monkul and Özhan, 2021; Williams et al., 2001) which are non-reactive and are do not decompose under microbial attack. They are available in different sizes and thicknesses in the form of rolls and can also be manufactured as per design specifications. Geosynthetics have changed the shape of soil mechanics and opened far more options, especially for slope stability and erosion control material (Yu et al., 2021). Geosynthetics are placed in or on the soil in layers and compacted so that the slope stability is improved (Kavazanjian et al., 2009) (Fig. 24.2). A separate drainage line is provided for water to move along so that the soil structure does not get affected. Although they are made to withstand the higher strain, geosynthetics cannot survive long under the presence of heat and cold cycles and tend to become brittle due to ageing (Monkul and Özhan, 2021). They then rupture and disintegrate.

A study on the environmental impact of geosynthetics showed that polymers are more prone to degradation under the influence of UV light as compared to seawater immersion (Scholz et al., 2021; Esiukova et al., 2019). Microplastics settle down on the sea bed and thus adversely affect marine aquaculture and corals. Another process for soil conservation used is the mulching films (Dong et al., 2013). They are a boon to farmlands as they help in creating a curing temperature in which a variety of plants can easily survive, they help retain moisture while allowing infiltration, control weed growth, and are relatively cheap (Chalise et al., 2019; Sith et al., 2019). In places with the issue of water stress, mulching helps crop productivity (Kumar and Sheela, 2021; Huang et al., 2020). In a study conducted by (Vinoth and Sheela, 2020), on the effect of plastic film mulching on farms, it was found that the distribution of plastic residue decreased with increasing depth. The exposure of soil to microplastics leads to the absorption of organic and heavy metal pollutants. These microplastics tend to spread in nearby land and water bodies from the area of contamination. Thus, enhancing pollution in the environment (Meng et al., 2020). As per the study by Steinmetz et al. (2016), "Plastic mulches have the potential to alter soil quality by shifting the edaphic biocoenosis, accelerate C/N metabolism eventually depleting soil organic matter stocks, increase soil water repellence and favor the release of greenhouse gases" (Steinmetz et al., 2016). In general, it can be said that the use of all traditional and advanced materials has some other disadvantages. The material that is focused more on strength parameters rarely undergoes long-term study of their environmental impact (Orts et al., 1999).

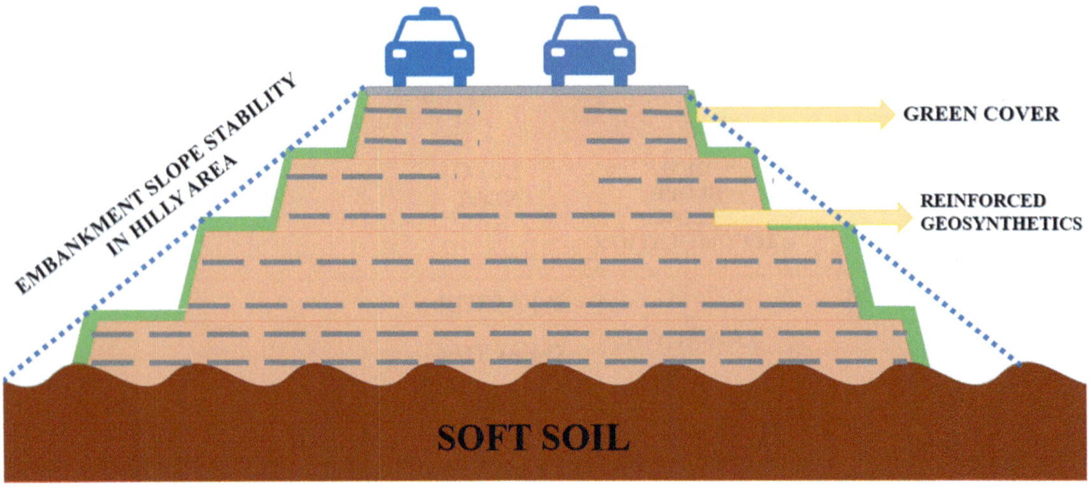

FIGURE 24.2 Use of geosynthetics in soil stability and erosion control in hilly areas.

Role of agriculture waste biopolymer in promoting environmental sustainability via green chemistry

Among other techniques, soil stabilization using chemical additives has been practiced for a long time (Zhai et al., 2018). Some common additives that are used are lime, cement, bitumen, fly ash, calcium chloride, sodium carbonate, sodium sulfate and their combination. However, most of these additives are unable to support the requirements of agricultural practice (Ahamad et al., 2018; Gómez-García et al., 2021). Additives like cement and agriculture lime have been reported to increase the carbon dioxide content in the atmosphere, increasing global warming (Alotaibi et al., 2022; Bahrami et al., 2016; Moorthy et al., 2020). Nonetheless, erosion mostly occurs in croplands and in this land, degradation process the top fertile soil is mostly lost (Amanambu et al., 2019; Dutal and Reis, 2020; Wang et al., 2021). This calls for a better eco-friendly source of soil stabilizers. Green chemistry is the new concept of minimizing or eliminating the usage of chemicals in products and processes for better pollution control and degradation of the quality of natural resources (Gómez-García et al., 2021; Scarlat et al., 2015). In the field of agriculture, concepts of green chemistry have promoted the application of biodegradable waste in innovative ways (Diaz-Sarachaga, 2021; Philippidis et al., 2019). The use of renewable biomass resources has increased the bio-friendly food products, which also promotes zero waste (SDG 12.5), increases socio-economic opportunities (SDG 8 and 1) and lowers adverse environmental impact (SDG 13 and 15) (Adewuyi and Awodumi, 2020; Anastas and Zimmerman, 2018; Kapović Solomun et al., 2018; Tóth et al., 2018). The roots of vegetation cover provide vertical and horizontal strengthening to surface soil and help in keeping the geomembrane layer in place. Similarly, the use of agriculture waste biopolymers (AWB) has gained superiority over other materials in soil conservation techniques (Lentz, 2015; Yu et al., 2021). AWBs are more eco-friendly, do not hamper soil quality, provide enrichment and strength, help in agriculture waste utilization, and promote a circular economy (Pandey and Kumari, 2021).

Types and characteristics of agriculture waste biopolymers

The disposal of agricultural waste is a common problem faced globally. If these wastes are left to dispose of naturally, they produce a tone of greenhouse gases such as methane, carbon dioxide, etc., which is a huge concern for the environment (Chen et al., 2018; Stringfellow et al., 2013; Upgupta and Singh, 2020; Vermeulen et al., 2012; Walia and Dick, 2018). In recent decades agricultural waste management has been considered crucial for both waste disposal and promoting the circular economy (concept of rethink, redesign, reduce, reuse, recycle) (Corrado and Sala, 2018; Gupta et al., 2021; Stefanakis, 2019). Agriculture waste is enriched with all sorts of organic compounds, and macromolecules (polysaccharides, polypeptides, polyaromatics, etc.) (Ramadhan and Handayani, 2020; Schnepp, 2013). Biopolymers can be defined as polymers made from natural compounds that are formed naturally in the life cycles of organisms and plants, mostly in their cell through complex metabolism (Chang et al., 2020; Kwon et al., 2019). Biopolymers are also considered as organic polymers naturally derived from agricultural biomass and compounds, microbes, fossil resources,

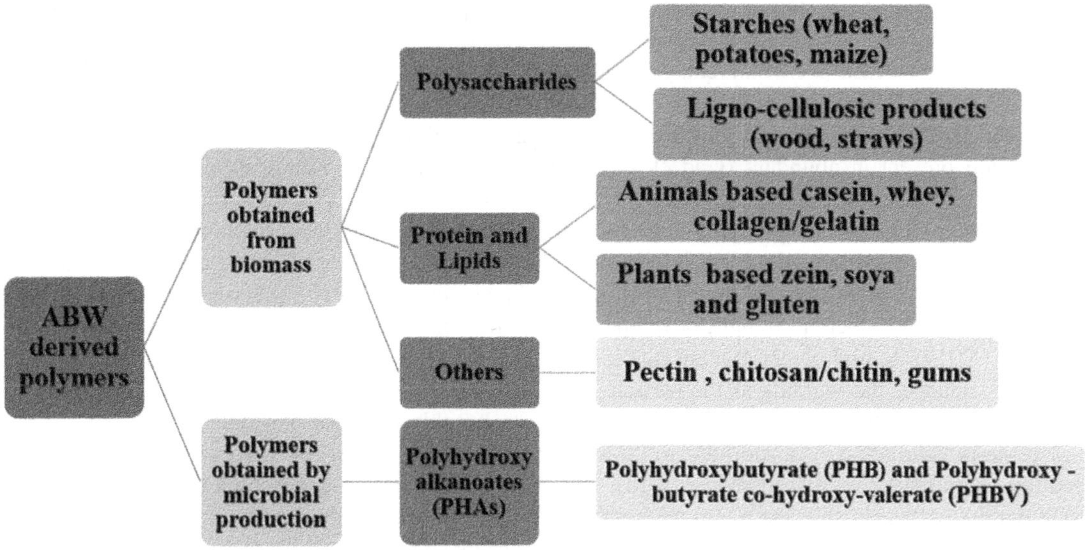

FIGURE 24.3 AWB extracted from agricultural biomass and microbial production.

and other agro-resource monomers (Fig. 24.3). The variation in the functional groups of biopolymers bestow them with varying properties such as bio-compatibility, enhanced porosity, water adsorption capacity, biodegradability, binding, modification properties (Erogbogbo et al., 2010). Hence, depending on their type and characteristics they can be extracted and used for different purposes. Additionally, biopolymers are cheap along with having greater surface area, functionality, and regenerative properties. Among various types of extractable biopolymers, one of the most popular one is polysaccharides which has diverse classification based on structure, source, origin, etc. (Joseph et al., 2020).

For instance, originating from plants consists of cellulose, starch, pectin, xylan, and gums; originating from animals consists of chitin, chitosan, and heparin; while originating from algal and microbial origin consists of alginate, agar, cellulose, dextran, curdlan, etc. (Sheibani et al., 2021; Banerjee et al., 2018). Out of these biopolymers, the wider application of gums (acacia, xanthan, gellan, guar, agar), chitosan and chitin, starch, casein, dextran are studied more actively for soil stabilization and in the civil construction industry. Guar gum has some advantageous features, like rapid hydration in cold water, and high viscosity at lower concentrations, as compared to xanthan gum for the same percentage of additive (Carpinteyro-Urban et al., 2012). The efficiency of guar gum is subjected to its concentration, temperature, viscosity, pH, and dispersion, although it requires a controlled hydration processing for at least 2 hours in practice for it to reach maximum viscosity (Torres and Carpinteyro-Urban, 2012). It has been found effective in treating the sandy soil and can also act as a barrier wall (Chen et al., 2013; Dehghan et al., 2019). Similarly different types of gum have their utility but their efficiency depends upon the type of soil, water content, concentration, and curing period (Chang et al., 2020). Many combinations of gums or other chemicals are also added to the polysaccharides to enhance their soil additive interaction (Chang et al., 2020). Chitin and chitosan are low-cost biopolymers obtained from skeletal polysaccharides or discarded crustacean shells (crabs, lobsters, shrimps, and insects) from aquaculture-based industries (Begum et al., 2021; Hataf et al., 2018). Shell waste is produced in mass and is estimated to be 1.44 million tonnes annually and is known for its effectiveness in erosion control, especially during irrigation (Rabea et al., 2003; Yusuf, 2017). Sandy soil coated with chitosan has low hydraulic conductivity, metal binding capacity, metal ion leach control, groundwater purification, and improves water quality (Chen et al., 2019). Under wet conditions, chitosan provides better interaction between particles in the early days while in a dry state loses its efficiency over time (Hataf et al., 2018). Cellulose is present in nature and is main constituent of the cell wall of plants (Jose et al., 2016; Ramos, 2011). Study has shown that cellulose can be used for strength improvement (both flexure and tensile strength) (Das et al., 2013; Mostafa et al., 2018).

Biopolymer-soil interaction mechanisms and impact on soil properties

The application of biopolymers for soil erosion stability is being researched frequently. The curing variation under different moisture content, microbial activity, concentration, etc., has been attempted frequently (Gao et al., 2018; Hataf et al., 2018). For a better bonding and understanding of the function of biopolymers with soil, it is crucial to understand the mechanism of their interaction (Benchaa et al., 2021; Chang et al., 2015; Latifi et al., 2016; Zhou et al., 2020). For soil strengthening, biopolymer additives are added in small quantities, and yet the results are significantly positive. Biopolymers have high surface area and charges thus promoting bonding between biopolymer and soil particles (Alminderej et al., 2021). The matrices formed are quite strong as the bonds formed are ionic and hydrogen bonds. Kaolinite (both negative and positive charge) (Anglade et al., 2022; Pal and Maiti, 2018) and montmorillonite (majorly negative charge) are present in abundant (Gad et al., 2021; Huo et al., 2021; Zhang et al., 2021). Kaolinite is non-swelling soils whereas montmorillonites (bonded with weak van der Waal forces) are swelling soils, that are in contact with water the soil starts to expand volumetrically (Essebaai et al., 2021). Clay absorbs biopolymers and forms direct links (ionic bond) between soil and the biopolymer matrix. Due to the presence of ionic bond,s the clay-biopolymer matrix has stronger interparticle connection and efficiency (Fatehi et al., 2021). The well-graded soil (having both coarse and fine soil) provides better strength development post-treatment as compared to pure clays (El Jazouli et al., 2020; Rawat et al., 2019). In high water content, the biopolymer soil bond seems weaker for most polysaccharides, but low water content enhances the bonding as the particle contact increases (Chen et al., 2019). The addition of chemicals can be used to improve the strengthening effects of biopolymers, but this will also be subject to the type of soil (Lentz, 2015). An increase in volumetric expansion due to the suction of water by hydrogels, lowers the strength bonds between particles and the overall stiffness of soil (Kaur, 2018). This reduces the shear strength by a tenth of soil strength in the dried state. In the case of sandy soil, the hydrogel form act as pore filler thus reducing the hydraulic conductivity of the soil. This makes biopolymers more efficient for grouting and barriers (Bai et al., 2019; Gao et al., 2018).

Application of AWB for erosion control

The use of biogenic excretion is far more appreciated than microbial cultivation for soil stabilization and strengthening (Issac and Kandasubramanian, 2021). The microbial process requires frequent injecting, curing time, more precipitation for growth and bonding, etc. Much research has shown that biopolymers have the potential to increase the slope stability and resistivity to erosion (Table 24.1) (Nguyen et al., 2021). Biopolymers, especially made from ABW (Xanthan gum, Chitosan, etc.) are better in controlling surface erosion, by multiple interactions such as bio-crusting, bio-aggregation, bio-cementation, etc. (Arab et al., 2021). Among other polysaccharides-biopolymers, xanthan gum is more effective in erosion control as it reduces the hydraulic conductivity of silts, sands, and silty sands through pore clogging. This improves the undrained shear strength of coarser soil as its liquid limit is enhanced (Kwon et al., 2019). Cellulose produces gelation which causes thick layering and bonding between particles, thus providing stable strengthening of soil (Jose et al., 2016; Maraveas, 2020). Similarly, curdlan also works on clogging potential in soil treatment, beta-glucan enhances the mechanical strength of the soil matrix, and agar gum and gellan gum provide thermoregulation and strength to cohesionless soil (Hataf et al., 2018). Chang et al. (2015) studied the effects of xanthan gum on soil strengthening and found that xanthan gum works better with soil at higher concentrations and was subjected to higher levels of the water content of the soil, type of soil, and mixing method. The strength enhancement occurs due to the strong attachment between additive polymer chains and clay soil which have high cation exchange capacity (Nascimento et al., 2021). Wade et al. (2021) studied the use of

TABLE 24.1 Biopolymers and their interaction with soil.

Biopolymer	Interaction with soil	Benefit to agriculture
Agar gum	Strengthening, bio-clogging, erosion reduction	High water absorption capacity, biodegradability, nutrient addition
Guar gum	Dust control, strengthening, bio-grouting	Increase soil moisture, water retention, biodegradability, nutrient addition
Gellan gum	Strengthening, bio-clogging, erosion reduction	Biodegradable, strong water retention capacity, low cost, controls nutrient leaching
Xanthan gum	Drilling mud thickener, strengthening	Increase water retention, reduce evaporation, increase percolation
Chitosan	Coagulants, heavy metal reduction in water	Pesticides, herbicides, insecticides, high productivity crop yield, fertilizer

alginate and guar gum spray to improve soil aggregation and soil crust formation against wind erosion. The spray increased the diameter of particles by 1.5 times and its mechanical strength. This reduced the vulnerability of soil to wind erosion (Pastore et al., 2021).

Conclusion

Soil erosion is one of the most problematic ecological issues faced globally. Irrespective of developed and developing country, global erosion cost billions of dollar loss every year. In the past few decades, the application of various techniques has been used to geotechnically modify soil strength and hydraulics to control erosion. The use of check dams, gabion walls, rubble machinery, geosynthetics, mulching, etc., have been practiced frequently to conserve soil against external forces like wind, water, etc. the soil gets weathered or eroded when the shear strength of external factors is greater than the shear strength of the soil. Additionally, if the shear strength of soil is genuinely low then it becomes susceptible to erosion easily. Compared to sandy soil (cohesionless soil) which is inert, clayey soils are far more cohesive and have better interparticle attraction. In the case of sandy soil, the pores are filled with hydrogel and the inert particles start to bond together. Overall, the application of biopolymers is effective but the efficiency and bond formation is subject to the type of soil, the concentration of biopolymers, water content, and curing period. The concentration and water content should be fixed before field application. The utilization of agricultural waste in soil strengthening against erosion would not only be an ecologically and economically better option but would also help in promoting a circular economy.

References

Adewuyi, A.O., Awodumi, O.B., 2020. Environmental pollution, energy import, and economic growth: evidence of sustainable growth in South Africa and Nigeria. Environmental Science and Pollution Research . Available from: https://doi.org/10.1007/s11356-020-11446-z.

Adnan, S., Aldefae, A.H., Humaish, W.H., 2021. Soil erosion and the influenced factors: a review article. IOP Conference Series: Materials Science and Engineering 1058 (1), 012041. Available from: https://doi.org/10.1088/1757-899x/1058/1/012041.

Ahamad, A., Madhav, S., Singh, P., Pandey, J., 2018. Assessment of groundwater quality with special emphasis on nitrate contamination in parts of Varanasi City, Uttar Pradesh, India. Applied Water Science 8 (4), 1–13. Available from: https://doi.org/10.1007/s13201-018-0759-xs.

Albaradeyia, I., Hani, A., Shahrour, I., 2011. WEPP and ANN models for simulating soil loss and runoff in a semi-arid Mediterranean region. Environmental Monitoring and Assessment 180 (1–4), 537–556. Available from: https://doi.org/10.1007/s10661-010-1804-x.

Alewell, C., Borrelli, P., Meusburger, K., Panagos, P., 2019. Using the USLE: chances, challenges and limitations of soil erosion modelling. International Soil and Water Conservation Research 7 (3), 203–225. Available from: https://doi.org/10.1016/j.iswcr.2019.05.004.

Almajed, A., Abbas, H., Almusallam, T., Al-Salloum, Y., 2021. Stabilization of sand using energy efficient materials under normal and extreme hot weathers. Journal of Cleaner Production 285. Available from: https://doi.org/10.1016/j.jclepro.2020.124914.

Alminderej, F.M., Ammar, C., El-Ghoul, Y., 2021. Functionalization, characterization and microbiological performance of new biocompatible cellulosic dressing grafted chitosan and Suaeda fruticosa polysaccharide extract. Cellulose 28 (15), 9821–9835. Available from: https://doi.org/10.1007/s10570-021-04155-4.

Alotaibi, E., Arab, M.G., Abdallah, M., Nassif, N., Omar, M., 2022. Life cycle assessment of biocemented sands using enzyme induced carbonate precipitation (EICP) for soil stabilization applications. Scientific Reports 12 (1). Available from: https://doi.org/10.1038/S41598-022-09723-7.

Amanambu, A.C., Li, L., Egbinola, C.N., Obarein, O.A., Mupenzi, C., Chen, D., 2019. Spatio-temporal variation in rainfall-runoff erosivity due to climate change in the Lower Niger Basin, West Africa. Catena 172 (August 2018), 324–334. Available from: https://doi.org/10.1016/j.catena.2018.09.003.

Anastas, P.T., Zimmerman, J.B., 2018. The United Nations sustainability goals: how can sustainable chemistry contribute? Current Opinion in Green and Sustainable Chemistry 13, 150–153. Available from: https://doi.org/10.1016/j.cogsc.2018.04.017.

Anglade, E., Aubert, J.E., Sellier, A., Papon, A., 2022. Physical and mechanical properties of clay–sand mixes to assess the performance of earth construction materials. Journal of Building Engineering 51 (December 2021), 1–15. Available from: https://doi.org/10.1016/j.jobe.2022.104229.

Arab, M.G., Alsodi, R., Shanablah, A., Kavazanjian, E., Zeiada, W., 2021. Evaluation of microstructural strength of bio-cemented sand crust using rheometry. Geotechnique Letters 11 (4), 306–312. Available from: https://doi.org/10.1680/JGELE.21.00030.

Bahrami, A., Soltani, N., Pech-Canul, M.I., Gutiérrez, C.A., 2016. Development of metal-matrix composites from industrial/agricultural waste materials and their derivatives, Critical Reviews in Environmental Science and Technology, 46. Taylor and Francis Inc, pp. 143–207Issue 2. Available from: https://doi.org/10.1080/10643389.2015.1077067.

Bai, J., Li, S., Jiang, Y., Liu, R., Li, Z., Li, W., 2019. An extension theoretical model for grouting effect evaluation in sand stratum of metro construction. KSCE Journal of Civil Engineering 23, 2349–2358. Available from: https://doi.org/10.1007/s12205-019-1128-5.

Banerjee, S., Thrall, P.H., Bissett, A., Van Der Heijden, M.G.A., Richardson, A.E., 2018. Linking microbial co-occurrences to soil ecological processes across a woodland-grassland ecotone. International Journal of Business Innovation and Research 17 (3), 8217–8230. Available from: https://doi.org/10.1002/ece3.4346.

Begum, S., Yuhana, N.Y., Md Saleh, N., Kamarudin, N.H.N., Sulong, A.B., 2021. Review of chitosan composite as a heavy metal adsorbent: material preparation and properties. Carbohydrate Polymers 259 (January), 117613. Available from: https://doi.org/10.1016/j.carbpol.2021.117613.

Benchaa, S., Gheriani, R., Achouri, A., Bouguettaia, H., Mechri, M.L., 2021. Structural characterizations of dune sand and construction sand of Sidi Slimane and Zaouia El Abidia areas in the Touggourt region in southeast Algeria. Arabian Journal of Geosciences 14 (22). Available from: https://doi.org/10.1007/s12517-021-08303-9.

Bhattacharyya, R., Ghosh, B.N., Mishra, P.K., Mandal, B., Rao, C.S., Sarkar, D., et al., 2015. Soil degradation in india: challenges and potential solutions. Sustainability (Switzerland) 7 (4), 3528–3570. Available from: https://doi.org/10.3390/su7043528.

Bhattacharya, R., Dolui, G., Das Chatterjee, N., 2019. Effect of instream sand mining on hydraulic variables of bedload transport and channel planform: an alluvial stream in South Bengal basin, India. Environmental Earth Sciences 78 (10), 1–24. Available from: https://doi.org/10.1007/s12665-019-8267-3.

Bojie, F., Academician, C., Jianbo, L., Guangyao, G., Shuai, W., Lei, J., et al., 2018. Foundation: National Natural Science Foundation of China, No.41390464; National Key Research and Development Program. Youth Innovation Promotion Association CAS, 28(1), 59–78. http://www.geogsci.com; http://www.springerlink.com/content/1009-637x.

Carpinteyro-Urban, S., Vaca, M., Torres, L.G., 2012. Can vegetal biopolymers work as coagulant-flocculant aids in the treatment of high-load cosmetic industrial wastewaters? Water, Air, and Soil Pollution 223 (8), 4925–4936. Available from: https://doi.org/10.1007/s11270-012-1247-9.

Chalise, D., Kumar, L., Kristiansen, P., 2019. Land degradation by soil erosion in Nepal: a review. Soil Systems 3 (1), 1–18. Available from: https://doi.org/10.3390/soilsystems3010012.

Chang, I., Im, J., Prasidhi, A.K., Cho, G.C., 2015. Effects of Xanthan gum biopolymer on soil strengthening. Construction and Building Materials 74 (x), 65–72. Available from: https://doi.org/10.1016/j.conbuildmat.2014.10.026.

Chang, I., Lee, M., Tran, A.T.P., Lee, S., Kwon, Y.M., Im, J., et al., 2020. Review on biopolymer-based soil treatment (BPST) technology in geotechnical engineering practices. Transportation Geotechnics 24 (June), 100385. Available from: https://doi.org/10.1016/j.trgeo.2020.100385.

Chen, R., Zhang, L., Budhu, M., 2013. Biopolymer stabilization of mine tailings. Journal of Geotechnical and Geoenvironmental Engineering 139 (10), 1802–1807. Available from: https://doi.org/10.1061/(asce)gt.1943-5606.0000902.

Chen, W., He, B., Nover, D., Duan, W., Luo, C., Zhao, K., et al., 2018. Spatiotemporal patterns and source attribution of nitrogen pollution in a typical headwater agricultural watershed in Southeastern China. Environmental Science and Pollution Research 25 (3), 2756–2773. Available from: https://doi.org/10.1007/s11356-017-0685-8.

Chen, C., Wu, L., Perdjon, M., Huang, X., Peng, Y., 2019. The drying effect on xanthan gum biopolymer treated sandy soil shear strength. Construction and Building Materials 197, 271–279. Available from: https://doi.org/10.1016/j.conbuildmat.2018.11.120.

Choudhary, J.S., Shukla, G., Prabhakar, C.S., Maurya, S., 2012. Assessment of local perceptions on climate change and coping strategies in Chotanagpur Plateau of Eastern India. Journal of Progressive Agriculture 3 (1), 8–15.

Corrado, S., Sala, S., 2018. Food waste accounting along global and European food supply chains: state of the art and outlook. Waste Management 79, 120–131. Available from: https://doi.org/10.1016/j.wasman.2018.07.032.

Das, S.P., Ravindran, R., Deka, D., Jawed, M., Das, D., Goyal, A., 2013. Bioethanol production from leafy biomass of mango (*Mangifera indica*) involving naturally isolated and recombinant enzymes. Preparative Biochemistry and Biotechnology 43 (7), 717–734. Available from: https://doi.org/10.1080/10826068.2013.773342.

De Baets, S., Poesen, J., 2010. Empirical models for predicting the erosion-reducing effects of plant roots during concentrated flow erosion. Geomorphology 118 (3–4), 425–432. Available from: https://doi.org/10.1016/j.geomorph.2010.02.011.

Deb, S., Sahu, S.S., 2010. Study of Mineralogy and Weathering of Two Soil Profiles of Undulating Plateau Area of Jumar Subwatershed, Jharkhand, India. 3(September), 295–301.

Dehghan, H., Tabarsa, A., Latifi, N., Bagheri, Y., 2019. Use of xanthan and guar gums in soil strengthening. Clean Technologies and Environmental Policy 21 (1), 155–165. Available from: https://doi.org/10.1007/s10098-018-1625-0.

Diaz-Sarachaga, J.M., 2021. Shortcomings in reporting contributions towards the sustainable development goals. Corporate Social Responsibility and Environmental Management 28 (4), 1299–1312. Available from: https://doi.org/10.1002/CSR.2129.

Dong, H., Liu, T., Li, Y., Liu, H., Wang, D., 2013. Effects of plastic film residue on cotton yield and soil physical and chemical properties in Xinjiang. Nongye Gongcheng Xuebao/Transactions of the Chinese Society of Agricultural Engineering 29 (8), 91–99. Available from: https://doi.org/10.3969/j.issn.1002-6819.2013.08.011.

Donovan, M., Monaghan, R., 2021. Impacts of grazing on ground cover, soil physical properties and soil loss via surface erosion: a novel geospatial modelling approach. Journal of Environmental Management 287 (March), 112206. Available from: https://doi.org/10.1016/j.jenvman.2021.112206.

Dutal, H., Reis, M., 2020. Determining the effects of land use on soil erodibility in the Mediterranean highland regions of Turkey: a case study of the Korsulu stream watershed. Environmental Monitoring and Assessment 192 (3). Available from: https://doi.org/10.1007/s10661-020-8155-z.

El Jazouli, A., Barakat, A., Khellouk, R., 2020. Geotechnical studies for Landslide susceptibility in the high basin of the Oum Er Rbia river (Morocco. Geology, Ecology, and Landscapes 00 (00), 1–8. Available from: https://doi.org/10.1080/24749508.2020.1743527.

Erogbogbo, F., Yong, K.T., Hu, R., Law, W.C., Ding, H., Chang, C.W., et al., 2010. Biocompatible magnetofluorescent probes: luminescent silicon quantum dots coupled with superparamagnetic iron(III) oxide. ACS Nano 4 (9), 5131–5138. Available from: https://doi.org/10.1021/nn101016f.

Esiukova, E.E., Chubarenko, B.V., Simon, F.G., 2019. Debris of geosynthetic materials on the shore of the South-Eastern Baltic (Kaliningrad Oblast, the Russian Federation). 2018 IEEE/OES Baltic International Symposium, BALTIC 2018, 3–8. Available from: https://doi.org/10.1109/BALTIC.2018.8634842.

Essebaai, H., Lgaz, H., Alrashdi, A.A., Habsaoui, A., Lebkiri, A., Marzak, S., et al., 2021. Green and eco-friendly montmorillonite clay for the removal of Cr(III) metal ion from aqueous environment: montmorillonite clay for the removal of Cr(III) metal ion from aqueous solutions. International Journal of Environmental Science and Technology, Iii . Available from: https://doi.org/10.1007/s13762-021-03303-4.

FAO, 2019. Soil erosion: the greatest challenge to sustainable soil management. Rome. 100 pp.

FAO, 2021. Food Waste Index Report 2021. In Unep.

FAO, 2015. Global food losses and food waste facts. Global Food Losses and Food Waste – Extent, Causes and Prevention. FAO, p. 4807.

FAO, ITPS, 2015. Status of the World's Soil Resources (SWSR)—Main Report. Food and Agriculture Organization of the United Nations and Intergovernmental Technical Panel on Soils, Rome, Italy. Available from: http://www.fao.org/3/a-i5199e.pdf.

Fatehi, H., Ong, D.E.L., Yu, J., Chang, I., 2021. Biopolymers as green binders for soil improvement in geotechnical applications: a review. Geosciences (Switzerland) 11 (7), 1–39. Available from: https://doi.org/10.3390/geosciences11070291.

Fenta, A.A., Yasuda, H., Shimizu, K., Haregeweyn, N., Negussie, A., 2016. Dynamics of soil erosion as influenced by watershed management practices: a case study of the agula watershed in the semi-arid highlands of Northern Ethiopia. Environmental Management 58 (5), 889–905. Available from: https://doi.org/10.1007/s00267-016-0757-4.

Filho, C., Cochrane, C., Norton, T.A., Caviglione, L.D., Johansson, J.H., 2001. Land degradation assessment: tools and techniques for measuring sediment load. 3rd International Land Degradation, 1–20.

Gad, Y.H., Ali, H.E., Hegazy, E.S.A., 2021. Radiation-induced improving mechanical and thermal properties of carboxymethyl cellulose/clay composite for application in removal of copper(II) ions from wastewater. Journal of Inorganic and Organometallic Polymers and Materials 31 (5), 2083–2094. Available from: https://doi.org/10.1007/s10904-020-01850-w.

Gao, X., Wang, X., Liu, X., 2018. New chemical grouting materials and rapid construction technology for inclined shaft penetrating Drift-Sand layer in coal mine. Advances in Materials Science and Engineering 2018. Available from: https://doi.org/10.1155/2018/2797419.

Gikas, G.D., Sylaios, G.K., Tsihrintzis, V.A., Konstantinou, I.K., Albanis, T., Boskidis, I., 2020. Comparative evaluation of river chemical status based on WFD methodology and CCME water quality index. Science of the Total Environment 745, 140849. Available from: https://doi.org/10.1016/j.scitotenv.2020.140849.

Gómez-García, R., Campos, D.A., Aguilar, C.N., Madureira, A.R., Pintado, M., 2021. Valorisation of food agro-industrial by-products: from the past to the present and perspectives. Journal of Environmental Management 299 (September). Available from: https://doi.org/10.1016/j.jenvman.2021.113571.

Gupta, A.K., Tomar, J.M.S., Kaushal, R., Kadam, D.M., Rathore, A.C., Mehta, H., et al., 2021. Aromatic plants based environmental sustainability with special reference to degraded land management. Journal of Applied Research on Medicinal and Aromatic Plants 22. Available from: https://doi.org/10.1016/j.jarmap.2021.100298.

Gyssels, G., Poesen, J., Bochet, E., Li, Y., 2005. Impact of plant roots on the resistance of soils to erosion by water: a review. Progress in Physical Geography 29 (2), 189–217. Available from: https://doi.org/10.1191/0309133305pp443ra.

Hataf, N., Ghadir, P., Ranjbar, N., 2018. Investigation of soil stabilization using chitosan biopolymer. Journal of Cleaner Production 170, 1493–1500. Available from: https://doi.org/10.1016/j.jclepro.2017.09.256.

Huang, Y., Liu, Q., Jia, W., Yan, C., Wang, J., 2020. Agricultural plastic mulching as a source of microplastics in the terrestrial environment. Environmental Pollution 260, 114096. Available from: https://doi.org/10.1016/j.envpol.2020.114096.

Huo, J., Min, X., Wang, Y., 2021. Zirconium-modified natural clays for phosphate removal: effect of clay minerals. Environmental Research 194 (December 2020), 110685. Available from: https://doi.org/10.1016/j.envres.2020.110685.

Imran, M., Mehmood, A., 2020. Analysis and mapping of present and future drivers of local urban climate using remote sensing: a case of Lahore, Pakistan. Arabian Journal of Geosciences 13 (6). Available from: https://doi.org/10.1007/s12517-020-5214-2.

IPBES, 2018. The IPBES Assessment Report on Land Degradation and Restoration. Secretariat of the Intergovernmental Science-Policy Platform on Biodiversity and Ecosystem Services, Bonn, Germany, p. 744. Available from: https://doi.org/10.5281/zenodo.3237392.

Issac, M.N., Kandasubramanian, B., 2021. Effect of microplastics in water and aquatic systems. Environmental Science and Pollution Research International . Available from: https://doi.org/10.1007/s11356-021-13184-2.

Jose, S., Salim, R., Ammayappan, L., 2016. An overview on production, properties, and value addition of pineapple leaf fibers (PALF), In Journal of Natural Fibers, 13. Taylor and Francis Inc, pp. 362–373. Issue 3. Available from: https://doi.org/10.1080/15440478.2015.1029194.

Joseph, A.A., Pardo-Vargas, A., Seeberger, P.H., 2020. Total synthesis of polysaccharides by automated glycan assembly. Journal of the American Chemical Society 142 (19), 8561–8564. Available from: https://doi.org/10.1021/jacs.0c00751.

Kapović Solomun, M., Barger, N., Cerda, A., Keesstra, S., Marković, M., 2018. Assessing land condition as a first step to achieving land degradation neutrality: a case study of the Republic of Srpska. Environmental Science and Policy 90 (July), 19–27. Available from: https://doi.org/10.1016/j.envsci.2018.09.014.

Kaur, M., 2018. Agricultural and industrial waste as soil stabilizer: an overview. International Journal of Pure and Applied Bioscience 6 (3), 631–636. Available from: https://doi.org/10.18782/2320-7051.6664.

Kavazanjian, E., Iglesias, E., Karatas, I., 2009. Biopolymer soil stabilization for wind erosion control. Proceedings of the 17th International Conference on Soil Mechanics and Geotechnical Engineering: The Academia and Practice of Geotechnical Engineering, 1, 881–884. Available from: https://doi.org/10.3233/978-1-60750-031-5-881.

Kumar, M.V., Sheela, A.M., 2021. Effect of plastic film mulching on the distribution of plastic residues in agricultural fields. Chemosphere 273, 128590. Available from: https://doi.org/10.1016/j.chemosphere.2020.128590.

Kumar, T., Jhariya, D.C., Pandey, H.K., 2020. Comparative study of different models for soil erosion and sediment yield in Pairi watershed, Chhattisgarh, India. Geocarto International 35 (11), 1245–1266. Available from: https://doi.org/10.1080/10106049.2019.1576779.

Kwon, Y.M., Chang, I., Lee, M., Cho, G.C., 2019. Geotechnical engineering behavior of biopolymer-treated soft marine soil. Geomechanics and Engineering 17 (5), 453–464. Available from: https://doi.org/10.12989/gae.2019.17.5.453.

Latifi, N., Horpibulsuk, S., Meehan, C.L., Majid, M.Z.A., Rashid, A.S.A., 2016. Xanthan gum biopolymer: an eco-friendly additive for stabilization of tropical organic peat. Environmental Earth Sciences 75 (9), 1–10. Available from: https://doi.org/10.1007/s12665-016-5643-0.

Lee, J.H., Heo, J.H., 2011. Evaluation of estimation methods for rainfall erosivity based on annual precipitation in Korea. Journal of Hydrology 409 (1–2), 30–48. Available from: https://doi.org/10.1016/j.jhydrol.2011.07.031.

Lentz, R.D., 2015. Polyacrylamide and biopolymer effects on flocculation, aggregate stability, and water seepage in a silt loam. Geoderma 241–242, 289–294. Available from: https://doi.org/10.1016/j.geoderma.2014.11.019.

Liu, F., Huang, K., Zheng, A., Xiao, F.S., Dai, S., 2018. Hydrophobic solid acids and their catalytic applications in green and sustainable chemistry. ACS Catalysis 8 (1), 372–391. Available from: https://doi.org/10.1021/acscatal.7b03369.

Maraveas, C., 2020. Production of sustainable and biodegradable polymers from agricultural waste. Polymers 12 (5). Available from: https://doi.org/10.3390/POLYM12051127.

Meng, F., Fan, T., Yang, X., Riksen, M., Xu, M., Geissen, V., 2020. Effects of plastic mulching on the accumulation and distribution of macro and micro plastics in soils of two farming systems in Northwest China. PeerJ 8, 1–20. Available from: https://doi.org/10.7717/peerj.10375.

Meshesha, D.T., Tsunekawa, A., Tsubo, M., 2012. Continuing land degradation: cause-effect in Ethiopia's Central Rift Valley. Land Degradation and Development 23 (2), 130–143. Available from: https://doi.org/10.1002/ldr.1061.

Mishra, H., Denis, D.M., Suryavanshi, S., Kumar, M., Srivastava, S.K., Denis, A.F., et al., 2017. Hydrological simulation of a small ungauged agricultural watershed Semrakalwana of Northern India. Applied Water Science 7 (6), 2803–2815. Available from: https://doi.org/10.1007/s13201-017-0531-7.

Monkul, M.M., Özhan, H.O., 2021. Microplastic contamination in soils: a review from geotechnical engineering view. Polymers 13 (23). Available from: https://doi.org/10.3390/polym13234129.

Moorthy, G.V., Thangaraj, P., Shanmugam, T., 2020. Strength aspects of high performance concrete by using waste material as replacement of fine aggregate in various percentages. IOP Conference Series: Materials Science and Engineering 955 (1). Available from: https://doi.org/10.1088/1757-899X/955/1/012047.

Moreno-Jiménez, E., Ochoa-Hueso, R., Plaza, C., Aceña-Heras, S., Flagmeier, M., Elouali, F.Z., et al., 2020. Biocrusts buffer against the accumulation of soil metallic nutrients induced by warming and rainfall reduction. Communications Biology 3 (1). Available from: https://doi.org/10.1038/s42003-020-1054-6.

Mostafa, N.A., Farag, A.A., Abo-dief, H.M., Tayeb, A.M., 2018. Production of biodegradable plastic from agricultural wastes. Arabian Journal of Chemistry 11 (4), 546–553. Available from: https://doi.org/10.1016/j.arabjc.2015.04.008.

Nabiollahi, K., Taghizadeh-Mehrjardi, R., Kerry, R., Moradian, S., 2017. Assessment of soil quality indices for salt-affected agricultural land in Kurdistan Province, Iran. Ecological Indicators 83 (August), 482–494. Available from: https://doi.org/10.1016/j.ecolind.2017.08.001.

Nascimento, C.M., de Sousa Mendes, W., Quiñonez Silvero, N.E., Poppiel, R.R., Sayão, V.M., Dotto, A.C., et al., 2021. Soil degradation index developed by multitemporal remote sensing images, climate variables, terrain and soil atributes. Journal of Environmental Management 277. Available from: https://doi.org/10.1016/j.jenvman.2020.111316.

Nguyen, V.P., Pham, D.H., Bui, P.D., Bui, N.T., Nguyen, D.V.A., 2021. Research the possibility of using sea sand in roadbed construction in vietnam. International Journal of Geomate 20 (77), 123–131. Available from: https://doi.org/10.21660/2020.77.47890.

Orts, W.J., Sojka, R.E., Glenn, G.M., Gross, R.A., 1999. Preventing soil erosion with polymer additives. Polymer News 24 (12), 406–413.

Pal, D., Maiti, S.K., 2018. Heavy metal speciation, leaching and toxicity status of a tropical rain-fed river Damodar, India. Environmental Geochemistry and Health 40 (6), 2303–2324. Available from: https://doi.org/10.1007/s10653-018-0097-9.

Panagos, P., Katsoyiannis, A., 2019. Soil erosion modelling: the new challenges as the result of policy developments in Europe. Environmental Research 172 (February), 470–474. Available from: https://doi.org/10.1016/j.envres.2019.02.043.

Pandey, J., Singh, R., 2017. Heavy metals in sediments of Ganga River: up- and downstream urban influences. Applied Water Science 7 (4), 1669–1678. Available from: https://doi.org/10.1007/s13201-015-0334-7.

Pandey, S., Kumari, N., 2021. Proceedings of advances in sustainable research for energy and environmental management. In: Mungray, A.K., Lad, V.N., Mungray, A., Chakraborty, M., Singh, J. (Eds.), First, 1. pp. 1–248. . Nandan Graphics and Library Services.

Pastore, G., Baird, T., Vermeesch, P., Resentini, A., Garzanti, E., 2021. Provenance and recycling of Sahara Desert sand. Earth-Science Reviews 216 (February), 103606. Available from: https://doi.org/10.1016/j.earscirev.2021.103606.

Patel, N., Kathwas, A.K., 2012. Assessment of spatio-temporal dynamics of soil erosional severity through geoinformatics. Geocarto International 27 (1), 3–16. Available from: https://doi.org/10.1080/10106049.2011.614359.

Philippidis, G., Sartori, M., Ferrari, E., M'Barek, R., 2019. Waste not, want not: a bio-economic impact assessment of household food waste reductions in the EU. Resources, Conservation and Recycling 146 (April), 514–522. Available from: https://doi.org/10.1016/j.resconrec.2019.04.016.

Poesen, J., 2018. Soil erosion in the Anthropocene: research needs. Earth Surface Processes and Landforms 43 (1), 64–84. Available from: https://doi.org/10.1002/esp.4250.

Prăvălie, R., Patriche, C., Borrelli, P., Panagos, P., Roşca, B., Dumitraşcu, M., Nita, I.A., Săvulescu, I., Birsan, M.V., Bandoc, G., 2021. Arable lands under the pressure of multiple land degradation processes. A global perspective. Environmental Research 194. Available from: https://doi.org/10.1016/j.envres.2020.110697.

Rabea, E.I., Badawy, M.E.T., Stevens, C.V., Smagghe, G., Steurbaut, W., 2003. Chitosan as antimicrobial agent: applications and mode of action. Biomacromolecules 4 (Issue 6), 1457–1465. Available from: https://doi.org/10.1021/bm034130m.

Rahmani, V., Kastens, J.H., de Noyelles, F., Jakubauskas, M.E., Martinko, E.A., Huggins, D.H., et al., 2018. Examining storage capacity loss and sedimentation rate of large reservoirs in the Central U.S. great plains. Water (Switzerland) 10 (2), 1–17. Available from: https://doi.org/10.3390/w10020190.

Ramadhan, M.O., Handayani, M.N., 2020. The potential of food waste as bioplastic material to promote environmental sustainability: a review. IOP Conference Series: Materials Science and Engineering 980 (1). Available from: https://doi.org/10.1088/1757-899X/980/1/012082.

Ramos, B.G.Z., 2011. Biopolymers employed in drug delivery. Biopolymers: Biomedical and Environmental Applications 559–573. Available from: https://doi.org/10.1002/9781118164792.CH19.

Rawat, K.S., Kumar, R., Singh, S.K., 2019. Topographical distribution of cobalt in different agro-climatic zones of Jharkhand state, India. Geology, Ecology, and Landscapes 3 (1), 14–21. Available from: https://doi.org/10.1080/24749508.2018.1481654.

Reddy, R.M., Reddy, A.G.S., Naveen, T., 2021. Impact assessment of artificial recharge structures on Kakatiya University Campus, Warangal Urban District, Telangana. Journal of the Geological Society of India 97 (4), 375–384. Available from: https://doi.org/10.1007/s12594-021-1695-1.

Ricci, G.F., Jeong, J., De Girolamo, A.M., Gentile, F., 2020. Effectiveness and feasibility of different management practices to reduce soil erosion in an agricultural watershed. Land Use Policy 90 (September 2019), 104306. Available from: https://doi.org/10.1016/j.landusepol.2019.104306.

Rostami, N., Karimi, H., Tavakoli, M., Omidipour, R., 2022. Short-term effect of oil-mulch on vegetation dynamics; Integration of ecological and remote sensing-based approaches. Land Degradation and Development 33 (2), 235–245. Available from: https://doi.org/10.1002/LDR.4140.

Sathya, A., Thampi, S.G., Chithra, N.R., 2021. Development of a framework for sand auditing of the Chaliyar River basin, Kerala, India using HEC-HMS and HEC-RAS model coupling. International Journal of River Basin Management 0 (0), 1–14. Available from: https://doi.org/10.1080/15715124.2021.1909604.

Scarlat, N., Dallemand, J.F., Monforti-Ferrario, F., Nita, V., 2015. The role of biomass and bioenergy in a future bioeconomy: policies and facts. Environmental Development 15, 3–34. Available from: https://doi.org/10.1016/j.envdev.2015.03.006.

Schnepp, Z., 2013. Biopolymers as a flexible resource for nanochemistry. Angewandte Chemie International Edition 52 (4), 1096–1108. Available from: https://doi.org/10.1002/ANIE.201206943.

Scholz, P., Putna-Nimane, I., Barda, I., Liepina-Leimane, I., Strode, E., Kileso, A., et al., 2021. Environmental impact of geosynthetics in coastal protection. Materials 14 (3), 1–13. Available from: https://doi.org/10.3390/ma14030634.

Sheibani, E., Hosseini, A., Sobhani Nasab, A., Adib, K., Ganjali, M.R., Pourmortazavi, S.M., et al., 2021. Application of polysaccharide biopolymers as natural adsorbent in sample preparation. Critical Reviews in Food Science and Nutrition . Available from: https://doi.org/10.1080/10408398.2021.1978385.

Singh, A.K., Mondal, G.C., Singh, T.B., Singh, S., Tewary, B.K., Sinha, A., 2012. Hydrogeochemical processes and quality assessment of groundwater in Dumka and Jamtara districts, Jharkhand, India. Environmental Earth Sciences 67 (8), 2175–2191. Available from: https://doi.org/10.1007/s12665-012-1658-3.

Singh, A.K., Giri, S., Chaturvedi, A., 2018. Fluvial geochemistry of Subarnarekha River basin, India. Journal of Earth System Science 127 (8), 1–22. Available from: https://doi.org/10.1007/s12040-018-1020-6.

Sith, R., Watanabe, A., Nakamura, T., Yamamoto, T., Nadaoka, K., 2019. Assessment of water quality and evaluation of best management practices in a small agricultural watershed adjacent to Coral Reef area in Japan. Agricultural Water Management 213, 659–673. Available from: https://doi.org/10.1016/j.agwat.2018.11.014.

Sran, D.S., Kukal, S.S., Singh, M.J., 2012. Run-off and sediment yield in relation to differential gully-plugging schemes in micro-catchments of Shiwaliks in the lower Himalayas. Archives of Agronomy and Soil Science 58 (11), 1317–1327. Available from: https://doi.org/10.1080/03650340.2011.577422.

Stefanakis, A.I., 2019. The role of constructedwetlands as green infrastructure for sustainable urban water management. Sustainability (Switzerland) 11 (24). Available from: https://doi.org/10.3390/su11246981.

Steinmetz, Z., Wollmann, C., Schaefer, M., Buchmann, C., David, J., Tröger, J., et al., 2016. Plastic mulching in agriculture. Trading short-term agronomic benefits for long-term soil degradation? Science of the Total Environment 550, 690–705. Available from: https://doi.org/10.1016/j.scitotenv.2016.01.153.

Stringfellow, W.T., Karpuzcu, M.E., Spier, C., Hanlon, J.S., Graham, J., 2013. Sizing mitigation wetlands in agricultural watersheds. Water Science and Technology 67 (1), 40–46. Available from: https://doi.org/10.2166/wst.2012.527.

Thakur, S., Maity, D., Mondal, I., Basumatary, G., Ghosh, P.B., Das, P., et al., 2020. Assessment of changes in land use, land cover, and land surface temperature in the mangrove forest of Sundarbans, northeast coast of India. Environment, Development and Sustainability 23 (2), 1917–1943. Available from: https://doi.org/10.1007/s10668-020-00656-7.

Thapa, P., 2020. Spatial estimation of soil erosion using RUSLE modeling: a case study of Dolakha district, Nepal. Environmental Systems Research 9 (1). Available from: https://doi.org/10.1186/s40068-020-00177-2.

Torres, L.G., Carpinteyro-Urban, S.L., 2012. Use of prosopis laevigata seed gum and opuntia ficus-indica mucilage for the treatment of municipal wastewaters by coagulation-flocculation. Natural Resources 03 (02), 35–41. Available from: https://doi.org/10.4236/nr.2012.32006.

Tóth, G., Hermann, T., da Silva, M.R., Montanarella, L., 2018. Monitoring soil for sustainable development and land degradation neutrality. Environmental Monitoring and Assessment 190 (2). Available from: https://doi.org/10.1007/s10661-017-6415-3.

Upgupta, S., Singh, P.K., 2020. Quantifying the dynamics and drivers of landscape change in an opencast coal mining area of Central India (East Bokaro, Jharkhand). Proceedings of the National Academy of Sciences India Section A - Physical Sciences 90 (3), 565–577. Available from: https://doi.org/10.1007/s40010-018-0589-0.

Ustaoğlu, F., Taş, B., Tepe, Y., Topaldemir, H., 2021. Comprehensive assessment of water quality and associated health risk by using physicochemical quality indices and multivariate analysis in Terme River, Turkey. Environmental Science and Pollution Research 28 (Issue 44), 62736–62754. Available from: https://doi.org/10.1007/s11356-021-15135-3.

Vasu, D., Tiwary, P., Chandran, P., Singh, S.K., 2019. Soil quality for sustainable agriculture. Nutrient Dynamics for Sustainable Crop Production 41–66. Available from: https://doi.org/10.1007/978-981-13-8660-2_2.

Vermeulen, S.J., Campbell, B.M., Ingram, J.S.I., 2012. Climate change and food systems. Annual Review of Environment and Resources 37, 195–222. Available from: https://doi.org/10.1146/annurev-environ-020411-130608.

Vigiak, O., Malagó, A., Bouraoui, F., Vanmaercke, M., Poesen, J., 2015. Adapting SWAT hillslope erosion model to predict sediment concentrations and yields in large Basins. Science of the Total Environment 538, 855–875. Available from: https://doi.org/10.1016/j.scitotenv.2015.08.095.

Wade, E., Zowada, R., Foudazi, R., 2021. Alginate and guar gum spray application for improving soil aggregation and soil crust integrity. Carbohydrate Polymer Technologies and Applications 2, 100114. Available from: https://doi.org/10.1016/j.carpta.2021.100114.

Walia, M.K., Dick, W.A., 2018. Selected soil physical properties and aggregate-associated carbon and nitrogen as influenced by gypsum, crop residue, and glucose. Geoderma 320 (April 2017), 67–73. Available from: https://doi.org/10.1016/j.geoderma.2018.01.022.

Wang, S.W., Munkhnasan, L., Lee, W.K., 2021. Land use and land cover change detection and prediction in Bhutan's high altitude city of Thimphu, using cellular automata and Markov chain. Environmental Challenges 2 (November 2020). Available from: https://doi.org/10.1016/j.envc.2020.100017.

Williams, M.A., Rice, C.W., Owensby, C.E., 2001. Nitrogen competition in a tallgrass prairie ecosystem exposed to elevated carbon dioxide. Soil Science Society of America Journal 65 (2), 340–346. Available from: https://doi.org/10.2136/sssaj2001.652340x.

Wolka, K., Mulder, J., Biazin, B., 2018. Effects of soil and water conservation techniques on crop yield, runoff and soil loss in Sub-Saharan Africa: a review. Agricultural Water Management 207 (January), 67–79. Available from: https://doi.org/10.1016/j.agwat.2018.05.016.

Wood, J.L., Harrison, S., Reinhardt, L., Taylor, F.E., 2020. Landslide databases for climate change detection and attribution. Geomorphology 355. Available from: https://doi.org/10.1016/j.geomorph.2020.107061.

Wu, H., Yao, C., Li, C., Miao, M., Zhong, Y., Lu, Y., et al., 2020a. Review of application and innovation of geotextiles in geotechnical engineering. Materials 13 (7), 1–21. Available from: https://doi.org/10.3390/MA13071774.

Wu, J., Li, P., Wang, D., Ren, X., Wei, M., 2020b. Statistical and multivariate statistical techniques to trace the sources and affecting factors of groundwater pollution in a rapidly growing city on the Chinese Loess Plateau. Human and Ecological Risk Assessment 26 (6), 1603–1621. Available from: https://doi.org/10.1080/10807039.2019.1594156.

Yu, C., Yang, Y., Wu, Z.X., Jiang, J.F., Liao, R.P., Deng, Y.F., 2021. Experimental study on the permeability and self-healing capacity of geosynthetic clay liners in heavy metal solutions. Geotextiles and Geomembranes 49 (2), 413–419. Available from: https://doi.org/10.1016/j.geotexmem.2020.10.012.

Yusuf, M., 2017. Handbook of Ecomaterials. Handbook of Ecomaterials, August. Available from: https://doi.org/10.1007/978-3-319-48281-1.

Zhai, Y., Wang, T., Zhu, Y., Peng, C., Wang, B., Li, X., et al., 2018. Production of fuel pellets via hydrothermal carbonization of food waste using molasses as a binder. Waste Management 77, 185–194. Available from: https://doi.org/10.1016/j.wasman.2018.05.022.

Zhang, T., Wang, W., Zhao, Y., Bai, H., Wen, T., Kang, S., et al., 2021. Removal of heavy metals and dyes by clay-based adsorbents: from natural clays to 1D and 2D nano-composites. Chemical Engineering Journal 420 (P2), 127574. Available from: https://doi.org/10.1016/j.cej.2020.127574.

Zhao, Y., Duan, X., Han, J., Yang, K., Xue, Y., 2018. The main influencing factors of soil mechanical characteristics of the gravity erosion environment in the dry-hot valley of Jinsha river. Open Chemistry 16 (1), 796–809. Available from: https://doi.org/10.1515/chem-2018-0086.

Zhou, B., Wang, J., Zhang, H., Shi, H., Fei, Y., Huang, S., et al., 2020. Microplastics in agricultural soils on the coastal plain of Hangzhou Bay, east China: multiple sources other than plastic mulching film. Journal of Hazardous Materials 388 (November), 121814. Available from: https://doi.org/10.1016/j.jhazmat.2019.121814.

Zhao, Y., Gao, G., Ding, G., Wang, L., Chen, Y., Zhao, Y., et al., 2022. Assessing the influencing factors of soil susceptibility to wind erosion: a wind tunnel experiment with a machine learning and model-agnostic interpretation approach. Catena 215. Available from: https://doi.org/10.1016/j.catena.2022.106324.

Chapter 25

Bioplastics: solution to a green environment and sustainability

Bisma Nisar[1], Heena Nisar Pahalvi[1], Aadil Gulzar[1], Sumaira Rashid[1], Lone Rafiya Majeed[2] and Azra N. Kamili[3]

[1]*Department of Environmental Sciences, University of Kashmir, Srinagar, Jammu and Kashmir, India,* [2]*Department of Environmental Science, Vivekananda Global University, Jaipur, Rajasthan, India,* [3]*Department of Botany, Central University of Kashmir, Ganderbal, Jammu and Kashmir, India*

Introduction

Plastic left-over forms a major portion of solid waste and creates many burning environmental issues and ecological imbalances. It accounts for almost 25% of the total volume of landfills (Lau et al., 2020). Notably, 92% of plastics in use are nonbiodegradable. Because of this, plastic takes hundreds or even thousands of years to disintegrate in landfills (Carroccio et al., 2021). Nowadays, bioplastic is becoming very popular as it contributes to sustainable development. It is biodegradable and has a wider range of disposal options. The term "bio-based" denotes the involvement of biomass as a source of raw material. Bioplastics have some characteristics like they have a more functionalized globular nature and a less mesh-like structure in the solid state (Di Bartolo et al., 2021). Besides, their molecular motion is high (Kwak and Lee, 2000), they have low melt viscosity (Hawker et al., 1995; Ball and McLeish, 1989; Roovers, 1994; Zhou and Larson, 2007), and are soluble in many solvents. Bioplastics degrade in two ways: they are naturally degraded by the aerobic process through composting and in another way, they have degraded aerobically in landfills. In both ways, the degradation by microorganisms releases the monomer units into the environment, which pose less impact on the landscapes (Yaradoddi et al., 2016). The present chapter provides a quantitative systematic review of scientific literature published over the last 30 years.

Types of bioplastics

Starch-based bioplastics

Starch-based bioplastics are made from the amalgamation of polymers of petrochemical plastic with decomposable polymeric starch materials (Manjunath and Sailaja, 2016). They can be acquired from maize, wheat, potato, etc. (Greene, 2014). When these bioplastics are disposed of, the microorganisms degrade the starch molecules, and the plastic polymer disintegrates (Peelman et al., 2013). The microbial fermentation of starch also yields bioplastics, for example, PLA and polyhydroxyalkanoates (PHA) (Thielen, 2014). The new generation bioplastics such as the bio-based polyolefin's polyethylene (PE), polyvinyl chloride (PVC), and the partly bio-based PE terephthalate (PET) are also manufactured from starch and other carbohydrates.

Cellulose-based bioplastics

Cellulosic plastics mostly come from two sources: wood pulp and cotton fibers. Organic esters of cellulose and regenerated cellulose are the two forms of cellulosic polymers (Harms, 2006). Cellulose esterification with organic acids produces organic esters of cellulose. Cellulose acetate butyrate, cellulose acetate propionate, and cellulose acetate are industrially important organic esters of cellulose. They are used in wrapping films, textile fibers, cigarette filters, and pharmaceuticals. The other type of cellulosic plastic, that is, cellulose regeneration, is synthesized through the chemical dissolution of cellulose and then restructured in the form of fibers or films (Thielen, 2014). Viscose silk, rayon, and viscose are a few examples of regenerated cellulose (Harms, 2006).

Protein-based bioplastics

Proteins from various sources, such as wheat gluten and casein, can be used to make bioplastics. They have the potential as a raw material for a variety of biodegradable polymers (Song et al., 2009). For instance, the body parts of the

Role of Green Chemistry in Ecosystem Restoration to Achieve Environmental Sustainability. DOI: https://doi.org/10.1016/B978-0-443-15291-7.00021-3

first Ford vehicle were constituted of soy plastic (Ralston et al., 2008). The plastics based on soy protein are fragile when exposed to water, and they are also costly. Soy protein is therefore amalgamated with other biodegradable polyesters to increase its water resilience and reduce its cost (Zhang et al., 2006).

PLA-based bioplastics

The most common bioplastics in the market today are PLA and polylactide plastics (Endres and Siebert, 2009). Via these PLA bioplastics, the problems found in starch-based bioplastics can be addressed. PLA and PLA-composites are currently available in granulated forms in a variety of grades for use in film production, molded parts, containers for drinks, dishes, flasks, and other common items (Lorcks, 2005). PLA-based plastics are increasingly being used in a variety of industries, including medicine, fashion, cosmetics, and domestic applications. Some other PLA-based plastics are door tread plates, dashboards, and other car parts (Lorcks, 2005).

Polyhydroxyalkanoates-based bioplastics

Utilizing microbial fermentation processes, PHA-based polymers are created from plant-based starch. PHA-based polymers have physical and chemical qualities that are the same as those of polyesters, PE, and polypropylene (Laycock et al., 2013). Some natural microbes make intercellular PHAs granules in the process of excessive carbon sources (Endres and Siebert, 2009). More than 75 taxa of prokaryotes and archaea manufacture PHA intracellularly (Reddy et al., 2013; Rehm, 2003). The various PHA groups usually used for plastic manufacture are PHB and polyhydroxy butyrate, polyhydroxy valerate (PHV), and poly-3-hydroxybutyrate-co-valerate (PHBV). Several large-scale companies produce PHA-based plastics like biopol and bionelle. The biopol is a PHB copolymer called PHBV, and the bionelle is a biodegradable plastic because of chemically synthesized PHA (Tokiwa et al., 2009). This is a nontoxic, renewable polymer (Forni et al., 1999) that may be utilized in a variety of items, including foodstuff wrapping, medical implants, farming, and so on (Suriyamongkol et al., 2007).

Classification of bioplastics

Based on the origin of bioplastics, they are divided into four major classes. These are further partitioned into two general categories polymers (category 1) and biopolyesters (categories 2 to 4) (Averous and Pollet, 2012) (Fig. 25.1).

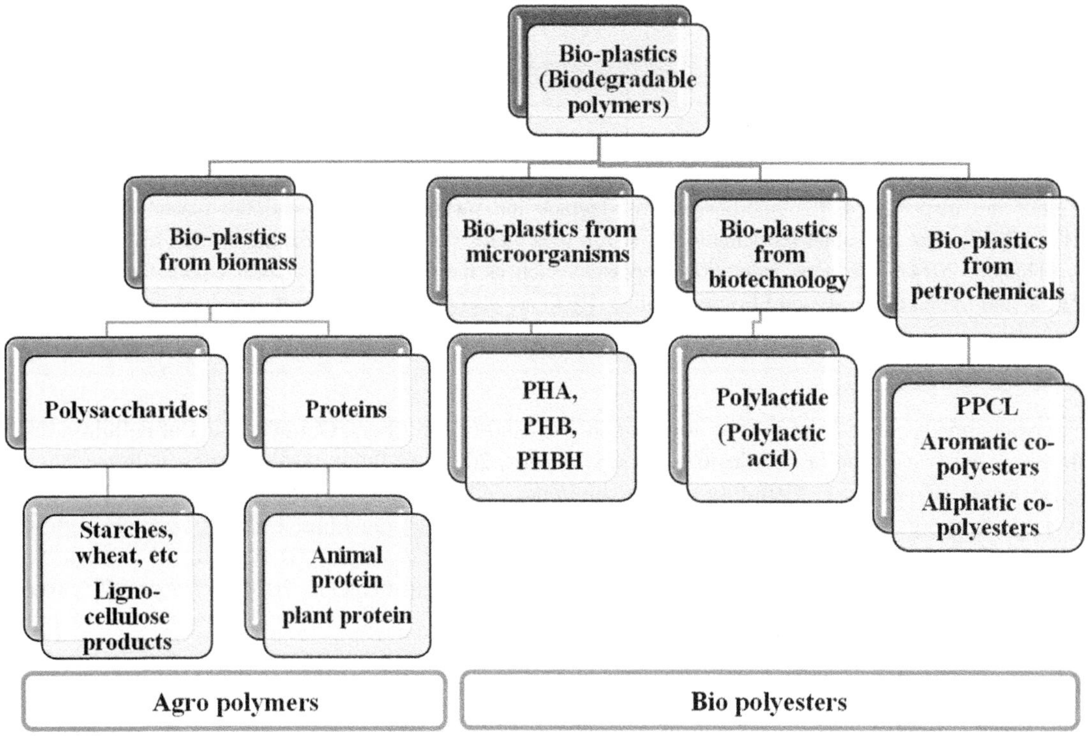

FIGURE 25.1 Bioplastic classification (Avérous and Pollet, 2012).

Applications of bioplastics

Packaging

One of the most common uses for bioplastic packaging is in shopping bags, which often may be used as a bag for compostable kitchen and garden waste and a bag for garbage. Certain uses include thermoformed inserts for fruit, vegetables, meat and egg trays, stuff for dairy goods, flasks, and nets for fruits and vegetables. There are jars and tubes for use in the cosmetics market. Packaging materials made from bioplastics (Table 25.1) with sound barriers and impenetrable to odors are now available (Lorcks, 2005). Covering paper and carton laminates with bioplastics results in modern packaging with intense moisture and fat or oil tolerance (Manjure, 2011).

Horticulture and agriculture

Mulch films composed of biodegradable plastic can be plowed in after use and do not need to be cleaned or disposed of as filthy plastic waste. Pots of plants and seed trays decompose in the soil and are no longer considered garbage.

TABLE 25.1 Current applications of bioplastics in the food industry (Kumar and Thakur, 2017).

Packaging application	Biopolymer	Company	References
Coffee and tea	Cardboard cups coated with PLA	KLM	Jager (2010)
Beverages	PLA cups	Mosburger (Japan)	Sudesh and Iwata (2008)
Fresh salads	PLA bowls	McDonald's	Haugaard et al. (2001)
Carbonated water, fresh juices, dairy drinks	PLA bottles	Biota, noble	Auras et al. (2004)
Freshly cut fruits, whole fruits, vegetables, bakery goods, salads	Rigid PLA trays and packs	Asda (retailer)	Koide and Shi (2007) Jager (2010)
Organic pretzels, potato chips	PLA bags	Snyder's of Hanover, PepsiCo's Frito-lay	Weston, 2012
Bread	Paper bags with PLA window	Delhaize (retailer)	Demirbas (2007)
Organic poultry	PLA bowls, absorb pads	Delhaize (retailer)	Demirbas (2007)
Starch-based Milk chocolates	Cornstarch trays	Cadbury Schweppes food group, Marks and Spencer	Website European bioplastics
Organic tomatoes	Corn-based packaging	Iper supermarkets (Italy), Coop Italia	website European bioplastics
Cellulose-based Kiwi	Bio-based trays wrapped with cellulose film	Wal-Mart	Blakistone and Sand (2007)
Potato chips	Metalized cellulose film	Boulder Canyon	Website European bioplastics
Organic pasta	Cellulose-based packaging	Birkel	Website European bioplastics
Sweets	Metalized cellulose film	Quality Street, Thornton	Website European bioplastics

Source: Data from Kumar, S., Thakur, K.S., 2017. Bioplastics-classification, production and their potential food applications. Journal of Hill Agriculture, 8(2), 118–129.

Plant trays for flowers and vegetables can be composted alongside kitchen and garden trash in the domestic compost heap (Lorcks, 2005). The film, woven fabric, and bioplastic networks keep back newly laid roadside banks and avoid soil erosion until plants stabilize them.

Medicine and personal care

For several years special bioplastics have been used in the area of medicine. These restorable bioplastics can be used for many tasks; for example, thermoplastic starch (TPS) is a substitute for gelatine as a pill and capsule material (Lorcks, 2005). PLA and its copolymers are used as surgical threads, as carriers for implanted active materials, or for the manufacture of absorbable implants like sutures, pins, or plates, which are destroyed by the metabolism and thus render another surgery unnecessary for removal. Some bioplastics have unique properties that make them a natural fit for hygiene products. These materials are often used as "breathing" biofilms for nappy liners, bed underlay, incontinence goods, feminine hygiene products, and disposable gloves because they enable water vapor to pass through them while remaining waterproof (Lorcks, 2005).

Consumer electronics

The Sony Walkman was the first electronic equipment in which bio-based plastic was used. PLA was first used in this country in 2002. In 2005, a cell phone with a casing composed of PLA and reinforced with Kenaf fibers was released. Many electrical products, from the computer mouse to keyboards to headphones, are now on the market, all with bio-based plastic housings or components.

Automobile manufacture

Henry Ford began experimenting with bioplastics based on wheat and soy for use in automobile manufacturing at the beginning of the last century in the United States. Today, Ford is experimenting with polyol based on up to 40% soya for polyurethane parts, particularly seats, headrests, and armrests. Toyota is yet another leader in this method. The spare wheel cover is made of PLA with kenaf support in the Prius car, which is presently one of the best eco-friendly cars in the world. Toyota was claimed to be launching a car with an inside surface cover made of roughly 80% bioplastic at the end of 2011. The "Sai" hybrid limousine traded in Japan has seat coverings and mats made of partly bio-based PET, among other items. A completely functioning steering wheel with an integrated airbag made from more than 50% renewable materials received the 7th "Bioplastics Award" in November 2012 (Gaumann, 2013).

Textiles

Textiles created from renewable materials are nearly as old as human beings (linen, cotton, etc.). Natural fabrics made from renewable materials now combine their "biological" origin with the technological characteristics of natural microfiber fabrics, such as good moisture transmission in particular, so that sweating is (almost) no longer a problem. Most biopolyesters are used for spinning fibers and producing textiles, chiefly PLA and terephthalate polytrimethylene (PTT).

Construction and housing

For the manufacture of upholstered furniture, biobased foams such as polyurethane are suitable; particle foams manufactured from PLA are accustomed for insulation of buildings. Patio decks and fascia cladding are just two examples of WPC (wood plastic composites). Water and sewage pipelines make great use of Bio-PE (and Bio-PVC). Paints and varnishes, linseed oil paints, wallpaper paste, and other products can all make use of bio-based functional polymers.

Waste management options for bioplastics

Many bioplastics are 100% compostable and usually biodegradable when put in a municipal composting facility in 180 days or less, while conventional plastics can take many years to undergo decomposition. As conventional plastics start to break down, they crumble into smaller and finer fragments that sometimes go to our water streams and finally enter into our food chain as animals consume the plastics. In comparison, compostable plastics are recycled back into the earth and converted to soil nutrients- thus closing the deadly loop (Reddy et al., 2013). The most appropriate alternative for disposal relies on the form of bio-based and biodegradable plastic:

Recycling

In European standards EN 13430 and EN 16848 (taken from ISO 18604), material recycling is characterized as the reprocessing of used items into a new product. An example of this is plastic which can be gathered, graded, and reprocessed into novel products after use. That recycling type is called mechanical recycling. One more method is chemical recycling, in which products are decomposed into monomers that can be used again for the manufacturing of polymers. One illustration is the Galactic Loopla cycle, where PLA is hydrolyzed to lactic acid (Looplife, 2016).

Mechanical recycling

Mechanical plastic recycling involves recovering plastic trash by mechanical processes (separating, grinding, washing, drying, re-granulating, and compounding) to yield materials that can be turned into novel goods. Unlike organic and chemical recycling, mechanical recycling looks to reuse plastics at the material level rather than at the building block level. The major share of bioplastics developed today is mechanically recyclable (European Bioplastics, 2017).

Organic recycling

Numerous industrial composting and anaerobic digestion techniques handle Municipal Organic Waste (Van den Oever et al., 2017). Industrial composting and anaerobic digestion, also called organic recycling, are the "circular" end-of-life solutions for biodegradable, compostable plastics, including wrapping items. These plastics biodegrade under clearly specified environments (e.g., temperature, humidity, and presence of micro-organisms) within a limited timeframe. In an industrial composting unit, the biodegradation cycle is called composting, transforming the bioplastic into carbon dioxide, water, and biomass. In anaerobic digestion, the process of biodegradation is called biomethanisation, and it also produces methane (biogas), which is used for energy and fuel production (European Bio-plastics, 2017).

Chemical recycling

Chemical recycling, also known as feedstock recycling, involves the recovery of the building blocks and monomers that a polymer or plastic is composed of. The raw materials produced by chemical processing can be utilized in new processes of polymerization and the manufacture of novel items. Chemical recycling suits well with the concept of a circular economy and there have been several active research projects in recent years (European Bio-plastics, 2017).

Incineration with energy recovery

Most of the plastics have equal or higher gross calorific values (GCV) than coal (Davis and Song, 2006). Thus, burning with energy recovery after all recyclable elements have been extracted is a potentially good choice. It is argued that when used again as a fuel in incineration, petrochemical carbon, which had earlier a high-value use, signifies a more eco-efficient alternative than direct burning of the oil (Miller, 2005). Energy recuperation by burning is considered an appropriate choice for all bioplastic polymers. Natural cellulose fiber and starch have comparatively lower GCV than coal but are comparable to wood, and thus also have considerable incineration value (Davis and Song, 2006). Furthermore, in the first place, the processing of fiber and starch materials requires considerably a smaller amount of energy (Patel et al., 2005), thereby contributing surely to the total energy balance in the life cycle.

Landfill

The landfill option for waste plastic is the least favoured in the waste hierarchy. Earlier it was chosen as it was very easy and inexpensive without essential separation, cleaning, or treatment. Landfilling of biodegradable products, including bioplastic polymers, and orchard and kitchen trash, poses a specific problem: methane (a greenhouse gas) can be generated under anaerobic situations (Hudgins, 1999). However, such a "landfill gas" can be captured and utilized as an energy source.

Advantages of bioplastics

The benefits of bioplastics over traditional plastics are numerous. Bioplastic goods have the key advantage that they are made from renewable resources rather than from fossil resources (Thielen, 2014). The use of renewable resources will help mitigate greenhouse gas emissions by reducing the carbon footprint (Demirbas, 2007; Gross and Kalra, 2002). Their carbon footprint is theoretically much smaller. It should be noted that a bioplastic carbon footprint relies vitally on how the plastic enduringly holds the carbon that the developing plant extracts from the environment. Plastic prepared from a biological source seizes the CO_2 that the plant absorbs during photosynthesis. This sequestration is reversed if the resultant bioplastic dissociates back into CO_2 and water. But a permanent bioplastic, designed to imitate PE or other traditional plastics, will forever store CO_2. Even if the plastic is regularly recycled, the CO_2 initially removed from the atmosphere remains sequestered. Bioplastic processing can produce about 80% less carbon dioxide than petrochemical plastics (Yu and Chen, 2008). The costs of making bioplastics are lower. Since plastics are prepared from $\sim$5% of the oil used by the world each year. With oil shortages, plastics manufacturing is rapidly exposed to fluctuating costs (Bezirhan and Ozsoy, 2014). The best-known benefit of biodegradable plastics lies in permanent litter reduction. Bioplastics can overcome the environmental issues arising from the inadequate handling of plastic waste (Chen, 2014). With the rise of global interest in the manufacturing of biofuels and bioplastics, crop prices such as maize have risen sharply, as countries across the globe are searching for substitutes to fossil fuels to safeguard the environment and accomplish energy security (Reddy et al., 2013).

Disadvantages of bioplastics

Bioplastics have a range of drawbacks that need to be considered. Uncontrolled and improper management of bioplastic waste often leads to littering and contamination of soil and water. Like traditional plastics, the littering of bioplastic wastes is often detrimental to wildlife. The dumping of bioplastic waste into a landfill can lead to the emission of greenhouse gases. Recycling biodegradable plastics with other types of plastics is not feasible. Cultivating crops to produce bioplastics can finally establish competition on cultivable land for food production (Reddy et al., 2013; ISWA, 2015).

Bioplastics and carbon cycle

The carbon cycle is the biogeochemical process of sharing carbon amongst the earth's spheres. The plant takes CO_2 as it grows, and when it biodegrades, it releases the carbon dioxide back into the atmosphere through a completely closed-loop process. Such natural resources undergo various polymer-forming processes. Polymers are transformed further into bioplastics. These plastics are utilized in different forms and degraded further. Carbon dioxide is released back into the atmosphere during biodegradation. Consequently, CO_2 removed from the environment is cycled back into the atmosphere through different phases, creating a closed loop (Reddy et al., 2013). GHG generation and its effect on climate change are significant features of plastics manufacturing. The Intergovernmental Panel on Climate Change (IPCC) estimates for stabilizing atmospheric GHG concentrations by 2050 at 450 ppm CO_2 require emission reductions of 80% compared to 1990 levels (Barker et al., 2007), a major challenge for all sectors of the economy (Williams et al., 2012). Implementing a national plan for a bio-based economy is a firm policy goal for many countries. Overall, emissions tend to be higher for fossil-based polymers, while those associated with bio-based polymers have negative and lower positive values (Jamshidian et al., 2010). But, in terms of carbon, we are still on a closed loop (Gontard et al., 2019). When we talk about a circular economy, considering the duration of a cycle is important; it must be consistent with human activity. This is the case with the process of biodegradation/photosynthesis, which is accomplished fairly fast (Gontard et al., 2019).

Bioplastics and environmental sustainability

Environmental sustainability is now a great concern for life forms due to the rising population, urbanization, and standard of living (AlQattan et al., 2018). Plastic garbage has become one of the most serious issues due to its widespread usage, difficulty in decomposition, and massive mass, all of which have severe consequences for disposal sites and aquatic ecosystems (de Moura et al., 2017). Over half of all plastic waste in 2015 was made of single-use plastic, which is frequently used for packaging and items that are only used once before being discarded (Giacovelli, 2018). For years, the world's reliance on petroleum-based synthetic polymers, or plastics as they are usually known, has made them one of the most appealing materials utilized for a wide range of purposes. Every year, more than 400 million tonnes of plastic are manufactured worldwide (Geyer et al., 2017). Due to its low price, ease of production, great endurance, low density, and aesthetic appearance, this record is commensurate with the material's features. In 2015, plastic packaging with single-use/disposable applications accounted for 36% of global plastic manufacturing. Food packaging, grocery bags, cutlery, bottles, straws, cups, and other products are among them. It also accounts for 47% of all plastic waste produced globally. Plastics are eventually recycled into other products, incinerated, landfilled, dumped, or strewn in the environment. By 2050, there will be roughly 12 billion tonnes of plastic garbage in landfills and the environment if current consumption trends remain unchanged and no environmentally friendly substitute materials are developed. Being a part of the natural ecosystem, plastic has billions of long-term detrimental consequences on the environment on both land and sea life (Geyer et al., 2017). Many different types of plastic, such as plastic bags, have been discovered to poison the respiratory tracts and stomachs of marine life. Animals like dolphins and turtles mistake plastic bags for food. Toxic compounds were absorbed during the transfer of the production process from inserting plastic into animal tissue to ultimately entering humans via the food chain (Jambeck et al., 2015). There are numerous environmental benefits to using bioplastic, including a potential reduction in carbon footprint and GHG emissions, lower manufacturing energy costs, reduced permanent litter, and considerably safer for the environment (Pilla, 2011). Bioplastics also have material advantages such as substantially better water vapor permeability than normal plastic, a less oily feel, improved printability, and being softer and more tactile. These nonpetrochemical, environmentally friendly polymers cut greenhouse gas emissions. They cut down on the use of fossil fuels and nonrenewable resources. In comparison to traditional plastic, the manufacturing process can save up to 65% of energy and produce less greenhouse gases. To meet the demands of a growing population, biodegradable polymers should be created and used on a big scale (Singh and Verma, 2020). Petroleum-based plastics can be substituted by bio-based polymers because of their biodegradability and renewability, reducing environmental hazards (Markus, 2014).

Future perspective

In general, bioplastics are marketed as a renewable alternative to traditional plastics. Bioplastic production, however, is the most challenging point because it can disrupt the potential sources of food. Using nonfood resources for this reason will ease up the situation. Those are referred to as second-generation bioplastics. Therefore, these must be produced by manufacturing methods like extrusion, compression, and injection molding. Future environmental issues and bioplastic effects have not been thoroughly explored and understood yet. Therefore, further research is needed to overcome the scarce resources available, boost resource productivity, and mitigate environmental issues (Sidek et al., 2019; Krzan, 2013).

Conclusion

Bioplastics are growing rapidly in many applications because of the obvious advantages. Such advantages would increase as the oil supply tightens. Their carbon footprint is also smaller than the equivalents made from oil. With increased consciousness of the environment and growing environmental consciousness among communities, those who invest in biodegradable plastic materials benefit as they have a head start. The many benefits of bioplastics, such as being 100% biodegradable, made from natural renewable resources, recycling, reusing, composting, or burning without generating noxious by-products, make them outstanding substitutes for petroleum-based plastics. Biopolymers reduce carbon dioxide production at the time of manufacturing and after disposal decompose to organic matter. Decomposable polymers may not be a one-stop solution to all environmental issues produced by traditional plastics. Still, it is a step on the right track as time is of the essence for producing biodegradable polymers.

Acknowledgments

The writers of this chapter are thankful to the Department of Environmental Science, University of Kashmir, Srinagar for sharing their facilities and those who made their literature available for the compilation of this manuscript.

References

AlQattan, N., Acheampong, M., Jaward, F.M., Ertem, F.C., Vijayakumar, N., Bello, T., 2018. Reviewing the potential of Waste-to-Energy (WTE) technologies for Sustainable Development Goal (SDG) numbers seven and eleven. Renewable Energy Focus (San Francisco, Calif.) 27, 97–110.

Auras, R., Harte, B., Selke, S., 2004. An overview of polylactides as packaging materials. Macromolecular Bioscience 4 (9), 835–864.

Avérous, L., Pollet, E., 2012. Environmental Silicate Nano-Biocomposites, Vol. 1. Springer, London, UK.

Ball, R.C., McLeish, T.C.B., 1989. Dynamic dilution and the viscosity of star-polymer melts. Macromolecules 22 (4), 1911–1913.

Barker, T., Bashmakov, I., Bernstein, L., Bogner, J., Bosch, P., Dave, R., et al., 2007. Technical summary. In Climate change 2007: mitigation. Contribution of Working Group III to the Fourth Assessment Report of the Intergovernmental Panel on Climate Change. Cambridge University Press.

Bezirhan, A.E., Ozsoy, H.D., 2014. Bio-plastics as a green material. International Conference on Green Infrastructure and Sustainable Socities/Cities 'GreInSus'-14. pp. 66–70.

Blakistone, B., Sand, C.K., 2007. Using sustainable packaging technologies to respond to consumer, retailer, and seafood industry needs. International Smoked Seafood Conference Proceedings 75–79.

Carroccio, S.C., Scarfato, P., Bruno, E., Aprea, P., Dintcheva, N.T., Filippone, G., 2021. Impact of nanoparticles on the environmental sustainability of polymer nanocomposites based on bioplastics or recycled plastics–a review of life-cycle assessment studies. Journal of Cleaner Production 130322.

Chen, Y.J., 2014. Bioplastics and their role in achieving global sustainability. Journal of Chemical and Pharmaceutical Research 6 (1), 226–231.

de Moura, I.G., de Sá, A.V., Abreu, A.S.L.M., Machado, A.V.A., 2017. Bioplastics from agro-wastes for food packaging applications. Food Packaging. Academic Press, pp. 223–263.

Demirbas, A., 2007. Biodegradable plastics from renewable resources. Energy Sources, Part A 29 (5), 419–424.

Di Bartolo, A., Infurna, G., Dintcheva, N.T., 2021. A review of bioplastics and their adoption in the circular economy. Polymers 13 (8), 1229.

Endres, H.J., Siebert-Raths, A., 2009. Technische Biopolymere. Hanser Verlag.

European Bioplastics, 2017. Available from: http://www.european-bioplastics.org (2017).

Forni, D., Wenk, C., Bee, G., 1999. Digestive utilization of novel biodegradable plastic in growing pigs. Annales de zootechnie 48 (3), 163–171.

Gaumann, U., 2013. Bioplastics magazine: biobased polymers for automotive safety components. Cellular Polymers 32 (4), 254–255.

Geyer, R., Jambeck, J.R., Law, K.L., 2017. Production, use, and fate of all plastics ever made. Science Advances 3 (7), e1700782.

Giacovelli, C., 2018. Single-Use Plastics: A Roadmap for Sustainability (rev. 2).

Gontard, N., Bruzaud, S. and Ghiglione, J.F., 2019. Biodegradable and compostable bioplastics.

Greene, J.P., 2014. Sustainable Plastics: Environmental Assessments of Biobased, Biodegradable, and Recycled Plastics. John Wiley & Sons.

Gross, R.A., Kalra, B., 2002. Biodegradable polymers for the environment. Science 297 (5582), 803–807.

Harms, H., 2006. March. Carbohydrates in the context of the wood biorefinery. EPNOE Workshop for Industry 23–24.

Haugaard, V.K., Udsen, A.M., Mortensen, G., Høegh, L., Petersen, K., Monahan, F., 2001. Potential food applications of biobased materials. An EU-concerted action project. Starch-Stärke 53 (5), 189–200.

Hawker, C.J., Farrington, P.J., Mackay, M.E., Wooley, K.L., Frechet, J.M., 1995. Molecular ball bearings: the unusual melt viscosity behavior of dendritic macromolecules. Journal of the American Chemical Society 117 (15), 4409–4410.

Hudgins, M., 1999, April. Aerobic landfill studies from the USA. In 1st International Conference on Solid Waste.

ISWA. 2015. Biodegradable Plastics-An Overview of the Compostability of Biodegradable Plastics and its Implications for the Collection and Treatment of Organic Wastes, International Solid Waste Association. Available from: http://www.iswa.org/index.

Jager, A., 2010. Ingeo™ polylactide een natuurlijke keus, presentation given at VMT conference: green packaging.

Jambeck, J.R., Geyer, R., Wilcox, C., Siegler, T.R., Perryman, M., Andrady, A., et al., 2015. Plastic waste inputs from land into the ocean. Science (New York, N.Y.) 347 (6223), 768–771.

Jamshidian, M., Tehrany, E.A., Imran, M., Jacquot, M., Desobry, S., 2010. Poly-lactic acid: production, applications, nanocomposites, and release studies. Comprehensive Reviews in Food Science and Food Safety 9 (5), 552–571.

Koide, S., Shi, J., 2007. Microbial and quality evaluation of green peppers stored in biodegradable film packaging. Food Control 18 (9), 1121–1125.

Krzan, A., 2013. Biodegradable Polymers and Plastics,"Innovative Value Chain Development for Sustainable Plastics in Central Europe." PLASTiCE.

Kumar, S., Thakur, K.S., 2017. Bioplastics-classification, production and their potential food applications. Journal of Hill Agriculture 8 (2), 118–129.

Kwak, S.Y., Lee, H.Y., 2000. Molecular relaxation and local motion of hyperbranched poly (ether ketone) s with reference to their linear counterpart. 1. Effect of degrees of branching. Macromolecules 33 (15), 5536–5543.

Lau, W.W., Shiran, Y., Bailey, R.M., Cook, E., Stuchtey, M.R., Koskella, J., et al., 2020. Evaluating scenarios toward zero plastic pollution. Science (New York, N.Y.) 369 (6510), 1455–1461.

Laycock, B., Halley, P., Pratt, S., Werker, A., Lant, P., 2013. The chemomechanical properties of microbial polyhydroxyalkanoates. Progress in polymer science 38 (3–4), 536–583.

Lörcks, J., 2005. Bioplastics, brochure of. *Fachagentur Nachwachsende Rohstoffe eV (FNR).*

Manjunath, L., Sailaja, R.R.N., 2016. Starch/polyethylene nanocomposites: mechanical, thermal, and biodegradability characteristics. Polymer Composites 37 (5), 1384–1395.

Manjure, S., 2011. PLA for paper coating. Bioplastics Magazine 6 (5), 34–37.

Markus, K., 2014. Everything you don't want to know about plastics. Swedish Society of Nature Conservation .

Miller, R., 2005. The landscape for biopolymers in packaging. Miller-Klein associates report. Summary and Full Report available from The National Non-Food Crops Centre. Heslington, York, UK.

Patel, M., Bastioli, C., Marini, L., Würdinger, E., 2005. Life-cycle assessment of bio-based polymers and natural fiber composites. Biopolymers Online: Biology Chemistry Biotechnology Applications 10.

Peelman, N., Ragaert, P., De Meulenaer, B., Adons, D., Peeters, R., Cardon, L., et al., 2013. Application of bioplastics for food packaging. Trends in Food Science & Technology 32 (2), 128–141.

Pilla, S. (Ed.), 2011. Handbook of Bioplastics and Biocomposites Engineering Applications. John Wiley & Sons.

Ralston, B.E., Osswald, T.A., 2008. The history of tomorrow's materials: protein-based biopolymers. Plastics Engineering 64 (2), 36–40.

Reddy, R.L., Reddy, V.S., Gupta, G.A., 2013. Study of bio-plastics as green and sustainable alternative to plastics. International Journal of Emerging Technology and Advanced Engineering 3 (5), 76–81.

Rehm, B.H., 2003. Polyester synthases: natural catalysts for plastics. Biochemical Journal 376 (1), 15–33.

Roovers, J., 1994. Concentration dependence of the relative viscosity of star polymers. Macromolecules 27 (19), 5359–5364.

Sidek, I.S., Draman, S.F.S., Abdullah, S.R.S., Anuar, N., 2019. Current development on bioplastics and its future prospects: an introductory review. INWASCON Technology Magazine 1, 03–08.

Singh, P., Verma, R., 2020. Bioplastics: a green approach toward sustainable environment. Environmental Microbiology and Biotechnology. Springer, Singapore, pp. 35–53.

Song, J.H., Murphy, R.J., Narayan, R., Davies, G.B.H., 2009. Biodegradable and compostable alternatives to conventional plastics. Philosophical Transactions of the Royal Society B: Biological Sciences 364 (1526), 2127–2139.

Sudesh, K., Iwata, T., 2008. Sustainability of biobased and biodegradable plastics. CLEAN–Soil, Air, Water 36 (5-6), 433–442.

Suriyamongkol, P., Weselake, R., Narine, S., Moloney, M., Shah, S., 2007. Biotechnological approaches for the production of polyhydroxyalkanoates in microorganisms and plants—a review. Biotechnology Advances 25 (2), 148–175.

Thielen, M., 2014. Bioplastics MAGAZINE. In: Bioplastics–Plants and crops, raw materials. Published in Berlin: Fachagentur Nachwachsende Rohstoffe eV (FNR). Order, (237).

Tokiwa, Y., Calabia, B.P., Ugwu, C.U., Aiba, S., 2009. Biodegradability of plastics. International Journal of Molecular Sciences 10 (9), 3722–3742.

Van den Oever, M., Molenveld, K., van der Zee, M., Bos, H., 2017. Bio-Based And Biodegradable Plastics: Facts and Figures: Focus on Food Packaging in the Netherlands (No. 1722). Wageningen Food & Biobased Research.

Weston, S., 2012. Packaging-assessing its environmental credentials.

Williams, J.H., DeBenedictis, A., Ghanadan, R., Mahone, A., Moore, J., Morrow III, W.R., et al., 2012. The technology path to deep greenhouse gas emissions cuts by 2050: the pivotal role of electricity. Science (New York, N.Y.) 335 (6064), 53–59.

Yaradoddi, J., Patil, V., Ganachari, S., Banapurmath, N., Hunashyal, A., Shettar, A., et al., 2016. Biodegradable plastic production from fruit waste material and its sustainable use for green applications. International Journal of Pharmaceutical Research and Allied Sciences 5 (4), 72–81.

Yu, J., Chen, L.X., 2008. The greenhouse gas emissions and fossil energy requirement of bioplastics from cradle to gate of a biomass refinery. Environmental Science & Technology 42 (18), 6961–6966.

Zhang, J., Jiang, L., Zhu, L., Jane, J.L., Mungara, P., 2006. Morphology and properties of soy protein and polylactide blends. Biomacromolecules 7 (5), 1551–1561.

Zhou, Q., Larson, R.G., 2007. Direct molecular dynamics simulation of branch point motion in asymmetric star polymer melts. Macromolecules 40 (9), 3443–3449.

Chapter 26

Resource recovery from the e-wastes through bioleaching

Mitun Sen, Nilothpal Sinha and Dipankar Ghosh
Microbial Engineering & Algal Biotechnology Laboratory, Department of Biosciences, JIS University, Agarpara, Kolkata, India

Introduction

E-waste, also known as electronic waste, is simply unwanted electronic equipment that is no longer functional and is nearing or at the end of its "useful life." The majority of electrical or electronics contain toxic materials (beryllium, cadmium, mercury, and lead), which poses harmful effect on human life and environments like our soil, water, air, and wildlife. When discarded e-waste is dug in a landfill, it becomes dissolved in microscopic traces into the sludge that permeates the landfill. In this context, it has been found that harmful chemicals leak into the earth underneath the landfill (Karimi-Maleh et al., 2021). In general, e-wastes consist of various metals and nonmetals components (30% organics, 30% ceramics, and 40% inorganics). Inorganic metals contain base metal (e.g., iron, aluminum, iron, tin, and copper), baronial metallic constituents (e.g., gold, palladium, and silver), heavy metallic residual components likely nickel, chromium, zinc, lead, cadmium, mercury, and beryllium, and rare-earth metals (e.g., platinum, gallium, and tantalum). E-waste is collected by around 17.4% and can be recycled. However, the rest of the e-waste materials (82.6%) have been traded through illegal markets and disposed of in municipal landfills (Herat and Agamuthu, 2012). Open dumping causes soil pollution that causes contamination in the food chain and groundwater pollution, which is directly used for domestic and irrigational purposes (Saha et al., 2021). Most e-wastes generate carcinogen-based health hazards, including damage to the kidney, nervous system, bones, and reproductive and endocrine system (Chen et al., 2011; Fowler and Weissberg, 1974; Bosshard et al., 1996; Coram and Rawlings, 2002; Bala and Goel, 2012). E-waste recycling treatment is the costliest process, which is why only a few developing countries can informally invest in e-waste management. However, several issues have evolved due to higher population density, improper knowledge of e-waste reutilization schemes, lack of e-waste management rules, and improper enforcement of existing legislation in a few developing nations (Rautela et al., 2021). In 2005, around two million tons of e-waste were generated in the nation of the US, but only 17%–18% of these two million tons of e-waste mass undergoes recycling. While the rest of the wastes (e.g., ~80%) have been disposed of by local landfilling (Bhutta et al., 2011). A survey report from MALT (Manufacturers' Association for Information Technology) announces that fifty thousand tons are being imported every year (Tsydenova and Bengtsson, 2011; Barbosa et al., 2005; Jomova et al., 2011; Sheng and Etsell, 2007). Bioleaching is the technique of extracting metal ions from ores or wastes using microorganisms. The direct bioleaching process predominantly uses minerals that can react to oxidation to establish a direct enzymatic strike towards metal separations from e-wastes using potential microbial assemblages. In the indirect bioleaching process, potent microbial regimes have not been involved in direct adjoining with metal ions that produce bioleaching agents that oxidize metals from e-waste biomass. In today's smart and productive world, the production of electronic materials is increasing, as a result, the amount of e-waste has been ameliorated (Karimi-Maleh et al., 2021). Acidophilic bacterial assemblages (e.g., *Acidothiobacillus ferrooxidans*, *Acidothiobacillus thiooxidans*, *Leptospirillum ferrooxidans*, and *Sulfolobus* sp.) play a decisive part in heavy metal bioleaching from waste. On the earth's surface, microbial communities are involved in decomposing diverse inorganic and organic substances. In bioleaching, microbes have shown their natural ability to transform solid metallic compounds into a soluble and extractable state. Three major groups of microorganisms take part in the bioleaching of heavy metals in e-waste bacterial autotrophs, heterotrophs, including heterotrophic fungal regimes. Moreover, chemolithotrophs are considered the utmost mineral recycling microbial communities that can break

Role of Green Chemistry in Ecosystem Restoration to Achieve Environmental Sustainability. DOI: https://doi.org/10.1016/B978-0-443-15291-7.00027-4

down iron and sulfur naturally following oxidative metabolic networks (Clark and Norris, 1996; Mishra and Rhee, 2010; Nagpal et al., 1993). The common underlying mechanisms in bioleaching are acidolysis, complexolysis, redoxolysis, and bioaccumulation (Chen et al., 2011). Mineral bio-oxidation methods have been involved in the employment of microorganisms that are trusted as a consortium of gram-negative bacteria (e.g., iron- and sulfur-oxidizing *A. ferrooxidans*) at temperatures of 40°C or less (Coram and Rawlings, 2002). Microorganisms extract metals from sulfide, iron-containing ores, and mineral aggregates (Lee et al., 2011). Though bioleaching is a time-consuming process it's cost-effective but does not show the highest metals recovery efficacies (Ren et al., 2009).

Methods of e-waste recycling

E-waste has been considered a variety of metals and nonmetals, including distinct and diversified characteristics. It consists of organics, ceramics, and inorganics metals (Kaya, 2016a). The e-waste resource recovery approach can provide employment and commercial prospects; its implementation might be slowed down by not having adequate infrastructures and effective scientific procedures (Lakshmi and Raj, 2017).

Conventional recycling approach

In a formal recycling procedure, there are three steps preprocessing, postprocessing, and disposal. Grouping and fractioning of metals are effective approaches for the preprocessing of e-waste (Ashiq et al., 2019). By heat, chemical, and metallurgical methods valuable elements from the homogenous group are extracted (Arya and Kumar, 2020). Some mechanical methods (e.g., shredding/crushing, magnetic separation, eddy current separation, ball milling, and gravity separation) play an important role in making dissociation between the metallic and non-metallic fractions of the wastes (Kaya, 2016b; Zheng et al., 2017; Chen et al., 2019; Awasthi et al., 2017). Different forms of acids (e.g., HNO_3, HCl, H_2SO_4, and aquilegia) are applied towards the hydrometallurgical process (preprocessing approach) to leach metallic residues from e-wastes (Ashiq et al., 2019). To track down valuable metals from e-wastes by sintering, smelting in a furnace, and incineration e-wastes need a molten metal bath at 1250°C (Cui and Zhang, 2008; Khaliq et al., 2014). The aforementioned process generates secondary pollutants as hazardous by-products. $CaCO_3$ is used to remove those secondary pollutants (Qin et al., 2020). The major problem comes when waste residues and acidic effluents are produced by postprocessing methods and are treated by landfill or incineration (Lucier and Gareau, 2019). It's a 4R eco-friendly technology; the 4Rs denote reduce, reuse, recycle, and recover (Debnath et al., 2016).

Informal recycling

Informal recycling or backyard recycling refers to the unscientific and unsophisticated methods used to recover resources from e-waste (Forti et al., 2020). Rag pickers in the informal recycling industry significantly lower the environmental impact of e-waste (Chi et al., 2011a). Temperature-exchange equipment, big equipment, screens, and information-technology equipment are the principal e-waste items handled by the informal sectors (Baldé et al., 2015). Postrecovery reduces unwanted materials. Those are deposited in neighboring sewers, rivers, lakes, or seas after being landfilled. Backyard e-waste recycling harms the environment by poisoning the soil eco-dynamics, eco-balance of waterbodies, and quality of air. Motherboards, PCBs, and wires are dipped in HCl, H_2SO_4, HNO_3, or aqua regia for a long treatment tenure (Jadhav and Hocheng, 2015). The copper-containing sludge has been subsequently drained off following the ion-exchange cycle as a combinatorial approach. Afterward, segregated copper was brought out and the rest effluent was discarded without additional treatment (Shibayama et al., 2010). By the same process gold and silver were also restored from e-wastes (Sivaraman, 2013; Chakraborty et al., 2018; Fu et al., 2016). Several problems exist in informal recycling by the toxin (produced by heavy metal interaction). Air pollution is caused by various reasons arsenic pollution in China by open burning of e-waste (Deng et al., 2006), cadmium pollution by mechanical treatment and open burning in China from e-waste recycling, copper pollution by mechanical treatment, crude recycling and open burning at China and India (Deng et al., 2006; Bi et al., 2010; Ha et al., 2009). In India chromium pollution occurs through open flaming and crude recycling (Ha et al., 2009) and nickel pollution by mechanical treatment and open burning (Bi et al., 2010). Soil pollution is caused by different activities like pollution of chromium in Delhi by e-waste dumping and landfilling, leaching processes, crude recycling, mechanical treatment, and open disposal (Salam and Varma, 2019; Wu et al., 2018; Pradhan and Kumar, 2014; Ackah, 2019).

Bioleaching of e-waste for resource recovery

Microbial bioleaching technologies may have been employed since ancient times, and copper was most likely recovered from groundwater by the Greeks and Romans more than 2000 years ago. Bacteria are primarily groomed to enrich the recovery of metals in aqueous state from solid mineral deposits and ore mines, but this has only been understood for around 50 years (Bosecker, 1997). In some cases, bioleaching is a more cost-effective procedure than chemical leaching. Acidophilic bacteria (e.g., *Acidithiobacillus ferrooxidans*, *A. thiooxidans*, *Leptospirillum ferrooxidans*, and *Sulfolobus* sp.) play an essential part in the bioleaching of heavy metals from wastes (Needhidasan et al., 2014). On the earth's surface, microorganisms are involved in the production and degradation of diverse inorganic and organic materials (Gaidajis et al., 2010). The three main categories of microorganisms engaged in metal bioleaching are autotrophic bacteria (e.g., *Thiobacillus* sp.), heterotrophic bacteria (e.g., *Pseudomonas* sp.; *Bacillus* sp.), and heterotrophic fungi (e.g., *Aspergillus* sp., *Penicillium* sp.). The most significant mineral-degrading microorganisms are chemolithotrophic iron- and sulfur-oxidizing type. The common mechanisms mixed up in bioleaching include acidolysis, complexolysis, radiolysis, and bioaccumulation. Mineral bio-oxidation processes involving microorganisms are made up of a consortium of gram-negative bacteria, including iron- and sulfur-oxidizing *A. ferrooxidans*, at temperatures below 40°C (Prabhu et al., 2014). Acidophilic bacteria thrive in low-pH environments, such as those with a pH of 1−2 (Gokul and Rajasekaran, 2019).

Impact of bioleaching on e-waste recycling

Bioleaching of disposed printed circuit boards

Printed circuit boards (PCBs) are the most widespread recycling target for the preservation of resources to establish a circular economy (Ilyas, 2010). The diversity of e-waste makes the recycling avenues staggeringly difficult. Although PCBs account for just 6% of electronic waste, they remain an essential component (Arshadi and Mousavi, 2015; Gu et al., 2017a). It is estimated that 0.5 million tonnes of discarded PCBs are disposed of each year (Gu et al., 2017b). Recycling of PCBs is most important due to their high precious metal content, large waste quantity, and short lifespan (e.g., ~4 years). The metal percentage accounts for approximately 28% of PCBs which includes copper, lead, nickel, platinum, gold, and silver (Arshadi and Mousavi, 2015; Xiang et al., 2010). Bioleaching has been reviewed as a commercially viable and low-cost method for extracting metals from low-grade ores (Yang et al., 2009; Bryan et al., 2015). Moreover, chemolithoautotrophic (mesophilic) bacteria or cyanogenic bacteria (*Chrotobacterium violaceum*) are predominantly utilized for PCB bioleaching (Ilyas et al., 2010). *C. violaceum* encodes six classes of HCN synthase enzymes to accelerate PCBs bioleaching. *C. violaceum* shows higher metabolic and enzymatic efficacy for PCB recycling and retrieves silver, gold, and copper as cyanide conjugates (Natarajan, 2018; Arshadi and Mousavi, 2014; Natarajan and Ting, 2014). PCB bioleaching has been performed in stirred tank reactors considering a semiindustrial process based on this technology and an economic study to certify the viability of the aforementioned bioprocess (Xia et al., 2017). By conducting a thorough and economically sustainable assessment for the recovery of metals (such as copper and gold) from PCBs. However, later on, pulsed plate bioreactor has also been employed to set forward the large-scale bioleaching of PCBs to extract copper (Jagannath et al., 2017). In another study, the efficiency of Cu leaching from PCBs (Gu et al., 2017b; Wang et al., 2016) has also been performed. The BRGM-KCC acidophilic consortium was employed in a bubble column with activated charcoal to investigate the biogenic ferric iron formation. Priya and Hait (2018) have proposed a hybrid bioleaching and hydrometallurgy technique in which citric acid was used as a chelating agent to boost PCB base metal recovery. A lower particular size and lowered density of pulp can establish higher gold recovery efficiencies (Arshadi and Mousavi, 2015). The amount of dissolved oxygen has a decisive influence on cyanogenic bacteria's metal recovery since oxygen is required as an electron acceptor in addition to being required for bacterial respiration (Liang et al., 2014). Mixed microbial consortiums of autotrophs and heterotrophs have shown higher metal removal efficacies than the axenic pure microbial cultures during the bioleaching process (Baniasadi et al., 2019). Moreover, copper mobilization from PCBs was enhanced when iron-oxidizing and sulfur-oxidizing chemolithoautotrophs coexist as a consortium (Arshadi and Mousavi, 2015). A combination of thermophilic microbe and chemolithoautotroph bacteria showed higher efficiency for PCB bioleaching (Ilyas et al., 2013; Ilyas et al., 2010).

Disposal of spent batteries through bioleaching

Electronic batteries have been categorized as likely primary non-rechargeable batteries and secondary rechargeable (Kim et al., 2016). It shows the primary category of battery, that is, ZMB, and the other type, that is, LBs, having nickel metal hydride and Ni-Cd batteries (Xin et al., 2012). Because of the presence of dangerous elements, the accumulation of wasted batteries in general generates significant difficulties for the ecology. Thus, technological advancements for the treatment and recycling of battery-based e-waste are of utmost necessity in the current global scenario (Zeng et al., 2012).

Lithium-ion batteries (LBs): Cathode materials are frequently the costliest component of a battery. The recycling of those e-waste must be maintained at all stages of manufacturing, usage, collection, recycling, and disposal (Bahaloo-Horeh and Mousavi, 2017). The cathode and anode rich in Co and Li are then combined, crushed, and sieved to the desired size (Niu et al., 2014; Zeng et al., 2012).

Nickel-cadmium-based batteries: Ni-Cd and nickel metal hydride are rechargeable major constituents in secondary batteries. It has tremendous industrial usage, especially in portable electronic devices including wireless communication and computer equipment (Kim et al., 2016). As a result, many EU directives have been adopted to gather them. For Ni-Cd battery recovery, many treatments based on pyrometallurgical, hydrometallurgical, or combination approaches are available (Velgosová et al., 2013).

Mechanism of bioleaching

Bioleaching is an environmentally benign approach where potent microorganisms grow on e-waste biomass surface and release potent enzymes that can degrade the e-waste to release value-added bioresources including metal ions to achieve environmental and industrial sustainability (Zhu et al., 2003; Zhao et al., 2008). Heterotrophic bacteria, fungi, and autotrophic bacteria can carry out bioleaching using e-waste complex materials (Ruan et al., 2014). Fungus (e.g., *Penicillium* sp. and *Aspergillus* sp.) may flourish bioleaching in higher pH-based microenvironments (Bas et al., 2013). Metal breakdown by fungal species is mediated by two major processes. The acidolysis process starts with metal ions exchanging hydrogen ions, followed by complexolysis-complex synthesis. A broad spectrum of the fungal regime including *Aspergillus niger* biosynthesizes organic acids (e.g., malic acid, gluconic acid, and oxalic acid) for metal recovery in WPCBs bioleaching (Faraji et al., 2018). Mesophilic heterotrophic bacteria, for example, *Thiobacillus* sp. and *Leptospirillum* sp. can carry out bioleaching. In comparison to mesophiles, moderate thermophiles had the highest efficiency in copper sulfide bioleaching (Marhual et al., 2008; Li et al., 2014). Mostly bioleaching takes place in stirred tanks (Ilyas and Lee, 2014; Hong and Valix, 2014). It is beneficial to mix acid along with the bioleaching of Cu produced from waste PCBs (Pandey, 2013). Copper bioleaching from PCB has been carried out through the involvement of *A. ferroxidans*. The connection of PCB particles with *A. ferroxidans* has been supposed to be favorable if and only if the van der Waals force was larger than the repulsive force (Kasper and Veit, 2018). When microorganisms adhere to mineral surfaces, they generate an exopolysaccharide (EPS) layer. The thiosulfate process is postulated for acid-insoluble metal sulfides such as molybdenite, pyrite, and polysulfide, whereas the polysulfide mechanism is used for oxidation and acid soluble metal sulfides such as chalcopyrite, gelane, and sphalerite (Zhu et al., 2011; Priya and Hait, 2017). *Pseudomonas* sp. and *C. violaceum* bacterial strains were shown to be capable of dissolving metals such as Ag and Au from WPCB by producing biochemicals such as cyanide. Bacteria provide cyanide when the conditions are met. In a high-pH solution, cyanide dissolves (Juju et al., 2015; Işıldar, 2019). The mechanism of bioleaching from e-waste to recover metals has been elaborately demonstrated in Fig. 26.1.

Factors influence bioleaching

Biotic factors that impact e-waste bioleaching

Microorganism nutritional type

Microbes are categorized into four according to their electron, carbon, and energy needs (Spellman and Drinan, 2014). Chemolithoautotrophs are known as extremophilic microbes having the metal-resistant capability and they capture oxygen from the air as the ultimate electron acceptor to oxidize metals into solution. Chemolithoautotrophs use ambient carbon dioxide as a carbon supply and inorganic hydrogen to contribute electrons to produce energy from reduced inorganic compounds (sulfides). Sulfur and iron oxidizers make up the majority of lithotroph's bioleaching members. *Acidothiobacillus* sp. (Yang et al., 2013; Işıldar, 2019) *Leptospirillum* sp., *Acidianus* sp., *Sulfolobus* sp., and other bacteria genera are good at bioleaching metals from their surfaces (Bosecker, 1997).

Size and growth of the inoculums

The potential of bioleaching on metallic residuals recoveries from e-wastes has been highly influenced by the microbial inoculum size, substrate consumption rate, and the metabolite pools within the reaction platform. It has been emphasized that increasing microbial inoculum size enhances the formation of organic acids which induce higher leaching of

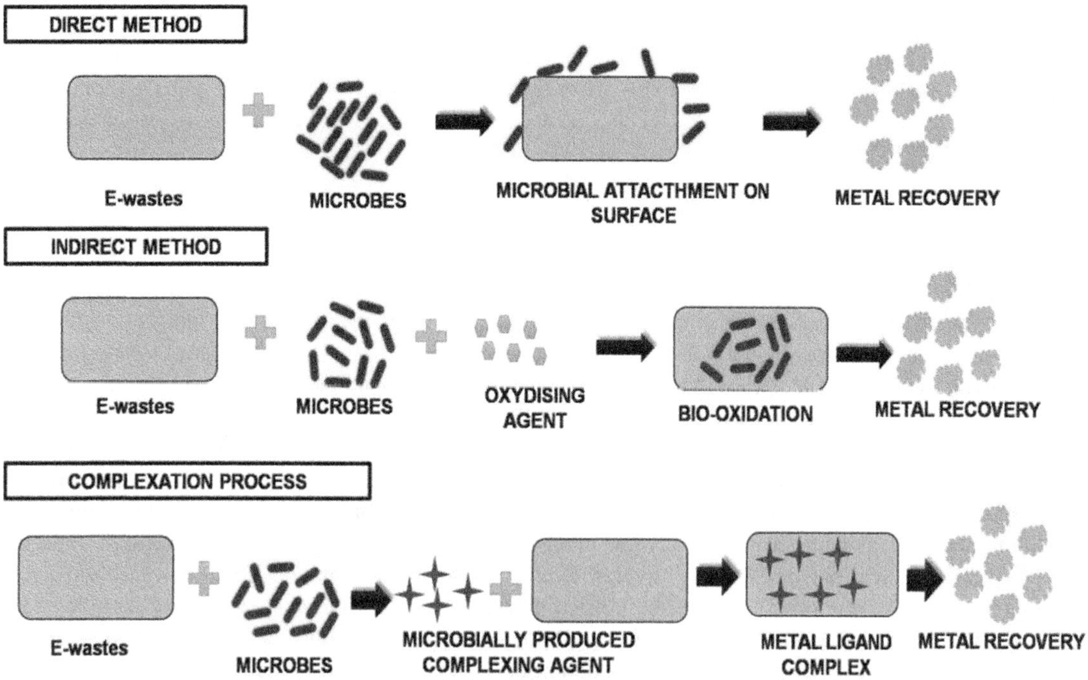

FIGURE 26.1 Mechanism of bioleaching from e-waste to recover various metals (Yaashikaa et al., 2022).

metals from e-wastes (Rasoulnia and Mousavi, 2016), alleviates heavy metal bioleaching (Chen and Cheng, 2019), and improves copper recovery from e-waste. When the inoculum size exceeds a particular threshold, the bioreactor becomes turbid and clogs, limiting the interaction between the solids and the cells, resulting in insufficient bioleaching time and inadequate metal bioleaching (Jagannath et al., 2017).

Abiotic factors that impact on e-waste bioleaching

Bioleaching physicochemical properties

The parameters of the organism, like pH, temperature, oxygen demand, and bioleaching medium composition, are determined by the leaching process system to be used in the process. The selected habitat's natural environment exclusively determines the initial pH and temperature that must be maintained during the bioprocess (Ilyas et al., 2007). The availability of protons is reduced at higher basal pH, which hinders the microbial oxidation process. The rate of leaching is related to the temperature, although it differs depending on whether the material is thermophile or mesophilic (Priya and Hait, 2017). The impact of Fe^{2+} content, pH (Xiang et al., 2010), temperature, carbon source, energy source, and additional supplemental optimization on bioleaching of metals from e-waste is observed in reports (Arshadi et al., 2016). When cyanogenic strains are used, supplementing the medium with cyanide synthesis precursors improves the bioleaching of essential metals from e-waste (Liu et al., 2016). In most bioleaching strains, oxygen serves as the terminal electron acceptor and participates in redox processes (Pradhan et al., 2008). The aeration rate in designed systems may be regulated to enable effective oxygen transfer for bioleaching (Foucher et al., 2003).

pH

One of the bioleaching environment's physicochemical factors is pH. The oxidation of iron and sulfides, as well as the dissolution of metals like zinc, nickel, and copper from sulfide-bearing minerals, are thought to be best at acidic pH, according to studies. Furthermore, aluminum dissolution from aluminous basalt rock is best achieved at a neutral pH (Mehta et al., 1979). The optimum pH value is required for leaching bacteria to flourish and is crucial for metal solubilizations. For the bacterial oxidation of ferrous iron and sulfide, a pH between 2.0 and 2.5 is ideal. *T. ferrooxidans* faces significant inhibition at pH levels below 2.0, although they can be acclimated to even lesser pH values by raising acid input (Tuovinen and Kelly, 1973).

Temperature

T. ferrooxidans can only oxidize ferrous iron and sulfide at temperatures between 28°C and 30°C. Metal extraction will be reduced at lower temperatures, however even at 4°C, bacterial solubilizations of copper, cobalt, nickel, and zinc are found (Ahonen and Tuovinen, 1989). At greater levels, thermophilic bacteria can be employed for leaching at temperatures between 50°C and 80°C (Brierley, 1978; Brierley, 1999).

Conclusion and future outlook

The bioleaching needs to be optimized, simulated, and ameliorated in such a way that it could be viable for industrial application on a large scale. Moreover, the current state of bioleaching for metal recovery from e-wastes suffers from the usage of high-cost feedstock for diverse microbial assemblages (bacteria, fungus, and algae) to support their growth and improve enzymatic activities for improving metal recovery efficacies in due course. Hence, the usage of alternative low-cost substrates for microbial growth and metabolism might boost up kinetics of bioleaching performance. The economic effectiveness of the heap bioleaching process might be improved by lowering the cost of specific microbial leaching operations. Moreover, bioleaching has lots of benefits toward resource recovery from e-waste likely a simple, cost-effective, and eco-friendly process. It has been evident that manganese bioleaching is that while heterotrophic bacteria may accomplish total manganese solubilization, using nutrients like glucose and other refined carbohydrates is more cost-effective. However, the usages of a diverse range of biocatalysts (having diverse microbial backgrounds) have also shown few positive outcomes. In a nutshell, microbial involvement in e-waste recycling includes mineral beneficiation, metal extraction, and environmental control. Gold, silver, copper, zinc, cobalt, nickel, and other minerals have all been identified as having exceptional mineral potential by potential firms. Therefore, a massive paradigm shift needs to be required to address current issues on microbes-based bioleaching approaches for value-added resource recovery from e-waste.

Acknowledgment

The authors are grateful to JIS University and JIS Group of Educational Initiative for their support.

References

Ackah, M., 2019. Soil elemental concentrations, accumulation index, non-carcinogenic and carcinogenic risks in functional areas of an informal e-waste recycling area in Accra, Ghana. Chemosphere 235, 908–917. Available from: https://doi.org/10.1016/j.chemosphere.2019.07.014.

Ahonen, L., Tuovinen, O.H., 1989. Effect of temperature on the microbiological leaching of sulfide ore material in percolators containing chalcopyrite, pentlandite, sphalerite and pyrrhotite as main minerals. Biotechnology Letters 11 (5), 331–336. Available from: https://doi.org/10.1007/bf01024513.

Arshadi, M., Mousavi, S.M., 2014. Simultaneous recovery of Ni and Cu from computer-printed circuit boards using bioleaching: statistical evaluation and optimization. Bioresource Technology 174, 233–242. Available from: https://doi.org/10.1016/j.biortech.2014.09.140.

Arshadi, M., Mousavi, S.M., 2015. Multi-objective optimization of heavy metals bioleaching from discarded mobile phone PCBs: simultaneous Cu and Ni recovery using *Acidithiobacillus ferrooxidans*. Separation and Purification Technology 147, 210–219. Available from: https://doi.org/10.1016/j.seppur.2015.04.020.

Arshadi, M., Mousavi, S.M., Rasoulnia, P., 2016. Enhancement of simultaneous gold and copper recovery from discarded mobile phone PCBs using Bacillus megaterium: RSM based optimization of effective factors and evaluation of their interactions. Waste Management (New York, N.Y.) 57, 158–167. Available from: https://doi.org/10.1016/j.wasman.2016.05.012.

Arya, S., Kumar, S., 2020. Bioleaching: urban mining option to curb the menace of E-waste challenge. Bioengineered 11 (1), 640–660. Available from: https://doi.org/10.1080/21655979.2020.1775988.

Ashiq, A., Kulkarni, J., Vithanage, M., 2019. Hydrometallurgical recovery of metals from E-waste. Electronic Waste Management and Treatment Technology. Butterworth-Heinemann, pp. 225–246. Available from: https://doi.org/10.1016/B978-0-12-816190-6.00010-8.

Awasthi, A.K., Zlamparet, G.I., Zeng, X., Li, J., 2017. Evaluating waste printed circuit boards recycling: opportunities and challenges, a mini review. Waste Management & Research 35 (4), 346–356. Available from: https://doi.org/10.1177/0734242X16682607.

Bahaloo-Horeh, N., Mousavi, S.M., 2017. Enhanced recovery of valuable metals from spent lithium-ion batteries through optimization of organic acids produced by Aspergillusniger. Waste Management (New York, N.Y.) 60, 666–679. Available from: https://doi.org/10.1016/j.wasman.2016.10.034.

Bala, S., Goel, S., 2012. A study of e-waste management in relation to awareness of college students. International Journal of Educational and Psychological Research 2, 31–35. Available from: https://ijepr.org/panel/assets/papers/13ij5.pdf.

Baldé, C.P. et al., 2015. The global e-waste monitor 2014, institute for the advanced study of sustainability (ias) and sustainable cycles (cycle). https://i.unu.edu/media/ias.unu.edu-en/news/7916/Global-E-waste-Monitor-2014-small.pdf.

Baniasadi, M., Vakilchap, F., Bahaloo-Horeh, N., Mousavi, S.M., Farnaud, S., 2019. Advances in bioleaching as a sustainable method for metal recovery from e-waste: a review. Journal of Industrial and Engineering Chemistry 76, 75–90.

Barbosa, F., Eduardo, T.S.J., Fernanda, G.R., 2005. A critical review of biomarkers used for monitoring human exposure to lead: advantages, limitations, and future needs. Environmental Health Perspectives 113 (12), 1669–1674. Available from: https://doi.org/10.1289/ehp.7917.

Bas, A.D., Deveci, H., Yazici, E.Y., 2013. Bioleaching of copper from low grade scrap TV circuit boards using mesophilic bacteria. Hydrometallurgy 138, 65–70. Available from: https://doi.org/10.1016/j.hydromet.2013.06.015.

Bhutta, K.S., Adnam, O., Xia Ozhe, Y., 2011. Electronic waste: a growing concern in today's environment. Economics Research International 2011, 8. Available from: https://doi.org/10.1155/2011/474230.

Bi, X., et al., 2010. The major components of particles emitted during recycling of waste printed circuit boards in a typical e-waste workshop of South China. Atmospheric Environment (Oxford, England: 1994) 44 (35), 4440–4445. Available from: https://doi.org/10.1016/j.atmosenv.2010.07.040.

Bosecker, K., 1997. Bioleaching: metal solubilization by microorganisms. FEMS Microbiology Reviews 20 (3–4), 591–604.

Bosshard, P.P., Bachofen, R., Brandl, H., 1996. Metal leaching of fly ash from municipal waste incineration by Aspergillusniger. Environmental Science & Technology 30 (10), 3066–3070. Available from: https://doi.org/10.1021/es960151v.

Brierley, C.L., 1999. Bacterial succession in bioheapleaching. Process Metallurgy 9, 91–97. Available from: https://doi.org/10.1016/S1572-4409(99)80008-X.

Brierley, J.A., 1978. Thermophilic iron-oxidizing bacteria found in copper leaching dumps. Applied and Environmental Microbiology 36 (3), 523–525. Available from: https://doi.org/10.1128/aem.36.3.523-525.1978.

Bryan, C.G., et al., 2015. The use of pyrite as a source of lixiviant in the bioleaching of electronic waste. Hydrometallurgy 152, 33–43. Available from: https://doi.org/10.1016/j.hydromet.2014.12.004.

Chakraborty, P., et al., 2018. PCBs and PCDD/Fs in soil from informal e-waste recycling sites and open dumpsites in India: levels, congener profiles and health risk assessment. The Science of the Total Environment 621, 930–938. Available from: https://doi.org/10.1016/j.scitotenv.2017.11.083.

Chen, A., Dietrich, K.N., Huo, X., Ho, S.M., 2011. Developmental neurotoxicants in e-waste: an emerging health concern. Environmental Health Perspectives 119 (4), 431–438. Available from: https://doi.org/10.1289/ehp.1002452.

Chen, S.-Y., Cheng, Y.-K., 2019. Effects of sulfur dosage and inoculum size on pilot-scale thermophilic bioleaching of heavy metals from sewage sludge. Chemosphere 234, 346–355. Available from: https://doi.org/10.1016/j.chemosphere.2019.06.084.

Chen, X., Zhu, J., Ruan, J., Tang, Y.T., Qiu, R.L., 2019. Debromination and decomposition mechanisms of phenolic resin molecules in ball milling with nano-zerovalent iron. ACS Sustainable Chemistry & Engineering 8 (1), 172–178. Available from: https://doi.org/10.1021/acssuschemeng.9b05071.

Chi, T.D., et al., 2011a. Bioleaching of gold and copper from waste mobile phone PCBs by using a cyanogenic bacterium. Minerals Engineering 24 (11), 1219–1222. Available from: https://doi.org/10.1016/j.mineng.2011.05.009.

Clark, D.A., Norris, P.R., 1996. *Acidimicrobium ferrooxidans* gen. nov., sp. nov.: mixed-culture ferrous iron oxidation with *Sulfobacillus* species. Microbiology (Reading, England) 142 (4), 785–790. Available from: https://doi.org/10.1099/00221287-142-4-785.

Coram, N.J., Rawlings, D.E., 2002. Molecular relationship between two groups of the genus *Leptospirillum* and the finding that *Leptospirillum ferriphilum* sp. nov.dominates South African commercial biooxidation tanks that operate at 40 C. Applied and Environmental Microbiology 68 (2), 838–845. Available from: https://doi.org/10.1128/AEM.68.2.838-845.2002.

Cui, J., Zhang, L., 2008. Metallurgical recovery of metals from electronic waste: a review. Journal of Hazardous Materials 158 (2-3), 228–256. Available from: https://doi.org/10.1016/j.jhazmat.2008.02.001.

Debnath, B., Roychoudhuri, R., Ghosh, S.K., 2016. E-waste management–a potential route to green computing. Procedia Environmental Sciences 35, 669–675. Available from: https://doi.org/10.1016/j.proenv.2016.07.063.

Deng, W.J., et al., 2006. Atmospheric levels and cytotoxicity of PAHs and heavy metals in TSP and PM2. 5 at an electronic waste recycling site in southeast China. Atmospheric Environment 40 (36), 6945–6955.

Faraji, F., et al., 2018. Fungal bioleaching of WPCBs using *Aspergillus niger*: observation, optimization and kinetics. Journal of Environmental Management 217, 775–787. Available from: https://doi.org/10.1016/j.jenvman.2018.04.043.

Forti, V., Balde, C.P., Kuehr, R., Bel, G., 2020. The global e-waste monitor 2020: quantities, flows and the circular economy potential, pp.120. https://collections.unu.edu/view/UNU:7737.

Foucher, S., et al., 2003. Evolution of the bacterial population during the batch bioleaching of a cobaltiferous pyrite in a suspended-solids bubble column and comparison with a mechanically agitated reactor. Hydrometallurgy 71 (1–2), 5–12. Available from: https://doi.org/10.1016/s0304-386x(03)00142-7.

Fowler, B.A., Weissberg, J.B., 1974. Arsine poisoning. The New England Journal of Medicine 291 (22), 1171–1174. Available from: https://doi.org/10.1056/NEJM197411282912207.

Fu, Z., et al., 2016. How PBDEs are transformed into dihydroxylated and dioxin metabolites catalyzed by the active center of cytochrome P450s: a DFT study. Environmental Science & Technology 50 (15), 8155–8163. Available from: https://doi.org/10.1021/acs.est.6b00524.

Gaidajis, G., et al., 2010. E-waste: environmental problems and current management. Journal of Engineering Science and Technology Review 3 (1), 193–199. Available from: https://doi.org/10.25103/jestr.031.32.

Gokul, V., Rajasekaran, T., 2019. Electronic waste management by biological leaching. Bulletin of Scientific Research 1 (1), 54–61. Available from: https://doi.org/10.34256/bsr1918.

Gu, W., et al., 2017a. Enhanced bioleaching efficiency of copper from waste printed circuit board driven by nitrogen-doped carbon nanotubes modified electrode. Chemical Engineering Journal (Lausanne, Switzerland: 1996) 324, 122–129. Available from: https://doi.org/10.1016/j.cej.2017.05.024.

Gu, W., et al., 2017b. Catalytic effect of graphene in bioleaching copper from waste printed circuit boards by *Acidithiobacillus ferrooxidans*. Hydrometallurgy 171, 172–178. Available from: https://doi.org/10.1016/j.hydromet.2017.05.012.

Ha, N.N., et al., 2009. Contamination by trace elements at e-waste recycling sites in Bangalore, India. Chemosphere 76 (1), 9–15. Available from: https://doi.org/10.1016/j.chemosphere.2009.02.056.

Herat, S., Agamuthu, P., 2012. E-waste: a problem or an opportunity? Review of issues, challenges and solutions in Asian countries. Waste Management & Research: The Journal of the International Solid Wastes and Public Cleansing Association, ISWA 30 (11), 1113–1129. Available from: https://doi.org/10.1177/0734242x12453378.

Hong, Y., Valix, M., 2014. Bioleaching of electronic waste using acidophilic sulfur oxidising bacteria. Journal of Cleaner Production 65, 465–472. Available from: https://doi.org/10.1016/j.jclepro.2013.08.043.

Ilyas, S., 2010. Bioleaching of metals from ores and electronic scrap (Doctoral dissertation). University of Agriculture, Faisalabad, Pakistan.

Ilyas, S., Lee, J.-C., 2014. Biometallurgical recovery of metals from waste electrical and electronic equipment: a review. ChemBioEng Reviews 1 (4), 148–169. Available from: https://doi.org/10.1002/cben.201400001.

Ilyas, S., et al., 2007. Bioleaching of metals from electronic scrap by moderately thermophilic acidophilic bacteria. Hydrometallurgy 88 (1–4), 180–188. Available from: https://doi.org/10.1016/j.hydromet.2007.04.007.

Ilyas, S., et al., 2010. Column bioleaching of metals from electronic scrap. Hydrometallurgy 101 (3–4), 135–140. Available from: https://doi.org/10.1016/j.hydromet.2009.12.007.

Ilyas, S., Lee, J.-C., Chi, R.-A., 2013. Bioleaching of metals from electronic scrap and its potential for commercial exploitation. Hydrometallurgy 131–132, 138–143. Available from: https://doi.org/10.1016/j.hydromet.2012.11.010.

Işıldar, A., 2019. Biotechnological strategies for the recovery of valuable and critical raw materials from waste electrical and electronic equipment (WEEE)-A review. Journal of Hazardous Materials 362, 467–481. Available from: https://doi.org/10.1016/j.jhazmat.2018.08.050.

Jadhav, U., Hocheng, H., 2015. Hydrometallurgical recovery of metals from large printed circuit board pieces. Scientific Reports 5 (1), 14574. Available from: https://doi.org/10.1038/srep14574.

Jagannath, A., Shetty, V., Saidutta, M.B., 2017. Bioleaching of copper from electronic waste using Acinetobacter sp. Cr B2 in a pulsed plate column operated in batch and sequential batch mode. Journal of Environmental Chemical Engineering 5 (2), 1599–1607. Available from: https://doi.org/10.1016/j.jece.2017.02.023.

Jomova, K., et al., 2011. Arsenic: toxicity, oxidative stress and human disease: toxicity of arsenic. Journal of Applied Toxicology: JAT 31 (2), 95–107. Available from: https://doi.org/10.1002/jat.1649.

Juju, R., et al., 2015. A novel designed bioreactor for recovering precious metals from waste printed circuit boards. Scientific Reports 5 (1), 13481. Available from: https://doi.org/10.1038/srep13481.

Karimi-Maleh, H., et al., 2021. Recent advances in using of chitosan-based adsorbents for removal of pharmaceutical contaminants: a review. Journal of Cleaner Production 291, 125880. Available from: https://doi.org/10.1016/j.jclepro.2021.125880.

Kasper, A.C., Veit, H.M., 2018. Gold recovery from printed circuit boards of mobile phones scraps using a leaching solution alternative to cyanide. Brazilian Journal of Chemical Engineering 35 (3), 931–942. Available from: https://doi.org/10.1590/0104-6632.20180353s20170291.

Kaya, M., 2016a. Recovery of metals and nonmetals from electronic waste by physical and chemical recycling processes. Waste Management (New York, N.Y.) 57, 64–90. Available from: https://doi.org/10.1016/j.wasman.2016.08.004.

Kaya, M., 2016b. Recovery of metals from electronic waste by physical and chemical recycling processes. International Journal of Chemical and Molecular Engineering 10 (2), 259–270. Available from: http://scholar.waset.org/1999.2/10003863.

Khaliq, A., Rhamdhani, M.A., Brooks, G., Masood, S., 2014. Metal extraction processes for electronic waste and existing industrial routes: a review and Australian perspective. Resources 3 (1), 152–179. Available from: https://doi.org/10.3390/resources3010152.

Kim, M.-J., et al., 2016. Bioleaching of spent Zn-Mn or Ni-Cd batteries by *Aspergillus* species. Waste Management (New York, N.Y.) 51, 168–173. Available from: https://doi.org/10.1016/j.wasman.2015.11.001.

Lakshmi, S., Raj, A., 2017. A review study of e-waste management in India. Asian Journal of Applied Science and Technology (AJAST) Volume 1, 33–36. Available from: https://ssrn.com/abstract = 3048625.

Lee, J.C., Yoo, J.M., Jeong, J.K., Jha, M.K., 2011. U.S. Patent No. 7,867,317. Washington, DC: U.S. Patent and Trademark Office.

Li, J., Guo, Z., Gao, J., 2014. Isothermal enriching perovskite phase from $CaO-TiO_2-SiO_2-Al_2O_3-MgO$ melt by supergravity. ISIJ International 54 (4), 743–749. Available from: https://doi.org/10.2355/isijinternational.54.743.

Liang, C.J., Li, J.Y., Ma, C.J., 2014. Review on cyanogenic bacteria for gold recovery from E-waste. Advanced Materials Research 878, 355–367. Available from: https://doi.org/10.4028/http://www.scientific.net/amr.878.355.

Liu, R., Li, J., Ge, Z., 2016. Review on *Chromobacterium violaceum* for gold bioleaching from e-waste. Procedia Environmental Sciences 31, 947–953. Available from: https://doi.org/10.1016/j.proenv.2016.02.119.

Lucier, C.A., Gareau, B.J., 2019. Electronic waste recycling and disposal: an overview. Assessment and Management of Radioactive and Electronic Wastes 1–12. Available from: https://doi.org/10.5772/intechopen.85983.

Marhual, N.P., et al., 2008. Differential bioleaching of copper by mesophilic and moderately thermophilic acidophilic consortium enriched from same copper mine water sample. Bioresource Technology 99 (17), 8331–8336. Available from: https://doi.org/10.1016/j.biortech.2008.03.003.

Mehta, A.P., Torma, A.E., Murr, L.E., 1979. Effect of environmental parameters on the efficiency of biodegradation of basalt rock by fungi. Biotechnology and Bioengineering 21 (5), 875–885. Available from: https://doi.org/10.1002/bit.260210510.

Mishra, D., Rhee, Y.H., 2010. Current research trends of microbiological leaching for metal recovery from industrial wastes. Current Research, Technology and Education Topics in Applied Microbiology and Microbial Biotechnology 2, 1289–1292. Available from: https://www.researchgate.net/profile/Dr-DebarajMishra/publication/267788229_Current_Research_Trends_of_Microbiological_Leaching_for_Metal_Recovery_from_Industrial_Wastes/links/551b30fc0cf2bb754078b87a/Current-Research-Trends-of-Microbiological-Leaching-for-Metal-Recovery-from-Industrial-Wastes.pdf.

Nagpal, S., Dahlstrom, D., Oolman, T., 1993. Effect of carbon dioxide concentration on the bioleaching of a pyrite-arsenopyrite ore concentrate. Biotechnology and Bioengineering 41 (4), 459–464. Available from: https://doi.org/10.1002/bit.260410409.

Natarajan, G., Ting, Y.-P., 2014. Pretreatment of e-waste and mutation of alkali-tolerant cyanogenic bacteria promote gold biorecovery. Bioresource Technology 152, 80–85. Available from: https://doi.org/10.1016/j.biortech.2013.10.108.

Natarajan, K.A., 2018. Biotechnology of Metals: Principles, Recovery Methods and Environmental Concerns. Elsevier. Available from: https://books.google.co.in/books?hl = en&lr = &id = uOJgDwAAQBAJ&oi = fnd&pg = PP1&dq = 98.%09Natarajan, + K. + A. + (2018) + Biotechnology + of + metals: + principles, + recovery + methods + and + environmental + concerns. + Elsevier. & ots = KeMza5S4Om&sig = 92RQNvhCudSVfilsSDL0XwuBFB8 &redir_esc = y#v = onepage&q = 98.%09Natarajan%2C%20K.%20A.%20(2018) %20Biotechnology%20 of %20metals%3A%20principles%2C%20recovery %20methods%20and%20environmental%20concerns.%20Elsevier.&f = false.

Needhidasan, S., Samuel, M., Chidambaram, R., 2014. Electronic waste-an emerging threat to the environment of urban India. Journal of Environmental Health Science and Engineering 12 (1), 1–9.

Niu, Z., et al., 2014. Process controls for improving bioleaching performance of both Li and Co from spent lithium-ion batteries at high pulp density and its thermodynamics and kinetics exploration. Chemosphere 109, 92–98. Available from: https://doi.org/10.1016/j.chemosphere.2014.02.059.

Pandey, B.D., 2013. Microbially assisted leaching of uranium - a review. Mineral Processing and Extractive Metallurgy Review 34, 81–113.

Prabhu, S.V., et al., 2014. Fungal and bacterial bioleaching studies on bauxite. International Journal of Bioprocess Technology 1 (1), 42–46.

Pradhan, J.K., Kumar, S., 2014. Informal e-waste recycling: environmental risk assessment of heavy metal contamination in Mandoli industrial area, Delhi, India. Environmental Science and Pollution Research International 21 (13), 7913–7928. Available from: https://doi.org/10.1007/s11356-014-2713-2.

Pradhan, N., Nathsarma, K.C., Rao, K.S., Sukla, L.B., Mishra, B.K., 2008. Heap bioleaching of chalcopyrite: a review. Minerals Engineering 21 (5), 355–365. Available from: https://doi.org/10.1016/j.mineng.2007.10.018.

Priya, A., Hait, S., 2017. Comparative assessment of metallurgical recovery of metals from electronic waste with special emphasis on bioleaching. Environmental Science and Pollution Research International 24 (8), 6989–7008. Available from: https://doi.org/10.1007/s11356-016-8313-6.

Priya, A., Hait, S., 2018. Extraction of metals from high grade waste printed circuit board by conventional and hybrid bioleaching using *Acidithiobacillus ferrooxidans*. Hydrometallurgy 177, 132–139. Available from: https://doi.org/10.1016/j.hydromet.2018.03.005.

Qin, B., Lin, M., Yao, Z., Zhu, J., Ruan, J., Tang, Y., et al., 2020. A novel approach of accurately rationing adsorbent for capturing pollutants via chemistry calculation: rationing the mass of CaCO3 to capture Br-containing substances in the pyrolysis of nonmetallic particles of waste printed circuit boards. Journal of Hazardous Materials 393, 122410. Available from: https://doi.org/10.1016/j.jhazmat.2020.122410.

Rasoulnia, P., Mousavi, S.M., 2016. Maximization of organic acids production by Aspergillusniger in a bubble column bioreactor for V and Ni recovery enhancement from power plant residual ash in spent-medium bioleaching experiments. Bioresource Technology 216, 729–736. Available from: https://doi.org/10.1016/j.biortech.2016.05.114.

Rautela, R., Arya, S., Vishwakarma, S., Lee, J., Kim, K.H., Kumar, S., 2021. E-waste management and its effects on the environment and human health. The Science of the Total Environment 773, 145623. Available from: https://doi.org/10.1016/j.scitotenv.2021.145623.

Ren, W.-X., et al., 2009. Biological leaching of heavy metals from a contaminated soil by Aspergillusniger. Journal of Hazardous Materials 167 (1–3), 164–169. Available from: https://doi.org/10.1016/j.jhazmat.2008.12.104.

Ruan, J., et al., 2014. A new strain for recovering precious metals from waste printed circuit boards. Waste Management (New York, N.Y.) 34 (5), 901–907. Available from: https://doi.org/10.1016/j.wasman.2014.02.014.

Saha, L., Kumar, V., Tiwari, J., Rawat, S., Singh, J., Bauddh, K., 2021. Electronic waste and their leachates impact on human health and environment: global ecological threat and management. Environmental Technology & Innovation 24, 102049. Available from: https://doi.org/10.1016/j.eti.2021.102049.

Salam, M.D., Varma, A., 2019. Toxic pollutants survey in soils of electronic waste-contaminated sites in Delhi NCR. Waste Management and Resource Efficiency. Springer, Singapore, pp. 841–851. Available from: https://doi.org/10.1007/978-981-10-7290-1_71.

Sheng, P.P., Etsell, T.H., 2007. Recovery of gold from computer circuit board scrap using aqua regia. Waste Management & Research: The Journal of the International Solid Wastes and Public Cleansing Association, ISWA 25 (4), 380–383. Available from: https://doi.org/10.1177/0734242X07076946.

Shibayama, A., et al., 2010. Treatment of smelting residue for arsenic removal and recovery of copper using pyro-hydrometallurgical process. Journal of Hazardous Materials 181 (1–3), 1016–1023. Available from: https://doi.org/10.1016/j.jhazmat.2010.05.116.

Sivaraman, S., 2013. E-waste management, disposal and its impacts on the environment. Universal Journal of Environmental Research & Technology 3 (5), 531–537. Available from: https://environmentaljournal.org/3-5/ujert-3-5-1.pdf.

Spellman, F.R., Drinan, J., 2014. Wastewater Stabilization Ponds. CRC Press, Boca Raton, International Standard Book Number-13: 978-1-4665-9319-0 (eBook - PDF).

Tsydenova, O., Bengtsson, M., 2011. Chemical hazards associated with treatment of waste electrical and electronic equipment. Waste Management (New York, N.Y.) 31 (1), 45–58. Available from: https://doi.org/10.1016/j.wasman.2010.08.014.

Tuovinen, O.H., Kelly, D.P., 1973. Studies on the growth of *Thiobacillus ferrooxidans*. ArchivfürMikrobiologie 88 (4), 285–298. Available from: https://doi.org/10.1007/BF00409941.

Velgosová, O., Kaduková, J., Marcinčáková, R., Palfy, P., Trpčevská, J., 2013. Influence of H2SO4 and ferric iron on Cd bioleaching from spent Ni-Cd batteries. Waste Management 33 (2), 456–461. Available from: https://doi.org/10.1016/j.wasman.2012.10.007.

Wang, S., et al., 2016. Enhanced bioleaching efficiency of metals from E-wastes driven by biochar. Journal of Hazardous Materials 320, 393–400. Available from: https://doi.org/10.1016/j.jhazmat.2016.08.054.

Wu, W., et al., 2018. Regional risk assessment of trace elements in farmland soils associated with improper e-waste recycling activities in Southern China. Journal of Geochemical Exploration 192, 112–119. Available from: https://doi.org/10.1016/j.gexplo.2018.06.009.

Xia, M.-C., et al., 2017. Recycling of metals from pretreated waste printed circuit boards effectively in stirred tank reactor by a moderately thermophilic culture. Journal of Bioscience and Bioengineering 123 (6), 714–721. Available from: https://doi.org/10.1016/j.jbiosc.2016.12.017.

Xiang, Y., et al., 2010. Bioleaching of copper from waste printed circuit boards by bacterial consortium enriched from acid mine drainage. Journal of Hazardous Materials 184 (1–3), 812–818. Available from: https://doi.org/10.1016/j.jhazmat.2010.08.113.

Xin, B., et al., 2012. Analysis of reasons for decline of bioleaching efficiency of spent Zn-Mn batteries at high pulp densities and exploration measure for improving performance. Bioresource Technology 112, 186–192. Available from: https://doi.org/10.1016/j.biortech.2012.02.133.

Yaashikaa, P.R., et al., 2022. A review on recent advancements in recovery of valuable and toxic metals from e-waste using bioleaching approach. Chemosphere 287 (Pt 2), 132230. Available from: https://doi.org/10.1016/j.chemosphere.2021.132230.

Yang, T., et al., 2009. Factors influencing bioleaching copper from waste printed circuit boards by *Acidithiobacillus ferrooxidans*. Hydrometallurgy 97 (1–2), 29–32. Available from: https://doi.org/10.1016/j.hydromet.2008.12.011.

Yang, Y., et al., 2013. Column bioleaching of low-grade copper ore by *Acidithiobacillus ferrooxidans* in pure and mixed cultures with a heterotrophic acidophile *Acidiphilium* sp. Hydrometallurgy 131–132, 93–98. Available from: https://doi.org/10.1016/j.hydromet.2012.09.003.

Zeng, G., et al., 2012. A copper-catalyzed bioleaching process for enhancement of cobalt dissolution from spent lithium-ion batteries. Journal of Hazardous Materials 199–200, 164–169. Available from: https://doi.org/10.1016/j.jhazmat.2011.10.063.

Zhao, L., Zhu, N.-W., Wang, X.-H., 2008. Comparison of bio-dissolution of spent Ni-Cd batteries by sewage sludge using ferrous ions and elemental sulfur as substrate. Chemosphere 70 (6), 974–981. Available from: https://doi.org/10.1016/j.chemosphere.2007.08.011.

Zheng, J., Ruan, J., Dong, L., Zhang, T., Huang, M., Xu, Z., 2017. Hollow aluminum particle in eddy current separation of recovering waste toner cartridges. ACS Sustainable Chemistry & Engineering 5 (1), 161–167. Available from: https://doi.org/10.1021/acssuschemeng.9b05071.

Zhu, N., et al., 2003. Recycling of spent nickel-cadmium batteries based on bioleaching process. Waste Management (New York, N.Y.) 23 (8), 703–708. Available from: https://doi.org/10.1016/S0956-053X(03)00068-0.

Zhu, N., et al., 2011. Bioleaching of metal concentrates of waste printed circuit boards by mixed culture of acidophilic bacteria. Journal of Hazardous Materials 192 (2), 614–619. Available from: https://doi.org/10.1016/j.jhazmat.2011.05.062.

Chapter 27

Hydrochar from agrowastes: a low-cost adsorbent for environmental application

Rishi Ranjan Pandey[1], Amit Ranjan Pandey[2], Arunima[1], Avnish Kumar Tripathi[1] and Jitendra Kumar[1]

[1]*Department of Chemistry, M. L. K. (P. G.) College, Balrampur, Uttar Pradesh, India,* [2]*Department of Applied Science and Humanities (APSH), Rajkiya Engineering College, Ambedkar Nagar, Uttar Pradesh, India*

Introduction

The vitality of water for the existence of life is undebatable. Although a copious amount of water is present on the earth's surface, the majority is undrinkable. The limited drinkable water resources are under stress due to the growing population. The uncontrolled release of harmful chemicals due to industrialization and day-to-day activities further worsens the quality of water (Järup, 2003; Lapworth et al., 2012). The harmful chemicals consist of organic and inorganic contaminants which are toxic and can disrupt normal physiochemical processes resulting in many water-borne diseases. Consequently, a growing concern for many countries including India is to provide safe and drinkable water to the rising population. Various water purification and recycling techniques such as reverse osmosis, foam floatation, electrochemical, ion exchange, chemical methods, etc. have been developed (Gebreeyessus, 2019; Chaplin, 2019; Greenlee et al., 2009; Gupta et al., 2012). The majority of these methods are very tedious, energy-demanding, and inadequate against ranging pollutants. Further, due to its costly nature, these methods are unaffordable by the large populace which necessitates the development of new technologies (Demirbas, 2008). Many materials like metal-oxides, clay, zeolites, activated carbon, etc. have been evaluated. Activated carbon displays excellent adsorption properties due to its highly porous structure and large surface area but the higher cost and regeneration difficulty impose limitations on its exclusive use (Kurniawan et al., 2006). Moreover, higher temperatures used in the synthesis result in the loss of beneficial functional groups of the precursors which necessitates additional steps of post-modification (Bhatnagar et al., 2013). The post-modification reintroduces chemical groups crucial for interaction with pollutants and also changes the porosity, thus controlling the overall performance of activated carbon as adsorbents (Heidarinejad et al., 2020). These limitations prompted the search for alternate adsorbents, accordingly, agricultural waste products have been explored as low-cost materials to develop effective, sustainable, and economic adsorbents for water remediation (Mashkoor and Nasar, 2021; Demirbas, 2008).

Agrowastes-a low-cost precursor for hydrochar

The term agrowaste encompasses a range of waste materials produced from agricultural activities which include manures, leaves, stems, barks, kitchen waste, etc. Currently, agrowastes are being employed as a precursor for the synthesis of biochars and hydrochars as low-cost novel adsorbent materials for water purification (Zhang et al., 2019). Biochars are eco-friendly, carbon-rich, porous solids produced by thermal treatment of organic waste like agrowaste, municipal wastes, etc. under oxygen-deprived conditions at high temperatures (350°C–700°C), while hydrochars are produced by hydrothermal carbonization at relatively lower temperature (130°C–250°C) under high pressures in a sealed vessel using water as solvent. The lignocellulosic agrowastes have several oxygen-rich functional groups due to cellulose, hemicelluloses, and lignin components (Zhang et al., 2020). Carbonization results in the collapse of their primary structure with significant densification and an increase in carbon content due to the formation of benzenoid moieties resulting in porous material with large surface area (Zhang et al., 2021). The rich chemical functionalities on the surface provide a tool for post-modification to tinker physicochemical properties. These attributes of bio/hydrochars render them suitable for adsorption through complexation, precipitation, ion exchange, electrostatic interactions, etc. (Fig. 27.1).

Role of Green Chemistry in Ecosystem Restoration to Achieve Environmental Sustainability. DOI: https://doi.org/10.1016/B978-0-443-15291-7.00013-4

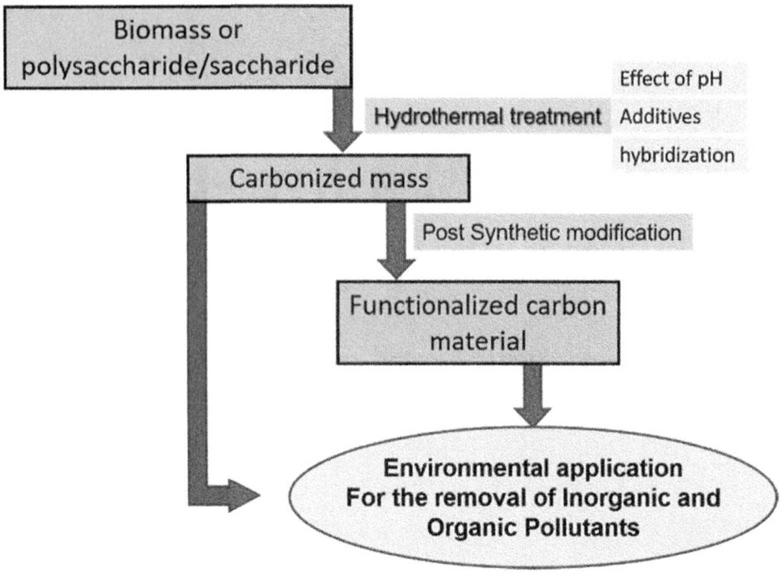

FIGURE 27.1 Preparation of hydrochar from agrowaste.

Hydrothermal technique for hydrochar production

Hydrothermal technology has emerged as a novel route for the synthesis of carbonized materials from agrowaste with potential applications (Titirici et al., 2012). Hydrothermal carbonization (HTC) mimics the natural process of peat or coal formation in the earth's womb under extreme pressure and temperature over thousands of years, but it accomplishes the same in a very short period. HTC is greener as the process is atom-economical, uses water as a reaction medium, and does not release any toxic gases or harmful chemicals into the environment (John and Omorogie, 2022). This process requires less energy input which can further be shortened by microwave irradiation offering an energy-efficient tool (Titirici et al., 2012; Hu et al., 2010). Hydrothermal conversion of agrowaste to value-added chemicals and fuels is another area of great interest (Cao et al., 2021). A range of agrowastes like shells, hulls, corn cobs, sawdusts, husk, straw, seeds, leaves, etc., have been employed for the generation of hydrochars (Zhang et al., 2020; Srivatsav et al., 2020; Enaime et al., 2020). A range of inorganic pollutants like heavy metal ions and various organic pollutants like dyes, herbicides, pharmaceuticals, polyaromatic hydrocarbons, etc. have been targeted for the removal of contaminated water using hydrochar (Román et al., 2018).

Characterization of hydrochar

The characterization of synthesized hydrochar is of fundamental importance to understand the mechanism of formation and its adsorption behavior for possible improvement in synthetic parameters and adsorption efficiency (Amin et al., 2016). First, hydrochars are characterized by elemental analysis, and using the C/H ratio extent of densification is known which is useful in studying the effect of reaction temperature and retention time for knowing the optimum reaction conditions. Generally, the C/H ratio of hydrochar increases whereas O/C ratios decrease compared to precursor agrowastes. The surface functionalities are determined by FTIR analysis which generally consists of hydroxy ($3400-3600\ cm^{-1}$), carbonyl groups ($1700\ cm^{-1}$), and aromatic rings ($1600-1510\ cm^{-1}$) (Xiao et al., 2018). The surface chemistry is also studied by measuring the point of zero charge from the concentrated dispersion of hydrochar at different pH. This information is critical in targeting anionic or cationic pollutants with complementary affinity concerning the hydrochar surface.

Metal ion removal by hydrochars

Certain metal ions are critical for the functioning and stability of many enzymes and proteins and thus, control vital biological phenomena like the growth and metabolism of living organisms (Moustakas, 2021). However, there are metal ions that adversely affect physiological processes and are considered toxic, for example, arsenic, lead, cadmium, mercury, etc (Mudgal et al., 2010). The wastewater from house-hold due to metal-based consumer products and many industries are laden with significantly high amount of toxic metal ions and remediation of wastewater are fundamental for environmental and health protection (Fu and Wang, 2011). Agrowaste hydrochars have been successfully employed as an economic adsorbent for metal ion removal in raw as well as modified forms (Babeker and Chen, 2021) and a few recent reports are summarized in Table 27.1.

The pre-treatment of agrowastes and post-modification steps significantly alter the chemical composition, surface charge, porosity, and increased oxygen functionality which contribute towards higher adsorption capacities. The addition of acids during HTC also affects the porosity and adsorption capacity (Khoshbouy et al., 2018). Chemical aging of hydrochar in acids has also been explored as a modification method to enhance the removal capacity of adsorbent material (Li et al., 2021).

Adsorption of organic pollutants

Organic pollutants, dyes, in particular, are extensively used by many industries such as textile, leather, food, and paper. According to a report, more than 7 lakh tons of dyes are produced annually worldwide, and approximately 20% of these dyes are discharged into water without any treatment (Mashkoor and Nasar, 2021). The dyes are persistent due to structural stability and are toxic, mutagenic, and carcinogenic. The dyes tend to lower sunlight penetration into the water causing abruption of photosynthetic phenomenon and toxicity to aquatic life. Smaller quantities of dyes impart significant colouration to the water rendering it undrinkable (Kadhom et al., 2020). Apart from dyes, other organic pollutants like pharmaceuticals, herbicides pesticides, aromatics, etc. end up in water bodies causing serious health concerns (Lapworth et al., 2012). Thus, it is imperative to develop low-cost and eco-friendly strategies to efficiently remove these organic pollutants. Although many technologies exist (Dai et al., 2019), adsorption using hydrochars from agrowastes has been successfully employed as a low-cost and eco-friendly alternative, and recent findings are given in Table 27.2.

Catalytic degradation of herbicides using hydrochar composite has also been investigated in which the degradation products are environmentally benign; offering a greener approach for the remediation of herbicides from water (Ding et al., 2023; Liu, Sun, et al., 2021). Pharmaceutics are also increasingly found in drinkable water and its overexposure has severe consequences. Hydrochars from agrowaste have been reported to successfully remove various antibiotics in the presence of pollutants to mimic the real system where metal ions and other pollutants can be present along with drugs (Minale et al., 2021; Guan et al., 2021). A recent study highlighted the role of different activating agents like H_2SO_4, KOH, and $KMnO_4$ for achieving different surface functionalities of oven cotton straw hydrochar which was evaluated for the removal of tetracycline and norfloxacin, the two most prescribed medicines. The $KMnO_4$-activated hydrochar displayed maximum adsorption capacity for tetracycline (58.09 mg/g) whereas H_2SO_4-activated hydrochar worked best for norfloxacin with an adsorption capacity of 49.64 mg/g (Cheng et al., 2021). Metal-organic framework incorporation has also been investigated for increasing the porosity of hydrochar-MOF composite and consequent adsorption of tetracycline (Jais et al., 2021). Apart from dyes, herbicides, and pharmaceuticals, other organic pollutants such as aromatics and other volatile organic compounds can also be removed using hydrochars and a few reports have been given in Table 27.2 (Yu et al., 2020).

Hydrochars as a practical approach to green chemistry

Green chemistry is defined as the "design of chemical products and processes to reduce or eliminate the use and generation of hazardous substances" (Anastas and Eghbali, 2010). The principles of green chemistry strive to tackle environmental and economic challenges sustainably through novel innovations. In this regard, there is a paradigm shift in utilizing agrowastes available locally as a greener alternative for the production of useful chemicals (Horváth and Anastas, 2007). The use of agrowaste for adsorbent generation is greener in its approach as the feedstocks are renewable, widely available, and have zero cost (John and Omorogie, 2022). The difficulty in managing this agrowaste from farand has prompted stubble burning; causing massive pollution which has become a serious environmental concern in many countries including India which necessitates finding novel applications of agrowastes. The hydrochar from agrowaste offers a unique and greener solution and opens up new avenues for its application while simultaneously providing economic value to the agrowaste precursors (Usman et al., 2019).

TABLE 27.1 Recent reports of agrowaste-based hydrochar adsorbents for a variety of metal ions.

Agrowaste for hydrochar	Modifications	Metal ions	Adsorption capacity	References
Pinewood	H_2O_2	Pb(II)	21 mg/g	Xia et al. (2019)
Sesame oil coke	H_2O_2	Pb(II)	24.5 mg/g	Khan et al. (2020)
Phoenix dactylifera palm leaves	H_2O_2	Pb(II)	74.5 mg/g	Hammud et al. (2021)
Saffron (*Crocus sativus*) petals	KOH and $ZnCl_2$	Pb(II)	89.5 mg/g	Rasam et al. (2021)
Sugarcane bagasse	KOH	Pb(II)	90.1 mg/g	Malool et al. (2021)
Sawdust Wheat straw Cornstalk	KOH	Cd(II),	30.4–40.7 mg/g	Sun et al. (2015)
Grape pomace	KOH	Pb(II)	137 mg/g	Petrović et al. (2016)
Canola straw	microwave	Pb(II)	78.2 mg/g	Nzediegwu et al. (2021)
Corn stower	H_3PO_4 Polyethylenimine Un-activated	Pb(II)	353.4 mg/g 214.0 mg/g 32.67 mg/g	Jiang et al. (2019)
Corn straw	KOH activation followed by Polyethylenimine	Zn(II) Cu(II)	207.6 mg/g 56.1 mg/g	He et al. (2021)
Banana peel (Dehydrated and fresh)	phosphoric acid	Pb(II)	359.0 mg/g 193.0 mg/g	Zhou et al. (2017)
Olive pomace	$FeCl_3$	As(V)	4.1 mg/g	Capobianco et al. (2020)
Sugarcane bagasse	N-doped magnetic hydrochar	Cd(II) Pb(II)	57.8 mg/g 59.9 mg/g	Khan et al. (2019)
Vinasse	Red mud	Pb(II)	200 mg/g	Kazak and Tor (2020)
Pinecone	Fe_3O_4 and grafting of sulfide containing group	Pb(II) Cd(II)	149.3 mg/g 62.4 mg/g	Zhang, Qu, et al. (2022)
Eucalyptus sawdust	KOH	Cr(VI)	45.8 mg/g	Zhang et al. (2018)
Walnut shells	Polyethylenimine	Cr(VI)	363 mg/g	Ghadikolaei et al. (2019)
Corn cobs	Polyethylenimine	Cr(VI) Ni(II)	33.6 mg/g 29.1 mg/g	Shi et al. (2018)
Bamboo sawdust	$ZnCl_2$ $AlCl_3$	Cr(VI)	14.0 mg/g 12.3 mg/g	Li, Zimmerman, et al. (2020)
Bamboo sawdust	Glycidyl methacrylate	Cr(VI)	424.0 mg/g	Zhang, Guo, et al. (2022)
Bamboo sawdust	Maleic anhydride	Cd(II) Methylene blue	1155.5 mg/g 90.0 mg/g	Li, Guo, et al. (2019)
Bamboo sawdust	Dithiocarbamate	Pb(II)	151.5 mg/g	Li, Guo, et al. (2020)
Microalgae		Cu(II)	19.2 mg/g	Saber et al. (2018)
Rice straw		Congo red berberine 2-naphthol Zn(II) Cu(II)	222.1 mg/g 174.0 mg/g 48.7 mg/g 112.8 mg/g 144.9 mg/g	Li, Tsend, et al. (2019)
Bamboo sawdust	Calcite-impregnation	Cu(II)	130.57	Chen et al. (2019)
Sawdust	Mg-Si-hydrochar composite	Cu(II),Zn(II), tetracycline	214.7 mg/g,227.3 mg/g,361.7 mg/g,	Deng et al. (2020)
Eucalyptus sawdust Corn straw Corncob	Microwave and KOH treatment	Cr(VI)	34.0 mg/g, 30.2 mg/g, 29.5 mg/g	Khushk et al. (2019)

TABLE 27.2 Recent reports of hydrochars for organic pollutants.

Agrowaste for hydrochar	Modification	Organic pollutant	Adsorption capacity	References
Coffee husk		Methylene blue	34.8 mg/g	Ronix et al. (2017)
Male oil palm flower		Methylene blue	42.9 mg/g	Said et al. (2020)
Kermes oak (*Quercus coccifera*)		Basic red-18	194.6 mg/g	Saleh et al. (2020)
Capsicum annuum chili seeds		Methylene blue	145 mg/g	Parra-Marfíl et al. (2020)
Acerola (*Malpighia emarginata* D.C.)		Methylene blue	6.7–85.4 mg/g	Nogueira et al. (2019)
Orange peel	Fe_3O_4	Brilliant black	10.49 mg/g	Çatlıoğlu et al. (2020)
Orange peel Grape skin		Methylene blue	66.2 mg/g 36.6 mg/g	Saha et al. (2020)
Coffee husk	KOH	Methylene blue	357.3 mg/g	Tran et al. (2020)
Bamboo sawdust		Congo red 2-Naphthol	90.5 mg/g 72.9 mg/g	Li, Meas, et al. (2018)
Olive-oil cake	Pristine NaOH	Methylene blue	24.2 mg/g 285.7 mg/g	Alshareef et al. (2021)
Camellia oleifera seed shells	KOH	Cr(VI) Methylene blue	307.2 and 493.0 mg/g	Guo et al. (2018)
Bamboo sawdust	HCl followed by NaOH NaOH	Methylene blue	655.7 mg/g 268.9 mg/g	Qian et al. (2018)
Hickory hull Peanut hulls	KOH H_3PO_4	Methylene blue	49.8 mg/g 187.3 mg/g	Fang et al. (2017)
Bamboo sawdust	Epichlorohydrin	Methyl orange	563 mg/g	Li, Wang, et al. (2018)
Bamboo sawdust	Polyaminocarboxylic acid	Cu(II) Methylene blue	140.6 mg/g 1238.6 mg/g	Li, Lv, et al. (2019)
Poplar sawdust	Polyethyleneimine	Congo red Cr(VI)	1019.2 and 287.7 mg/g	Wu et al. (2022)
Rice straw	Quaternary ammonium group	Methyl orange	49 mg/g	Peng et al. (2019)
Coffee husk	Fe_3O_4	Methylene blue	78 mg/g	Krishna Murthy et al. (2020)
Bamboo sawdust	Fe_3O_4 incorporated amine functionalized	Methylene blue Methyl orange	148.8 mg/g 202.0 mg/g	Liu, Qian, et al. (2021)
Walnut shells	Silica composite	Methylene blue	434.7 mg/g	Duan et al. (2018)
Pinewood	Oxone	Pb(II) Methylene blue	46.7 mg/g 86.7 mg/g	Madduri et al. (2020)
Pinewood	Ozone followed by polyethylenimine	Remazol brilliant blue R Reactive black 5	218.3 mg/g 182.7 mg/g	Elsayed et al. (2022)
Pomegranate peels	Microwave irradiation	Methylene blue	194.9 mg/g	Hessien (2022)
Cornstalk		Mesotrione	3.2 mg/g	Zhang, Zhou, et al. (2021)

(*Continued*)

TABLE 27.2 (Continued)

Agrowaste for hydrochar	Modification	Organic pollutant	Adsorption capacity	References
Rice husk	Microwave Irradiation KOH	Atrazine	4.06 mg/g	Phan et al. (2022)
Prunus serrulata bark		Atrazine	63.3 mg/g	Netto et al. (2022)
Lettuce, taro and watermelon peels	Peroxymonosulfate degradation	2,4-Dichlorophenoxy acetic acid (2,4-D)	73.5–88.4 mg/g	Liu, Sun, et al. (2021)
Pomegranate	Fe_3O_4	2,4-D	101.1 mg/g	Sayğılı and Sayğılı (2022)
Coffee waste		Sulfadiazine sulfamethoxazole	82.2 μg/g 85.7 μg/g	Zhang, Zhang, et al. (2020)
Cotton straw	$KMnO_4$	Tetracycline	58.0 mg/g	Cheng et al. (2021)
Sugarcane bagasse	H_2SO_4	Norfloxacin	49.6 mg/g	Jais et al. (2021)
	NiFe-bimetallic metal organic framework	Tetracycline	568.1 mg/g	
Saccharum ravennae		Diclofenac, Ibuprofen, Naproxen	191.4–230.0 mg/g	Sachan and Das (2021)
Saccarum oficinarum	At pH 9 At pH 3	Brilliant green Sulfadiazine	62.0–103.4 mg/g	Qiao et al. (2022)
Pine fruit shell	NaOH	Bisphenol A	708 mg/g	de Lima et al. (2021)
Sawdust	Sulfamic acid	Benzotriazole Cu(II)	159.9 mg/g 298.8 mg/g	Deng et al. (2019)
Alfalfa (*Medicago sativa*)	Fe_3O_4 and β-cyclodextrin grafting	Anthracene Cd(II)	60.2 mg/g 47.3 mg/g	Qu et al. (2021)

Conclusion

Agrowaste-based hydrochars are an excellent alternative for water remediation apart from finding novel applications in catalysis, energy storage, biofuels, chemicals, etc. The utilization of abundant and zero-value agrowastes and low energy input in HTC significantly reduces the cost of hydrochars, making them an economically viable and eco-friendly option for adsorbents. Significant scientific contributions from various groups have enabled a better understanding of converting biomass into functional materials via HTC, yet there are significant challenges that need to be overcome. Since adsorbent properties are highly dependent on precursor agrowastes, exhaustive and comparative investigations need to be performed on different precursors to find ideal raw materials. These studies will furnish a better understanding of common structural and technical parameters in obtaining superior adsorbents. Recent development has seen a significant rise in hetero atom doping strategies to enhance the performance of the synthesized material, which often results in reduced pore volume. A thorough investigation is needed to achieve reliable process parameters to scale up the synthesis on an industrial scale.

References

Alshareef, Shareefa Ahmed, Otero, Marta, Alanazi, Hamdah S., Siddiqui, Masoom Raza, Khan, Moonis Ali, Alothman, Zeid A., 2021. Upcycling olive oil cake through wet torrefaction to produce hydrochar for water decontamination. Chemical Engineering Research and Design 170, 13–22.

Amin, Farrukh Raza, Huang, Yan, He, Yanfeng, Zhang, Ruihong, Liu, Guangqing, Chen, Chang, 2016. Biochar applications and modern techniques for characterization. Clean Technologies and Environmental Policy 18, 1457–1473.

Anastas, Paul, Eghbali, Nicolas, 2010. Green chemistry: principles and practice. Chemical Society Reviews 39, 301–312.

Babeker, Tawasul Mohammed Ali, Chen, Quanyuan, 2021. Heavy metal removal from wastewater by adsorption with hydrochar derived from biomass: current applications and research trends. Current Pollution Reports 7, 54–71.

Bhatnagar, Amit, Hogland, William, Marques, Marcia, Sillanpää, Mika, 2013. An overview of the modification methods of activated carbon for its water treatment applications. Chemical Engineering Journal 219, 499–511.

Cao, Yang, He, Mingjing, Dutta, Shanta, Luo, Gang, Zhang, Shicheng, Tsang, Daniel C.W., 2021. Hydrothermal carbonization and liquefaction for sustainable production of hydrochar and aromatics. Renewable and Sustainable Energy Reviews 152, 111722.

Capobianco, Laura, Caprio, Fabrizio Di, Altimari, Pietro, Astolfi, Maria Luisa, Pagnanelli, Francesca, 2020. Production of an iron-coated adsorbent for arsenic removal by hydrothermal carbonization of olive pomace: effect of the feedwater pH. Journal of environmental management 273, 111164.

Çatlıoğlu, F.N., Akay, S., Gözmen, B., Turunc, E., Anastopoulos, I., Kayan, B., et al., 2020. 'Fe-modified hydrochar from orange peel as adsorbent of food colorant Brilliant Black: process optimization and kinetic studies. International Journal of Environmental Science and Technology 17, 1975–1990.

Chaplin, Brian P., 2019. The prospect of electrochemical technologies advancing worldwide water treatment. Accounts of chemical research 52, 596–604.

Chen, Xiao-qin, Li, Bing, Shen, Yang, Jian-Zhong Guo, 2019. Facile synthesis of calcite-impregnated hydrochar with high sorption capacity for Cu (II) from aqueous solution. ACS Omega 4, 15022–15029.

Cheng, Long, Ji, Yuanhui, Shao, Qing, 2021. Facile modification of hydrochar derived from cotton straw with excellent sorption performance for antibiotics: coupling DFT simulations with experiments. Science of the Total Environment 760, 144124.

Dai, Yingjie, Zhang, Naixin, Xing, Chuanming, Cui, Qingxia, Sun, Qiya, 2019. The adsorption, regeneration and engineering applications of biochar for removal organic pollutants: a review. Chemosphere 223, 12–27.

de Lima, Hugo Henrique Carline, Llop, Maria Eugênia Grego, Maniezzo, Rogério dos Santos, Moisés, Murilo Pereira, Janeiro, Vanderly, Arroyo, Pedro Augusto, et al., 2021. Enhanced removal of bisphenol A using pine-fruit shell-derived hydrochars: adsorption mechanisms and reusability. Journal of Hazardous Materials 416, 126167.

Demirbas, Ayhan, 2008. Heavy metal adsorption onto agro-based waste materials: a review. Journal of Hazardous Materials 157, 220–229.

Deng, Jiaqin, Li, Xiaodong, Wei, Xue, Liu, Yunguo, Liang, Jie, Song, Biao, et al., 2020. Hybrid silicate-hydrochar composite for highly efficient removal of heavy metal and antibiotics: coadsorption and mechanism. Chemical Engineering Journal 387, 124097.

Deng, Jiaqin, Li, Xiaodong, Wei, Xue, Liu, Yunguo, Liang, Jie, Tang, Ning, et al., 2019. Sulfamic acid modified hydrochar derived from sawdust for removal of benzotriazole and Cu(II) from aqueous solution: adsorption behavior and mechanism. Bioresource Technology 290, 121765.

Ding, Chunxia, Zeng, Guangyong, Tao, Yaping, Long, Xiuyu, Gong, Daoxin, Zhou, Nan, et al., 2023. Environmental-friendly hydrochar-montmorillonite composite for efficient catalytic degradation of dicamba and alleviating its damage to crops. Science of the Total Environment 856, 158917.

Duan, Xinhui, Hong, Wei, Srinivasakannan, C., Wang, Xin, 2018. Hydrochar silicate composite sorbent via simple hydrothermal carbonization and its application to methylene blue removal. Materials Research Express 6, 035601.

Elsayed, Islam, Madduri, Sunith, El-Giar, Emad M., Hassan, El. Barbary, 2022. Effective removal of anionic dyes from aqueous solutions by novel polyethylenimine-ozone oxidized hydrochar (PEI-OzHC) adsorbent. Arabian Journal of Chemistry 15, 103757.

Enaime, Ghizlane, Baçaoui, Abdelaziz, Yaacoubi, Abdelrani, Lübken, Manfred, 2020. Biochar for wastewater treatment-conversion technologies and applications. Applied Sciences 10, 3492.

Fang, June, Gao, Bin, Mosa, Ahmed, Zhan, Lu, 2017. Chemical activation of hickory and peanut hull hydrochars for removal of lead and methylene blue from aqueous solutions. Chemical Speciation & Bioavailability 29, 197–204.

Fu, Fenglian, Wang, Qi, 2011. Removal of heavy metal ions from wastewaters: a review. Journal of Environmental Management 92, 407–418.

Gebreeyessus, Getachew Dagnew, 2019. Status of hybrid membrane–ion-exchange systems for desalination: a comprehensive review. Applied Water Science 9, 1–14.

Ghadikolaei, Nazanin Farmanbordar, Kowsari, Elaheh, Balou, Salar, Moradi, Abbas, Taromi, Faramarz Afshar, 2019. Preparation of porous biomass-derived hydrothermal carbon modified with terminal amino hyperbranched polymer for prominent Cr(VI) removal from water. Bioresource Technology 288, 121545.

Greenlee, Lauren F., Lawler, Desmond F., Freeman, Benny D., Marrot, Benoit, Moulin, Philippe, 2009. Reverse osmosis desalination: water sources, technology, and today's challenges. Water Research 43, 2317–2348.

Guan, Junjie, Liu, Yuyan, Jing, Fanqi, Ye, Rong, Chen, Jiawei, 2021. Contrasting impacts of chemical and physical ageing on hydrochar properties and sorption of norfloxacin with coexisting Cu2 + . Science of the Total Environment 772, 145502.

Guo, Huiqin, Bi, Chenyang, Zeng, Chuncheng, Ma, Wentian, Yan, Liushui, Li, Kexin, et al., 2018. Camellia oleifera seed shell carbon as an efficient renewable bio-adsorbent for the adsorption removal of hexavalent chromium and methylene blue from aqueous solution. Journal of Molecular Liquids 249, 629–636.

Gupta, Kumar, Vinod, Ali, Imran, Saleh, Tawfik A., Nayak, Arunima, Agarwal, Shilpi, 2012. Chemical treatment technologies for waste-water recycling-an overview. RSC Advances 2, 6380–6388.

Hammud, Hassan H., Karnati, Ranjith Kumar, Shafee, Maryam Al, Fawaz, Youssef, Holail, Hanafy, 2021. Activated hydrochar from palm leaves as efficient lead adsorbent. Chemical Engineering Communications 208, 197–209.

He, Xinyue, Zhang, Tao, Xue, Qing, Zhou, Yalan, Wang, Hailong, Bolan, Nanthi S., et al., 2021. Enhanced adsorption of Cu(II) and Zn(II) from aqueous solution by polyethyleneimine modified straw hydrochar. Science of the Total Environment 778, 146116.

Heidarinejad, Zoha, Dehghani, Mohammad Hadi, Heidari, Mohsen, Javedan, Gholamali, Ali, Imran, Sillanpää, Mika, 2020. Methods for preparation and activation of activated carbon: a review. Environmental Chemistry Letters 18, 393–415.

Hessien, Manal, 2022. Microwave-assisted hydrothermal carbonization of pomegranate peels into hydrochar for environmental applications. Energies 15, 3629.

Horváth, István T., Anastas, Paul T., 2007. Innovations and green chemistry. Chemical Reviews 107, 2169–2173.

Hu, Bo, Wang, Kan, Wu, Liheng, Yu, Shu-Hong, Antonietti, Markus, Titirici, Maria-Magdalena, 2010. Engineering carbon materials from the hydrothermal carbonization process of biomass. Advanced Materials 22, 813–828.

Jais, Farahin Mohd, Ibrahim, Shaliza, Chee, Ching Yern, Ismail, Zubaidah, 2021. Solvothermal growth of the bimetal organic framework (NiFe-MOF) on sugarcane bagasse hydrochar for the removal of dye and antibiotic. Journal of Environmental Chemical Engineering 9, 106367.

Järup, Lars, 2003. Hazards of heavy metal contamination. British Medical Bulletin 68, 167–182.

Jiang, Qun, Xie, Weiling, Han, Siyue, Wang, Yifan, Zhang, Ying, 2019. Enhanced adsorption of Pb (II) onto modified hydrochar by polyethyleneimine or H3PO4: an analysis of surface property and interface mechanism. Colloids and Surfaces A: Physicochemical and Engineering Aspects 583, 123962.

John, Kingsley I., Omorogie, Martins O., 2022. Biomass-based hydrothermal carbons for catalysis and environmental cleanup: a review. Green Chemistry Letters and Reviews 15, 162–186.

Kadhom, Mohammed, Albayati, Noor, Alalwan, Hayder, Al-Furaiji, Mustafa, 2020. Removal of dyes by agricultural waste. Sustainable Chemistry and Pharmacy 16, 100259.

Kazak, Omer, Tor, Ali, 2020. In situ preparation of magnetic hydrochar by co-hydrothermal treatment of waste vinasse with red mud and its adsorption property for Pb(II) in aqueous solution. Journal of Hazardous Materials 393122391.

Khan, Moonis Ali, Alqadami, Ayoub Abdullah, Otero, Marta, Siddiqui, Masoom Raza, Alothman, Zeid Abdullah, Alsohaimi, Ibrahim, et al., 2019. Heteroatom-doped magnetic hydrochar to remove post-transition and transition metals from water: synthesis, characterization, and adsorption studies. Chemosphere 218, 1089–1099.

Khan, Moonis Ali, Alqadami, Ayoub Abdullah, Wabaidur, Saikh Mohammad, Siddiqui, Masoom Raza, Jeon, Byong-Hun, Alshareef, Shareefa Ahmed, et al., 2020. Oil industry waste based non-magnetic and magnetic hydrochar to sequester potentially toxic post-transition metal ions from water. Journal of Hazardous Materials 400, 123247.

Khoshbouy, Reza, Rico Lejiu, Fumitake Takahashi, Kunio Yoshikawa, 2018. Effect of acid-assisted hydrothermal carbonization (HTC) process of tree branches using nitric acid on cadmium adsorption. In *E3S Web of Conferences*, 03033. EDP Sciences.

Khushk, Shujauddin, Lei Zhang, Abdul Majeed Pirzada, Muhammad Irfan, Aimin Li. 2019. Cr (VI) heavy metal adsorption from aqueous solution by KOH treated hydrochar derived from agricultural wastes. In *AIP Conference Proceedings*, 020003. AIP Publishing LLC.

Krishna Murthy, T.P., Gowrishankar, B.S., Hari Krishna, R., Chandraprabha, M.N., Mathew, Blessy Baby, 2020. Magnetic modification of coffee husk hydrochar for adsorptive removal of methylene blue: isotherms, kinetics and thermodynamic studies. Environmental Chemistry and Ecotoxicology 2, 205–212.

Kurniawan, T.A., Chan, G.Y.S., h Lo, W., Babel, S., 2006. Comparisons of low-cost adsorbents for treating wastewaters laden with heavy metals. Science of the Total Environment 366, 409–426.

Lapworth, D.J., Baran, Nicole, Stuart, M.E., Ward, R.S., 2012. Emerging organic contaminants in groundwater: a review of sources, fate and occurrence. Environmental Pollution 163, 287–303.

Li, Bing, Guo, Jian-Zhong, Liu, Jia-Lin, Fang, Ling, Lv, Jian-Quan, Lv, Kangle, 2020. Removal of aqueous-phase lead ions by dithiocarbamate-modified hydrochar. Science of the Total Environment 714, 136897.

Li, Bing, Guo, Jianzhong, Lv, Kangle, Fan, Jiajie, 2019. Adsorption of methylene blue and Cd (II) onto maleylated modified hydrochar from water. Environmental Pollution 254, 113014.

Li, Bing, Lv, Jian-Quan, Guo, Jian-Zhong, Fu, Shen-Yuan, Guo, Ming, Yang, Ping, 2019. The polyaminocarboxylated modified hydrochar for efficient capturing methylene blue and Cu(II) from water. Bioresource Technology 275, 360–367.

Li, Bing, Wang, Qian, Guo, Jian-Zhong, Huan, Wei-Wei, Liu, Li, 2018. Sorption of methyl orange from aqueous solution by protonated amine modified hydrochar. Bioresource Technology 268, 454–459.

Li, Detian, Cui, Hongbiao, Cheng, Yueqin, Xue, Lihong, Wang, Bingyu, He, Huayong, et al., 2021. Chemical aging of hydrochar improves the Cd2 + adsorption capacity from aqueous solution. Environmental Pollution 287, 117562.

Li, Feiyue, Zimmerman, Andrew R., Hu, Xin, Gao, Bin, 2020. Removal of aqueous Cr (VI) by Zn-and Al-modified hydrochar. Chemosphere 260, 127610.

Li, Yin, Meas, Arun, Shan, Shengdao, Yang, Ruiqin, Gai, Xikun, Wang, Hongpeng, et al., 2018. Hydrochars from bamboo sawdust through acid assisted and two-stage hydrothermal carbonization for removal of two organics from aqueous solution. Bioresource Technology 261, 257–264.

Li, Yin, Tsend, Nyamkhand, Li, Ti.Kai, Liu, Heyang, Yang, Ruiqin, Gai, Xikun, et al., 2019. Microwave assisted hydrothermal preparation of rice straw hydrochars for adsorption of organics and heavy metals. Bioresource Technology 273, 136–143.

Liu, Jia-Lin, Qian, Wei-Cong, Guo, Jian-Zhong, Shen, Yan, Li, Bing, 2021. Selective removal of anionic and cationic dyes by magnetic Fe3O4-loaded amine-modified hydrochar. Bioresource Technology 320, 124374.

Liu, Yuyan, Sun, Yuqing, Wan, Zhonghao, Jing, Fanqi, Li, Zhixiong, Chen, Jiawei, et al., 2021. Tailored design of food waste hydrochar for efficient adsorption and catalytic degradation of refractory organic contaminant. Journal of Cleaner Production 310, 127482.

Madduri, Sunith, Elsayed, Islam, Hassan, El. Barbary, 2020. Novel oxone treated hydrochar for the removal of Pb(II) and methylene blue (MB) dye from aqueous solutions. Chemosphere 260, 127683.

Malool, Mohamad Ebrahim, Keshavarz Moraveji, Mostafa, Shayegan, Jalal, 2021. Optimized production, Pb (II) adsorption and characterization of alkali modified hydrochar from sugarcane bagasse. Scientific Reports 11, 1–17.

Mashkoor, Fouzia, Nasar, Abu, 2021. Environmental application of agro-waste derived materials for the treatment of dye-polluted water: a review. Current Analytical Chemistry 17, 904–916.

Minale, Mengist, Guadie, Awoke, Liu, Yiyang, Meng, Yuan, Wang, Xuejiang, Zhao, Jianfu, 2021. Insight into adsorption behavior of chlortetracycline and cadmium in aqueous solution using a novel manganese ferrite loaded swine manure hydrochar. Journal of Environmental Chemical Engineering 9, 106785.

Moustakas, Michael, 2021. The role of metal ions in biology, biochemistry and medicine. Materials 14, 549.

Mudgal, Varsha, Madaan, Nidhi, Mudgal, Anurag, Singh, R.B., Mishra, Sanjay, 2010. Effect of toxic metals on human health. The Open Nutraceuticals Journal 3, 94–99.

Netto, Matias S., Georgin, Jordana, Franco, Dison S.P., Mallmann, Evandro S., Foletto, Edson Luiz, Godinho, Marcelo, et al., 2022. Effective adsorptive removal of atrazine herbicide in river waters by a novel hydrochar derived from Prunus serrulata bark. Environmental Science and Pollution Research 29, 3672–3685.

Nogueira, Geraldo D.R., Duarte, Claudio R., Barrozo, Marcos A.S., 2019. Hydrothermal carbonization of acerola (*Malphigia emarginata* D.C.) wastes and its application as an adsorbent. Waste Management 95, 466–475.

Nzediegwu, Christopher, Anne Naeth, M., Chang, Scott X., 2021. Lead(II) adsorption on microwave-pyrolyzed biochars and hydrochars depends on feedstock type and production temperature. Journal of Hazardous Materials 412, 125255.

Parra-Marfíl, Adriana, Ocampo-Pérez, Raúl, Collins-Martínez, Virginia H., Flores-Vélez, Luisa Ma., Gonzalez-Garcia, Raúl, Medellín-Castillo, Nahúm A., Labrada-Delgado, Gladis J., 2020. Synthesis and characterization of hydrochar from industrial Capsicum annuum seeds and its application for the adsorptive removal of methylene blue from water, Environmental Research, 184, 109334.

Peng, Xiong, Yan, Zongcheng, Cheng, Xi, Li, Yu, Wang, Aili, Chen, Li, 2019. Quaternary ammonium-functionalized rice straw hydrochar as efficient adsorbents for methyl orange removal from aqueous solution. Clean Technologies and Environmental Policy 21, 1269–1279.

Petrović, Jelena T., Stojanović, Mirjana D., Milojković, Jelena V., Petrović, Marija S., Šoštarić, Tatjana D., Laušević, Mila D., et al., 2016. Alkali modified hydrochar of grape pomace as a perspective adsorbent of Pb2 + from aqueous solution. Journal of Environmental Management 182, 292–300.

Phan, Kim Anh, Phihusut, Doungkamol, Tuntiwiwattanapun, Nattapong, 2022. Preparation of rice husk hydrochar as an atrazine adsorbent: optimization, characterization, and adsorption mechanisms. Journal of Environmental Chemical Engineering 10, 107575.

Qian, Wei-Cong, Luo, Xi-Ping, Wang, Xing, Guo, Ming, Li, Bing, 2018. Removal of methylene blue from aqueous solution by modified bamboo hydrochar. Ecotoxicology and Environmental Safety 157, 300–306.

Qiao, Xu, Xia, Yuhan, Su, Xuhan, Wang, Bingkun, Chen, Guang, Chen, Huan, 2022. Preparation of biomass carbon material based on fulvic acid and its application in dye and antibiotic treatments. Colloids and Surfaces A: Physicochemical and Engineering Aspects 638, 128302.

Qu, Jianhua, Wang, Siqi, Wang, Yihui, Tian, Xue, Jiang, Zhao, Tao, Yue, et al., 2021. Removal of Cd(II) and anthracene from water by β-cyclodextrin functionalized magnetic hydrochar: performance, mechanism and recovery. Bioresource Technology 337, 125428.

Rasam, Sajjad, Moraveji, Mostafa Keshavarz, Soria-Verdugo, Antonio, Salimi, Ali, 2021. Synthesis, characterization and absorbability of *Crocus sativus* petals hydrothermal carbonized hydrochar and activated hydrochar. Chemical Engineering and Processing - Process Intensification 159, 108236.

Román, Silvia, Libra, Judy, Berge, Nicole, Sabio, Eduardo, Ro, Kyoung, Li, Liang, et al., 2018. Hydrothermal carbonization: modeling, final properties design and applications: a review. Energies 11.

Ronix, Amanda, Pezoti, Osvaldo, Souza, Lucas S., Souza, Isis P.A.F., Bedin, Karen C., Souza, Patrícia S.C., et al., 2017. Hydrothermal carbonization of coffee husk: optimization of experimental parameters and adsorption of methylene blue dye. Journal of Environmental Chemical Engineering 5, 4841–4849.

Saber, Mohammad, Takahashi, Fumitake, Yoshikawa, Kunio, 2018. Characterization and application of microalgae hydrochar as a low-cost adsorbent for Cu(II) ion removal from aqueous solutions. Environmental Science and Pollution Research 25, 32721–32734.

Sachan, Deepa, Das, Gopal, 2021. Selective adsorption of drug micropollutants from synthetic wastewater using hydrochar derived from carbonisation of unused leaves. International Journal of Environmental Analytical Chemistry 1–20.

Saha, Nepu, Volpe, Maurizio, Fiori, Luca, Volpe, Roberto, Messineo, Antonio, Reza, M.T., 2020. Cationic dye adsorption on hydrochars of winery and citrus juice industries residues: performance, mechanism, and thermodynamics. Energies 13.

Said, Ahmad, Tekasakul, Surajit, Phoungthong, Khamphe, 2020. Investigation of hydrochar derived from male oil palm flower: characteristics and application for dye removal. Polish Journal of Environmental Studies 29, 807–815.

Saleh, Mohammed, Bilici, Zeynep, Ozay, Yasin, Yabalak, Erdal, Yalvac, Mutlu, Dizge, Nadir, 2020. Green synthesis of *Quercus coccifera* hydrochar in subcritical water medium and evaluation of its adsorption performance for BR18 dye. Water Science and Technology 83, 701–714.

Sayğılı, G.ülbahar Akkaya, Sayğılı, Hasan, 2022. Fabrication of a magnetic hydrochar composite via an in situ one-pot hydrocarbonization strategy for efficient herbicide removal. Diamond and Related Materials 128, 109302.

Shi, Yuanji, Zhang, Tao, Ren, Hongqiang, Kruse, Andrea, Cui, Ruofan, 2018. Polyethylene imine modified hydrochar adsorption for chromium (VI) and nickel (II) removal from aqueous solution. Bioresource Technology 247, 370–379.

Srivatsav, Prithvi, Bhargav, Bhaskar Sriharsha, Shanmugasundaram, Vignesh, Arun, Jayaseelan, Gopinath, Kannappan Panchamoorthy, Bhatnagar, Amit, 2020. Biochar as an eco-friendly and economical adsorbent for the removal of colorants (dyes) from aqueous environment: a review. Water 12, 3561.

Sun, Kejing, Tang, Jingchun, Gong, Yanyan, Hairong Zhang, 2015. Characterization of potassium hydroxide (KOH) modified hydrochars from different feedstocks for enhanced removal of heavy metals from water. Environmental Science and Pollution Research 22, 16640–16651.

Titirici, Maria-Magdalena, White, Robin J., Falco, Camillo, Sevilla, Marta, 2012. Black perspectives for a green future: hydrothermal carbons for environment protection and energy storage. Energy & Environmental Science 5, 6796–6822.

Tran, Thi Hien, Le, Anh Hoang, Pham, Thien Huu, Nguyen, Dinh Thanh, Chang, Soon Woong, Chung, Woo Jin, et al., 2020. Adsorption isotherms and kinetic modelling of methylene blue dye onto a carbonaceous hydrochar adsorbent derived from coffee husk waste. Science of the Total Environment 725, 138325.

Usman, Muhammad, Chen, Huihui, Chen, Kaifei, Ren, Shuang, Clark, James H., Fan, Jiajun, et al., 2019. Characterization and utilization of aqueous products from hydrothermal conversion of biomass for bio-oil and hydro-char production: a review. Green Chemistry 21, 1553–1572.

Wu, Zhengde, Zhao, Chen, Zeng, Wengen, Wang, Ximo, Liu, Chuanfu, Yu, Zhongye, et al., 2022. Ultra-high selective removal of CR and Cr(VI) from aqueous solutions using polyethyleneimine functionalized magnetic hydrochar: application strategy and mechanisms insight. Chemical Engineering Journal 448, 137464.

Xia, Yu, Yang, Tianxue, Zhu, Nengmin, Li, Dong, Chen, Zeliang, Lang, Qianqian, et al., 2019. Enhanced adsorption of Pb (II) onto modified hydrochar: modeling and mechanism analysis. Bioresource Technology 288, 121593.

Xiao, Kangxin, Liu, Huan, Li, Yang, Yi, Linlin, Zhang, Xiuju, Hu, Hongyun, et al., 2018. Correlations between hydrochar properties and chemical constitution of orange peel waste during hydrothermal carbonization. Bioresource Technology 265, 432–436.

Yu, Xinning, Liu, Shaojun, Lin, Guoxin, Yang, Yang, Zhang, Shuo, Zhao, Haitao, et al., 2020. KOH-activated hydrochar with engineered porosity as sustainable adsorbent for volatile organic compounds. Colloids and Surfaces A: Physicochemical and Engineering Aspects 588, 124372.

Zhang, Xiaojuan, Zhang, Lei, Li, Aimin, 2018. Eucalyptus sawdust derived biochar generated by combining the hydrothermal carbonization and low concentration KOH modification for hexavalent chromium removal. Journal of Environmental Management 206, 989–998.

Zhang, Xinbo, Zhang, Yongchao, Ngo, Huu Hao, Guo, Wenshan, Wen, Haitao, Zhang, Dan, et al., 2020. Characterization and sulfonamide antibiotics adsorption capacity of spent coffee grounds based biochar and hydrochar. Science of the Total Environment 716, 137015.

Zhang, Xingguang, Wang, Yang, Cai, Junmeng, Wilson, Karen, Lee, Adam F., 2020. Bio/hydrochar sorbents for environmental remediation. Energy & Environmental Materials 3, 453–468.

Zhang, Ying, Qu, Jianhua, Yuan, Yihang, Song, Haijiao, Liu, Yang, Wang, Siqi, et al., 2022. Simultaneous scavenging of Cd(II) and Pb(II) from water by sulfide-modified magnetic pinecone-derived hydrochar. Journal of Cleaner Production 341, 130758.

Zhang, Yu-Nan, Guo, Jian-Zhong, Wu, Chunzheng, Huan, Wei-Wei, Chen, Lin, Li, Bing, 2022. Enhanced removal of Cr(VI) by cation functionalized bamboo hydrochar. Bioresource Technology 347, 126703.

Zhang, Zhikun, Zhu, Zongyuan, Shen, Boxiong, Liu, Lina, 2019. Insights into biochar and hydrochar production and applications: a review. Energy 171, 581–598.

Zhang, Zhiming, Yang, Jiantao, Qian, Jianqiang, Zhao, Yong, Wang, Tengfei, Zhai, Yunbo, 2021. Biowaste hydrothermal carbonization for hydrochar valorization: skeleton structure, conversion pathways and clean biofuel applications. Bioresource Technology 324, 124686.

Zhang, Zhongqing, Zhou, Mengmeng, Liu, Jinhua, Li, Jiahao, Yang, Jingmin, Chang, Haibo, 2021. Preparation and characterization of cornstalk microspheric hydrochar and adsorption mechanism of mesotrione. Royal Society Open Science 8, 202209.

Zhou, Nan, Chen, Honggang, Xi, Junting, Yao, Denghui, Zhou, Zhi, Tian, Yun, et al., 2017. Biochars with excellent Pb(II) adsorption property produced from fresh and dehydrated banana peels via hydrothermal carbonization. Bioresource Technology 232, 204–210.

Chapter 28

Sustainable food processing waste management for environmental protection

Matthew C. Ogwu[1], Cherubala N. Malikia[2], Aidan Stansfield[1], Alyson D. Gonzalez-Torres[3] and Sylvester C. Izah[4]

[1]*Goodnight Family Department of Sustainable Development, Appalachian State University, Boone, NC, United States,* [2]*Department of Geography and Regional Planning, College of Arts and Sciences, Appalachian State University, Boone, NC, United States,* [3]*Department of Biology, College of Arts and Sciences, Appalachian State University, Boone, NC, United States,* [4]*Department of Microbiology, Faculty of Science, Bayelsa Medical University, Yenagoa, Bayelsa State, Nigeria*

Introduction

Food is a major substance needed for human growth, development, and survival (Izah et al., 2017a; Ogwu, 2019a). Foods are obtained from mainly living forms, including animals, plants, and microorganisms (Izah et al., 2015; Aiwansoba et al., 2019; Ogwu et al., 2019). Therefore, agricultural activities (growing of plants and rearing of animals) are crucial for human survival (Ogwu et al., 2018; Ogwu, 2019b). But due to inadequate processing facilities, a substantial amount of food produced ends up as waste, especially in developing nations (Chime et al., 2018; Ogwu, 2019c; Omorotionmwan et al., 2019). According to Nair et al. (2017), around 1.3 billion tonnes of food is lost or wasted every year. Rising food waste volumes worldwide have created an environmental pollution problem. This is because, during food processing, three major waste streams are generated, viz., solid, liquid, and gaseous emissions (Ohimain et al., 2013a). The type of waste generated depends on the food that is being processed. For instance, during palm oil production from palm fruits, three major waste streams are produced, including solid (empty fruit bunch, chaff, palm kernel shell, and palm press fiber), liquid (palm oil mill effluents), and gaseous emissions (including particulates, carbon monoxide, volatile organic compounds, etc.) (Ohimain et al., 2013a,b; Ohimain and Izah, 2013a,b, 2014a). Similarly, during cassava processing into cassava flake, cassava mill effluents, cassava peels, and gaseous emissions are some of the waste streams generated (Izah et al., 2017b,c,d,e). Also, the waste generation in the palm oil mill industry includes solid and liquid wastes (Ohimain et al., 2013b; Ohimain and Izah, 2013b). Furthermore, food wastes may be classified based on the point of generation, including kitchen waste, market waste, and pre and postconsumption food waste. It can also be classified based on the edibility of the food, including edible, inedible, and based on biodegradability, viz., compostable and noncompostable. Studies have shown that food processing wastes (FPW) can be utilized in the production of energy, and in the processes, the attendant environmental pollution associated with the wastes can be avoided (Izah et al., 2016). FPWs are heterogeneous biodegradable biochemical by-products from food processing processes that can be potentially reused. FPW contributes to postharvest and other losses within the food systems and alone is estimated to reach gigatons in 2050. In the field of bioenergy, food wastes have been widely studied as potential feedstocks. As such, it is useful in the production of solid, liquid, and gaseous fuel (Ohimain and Izah, 2015, 2014a,b; Izah and Ohimain, 2015; Izah, 2019) and bioelectricity (Izah, 2019). Bioenergy can be produced from FPW via food waste by diverse biological, thermal, and chemical transformations (Kaur et al., 2021) as well as microbial catalysis. Furthermore, FPWs have been found valuable in the production of feeds (Izah, 2018a) and biofertilizers. The availability, collection, chemical complexity, and amenability of various treatment techniques all influence its ability to be treated and recycled into value-added products (Kaur et al., 2021). This chapter looks at the characteristics of communal, regional, and global agro-food systems; food waste management for environmental challenges and protection; food

Role of Green Chemistry in Ecosystem Restoration to Achieve Environmental Sustainability. DOI: https://doi.org/10.1016/B978-0-443-15291-7.00010-9

processing processes, wastes, and waste streams; and sustainable technologies for minimizing FPWs. These considerations are important considering the increasing human population, FPWs, and associated problems associated as environmental pollution because of the high chemical demand and volume, economic loss, food insecurity, health risks, scarcity of land, etc. (Evivie et al., 2020; Izah et al., 2022). Besides reducing these negative impacts, knowledge of food processing processes and waste streams would drive the incorporation of green chemistry for the reutilization of by-products and promote sustainability in the food processing sector (Chime et al., 2016; Ishangulyyev et al., 2019; Ogwu et al., 2019; Dueñas and García-Estévez, 2020; Banerjee et al., 2017; Chang et al., 2022).

Communal, regional, and global agro-food systems

Agricultural food processing is the transformation of raw materials by physical, biological, chemical, or other means into food. This also includes activities common to manufacturing such as transportation, testing/analysis, packaging, and storage. A substantial amount of the global population is directly or indirectly engaged in the transformation of raw agricultural materials into products for intermediate or final consumption as food. In the past, energy comprised only 3% of total production costs (Sandberg and Soderstrom, 2003). Recent statistics show that the agri-food value chain consumes 30% of the world's available energy (FAO, 2021), with energy being consumed at every stage of the food supply chain, heavily relying on non-renewable resources such as fossil fuels. This consumption of energy has resulted in GHG emissions. Around 20% of GHG emissions originate from food production; 31% of land-use (FAO, 2021). Between farm and retail, about 14% of this energy is lost, globally. Not only are our food processing systems costly in terms of energy and environmental pollution via greenhouse gas emissions, but they also harm biodiversity, compete and add pressure on an already scarce freshwater as well and cause alterations in ecosystem integrity through loss of functionality (Ogwu, 2020; Ogwu and Osawaru, 2022; Ogwu et al., 2018, 2022; Ikhajiagbe et al., 2022a,b). A food supply chain describes the process of food moving from a farmed product to one's plate. Processors in the food supply chain, by default, include producers, processors, distributors, and retailers. At the global level, food and agriculture production dates back to the end of World War II with population and economic growth, as well as technological shifts in the ways food was produced. Global food production yields have been on the increase since the 1950s with more food produced per person than ever recorded (Gladek et al., 2017). Although this large-scale production was necessary to avoid food shortages, impacts on the environment have shown the downside of these intensification practices. In a community food system, food production, processing, distribution, and consumption are often integrated to enhance the environmental, economic, social, and nutritional health value within the system. In a communal food system, it is more likely and common for consumers to purchase food directly from those who grew it because there is a higher percentage of food grown within the region. This improves communal reliance, as well as keeps consumer food dollars circulating in the community (Sirrine, 2022).

Food waste management for environmental protection

Food waste is simply food that is fit for consumption but is discarded by choice or because it has been left to spoil or expire, with food referring to processed, semi-processed, or raw edible products going to human consumption (Martin-Rios, 2020). The European Union goes a step further with its, food use for social innovation by optimizing waste prevention strategies program, stating that, food waste is any food, and inedible parts of food, removed from the food supply chain to be recovered or disposed of. Furthermore, the terminology around food waste needs to be addressed. For instance, the Environmental Protection Agency (EPA), uses the term *wasted food* instead of *food waste* for food that was not used for its intended purpose as it conveys that a valuable resource is being wasted, whereas food waste implies that the food no longer has value and needs to be managed as waste (EPA, 2022). Characterization of food waste is yet another factor of food waste management that will need to be standardized, while still allowing for local applicability. A food waste characterization study performed in Hong Kong highlighted that food waste from different sources or generation stages may have different compositional characteristics and therefore can be recycled into different products. According to Ho and Chu (2019), analyzing waste source/type would then allow for a better understanding of waste chemical composition, therefore permitting better decision-making concerning repurposing food waste. Another food characterization study was conducted by the Minnesota Pollution Control Agency (Minnesota Pollution Control Agency (MPCA), 2021), looking at food waste generation and composition in 2019/2020. Differently from the Hong Kong study, MPCA characterized food waste based on the different forms of foods. The categories and descriptions according to MCPA are presented in Table 28.1.

TABLE 28.1 Waste characterization methods are diverse, with each method serving different purposes and having varied results.

Category	Definition	Action
Inedible food		
Food scraps	The leftover portion of food after consumption whether edible or inedible parts	The measure of food that is ideal for composting and processing for reuse
Wasted edible food		
Food, no packaging	Unconsumed food that has past the shelf life date and includes anything spoiled, moldy, half-eaten, and generally unfit for human consumption	A measure of food stream that could have been consumed but was not
Opened or expired packaged	Food in compromised packages or opened food containers with edible or inedible food within	Food composting would require removal from the package or container
Rescuable whole or unopened packaged food	Food that is fit for human consumption and meets the criteria for donation. Often limited to pre-consumed surplus food but may include inedible food	Measures the portion of food that would be eligible for donation
Compostable paper and packaging		
Compostable paper products	Paper towels, napkins, and unlined paper bags only	An additional compostable stream that can be readily diverted from landfill or incineration to an industrial-sized composting facility
Noncompostable packaging		
Packaging of opened or expired packaged food	Packaging removed from opened or expired depackaged food	Measures the percentage of food that would require individuals or a machine to remove packaging before composting or digesting
Packaging of rescuable whole or unopened packaged food	Packaging removed from rescuable depackaged food	Measures the percentage of food that would require individuals or a machine to remove packaging before composting or digesting
Other materials		
Liquid waste food that is generally unsuitable for backyard	Liquid captured from emptying beverage containers	Food that is generally unsuitable for backyard composting or curbside composting programs could be utilized in an anaerobic digestion system
Remaining sample material		Recyclables, refuse, yard waste

Source: Adapted from Minnesota Pollution Control Agency (MPCA), 2021. 2019/2020 Food Waste Generation And Composition Study Analysis. Minnesota Pollution Control Agency, USA. 44 p.

Independent of food waste definitions and characterization methods, food waste is experienced globally. In 2011, the FAO estimated that about one-third of all food produced for human consumption worldwide is lost or wasted. In 2015, statistics remained the same. However, there are no accurate estimates of the full extent of food waste loss (FWL). Studies estimate that FLW is roughly 30% of all food produced globally amounting to 1.3 billion tonnes per year (Nair et al., 2017; World Bank, 2018; FAO, 2021). In terms of sectors of food waste production, around 14% of food produced is lost between harvest and retail, while an estimated 17% of total global food production is wasted with 11% accounted for by households, 5% in the food service, and 2% in retail. According to the United States Department of Agriculture, between 30% and 40% of the food supply is wasted. In the European Union, 20% of food produced is either lost or wasted. Data obtained from the World Bank detail that for low-income countries, FWL occurs at the production, postharvest handling, storage, and processing stages predominantly due to managerial and technical limitations (Chime et al., 2015, 2016; Ikhajiagbe et al., 2021). Conversely, FWL in middle to high-income countries occurs mainly at the distribution and consumption stages of food processing, but can also occur in earlier stages such as agricultural subsidies which lead to the overproduction of farm crops (World Bank, 2018).

Food processing processes

Food can be minimally processed which can include food that has been cooked through various heating methods to highly processed food that can contain high amounts of additives and preservatives. The earliest evidence of processed food goes back about 2,000,000 years ago, although recent studies have shown it could've possibly been even more than that (Huebbe and Rimbach, 2020). The process of salting helped keep food fresh for longer periods (Huebbe and Rimbach, 2020). As humans started to develop more technology the access to machines that contribute to food production and preservation became more accessible. Modern-day practices like ready-to-eat meals can be linked to moments in history when people needed food faster and easier (Izah et al., 2017e). The biggest change was seen during wars and the Industrial Revolution. These were both periods that had difficulty processing food to meet demand. After wars like World War II, there was also an increase in women in the workforce. Easy-to-cook meals became a staple for the working middle class as a way to make food because it was highly marketed to them. As more methods have developed they have lowered the price of food by making it abundant. Food processing today is focused on extending the shelf-life of products, increasing the nutritional value of certain foods, increasing crop yield, and giving us a higher variety of food product options (Bleiweiss-Sande et al., 2019). The words genetically modified organisms (GMO) have been thrown around for many years in various articles arguing that they are dangerous or safe to consume (Sanders, 1999; Ogwu et al., 2019). While there is plenty of research backing up the safety of GMOs one thing is for certain, GMOs are one of the food processing methods that keep the cost of food low and accessible to many people (Ronald, 2011). The technology used in GMO production will be important for our future as our climate changes. A method such as salting is not as effective as cooling or putting the products in an airtight container but it helps maintain products like meats fresh for longer periods. From this method though came the inspiration to put additives in food products to make them last longer (Capozzi et al., 2021). Every technological use of food processing has to go through an extensive amount of work to be considered safe (Bleiweiss-Sande et al., 2019). Cooling preservation increased the distance of trade and the variety of products that could be traded between countries (Misra et al., 2017). Today refrigerators are seen in almost every home and they have been increasing in size in the last couple of decades (Sproesser et al., 2019). In the contraposition of cooling came the process of heating. Heating food can be done in various methods to cook meat and crops. Heating raw meat helps decrease the amount of foodborne diseases and makes food easier to eat. It also contributes to preservation methods like using airtight containers. Some of the first methods for putting food in cans included putting the cans in boiling water, followed by cooling. This created a space for food to extend its shelf life while having a minimal probability of a consumer getting food poisoning. While three states of matter are generally recognized there may be four states. Plasma is the fourth state of matter that has been popular with processing foods. As the discovery of plasma developed starting during the mid-19th century it was found that plasma can interact with delicate biological material. The first instance plasma was used for human consumption was through the use of water treatment in various European countries in the 1950s. During the period this was being used it was found that plasma helped with the decontamination of water and it could be applied to various other food products by using the plasma alongside a combination of gasses (Misra et al., 2017). Using pressurized states of matter, like fluid, has been highly popular in food processing (Misra et al., 2017). Just as important as the preservation of food products is the prevention of foodborne diseases. Food industries must ensure the consumer is safe by preventing microbiological hazards. In poorer countries that don't have access to current food processing technology people of all ages get sick or even die from various pathogens in their food and water systems (Sanders, 1999). The longevity of lifespan in humans has increased throughout the decades. Problems like overconsumption and poor diets now have health on a decline (Poti et al., 2017). Food can be minimally processed and this usually refers to food that has been cooked using a traditional diet which includes home-cooked meals and traditional eating habits (Sproesser et al., 2019). Unfortunately, the most accessible food in the United States is ultra-processed food which involves food that has a high amount of additives, and preservatives and may lack nutritional value (Elizabeth et al., 2020). Most of these ultra-processed foods come in the form of ready-to-eat meals that are usually cheaper than healthier options. Most of these products are targeted at the working middle class as an accessible alternative. Companies benefit from the convenience that these foods provide to their customers (Poti et al., 2017). Highly marketed food items in the US usually encompass fried foods and sweets. The convenience of being able to purchase a meal on the go from any fast-food chain is a staple in American culture. Due to these meals being cheap and in most corners of the nation people purchase these meals regularly. Ultra-processed foods can be linked to various health problems like obesity, cardiovascular disease (CVD), and depression (Elizabeth et al., 2020; Bleiweiss-Sande et al., 2019). Pre-prepared meals are highly popular in modern diets for being convenient (Sproesser et al., 2019). Not only are these products bad in excess but to produce and maintain these foods they have to go through an extensive process which has led to an increase in pollutants in our environment (Capozzi et al., 2021). Products like meat that are part of modern diets can be harmful to the environment. Livestock takes up about 80% of the world's agricultural lands. Simply reducing the amount of meat we consume every day can not only help the environment but contribute to a healthier lifestyle (Ritchie, 2017).

Food processing wastes and waste streams

Food waste is one of the problems causing climate change today. Overconsumption and our poor food waste management have caused an excess of perfectly good food to be thrown out. Food that would otherwise feed hundreds of people who need it most gets added to piles in landfills. While climate change is caused by many factors that are brought on by humans, food processing and waste management can be one of the easiest solutions for us to start with. While some problems have more complicated situations the way we treat our food can be a fix that at an individual level can have a big impact (Cotruvo, 2018). Organic materials like food waste would highly be reduced in various ways. Just because it's at a landfill does not mean it has no use. Food waste can contain micro-and macro-nutrients that can be used to produce alcohols that will be used for industrial purposes. This can be applied to things like feeding livestock, and biofuels and reducing the use of water (Hedge et al., 2018). The landfill can be pre-treated to be put through a process of fermentation or extraction. The material that is obtained can then be used to produce biopolymers which have various uses in industries like creating bioplastics (Ranganathan et al., 2020).

Environmental challenges associated with food processing wastes

The unsustainable environmental impacts of FPW led the UN SDG to reduce these avoidable losses by 50% by 2030 (Ishangulyyev et al., 2019; Read et al., 2020). This supports the view of Cooper et al. (2018) that all hands need to be on deck to minimize food waste and loss and the attendant environmental impact. Necessitating management through characterization and reutilization is the many million tons of food that are lost and wasted during food processing, which have been shown to create or exacerbate environmental problems like climate change as well as economic and food insecurity challenges (Ogwu, 2019b; Dueñas and García-Estévez, 2020; Georganas et al., 2020). Food processing activities may give off ammonia, volatile organic compounds, methane, nitrous oxide, sulfur, and particulate matter into the environment (Shaw et al., 2007; Aneja et al., 2009; Ohimain et al., 2013a). These would require measuring and monitoring to determine the environmental cost and design a reduction plan (Kummu et al., 2017; Cattaneo et al., 2021). A key determinant of a load of these toxic materials from food processing activities is infrastructure, quality or safety standards, requirements, environmental factors, material input (food resources and sources), energy source and efficiency, and processing stage produced. The United States Environmental Protection Agency's "Waste Reduction Model" is useful for estimating the baseline impacts and sustainability of alternative practices like source reduction, anaerobic digestion, recycling, combustion, composting, and landfilling of waste from food processing activities. Besides the load, another consideration is the multiple stages and processes associated with the life cycle of these waste products from food processing (Tonini et al., 2018). Although the environmental impacts of FPW may be diverse based on the type and class of food, meat, and fish are considered the single largest contributors (Cooper et al., 2018). Food processing consumes a significant amount of freshwater and fossil fuels both of which have effects on global climate change (Hall et al., 2009). In a way, the environmental impact of FPW is a trade-off of the modern food value chain. Reducing the environmental impact of FPW and loss will yield consumption and production benefits. The work of Brennan and Browne (2021) established the FPW and lost the causative link between environmental health and public health. They further identified interdisciplinary intervention measures for environmental sustainability. Using a different model Read et al. (2020) showed that intervention methods should target the following sectors—food service, food processing, food distribution, food policy, and household consumption. Change in people's behavior through public education would likely have the most impact on reducing the environmental impacts of FPWs from different waste streams alongside the appropriate formulation of products, packaging, portioning, and utilization methods (Cooper et al., 2018).

Sustainable technologies for reducing food processing wastes

Industrialized agriculture and food production have never been more prevalent as the human population continues to increase exponentially. Industrial agriculture produces foodstuffs quickly and in large quantities, and tremendous amounts of food are wasted along the supply chain. Food waste management policies are often derived from environmental goals but are constrained by budgets and executive ability. As a result, governments must prioritize food waste management strategies based on cost and accessibility. There is an inverse relationship between cost and environmental effects, as prevention strategies are more costly than intervening later in the supply chain. Because of this, many governments cannot deploy novel prevention strategies at a large scale and resort to less sustainable mitigation methods (Cristobal et al., 2018). For instance, in 2009, 30% of Taiwan's municipal landfill waste was food scraps. Industrial practices contributing to this overproduction, and subsequent waste, have adverse effects on the environment. Food production uses 70% of yearly freshwater resources and 2000–5000 liters are used to produce enough food for one person. The livestock industry alone accounts for 18% of yearly global GHG emissions; beef and pork production emits 30 times more GHG than plant production. Inadequate farming practices such as overgrazing and excess nitrogen contribute to soil depletion. Soil degradation exacerbates food insecurity in developing nations. Sub-Saharan Africa's nitrogen balance deficit increased from -22 ha^{-1}, year in 1983, to -26 in 2000 (Lin et al., 2009). This reduces the amount of arable land in a region that already faces extreme food insecurity. Recycling food waste as pig feed is a promising mediation technique originating from South Korea. Salemdeeb et al. (2017) found recycling municipal food waste as pig feed to be less expensive than traditional pig feed, and more environmentally favorable than composting and landfills. Biorefinery via anaerobic digestion presents another pathway to sustainable food waste management. Combining food and wastewater treatment results in higher biogas yields, increased phosphorus recovery, as well as increased recovery of biodegradable polymers (Cecchi and Cavinato, 2019). Adjacent to biogas production via anaerobic digestion is the option to recover desirable molecules from food waste byproducts (Ohimain and Izah, 2017, 2014b). This waste reduction measure involves sorting food waste from both industrial and agricultural food production, drying, removal of harmful material, and extracting molecules (Torres-Leon et al., 2018). These measures represent innovations in food waste management, however, these technologies do not address the disproportionate impact that industrial food production and waste have on developing nations. The improvement of infrastructure in developing nations is a key challenge to a globally sustainable food system. Overall, large households with poor food access are much more likely to face food insecurity than large families with adequate food access (Nyangasa et al., 2019). Lack of infrastructure and purchasing power in developing countries obstruct their ability to deploy sustainable agricultural technologies. Top-down strategies disregarding environmental impact and food security are the cause of the tremendous amounts of yearly food waste. Locally-derived solutions to food production are the best defense against food insecurity, as these solutions are specific to the local environment and often do not rely on industrial agriculture.

Conclusion

Despite their potential value, FPWs continue to create multidimensional problems because of their composition and characteristics vis a vis environmental and economic impacts. Several food processing industries around the world, however, have been working to reduce the impact of these food wastes by recycling products, reusing processing wastes, and reducing waste generation. Furthermore, in many developed nations, landfill is the main method of disposing of waste, while in developing nations, food wastes are still dumped in open dumpsites, and the residents of coastal communities also dispose of their wastes on surface water bodies. Food wastes can be sustainably disposed of through biotechnological advances; thus, animal feed production, biofertilizer, and bioenergy production. Also, food processing waste can be reused by other industries.

References

Aiwansoba, R.O., Ogwu, M.C., Osawaru, M.E., 2019. Assessing the relatedness of *Abelmoschus* accessions using morphological characters. Journal of Tropical Biology and Conservation 16, 193–207.

Aneja, V.P., Schlesinger, W.H., Erisman, J.W., 2009. Effects of agriculture upon the air quality and climate: research, policy, and regulations. Environmental Science and Technology 43 (12), 4234–4240.

Banerjee, J., Singh, R., Vijayaraghavan, R., MacFarlane, D., Patti, A.F., Arora, A., 2017. Bioactives from fruit processing wastes: green approaches to valuable chemicals. Food Chemistry 225, 10–22.

Bleiweiss-Sande, R., Chui, K., Evans, E.W., Goldberg, J., Amin, S., Sacheck, J., 2019. Robustness of food processing classification systems. Nutrients 11 (6), 1344.

Brennan, A., Browne, S., 2021. Food waste and nutrition quality in the context of public health: a scoping review. International Journal of Environmental Research and Public Health 18 (10), 5379.

Capozzi, F., Magkos, F., Fava, F., Milani, G.P., Agostoni, C., Astrup, A., et al., 2021. A multidisciplinary perspective of ultra-processed foods and associated food processing technologies: a view of the sustainable road ahead. Nutrients 13 (11), 3948.

Cattaneo, A., Sánchez, M.V., Torero, M., Vos, R., 2021. Reducing food loss and waste: Five challenges for policy and research. Food Policy 98, 101974.

Cecchi, F., Cavinato, C., 2019. Smart approaches to food waste final disposal. International Journal of Environmental Research and Public Health 16, 2860.

Chang, Y., Shi, X., He, F., Wu, T., Jiang, L., Normakhamatov, N., et al., 2022. Valorization of food processing waste to produce valuable polyphenolics. Journal of Agricultural and Food Chemistry 70 (29), 8855–8870. Available from: https://doi.org/10.1021/acs.jafc.2c02655.

Chime, A.O., Aiwansoba, R.O., Ogwu, M.C., 2018. Pathological status of plant germplasm and sustainable crop production and conservation. Journal of Energy and Natural Resources Management 1, 17–21.

Chime, A.O., Aiwansoba, R.O., Danagogo, S.J., Egharevba, I.I., Osawaru, M.E., Ogwu, M.C., 2016. Effects of lining with leaves of *Triclisia dictyophylla* on the fungal composition of *Cola nitida* during storage. Journal of Industrial Research and Technology 5 (2), 128–136.

Chime, A.O., Osawaru, M.E., Ogwu, M.C., Ogunlowo, E.O., 2015. Preservation of fresh tomato fruits (*Solanum lycopersicon* L.) using leaf extracts of bitter melon (*Momordica charantia* L., Cucurbitaceae). Journal of Phytopathology and Plant Health 3, 29–37.

Cooper, K.A., Quested, T.E., Lanctuit, H., Zimmermann, D., Espinoza-Orias, N., Roulin, A., 2018. Nutrition in the bin: a nutritional and environmental assessment of food wasted in the UK. Frontiers in Nutrition 5, 19.

Cotruvo, J., 2018. Wastewater treatment challenges in food processing and agriculture. Available online: http://www.watertechonline.com/wastewater/article/15550688/wastewater-treatment-challenges-in-food-processing-and-agriculture (accessed 15.05.22).

Cristobal, J., Castellani, V., Manfredi, S., Sala, S., 2018. Prioritizing and optimizing sustainable measures for food waste prevention and management. Waste Management (New York, N.Y.) 72, 3–16.

Dueñas, M., García-Estévez, I., 2020. Agricultural and food waste: analysis, characterization and extraction of bioactive compounds and their possible utilization. Foods (Basel, Switzerland) 9 (6), 817.

Elizabeth, L., Machado, P., Zinöcker, M., Baker, P., Lawrence, M., 2020. Ultra-processed foods and health outcomes: a narrative review. Nutrients 12 (7), 1955.

Evivie, S.E., Ogwu, M.C., Ebabhamiegbebho, P.A., Abel, E.S., Imaren, J.O., Igene, J.O., 2020. Packaging and the Nigerian food industry: challenges and opportunities. In: Ogunlade, C.A., Adeleke, K.M., Oladejo, M.T. (Eds.), Food Technology and Culture in Africa. Reamsworth Publishing, Ibadan, Nigeria, pp. 28–99.

FAO, 2021. Sustainable food systems concept and framework. Rome. 45 p. https://www.fao.org/fao-stories/article/en/c/1412108/.

Georganas, A., Giamouri, E., Pappas, A.C., Papadomichelakis, G., Galliou, F., Manios, T., et al., 2020. Bioactive compounds in food waste: a review on the transformation of food waste to animal feed. Foods (Basel, Switzerland) 9 (3), 291. Available from: https://doi.org/10.3390/foods9030291.

Gladek, E., Roemers, G., Sabag Munos, oscar, Kennedy, E., Fraser, M., et al., 2017. The global food system: trends, impacts, and solutions. Metabolic . Available online. Available from: https://www.metabolic.nl/publication/global-food-system-an-nalysis/ (retrieved 14.05.22).

Hall, K.D., Guo, J., Dore, M., Chow, C.C., 2009. The progressive increase of food waste in America and its environmental impact. PLoS One 4 (11), e7940. Available from: https://doi.org/10.1371/journal.pone.0007940.

Ho, K.S., Chu, L.M., 2019. Characterization of food waste from different sources in Hong Kong. Journal of the Air and Waste Management Association 69 (3), 277–288.

Huebbe, P., Rimbach, G., 2020. Historical reflection of food processing and the role of legumes as part of a healthy balanced diet. MDPI Foods 9 (8), 1058.

Ikhajiagbe, B., Aituae, W., Ogwu, M.C., 2022a. Morpho-physiological assessment of oil palm (*Elaeis guineensis* Jacq.) seedlings exposed to simulated drought conditions. Journal of Oil Palm Research 34 (1), 26–34. Available from: https://doi.org/10.21894/jopr.202100018.

Ikhajiagbe, B., Ogwu, M.C., Ekhator, P.O., Omoayena, I.I., 2022b. Implications of soil nitrogen enhancement on the yield performance of soybean (*Glycine max*) in cadmium-polluted soil. Journal of Pollution Effects and Control 10 (2), 1–12. Available from: https://doi.org/10.1101/2021.11.22.469589. #: 1000334.

Ikhajiagbe, B., Ogwu, M.C., Lato, N.F., 2021. Growth and yield response of soybean (*Glycine max* [L.] Merr.) accessions after exposure to cadmium. VEGETOS – International Journal of Plant Research and Biotechnology . Available from: https://doi.org/10.1007/s.42535-021-00189-y.

Ishangulyyev, R., Kim, S., Lee, S.H., 2019. Understanding food loss and waste-why are we losing and wasting food? Foods (Basel, Switzerland) 8 (8), 297. Available from: https://doi.org/10.3390/foods8080297.

Izah, S.C., Ovuru, K.F., Ogwu, M.C., 2022. Lassa fever in Nigeria: social and ecological risk factors exacerbating transmission and sustainable management strategies. International Journal of Tropical Diseases 5 (2), 065. Available from: https://doi.org/10.23937/2643-461x/1710065.

Izah, S.C., 2019. Cassava mill effluents recycling through bioenergy production: a review. Environmental Analyses and Ecological Studies 5 (4), EAES.000618.2019.

Izah, S.C., 2018a. Feed potentials of *Saccharomyces cerevisiae* biomass cultivated in palm oil and cassava mill effluents. Journal of Bacteriology and Mycology Open Access 6 (5), 287–293.

Izah, S.C., Inyang, I.R., Angaye, T.C.N., Okowa, I.P., 2017a. A review of heavy metal concentration and potential health implications in beverages consumed in Nigeria. Toxics 5 (1), 1–15. Available from: https://doi.org/10.3390/toxics5010001.

Izah, S.C., Bassey, S.E., Ohimain, E.I., 2017b. Assessment of pollution load indices of heavy metals in cassava mill effluents contaminated soil: a case study of small-scale cassava processing mills in a rural community of the Niger Delta region of Nigeria. Bioscience Methods 8 (1), 1–17.

Izah, S.C., Bassey, S.E., Ohimain, E.I., 2017c. Assessment of heavy metal in cassava mill effluent contaminated soil in a rural community in the Niger Delta region of Nigeria. EC Pharmacology and Toxicology 4 (5), 186–201.

Izah, S.C., Bassey, S.E., Ohimain, E.I., 2017d. Changes in the treatment of some physico-chemical properties of cassava mill effluents using *Saccharomyces cerevisiae*. Toxic 5 (4), 28. Available from: https://doi.org/10.3390/toxics5040028. PMID:29051460.

Izah, S.C., Bassey, S.E., Ohimain, E.I., 2017e. Removal of heavy metals in cassava mill effluents with *Saccharomyces cerevisiae* isolated from palm wine. MOJ Toxicology 3 (4), 00057.

Izah, S.C., Angaye, T.C.N., Ohimain, E.I., 2016. Environmental impacts of oil palm processing in Nigeria. Biotechnological Research 2 (3), 132–141.

Izah, S.C., Aseiba, E.R., Orutugu, L.A., 2015. Microbial quality of polythene packaged sliced fruits sold in major markets of Yenagoa Metropolis, Nigeria. Point Journal of Botany and Microbiology Research 1 (3), 30–36.

Izah, S.C., Ohimain, E.I., 2015. Comparison of traditional and semi-mechanized palm oil processing approaches in Nigeria; Implications on biodiesel production. Journal of Environmental Treatment Techniques 3 (2), 82–87.

Kaur, G., Lasaridi, K., Wong, J., 2021. Sustainable food waste management: an introduction. Current Developments in Biotechnology and Bioengineering. Elsevier, pp. 1–10.

Kummu, M., Fader, M., Gerten, D., Guillaume, J.H., Jalava, M., Jägermeyr, J., et al., 2017. Bringing it all together: linking measures to secure nations' food supply. Current Opinion in Environmental Sustainability 29, 98–117.

Lin, A.Y.-C., Huang, S.T.-Y., Wahlqvist, M.L., 2009. Waste management to improve food safety and security for health advancement. Asia Pacific Journal of Clinical Nutrition 18 (4), 538–545.

Martin-Rios, C., 2020. Food Waste Management Innovations in the foodservice industry. Hospitality News & Business Insights by EHL. Available online. Available from: https://hospitalityinsights.ehl.edu/food-waste-management-innovations (retrieved 14.05.22).

Minnesota Pollution Control Agency (MPCA), 2021. 2019/2020 Food Waste Generation And Composition Study Analysis. Minnesota Pollution Control Agency, USA. 44 p.

Misra, N.N., Koubaa, M., Roohinejad, S., Juliano, P., Alpas, H., Inácio, R.S., et al., 2017. Landmarks in the historical development of twenty first century food processing technologies. Food Research International 97, 318–339. Available from: https://doi.org/10.1016/j.foodres.2017.05.001.

Nair, N.R., Sekhar, V.C., Nampoothiri, K.M., Pandey, A., 2017. Food wastes properties. Current Developments in Biotechnology and Bioengineering. Elsevier, Amsterdam, pp. 157–190.

Nyangasa, M.A., Buck, C., Kelm, S., Sheikh, M., Hebestreit, A., 2019. Exploring food access and sociodemographic correlates of food consumption and food insecurity in zanzibari households. International Journal of Environmental Research and Public Health 16, 1557.

Ogwu, M.C., 2020. Value of *Amaranthus* [L.] species in Nigeria. In: Waisundara, V. (Ed.), Nutritional Value of Amaranth. IntechOpen, UK, pp. 1–21.

Ogwu, M.C., Osawaru, M.E., 2022. Traditional methods of plant conservation for sustainable utilization and development. In: Chibueze Izah, S. (Ed.), Biodiversity in Africa: Potentials, Threats and Conservation. Sustainable Development and Biodiversity, Vol. 29. Springer, Singapore, pp. 451–472. Available from: http://doi.org/10.1007/978-981-19-3326-417.

Ogwu, M.C., Ahana, C.M., Osawaru, M.E., 2018. Sustainable food production in Nigeria: a case study for Bambara groundnut (*Vigna subterranean* (L.) Verdc. Fabaceae). Journal of Energy and Natural Resources Management 1, 68–77.

Ogwu, M.C., Chime, A.O., Aiwansoba, R.O., Emere, A.O., 2019. Effects of storage methods and duration on the microbial composition and load of tomato (*Solanum lycopersicum* [L.], Solanaceae) fruits. Bitlis Eren University Journal of Science and Technology 9 (1), 1–7. Available from: https://dergipark.org.tr/download/article-file/746056.

Ogwu, M.C., Izah, S.C., Iyiola, A.O., 2022. An overview of the potentials, threats and conservation of biodiversity in Africa. In: Chibueze Izah, S. (Ed.), Biodiversity in Africa: Potentials, Threats and Conservation. Sustainable Development and Biodiversity, Vol. 29. Springer, Singapore, pp. 3–20. Available from: http://doi.org/10.1007/978-981-19-3326-41.

Ogwu, M.C., 2019a. Lifelong consumption of plant-based GM foods: is it safe? In: Papadopoulou, P., Misseyanni, A., Marouli, C. (Eds.), Environmental Exposures and Human Health Challenges. IGI Global, Pennsylvania, USA, pp. 158–176. Available from: http://doi.org/10.4018/978-1-5225-7635-8.ch008.

Ogwu, M.C., 2019b. Towards sustainable development in Africa: the challenge of urbanization and climate change adaptation. In: Cobbinah, P.B., Addaney, M. (Eds.), The Geography of Climate Change Adaptation in Urban Africa. Springer Nature, Switzerland, pp. 29–55.

Ogwu, M.C., 2019c. Understanding the composition of food waste: an "-omics" approach to food waste management. In: Gunjal, A.P., Waghmode, M.S., Patil, N.N., Bhatt, P. (Eds.), Global Initiatives for Waste Reduction and Cutting Food Loss. IGI Global, Pennsylvania, USA, pp. 212–236.

Ohimain, E.I., Izah, S.C., 2017. A review of biogas production from palm oil mill effluents using different configurations of bioreactors. Renewable & Sustainable Energy Reviews 70, 242–253.

Ohimain, E.I., Izah, S.C., 2015. Estimation of potential biohydrogen from palm oil mills' effluent in Nigeria using different microorganisms under light independent fermentation. Journal of Environmental Treatment Techniques 3 (2), 97–104.

Ohimain, E.I., Izah, S.C., 2014a. Potential of biogas production from palm oil mills' effluent in Nigeria. Sky Journal of Soil Sciences and Environmental Management 3 (5), 50–58.

Ohimain, E.I., Izah, S.C., 2014b. Estimation of potential electrical energy and currency equivalent from un-tapped palm oil mill effluents in Nigeria. International Journal of Farming and Allied Science 3 (8), 855–862.

Ohimain, E.I., Izah, S.C., 2013a. Gaseous emissions from a semi-mechanized oil palm processing mill in Bayelsa state, Nigeria. Continental Journal of Water, Air and Soil Pollution 4 (1), 15–25.

Ohimain, E.I., Izah, S.C., 2013b. Water minimization and optimization by small-scale palm oil mill in Niger Delta, Nigeria. Journal of Water Research 135, 190–198.

Ohimain, E.I., Izah, S.C., Abah, S.O., 2013a. Air quality impacts of smallholder oil palm processing in Nigeria. Journal of Environmental Protection 4, 83–98.

Ohimain, E.I., Izah, S.C., Obieze, F.A.U., 2013b. Material-mass balance of smallholder oil palm processing in the Niger Delta, Nigeria. Advance Journal of Food Science and Technology 5 (3), 289–294.

Omorotionmwan, F.O.-O., Ogwu, H.I., Ogwu, M.C., 2019. Antibacterial characterization and bacteria composition of pineapple (*Ananas comosus* [Linn.] Merr.) peel and pulp. Food and Health 5 (1), 1–11.

Poti, J.M., Braga, B., Qin, B., 2017. Ultra-processed food intake and obesity: what really matters for health-processing or nutrient content? Current Obesity Reports 6 (4), 420–431. Available from: https://doi.org/10.1007/s13679-017-0285-4.

Ranganathan, S., Dutta, S., Moses, J.A., Anandharamakrishnan, C., 2020. Utilization of food waste streams for the production of biopolymers. Heliyon 6 (9), e04891.

Read, Q.D., Brown, S., Cuéllar, A.D., Finn, S.M., Gephart, J.A., Marston, L.T., et al., 2020. Assessing the environmental impacts of halving food loss and waste along the food supply chain. The Science of the Total Environment 712, 136255.

Ritchie, H., 2017. How much of the world's land would we need in order to feed the global population with the average diet of a given country? Available online: https://ourworldindata.org/agricultural-land-by-global-diets (accessed 19.05.22).

Ronald, P., 2011. Plant genetics, sustainable agriculture and global food security. Genetics 188 (1), 11–20. Available from: https://doi.org/10.1534/genetics.111.128553.

Salemdeeb, R., z uErmgassen, E.K.H.J., Kim, M.H., Balmford, A., Al-Tabbaa, A., 2017. Environmental and health impacts of using food waste as animal feed: a comparative analysis of food waste management options. Journal of Cleaner Production 140, 871–880. Available from: https://doi.org/10.1016/j.jclepro.2016.05.049.

Sandberg, P., Soderstrom, M., 2003. Industrial energy efficiency: the need for investment decision support from a manager perspective. Journal of Energy Policy 31, 1629–1634.

Sanders, T.A., 1999. Food production and food safety. BMJ (Clinical Research Ed.) 318 (7199), 1689–1693.

Shaw, S.L., Mitloehner, F.M., Jackson, W., DePeters, E.J., Fadel, J.G., Robinson, P.H., et al., 2007. Volatile organic compound emissions from dairy cows and their waste as measured by proton-transfer- reaction mass spectrometry. Environmental Science and Technology 41, 1310–1316.

Sirrine, R., 2022. How is a community food system different than the globalized food system? MSU Extension Retrieved 14.05.22 from. Available from: https://www.canr.msu.edu/news/how_is_a_community_food_system_different_than_the_globalized_food_system.

Sproesser, G., Ruby, M.B., Arbit, N., Akotia, C.S., Alvarenga, M.D.S., Bhangaokar, R., et al., 2019. Understanding traditional and modern eating: the TEP10 framework. BMC Public Health 19 (1), 1606. Available from: https://doi.org/10.1186/s12889-019-7844-4.

Tonini, D., Albizzati, P.F., Astrup, T.F., 2018. Environmental impacts of food waste: learnings and challenges from a case study on UK. Waste Management (New York, N.Y.) 76, 744–766. Available from: https://doi.org/10.1016/j.wasman.2018.03.032.

Torres-Leon, C., Ramirez-Guzman, N., Londono-Hernandez, L., Martinez-Medina, G.A., Diaz-Herrera, R., Navarro-Macias, V., et al., 2018. Food waste and byproducts: an opportunity to minimize malnutrition and hunger in developing countries. Frontiers in Sustainable Food Systems 2, 52. Available from: https://doi.org/10.3389/fsufs.2018.00052.

World Bank, 2018. What a waste 2.0. Global food loss and waste. Retrieved 14.05.22 from https://datatopics.worldbank.org/what-a-waste/global_food_loss_and_waste.html.

Chapter 29

Sequestration of pesticide residues using biofabricated nanomaterials: challenges and future prospects

Usman Lawal Usman[1,2], Bharat Kumar Allam[3], Sushmita Banerjee[1] and Nakshatra Bahadur Singh[4]

[1]*Department of Environmental Sciences, Sharda University, Greater Noida, Uttar Pradesh, India,* [2]*Department of Biology, Umaru Musa Yar'adua University, Katsina, Nigeria,* [3]*Department of Chemistry, Rajiv Gandhi University (A Central University), Doimukh, Arunachal Pradesh, India,* [4]*Department of Chemistry and Biochemistry, Sharda University, Greater Noida, Uttar Pradesh, India*

Introduction

The decline in potable drinking water is increasing globally, this is because of an upsurge in non-point and point origins of pollution that empties into the aquatic bodies. Among the major causes of pollution comprise; industrial wastes; waste generated from industries forms the main cause of all pollutants in the water, and the majority of industries significantly contribute to water pollution (Wang et al., 2021). Agricultural wastes: waste emerges from crop cultivation (plowing) and animals which forms sediment and other pollutants in the water bodies such as fertilizers, and pesticides (Dwivedi and Shikha, 2016). Consistent use of pesticides in modern agricultural practice leads to an increase in surface and groundwater contamination (Al-Maliki et al., 2021). Domestic Wastes; are waste generated from households. Domestic waste is mainly from sewage or leakages of a septic tank that is discharged into water bodies. Similarly, plant nutrients like nitrogen and phosphorous from human waste and used fertilizers in household lawns and gardens end up in water bodies (Liu et al., 2021). Nowadays, a lot of garbage is dumped into various water bodies, for example, rivers, streams, lakes, etc. hence rendering the water bodies to harbor bottles, cans, plastics, and many household products (Raja and Venkatesan, 2010), which can adversely affect the quality of the water and the aquatic biota at large. The unauthorized use of pesticides to protect agricultural produce is one of the major problems affecting developing countries. A small quantity ($\sim 10\%$) of pesticides used in farmland are adsorbed by the target plant and large volumes are spread off into the soil, air, and water thereby affecting their quality based on their shelf life (Mulvey and Cresswell, 2020). Pesticides are known to be very toxic, carcinogenic, suicidal, and homicidal agents (Bonvoisin et al., 2020). Hence, they have been categorized as prominent hazardous materials by the European Commission; Water Framework Directive. However, despite the strict regulation on the application of these pesticides, they can still be detected in some natural water bodies including both surface and ground waters (Sasakova et al., 2018). Concerning the water quality meant for consumption by humans, the 98/83/EC; Drinking Water Directive prescribes 0.1 and 0.5 μg/L as maximum limits for a single pesticide and the total of multiple pesticides respectively (European Commission, 2019). Pesticides are classified either based on their chemical composition or used targets (herbicides, fungicides, and insecticides). Thus, the major group of pesticides comprises organophosphate pesticides (OPs), organochlorine pesticides (OCs), carbamate pesticides, substituted urea compounds (SU), bio-pesticides, and miscellaneous pesticides (Kafilzadeh, 2015; Kaushal et al., 2021; Rani and Shanker, 2018; Rani et al., 2021).

Pesticides in the environment

Pesticides have captured the attention of the whole world due to their toxicity, tumorogenic, carcinogenic, and mutagenic (Rocha and Grisolia, 2019). They affect different organisms across various strata of the environment through misuse or spillage (Broughton, 2005). About 51% of foods produced are adulterated with the residue of pesticides (Gupta, 2004). Similarly, Calabrese (1982) reported contamination of milk with pesticide residue from different sources in the USA. The persistent use of OP and OC made it biodegradable resistant or gradually degraded to more complex

Role of Green Chemistry in Ecosystem Restoration to Achieve Environmental Sustainability. DOI: https://doi.org/10.1016/B978-0-443-15291-7.00008-0

pesticides (Nagata et al., 1999). Kafilzadeh (2015) reported the presence of dichlorodiphenyl dichloroethylene (DDE) in Lake Tashk, Iran in fish, water, and sediment. Kannan et al. (1997) described a high concentration of chlordane compound in seafood within Australia. Similarly, high dichlorodiphenyltrichloroethane (DDT) and hexachlorocyclohexane (HCH) concentrations in Chinese foodstuffs were reported by Li et al. (1998). Thus, HCH and DDT productions were barred in China in 1983 (Wolfe et al., 1984; Chen and Gao, 1993).

Pesticides toxicity in water and the environment

Pesticides are known to be toxic, especially OCs and OPs which are carcinogenic with great persistent in water and soil, and lower degradation potential thereby having more stability toward heat, microbial attack, and oxidizing substances (Cortes and Hites, 2000; Barcelo and Hennion, 2000). The surface and groundwater along with effluents containing below 1 ppm concentration of pesticides are considered to be polluted water (Gilliom, 2001). OCs were detected in foodstuffs and water with a range of about 0.027−0.045 and 0.068−0.098 ppb respectively, along with a trace in human breast milk (Shoiful et al., 2013; Tang et al., 2013). Occasionally, the metabolites from pesticides are considered to be more toxic than the pesticides (Gupta et al., 2012). Matsumura (1985) reported that the organophosphate persistent pesticide metabolites are more noxious to animals and plants compared to the pesticide compound (Vasilić et al., 1992). The degradation of DDT and dieldrin which comprises DDE, dichlorodiphenyl dichloroethane, and aldrin, respectively, is known to be highly toxic (Tian et al., 2009a,b). Thus, the 91/414/EEC European Directive and its succeeding modifications in 1997 compel that environmental data must be provided before the launching of any new pesticide on the market (Barcelo and Hennion, 2000).

Influence of pesticides on human health and ecosystem

Inappropriate usage of pesticides can end up in the remnant part of the food chain residues which can result in pollution in the soil, water as well and air prompting the entire ecosystem (Ahmed et al., 2017). There are various ways through which humans can be exposed to pesticides, this includes skin contact, breathing, and digestion (Paul et al., 2018). The pesticide's impact on human is highlighted in Table 29.1. Some of the instantaneous pesticide's health effects are nausea, diarrhea, dizziness, and exasperation of the throat, skin, and nose resulting in stinging, burning, itching, rashes, and blisters. The chronic effects comprise cancer, damage to the central nervous system, infertility, defects in birth, and damage to the lungs, liver, kidneys, etc. (Sun et al., 2020).

Methods of pesticide degradation

Different techniques are employed in the degradation of pesticides due to their toxic and hazardous effects on plants, animals, and the environment. These techniques include; aerobic degradation (Rajashekara Murthy et al., 2010), extraction of fluid, bioremediation (Lagadec et al., 2000), nano-filtration membranes (Ahmad et al., 2008), biological oxidation, electrochemical and advanced oxidation processes (Ballesteros Martín et al., 2008b; Bandala et al., 2002), adsorption (Danish et al., 2010) and extraction in solid phase (Masselon et al., 1996). Activated carbon (AC) is also widely used for the elimination of pesticides, this is due to their porosity and larger surface area (Kitous et al., 2009)

TABLE 29.1 Influence of different pesticides on human health.

Pesticides	Influence on human	References
Fumigants	Symptoms comprise itching (eyes and skin), burning, irritation of the respiratory tract, nose bleeding, and coughing	Preisser et al. (2011)
Carbamates and OPs	Signs consist of dizziness, chest pain, diarrhea, headaches, pain in the muscles, nausea, confusion, and vomiting	Gupta et al. (2012)
Pyrethroids	They are known as carcinogens and excitant nerve poison	Chrustek et al. (2018)
OCs	Even though these pesticides are usually non-immediate as toxic compared to organophosphates or carbamates, however, they are very hazardous and they are persevering in the ecosystem and tend to bioaccumulate in the tissues via the food chain	Sun et al. (2020)

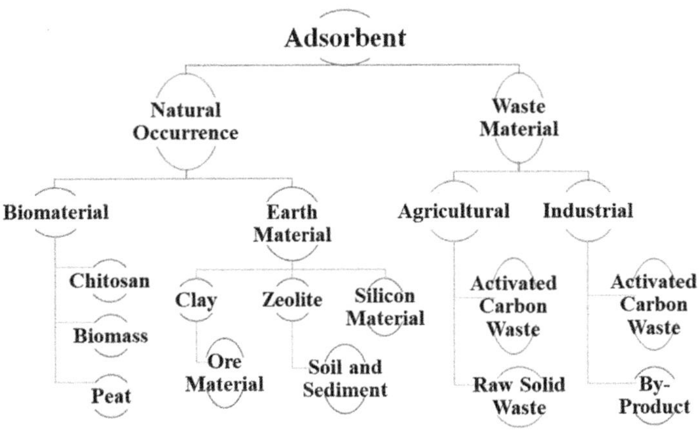

FIGURE 29.1 Adsorbent materials for pesticide removal.

alongside their removal capacity which is largely based on the mode of synthesis ranging between 170 and 422 mg/g as reported by (Ayranci and Hoda, 2005; Hameed et al., 2009), this resulted in the other alternative cost-effective and efficient adsorbent. Plants are found to be highly effective in the removal of various pollutants via their chemical processes, this is due to their possession of both acidic and hydroxyl group compounds (Usman et al., 2022). Similarly, based on their functions they have a higher removal ability compared to the activated carbon and raw materials (Ali et al., 2020; Aouada et al., 2009). Similar to the waste from agricultural products, the effluent from industries such as carbon slurry, fly ash, and sludge was reported to function as good materials for the elimination of noxious pesticides like metolachlor, metribuzin, atrazine, etc. (Sharma et al., 2019). Organic contaminant from industries and pesticides have been effectively immobilized with the use of cost-effective organic material such as natural and organoclay composite adsorbent and crystalline zeolites which has a surface area and void volume of about 903–1132 m^2/g and 18%–50% respectively (Ibrahim and Jbara, 2009). Adsorption techniques using different adsorbents as depicted in Fig. 29.1 were reported to be the most effective, simple, flexible, and low-cost methods used for pesticide degradation (Foo and Hameed, 2010).

The advanced oxidation processes (AOPs) involve the generating the free radical species that are highly reactive, which react with organic pollutants to form non-toxic byproducts and mineral salts (Bandala et al., 2002). The free radical mechanisms occur by abstraction of hydrogen or via the addition of electrophilic to double bonds which ensure the generation of free radicals (R•) from the contaminants, thereby forming peroxy radicals that initiate the oxidative degradation chain reactions which in the end lead to the mineralization of the organic RH + HO• or SO_4•—to HR• + H_2O (Blanco, 2003; Bolton and Cater, 2018). However, heterogeneous photocatalytic process functions with oxides utilization such as ZrO_2 ZnO, TiO_2, etc., and sulfides of metals example, FeS, ZnS, etc. which serve as semiconductor catalysts when exposed to UV or solar radiation (Robertson, 1996). In the last two decades, metallic iron which acts as moderately reducing reagents and zero-valent iron (FeO) has been extensively explored by different researchers and various agencies of government largely due to their cost effectiveness and non-toxicity (Gillham and O'Hannesin, 1994). Synthesis of FeO comprises corrosion or electrochemical reaction in which iron oxidized from contact with water and oxygen thereby making the pollutants accept free electrons from oxidation and reduction of iron based on their stoichiometry (Rani and Shanker, 2018). Recently, the metallic nanoscale has been widely exploited in the form of nanocomposites and nano-biocomposites (Avérous and Pollet, 2012; Rhim et al., 2013).

OC pesticides degradation

The development of different advanced and ground-breaking technologies for the removal of organochlorine pesticides into eco-friendly compounds is increasing worldwide. Rani et al. (2017) described the use of photocatalysts (TiO_2) and redox reactions (FeO) to degrade OC pesticides (Table 29.2).

OPs pesticides degradation

Generally, the effective degradation of organic pollutants cannot be achieved via conventional approaches like physical and chemical treatment, digestion, anaerobic, activated sludge, etc. (Galindo et al., 2001; Kuo and Ho, 2001). Different forms of the photocatalytic degradation process of organophosphate pesticides were reviewed by Sud and Kaur (2012). Photocatalysis using heterogeneous composites, such as TiO_2-doped composite with nZVI and different metals, was used for the degradation of organophosphates pesticides such as parathion, malathion, methamidophos, etc. (Table 29.3).

TABLE 29.2 OC pesticide degradation using nanomaterials.

Pesticides	Nanomaterials	Degradation	Techniques	References
Endrin, aldrin and lindane	Fe_3O_4 @ polystyrene	33.5, 24.7, and 10.2 mg/g	Adsorption	Lan et al. (2014)
Aldrin	TiO_2 (25 mg)	100% degraded in water to chlordane, dieldrin, and 12-hydroxy-dieldrin	Photodegradation	Bandala et al. (2002)
Dicofol	TiO_2 (0.25 mg/mL)	100% degraded in 2 h	Photocatalytic removal (UV)	Shen et al. (2007)
DDT	Fe_3O_4-Hexagonal Mesoporous Silica (HMS)	2.77 μg/mg of DDT was absorbed	Adsorption	Tian et al. (2009a,b)
DDT	Fe_3O_4	92% and 22.4% degraded in water and soil, respectively	Reductive removal	El-Temsah et al. (2016)
DDT (150 mL of 100 μg/mL)	Bimetallic Ni/Fe (0.05 g/L)	Effectively degraded at 4 h in weakly acid or alkaline medium	Redox removal	Tian et al. (2009a,b)
2 and 3 chlorophenoxy acetic acid	ZnO/γ-Fe_2O_3 (0.20 g/L)	55.2% and 57.0% degradation	Photocatalytic removal	Abdullah et al. (2013)
Endosulfan (1140 μg/kg)	Fe_3O_4 (1000 mg/kg of soil)	20% and 8% endosulfan removal from soil by Tulsi and Chittaratha, respectively in 28 days	Adsorption, reductive degradation	Pillai and Kottekottil (2016)
Endosulfan	Copper	Annihilates in solution	Adsorption	Mitra (2013)
Lindane (5×10^{-4} mmol/L)	P-25 Degussa, TiO_2 Anatase, TiO_2 N-doped	Effectively 100% degradation using visible light	Photodegradation	Senthilnathan and Philip (2010)
Lindane	TiO_2-N	Efficient photocatalytic activity in visible light	Photocatalyst	Senthilnathan and Philip (2010)
Lindane	Ag-reduced graphene oxide	99.90%	Degradation	Gupta et al. (2015)
Lindane (7.5)	0.015 g/L Zero-valent iron (Fe_3O_4)	Metabolize to chlorides and benzene in 24 h	Reductive degradation	Elliott et al. (2009)
Lindane	0.5 g/L Fe-Pd Bimetallic	100% and 65% degraded in reduction anaerobic and aerobic Fenton-like process, respectively	Redox degradation	Joo and Zhao (2008)
Lindane, aldrin, dieldrin, and endrin	Fe_3O_4@ Polystyrene	Adsorption capacity 10.2, 24.7, 21.3, and 33.5 mg/g, respectively	Adsorption	Lan et al. (2014)
Lindane	FeS-biopolymers stabilized	94% of degradation in 8 h	Oxidative degradation	Paknikar et al. (2005)
Lindane, DDT and aldrin	TiO_2-Ag/0.5, 1.0, and 2.0 g Fe-based compounds	79%, 81%, and 100%, after reaction rate of 12 h constants 0.0269, 0.0327, and 0.1381, respectively	Redox removal	Shoiful et al. (2016)
Organochlorine pesticides (α-hexachlorobenzene (BHC) and dicofol)		High surface area and lower electron-hole recombination	Photocatalyst	Guo et al. (2009)
δ-, γ-, β-, α-BHC, cypermethrin and dicofol (20 μg each)	TiO_2 coated films (2.24 mg/cm^2)	Degradation within 10 min of 0.4%, 9%, 0.2%, 44%, 32%, and 31%, respectively	Photocatalytic removal (UV)	Yu et al. (2007)
4-Chloro phenol	0.05 g C/ZnO/Cd-S	98.0% in 120 min	Photocatalytic removal	Lavand and Malghe (2015)

TABLE 29.3 OP pesticides degradation using nanomaterials.

Pesticides	Nanomaterials	Degradation	Techniques	References
Acephate	TiO_2 (4 mg/L)	100% mineralization	Photocatalysis	Han et al. (2009)
Chlorpyrifos	Ag and Au modified on neutral alumina	100% degradation in 3 h using supported Ag	Adsorption	Bootharaju and Pradeep (2012)
Chlorpyrifos	Co-Fe_2O_4@TiO_{2-0}.4 g/L reduced graphene	89.9% in 60 min	Photocatalysis	Gupta et al. (2015)
Chlorpyrifos, cypermethrin, chlorothalonil	5 g/L activated carbon decorated Fe-granular	100% degraded within 1 min in 100 mg/L H_2O_2	Oxidation	Affam et al. (2016)
Dichlorvos	Chitosan, polylactic acid, and gum ghatti-mediated CuO-MMT (0.5–3.0 g/L)	93.4%, 63.3%, and 87.8% degradation, respectively	Adsorption	Sahithya et al. (2015)
Malathion (10 ppm)	Au/ZnO, TiO_2 and ZnO (10^{-5}, 3×10^{-5} and 10^{-4} M)	30.0% removal within 30 min	Photocatalysis	Fouad and Mohamed (2012)
Monocrotophos	ZnO-modified Zeolite	100% removal	Photocatalytic degradation	Anandan et al. (2006)
Monocrotophos	ZnO supported on zeolite (1 or 5 g/L)	100% removal within 4 h	Photocatalysis	Tomašević et al. (2010)
Monocrotophos, quinalphos (20 each)	TiO_2/ZnO (0.1 g)	96% and 84.2% removal, respectively	Photocatalysis	Kaur et al. (2019)
Malathion	Fe_3O_4 (0.1 g/kg)	Within 8 min oxidized in soil	Oxidation	Singhal et al. (2012)
Monocrotophos, dichlorv	TiO_2-zeolite (50 mg)	100% removal	Photocatalysis	Gomez et al. (2015)
Malathion	TiO_2 (anatase)/WO_3 (2 wt. %) (125 or 250 mg)	100% removal after 2 h	Photocatalysis	Ramos-Delgado et al. (2013)
Malathion	CuO-chitosan (0.02 g)	99.9%, removal	Adsorption	Jaiswal et al. (2012)
Terbufos	TiO_2 (50 mg)	~99% removal in 90 min, rate constant $k = 0.0854$/min	Photocatalysis	Wu et al. (2009)
Parathion methyl (PM) (100 μL of 10,000 mg /L)	Titania-iron mixed oxides (50 mg in 400 μL)	<70% maximum degradation with Fe:Ti ratio 1:0.25 and 0.25:1	Nucleophilic substitution SN^2	Henych et al. (2015)
Phoxim	La-modified TiO_2 (0.5 g/L)	100% mineralization afterward 4-h	Photocatalysis	Dai et al. (2009)
PM (200 μL of 5000 mg/ L)	MnO_2 (IV) (50 mg in 300 μL)	90% within 2 h in a nonpolar organic solvent	Nucleophilic substitution SN^2	Šťastný et al. (2016)
PM, chlorpyrifos, dichlofenthion, fenchlorphos, prothiofos (20 mg)	1 g CeO_2	90% removal per min, P-O-aryl separation of the bond	Nucleophilic substitution SN^2	Janos et al. (2014)
PM (100 μL of 10,000 mg/L)	Titania-ceria oxides mixture (50 mg in 400 μL)	Degradation ability:Nano-ceria> nano-titania; highest degradation at Ti:Ce 2:8 and 1:1 molar ratio	Nucleophilic substitution SN^2	Šťastný et al. (2016)

(Continued)

TABLE 29.3 (Continued)

Pesticides	Nanomaterials	Degradation	Techniques	References
Organophosphorous pesticides	Fe_3O_4-C18	80% typical recovery for organophosphorous pesticides was recorded	Adsorption	Shen et al. (2007)
Malathion	TiO_2@ Au–Pd	Effective photo-generated degradation	Photocatalysis	Mamdouh Fouad et al. (2011)
Chlorpyrifos	TiO_2-V, Mo, Th	High surface area, lower electron-hole recombination	Photocatalysis	Devi et al. (2009)
Dimethyl methyl phosphonate (DMMP)	Al_2O_3-cerium oxide	775 μg/g of DMMP was adsorbed under 25°C	Adsorption	Mitchell et al. (2004)

Substituted urea pesticide degradation

The toxicity effect of substituted urea pesticides as well as chronic and long-term exposure has been reported throughout the food chain despite their low presence in water and soil (Zhou et al., 2009). Degradation of substituted urea pesticides using microbes and chemical hydrolysis in the soil is largely influenced by the activities of the microbes, pH, and moisture content (Saha and Kulshrestha, 2008).

Miscellaneous pesticides degradation

Effective degradation of several pesticides and pesticide residue has been emerging through the use of heterogeneous photocatalysis comprising different nanomaterials (Table 29.4). Carbon nanotubes (CNTs) are considered highly effective adsorbents due to their distinctive chemical orientation and fascinating physical characteristics such as high reactivity and greater surface area which aid in forming a stronger interaction between the carbon nanotubes and organic chemicals (Liang et al., 2015).

Enzymatic degradation of pesticides using immobilized laccases

The current investigation on the enzymatic removal of different pesticides using immobilized laccases is depicted in Table 29.5. To enhance the dichlorophenol biodegradation, *Coriolis laccase* was immobilized by Vidal on the synthetic nanostructured mesoporous silica-form (MSUF) via a physical absorption process for the fabrication of the nanomaterial's hybrid (Laccase-MSUF). The efficient adsorption of the nanocomposites by the enzymatic process (7 μM/μg MSUF) reveals the stabilization of the enzymes across the inner MSUF pores can be altered by electrostatic interaction (Lingamdinne et al., 2019; Xiaochu et al., 2021). The typical fabricated hybrid MSU-F laccase nanocomposites can effectively oxidize various pesticides into dichlorophenol, thereby resulting in a substantial decrease in the apoptosis acute genotoxicity. In addition, afterward, the immobilization of the laccase, MSUF yielded a potential shielding effect. Consequently, because of their high load capacity, the mesoporous nanomaterials can be distended to a few grams (Das et al., 2017; Vera et al., 2020). Similarly, laccase immobilization of *Myceliophthora thermophila* was conducted recently on the polyglycidyl methacrylate monodisperse microspheres, which was used biologically to degrade azinphos-methyl (Masjoudi et al., 2021).

Challenges and future prospects

The world's population is rapidly growing; hence pesticides have been included in contemporary life to supply the food needs of this expanding population. The targeted pests have developed acquired resistance as a result of the overuse of pesticides. Pesticide overuse has cleared out the normally susceptible population, but it has given resistant pests a selective advantage, which has increased their population. The development of resistance to significant crop pests, livestock parasites, household pests, and vector-borne diseases has made control very difficult. Pest resurgence, often known as the reappearance of significant pest populations following the administration of a pesticide, is another problem with pesticides. The use of pesticides at less lethal dosages has been linked to the killing of natural enemies, an increase in

TABLE 29.4 Miscellaneous pesticide degradation using nanomaterials.

Pesticides	Nanomaterials	Degradation	Techniques	References
Neonicotinoids pesticides				
Atrazine (2 g)	1:1 ratio of Ag-chitosan	98% degradation in 65 min	Adsorption	Saifuddin et al. (2011)
Acetamiprid	TiO_2	Degraded efficiently in alkaline pH	Photocatalysis	Khan et al. (2012)
Atrazine	Carbon nanotubes	>90 degradation	Adsorption	Yan et al. (2008)
Clothianidin, thiamethoxam, and imidacloprid	Immobilized TiO_2	Oxidized in 2 h by $k = 0.021$, 0.019, and 0.035/min, respectively	Photocatalysis (UV)	Žabar et al. (2012)
Thiamethoxam and imidacloprid	0.06 mol/L Fe/Fe_3O_4 (with/or without H_2O_2)	> 90% degradation within 30 min especially in acidic pH	Redox removal	de Urzedo et al. (2009)
Thiamethoxam and imidacloprid	Fe, Sn, Zn-zero-valent metals	>90% degradation after 30 min through initial reduction of NO_2 − > NH2	Ultrasonication	Ballesteros Martín et al. (2008a)
Phosphamidon, isoproctum, and imidacloprid	TiO_2-modified porous nanosilica	Degraded within 120, 240, and 90 min, respectively at 7.0 pH	Photocatalysis (UV)	Sharma et al. (2009)
Thiamethoxam (3.4 mM)	1 g/L of TiO_2	Degraded effectively	Photocatalysis	Mir et al. (2013)
Pyridazinone pesticides				
Chloridazon and metribuzin	1 g/L of TiO_2 ()	Chloridazon >metribuzin in acidic pH while metribuzin >chloridazon in alkaline pH	Photocatalysis	Khan et al. (2012)
Chloridazon	Core-shell-Fe_3O_4@Au and FeO_4	>90% degradation	Photocatalysis	Fouad and Mohamed (2012)
Chloridazon	Au/TiO_2 (3 × 10^5, $10^{-5}$and 10^{-4} M)	50% in 30 min	Photocatalysis	Fouad and Mohamed (2012)
Azoxystrobin, primicarb, kresoxim- methyl, propyzamide, tebuconazole, triadimenol, hexaconazole and pyrimethanil	ZnO@$Na_2S_2O_8$	Decrease in reaction time and increase the degradation rate	Photocatalyst	Navarro et al. (2009)
Permethrin (25 mL, 0.1 mg/L) (pyrethroid pesticide)	ZnO@Chitosan (0.5 g)	99% degradation	Adsorption	Moradi Dehaghi et al. (2014)
Carbamate pesticides				
Methomyl	Fe-zeolite (1 or 5 g/L)	100% degradation in 4 h	Photocatalysis	Tomašević et al. (2010)
Diazinon	Al_2O_3 and MgO, activated carbon	The maximum adsorption rate was 19.36 μL/min	Adsorption	Behnam et al. (2013)
Diazinon and fenitrothion	Al_2O_3	57% and 90% effective removal with 0.28 and 0.32 mg/mL initial dosage, respectively	Adsorption	Armaghan and Amini (2012)

(Continued)

TABLE 29.4 (Continued)

Pesticides	Nanomaterials	Degradation	Techniques	References
Methomyl	TiO_2-$CdSO_4$	Recombination and capturing from lower and strong electron hole, respectively	Photocatalyst	Barakat et al. (2014)
Vitavax	LFCOs and LFCO1 NPs	139 and 155 mg/g adsorption capacity, respectively	Adsorption	Tavakkoli and Yazdanbakhsh (2013)
Permethrin	Zinc oxide-chitosan	99% degradation at room temperature and 7.0 pH	Adsorption	Moradi Dehaghi et al. (2014)
Isoproturon	TiO_2@ C, N, and S	Maximum absorption within the range of visible light, wide surface area, and recombination from lower electron hole	Photocatalyst	Anil Kumar Reddy et al. (2010)

TABLE 29.5 Enzymatic degradation of pesticides at different time intervals.

Pesticide type	Enzyme source	Enzyme quantity	Redox mediator	Degradation (%)	Time	References
Atrazine	*Trametes versicolor*	0.05 Um/L	1-Hydroxybenzotriazole	90%	8 d	Briceño et al. (2018)
Chlorothalonil	*T. versicolor*	0.05 Um/L	Acetosyringone	90%	8 d	Kupski et al. (2019)
Chlorpyrifos	*Pseudomonas* sp.	210 nkat/ML	No or sinapinic acid	83.25/80.56%	6 h	Aswathi et al. (2019)
Chlorpyrifos	*T. versicolor*	0.05 Um/L	Vanillin	90%	8 d	Hu et al. (2020)
Chlorpyrifos	*T. versicolor*	0.05 Um/L	Vanillin	90%	8 d	Murillo-Zamora et al. (2017)
Dichlorophos	*Pseudomonas* sp.	210 nkat/ML	No or sinapinic acid	45.99/45.82%	6 h	Shabbir et al. (2018)
Glyphosate	*T. versicolor*	0.15 Um/L	ABTS	40.90%	24 h	Ambreen and Yasmin (2020)
Glyphosate	*T. versicolor*	0.15 Um/L	ABTS + Mn^{2+} + Tween 80	90.10%	24 h	Shabbir et al. (2018)
Glyphosate	*T. versicolor*	0.15 U/mL	Mn^{2+} + Tween 80	62.80%	24 h	Góngora-Echeverría et al. (2020)
Isoproturon	*T. versicolor*	0.05 UmL^{-1}	Violuric acid	100%	24 h	Bayramoglu and Arica (2019)
Isoproturon	*T. versicolor*	0.30 Um/L	HBT	68.20%	24 h	Lizano-Fallas et al. (2017)
Isoproturon	*T. versicolor*	0.30 UmL^{-1}	HBT	100%	24 h	Castro-Gutiérrez et al. (2019)
Monocrotophos	*Pseudomonas* sp.	210 nkat/mL	No or sinapinic acid	75.45/78.82%	6 h	Briceño et al. (2018)
Pyrimethanil	*T. versicolor*	0.05 Um/L	Violuric acid	100%	24 h	Cambronero-Heinrichs et al. (2018)
Profenovos	*Pseudomonas* sp.	210 nkat/mL	No or sinapinic acid	81.84/82.15%	6 h	Beltrán-Flores et al. (2020)

reproductivity, and the feeding of insect pests. Secondary pests can occasionally turn into primary pests as a result of the elimination of the primary pests. Earthworms and other nontarget creatures like pollinators have a significant impact on agricultural productivity. The majority of pesticides are nonbiodegradable and highly toxic, leading to bioaccumulation and biomagnification, and causing air, water, and soil pollution. In addition to upsetting the ecosystem's delicate balance, pesticide use has substantial negative effects on the environment and human health. The prospect of removing pesticides from the environment involves the use of engineered nanomaterials fabricated with biological organisms such as bacteria, fungi, enzymes, plants, etc. It has been reported for several pesticides, but further research is needed to broaden the chemical range it can remediate and to understand the mechanism underlying the remediation process.

Conclusions

The biogenic molecules such as polysaccharides, lipids proteins, etc. serve both as reducing or chelating and capping agents during the synthesis of the composites, hence the resultant composite is prevented from further reaction and accumulation thereby enhancing their longevity and stability. Most of the highlighted nanomaterials can effectively remove more than 80% of the pesticides within a short period. However, OCs and OPs are commonly removed using nanomaterial but substituted urea compounds and carbamate pesticides which are referred to as the future contaminants are yet to be explored extensively by the nanomaterials-mediated composites. Therefore, studies on the various techniques especially the biogenic-mediated synthesis of biopolymer-based and nano-biocomposites will be the solution to the proliferation of different contaminants with their metabolites from various pesticide pollutants in the environment.

References

Abdullah, A.H., Mun, L.K., Zainal, Z., Hussein, M.Z., 2013. Photodegradation of Chlorophenoxyacetic acids by ZnO/r-Fe2O3 nanocatalysts: a comparative study. International Journal of Chemistry . Available from: https://doi.org/10.5539/ijc.v5n4p56.

Affam, A.C., Chaudhuri, M., Kutty, S.R.M., Muda, K., 2016. Degradation of chlorpyrifos, cypermethrin and chlorothalonil pesticides in aqueous solution by FeGAC/H2O2 process. Desalination and Water Treatment . Available from: https://doi.org/10.1080/19443994.2014.1001441.

Ahmad, A.L., Tan, L.S., Shukor, S.R.A., 2008. Dimethoate and atrazine retention from aqueous solution by nanofiltration membranes. Journal of Hazardous Materials . Available from: https://doi.org/10.1016/j.jhazmat.2007.05.047.

Ahmed, M.B., Zhou, J.L., Ngo, H.H., Guo, W., Thomaidis, N.S., Xu, J., 2017. Progress in the biological and chemical treatment technologies for emerging contaminant removal from wastewater: a critical review. Journal of Hazardous Materials . Available from: https://doi.org/10.1016/j.jhazmat.2016.04.045.

Ali, N., Khan, A., Bilal, M., Malik, S., Badshah, S., Iqbal, H.M.N., 2020. Chitosan-based bio-composite modified with thiocarbamate moiety for decontamination of cations from the aqueous media. Molecules (Basel, Switzerland) . Available from: https://doi.org/10.3390/molecules25010226.

Al-Maliki, L.A., Farhan, S.L., Jasim, I.A., Al-Mamoori, S.K., Al-Ansari, N., 2021. Perceptions about water pollution among university students: a case study from Iraq. Cogent Engineering . Available from: https://doi.org/10.1080/23311916.2021.1895473.

Ambreen, S., Yasmin, A., 2020. Novel metabolites of triazophos formed during degradation by bacterial strains *Pseudomonas kilonensis* MB490, *Pseudomonas kilonensis* MB498 and pseudomonas sp. MB504 isolated from cotton fields. Journal of Environmental Science and Health - Part B Pesticides, Food Contaminants, and Agricultural Wastes . Available from: https://doi.org/10.1080/03601234.2020.1823171.

Anandan, S., Vinu, A., Venkatachalam, N., Arabindoo, B., Murugesan, V., 2006. Photocatalytic activity of ZnO impregnated Hβ and mechanical mix of ZnO/Hβ in the degradation of monocrotophos in aqueous solution. Journal of Molecular Catalysis A: Chemical . Available from: https://doi.org/10.1016/j.molcata.2006.05.012.

Anil Kumar Reddy, P., Venkata Laxma Reddy, P., Maitrey Sharma, V., Srinivas, B., Kumari, V.D., Subrahmanyam, M., 2010. Photocatalytic degradation of isoproturon pesticide on C, N and S doped TiO2. Journal of Water Resource and Protection . Available from: https://doi.org/10.4236/jwarp.2010.23027.

Aouada, F.A., Pan, Z., Orts, W.J., Mattoso, H.C.L., 2009. Removal of paraquat pesticide from aqueous solutions using a novel adsorbent material based on polyacrylamide and methylcellulose hydrogels. Journal of Applied Polymer Science . Available from: https://doi.org/10.1002/app.30339.

Armaghan, M., Amini, M.M., 2012. Adsorption of diazinon and fenitrothion on nanocrystalline alumina from non-polar solvent1. Colloid Journal . Available from: https://doi.org/10.1134/S1061933X12040047.

Aswathi, A., Pandey, A., Sukumaran, R.K., 2019. Rapid degradation of the organophosphate pesticide – chlorpyrifos by a novel strain of Pseudomonas nitroreducens AR-3. Bioresource Technology . Available from: https://doi.org/10.1016/j.biortech.2019.122025.

Avérous, L., Pollet, E., 2012. Environmental silicate nano-biocomposites. Green Energy and Technology . Available from: https://doi.org/10.1007/978-1-4471-4108-2.

Ayranci, E., Hoda, N., 2005. Adsorption kinetics and isotherms of pesticides onto activated carbon-cloth. Chemosphere . Available from: https://doi.org/10.1016/j.chemosphere.2005.02.040.

Ballesteros Martín, M.M, et al., 2008. Combined photo-Fenton and biological oxidation for pesticide degradation: Effect of photo-treated intermediates on biodegradation kinetics. Chemosphere .

Ballesteros Martín, M.M., Sánchez Pérez, J.A., García Sánchez, J.L., Montes de Oca, L., Casas López, J.L., Oller, I., et al., 2008a. Degradation of alachlor and pyrimethanil by combined photo-Fenton and biological oxidation. Journal of Hazardous Materials . Available from: https://doi.org/10.1016/j.jhazmat.2007.11.069.

Bandala, E.R., Gelover, S., Leal, M.T., Arancibia-Bulnes, C., Jimenez, A., Estrada, C.A., 2002. Solar photocatalytic degradation of Aldrin. Catalysis Today . Available from: https://doi.org/10.1016/S0920-5861(02)00218-3.

Barakat, N.A.M., Nassar, M.M., Farrag, T.E., Mahmoud, M.S., 2014. Effective photodegradation of methomyl pesticide in concentrated solutions by novel enhancement of the photocatalytic activity of TiO2 using CdSO4 nanoparticles. Environmental Science and Pollution Research . Available from: https://doi.org/10.1007/s11356-013-2027-9.

Barcelo, D., Hennion, M., 2000. Trace Determination of Pesticides and Their Degradation Products in Water. *Elsevier*.

Bayramoglu, G., Arica, M.Y., 2019. Biodegradation of methylene blue and carbaryl by *Trametes versicolor* laccase preparations in the presence of a mediator compound. Journal of Macromolecular Science, Part A: Pure and Applied Chemistry . Available from: https://doi.org/10.1080/10601325.2019.1565549.

Behnam, R., Morshed, M., Tavanai, H., Ghiaci, M., 2013. Destructive adsorption of diazinon pesticide by activated carbon nanofibers containing Al2O3 and MgO nanoparticles. Bulletin of Environmental Contamination and Toxicology . Available from: https://doi.org/10.1007/s00128-013-1064-x.

Beltrán-Flores, E., Torán, J., Caminal, G., Blánquez, P., Sarrà, M., 2020. The removal of diuron from agricultural wastewaters by *Trametes versicolor* immobilized on pinewood in simple channel reactors. Science of the Total Environment . Available from: https://doi.org/10.1016/j.scitotenv.2020.138414.

Blanco, J., 2003. Development of CPC solar collectors for application to photochemical degradation of persistent pollutants in water Editorial CIEMAT, Madrid. (In Spanish).

Bolton, J.R., Cater, S.R., 2018. Homogeneous photodegradation of pollutants in contaminated water: an introduction. Aquatic and Surface Photochemistry . Available from: https://doi.org/10.1201/9781351069847.

Bonvoisin, T., Utyasheva, L., Knipe, D., Gunnell, D., Eddleston, M., 2020. Suicide by pesticide poisoning in India: a review of pesticide regulations and their impact on suicide trends. BMC Public Health . Available from: https://doi.org/10.1186/s12889-020-8339-z.

Bootharaju, M.S., Pradeep, T., 2012. Understanding the degradation pathway of the pesticide, chlorpyrifos by noble metal nanoparticles. Langmuir: The ACS Journal of Surfaces and Colloids . Available from: https://doi.org/10.1021/la2050515.

Briceño, G., Vergara, K., Schalchli, H., Palma, G., Tortella, G., Fuentes, M.S., et al., 2018. Organophosphorus pesticide mixture removal from environmental matrices by a soil Streptomyces mixed culture. Environmental Science and Pollution Research . Available from: https://doi.org/10.1007/s11356-017-9790-y.

Broughton, E., 2005. The Bhopal disaster and its aftermath: a review. Environmental Health: A Global Access Science Source . Available from: https://doi.org/10.1186/1476-069X-4-6.

Calabrese, E.J., 1982. Human breast milk contamination in the United States and Canada by chlorinated hydrocarbon insecticides and industrial pollutants: current status. Journal of the American College of Toxicology . Available from: https://doi.org/10.3109/10915818209018020.

Cambronero-Heinrichs, J.C., Masís-Mora, M., Quirós-Fournier, J.P., Lizano-Fallas, V., Mata-Araya, I., Rodríguez-Rodríguez, C.E., 2018. Removal of herbicides in a biopurification system is not negatively affected by oxytetracycline or fungally pretreated oxytetracycline. Chemosphere . Available from: https://doi.org/10.1016/j.chemosphere.2018.01.122.

Castro-Gutiérrez, V., Masís-Mora, M., Carazo-Rojas, E., Mora-López, M., Rodríguez-Rodríguez, C.E., 2019. Fungal and bacterial co-bioaugmentation of a pesticide-degrading biomixture: pesticide removal and community structure variations during different treatments. Water, Air, and Soil Pollution . Available from: https://doi.org/10.1007/s11270-019-4282-y.

Chen, J., Gao, J., 1993. The Chinese total diet study in 1990. Part I. Chemical contaminants. Journal of AOAC International . Available from: https://doi.org/10.1093/jaoac/76.6.1193.

Chrustek, A., Hołyńska-Iwan, I., Dziembowska, I., Bogusiewicz, J., Wróblewski, M., Cwynar, A., et al., 2018. Current research on the safety of pyrethroids used as insecticides. Medicina (Lithuania) . Available from: https://doi.org/10.3390/medicina54040061.

Cortes, D.R., Hites, R.A., 2000. Detection of statistically significant trends in atmospheric concentrations of semivolatile compounds. Environmental Science and Technology . Available from: https://doi.org/10.1021/es990466l.

Dai, K., Peng, T., Chen, H., Liu, J., Zan, L., 2009. Photocatalytic degradation of commercial phoxim over la-doped TiO2 nanoparticles in aqueous suspension. Environmental Science and Technology . Available from: https://doi.org/10.1021/es802724q.

Danish, M., Sulaiman, O., Rafatullah, M., Hashim, R., Ahmad, A., 2010. Kinetics for the removal of paraquat dichloride from aqueous solution by activated date (phoenix dactylifera) stone carbon. Journal of Dispersion Science and Technology . Available from: https://doi.org/10.1080/01932690903167368.

Das, A., Singh, J., Yogalakshmi, K.N., 2017. Laccase immobilized magnetic iron nanoparticles: fabrication and its performance evaluation in chlorpyrifos degradation. International Biodeterioration and Biodegradation . Available from: https://doi.org/10.1016/j.ibiod.2017.01.007.

de Urzedo, A.P.F.M., Nascentes, C.C., Augusti, R., 2009. Degradation of the insecticides thiamethoxam and imidacloprid in aqueous solution as promoted by an innovative Feo/Fe3O4 composite. Journal of the Brazilian Chemical Society . Available from: https://doi.org/10.1590/s0103-50532009000100010.

Devi, L.G., Murthy, B.N., Kumar, S.G., 2009. Photocatalytic activity of V5 + , Mo6 + and Th4 + doped polycrystalline TiO2 for the degradation of chlorpyrifos under UV/solar light. Journal of Molecular Catalysis A: Chemical . Available from: https://doi.org/10.1016/j.molcata.2009.04.007.

Dwivedi, S., Shikha, D., 2016. Water pollution: causes, effects and control. Biochemical and Cellular Archives .

Elliott, D.W, et al., 2009. Degradation of Lindane by Zero-Valent Iron Nanoparticles. Journal of Environmental Engineering .

El-Temsah, Y.S., Sevcu, A., Bobcikova, K., Cernik, M., Joner, E.J., 2016. DDT degradation efficiency and ecotoxicological effects of two types of nano-sized zero-valent iron (nZVI) in water and soil. Chemosphere . Available from: https://doi.org/10.1016/j.chemosphere.2015.10.122.

European Commission, 2019. Analytical Quality Control and Method Validation for Pesticide Residues Analysis in Food and Feed (SANTE/12682/2019). Sante/12682/2019.

Foo, K.Y., Hameed, B.H., 2010. Detoxification of pesticide waste via activated carbon adsorption process. Journal of Hazardous Materials . Available from: https://doi.org/10.1016/j.jhazmat.2009.10.014.

Fouad, D.M., Mohamed, M.B., 2012. Comparative study of the photocatalytic activity of semiconductor nanostructures and their hybrid metal nanocomposites on the photodegradation of malathion. Journal of Nanomaterials . Available from: https://doi.org/10.1155/2012/524123.

Galindo, C., Jacques, P., Kalt, A., 2001. Photooxidation of the phenylazonaphthol AO20 on TIO2: kinetic and mechanistic investigations. Chemosphere . Available from: https://doi.org/10.1016/S0045-6535(01)00118-7.

Gillham, R.W., O'Hannesin, S.F., 1994. Enhanced degradation of halogenated aliphatics by zero-valent iron. Groundwater . Available from: https://doi.org/10.1111/j.1745-6584.1994.tb00935.x.

Gilliom, R.J., 2001. Pesticides in the hydrologic system - what do we know and what's next? Hydrological Processes . Available from: https://doi.org/10.1002/hyp.501.

Gomez, S., Marchena, C.L., Renzini, M.S., Pizzio, L., Pierella, L., 2015. In situ generated TiO2 over zeolitic supports as reusable photocatalysts for the degradation of dichlorvos. Applied Catalysis B: Environmental . Available from: https://doi.org/10.1016/j.apcatb.2014.06.047.

Góngora-Echeverría, V.R., García-Escalante, R., Rojas-Herrera, R., Giácoman-Vallejos, G., Ponce-Caballero, C., 2020. Pesticide bioremediation in liquid media using a microbial consortium and bacteria-pure strains isolated from a biomixture used in agricultural areas. Ecotoxicology and Environmental Safety . Available from: https://doi.org/10.1016/j.ecoenv.2020.110734.

Guo, G., Yu, B., Yu, P., Chen, X., 2009. Synthesis and photocatalytic applications of Ag/TiO2-nanotubes. Talanta . Available from: https://doi.org/10.1016/j.talanta.2009.04.036.

Gupta, P.K., 2004. Pesticide exposure - Indian scene. Toxicology . Available from: https://doi.org/10.1016/j.tox.2004.01.021.

Gupta, B., Rani, M., Kumar, R., 2012. Degradation of thiram in water, soil and plants: a study by high-performance liquid chromatography. Biomedical Chromatography . Available from: https://doi.org/10.1002/bmc.1627.

Gupta, V.K., Eren, T., Atar, N., Yola, M.L., Parlak, C., Karimi-Maleh, H., 2015. CoFe2O4@TiO2 decorated reduced graphene oxide nanocomposite for photocatalytic degradation of chlorpyrifos. Journal of Molecular Liquids . Available from: https://doi.org/10.1016/j.molliq.2015.04.032.

Hameed, B.H., Salman, J.M., Ahmad, A.L., 2009. Adsorption isotherm and kinetic modeling of 2,4-D pesticide on activated carbon derived from date stones. Journal of Hazardous Materials . Available from: https://doi.org/10.1016/j.jhazmat.2008.06.069.

Han, S., et al., 2009. Photocatalytic decomposition of acephate in irradiated TiO_2 suspensions. Journal of Hazardous Materials .

Henych, J., Štengl, V., Slušná, M., Matys Grygar, T., Janoš, P., Kuráň, P., et al., 2015. Degradation of organophosphorus pesticide parathion methyl on nanostructured titania-iron mixed oxides. Applied Surface Science . Available from: https://doi.org/10.1016/j.apsusc.2015.02.181.

Hu, K., Peris, A., Torán, J., Eljarrat, E., Sarrà, M., Blánquez, P., et al., 2020. Exploring the degradation capability of *Trametes versicolor* on selected hydrophobic pesticides through setting sights simultaneously on culture broth and biological matrix. Chemosphere . Available from: https://doi.org/10.1016/j.chemosphere.2020.126293.

Ibrahim, K.M., Jbara, H.A., 2009. Removal of paraquat from synthetic wastewater using phillipsite-faujasite tuff from Jordan. Journal of Hazardous Materials . Available from: https://doi.org/10.1016/j.jhazmat.2008.06.109.

Jaiswal, M., Chauhan, D., Sankararamakrishnan, N., 2012. Copper chitosan nanocomposite: synthesis, characterization, and application in removal of organophosphorous pesticide from agricultural runoff. Environmental Science and Pollution Research . Available from: https://doi.org/10.1007/s11356-011-0699-6.

Janos, P., Kuran, P., Kormunda, M., Stengl, V., Grygar, T.M., Dosek, M., et al., 2014. Cerium dioxide as a new reactive sorbent for fast degradation of parathion methyl and some other organophosphates. Journal of Rare Earths . Available from: https://doi.org/10.1016/S1002-0721(14)60079-X.

Joo, S.H., Zhao, D., 2008. Destruction of lindane and atrazine using stabilized iron nanoparticles under aerobic and anaerobic conditions: effects of catalyst and stabilizer. Chemosphere . Available from: https://doi.org/10.1016/j.chemosphere.2007.06.070.

Kafilzadeh, F., 2015. Assessment of organochlorine pesticide residues in water, sediments and fish from Lake Tashk, Iran. Achievements in the Life Sciences . Available from: https://doi.org/10.1016/j.als.2015.12.003.

Kannan, K., Tanabe, S., Giesy, J.P., Tatsukawa, R., 1997. Organochlorine pesticides and polychlorinated biphenyls in foodstuffs from Asian and Oceanic countries. Reviews of Environmental Contamination and Toxicology . Available from: https://doi.org/10.1007/978-1-4612-1964-4_1.

Kaur, R., et al., 2019. Toxicity and degradation of the insecticide monocrotophos. Environmental Chemistry Letters .

Kaushal, J., Khatri, M., Arya, S.K., 2021. A treatise on Organophosphate pesticide pollution: current strategies and advancements in their environmental degradation and elimination. Ecotoxicology and Environmental Safety . Available from: https://doi.org/10.1016/j.ecoenv.2020.111483.

Khan, A., Mir, N.A., Faisal, M., Muneer, M., 2012. Titanium dioxide-mediated photcatalysed degradation of two herbicide derivatives chloridazon and metribuzin in aqueous suspensions. International Journal of Chemical Engineering . Available from: https://doi.org/10.1155/2012/850468.

Kitous, O., Cheikh, A., Lounici, H., Grib, H., Pauss, A., Mameri, N., 2009. Application of the electrosorption technique to remove Metribuzin pesticide. Journal of Hazardous Materials . Available from: https://doi.org/10.1016/j.jhazmat.2008.04.091.

Kuo, W.S., Ho, P.H., 2001. Solar photocatalytic decolorization of methylene blue in water. Chemosphere . Available from: https://doi.org/10.1016/S0045-6535(01)00008-X.

Kupski, L., Salcedo, G.M., Caldas, S.S., de Souza, T.D., Furlong, E.B., Primel, E.G., 2019. Optimization of a laccase-mediator system with natural redox-mediating compounds for pesticide removal. Environmental Science and Pollution Research . Available from: https://doi.org/10.1007/s11356-018-4010-y.

Lagadec, A.J.M., Miller, D.J., Lilke, A.V., Hawthorne, S.B., 2000. Pilot-scale subcritical water remediation of polycyclic aromatic hydrocarbon- and pesticide-contaminated soil. Environmental Science and Technology . Available from: https://doi.org/10.1021/es990722u.

Lan, J., Cheng, Y., Zhao, Z., 2014. Effective organochlorine pesticides removal from aqueous systems by magnetic nanospheres coated with polystyrene. Journal Wuhan University of Technology, Materials Science Edition . Available from: https://doi.org/10.1007/s11595-014-0887-6.

Lavand, A.B., Malghe, Y.S., 2015. Visible light photocatalytic degradation of 4-chlorophenol using C/ZnO/CdS nanocomposite. Journal of Saudi Chemical Society . Available from: https://doi.org/10.1016/j.jscs.2015.07.001.

Li, Y.F., Cai, D.J., Singh, A., 1998. Technical hexachlorocyclohexane use trends in China and their impact on the environment. Archives of Environmental Contamination and Toxicology . Available from: https://doi.org/10.1007/s002449900432.

Liang, J., Liu, J., Yuan, X., Dong, H., Zeng, G., Wu, H., et al., 2015. Facile synthesis of alumina-decorated multi-walled carbon nanotubes for simultaneous adsorption of cadmium ion and trichloroethylene. Chemical Engineering Journal . Available from: https://doi.org/10.1016/j.cej.2015.03.069.

Lingamdinne, L.P., Koduru, J.R., Karri, R.R., 2019. A comprehensive review of applications of magnetic graphene oxide based nanocomposites for sustainable water purification. Journal of Environmental Management . Available from: https://doi.org/10.1016/j.jenvman.2018.10.063.

Liu, H., Chen, J., Zhang, L., Sun, K., Cao, W., 2021. Simulation effects of clean water corridor technology on the control of non-point source pollution in the Paihe River basin, Chaohu lake. Environmental Science and Pollution Research . Available from: https://doi.org/10.1007/s11356-020-12274-x.

Lizano-Fallas, V., Masís-Mora, M., Espinoza-Villalobos, D., Lizano-Brenes, M., Rodríguez-Rodríguez, C.E., 2017. Removal of pesticides and ecotoxicological changes during the simultaneous treatment of triazines and chlorpyrifos in biomixtures. Chemosphere . Available from: https://doi.org/10.1016/j.chemosphere.2017.04.147.

Mamdouh Fouad, D., et al., 2011. Studies on the photo-catalytic activity of semiconductor nanostructures and their gold core−shell on the photodegradation of malathion. Nanotechnology .

Masjoudi, M., Golgoli, M., Ghobadi Nejad, Z., Sadeghzadeh, S., Borghei, S.M., 2021. Pharmaceuticals removal by immobilized laccase on polyvinylidene fluoride nanocomposite with multi-walled carbon nanotubes. Chemosphere . Available from: https://doi.org/10.1016/j.chemosphere.2020.128043.

Masselon, C., Krier, G., Muller, J.F., Nélieu, S., Einhorn, J., 1996. Laser desorption Fourier transform ion cyclotron resonance mass spectrometry of selected pesticides extracted on C18 silica solid-phase extraction membranes. Analyst . Available from: https://doi.org/10.1039/AN9962101429.

Matsumura, F., 1985. Toxicology of insecticides. Toxicology of Insecticides . Available from: https://doi.org/10.1007/978-1-4613-2491-1.

Mir, N.A., Khan, A., Muneer, M., Vijayalakhsmi, S., 2013. Photocatalytic degradation of a widely used insecticide Thiamethoxam in aqueous suspension of TiO2: adsorption, kinetics, product analysis and toxicity assessment. Science of the Total Environment . Available from: https://doi.org/10.1016/j.scitotenv.2013.04.041.

Mitchell, M.B., Sheinker, V.N., Cox, W.W., Gatimu, E.N., Tesfamichael, A.B., 2004. The room temperature decomposition mechanism of dimethyl methylphosphonate (DMMP) on alumina-supported cerium oxide - participation of nano-sized cerium oxide domains. Journal of Physical Chemistry B . Available from: https://doi.org/10.1021/jp035590c.

Mitra, D., 2013. Remediation of pesticide endosulfan in solution by ionizing radiation, advanced oxidation process and copper nano particle interaction a comparative studies using GC-MS analysis. IOSR Journal Of Environmental Science, Toxicology And Food Technology . Available from: https://doi.org/10.9790/2402-0720811.

Moradi Dehaghi, S., Rahmanifar, B., Moradi, A.M., Azar, P.A., 2014. Removal of permethrin pesticide from water by chitosan-zinc oxide nanoparticles composite as an adsorbent. Journal of Saudi Chemical Society . Available from: https://doi.org/10.1016/j.jscs.2014.01.004.

Mulvey, J., Cresswell, J.E., 2020. Time-dependent effects on bumble bees of dietary exposures to farmland insecticides (imidacloprid, thiamethoxam and fipronil). Pest Management Science . Available from: https://doi.org/10.1002/ps.5838.

Murillo-Zamora, S., Castro-Gutiérrez, V., Masís-Mora, M., Lizano-Fallas, V., Rodríguez-Rodríguez, C.E., 2017. Elimination of fungicides in biopurification systems: effect of fungal bioaugmentation on removal performance and microbial community structure. Chemosphere . Available from: https://doi.org/10.1016/j.chemosphere.2017.07.162.

Nagata, Y., Futamura, A., Miyauchi, K., Takagi, M., 1999. Two different types of dehalogenases, LinA and LinB, involved in γ-hexachlorocyclohexane degradation in Sphingomonas paucimobilis UT26 are localized in the periplasmic space without molecular processing. Journal of Bacteriology . Available from: https://doi.org/10.1128/jb.181.17.5409-5413.1999.

Navarro, S., Fenoll, J., Vela, N., Ruiz, E., Navarro, G., 2009. Photocatalytic degradation of eight pesticides in leaching water by use of ZnO under natural sunlight. Journal of Hazardous Materials . Available from: https://doi.org/10.1016/j.jhazmat.2009.07.137.

Paknikar, K.M., Nagpal, V., Pethkar, A.V., Rajwade, J.M., 2005. Degradation of lindane from aqueous solutions using iron sulfide nanoparticles stabilized by biopolymers. Science and Technology of Advanced Materials . Available from: https://doi.org/10.1016/j.stam.2005.02.016.

Paul, K.C., Chuang, Y.H., Cockburn, M., Bronstein, J.M., Horvath, S., Ritz, B., 2018. Organophosphate pesticide exposure and differential genome-wide DNA methylation. Science of the Total Environment . Available from: https://doi.org/10.1016/j.scitotenv.2018.07.143.

Pillai, H.P.S., Kottekottil, J., 2016. Nano-phytotechnological remediation of endosulfan using zero valent iron nanoparticles. Journal of Environmental Protection . Available from: https://doi.org/10.4236/jep.2016.75066.

Preisser, A.M., Budnik, L.T., Hampel, E., Baur, X., 2011. Surprises perilous: toxic health hazards for employees unloading fumigated shipping containers. Science of the Total Environment . Available from: https://doi.org/10.1016/j.scitotenv.2011.04.053.

Raja, G., Venkatesan, P., 2010. Assessment of groundwater pollution and its impact in and around Punnam area of Karur District, Tamilnadu, India. E-Journal of Chemistry . Available from: https://doi.org/10.1155/2010/782134.

Rajashekara Murthy, H.M., Thakur, M.S., Manonmani, H.K., 2010. Degradation of technical grade hexachlorocyclohexane in soil slurry by a defined microbial consortium. International Journal of Environmental Research . Available from: https://doi.org/10.22059/ijer.2010.232.

Ramos-Delgado, N.A., et al., 2013. Synthesis by sol−gel of WO_3/TiO_2 for solar photocatalytic degradation of malathion pesticide. Catalysis Today .

Rani, M., Shanker, U., 2018. Advanced treatment technologies. Handbook of Environmental Materials Management . Available from: https://doi.org/10.1007/978-3-319-58538-3_33-1.

Rani, M., Shanker, U., Jassal, V., 2017. Recent strategies for removal and degradation of persistent and toxic organochlorine pesticides using nanoparticles: a review. Journal of Environmental Management . Available from: https://doi.org/10.1016/j.jenvman.2016.12.068.

Rani, L., Thapa, K., Kanojia, N., Sharma, N., Singh, S., Grewal, A.S., et al., 2021. An extensive review on the consequences of chemical pesticides on human health and environment. Journal of Cleaner Production 283, 124657.

Rhim, J.W., Park, H.M., Ha, C.S., 2013. Bio-nanocomposites for food packaging applications. Progress in Polymer Science . Available from: https://doi.org/10.1016/j.progpolymsci.2013.05.008.

Robertson, P.K.J., 1996. Semiconductor photocatalysis: an environmentally acceptable alternative production technique and effluent treatment process. Journal of Cleaner Production . Available from: https://doi.org/10.1016/S0959-6526(96)00044-3.

Rocha, G.M., Grisolia, C.K., 2019. Why pesticides with mutagenic, carcinogenic and reproductive risks are registered in Brazil. Developing World Bioethics . Available from: https://doi.org/10.1111/dewb.12211.

Saha, S., Kulshrestha, G., 2008. Hydrolysis kinetics of the sulfonylurea herbicide Sulfosulfuron. International Journal of Environmental Analytical Chemistry . Available from: https://doi.org/10.1080/03067310802124278.

Sahithya, K., Das, D., Das, N., 2015. Effective removal of dichlorvos from aqueous solution using biopolymer modified MMT-CuO composites: equilibrium, kinetic and thermodynamic studies. Journal of Molecular Liquids . Available from: https://doi.org/10.1016/j.molliq.2015.08.013.

Saifuddin, N., Nian, C.Y., Zhan, L.W., Ning, K.X., 2011. Chitosan-silver nanoparticles composite as point-of-use drinking water filtration system for household to remove pesticides in water. Asian Journal of Biochemistry . Available from: https://doi.org/10.3923/ajb.2011.142.159.

Sasakova, N., Gregova, G., Takacova, D., Mojzisova, J., Papajova, I., Venglovsky, J., et al., 2018. Pollution of surface and ground water by sources related to agricultural activities. Frontiers in Sustainable Food Systems . Available from: https://doi.org/10.3389/fsufs.2018.00042.

Senthilnathan, J., Philip, L., 2010. Photocatalytic degradation of lindane under UV and visible light using N-doped TiO2. Chemical Engineering Journal . Available from: https://doi.org/10.1016/j.cej.2010.04.034.

Shabbir, M., Singh, M., Maiti, S., Kumar, S., Saha, S.K., 2018. Removal enactment of organo-phosphorous pesticide using bacteria isolated from domestic sewage. Bioresource Technology . Available from: https://doi.org/10.1016/j.biortech.2018.04.122.

Sharma, V.K., Yngard, R.A., Lin, Y., 2009. Silver nanoparticles: green synthesis and their antimicrobial activities. Advances in Colloid and Interface Science . Available from: https://doi.org/10.1016/j.cis.2008.09.002.

Sharma, G., Kumar, A., Sharma, S., Naushad, M., Prakash Dwivedi, R., ALOthman, Z.A., et al., 2019. Novel development of nanoparticles to bimetallic nanoparticles and their composites: a review. Journal of King Saud University - Science . Available from: https://doi.org/10.1016/j.jksus.2017.06.012.

Shen, H.Y., Zhu, Y., Wen, X.E., Zhuang, Y.M., 2007. Preparation of Fe3O4-C18 nano-magnetic composite materials and their cleanup properties for organophosphorous pesticides. Analytical and Bioanalytical Chemistry . Available from: https://doi.org/10.1007/s00216-006-1082-1.

Shoiful, A., Fujita, H., Watanabe, I., Honda, K., 2013. Concentrations of organochlorine pesticides (OCPs) residues in foodstuffs collected from traditional markets in Indonesia. Chemosphere . Available from: https://doi.org/10.1016/j.chemosphere.2012.10.022.

Shoiful, A., Ueda, Y., Nugroho, R., Honda, K., 2016. Degradation of organochlorine pesticides (OCPs) in water by iron (Fe)-based materials. Journal of Water Process Engineering . Available from: https://doi.org/10.1016/j.jwpe.2016.02.011.

Singhal, R.K., Gangadhar, B., Basu, H., Manisha, V., Naidu, G.R.K., Reddy, A.V.R., 2012. Remediation of malathion contaminated soil using zero valent iron nano-particles. American Journal of Analytical Chemistry . Available from: https://doi.org/10.4236/ajac.2012.31011.

Šťastný, M., Štengl, V., Henych, J., Tolasz, J., Vomáčka, P., Ederer, J., 2016. Mesoporous manganese oxide for the degradation of organophosphates pesticides. Journal of Materials Science . Available from: https://doi.org/10.1007/s10853-015-9577-9.

Sud, D., Kaur, P., 2012. Heterogeneous photocatalytic degradation of selected organophosphate pesticides: a review. Critical Reviews in Environmental Science and Technology . Available from: https://doi.org/10.1080/10643389.2011.574184.

Sun, J., Covaci, A., Bustnes, J.O., Jaspers, V.L.B., Helander, B., Bårdsen, B.J., et al., 2020. Temporal trends of legacy organochlorines in different white-tailed eagle (*Haliaeetus albicilla*) subpopulations: a retrospective investigation using archived feathers. Environment International . Available from: https://doi.org/10.1016/j.envint.2020.105618.

Tang, Z., Huang, Q., Yang, Y., Zhu, X., Fu, H., 2013. Organochlorine pesticides in the lower reaches of Yangtze River: occurrence, ecological risk and temporal trends. Ecotoxicology and Environmental Safety . Available from: https://doi.org/10.1016/j.ecoenv.2012.10.001.

Tavakkoli, H., Yazdanbakhsh, M., 2013. Fabrication of two perovskite-type oxide nanoparticles as the new adsorbents in efficient removal of a pesticide from aqueous solutions: kinetic, thermodynamic, and adsorption studies. Microporous and Mesoporous Materials . Available from: https://doi.org/10.1016/j.micromeso.2013.03.043.

Tian, H., Li, J., Mu, Z., Li, L., Hao, Z., 2009a. Effect of pH on DDT degradation in aqueous solution using bimetallic Ni/Fe nanoparticles. Separation and Purification Technology . Available from: https://doi.org/10.1016/j.seppur.2008.11.018.

Tian, H., Li, J., Shen, Q., Wang, H., Hao, Z., Zou, L., et al., 2009b. Using shell-tunable mesoporous Fe3O4@HMS and magnetic separation to remove DDT from aqueous media. Journal of Hazardous Materials . Available from: https://doi.org/10.1016/j.jhazmat.2009.06.029.

Tomašević, A., Kiss, E., Petrović, S., Mijin, D., 2010. Study on the photocatalytic degradation of insecticide methomyl in water. Desalination . Available from: https://doi.org/10.1016/j.desal.2010.06.019.

Usman, U.L., Kumar Allam, B., Bahadur Singh, N., Banerjee, S., 2022. Adsorptive removal of Cr(VI) from wastewater by hexagonal boron nitride-magnetite nanocomposites: kinetics, mechanism and LCA analysis. Journal of Molecular Liquids 354, 118833. Available from: https://doi.org/10.1016/j.molliq.2022.118833.

Vasilić, Ž., Drevenkar, V., Rumenjak, V., Štengl, B., Fröbe, Z., 1992. Urinary excretion of diethylphosphorus metabolites in persons poisoned by quinalphos or chlorpyrifos. Archives of Environmental Contamination and Toxicology . Available from: https://doi.org/10.1007/BF00212552.

Vera, M., Nyanhongo, G.S., Guebitz, G.M., Rivas, B.L., 2020. Polymeric microspheres as support to co-immobilized Agaricus bisporus and *Trametes versicolor* laccases and their application in diazinon degradation. Arabian Journal of Chemistry . Available from: https://doi.org/10.1016/j.arabjc.2019.07.003.

Wang, B., Wang, Y., Wang, S., 2021. Improved water pollution index for determining spatiotemporal water quality dynamics: case study in the Erdao Songhua River Basin, China. Ecological Indicators . Available from: https://doi.org/10.1016/j.ecolind.2021.107931.

Wolfe, D.A., Champ, M.A., Cross, F.A., Kester, D.R., Park, P.K., Swanson, R.L., 1984. Marine pollution research facilities in the People's Republic of China. Marine Pollution Bulletin . Available from: https://doi.org/10.1016/0025-326X(84)90288-1.

Wu, Ren-Jang, et al., 2009. Titanium dioxide-mediated heterogeneous photocatalytic degradation of terbufos: Parameter study and reaction pathways. Journal of Hazardous Materials .

Xiaochu, C., Zhou, Q., Liu, F., Peng, Q., Bian, Y., 2021. Performance and kinetic of pesticide residues removal by microporous starch immobilized laccase in a combined adsorption and biotransformation process. Environmental Technology and Innovation . Available from: https://doi.org/10.1016/j.eti.2020.101235.

Yan, X.M., Shi, B.Y., Lu, J.J., Feng, C.H., Wang, D.S., Tang, H.X., 2008. Adsorption and desorption of atrazine on carbon nanotubes. Journal of Colloid and Interface Science . Available from: https://doi.org/10.1016/j.jcis.2008.01.047.

Yu, B., et al., 2007. Investigation of the photocatalytic degradation of organochlorine pesticides on a nano-TiO_2 coated film. Talanta .

Žabar, R., Komel, T., Fabjan, J., Kralj, M.B., Trebše, P., 2012. Photocatalytic degradation with immobilised TiO2 of three selected neonicotinoid insecticides: imidacloprid, thiamethoxam and clothianidin. Chemosphere . Available from: https://doi.org/10.1016/j.chemosphere.2012.04.039.

Zhou, J., Liu, R., Song, G., Zhang, M., 2009. Determination of carbamate and benzoylurea insecticides in peach juice drink by floated organic drop microextraction-high performance liquid chromatography. Analytical Letters . Available from: https://doi.org/10.1080/00032710903060677.

Chapter 30

Nitrogen cycle and its effect on phytoplankton community structure

Tiina Nõges[1], Mark J. McCarthy[1,2], Peter Nõges[1] and Rahul Kumar[1]
[1]*Chair of Hydrobiology and Fishery, Institute of Agricultural and Environmental Sciences, Estonian University of Life Sciences (EMÜ), Fr. R. Kreutzwaldi 1, Tartu, Estonia,* [2]*Department of Earth and Environmental Sciences, Wright State University, Dayton, OH, United States*

Introduction

One of the most significant biogeochemical cycles on Earth's surface is the nitrogen cycle, as all living beings are dependent on nitrogen (Lehnert et al., 2018). Nitrogen is the most bountiful component present in our atmosphere but less than 2% is available to the organisms (Galloway, 1998) and is of foremost significance as it is extremely critical for the existence of life. Nitrogen is found everywhere including plants, creatures, soil, water, and air. In addition to N_2 and NH_3, nitrogen exists in various forms, including organic nitrogen and inorganic nitrogen. Appropriately, nitrogen goes through different changes in the biological framework, beginning from one form to another form as living creatures use it for development, presence, and energy. The tremendous changes in nitrogen are nitrogen fixation, nitrification, denitrification, anammox, and ammonification (Fig. 30.1). The difference in nitrogen in its various oxidation states is a key to helpfulness in the biosphere and is significantly dependent upon microbial activities (Suddick and Davidson, 2012). Natural mechanisms are responsible for the availability and contribution of around 120 megatons of nitrogen each year to the biosphere (Smil, 2001), generally because of microorganism's nitrogen-fixing process in relationship with leguminous crops, such as alfalfa, clover, beans, peas, lentils, and lupins. Human beings have significantly expanded these cycles, utilizing the modern Haber-Bosch process, where N2 and H2 are changed into ammonia at very high temperature and pressure, following the chemical reaction: N2 + 3H2 → 2NH3. The Haber-Bosch process contributes fundamentally a comparative proportion of fixed nitrogen to the biosphere as do normal cycles, with around 100

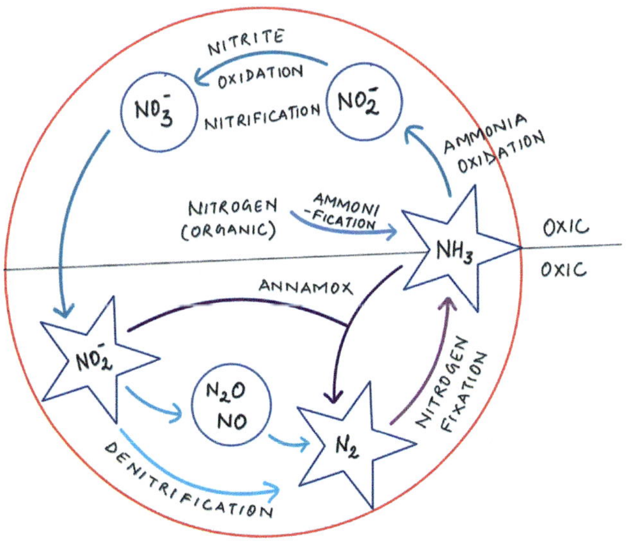

FIGURE 30.1 Significant changes in the nitrogen cycle. From: *Bernhard (2010).*

Role of Green Chemistry in Ecosystem Restoration to Achieve Environmental Sustainability. DOI: https://doi.org/10.1016/B978-0-443-15291-7.00029-8

million tons of nitrogen fertilizer each year (Fowler et al., 2013). This process is liable for around 2%−3% of worldwide petroleum product utilization (Smil, 2001). Reactive nitrogen (Nr) is made from N2 by three important and natural activities that include biological fixation of nitrogen, burning of biomass, and lightning (Fowler et al., 2013). Nitrogen is an enhancement of vital significance for the presence, development, and appropriate working of living beings including plants, animals, and human beings, and a principal part of food security (Hatfield and Follett, 2008). Humanity has looked for various methods of expanding harvest creation to give food to an enormous population for survival and growth. This has prompted the development and production of synthetic fertilizers subject to the Haber-Bosch process (Erisman et al., 2008). At this point, nitrogen gives many benefits to society, especially to farming and industry (Stoumann Jensen et al., 2011). Even though the current obsession with N_2 has a significant impact on food production, it has negative effects on the environment and human health. The current book chapter discusses the importance of the nitrogen cycle and its impact on phytoplankton community structure along various gradients including climate, catchment, and trophic status.

Origin and conversion of nitrogen in water

In freshwaters, nitrogen can be found in a variety of forms, including dissolved molecular N_2, a huge range of organic molecules ranging from amino acids and amines to proteins and refractory humic substances with low nitrogen content, ammonia (NH_4^+), nitrite (NO_2^-), and nitrate (NO_3^-) (Liu et al., 2011). Nitrogen comes from a variety of sources, including (1) precipitation that falls straightforwardly on the lake's surface, (2) fixation of nitrogen in the water and silt, and (3) drainage of surface water and groundwater. Nitrogen from precipitation and direct bulk (particulate) fallout varies dramatically depending on local climatic circumstances, wind patterns, and lake location about industrial and agricultural outputs. Nitrogen losses occur by the (1) outflow of the effluent from the basin, (2) decrease of NO_3^- to N_2 by bacterial denitrification with a resulting return of N_2 to the environment, and (3) never-ending sedimentation loss of organic and inorganic nitrogen-containing mixtures to the residue.

Nitrogen in precipitation and fallout

Nitrogen can enter a drainage basin in a variety of forms, including dissolved N_2, nitric acid, NH_4^+, and NO_3^-, as well as NH_4^+ adsorbed to inorganic particulate matter and organic molecules, which can be dissolved or particulate (Wetzel, 2001). The amount of nitrogen input per area of land or water has no direct relationship with the volume of rainfall or snowfall (Tahir et al., 2021). Dry fallout can contain up to ten times the amount of nutrients present in rainwater (Fish, 1976). Even though snow makes just a modest percentage of total nitrogen influx, its nitrogen content is often much larger than that of rain, and it can account for as much as half of a lake's annual nitrogen intake (Davidson, 2012). N_2-fixing bacteria are found in modest numbers in rainwater, but their contribution to overall nitrogen input is likely insignificant in contrast to other sources (Yano et al., 2017). Most of the combined nitrogen in the atmosphere is ammonia in non-polluted areas, with much of it coming from the decomposition of terrestrial organic matter (Bernhard, 2010). Dust-borne ammonia can be converted to nitrate in the atmosphere, resulting in precipitation containing both ammonia and nitrate (Aczel, 2019).

Fixation of molecular nitrogen

Although N_2 is not especially soluble in water, during periods of lake circulation, N_2 is frequently saturated in terms of surface temperature and pressure. Because of the higher solubility at cooler temperatures, the highest concentrations are found in the winter. The heating of the epilimnion reduces N_2 solubility during summer thermal stratification, whereas N_2 concentrations in the metalimnion and much of the hypolimnion may be slightly supersaturated maintaining a higher gas concentration despite the heating in these water strata (Giling et al., 2017). Reduction in the N_2 content in the lower hypolimnion over the sediments is most probably connected to the bacterial fixation of N_2 in productive lakes (Özkundakci et al., 2011).

Nitrogen fixation: cyanobacteria and other photosynthetic bacteria

The presence of heterocyst-bearing cyanobacteria in the open waters of lakes has been directly linked to nitrogen fixation (Kumar et al., 2010). Heterocysts are specialized cells that appear singly in most filamentous cyanobacteria, except the *Oscillatoriaceae*, and are the only source of nitrogen fixation in heterocyst-forming cyanobacteria growing aerobically (Kumar et al., 2010). Since the process requires reducing power and adenosine triphosphate (ATP), the two of which are delivered during the process of photosynthesis, nitrogen fixation is essentially light-dependent. Dinitrogenase reductase provides electrons for the reduction of N_2 in a highly endergonic reaction. Nitrogen-fixing photosynthetic bacteria can fix a small amount of N_2 in the dark, but only at a fraction of their highest daytime rates (Severin and Stal, 2008). In full sunlight, N_2 fixation is frequently inhibited at the surface reaches a maximum depth below the surface, and then declines rapidly, nearly exponentially, as depth and concomitant light attenuation increase (Piazena et al., 2002). Photosynthetic bacteria fix N_2 at the same time that they emit molecular H_2 via a non-cyclic electron flow caused by photophosphorylation. An external H-donor, such as thiosulfate, is the source of these electrons. The order Rhodospirillales contains three genera of photosynthetic bacteria that create bacteriochlorophyll and perform anoxygenic photosynthesis (Imhoff et al., 2019).

Nitrogen fixation: wetland sources

Lakes and streams are frequently flanked by dense stands of shrub-like plants that fix nitrogen from the atmosphere, in addition to nitrogen inputs from drainage discharge from terrestrial sources. Nearby lakes can be contaminated by leachate from direct leaf fall or released during the decomposition of foliage (Duan et al., 2014). N_2-fixation rates by cyanobacteria and heterotrophic bacteria can add considerable amounts of nitrogen to wetlands. Microbes associated with the aerobic layer of wetland HydroSoils, anaerobic regions such as the rhizosphere of non-nodulated plants, and floating mats of particulate detritus and related microorganisms all fix nitrogen to wetlands (Wetzel, 2001). On the flood plain of a tropical riverine lake, nitrogen fixation by associated cyanobacteria and heterotrophic bacteria produced a considerable but small (5%–10%) share of total nitrogen imports (Doyle and Fisher, 1994). Azolla is a small (1–3 cm) aquatic fern that floats freely on the water surface of many natural and manmade wetlands across the world (Wang et al., 2019). *Anabaena azollae*, a symbiotic heterocyst-forming, N_2-fixing cyanobacterium, is found in all Azolla species (Peters and Meeks, 1989).

Inorganic and organic nitrogen

In addition to atmospheric nitrogen in the form of precipitation, dry fallout, and N_2 fixation, influents, both from surface land drainage and groundwater sources are a key source of nitrogen income to lakes (Højberg et al., 2017). Inorganic dissolved and organic (both dissolved and particulate) chemicals in effluents moving out of lake basins are a noticeable loss (Liu et al., 2021). Volatilization of chemicals from the water surface, such as ammonia at high pH and N_2 produced during microbial denitrification of nitrate, can also cause nitrogen loss (Aczel, 2019).

Ammonia

The biological dissimilation of nitrate produces ammonia. However, much of the ammonia produced by heterotrophic bacteria during the decomposition of organic material comes from the breakdown of proteins, amino acids, urea, and other organic molecules that contain nitrogen (Tarazona and Ramos-Peralonso, 2014).

Nitrification

The nitrification process changes ammonia to nitrite and afterward to nitrate. In the global nitrogen cycle, it is the second most significant stage. Nitrification is usually carried out by prokaryotic organisms in the presence of oxygen or under aerobic circumstances. The ammonia-oxidizers first convert NH_4^+ to NO_2^-, thereafter convert into NO_3^-. The nitrite-oxidizing bacteria of the genera *Nitrospira*, *Nitrobacter*, and *Nitrococcus* have performed this process.

$$NH_3 + O_2 + 2e^- \rightarrow NH_2OH + H_2O$$

$$NH_2OH + H_2O \rightarrow NO_2{}^- + 5H^+ + 4e^-$$

Nitrification is inhibited by certain dissolved organic compounds, like tannins (Hockenbury and Grady, 1977).

Nitrate reduction and denitrification

Nitrate is broken down by the enzyme nitrate reductase into nitrite during denitrification, which is a microbial-dependent process. Nitric oxide is converted into nitrous oxide by nitric oxide reductase. Lastly, dinitrogen is produced when nitrous oxide is broken down by nitrous oxide reductase (Vázquez-Torres and Bäumler, 2016). A redox reaction can be used to describe the entire denitrification process:

$$2NO_3^- + 10e^- + 12H^+ \rightarrow N_2 + 6H_2O$$

Because some species (such as *Escherichia coli*) only generate nitrate reductase, they can only complete the initial reduction, resulting in nitrite buildup. Others, such as *Paracoccus denitrificans* and *Pseudomonas stutzeri*, eliminate nitrate. Since certain intermediates of denitrification are important ozone-harming substances that join with the sun and ozone gas to shape nitric acid, a part of acidic rain or precipitation, complete denitrification is an environmentally critical interaction.

Dissolved organic nitrogen

Broken-up organic nitrogen compounds in lakes and streams come from photosynthetic organic entities and creature discharge of nitrogenous waste, although organic nitrogen can also come through soil leachate, sewage discharge, and atmospheric deposition (Anderson et al., 2002). Proteins, free amino acids, amino sugars, and nucleic acids (RNA and DNA) are the crucial sources of dissolved organic nitrogen (DON) compounds available in freshwater, yet discharge items like urea and methylamines are additionally normal (Glibert et al., 2005a,b). Furthermore, several DON compounds have been discovered in humic materials dissolved in freshwater, while their chemical structure is unknown to the scientific community. Bacteria and some microalgae benefit from DON as these microorganisms use it as a source of nitrogen and carbon.

Carbon-to-nitrogen ratios

Organic matter has a carbon concentration that is at least one order of magnitude larger than nitrogen. Decomposition and mineralization of complex combinations of organic molecules in particulate and dissolved phases result in inorganic carbon, principally CO_2 and inorganic nitrogen compounds. Fungi and bacteria remove proportionately more nitrogen than carbon through their proteolytic metabolism. The pace of decomposition slows as the leftover organic components become more chemically resistant, and the selective removal of nitrogen by bacteria results in a net increase in the C:N ratios. In lakes, there are considerable seasonal and regional changes in particle organic matter, however, most of the organic matter is in the dissolved fraction, and dissolved organic matter concentrations are extremely consistent. Lakes can be classified according to the amount of organic carbon and dissolved organic matter they contain. With increased organic carbon content, there is a distinct drop in organic nitrogen, resulting in higher C:N ratios. Wetland and littoral material dominate the dissolved organic matter pool of lakes, streams, and particularly bogs and bog lakes. Before and during delivery to the lake, organic matter from terrestrial and marsh areas undergoes variable degrees of decomposition, during which much of the organic nitrogen can be consumed. As a result, allochthonous organic matter includes about 6% crude protein and a C:N ratio of 45:1 to 50:1 in general. The dissolved organic matter comprises a large percentage of humic acid molecules with low nitrogen content, which gives the water a stained brown appearance (Schnitzer and Khan, 1972).

Nitrogen as a nutrient in freshwater ecosystem

Nitrogen plays a crucial role in freshwater ecosystems and is a crucial supplement that is required to support the essential production of green algae and higher plants. Nitrogen likewise accepts a section in declining the food web structure and relative usefulness of a water body through microbial, algal, and nitrogen usage by plants such as DIN and DON. Any change in the rate at which nitrogen is delivered to an aquatic body will result in changes in the productivity of the water body and its microbial metabolism. These changes will also have discretionary effects on the local species organization and relative abundance of microbial, plant, and animal species, as well as the structure and balance of the amphibian food web.

Consequences of increased nitrogen loads in lakes

Toxic effects

Particularly two types of Nr to be specific nitrite and ammonia were demonstrated to be poisonous to various living entities (Lucassen et al., 2003). In the aquatic environment, poisonous impacts of NO_2^- and NH_4^+ can be seen all through the food chain. Poisonous impacts are essentially seen as brief-time frame reactions to high NO_2^- and NH_4^+ loads. However, life forms are being exposed to unnaturally high NO_2^- and NH_4^+ concentrations for an extended period due to the persistent environmental deposition of reactive nitrogen in a water body.

Water acidification

In undisturbed lake catchments, most nitrogen transfer is in the natural or organic nitrogen form (Kortelainen et al., 2006). When there are no direct catchment sources, a higher concentration of nitrate can indicate anthropogenic nitrogen deposition in undeveloped lake catchments (intensive agriculture, factory, or metropolitan regions). Information gathered from previous research works proposes that NO_3^- concentrations in acid-delicate lakes enhanced from their exceptionally low levels during the 1970s and 1980s (Henriksen and Brakke, 1988). Unfortunate results of acid deposition can occur at the primary producers' level (Lewandowska et al., 2012).

Species loss

In an aquatic ecosystem, a relationship between species' lavishness and nitrogen load could be taken into account. In a tremendous report on 60 significant water bodies (freshwater lakes) that were situated in Poland and the United Kingdom, James et al. (2005) recorded an association between the lavishness of reduced species of macrophyte and enhanced loads of nitrate. The primary driver behind the downfall of species can incorporate the advancement of nitrophilous (species with a high affinity to nitrogen), which can appropriately outcompete different species (Stevens et al., 2004).

Sensitivity to other stressors

Increasing the nitrogen loads in water bodies can enhance the number of stressors or expand the vulnerability to the existing stressors (Pardo et al., 2011). By enhancing the amount of one supplement in an environment, while leftover supplements stay unaffected, the extent between supplements changes, resulting in far-reaching ramifications for biological systems. The phytoplankton's development and growth depend on the essential nutrients, that is, nitrogen and phosphorus whereas, Si is a significant supplement for diatoms and some *Chrysophyceae* species. Although the primary or basic level producers are fundamentally versatile concerning supplement proportions, there are ideal circumstances for development. The purported Redfield Ratio describes the best molar essential stoichiometry of phytoplankton as C:N:P = 106:16:1.

Possible effects of nitrogen enrichment on phytoplankton community structure

It is of paramount importance to understand the procedures that affect the variety and abundance of phytoplankton, with respect to the progression of phytoplankton species (Smayda, 1980) and futile fracas (Tett et al., 2007). According to Howarth and Marino (2006), nitrogen—such as nitrate, nitrite, and ammonia—has traditionally been regarded as the essential supplemental component for the growth, development, and biomass of phytoplankton in water. As a result, it may play a significant role in determining the structure and diversity of phytoplankton communities: alterations in concentration (Gowen et al., 2012), the degree to which various nutrients are utilized, and the variety of chemical structures (Davidson et al., 2012) of nitrogen can alter the structure and diversity of phytoplankton communities (Gowen et al., 2012; Glibert et al., 2016), and possibly spread the growth and development of dangerous and poisonous species (Glibert et al., 2005a,b) or altering the flux of carbon exports. The expanding proportion of supplements (nitrogen and phosphorus) input may lead to an alteration of the phytoplankton community structure (Zhu et al., 2013).

Variable responses of phytoplankton to nitrogen addition in lakes

An expanding nitrogen supply is of organic significance since it prompts changes in the accessible supplement proportions, and in the long run, to excess nitrogen conditions and phosphorus insufficiency in aquatic and terrestrial ecosystems (Elser et al., 2009a,b). Nitrogen load enhancement is the most serious concern for biological systems or ecosystems in which the basic level producers are restricted by nitrogen, and accordingly, outcomes of enhancing biomass and primary production might emerge through additional fertilization (Elser et al., 2009b; Hessen, 2013). In addition, autotrophs, like phytoplankton, have the adaptability to adjust their elemental (stoichiometric) structure as per the encompassing conditions (Klausmeier et al., 2004). Recently, it was demonstrated that for prevalently nitrogen-confined freshwater bodies, in an inclination of atmospheric nitrogen deposition and the greater availability of dissolved N:P extents can increase the amount of seston biomass N:P ratios (Elser et al., 2009a; Hessen, 2013).

Effects of nitrate and ammonium supply rates on phytoplankton community dynamics

It has concurred that the management of nutrients should incorporate nitrate and ammonia control (Lewis et al., 2011) displays an exceptionally significant role in terms of bioavailability primary production (Peierls and Paerl, 1997). The consideration of both the nitrogen (NO_3^- and NH_4^+) compounds seems to be clear in fundamentally nitrogen-restricted frameworks, where nitrogen is inadequately accessible and since nitrogen is attempted to be globally an essential limiting macronutrient (Elser et al., 2007). Nonetheless, notwithstanding supplement limitation conditions likewise according to the energetical perspective qualitative contrasts in the supply of NO_3^- and NH_4^+ compounds ought to affect essential production in biological systems. NH4 + can be taken up straight by basic producers and doesn't require costly catalyst action of NO_3^- reduction (Falkowski, 1975). In addition, phytoplankton physiological changes can stifle the NO_3^- reductase synthesis under higher concentrations ($>0.5-1$ μmol/L) of NH_4^+ (Eppley et al., 1969).

Effects of ammonium on algae and their transfer to higher trophic levels

The consistent contribution of ammonium can aggregate and cause an enduring addition of ammonium focuses in environments (Glibert et al., 2014). Bergström and Jansson (2006) showed a connection between the dissolved inorganic nitrogen (DIN) load of a lake and the measure of environmental Nr deposition. A consistent contribution of nitrogen will consequently result in enhanced exposure of organisms living in water to high concentrations of ammonia (Glibert et al., 2016; Ip et al., 2005; Glibert et al., 2016). Adverse consequences of a huge concentration of ammonium on algal growth were regularly noticed (Collos and Harrison, 2014). Dai, Shang and Qiu (2012) analyzed the reaction of various species of algae to rising ammonium concentrations (NH_4^+). They determined the EC50 value for 18 species of algae addressing the concentrations of ammonia for which a decline of the functional PSII quantum yield by half contrasted with the control was noticed. These EC50 values changed to a great extent between algae, beginning from 0.27 mM NH_4^+ for *Tabellaria* sp. reaching up to the highest dose of 18.25 mM for *Chlamydomonas microsphaera*. However, ammonia (NH_4^+) may have indirect impacts on consumers through base-up driven impacts when their feeding pattern was disclosed to a raised quantity of NH_4^+.

Effects of nitrogen and phosphorus on phytoplankton biomass and composition

Supplement enrichment typically invigorates the growth and development of phytoplankton in lakes. A few phytoplankton genera including *Microcystis*, *Anabaena*, *Nostoc*, and *Aphanizomenon*, usually break out and stand steadily, prompting issues with hypoxia, poisons, and changes in the biological community structure (Chen et al., 2008). Understanding the connections between the concentration of supplements and the biomass of algae is significant in the endeavors for the administration or management of eutrophication. Phosphorus and Nitrogen are regularly treated as the chief restricting supplements for the growth and production of freshwater algae because of their short stockpile compared to cell growth requirements. Sakamoto (1966) recommended that the biomass of phytoplankton was reliant upon total phosphorus when TN:TP (Total Nitrogen: Total Phosphorus) by weight was >17, on total (absolute) nitrogen (TN) when TN:TP was <10 and both on TN and TP when the proportion of TN to TP was within the limit of 10–17. The significance of TN:TP proportion for the overall extent of cyanobacteria in phytoplankton where TN:TP by weight was <29 has been underlined by Smith and Kalff (1982). Paerl et al. (2001a,b) suggested that the "N:P rule" was not appropriate for highly eutrophic systems especially when the loadings of N and P surpass the assimilative limit of the phytoplankton. Extensive analysis of the Yangtze shallow lakes showed that the N:P proportions shifted significantly among various lakes that are eutrophic (Wang et al., 2008). In this manner, they recommended that the TN:TP proportion was not a suitable index to decide if nitrogen or phosphorus is the restricting supplement in a lake. Lv et al. (2011) performed a research activity to examine the phytoplankton community structure and biomass in 15 metropolitan lakes that were eutrophic alongside the impacts of primary environmental variables, including the concentration of nutrients and the proportion of nitrogen to phosphorus on the phytoplankton community structure. Due to high convergences of total phosphorus and total nitrogen, the research group noted that many metropolitan lakes were hypertrophic with average concentrations of 490 and 5380 mg/m^3. The phytoplankton community was overwhelmed by *Microcystis aeruginosa* and *Euglena caudate* during the summer season and *Cryptomonas ovata* and *Cyclotella meneghiniana* during the winter season. During the summer season, the phytoplankton's average biomass reached up to 456.87 mg/L in the summer season and the yearly level was 189.24 mg/L.

Effects of catchment land use and trophic state on phytoplankton composition

Rising stress on water assets through human actions has brought about water quality degradation in many regions across the globe (Vörösmarty et al., 2000). Low water quality is regularly connected with an enhanced trophic state and the existence of toxic algal blooms (Paerl et al., 2001a,b). The trophic status of a lentic water body can be evaluated from the phytoplankton efficiency (Carlson, 1977; Bellinger and Sigee, 2010). Trophic status indicators have been formed at the phytoplankton species level (Reynolds, 1990) and the levels of phyla and order (Interlandi and Kilham, 2001). Phytoplankton structure is impacted not just by the trophic condition of an aquatic body but also by different variables (Reynolds, 1998), like mixing regime (Ptacnik et al., 2008; Ryan et al., 2006), while species variation and abundance give extra measurements that have been related to the trophic level 2006, temperature (Paerl and Huisman, 2008; Brookes and Carey, 2011), depth of water (Jeppesen et al., 2005), euphotic depth to mixing depth ratio, amount of dissolved oxygen (Dokulil and Teubner, 2000) and proportions of macronutrients (Hamilton et al., 2006). The trophic condition of lakes has been associated with the usage of catchment land area however, direct connections between phytoplankton taxa and land use are very limited. Paul et al. (2012) estimated the composition of phytoplankton, addressed by the relative cell richness of phyla for 11 lakes located in New Zealand over 4 years. Paul et al. (2012) observed that trophic level was negatively correlated with local forest region whereas, positively correlated with pasture land and urban region.

Conclusion

The nitrogen cycle is loaded with complex chemistry, surprising enzymes, multimer models, multi-component dynamic locales, and advantageous reactions—science that excites and astounds simultaneously. Most studies uncovered the significance of mixotroph species for noticed impacts of raising nitrogen loads and changes in nitrogen source proportions. Hence a significant impact of changing atmospheric nitrogen input is the support of the microbial loop. A few studies recommended that the enrichment of nitrogen additionally influences phytoplankton dynamics in lakes with an assumed shortage of phosphorus. The primary initiator of the dynamics noticed was ammonia. Higher ammonia proportions might expand the concentration of chlorophyll-a and influence the growth and development of specific phytoplankton groups, along with the phytoplankton resources use effectiveness. Such a change might also promote shifts towards higher extents of green algae, particularly flagellated species.

References

Aczel, M., 2019. What is the nitrogen cycle and why is it key to life? Frontiers for Young Minds 7, 41. Available from: https://doi.org/10.3389/frym.2019.00041.

Anderson, D., Glibert, P., Burkholder, J., 2002. Harmful algal blooms and eutrophication: nutrient sources, composition, and consequences. Estuaries 25, 704–726. Available from: https://doi.org/10.1007/BF02804901.

Bellinger, E.G., Sigee, D.C., 2010. Freshwater Algae: Identification and Use as Bioindicators. *John Wiley & Sons*, Chichester, UK.

Bergström, A.K., Jansson, M., 2006. Atmospheric nitrogen deposition has caused nitrogen enrichment and eutrophication of lakes in the northern hemisphere. Global Change Biology 12 (4), 635–643.

Bernhard, A., 2010. The nitrogen cycle: processes, players, and human impact. Nature Education Knowledge 3 (10), 1–25.

Brookes, J.D., Carey, C.C., 2011. Resilience to blooms. Science (New York, N.Y.) 334 (46), 46–47.

Carlson, R.E., 1977. A trophic state index for lakes. Limnology and Oceanography 22, 361–369.

Chen, W., Song, L.R., Peng, L., Wan, N., Zhang, X.M., Gan, N.Q., 2008. Reduction in microcystin concentrations in large and shallow lakes: water and sediment-interface contributions. Water Research 42, 763–773.

Collos, Y., Harrison, P.J., 2014. Acclimation and toxicity of high ammonium concentrations to unicellular algae. Marine Pollution Bulletin 80 (1–2), 8–23. Available from: https://doi.org/10.1016/j.marpolbul.2014.01.006.

Dai, G.Z., Shang, J.L., Qiu, B.S., 2012. Ammonia may play an important role in the succession of cyanobacterial blooms and the distribution of common algal species in shallow freshwater lakes. Global Change Biology 18 (5), 1571–1581.

Davidson, E., 2012. The role of nitrogen in climate change and the impacts of nitrogen-climate interactions on terrestrial and aquatic ecosystems, agriculture, and human health in the United States. *97th ESA Annual Convention*.

Davidson, E.A., Townsend, A.R., Suddick, E.C., 2012. The Role of Nitrogen in Climate Change and the Impacts of Nitrogen-Climate Interactions on Terrestrial and Aquatic Ecosystems, Agriculture, and Human Health in the United States: A Technical Report Submitted to the US National

Climate Assessment. North American Nitrogen Center of the International Nitrogen Initiative (NANC-INI), Woods Hole Research Center, Falmouth, USA.

Dokulil, M.T., Teubner, K., 2000. Cyanobacterial dominance in lakes. Hydrobiologia 438, 1–12.

Doyle, R.D., Fisher, T.R., 1994. Nitrogen fixation by periphyton and plankton on the Amazon floodplain at Lake Calado. Biogeochemistry 26, 41–66. Available from: https://doi.org/10.1007/BF02180403.

Duan, S., Delaney-Newcomb, K., Kaushal, S.S., Findlay, S.E.G., Belt, K.T., 2014. Potential effects of leaf litter on water quality in urban watersheds. Biogeochemistry 121, 61–80. Available from: http://www.jstor.org/stable/24717568.

Elser, J.J., Bracken, M.E.S., Cleland, E.E., Gruner, D.S., Harpole, W.S., Hillebrand, H., et al., 2007. Global analysis of nitrogen and phosphorus limitation of primary producers in freshwater, marine and terrestrial ecosystems. Ecology Letters 10 (12), 1135–1142.

Elser, J.J., Andersen, T., Baron, J.S., Bergström, A.K., Jansson, M., Kyle, M., et al., 2009a. Shifts in lake N:P stoichiometry and nutrient limitation driven by atmospheric nitrogen deposition. Science (New York, N.Y.) 326 (5954), 835–837.

Elser, J.J., Kyle, M., Steger, L., Nydick, K.R., Baron, J.S., 2009b. Nutrient availability and phytoplankton nutrient limitation across a gradient of atmospheric nitrogen deposition. Ecology 90 (11), 3062–3073.

Eppley, R.W., Coatsworth, J.L., Solorzano, L., 1969. Studies of nitrate reductase in marine phytoplankton. Limnology and Oceanography 14, 194–205.

Erisman, J., Sutton, M., Galloway, J., Klimont, Z., Winiwarter, W., 2008. How a century of ammonia synthesis changed the world. Nature Geosciences 1, 636–639. Available from: https://doi.org/10.1038/ngeo325.

Falkowski, P.G., 1975. Nitrate uptake in marine phytoplankton: comparison constants from seven species. Limnology and Oceanography 20, 412–417.

Fish, G.R., 1976. The fallout of nitrogen and phosphorus compounds from the atmosphere at Ngapuna, near Rotorua, New Zealand. Journal of Hydrology (New Zealand) 15 (1), 27–34.

Fowler, D., Coyle, M., Skiba, U., Sutton, M.A., Cape, J.N., Reis, S., et al., 2013. The global nitrogen cycle in the twenty-first century. Philosophical Transactions of the Royal Society of London. Series B, Biological Sciences 368 (1621), 20130164. Available from: https://doi.org/10.1098/rstb.2013.0164.

Galloway, J.N., 1998. The global nitrogen cycle: changes and consequences. Environmental Pollution 102 (1), 15–24. Available from: https://doi.org/10.1016/S0269-7491(98)80010-9.

Giling, D., Staehr, P., Grossart, H.P., Andersen, M., Boehrer, B., Escot, C., et al., 2017. Delving deeper: metabolic processes in the metalimnion of stratified lakes: Metalimnetic metabolism in lakes. Limnology and Oceanography 62, 1–19. Available from: https://doi.org/10.1002/lno.10504.

Glibert, P., Seitzinger, S., Heil, C., Burkholder, J., Parrow, M., Codispoti, L., et al., 2005a. The role of eutrophication in the global proliferation of harmful algal blooms. Oceanography 18, 198–209.

Glibert, P.M., Anderson, D.M., Gentien, P., Granéli, E., Sellner, K.G., 2005b. The global, complex phenomena of harmful algal blooms. Oceanography 18 (2), 136–147. Available from: https://doi.org/10.5670/oceanog.2005.49.

Glibert, P.M., Hinkle, D.C., Sturgis, B., Jesien, R.V., 2014. Eutrophication of a Maryland/Virginia coastal lagoon: a tipping point, ecosystem changes, and potential causes. Estuaries and Coasts 37 (1), 128–146.

Glibert, P.M., Wilkerson, F.P., Dugdale, R.C., Raven, J.A., Dupont, C.L., Leavitt, R.R., et al., 2016. Pluses and minuses of ammonium and nitrate uptake and assimilation by phytoplankton and implications for productivity and community composition, with emphasis on nitrogen-enriched conditions. Limnology and Oceanography 61, 165–197. Available from: https://doi.org/10.1002/lno.10203.

Gowen, R.J., Tett, P., Bresnan, E., Davidson, K., McKinney, A., Harrison, P.J., et al., 2012. Anthropogenic nutrient enrichment and blooms of harmful phytoplankton. In: Barnes, M. (Ed.), Oceanography and Marine Biology, Oceanography and Marine Biology—An Annual Review. CRC Press, Boca Raton, pp. 65–126.

Hamilton, D.P., Douglas, G.B., Adeney, J.A., Radke, L.C., 2006. Seasonal changes in major ions, nutrients and Chlorophyll-α at two sites in the Swan River estuary, Western Australia. Marine & Freshwater Research 57, 803–815.

Hatfield, J.L., Follett, R.F., 2008. Nitrogen in the Environment: Sources, Problems, and Management. Elsevier, Amsterdam, The Netherlands, pp.

Henriksen, A., Brakke, D.F., 1988. Increasing contributions of nitrogen to the acidity of surface waters in Norway. Water, Air, and Soil Pollution 42, 183–201.

Hessen, D.O., 2013. Inorganic nitrogen deposition and its impacts on N:P-ratios and lake productivity. Water 5, 327–341.

Hockenbury, M.R., Grady, C.P.L., 1977. Inhibition of nitrification-effects of selected organic compounds. Journal of Water Pollution Control Federation 49 (5), 768–777.

Højberg, A.L., Hansen, A.L., Wachniew, P., Żurek, A.J., Virtanen, S., Arustiene, J., et al., 2017. Review and assessment of nitrate reduction in groundwater in the Baltic Sea Basin. Journal of Hydrology: Regional Studies 12, 50–68. Available from: https://doi.org/10.1016/j.ejrh.2017.04.001.

Howarth, R.W., Marino, R., 2006. Nitrogen as the limiting nutrient for eutrophication in coastal marine ecosystems: evolving views over three decades. Limnology and Oceanography 51, 364–376.

Imhoff, J.F., Rahn, T., Künzel, S., Neulinger, S.C., 2019. Phylogeny of anoxygenic photosynthesis based on sequences of photosynthetic reaction center proteins and a key enzyme in bacteriochlorophyll biosynthesis, the chlorophyllide reductase. Microorganisms 7, 576. Available from: https://doi.org/10.3390/microorganisms7110576.

Interlandi, S.J., Kilham, S.S., 2001. Limiting resources and the regulation of diversity in phytoplankton. Ecology 82, 1270–1282.

Ip, Y.K., Leong, M.W., Sim, M.Y., Goh, G.S., Wong, W.P., Chew, S.F., 2005. Chronic and acute ammonia toxicity in mudskippers, *Periophthalmodon schlosseri* and *Boleophthalmus boddaerti*: brain ammonia and glutamine contents, and effects of methionine sulfoximine and MK801. Journal of Experimental Biology 208 (10), 1993–2004.

James, C., Fisher, J., Russell, V., Collings, S., Moss, B., 2005. Nitrate availability and hydrophyte species richness in shallow lakes. Freshwater Biology 50, 1049–1063. Available from: https://doi.org/10.1111/j.1365-2427.2005.01375.x.

Jeppesen, E., Søndergaard, M., Jensen, J.P., Havens, K.E., Anneville, O., Carvalho, L., et al., 2005. Lake responses to reduced nutrient loading – an analysis of contemporary long-term data from 35 case studies. Freshwater Biology 50, 1747–1771.

Klausmeier, C.A., Litchman, E., Daufresne, T., Levin, S.A., 2004. Optimal nitrogen-to-phosphorus stoichiometry of phytoplankton. Nature 429, 171–174.

Kortelainen, P., Mattsson, T., Finer, L., Ahtiainen, M., Saukkonen, S., Sallantaus, T., 2006. Controls on the export of C, N, P and Fe from undisturbed boreal catchments, Finland. Aquatic Sciences – Research Across Boundaries 68, 453–468. Available from: https://doi.org/10.1007/s00027-006-0833-6.

Kumar, K., Mella-Herrera, R.A., Golden, J.W., 2010. Cyanobacterial heterocysts. Cold Spring Harbor Perspectives in Biology 2 (4), a000315. Available from: https://doi.org/10.1101/cshperspect.a000315.

Lehnert, N., Dong, H.T., Harland, J.B., Hunt, A.P., White, C.J., 2018. Reversing nitrogen fixation. Nature Reviews Chemistry 2, 278–289. Available from: https://doi.org/10.1038/s41570-018-0041-7.

Lewandowska, A.M., Breithaupt, P., Hillebrand, H., Hoppe, H.G., Jürgens, K., Sommer, U., 2012. Responses of primary productivity to increased temperature and phytoplankton diversity. Journal of Sea Research 72, 87–93. Available from: https://doi.org/10.1016/j.seares.2011.10.003.

Lewis, W.M., Wurtsbaugh, W.A., Paerl, H.W., 2011. Rationale for control of anthropogenic nitrogen and phosphorus to reduce eutrophication of inland waters. Environmental Science & Technology 45, 10300–10305.

Liu, H., Joonseon, J., Holly, G., Scott, S., David, S., 2011. Algal uptake of hydrophobic and hydrophilic dissolved organic nitrogen in effluent from biological nutrient removal municipal wastewater treatment systems. Environmental Science & Technology 46, 713–721. Available from: https://doi.org/10.1021/es203085y.

Liu, K., Xiao, X., Zhang, D., Ding, Y., Li, L., Zhao, M., 2021. Quantitative estimates of organic carbon contributions to the river-estuary-marine system in the Jiaozhou Bay, China. Ecological Indicators 129, 107929. Available from: https://doi.org/10.1016/j.ecolind.2021.107929.

Lucassen, E.C.H.E., Bobbink, R., Smolders, A.J.P., Van der Ven, P.J.M., Lamers, L.P.M., Roelofs, J.G.M., 2003. Interactive effects of low pH and high ammonium levels responsible for the decline of *Cirsium dissectum* (L.) Hill. Plant Ecology 165, 45–52.

Lv, J., Wun, H., Chen, M., 2011. Effects of nitrogen and phosphorus on phytoplankton composition and biomass in 15 subtropical, urban shallow lakes in Wuhan, China. Limnologica 41, 48–56. Available from: https://doi.org/10.1016/j.limno.2010.03.003.

Özkundakci, D., Hamilton, D., Gibbs, M., 2011. Hypolimnetic phosphorus and nitrogen dynamics in a small, eutrophic lake with a seasonally anoxic hypolimnion. Hydrobiologia 661, 5–20. Available from: https://doi.org/10.1007/s10750-010-0358-9.

Paerl, H., Fulton, R., Moisander, P., Dyble, J., 2001a. Harmful freshwater algal blooms with an emphasis on cyanobacteria. The Scientific World 1, 76–113. Available from: https://doi.org/10.1100/tsw.2001.16.

Paerl, H.W., Fulton, R.S., Moisander, P.H., Dyble, J., 2001b. Harmful freshwater algal blooms, with an emphasis on cyanobacterial. Scientific World Journal 1, 76–113.

Paerl, H.W., Huisman, J., 2008. Blooms like it hot. Science (New York, N.Y.) 320, 57–58.

Pardo, L.H., Robin-Abbott, M.J., Driscoll, C.T., 2011. Assessment of nitrogen deposition effects and empirical critical loads of nitrogen for ecoregions of the United States. Gen. Tech. Rep. NRS-80. U.S. Department of Agriculture, Forest Service, Northern Research Station, Newtown Square, PA, p. 291. Available from: https://doi.org/10.2737/NRS-GTR-80.

Paul, W.J., Hamilton, D.P., Ostrovsky, I., Miller, S.D., Zhang, A., Muraoka, K., 2012. Catchment land use and trophic state impacts on phytoplankton composition: a case study from the Rotorua lakes' district, New Zealand. Hydrobiologia 698, 133–146. Available from: https://doi.org/10.1007/s10750-012-1147-4.

Peierls, B.L., Paerl, H.W., 1997. Bioavailability of atmospheric organic nitrogen deposition to coastal phytoplankton. Limnology and Ocenography 42 (8), 1819–1823. Available from: https://doi.org/10.4319/lo.1997.42.8.1819.

Peters, G.A., Meeks, J.C., 1989. The Azolla-Anabaena symbiosis: basic biology. Annual Review of Plant Physiology and Plant Molecular Biology 40 (1), 193–210.

Piazena, H., Perez-Rodriguez, E., Häder, D., Figueroa, F.L., 2002. Penetration of solar radiation into the water column of the central subtropical Atlantic Ocean - optical properties and possible biological consequences. Deep Sea Research Part II: Topical Studies in Oceanography 49, 3513–3528. Available from: https://doi.org/10.1016/S0967-0645(02)00093-0.

Ptacnik, R., Lepisto, L., Willen, E., Brettum, P., Andersen, T., Rekolainen, S., et al., 2008. Quantitative responses of lake phytoplankton to eutrophication in Northern Europe. Aquatic Ecology 42, 227–236.

Reynolds, C.S., 1990. Temporal scales of variability in pelagic environments and the response of phytoplankton. Freshwater Biology 23, 25–53.

Reynolds, C.S., 1998. What factors influence the species composition of phytoplankton in lakes of different trophic status? Hydrobiologia 369, 11–26. Available from: https://doi.org/10.1023/A:1017062213207.

Ryan, E.F., Duggan, I.C., Hamilton, D.P., Burger, D.F., 2006. Phytoplankton assemblages in North Island lakes in New Zealand: is trophic state, mixing, or light climate more important? New Zealand Journal of Marine and Freshwater Research 40, 389–398.

Sakamoto, M., 1966. Primary production by phytoplankton community in some Japanese lakes and its dependence on lake depth. Archiv fur Hydrobiologie 62, 1–28.

Schnitzer, M., Khan, S.U., 1972. Humic Substances in the Environment. Marcel Dekker, Inc, New York.

Severin, I., Stal, L., 2008. Light dependency of nitrogen fixation in a coastal cyanobacterial mat. ISME Journal 2, 1077–1088. Available from: https://doi.org/10.1038/ismej.2008.63.

Smayda, T.J., 1980. Phytoplankton species succession. In: Morris, I. (Ed.), The Physiological Ecology of Phytoplankton. Blackwell, Oxford, pp. 493–570.

Smil, V., 2001. Enriching the Earth: Fritz Haber, *Carl Bosch, and the Transformation of World Food Production'*. The MIT Press, Cambridge, USA.

Smith, R.E.H., Kalff, J., 1982. Size-dependent phosphorus uptake kinetics and cell quota in phytoplankton. Journal of Phycology 18, 275–284.

Stevens, C.J., Dise, N.B., Mountford, J.O., Gowing, D.J., 2004. Impact of nitrogen deposition on the species richness of grasslands. Science (New York, N.Y.) 303 (5665), 1876–1879. Available from: https://doi.org/10.1126/science.1094678.

Stoumann Jensen, L., Schjoerring, J.K., van der Hoek, K.W., Dammgaard-Poulsen, H., Zevenbergen, J.F., Pallière, C., et al., 2011. Benefits of nitrogen for food, fibre and industrial production. In: Sutton, M.A., Howard, C.M., Erismans, J.W., Billen, G., Bleeker, A., Grennfelt, P., van Grinsven, H., Grizzetti, B. (Eds.), The European Nitrogen Assessment: Sources. Effects and Policy Perspectives Cambridge University Press.

Suddick, E.C., Davidson, E.A., 2012. The role of nitrogen in climate change and the impacts of nitrogen-climate interactions on terrestrial and aquatic ecosystems, agriculture, and human health in the United States: a technical report submitted to the US National Climate Assessment, North American Nitrogen Center of the International Nitrogen Initiative (NANC-INI), Woods Hole Research Center, 149 Woods Hole Road, Falmouth, MA, 02540-1644 USA.

Tahir, N., Li, J., Ma, Y., Ullah, A., Zhu, P., Peng, C., et al., 2021. 20 Years nitrogen dynamics study by using APSIM nitrogen model simulation for sustainable management in Jilin China. Scientific Reports 11, 17505. Available from: https://doi.org/10.1038/s41598-021-96386-5.

Tarazona, J.V., Ramos-Peralonso, M.J., 2014. Ecotoxicology terrestrial. In: Wexler, P. (Ed.), Encyclopedia of Toxicology, third ed. Academic Press, pp. 301–305. Available from: https://doi.org/10.1016/B978-0-12-386454-3.00499-1.

Tett, P., Gowen, R., Mills, D., Fernandes, T., Gilpin, L., Huxham, M., et al., 2007. Defining and detecting undesirable disturbance in the context of marine eutrophication. Marine Pollution Bulletin 55, 282–297.

Vázquez-Torres, A., Bäumler, A.J., 2016. Nitrate, nitrite and nitric oxide reductases: from the last universal common ancestor to modern bacterial pathogens. Current Opinion in Microbiology 29, 1–8. Available from: https://doi.org/10.1016/j.mib.2015.09.002.

Vörösmarty, C.J., Green, P., Salisbury, J., Richard, B., 2000. Global water resources: vulnerability from climate change and population growth. Science (New York, N.Y.) 289, 284–288.

Wang, H.J., Liang, X.M., Jiang, P.H., Wang, J., Wu, S.K., Wang, H.Z., 2008. TN:TP ratio and planktivorous fish do not affect nutrient-chlorophyll relationships in shallow lakes. Freshwater Biology 53, 935–944.

Wang, H., Wang, F., Wang, C., Han, Y., 2019. Effects of floating Azolla on phosphorus fluxes and recovery from former agricultural lands in wetland microcosms. Soil Science and Plant Nutrition 65 (1), 90–99. Available from: https://doi.org/10.1080/00380768.2018.1536387.

Wetzel, R.G., 2001. '*Limnology: Lake and River Ecosystems*'. Elsevier Publication, pp. 205–235.

Yano, K., Ohmori, Y., Fujiwara, T., 2017. Chapter 17 - Introduction to GWAS and MutMap for identification of genes/QTL using next-generation sequencing. In: Hossain, M.A., Kamiya, T., Burritt, D.J., Phan Tran, L.-S., Fujiwara, T. (Eds.), Plant Macronutrient Use Efficiency. Academic Press, pp. 307–321. Available from: https://doi.org/10.1016/B978-0-12-811308-0.00017-X.

Zhu, X., Huang, W., Zeng, J., Jiang, Z., Liu, J., Xu, X., et al., 2013. Effects of nitrogen and phosphorus ratios on phytoplankton community structure in winter. Chinese Journal of Applied Environmental Biology 19, 293–299. Available from: https://doi.org/10.3724/SP.J.1145.2013.00293.

Chapter 31

Mercury phytovolatilization: an overview of the mechanism and mitigation

Khushbu Sharma, Priyanka Devi, Shipa Rani Dey and Prasann Kumar
Department of Agronomy, School of Agriculture, Lovely Professional University, Phagwara, Punjab, India

Introduction

Despite its persistence in the environment, mercury continues to pose a serious threat to ecosystems as a result of its long-term presence in the environment. It is without a doubt that mercury is one of the most toxic elements on earth, and when it is in the presence of living organisms. As a result, there is a risk of death. It has been shown that when the concentration of mercury in drinking water is higher than the critical limit of 1 g/L (drinking water), it directly affects the nervous system, the immune system, and the digestive system of a human being. There is a naturally occurring element called mercury (Hg) that is spread through the weathering of rocks. It is exposed to the environment through mining, it is carried through the air by wind, and it is leached into groundwater, which contributes to further pollution. Several anthropogenic activities are also involved in the contamination of the environment with heavy metals, such as the dumping of industrial wastes into rivers, the overmining of contaminated sites, and the regular practice of outdoor activities in a contaminated environment (Campos et al., 2019; Xu et al., 2015). There are many toxicological effects caused by mercury and some of these effects have been associated with neurodegenerative diseases such as Amyotrophic lateral sclerosis, Alzheimer's disease, and Parkinson's disease (Mutter et al., 2004). It has been reported that elements of mercury, as well as inorganic mercury compounds, are capable of damaging the immune system and kidneys (Holmes et al., 2009), while meta mercury (MeHg) has been reported to pose a threat to the cardiovascular and nervous systems (Hogberg et al., 2010). As a result of methylation, mercury is converted into a more lipophilic form that enters the food chain as products like seafood are ingested. In Japan, the leading cause of human health concerns is the presence of MeHg in the food chain. MeHg is capable of biomagnifying in the food chain, and it has been associated with Minamata disease outbreaks. There is a possibility that MeHg can be taken up by aquatic organisms and accumulated in fish later on. The population that consumes a large amount of fish regularly in areas of mercury contamination is therefore at a much greater risk of MeHg poisoning (Harada et al., 2001). As a result of recent research, it has been found that rice can also accumulate MeHg from soils that have been contaminated with mercury in mercury mining areas (Zhang et al., 2010). In areas of China where mercury mining is widespread (Zhang et al., 2010) where fish is not the major source of protein, rice has now been identified as a major pathway for MeHg exposure (Feng et al., 2008).

Mobility of mercury

There are a variety of types of plants that contain mercury, and it is transported through the water to various organs of plants that contain mercury. Despite this, most of the mercury absorbed by plants is accumulated in the roots, with only a small amount translocated to the shoots after absorption. Moreover, even at low concentrations of mercury used in agriculture, mercury can cause a number of damages to plants, including growth retardation, suppression of photosynthesis, formation of reactive oxygen species (Kim et al., 2017), lipid membrane oxidation, and DNA and protein damage in plants (Malar et al., 2015). To combat the oxidative damage caused by heavy metals, plants have evolved a number of defense strategies. In general, the industrial sector is responsible for 27% of the emissions of gaseous elemental mercury globally (Zhang et al., 2016). According to Gworek et al. (2017), the European Union's member states

Role of Green Chemistry in Ecosystem Restoration to Achieve Environmental Sustainability. DOI: https://doi.org/10.1016/B978-0-443-15291-7.00031-6

produced 20% of the world's total mercury emissions in 2016. It is also pertinent to note that industrial wastes such as chlorine and sodium hydroxide contribute to the increase in the amount of mercury pollution in the atmosphere. A small portion of anthropogenic global Hg emissions came from chlor-alkali-related sources in the last decade, which accounted for 18.5–210, 8 t/year, or 2.7% of the total anthropogenic Hg emissions. To eliminate hazardous heavy metals from the environment, various strategies are being implemented.

The interface of mercury phytovolatilization

A well-behaved height profile of atmospheric mercury concentrations (air and soil air) is visible in the boundary layer around the soil/air interface as a result of the phtytovolatilization process. Using a comprehensive 9-point vertical gradient to characterize this profile, it was found that it perfectly matches the mathematical description (Hg) = 270.5 above ground level. Theoretically, this power function connection is examined, and it is supported by comparisons to a variety of atmospheric mercury samples that have been published in high-profile scientific journals. During the biogeochemical cycle of mercury, wetlands play an important role in the turnover of mercury from soils. There are a number of mechanisms that provide defense against oxidative stress, including enzymatic and nonenzymatic antioxidants, such as glutathione, phytochelatin salicylic acids, ascorbic acid (Sobrino-Plata et al., 2009), selenium, proline, and tocopherols (Fig. 31.1). There has been a significant increase in the attention paid to risk assessment and food security in the past few years (Natasha et al., 2022). Further, risk assessments play an important role in predicting the potential pathways through which metals may enter the food chain as well as the associated health risks (Antoniadis et al., 2019). It is clear that the consumption of Hg-contaminated foods is a significant route for human exposure to Hg, and the presence of Hg in edible parts of plants may pose a threat to human health. A total of more

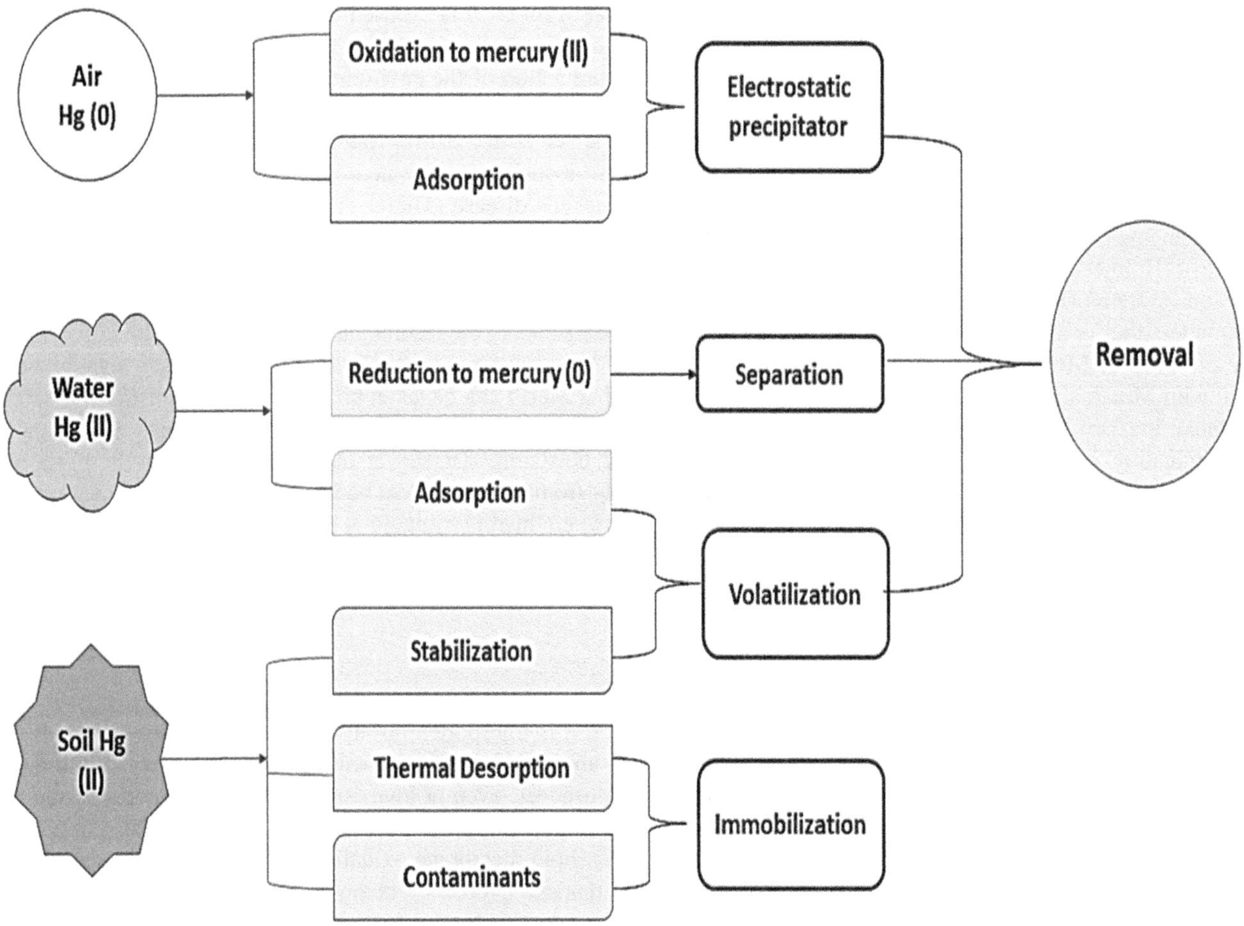

FIGURE 31.1 The interface of mercury through air, water and soil.

than 100 plant species have been studied in the last two decades to find the most effective phytoaccumulator of mercury. As a result of the time and cost involved, perennial plants are preferred for the remediation process. Preproof phytoremediation is applied to soils that are contaminated with mercury. Although a few plant species (such as *Eremochloa ciliaris* and *Erato polymnioides*) have been identified as "possible Hg hyperaccumulators" (Chamba et al., 2017), there has not yet been such identification of an Hg hyperaccumulator. In several plants, including *Aster ageratoides*, the concentrations of bio-accumulated mercury in the green, or shoot, parts of the plant are greater than those found in the roots of those plants (Qian et al., 2018).

Phytovolatilization of mercury through different plant species

In soils that have been contaminated with mercury, because of the high level of biomass present in plant shoots, a process of phytovolatilization can occur. This process is a result of mercury concentrations being frequently removed from the soil through this process due to the high level of biomass in plant shoots. This is because it is critical to take into account that the shoots of plants contain more biomass, which can assist in the phytoremediation of Hg-contaminated soils, and they should also be removed regularly to facilitate this process. This study examined several experimental plant species that were projected to grow in Hg-contaminated soils and found that their biomass of roots and shoots decreased as a result of the toxicity caused by subterranean exposure to Hg (Marrugo-Negrete et al., 2015). Compared with potting mixes without the addition of Hg, Triandra grew more vigorously in a potting mix that contained Hg, whereas, in a potting mix that contained Hg spikes, Triandra grew less aggressively.

The drawback of phytovolatilization

There are a number of drawbacks to phytoremediation technology, including its inability to deal with extremely polluted soils, Hg interfering with plant metabolic processes and limiting plant development, and a lengthy remediation time. There have been numerous studies that have used these methods to address the difficulties raised during phytoremediation of a heavy metal contaminated site as a result of this. Chemical-aided procedures are commonly used as a means of boosting the efficacy of phytoremediation to achieve greater results. A total of five types of mercury are known: exchangeable (EX), carbonate-bound (CB), Fe−Mn oxides-bound (OX), organic material-bound (OM), and residual (RS). EX $>$ CB $>$ OX $>$ OM $>$ RS is the five components with the highest level of bioavailability (Su et al., 2021). There is evidence that nonbioavailable substances can be transformed by using chemical accelerators. Although excessive amounts of iodide may be damaging to willow plants, Wang and Greger demonstrated that by adding potassium iodide (KI) to soils contaminated with heavy metals, the bioavailability of Hg may be improved, thereby promoting phytoextraction from the soils. Wang and Greger (2006), however, found that the addition of potassium iodide (KI) to the willow was not associated with an improvement in the low TF of Hg by the willow. Phytoremediation was significantly improved when compost and potassium iodide were added simultaneously to the soil during the process of phytoremediation. Due to its association with a less bioavailable amount of mercury in soils than sodium thiosulfate (TS), ammonium thiosulfate (TS) may be more effective than sodium TS. Researchers have also suggested that (NH4)2S2O3 may enhance total mercury (THg) solubility and boost the phytoextraction capability of a number of plants, including *Chenopodium glaucum* L., *Poa annua*, *Brassica juncea*, and *Helianthus annuus* toward Hg uptake and translocation (Pedron et al., 2013). Also, sodium TS (Na2S2O3) has been used to promote the phytoextraction process by increasing the total Hg accumulations in the roots of *Lepidium sativum* L. and *Oxalis corniculata* L. It was shown that the accumulation of Hg in the roots was caused by leakage of potassium (K) which was concentrated in the epidermis of the plant as a result of the treatment with sodium TS. As a result of adding thiosulphate to soils, mercury was found to be sequestered in the epidermis of roots, while it was transferred to the xylems of roots and the vascular bundles of stems (Wang et al., 2018). TEM-EDX spectroscopy was used to detect the transport of mercury in the treated roots in a manner comparable to sulfur transport observed in the treated roots (Wang et al., 2018). As a result of the addition of hydrochloric acid (HCl) to Hg-contaminated soils, there was a significant increase in plant biomass and Hg accumulation, whereas the addition of EDTA led to a huge reduction in plant biomass and a large increase in Hg accumulation in plant tissues. A recent study conducted by Li and colleagues in 2019 found that the application of sulfur fertilizer to the roots, straw, and grains of rice had several beneficial effects. Firstly, it enhanced the absorption of methyl mercury (MeHg) and inorganic mercury (IHg), and secondly, it improved the mobility of mercury (Hg). This improvement was achieved by speeding up the conversion of mercuric sulfide (HgS) to RS-Hg-SR (Hg(GS)2) and mercury dichloride. Moreover, the addition of sulfur may also play a role in the synthesis of thiomolecules, which are crucial for the detoxification of mercury (Hg). A study was conducted to examine the impact of adding thioligands to *B. juncea* on its Hg buildup and volatilization. In the study, It is found that the accumulation of Hg in the plant was enhanced without affecting the volatilization of Hg, indicating that these two processes are not interdependent. Furthermore, studies have shown that cytokinin and ammonium TS have synergistic effects on Hg absorption and translocation in *H. annuus* and *B. juncea* (Cassina et al., 2012). There has been an attempt to increase the metal uptake ability of *Solanum nigrum* L. by using attapulgite (MA2) and biochar (MB2) materials. It has been proven that both MA2 and MB2 amendments increased the plant length and fresh weight of *Solanum nigrum* L. The results indicate that these amendments have the potential to reduce Hg phytotoxicity (Li) when compared to the nonamended treatment (Table 31.1).

Volatilization of mercury in vascular plants

Plants have three major defense mechanisms to deal with stress: the ability to avoid stress, the ability to resist stress, and the ability to tolerate stress. As a result of avoidance, plants are unable to absorb metal ions from the soil. As a result, the metal ions accumulate within the root cells through sequestration and binding. Additionally, the translocation of metal ions in leaves and shoots is reduced as a result of this treatment. Mercury is one of the metal ions that can be excluded by some terrestrial plants and some aquatic plants. In addition to excreting some exudates, terrestrial plants are also responsible for excreting some exudates such as citrates, oxalates, etc. These exudates can be used to evaluate the rhizosphere's metal concentration. Because these excavates act as chelates and bind to metal ions within the soil. This contributes to the bioaccumulation of metal ions within the soil, which results in a reduction in the mobility of the metal ions. When it comes to mercury uptake, the pathway was controlled by the root cells in the case of mercury uptake. In the root cell, the transporter channels within the network of transporters play an imperative role in the transport of mercury as well. The uptake of mercury by rice has been investigated as an active transport process. It seems that the uptake of mercury and cadmium in terms of their chemical properties was somehow similar in terms of their

TABLE 31.1 Mercury accumulation.

Chemical compound	Pant species	Accumulation site	References
(NH4)2S2O3	*Chenopodium glaucum, Poa annua, Brassica juncea Helianthus annuus*	(Hg accumulation in roots, stems, and leaves)	Pedron et al. (2013)
Cytokinin	*B. juncea, H. annuus*	Hg accumulation in roots	Cassina et al. (2012)
Na2SO3	*Lepidium sativum* L., *B. juncea*	(Hg accumulation in roots)	
Potassium Iodide	*Willow, B. juncea*	(Hg accumulation in soils)	Wang and Greger (2006)
	Cyrtomium macrophyllum	Accumulated 36 mg/kg H	
	Chelating resin	Accumulated 78 mg/kg Hg	
	Leontodon taraxacoides		
EDTA	*Lupinus albus* *B. juncea*	(Hg accumulation in plants)	
Fertilizers Sulfur	*Oryza sativa* L.	(MeHg accumulation in grains)	Li et al. (2019)
Compost	*L. sativum* L.	(Hg accumulation in soils)	Wang and Greger (2006)
Algae and fungi-based remediation	*Ulva lactuca*	Due to several functional groups, significantly accumulated Hg	
	Lindgomycetaceae P87		
	Enterobacter cloacae		
	Pseudomonas putida SP-1	89% removal of Hg, volatilized 100% of total H	

uptake. Depending on how much mercury is present in the solution, it might be possible that the permeability of the metals through the transmembrane is influenced by the amount of mercury present in the solution. When Cd and Hg are present in the membrane, the membrane will be destroyed as a result of the presence of these toxic elements. As a result, metal ions can more easily pass through the membrane into the cells and get into the cells. For heavy metal ions to pass through the body through specific protein channels, they need to pass through a specific protein channel.

Phytoremediation in root cell

Daily, it is becoming increasingly evident that mercury-contaminated sites are multiplying across the country. Last but not least, it has been found in reports that there are over 3,000 sites in the world that are contaminated with mercury. The number of sites may increase over time. Involuntary contamination of sites caused by industry involvement has resulted in the extension of an area that covers as much as 50 km as a result of the contamination. Due to mercury's solubility in water, it is more mobile and therefore more hazardous due to its increased mobility, which is due to its higher solubility in water (Table 31.2).

Gaseous transport of mercury through VOCs

When the VOCs are introduced into the plant, they are transported throughout the plant through mechanisms that are dependent on the physiology of the plant or the characteristics of mercury after the VOCs have been introduced into the plant. As a result of advection with the transpiration stream, many plants and VOCs are transported to the surface tissues through advection as a result of advection. It has been found that when direct phytovolatilization fluxes in these systems are associated with transpiration rates, there will be periods of low direct phytovolatilization at night and periods of high direct phytovolatilization during the day. Carbon dioxide, water vapor, oxygen, ethylene, and certain signaling molecules are only a small percentage of some of the gases that plants have evolved to be able to regulate, including carbon dioxide, water vapor, oxygen, and ethylene. After being taken up by plants, there is evidence to

TABLE 31.2 Screening of plant species in Hg accumulation from an Hg-contaminated site in the vicinity of a chloralkali plant in June 2002.

Plants	Hg in plants, μg/(g DW)		Hg in soil, μg/(g DW)
	Shoots	Roots	
Equisetum arvense	0.09	0.21	3.62
Cirsium arvense	1.84	6.4	30.3
Salix cinerea	0.26	–	3.2
Potentilla anserina	0.41	–	10.6
Plantago major	0.78	1.78	14.9
Chamomoilla suvaveolens	0.65	–	10.6
Aegopodium podagraria	0.24	–	3.2
Vicia cracca	0.22	–	16.4
Epilobium angustifolium	0.34	0.98	16.4
Populus tremula	0.88	1.98	22.5
Calamagrostis arrundinacea	0.75	2.23	50.0
Betulaceae pendula	0.24	–	16.4
Trifolium medium	0.46	–	16.4
Plantaginaceae spp.	0.86	–	50.0
Rumex spp.	0.72	–	16.4

suggest that VOCs are distributed throughout them using mechanisms that are dependent on the plant's physiology or characteristics of mercury after they are absorbed by the plant (Nagajyoti et al., 2010). As a result of advection through transpiration streams, it is possible to transport a large number of plants and VOCs to aboveground tissues. There is a direct relation between direct phytovolatilization fluxes and transpiration rates in these systems, which results in periods of low direct phytovolatilization during the night and high direct phytovolatilization during the day in these systems. Moreover, there is the possibility of transporting substances to the aboveground tissues of plants as well. During the life cycle of a plant, VOCs spread through water and air gaps, resulting in evaporation and diffusion into trunks and stems, where they can cause damage to the plant. There are some cases where diffusion is the primary mode of transport for VOCs in plants in some cases. Many places around the world are home to wild plants, and they can be found in a variety of environments.

Hg absorption by plants

It is common for Hg to be added to the soil through fertilizers, lime, sludge, and manures, resulting in soil contamination. There is no linear relationship between soil levels of mercury and plant uptake which is influenced by a number of factors (e.g., the cation-exchange capacity of the soil, soil pH, soil aeration, and the type of plant). The absorption of water may be reduced when the soil has a high pH and/or a lot of lime and salts in it (Clarkson and Magos, 2006). Among the 45 plant groups, there are at least 45 species that accumulate metals in their cells. In the majority of cases, plants that absorb Hg tend to store it in their roots, but some of them are even able to store significant quantities of it in their shoots either by translocating it from their roots or by absorbing it directly from the vapor. In Suszcynsky and Shann's research, it was found that plants that are exposed to HgO can absorb and store that metal in their shoots, but not in their roots. In general, it is estimated that 27% of the emissions of gaseous elemental mercury are required to effectively immobilize lead-contaminated materials. By entering plant cells through the same uptake pathway as micronutrients, toxic metal ions are hypothesized to compete with micronutrients and compete for uptake in plant cells.

Conclusion

There is a greater tendency for Hg(O) mercury species to become volatile because HgO mercury species are more susceptible to evaporation from the soil. There is generally accepted information that mercury evaporating from the soil in both nonbackground and background areas is primarily in the form of Hg(0) vapor and dimethyl mercury vapor, probably with minor contributions of monomethyl mercury and soluble Hg(II)-salts contributing to the evaporation. By co-distilling evaporating water with some of the volatile and nonvolatile mercury species in the soil, some of these species can be removed from the soil by evaporation. There is a considerable difference in the volatility of the evaporating species from one another, with elemental mercury and dimethyl mercury being by far the most volatile. As for the other mercury species, they are classified as "nonvolatile," with air/water distribution constants that are at least 500 times slower than the ones for the volatile mercury species.

References

Antoniadis, V., Golia, E.E., Liu, Y.T., Wang, S.L., Shaheen, S.M., Rinklebe, J., 2019. Soil and maize contamination by trace elements and associated health risk assessment in the industrial area of Volos, Greece. Environment International 124, 79–88.

Campos, C., Nobre, T., Goss, M.J., Faria, J., Barrulas, P., Carvalho, M., 2019. Transcriptome analysis of wheat roots reveals a differential regulation of stress responses related to arbuscular mycorrhizal fungi and soil disturbance. Biology 8 (4), 93.

Cassina, L., Tassi, E., Pedron, F., Petruzzelli, G., Ambrosini, P., Barbafieri, M., 2012. Using a plant hormone and a thioligand to improve phytoremediation of Hg-contaminated soil from a petrochemical plant. Journal of Hazardous Materials 231, 36–42.

Chamba, I., Rosado, D., Kalinhoff, C., Thangaswamy, S., Sánchez-Rodríguez, A., Gazquez, M.J., 2017. Erato polymnioides – a novel Hg hyperaccumulator plant in Ecuadorian rainforest acid soils with potential of microbe-associated phytoremediation. Chemosphere 188, 633–641.

Clarkson, T.W., Magos, L., 2006. The toxicology of mercury and its chemical compounds. Critical Reviews in Toxicology 36 (8), 609–662.

Feng, X., Li, P., Qiu, G., Wang, S., Li, G., Shang, L., et al., 2008. Human exposure to methylmercury through rice intake in mercury mining areas, Guizhou Province, China. Environmental Science & Technology 42 (1), 326–332.

Gworek, B., Dmuchowski, W., Baczewska, A.H., Brągoszewska, P., Bemowska-Kałabun, O., Wrzosek-Jakubowska, J., 2017. Air contamination by mercury, emissions and transformations—a review. Water, Air, & Soil Pollution 228 (4), 1–31.

Harada, M., Nakanishi, J., Yasoda, E., Maria da Conceicâo, N.P., Oikawa, T., de Assis Guimarâes, G., et al., 2001. Mercury pollution in the Tapajos River basin, Amazon: mercury level of head hair and health effects. Environment International 27 (4), 285–290.

Hogberg, H.T., Kinsner-Ovaskainen, A., Coecke, S., Hartung, T., Bal-Price, A.K., 2010. mRNA expression is a relevant tool to identify developmental neurotoxicants using an in vitro approach. Toxicological Sciences 113 (1), 95–115.

Holmes, P., James, K.A.F., Levy, L.S., 2009. Is low-level environmental mercury exposure of concern to human health? Science of the total environment 408 (2), 171–182.

Kim, Y.H., Khan, A.L., Waqas, M., Lee, I.J., 2017. Silicon regulates antioxidant activities of crop plants under abiotic-induced oxidative stress: a review. Frontiers in Plant Science 8, 510.

Li, C., Zhou, K., Qin, W., Tian, C., Qi, M., Yan, X., et al., 2019. A review on heavy metals contamination in soil: effects, sources, and remediation techniques. Soil and Sediment Contamination: An International Journal 28 (4), 380–394.

Malar, S., Sahi, S.V., Favas, P.J., Venkatachalam, P., 2015. Mercury heavy-metal-induced physiochemical changes and genotoxic alterations in water hyacinths [Eichhornia crassipes (Mart.)]. Environmental Science and Pollution Research 22 (6), 4597–4608.

Marrugo-Negrete, J., Durango-Hernández, J., Pinedo-Hernández, J., Olivero-Verbel, J., Díez, S., 2015. Phytoremediation of mercury-contaminated soils by Jatropha curcas. Chemosphere 127, 58–63.

Mutter, J., Naumann, J., Sadaghiani, C., Walach, H., Drasch, G., 2004. Amalgam studies: disregarding basic principles of mercury toxicity. International Journal of Hygiene and Environmental Health 207 (4), 391–397.

Nagajyoti, P.C., Lee, K.D., Sreekanth, T.V.M., 2010. Heavy metals, occurrence and toxicity for plants: a review. Environmental chemistry letters 8 (3), 199–216.

Natasha, N., Shahid, M., Khalid, S., Bibi, I., Naeem, M.A., Niazi, N.K., et al., 2022. Influence of biochar on trace element uptake, toxicity and detoxification in plants and associated health risks: a critical review. Critical Reviews in Environmental Science and Technology 52 (16), 2803–2843.

Pedron, F., Petruzzelli, G., Barbafieri, M., Tassi, E., 2013. Remediation of a mercury-contaminated industrial soil using bioavailable contaminant stripping. Pedosphere 23 (1), 104–110.

Qian, X., Wu, Y., Zhou, H., Xu, X., Xu, Z., Shang, L., et al., 2018. Total mercury and methylmercury accumulation in wild plants grown at wastelands composed of mine tailings: Insights into potential candidates for phytoremediation. Environmental Pollution 239, 757–767.

Sobrino-Plata, J., Ortega-Villasante, C., Flores-Cáceres, M.L., Escobar, C., Del Campo, F.F., Hernández, L.E., 2009. Differential alterations of antioxidant defenses as bioindicators of mercury and cadmium toxicity in alfalfa. Chemosphere 77 (7), 946–954.

Su, Y., Kwong, R.W., Tang, W., Yang, Y., Zhong, H., 2021. Straw return enhances the risks of metals in soil? Ecotoxicology and Environmental Safety 207, 111201.

Wang, Y., Greger, M., 2006. Use of iodide to enhance the phytoextraction of mercury-contaminated soil. Science of the Total Environment 368 (1), 30–39.

Wang, J., Anderson, C.W., Xing, Y., Fan, Y., Xia, J., Shaheen, S.M., et al., 2018. Thiosulphate-induced phytoextraction of mercury in Brassica juncea: spectroscopic investigations to define a mechanism for Hg uptake. Environmental Pollution 242, 986–993.

Xu, J., Bravo, A.G., Lagerkvist, A., Bertilsson, S., Sjöblom, R., Kumpiene, J., 2015. Sources and remediation techniques for mercury contaminated soil. Environment International 74, 42–53.

Zhang, H., Feng, X., Larssen, T., Qiu, G., Vogt, R.D., 2010. In inland China, rice, rather than fish, is the major pathway for methylmercury exposure. Environmental Health Perspectives 118 (9), 1183–1188.

Zhang, L., Wang, S., Wu, Q., Wang, F., Lin, C.J., Zhang, L., et al., 2016. Mercury transformation and speciation in flue gases from anthropogenic emission sources: a critical review. Atmospheric Chemistry and Physics 16 (4), 2417–2433.

Chapter 32

Plant growth-promoting rhizobacteria for sustainable agriculture: recent progress and challenges

Manzari Kushwaha[1], Anuradha Mishra[1], Shiv Shankar[2], Divya Goel[2], Sarita Joshi[2] and Siya Ram[3]

[1]*Department of Applied Chemistry, School of Vocational Studies and Applied Sciences (SoVSAS), Gautam Buddha University (GBU), Government of Uttar Pradesh, Greater Noida, Uttar Pradesh, India,* [2]*Department of Environmental Science, School of Vocational Studies and Applied Sciences (SoVSAS), Gautam Buddha University (GBU), Government of Uttar Pradesh, Greater Noida, Uttar Pradesh, India,* [3]*Life Sciences Discipline, School of Sciences, Indira Gandhi National Open University (IGNOU), Maidan Garhi, Delhi, India*

Introduction

Worldwide, the twenty-first century has seen a significant increase in human population. Food is a basic human requirement that plays an important role in sustaining the growing human population and the well-being and progress of society. While feeding an ever-increasing human population, maintaining the quality of food to supplement a diet is proving to be a considerable challenge. Crop productivity, on the other hand, is declining, which might be a reason for more concern in the near future. Farmers employ chemical fertilizers, insecticides, pesticides, and other chemicals indiscriminately to boost agricultural productivity and to cater to the need for food grains demands from the increasing population. The soil ecology and human health both have been adversely affected by the extensive use of agrochemicals. These agrochemicals tend to bioaccumulate at different trophic levels of the food chain thereby causing adverse impacts on the ecosystem and human health. Environmental stressors on crops are anticipated to increase due to the effects of climate change worldwide (Pachauri et al., 2014). At present, agricultural policies are encouraging sustainable agricultural practices with an emphasis on the utilization of advantageous soil microorganisms present in the rhizospheric zone with a broad range of characteristics that stimulate plant growth to mitigate the negative effects of stress on crop plants (Basu et al., 2021). The surrounding of the plant root is known as the rhizosphere which is crucial for the plant's growth and development (Mendes et al., 2013). Among the numerous carbon-rich metabolites secreted by plant roots are vitamins, carbohydrates, organic acids, and amino acids. These chemicals serve as signals to entice microbial communities and stimulate their growth (Meena et al., 2020; Zehra et al., 2021). The term "rhizo-microbiome" refers to the entire microbial community present in the rhizosphere, which is different from the microbial community of the adjacent soil ecosystem (Kumar and Dubey, 2020). Depending on the interaction of plant growth-promoting rhizobacteria (PGPR) with plant roots, they can be classified as extracellular (ePGPR) or intracellular (iPGPR) species (Martínez-Viveros et al., 2010). PGPRs promote plant growth by producing phytohormones, symbiotic nitrogen fixation, phosphate solubilization, and secretion of volatile organic molecules (Meena et al., 2017; Chandran et al., 2020). In light of the aforesaid context, this chapter aims to highlight various aspects of PGPR, such as the role of PGPR in sustainable agriculture, the underlying mechanisms involved in growth promotion, the use of these rhizobacteria under various stress conditions, and challenges in PGPR applications.

Occurrence of rhizobacteria

Rhizobacteria that promote plant growth are categorized as either ePGPR or iPGPR. While ePGPRs dwell in the rhizosphere, on the rhizoplane, or in the gaps between the cells of the root cortex, on the other hand, iPGPRs are mostly found inside the specialized nodular structures of the root cells. The widely explored rhizobacterial genera include

Role of Green Chemistry in Ecosystem Restoration to Achieve Environmental Sustainability. DOI: https://doi.org/10.1016/B978-0-443-15291-7.00016-X

Bacillus, Burkholderia, Agrobacterium, Chromobacterium, Azotobacter, Azospirillum, Micrococcus, Caulobacter, Erwinia, Pseudomonas, Arthrobacter, Flavobacterium, and *Serratia*. To fix atmospheric nitrogen, endophytic bacteria such as *Mesorhizobium, Rhizobium Allorhizobium, Frankia* species and *Bradyrhizobium*, can be used. Key plant species that co-exist in symbiotic partnerships with these rhizobial critters include *Cajanus cajan, Cercis canadensis, Acacia* sp., *Arachis hypogaea, Lens culinaris, Lotus corniculatus, Cicer arietinum, Glycine max, Pisum sativum* (Sousa et al., 2008). Several bacteria *viz., Streptosporangium* sp., *Thermobifida* sp., *Micromonospora* sp., and *Streptomyces* sp., have been found to colonize the plant rhizosphere. The occurrence of these microbes in the rhizosphere indicates their importance as biocontrol agents against fungal pathogens (Franco-Correa et al., 2010). For rhizospheric bacteria, all of these secretions function as chemoattractants. PGPR can reach the root and establish an endophytic interaction with it (Wozniak et al., 2019; Papik et al., 2020).

Mode of action of PGPR

By generating and secreting a number of chemical regulators near the rhizosphere, PGPRs are directly or indirectly involved in promoting plant growth and development (Panchami and Karthikeyan, 2017; Vacheron et al., 2014).

Direct mechanism

The term "direct mechanisms" refers to the application of bacterial features that directly promote plant development. Different types of phytohormones are produced by Rhizobacteria which include abscisic acid, gibberellins, ethylene, and auxins which facilitate cell enlargement and root enlargement as well as extension (Kaur et al., 2016). Rhizobacteria play a crucial role in the production of indole-3-acetic acid, it is required by plants for cell division, elongation, and differentiation processes (Méndez-Gómez et al., 2016). Increased growth rate is also observed in the case of the plants that are treated with cytokinin-producing rhizobacteria (Jha and Saraf, 2015). Gibberellins play a crucial role in stem elongation, flowering process, seed germination, and fruit setting in higher plants (Saleem et al., 2015). Organic phosphate is also solubilized by *Rhizobacteria* through some organic acids (e.g., 2-ketogluconic acid, citric acid, oxalic acid, lactic acid, tartaric acid, malic acid, succinic acid, and acetic acid) (Stephen et al., 2015).

Indirect mechanism

Indirect mechanisms are bacterial characteristics that prevent one or more pathogenic fungi or bacteria from performing their functions. Indirect strategies include ACC deaminase, antibiotics, induced systemic resistance, cell wall-degrading hydrogen cyanide enzymes, competition, quorum quenching, and siderophores. *Rhizobacteria* play a very vital role in developing resistance to any kind of biotic stress as well as protecting the plant from any kind of abiotic stress such as unfavorable environmental conditions (Glick, 2014). PGPR plays a crucial role in the synthesis of biosurfactants and antibiotics. They play a very crucial role in the suppression of harmful microorganisms in the rhizosphere region for example proteases, chitinases, and cellulases (Patel, 2015). Induced systematic resistance (ISR) uses certain compounds to suppress the growth of pathogens. Many researchers have reported that antibiotic production, siderophores, and bacterial structural components particularly liposaccharides and flagella are the key determinants of ISR (Naureen et al., 2015). PGPR also produces some biochemicals such as N-alkylated benzylamine derivative, dimethyl disulfide, and volatile organic compounds which are proved to be very beneficial in triggering ISR (Kumar et al., 2015).

Role of PGPR in sustainable agriculture

PGPR as biofertilizers

Nitrogen fixation

The key nutrient that plants need for growth and productivity is nitrogen (N). Plants alone can't directly utilize 78% of nitrogen present in the atmosphere. They get the nitrogen from the soil which is fixed by microorganisms. Biologic nitrogen fixation (BNF) utilizes an enzyme system called nitrogenase to transform atmospheric N_2 into forms that plants can use (Zhang et al., 2021). Nitrogen may be physiologically fixed by nitrogen-fixing bacteria at room temperature (Raymond et al., 2004).

Phosphate solubilization

The second-most crucial nutrient for plant growth after nitrogen, is phosphorus, which is widely present in soils in both organic and inorganic forms (Bhattacharyya and Jha, 2012). Insoluble phosphorus can be found as apatite and inositol phosphate (soil phytate), phosphomonoester, and phosphodiester, which are organic forms of the mineral (Glick, 2012). To alleviate soil P deficiency, phosphatic fertilizers are commonly applied to soil. Only a small amount of phosphate fertilizers is absorbed by plants; the remainder quickly degrades in the soil into insoluble complexes (Amy et al., 2022).

Potassium solubilization

The presence of potassium is necessary for the activity of enzymes that are involved in photosynthesis and protein synthesis. Low potassium level in plants results in slow development, fewer seeds, and a weak root system. There must be an alternative indigenous supply of potassium to sustain the soil's potassium status and plant uptake (Kumar and Dubey, 2012). Several bacteria and fungi have been demonstrated to release potassium from soil's potassium-bearing minerals, including Ferrooxidans species, *Bacillus mucilaginous* species, *Acidothiobacter* species, *Bacillus edaphic*, *Paenibacillus* species, and *Burkholderia* species (Liu et al., 2012). For this reason, biofertilizers like potassium solubilizing PGPR can help minimize pesticide use while also increasing long-term crop yields (Bakhshandeh et al., 2017).

Production of siderophores

Bacteria frequently produce siderophores. The siderophore releases Fe^{2+} into the cell through a gating mechanism that connects the inner and outer membranes when the Fe^{3+} siderophore complex present on the membrane of rhizobacteria is reduced to Fe^{2+}. The siderophore may be disposed of or recycled throughout the reduction process. When there is an iron deficiency, siderophores function as solubilizers for it. When combined with other potentially dangerous heavy metals and radionuclides, siderophores produce stable complexes that can be damaging to the environment. The amount of soluble metal in a substance increases whenever a siderophore binds to a metal. Siderophore-assisted uptake of iron has been accomplished by the inoculation of plants with siderophore-producing rhizobacteria (Rajkumar et al., 2010). Rhizospheric bacteria are the source of the iron needed by the oat plant (Crowley and Kraemer, 2007). *Pseudomonas fluorescens* C7 Fe-pyoverdine complex was also absorbed by *Arabidopsis thaliana*. It results in an increase in iron levels in plant tissues favoring good plant growth (Vansuyt et al., 2007). After 45 days, the chlorotic symptoms of strain GRP3-infected plants decreased while Chlorophyll a & b, and iron levels were found to be increased in *Vigna radiata* (Sharma et al., 2003).

Zinc solubilization

A minute quantity of Zn is required by plants for proper development and reproduction in their tissues (Singh et al., 2005). A high pH, high biomass and bicarbonate content, a high magnesium-to-calcium ratio, and a surplus of phosphorus and iron in the soil are just a few of the variables that might result in Zn deficiency (Li et al., 2016). According to the results from a 7-day study, 96% of the Zn applied was transformed into a variety of insoluble forms that only partially met the plant's needs within that period (Saravanan et al., 2004). According to these studies, a Zn shortfall in the crop can be attributed to the degradation of soluble Zn fertilizers into insoluble forms that plants cannot absorb. Microbes that solubilize complex Zn in soil might be a feasible option for meeting plant Zn needs. Zn fortification of grains is facilitated by the solubilization of the fixed form of Zn and enhanced Zn absorption by PGPR (Barbagelata and Mallarino, 2013).

PGPR as biopesticides/biocontrol agents

By reducing the negative effects of infections on plant growth and development through the use of biocontrol agents, PGPR indirectly promotes plant growth and development. Müller et al. (2016) described the soil as a diverse ecosystem containing bacteria, fungi, protists, and mammals. Numerous plant-associated soil/root microorganisms have developed competitive, exploitative, or neutral interactions with plants; the plant and the photomicrobiome are part of one another (Lyu et al., 2020). Using beneficial rhizobacteria to reduce pathogenic effects and boost plant growth is a new concept being explored by scientists (Qiao et al., 2017). The secretion of metabolites/chemicals by the microbes for the control of disease is plant-specific (Lommen et al., 2019; Goh et al., 2013). The phytomicrobiome contains the most beneficial microorganisms in the form of bacteria. In the rhizosphere, the gaps between the root cortex cells, the root surface, and the root cells themselves may all contain these bacteria

(Inui Kishi et al., 2017). *Bacillus subtilis*, *Pseudomonas stutzeri*, and *Bacillus amyloliquefaciens* successfully colonize the roots of cucumber plants and significantly reduce the pathogen *Phytophthora capsici* (Islam et al., 2016). Post-harvest *B. subtilis* treatment has been shown to effectively suppress fruit infections caused by *Penicillium* sp. and *Rhizopus stolonifer* (Punja et al., 2016). *Fusarium oxysporum*-induced Fusarium wilt disease is well suppressed by *B. amyloliquefaciens* isolates grown in greenhouses (Gowtham et al., 2016; Goswami and Deka, 2020). It's reasonable to assume that PGPR, if they've been around for a long time, will have a positive impact on the host plant in the form of lower carbon emissions and increased niche space (Fan et al., 2020).

PGPR as phytostimulators

Plant growth regulators/phytostimulators are organic chemicals that, even at low levels, stimulate, inhibit, or alter plant growth and development (Damam et al., 2016). Ironically, some microorganisms, such as PGPR in plants, can also promote the synthesis of these phytostimulators (Table 32.1).

PGPR in environmental remediation

PGPR has been proven to play a role in environmental restoration (Satyanarayana et al., 2012). The detoxification of organic pollutants via microbe-mediated phytoremediation has been investigated in lab, greenhouse, and field settings (Ukaegbu-Obi and Mbakwem-Aniebo, 2014). PGPR siderophore can be used to absorb, chromium, gallium, and thorium metals (Kazy et al., 2009). Rhizoremediation of herbicides and pesticides improved soil quality while also shielding plants from the adverse effects of these chemicals (Kuiper et al., 2004). The role of different PGPRs has been highlighted in Table 32.2.

PGPR for stress reduction

Bacillus thuringiensis produces thuricin 17 under water stress, which enhances root abscisic acid, total nitrogen content, and length of nodules in soybean (Prudent et al., 2015). Microorganisms that benefit plants can also help

TABLE 32.1 Prominent PGPR strains as phyto-stimulator.

Phytohormone	Host	PGPR	References
Gibberellin	*Tephrosia apollinea*	*Sphingomonas* sp. LK11	Khan et al. (2014)
	Chickpea	*Bacillus cereus*	Baliyan et al. (2021)
	Alder	*Bacillus* sp.	Gutierrez-Manero et al. (2001)
Cytokinin	Arabidopsis	*Pseudomonas fluorescens* G20–18	Großkinsky et al. (2016)
	Solanum lycopersicum	*P. fluorescens* G20–18	Mekureyaw et al. (2022)
	Maize	*Bacillus toyonensis*	Zerrouk et al. (2020)
	Wheat	*Bacillus subtilis* IB-22	Kudoyarova et al. (2014)
IAA	Rice	*Aeromonas veronii*	Mehnaz et al. (2001)
	Cassia tora	*Pseudomonas*	Kumar et al. (2015)
	Turmeric	*Pseudomonas, Bacillus*	Kumar et al. (2016)
	–	*Bacillus pumilus* (E11), *P. fluorescens* (A5 and E9)	Ait Bessai et al. (2022)
	Abrus precatorius L.	*Enterobacter* spp. *(Gammaproteobacteria)*	Kumar Ghosh et al. (2015)
	Duckweed	*Herbaspirillum* strain	Gilbert et al. (2018)
	Withania somnifera (L.)	*Agrobacterium tumefaciens* AS38, *Pseudomonas aeruginosa* AS36, and *B. pumilus* AS02,	Mukherjee et al. (2017)
	Ophioglossum reticulatum L	*Bacillus* OPR 7	Mukherjee et al. (2017)

TABLE 32.2 List of some important PGPR strains employed in environmental clean-up.

PGPR	Crop	Heavy metal	References
Azotobacter chrocooccum HKN-5, *Bacillus megaterium* HKP-1	*Brassica juncea*	Pb and Zn	Wu et al. (2006)
Achromobacter xylosixidaus Ax 10	*B. juncea*	Cu	Ma et al. (2009)
Bacillus subtilis Sj-101	*B. juncea*	Ni	Zaidi et al. (2009)
Rhizobacterium sp. D14	*Populus deltoides*	As	Wang et al. (2011)
Khuyvera ascorbate SUD 165, SUD 165/26	*B. juncea, Lyopersicon esculentum*	Ni, Pb, and Zn	Burd et al. (2000)
Bacillus sp. PSB 10	*Cicer arietinum*	Cr	Wani and Khan (2010)
Mesohizibium huakuii B3	*Asreagalus sinicus*	Cd	Sriprang et al. (2003)
Brevibacillus B-1	*Trifolium repens*	Zn	Vivas et al. (2006)
Bradyrhizobium sp. RM8	*Brassica napus*	Ni and Zn	Wani et al. (2007)
Bacillus cereus SRA 10, *Pseudomonas* sp. SRA 2, SRA 1	*Brassica oxyrrhina, B. juncea*	Ni	Ma et al. (2009)
Serratia SY5	*Zea mays*	Cu and Cd	Koo and Cho (2009)
Pseudomonas jesenii M15, *Pseudomonas* sp. M6	*Ricinus communis*	Ni, Cu and Zn	Rajkumar and Freitas (2008)
Bacillus sp. RJ 31, *Pseudomonas* sp. RJ10	*B. napus*	Cd	Sheng and Xia (2006)
Sinorhizobium sp. Pb002	*B. juncea*	Pb	Di Gregorio et al. (2006)
Pseudomonas sp. Z5	*Oryza sativa*	Zn	
Serratia sp.	*Triticum aestivum*	Zn and P	Abaid-Ullah et al. (2011)

them survive floods. In wet conditions, the root elongation of rice seedlings treated with *P. fluorescens* REN1 ACC deaminase-producing strain increased (Etesami et al., 2014). ACC deaminase can be used to alleviate salt stress, photosynthesis, electron transport, and ion homeostasis in plants that have been infected with *Variovorax paradoxus* 5C-2. This bacterium produces the enzyme for ACC, which is critical to the plant's capacity to withstand salt stress (Wang et al., 2016). A greater chlorophyll content was found in the *B. amyloliquefaciens* SQR9-infected maize seedlings as compared to the control (Bharti et al., 2016). Crop failure can occur in the worst-case scenario if the crops are exposed to extreme cold or heat (Cheng, 2014). *Capsicum annum* (*Capsicum sativum*) development is encouraged by *Serratia nematodiphila* (PGPR), which produces gibberellin (Kang et al., 2015). The use of *Pseudomonada* spp. on tomato (*Solanum lycopersicum*) plants increased the expression of cold adaptation genes and antioxidant activity (Subramanian and Smith, 2015).

Biodegradation

The breakdown of organic molecules into microbial biomass and other simpler compounds, which eventually release water and CO_2 into the environment, is known as biodegradation (Ramana and Singh, 2000). A vast variety of substances, including hydrocarbons (oil), radionuclides, polychlorinated biphenyls (PCBs), and polyaromatic hydrocarbons (PAHs) can be broken down, converted, or accumulated by soil microbes due to their extensive catabolic diversity. Cytochrome P450 monooxygenases, laccases, peroxidases, dioxygenases, phosphatases, nitroreductases, and dehalogenases are a few examples of pollutant-degrading enzymes produced by PGPR. They also take part in the direct or cooperative metabolism of organic pollutants. The breakdown or removal of petroleum-contaminated soils has been reported using the bacteria *Pseudomonas* and *Brevibacillus*, which encourage plant growth (Zhuang et al., 2007).

Limitations in applications of PGPR

Farmers' attitude

Due to the efficacy and speed with which chemical fertilizers and pesticides work, farmers and producers have developed a strong faith in them, making it difficult to introduce and/or substitute creative alternatives (Moser et al., 2008). Transferring these techniques to the field, specifically microbe-based green technologies may be problematic since it depends on farmers' attitudes and willingness to take risks in their fields by implementing new tactics. The absence of empirical data demonstrating the efficacy of microbe-based therapy in the field is the first barrier to acceptability for farmers (Moser et al., 2008). Second, agrochemicals have been demonstrated to be more effective and prompter than PGPR strains in terms of nutrient accessibility and pathogen control in modest dosages, restricting the usage of PGPR and other microbial products.

Microbe crop specificity

Employing PGPR in the field might be challenging due to their unpredictable behavior in contrast to controlled environmental conditions. Even after completing successful and effective experiments in labs, predicting an organism's behavior in natural situations is extremely challenging (Mushtaq, 2020). The viability of the planned inoculant and root colonization are two additional problems that PGPR developers typically encounter. To populate the host and produce an effective reaction, the inoculant has to colonize it at a specific population density (Mcnear, 2013). The microbial consortium has its limitations, they can interact synergistically and/or be more promising in the rhizosphere than individual strains because they support other beneficial members of the plant microbial flora. This might support host plants indirectly by assisting the colonization of new bacteria by manufacturing certain chemicals, or it could function as a signaling component, enhancing the plant-microbe link even further (Bender et al., 2016).

Issues of law and regulation

One of the obstacles to the manufacturing and distribution of goods derived from microbes is the application of risk assessment and testing regulations by nations and occasionally by local governments to prevent the production and sale of lethal or hazardous organisms (Ahmad et al., 2018). In the case of bio-based products, regulatory processes follow a highly sophisticated and lengthy methodology established by each country's regulatory and health agencies. If a company wants to sell its goods in several different areas or nations, it might be challenging for them to comply with regulatory standards in each country and its sub-jurisdictions (Basu et al., 2021).

Conclusions and future prospects

Consequently, PGPRs are now being used to combat nutrient deficiency by forming mutualistic relationships with host plants that are connected to nutrient absorption, increased stress resistance, and regulation of plant growth. Root exudates, which are released into the rhizosphere by host plants as a reduced carbon supply for phytomicrobiome members, also aid in providing a stable environment for microbe development. As a result, using PGPR to increase yield and improve agricultural plant resistance to varied situations that challenge crop growth, development, and production is an accessible and under-utilized method. Through a variety of techniques and procedures, PGPR has increased agricultural production. To improve PGPR, it is vital to apply contemporary methods and approaches. Additional research should focus on creating microbial communities, choosing appropriate rhizosphere microorganisms, and transdisciplinary research that blends biotechnology and nanotechnology applications.

References

Abaid-Ullah, M., Hassan M.N., Nawaz, M.K., Hafeez, F.Y., 2011. Biofortification of wheat (*Triticum aestivum* L.) through Zn mobilizing PGPR. In Proceedings of International Science Conference "Prospects and Challenges to Sustainable Agriculture." Azad Jammu and Kashmir University, Pakistan, 298.

Ahmad, T., Kumar, A., Pandey, D., Prasad, B., 2018. Correlation and path coefficient analysis for yield and its attributing traits in bread wheat (Triticum aestivum L. em Thell). Journal of Applied and Natural Science 10 (4), 1078–1084. Available from: https://doi.org/10.31018/jans.v10i4.1867.

Ait Bessai, S., Bensidhoum, L., Nabti, E., 2022. Optimization of IAA production by telluric bacteria isolated from northern Algeria. Biocatalysis and Agricultural Biotechnology 41, 102319.

Amy, C., Avice, J.C., Laval, K., Bressan, M., 2022. Are native phosphate solubilizing bacteria a relevant alternative to mineral fertilizations for crops? Part I. when rhizobacteria meet plant P requirements. Rhizosphere 21, 100476. Available from: https://doi.org/10.1016/j.rhisph.2022.1004764.

Bakhshandeh, E., Pirdashti, H., Lendeh, K.S., 2017. Phosphate and potassium-solubilizing bacteria effect on the growth of rice. Ecological Engineering 103, 164–169.

Baliyan, N., Dhiman, S., Dheeman, S., et al., 2021. Optimization of gibberellic acid production in endophytic *Bacillus cereus* using response surface methodology and its use as plant growth regulator in chickpea. Journal of Plant Growth Regulation 41 (1).

Barbagelata, P.A., Mallarino, A.P., 2013. Field correlation of potassium soil test methods based on dried and field-moist soil samples for corn and soybean. Soil Science Society of America Journal 77 (1), 318–327.

Basu, A., Prasad, P., Das, S.N., Kalam, S., Sayyed, R.Z., Reddy, M.S., et al., 2021. Plant growth promoting rhizobacteria (PGPR) as green bioinoculants: recent developments, constraints, and prospects. Sustainability 13 (3), 1140.

Bender, S.F., Wagg, C., Van Der Heijden, M.G., 2016. An underground revolution: biodiversity and soil ecological engineering for agricultural sustainability. Trends in Ecology & Evolution 31 (6), 440–452.

Bharti, N., Pandey, S.S., Barnawal, D., Patel, V.K., Kalra, A., 2016. Plant growth promoting rhizobacteria *Dietzia natronolimnaea* modulates the expression of stress responsive genes providing protection of wheat from salinity stress. Scientific Reports 6, 34768.

Bhattacharyya, P.N., Jha, D.K., 2012. Plant growth-promoting rhizobacteria (PGPR): emergence in agriculture. World Journal of Microbiology and Biotechnology 28 (4), 1327–1350.

Burd, G.I., Dixon, D.G., Glick, B.R., 2000. Plant growth-promoting bacteria that decrease heavy metal toxicity in plants. Canadian Journal of Microbiology 46 (3), 237–245.

Chandran, H., Meena, M., Sharma, K., 2020. Microbial biodiversity and bioremediation assessment through omics approaches. Frontiers in Environmental Science 1, 570326.

Cheng, Z.-M., 2014. Introduction to the special issue: stress biology of specialty crops. Didcot: Taylor & Francis. Critical Reviews in Plant Sciences 33 (2–3), 90–91.

Crowley, D.E., Kraemer, S.M., 2007. Function of siderophores in the plant rhizosphere. In: Pinton, R., et al., (Eds.), The Rhizosphere, Biochemistry and Organic Substances at the Soil-Plant Interface, 26. *CRC Press*, pp. 73–109. (1).

Damam, M., Kaloori, K., Gaddam, B., Kausar, R., 2016. Plant growth promoting substances (phytohormones) are produced by rhizobacterial strains isolated from the rhizosphere of medicinal plants. International Journal of Pharmaceutical Sciences Review and Research 37 (1), 130–131.

Di Gregorio, S., et al., 2006. Combined application of Triton X-100 and Sinorhizobium sp Pb002 inoculum for the improvement of lead phytoextraction by *Brassica juncea* in EDTA amended soil. Chemosphere 63 (2), 293–299.

Etesami, H., Mirseyed Hosseini, H., Alikhani, H.A., 2014. Bacterial biosynthesis of 1-aminocyclopropane-1-caboxylate (ACC) deaminase, a useful trait to elongation and endophytic colonization of the roots of rice under constant flooded conditions. Physiology and Molecular Biology of Plants 20 (4), 425–434.

Fan, D., Subramanian, S., Smith, D.L., 2020. Plant endophytes promote growth and alleviate salt stress in *Arabidopsis thaliana*. Scientific Reports 10 (1), 12740.

Franco-Correa, M., Quintana, A., Duque, C., Suarez, C., Rodríguez, M.X., Barea, J.M., 2010. Evaluation of actinomycete strains for key traits related with plant growth promotion and mycorrhiza. Applied Soil Ecology 45 (3), 209–217.

Gilbert, S., Xu, J., Acosta, K., Poulev, A., Lebeis, S., Lam, E., 2018. Bacterial production of indole related compounds reveals their role in association between duckweeds and endophytes. Frontiers in Chemistry 6, 265.

Glick, B.R., 2014. Bacteria with ACC deaminase can promote plant growth and help to feed the world. Microbiological Research 169 (1), 30–39.

Glick, B.R., 2012. Plant growth-promoting bacteria: mechanisms and applications Hindawi Publishing Corporation. Scientifica 963401.

Goh, C.H., Veliz Vallejos, D.F., Nicotra, A.B., Mathesius, U., 2013. The impact of beneficial plant-associated microbes on plant phenotypic plasticity. Journal of Chemical Ecology 39 (7), 826–839.

Goswami, M., Deka, S., 2020. Plant growth-promoting rhizobacteria—alleviators of abiotic stresses in soil: a review. Pedosphere 30 (1), 40–61.

Gowtham, H.G., Hariprasad, P., Nayak, S.C., Niranjana, S.R., 2016. Application of rhizobacteria antagonistic to *Fusarium oxysporum* f. Sp. lycopersici for the management of Fusarium wilt in tomato. Rhizosphere 2 (2), 72–74.

Großkinsky, D., Tafner, R., Moreno, M., et al., 2016. Cytokinin production by *Pseudomonas fluorescens* G20-18 determines biocontrol activity against *Pseudomonas syringae* in *Arabidopsis*. Scientific Reports 623310.

Gutierrez-Manero, F.J., Ramos-Solano, B., Probanza, A., Mehouachi, J., Tadeo, F.R., Talon, M., 2001. The plant-growth-promoting rhizobacteria *Bacillus pumilus* and *Bacillus licheniformis* produce high amounts of physiologically active gibberellins. Physiologia Plantarum 111 (2), 206–211.

Inui Kishi, R.N., Galdiano Júnior, R.F., Val-Moraes, S.P., Kishi, L.T., 2017. Soil microbiome and their effects on nutrient management for plants. In: Kumar, V., Kumar, M., Sharma, S., Prasad, R. (Eds.), Probiotics in Agroecosystem. Springer, Singapore.

Islam, S., Akanda, A.M., Prova, A., Islam, M.T., Hossain, M.M., 2016. Isolation and identification of plant growth promoting rhizobacteria from cucumber rhizosphere and their effect on plant growth promotion and disease suppression. Frontiers in Microbiology 6, 1360.

Jha, C.K., Saraf, M., 2015. Plant growth promoting rhizobacteria (PGPR): a review. Journal of Agricultural Research and Development 5, 108–119.

Kang, S.M., Khan, A.L., Waqas, M., You, Y.H., Hamayun, M., Joo, G.J., et al., 2015. Gibberellin-producing *Serratia nematodiphila* PEJ1011 ameliorates low temperature stress in *Capsicum annuum* L. European Journal of Soil Biology 68, 85–93.

Kaur, H., Kaur, J., Gera, R., 2016. Plant growth promoting rhizobacteria: a boon to agriculture. International Journal of Cell Science and Biotechnology 5, 317–322.

Kazy, S.K., D'Souza, S.F., PinakiSar, P., 2009. Uranium and thorium sequestration by a Pseudomonas sp.: mechanism and chemical characterization. Journal of Hazardous Materials 163 (1), 65–72.

Khan, A., Sharif, M., Ali, A., Shah, S.N.M., Mian, I.A., Wahid, F., et al., 2014. Potential of AM fungi in phytoremediation of heavy metals and effect on yield of wheat crop. American Journal of Plant Sciences 5, 1578–1586.

Koo, S.Y., Cho, K.S., 2009. Isolation and characterization of a plant growth-promoting rhizobacterium, Serratia sp. SY5. Journal of Microbiology and Biotechnology 19 (11), 1431–1438.

Kudoyarova, G.R., Melentiev, A.I., Martynenko, E.V., Timergalina, L.N., Arkhipova, T.N., Shendel, G.V., et al., 2014. Cytokinin producing bacteria stimulate amino acid deposition by wheat roots. Plant Physiology and Biochemistry 83, 285–291.

Kuiper, I., Lagendijk, E.L., Bloemberg, G.V., Lugtenberg, B.J., 2004. Rhizoremediation: a beneficial plant-microbe interaction. Molecular Plant-Microbe Interactions 17, 6–15.

Kumar Ghosh, P., Kumar Sen, S., Kanti Maiti, T., 2015. Production and metabolism of IAA by Enterobacter spp. (*Gammaproteobacteria*) isolated from root nodules of a legume *Abrus precatorius* L. Biocatalysis and Agricultural Biotechnology 4, 296–303.

Kumar, A., Dubey, A., 2020. Rhizosphere microbiome: engineering bacterial competitiveness for enhancing crop production. Journal of Advanced Research 24, 337–352.

Kumar, A., Gond, S.K., Mishra, A., Sharma, V.K., Verma, S.K., Singh, D.K., et al., 2015. Salicylic acid and its role in systemic resistance induced by *Pseudomonas fluorescens* to early blight disease of tomato. VEGETOS: An International Journal of Plant Research & Biotechnology 28 (3), 12–19.

Kumar, A., Vandana, R.S., Singh, M., Pandey, K.D., 2016. Plant growth promoting rhizobacteria (PGPR). A promising approach for disease management. Microbes and Environmental Management. Studium Press, New Delhi, pp. 195–209.

Kumar, P., Dubey, R.C., 2012. Plant growth promoting rhizobacteria for biocontrol of phytopathogens and yield enhancement of *Phaseolus vulgaris* L. Journal of Current Perspectives in Applied Microbiology 1, 6–38.

Li, M., Ahammed, G.J., Li, C., Bao, X., Yu, J., Huang, C., et al., 2016. Brassinosteroid ameliorates zinc oxide nanoparticles-induced oxidative stress by improving antioxidant potential and redox homeostasis in tomato seedling. Frontiers in Plant Science 7, 615.

Liu, D., Lian, B., Dong, H., 2012. Isolation of Paenibacillus sp. and assessment of its potential for enhancing mineral weathering. Geomicrobiology Journal 29 (5), 413–414.

Lommen, S.T.E., Koops, K.G., Cornelder, B.A., de Jong, P.W., Brakefield, P.M., 2019. Genetics and selective breeding of variation in wing truncation in a flightless aphid control agent. Entomologia Experimentalis et Applicata 167, 636–645.

Lyu, D., Backer, R., Subramanian, S., Smith, D., 2020. Phytomicrobiome coordination signals hold potential for climate change-resilient agriculture. Frontiers in Plant Science 11, 634.

Ma, Y., Rajkumar, M., Freitas, H., 2009. Improvement of plant growth and nickel uptake by nickel resistant-plant-growth promoting bacteria. Journal of Hazardous Materials 166 (2), 1154–1161.

Martínez-Viveros, O., Jorquera, M.A., Crowley, D.E., Gajardo, G., Mora, M.L., 2010. Mechanisms and practical considerations involved in plant growth promotion by rhizobacteria. Journal of Soil Science and Plant Nutrition 10, 293–319.

Mcnear, D.H., 2013. The rhizosphere-roots, soil and everything in between. Nature Education Knowledge 4 (3), 1.

Meena, M., Swapnil, P., Divyanshu, K., Kumar, S., Harish, Tripathi, Y.N., et al., 2020. PGPR-mediated induction of systemic resistance and physiochemical alterations in plants against the pathogens. Journal of Current Perspectives in Basic Microbiology 60 (10), 828–861.

Meena, V.S., Mishra, P.K., Bisht, J.K., Pattanayak, A., 2017. Agriculturally Important Microbes for Sustainable Agriculture, Plant-Soil-Microbe Nexus, 1. Springer, Singapore.

Mehnaz, S., Mirza, M.S., Haurat, J., Bally, R., Normand, P., Bano, A., et al., 2001. Isolation and 16S rRNA sequence analysis of the beneficial bacteria from the rhizosphere of rice. Canadian Journal of Microbiology 47 (2), 110–117.

Mekureyaw, M.F., Pandey, C., Hennessy, R.C., Nicolaisen, M.H., Liu, F., Nybroe, O., et al., 2022. The cytokinin-producing plant beneficial bacterium *Pseudomonas fluorescens* G20-18 primes tomato (*Solanum lycopersicum*) for enhanced drought stress responses. Journal of Plant Physiology 270, 153629.

Mendes, R., Garbeva, P., Raaijmakers, J.M., 2013. The rhizosphere microbiome: significance of plant beneficial, plant pathogenic, and human pathogenic microorganisms. FEMS Microbiology Reviews 37 (5), 634–663.

Méndez-Gómez, M., Castro-Mercado, E., Alexandre, G., García-Pineda, E., 2016. Superoxide anion production in the interaction of wheat roots and rhizobacteria *Azospirillum brasilense* Sp245. Plant and Soil 400 (1–2), 55–65.

Moser, R., Pertot, I., Elad, Y., Raffaelli, R., 2008. Farmers' attitudes toward the use of biocontrol agents in IPM strawberry production in three countries. Biological Control 47 (2), 125–132.

Mukherjee, A., Bhattacharjee, P., Das, R., Pal, A., K. Paul, A., 2017. Endophytic bacteria with plant growth promoting abilities from *Ophioglossum reticulatum* L. AIMS Microbiology 3 (3), 596–612.

Müller, D.B., Vogel, C., Bai, Y., Vorholt, J.A., 2016. The plant microbiota: systems-level insights and perspectives. Annual Review of Genetics 50, 211–234.

Mushtaq, Z., 2020. PGPR: present role, mechanism of action and future prospects along bottlenecks in commercialization. EQA - International Journal of Environmental Quality 41, 9–15.

Naureen, Z., Hafeez, F.Y., Hussain, J., Al Harrasi, A., Bouqellah, N., Roberts, M.R., 2015. Suppression of incidence of *Rhizoctonia solani* in rice by siderophore producing rhizobacterial strains based on competition for iron. European Scientific Journal 11 (3).

Pachauri, R.K., Allen, M.R., Barros, V.R., Broome, J., Cramer, W., Christ, R., 2014. Climate change 2014: synthesis report. In: Pachauri R., Meyer L. (Eds.), Contribution of Working Groups I, II and III to the Fifth Assessment Report of the Intergovernmental Panel on Climate Change. https://www.ipcc.ch/site/assets/uploads/2018/05/SYR_AR5_FINAL_full_wcover.pdf.

Panchami, P., Karthikeyan, S., 2017. Isolation and characterization of phosphorus solubilizing micro-organisms from cardamom rhizosphere. International Journal of Environmental Science and Technology 6 (4), 2358–2368.

Papik, J., Folkmanova, M., Polivkova, M., Suman, J., Uhlik, O., 2020. The invisible life inside plants: deciphering the riddles of endophytic bacterial diversity. Biotechnology Advances 44, 107614.

Patel, D.K., 2015. Biochemical and molecular approaches for developing phosphate solubilizing plant growth promoting rhizobacteria PGPR. European Journal of Biological Research 8 (4), 191–213.

Prudent, M., Salon, C., Souleimanov, A., Emery, R.J.N., Smith, D.L., 2015. Soybean is less impacted by water stress using *Bradyrhizobium japonicum* and thuricin-17 from *Bacillus thuringiensis*. Agronomy for Sustainable Development 35, 749–757.

Punja, Z.K., Rodriguez, G., Tirajoh, A., 2016. Effects of *Bacillus subtilis* strain 713 and storage temperatures on post-harvest disease development on greenhouse tomatoes. Crop Protection 84, 98–104.

Qiao, J., Yu, X., Liang, X., Liu, Y., Borriss, R., Liu, Y., 2017. Addition of plant-growth-promoting *Bacillus subtilis* pts-394 on tomato rhizosphere has no durable impact on composition of root microbiome. BMC Microbiology 17, 131.

Rajkumar, M., Ae, N., Prasad, M.N., et al., 2010. Potential of siderophore–producing bacteria for improving heavy metal phytoextraction. Trends in Biotechnology 28 (3), 142–149.

Rajkumar, M., Freitas, H., 2008. Influence of metal resistant-plant growth-promoting bacteria on the growth of *Ricinus communis* in soil contaminated with heavy metals. Chemosphere 71 (5), 834–842.

Ramana, K.V., Singh, L., 2000. Microbial degradation of organic waste at low temperature. Defence Science Journal 50 (4), 371–382.

Raymond, J., Siefert, J.L., Staples, C.R., Blankenship, R.E., 2004. The natural history of nitrogen fixation. Molecular Biology and Evolution 21 (3), 541–554.

Saleem, M., Zamir, M.S.I., Haq, I., Irshad, M.Z., Khan, M.K., Asim, M., et al., 2015. Yield and quality of forage oat (*Avena sativa* L.) cultivars as affected by seed inoculation with nitrogenous strains. American Journal of Plant Science 6 (19), 3251–3259.

Saravanan, V.S., Subramoniam, S.R., Raj, S.A., 2004. Assessing in vitro solubilization potential of different zinc solubilizing bacterial (zsb) *isolates*. Brazilian Journal of Microbiology 35 (1–2), 121–125.

Satyanarayana, T.J., Narain, B., Prakash, A., 2012. Microorganisms in Sustainable Agriculture and Biotechnology. Springer, Dordrecht, *Netherlands*, 829.

Sharma, A., Johri, B.N., Sharma, A.K., Glick, B.R., 2003. Plant growth-promoting bacterium Pseudomonas sp. strain GRP3 influences iron acquisition in mung bean (*Vigna radiata* L. Wilzeck). Soil Biology & Biochemistry 35, 887–894.

Sheng, X.F., Xia, J.J., 2006. Improvement of rape (*Brassica napus*) plant growth and cadmium uptake by cadmium-resistant bacteria. Chemosphere 64, 1036–1042.

Singh, B., Natesan, S.K.A., Singh, B.K., Usha, K., 2005. Improving zinc efficiency of cereals under zinc deficiency. Current Science 88 (1), 36.

Sousa, C.S., Soares, A.C.F., Garrido, M.S., 2008. Characterization of streptomycetes with potential to promote plant growth and biocontrol. Scientia Agricola 65 (1), 50–55.

Sriprang, R., Hayashi, M., Ono, H., Takagi, M., Hirata, K., Murooka, Y., 2003. Enhanced accumulation of Cd21 by a Mesorhizobium sp. transformed with a gene from *Arabidopsis thaliana* coding for phytochelatin synthase. Applied and Environmental Microbiology 69 (3), 1791–1796.

Stephen, J., Shabanamol, S., Rishad, K.S., Jisha, M.S., 2015. Growth enhancement of rice (*Oryza sativa*) by phosphate solubilizing Gluconacetobacter sp. (MTCC 8368) and Burkholderia sp. (MTCC 8369) under greenhouse conditions*3* Biotech 5 (5), 831–837.

Subramanian, S., Smith, D.L., 2015. Bacteriocins from the rhizosphere microbiome - from an agriculture perspective. Frontiers in Plant Science 6, 909.

Ukaegbu-Obi, K.M., Mbakwem-Aniebo, C.C., 2014. The synergistic approach of plants and rhizobacteria in crude oil contaminated soil remediation. European Journal of Environmental and Safety Sciences 2, 65–70.

Vacheron, J., Desbrosses, G., Bouffaud, M.-L., Touraine, B., Prigent-Combaret, C., 2014. Plant growth-promoting rhizobacteria and root system functioning. Ecophysiology of Root Systems-Environment Interactions 166.

Vansuyt, G., Robin, A., Briat, J.F., Curie, C., Lemanceau, P., 2007. Iron acquisition from Fe-pyoverdine by *Arabidopsis thaliana*. Molecular Plant-Microbe Interactions 20, 441–447.

Vivas, A., Biro, B., Ruiz-Lozano, J.M., Barea, J.M., Azcon, R., 2006. Two bacterial strains isolated from a Zn-polluted soil enhance plant growth and mycorrhizal efficiency under Zn-toxicity. Chemosphere 62 (9), 1523–1533.

Wang, Q.Y., Dodd, I.C., Belimov, A.A., Jiang, F., 2016. Rhizosphere bacteria containing 1-aminocyclopropane-1-carboxylate deaminase increase growth and photosynthesis of pea plants under salt stress by limiting Na + accumulation. Functional Plant Biology 43, 161–172.

Wang, Q., Xiong, D., Zhao, P., Yu, X., Tu, B., Wang, G., 2011. Effect of applying an arsenic-resistant and plant growth-promoting rhizobacterium to enhance soil arsenic phytoremediation by *Populus deltoides* LH05–17. Journal of Applied Microbiology 111, 1065–1074.

Wani, P.A., Khan, M.S., Zaidi, A., 2007. Synergistic effects of the inoculation with nitrogen-fixing and phosphate-solubilizing rhizobacteria on the performance of field-grown chickpea. Journal of Plant Nutrition and Soil Science 170, 283–287.

Wani, P.A., Khan, M.S., 2010. Bacillus species enhance growth parameters of chickpea (*Cicer arietinum* L.) in chromium stressed soils. Food and Chemical Toxicology: An International Journal Published for the British Industrial Biological Research Association 48 (11), 3262–3267.

Wozniak, M., Gałazka, A., Tyśkiewicz, R., Jaroszuk-Sciseł, J., 2019. Endophytic bacteria potentially promote plant growth by synthesizing different metabolites and their phenotypic/physiological profiles in the biolog GEN III MicroPlateTM Test. International Journal of Molecular Sciences 20, 5283.

Wu, C.H., Wood, T.K., Mulchandani, A., Chen, W., 2006. Engineering plant-microbe symbiosis for rhizoremediation of heavy metals. Applied and Environmental Microbiology 72 (2), 1129–1134.

Zaidi, A., Khan, M.S., Ahemad, M., Oves, M., 2009. Plant growth promotion by phosphate solubilizing bacteria. Acta Microbiologica et Immunologica Hungarica 56, 263–284.

Zehra, A., Raytekar, N.A., Meena, M., Swapnil, P., 2021. Efficiency of microbial bio-agents as elicitors in plant defense mechanism under biotic stress: a review. Current Research in Microbial Sciences 2, 100054.

Zerrouk, I.Z., Rahmoune, B., Auer, S., Rösler, S., Lin, T., Baluska, F., Ludwig-Müller, J., 2020. Growth and aluminum tolerance of maize roots mediated by auxin- and cytokinin-producing *Bacillus toyonensis* requires polar auxin transport. Environmental and Experimental Botany 104064.

Zhang, J., Cook, J., Nearing, J.T., Zhang, J., Raudonis, R., Glick, B.R., Langille, M.G.I., Cheng, Z., 2021. Harnessing the plant microbiome to promote the growth of agricultural crops. Microbiological Research 245, 126690. Available from: https://doi.org/10.1016/j.micres.2020.1266903.

Zhuang, X., Chen, J., Shim, H., Bai, Z., 2007. New advances in plant growth-promoting rhizobacteria for bioremediation. Environment International 33 (3), 406–413. Available from: https://doi.org/10.1016/j.envint.2006.12.0057.

Chapter 33

Perspectives of crop improvement techniques and green chemistry towards sustainable agriculture

Nishish Mishra[1], **Arpita Ghosh**[2] **and Gargi Biswas**[3]

[1]*Department of Biotechnology, National Institute of Technology Durgapur, Durgapur, West Bengal, India,* [2]*Indian Institute of Management Sirmaur, Paonta Sahib, Himachal Pradesh, India,* [3]*Waste Management (I) Division, Central Pollution Control Board, Delhi, India*

Introduction

Food-water scarcity has become a global concern to feed the continuously growing population. The main obstacle to food-producing agriculture is climate change and drought, which continues to have a big impact in both developed and developing nations. United Nations research has estimated global population growth of almost 50% from 2000 to 2050 (Henchion et al., 2017), and such an increased population of human beings may be a burden for the need for healthy food requirements against individual demand (Roy et al., 2022). The increasing population of human beings has led to the demand for food availability to overcome the situation of starvation and malnourishment (Khan et al., 2022; Adegboye et al., 2021; Devaux et al., 2021). To supply enough food to the growing population, toxic chemical fertilizers are being used to enhance yield production across the world. Still, the major drawback of using these chemical-oriented products in a field is environmental pollution and the degradation of the field quality, which also causes human health issues. An increment of 60% production rate would be required to fulfill the need for food hunger among the 10 billion populations of humans (Hasegawa et al., 2021). Hence, the improvement in crop productivity and its sustainability is essential for the rapid urbanization of the entire world. The crop improvement "Jethobudho" is an aromatic rice landrace of Nepal's Pokhara Valley middle hills (Strait et al., 2012). The structural variants (SVs) are the critical substantial crop improvement and domestication characteristics. For capturing 238,490 SVs in different tomato traits, long-read nanopore sequencing is required. From wild ancestors, genetic variation is the cause of phenotype variation in crop plants. Genetic variation has been used to learn how any selected genes and variants were responsible for quantitative trait variation (Alonge et al., 2020; Yuan et al., 2021). Researchers have collected over 800 tomato successions, and hereafter, the SVs collector algorithm has expressed on it to select 51 different modern samples optimally. The study of tomato research and breeding communities has separately selected 49 wild species of tomato (Alonge et al., 2020). After that, the last collection of 100 accessions identified as phylogenetic diversity extended over the closest wild relatives of domesticated tomato (*Solanum cheesmaniae* [CHE], *Solanum pimpinellifolium* [SP], and *Solanum galapagense* [GAL]). The objective of the present study was to discuss the techniques of crop improvement of different countries. Moreover, bibliometric analysis was performed to understand the stress on the research in this domain all over the world.

Findings and critical discussion on crop improvement

Following are some of the techniques and ideas introduced by scientists and researchers to improve crops through genome editing, phenotype roots, nanoparticles, interspecific hybridization, metabolic, seed-coating technology, the significance of hydrogels, etc. The genetic variation introduced the method of genome editing in agriculture for crop improvement. TALENs (transcription activator-like effector nucleases) and ZFNs (zink-finger nucleases) have been used in genome editing for around two decades. CRISPR (clustered regularly interspaced short palindromic repeats) and CAS (CRISPR-associated protein 9) have been launched as modified systems for studying engendering crop genomes. TALEN has been widely used for genome editing in most plants, including nicotiana, brachypodium, arabidopsis, tomato, potato, barley, rice, maize, soya bean, rapeseed, flax, and wheat. Rice was the first crop to be improved using TALEN-mediated genome editing (Zhang et al., 2018). Table 33.1 shows the summary of crop improvement techniques from recent literature.

Role of Green Chemistry in Ecosystem Restoration to Achieve Environmental Sustainability. DOI: https://doi.org/10.1016/B978-0-443-15291-7.00025-0

CRISPR technology across central dogma

The genetic information is transferred from DNA through RNA to proteins, which is known as the central dogma (CD) of molecular biology. CD processes are responsible for maintaining the genetic coherence for the entire life cycle in the plants and the transmission of genetic information to the next generation. CRISPR technology is getting more popular than TALEN and ZFNs because of their beneficial characteristics (simple design, more efficiency, less cost, reversibility, target accessibility in a wide range, and precision editing). In the beginning, the CRISPR system was widely used to target the genomic DNA in yeast, bacteria, animals, and plants for either gene knock-in experiments or gene knockout. However, with the expansion of this CRISPR technology, the genome editing toolbox of CRISPR allows the transferring of genetic information across the CD. Hence, CRISPR-based techniques have been used in gene expression for crop improvement through CD (Pramanik et al., 2021). The genome sequence has been targeted for using the CRISPR/Cas system without making DSBs in the DNA by supplying the cargo (Fusion of inhibitory molecules with a catalytic diminished Cas enzyme). A broad spectrum of CRISPR-mediated equipment is available for engineering CD processes. Base editing is reached up to the adoption of Cas-fused enzymes that instantly make accurate DNA or RNA point mutations without any deletion in the nucleic acid sequences. CRISPR/Cas system can target the same or different chromosomes simultaneously. CRISPR-developed chromosomal edits with stable inheritance have been obtained in plants, animals, and yeast (Tan et al., 2022; Gaj, 2021; Moses and Kaur, 2019; Zhang, 2019).

TABLE 33.1 Recent literature on crop improvement techniques.

Aim	Techniques	Place of study	Crops improved	References
Crop improvement by genome editing	Genome editing by the TALEN method	China	Rapeseed, soybean, maize, tomato	Zhang et al. (2018)
Crop improvement by phenotypic roots	Phenotyping	India, United States	Wheat, maize, barley, rice, pea	Tracy et al. (2020)
Crop improvement by nanoparticles	Nanometal treatment	Hungary	Tomato, wheat	Kolbert et al. (2022)
Crop improvement by interspecific hybridization	Interspecific hybridization	Germany, Chile	Brassica crops	Katche et al. (2019)
Crop improvement by multiomics	Through metabolites	Bulgaria, Italy, Germany,	Dwarf rice, melon, strawberry, soybean, maize, wheat	Scossa et al. (2021)
Crop improvement by Genome editing	CRISPR/Cas9 system	Japan, United States	Rice, maize, tomato, potato, rapeseed, flax, cucumber, cotton, oranges, apple, barley, and watermelon	Ahmar et al. (2020)
Crop improvement by In vitro ploidy Manipulation	Uses of antimitotic agents	United States	Actinidiaceae, amarylidaceae, brassicaceae, cucurbitaceae, ericaceae, iridaceae	Touchell et al. (2020)
Crop improvement through Seed technology	Seed-coating technology	China, Pakistan, Saudi Arabia	Wheat, cowpea, chickpea, broccoli, cucumber, chili, artichoke, radish, etc.	Afzal et al. (2020)
Crop improvement by an insect pollinator	Insect pollination	Australia	Almonds, apples, cranberries, melons, broccoli, cherries, blueberries	Tanda (2021)
Crop improvement by Microbial technology	Uses of biofertilizer and biopesticides	India, United States, Brazil, France, Argentina	Rice, wheat, maize	Bhatt and Nailwal (2018)
Crop improvement by interspecific hybridization	Interspecific hybridization	South Carolina	Cucumis	

Crop improvement through interspecific hybridization

Interspecific hybridization is a widely used biotechnical tool for crop improvement for the yield of new species or the carrier of genetic information between the species. It is utilized to combine the beneficial traits of two different species or to transfer significant traits from wild to cultivated species (Lidder and Sonnino, 2012). The chromosome number is then doubled during a subsequent culture to produce haploid plants from diploid plants by using chemicals, such as colchicine, or other in vitro procedures. Colchicine is additionally used to produce functional pollen by artificially increasing the number of chromosomes in sterile wild hybrid plants. Two models for interspecific hybridization's role in plants' future development are genome polyploidization and stability and gene flow with ploidy maintenance (Zhang et al., 2016). One of the successful examples of crop improvement through interspecific hybridization is in BRASSICA crops (Katche et al., 2019). Interspecific hybridization has also been performed in Cucurbitaceae which are produced in various genera including, Cucumis, Citrullus, and Cucurbita. In the view of genus Cucumis, a cross has been done between the *C. dipsacus* E. ex S. and C. Anguria L. given the result of the amphidiploid observed by the researchers. In the Cucurbitaceae family, the only species Cucurbita has interspecific hybridization has been given the effective result for the utilization of the crop improvement. Cucumber and Melons are the two economical species of the Cucumis. Cucumber has relatively limited genetic variation and thus strives for interspecific hybrids to become expository and meaningful. *Cucumis hystrix* species has a vital role in the study of phylogenetic relationships. Compared to African species, Asian species, such as melon (*Cucumis melo*, $n = 12$), cucumber (*Cucumis sativus*, $n = 7$), and wild cucumber (*Cucumis hystrix*, $n = 12$), show highly scrambled genomes that brought on by large-scale inversions, centromere relocation, and rearrangements resembling chromothripsis (Chen and Adhelberg, 2020; Zhao et al., 2021; Swamy, 2017).

Crop improvement through in vitro ploidy manipulation

Antimitotic agents, such as oryzalin, colchicine, and trifluralin, are responsible for the commonly induced way of in vitro ploidy manipulation. Some of the significant contributions of polyploidy in terms of improved traits such as better resistive nature to environmental stresses, thicker and dark-colored leaves, more substantial and long-lasting flowers, enhanced vigor, resistive spirit towards pests and pathogens, enhanced production of metabolites, and perhaps regain fertility in sterile wide hybrids because of chromosome doubling of the various species. Plant tissue culture systems have some practical limitations for many taxa, mainly in woody plants, and many of the genera have not been raised successfully through tissue culture (Touchell et al., 2020). The most known way for the in vitro application of antimitotic inhibitors has been followed through the treatment in the liquid media, with the recovering tissue on a propagation medium, polyploids developed by the solidification of media by the use of antimitotic agents, in the similar composition the antimitotic agents are mixed with the media to impact chromosome doubling (Touchell et al., 2020).

Crop improvement from phenotype roots

With constantly evolving genomics technologies, significant progress has been achieved in deciphering and comprehending crop genomes. An excellent field phenotyping of shoots, roots, and yield products is essential for the validation of trait values in target field conditions. For root phenotypes and germplasm origin head-on in the field, the techniques are accessible today to choose genotypes with preferred root attributes that are meddling, barely invasive, or indirect via soil or shoots (Tracy et al., 2020). Coring and "Shovelomics" are generally adapted invasive methods. Coring, in field core-break, Bayesian hierarchical nonlinear mixed modeling, and field automated imaging of core faces add in soil over soil depths, that can be nursed as a single genetic activity. The traits on the washed "root crown" were measured by Shovelomics. On various species and soil conditions, growth, angle, and root number were scaled. Root calculator for Shovelomics trait, multiangle imaging program, and digital clarifying of root traits (Tracy et al., 2020; Mir et al., 2019; Cobb et al., 2013; Paez-Garcia et al., 2015).

Seed-coating technology

Modern seed-coating technology's objective is to equally distribute a range of active compounds (ingredients) on agricultural seeds at the right dosages to facilitate sowing and enhance crop performance. In the 1800s as a soak, copper sulfate was applied to cereals which obtained an effective and meaningful seed treatment (Sohail et al., 2022). Crop science describes commercial batch treaters to measure the weight of the seed and apply the exact amount of seed treatment products one batch at a time, while continuous flow treater with the seed treatment and coating material treats the calculated amount of seed at a given flow rate (Afzal et al., 2020; Pedrini et al., 2017).

Uses of nano forms of essential metals in crop improvement

The protection of crops from external elements like viruses, bacteria, fungi, and oomycetes has been cured by nano-pesticides such as CuO, ZnO, nFe2O3, nFe3O4, and NiO. The plant needs perfect availability of at least 14 macronutrients (N, P, K, S, Ca, Mg) and micronutrients (Mn, Zn, Cu B, Mo, Si, Ni, and Cl). Plant breeding, which has improved capacities to adapt to nutrient-rich circumstances, is the optimum option for the growth of plant genotypes (Kolbert et al., 2022). There are various factors for getting the effect of essential nanometals their morphology, hydrodynamic size, crystal structure, and charge on the surface are the reasons for the diversified effect on the plants. A study expressed a treatment on the soybean seeds with nanometals like (Cu, Fe, or Co NPS) moderately prevailed upon germination, when it is used on the ground level many new changes were observed such as increment in chlorophyll content, pods number per plant, nodules per root, the weight of pod and 1000-gain weight. Silicon nanoparticles act as a unique kind of pesticide and fertilizer because of their mesoporous nature which provides a wide surface area, due to this use it defines concepts like nano-fertilizer and nano-pesticides that may help in crop improvement.

Role of biofertilizers and biopesticides in sustainable agriculture

Fungi, algae, and bacteria are the ingredients of the biofertilizers that support the growth of a plant and increase crop production. It is also helpful in improving soil quality by fixing phosphate and nitrogen. In comparison to chemical fertilizers, biofertilizers are inexpensive (Seenivasagan and Babalola, 2021; Kumar et al., 2017). Its continuous use on the land for 3–4 years can hold fertility. The components of biofertilizers, which encourage plant development and boost agricultural yield, include fungi, algae, and bacteria. Fixing the phosphate and nitrogen also aids in enhancing the quality of the soil. Biopesticides can be defined as naturally occurring substances which is beneficial for controlling pests in plants by using nonchemical methods hence, they are eco-friendly too. They can be derived from plants (*Azadirachta, chrysanthemum*), animals (nematodes), and microorganisms (*Bacillus*, *Thuringiensis*, *Trichoderma*, etc.). Biopesticides are generally less toxic than chemical-based pesticides (Strait et al., 2012).

Importance of super absorbent hydrogels in agriculture

Due to the increasing drought conditions, crop production is affected and such conditions can lead to lipid peri-oxidation and oxidative stress in the plants. Because of this problem, researchers have found the use of super adsorbent hydrogels which have the maximum water-holding capacity for which moisture will be trapped in soil for a long time and could be also used as fertilizers. The use of nontoxic hydrogels will be beneficial for improving soil quality. The use of such hydrogels in agriculture will be beneficial for irrigation in low-rainfall areas (Sitnikova et al., 2019). Another use of green chemistry is to modify natural polymers to create hydrogels for agricultural uses (Abdel-Raouf et al., 2018). Some of the most common hydrogels practiced in agriculture are Agrosoak; Stockosorb; Pusa hydrogel, Jalshakti, and Soil Moist (Dar and Ram, 2016). In some cases, the combination of both the physical and chemical hydrogels gives the resultant dual-network hydrogels. Dual-network has more capacity to hold a large amount of liquid and it is also more sensitive in different pH ranges (Qamruzzaman et al., 2021). The greatest benefit of hydrogels is their ability to satisfy the need for water during drought conditions, which helps to preserve agricultural yield (Neethu et al., 2018).

Role of green chemistry in crop improvement

As the focus of consumers moves to secure a sustainable and reliable food supply, the agrochemical industry will require a second "green revolution" based on green chemistry principles to continue manufacturing products pertinent to agricultural operations. Because feedstock can reduce the output of agricultural waste, research, and development of sustainable technologies for biomass conversion have greatly grown. These processes strictly follow the principles of Green Chemistry (Perlatti et al., 2014). The use of solvent-free or environmentally friendly recyclable solvent systems is emphasized in green chemistry (Bhandari and Kasana, 2018). Green chemistry is an area of sustainable chemistry focused on such a chemical reaction that minimizes the generation of hazardous substances. Hence, it has become the new paradigm in agriculture today. Agro-chemists have developed the fraction of nutrients in fertilizers absorbed by plants to supply nutrients in a form that is available to plants, which helps to greatly reduce fertilizer application rates and prevent runoff to surrounding surface water (Bejarano et al., 2014).

Bibliometric analysis

The Scopus database, one of the most trustworthy scientific databases, has also been chosen as the primary data source for this study. In this study, the phrases "food security," "sustainability," "crop enhancement," and "agricultural" were used to search articles from the Scopus database. The search period included the years 1960 to 2022. A maximum of 813 articles were taken into an Excel file (.csv). VOSviewer, an open-access and free bibliometric program was used to evaluate the data. The Scopus file was imported into VOSviewer, data filtration was initiated to analyze mostly occurred keywords, authors, countries, etc. Fig. 33.1 rationally depicts how bibliometric analysis steps were conducted. To begin the study, a list of keywords connected to direct food security and other associated terms is provided (Fig. 33.2). For the country analysis, the minimum number of documents for a country was kept at 1, and the minimum number of citations was kept at 0. Among 114 countries, all 114 countries met the threshold and were selected for

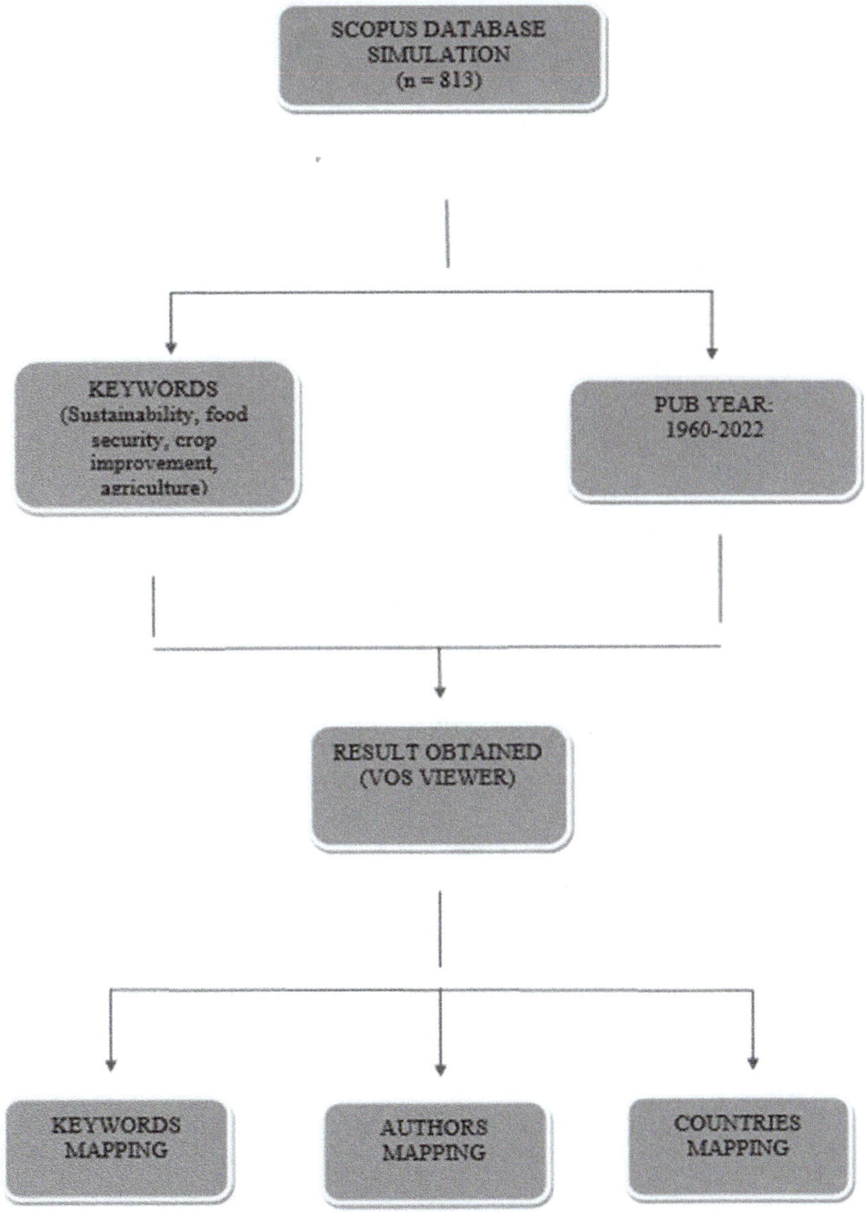

FIGURE 33.1 Flowchart of bibliometric method.

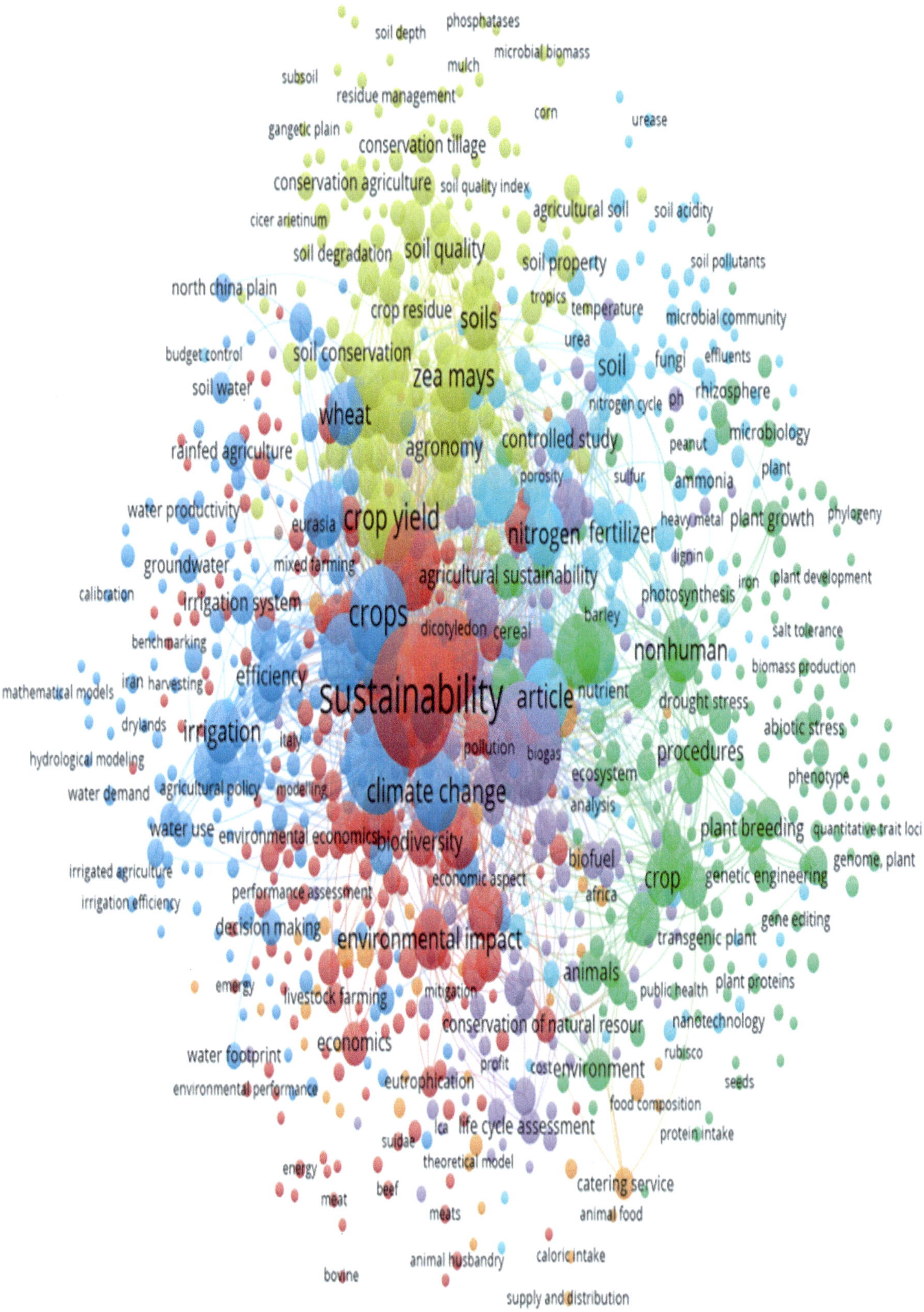

FIGURE 33.2 Keywords mapping through network visualization.

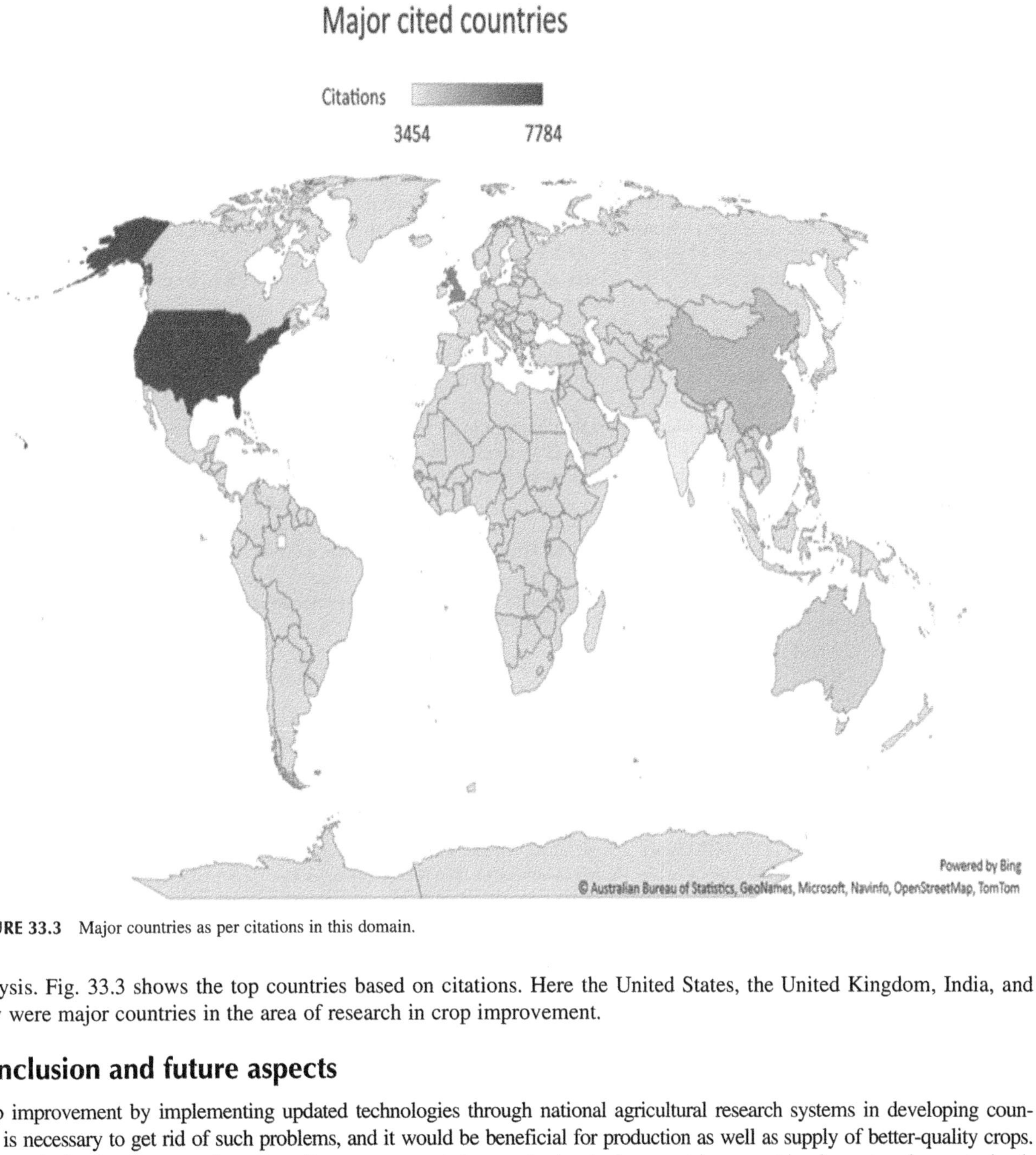

FIGURE 33.3 Major countries as per citations in this domain.

analysis. Fig. 33.3 shows the top countries based on citations. Here the United States, the United Kingdom, India, and Italy were major countries in the area of research in crop improvement.

Conclusion and future aspects

Crop improvement by implementing updated technologies through national agricultural research systems in developing countries is necessary to get rid of such problems, and it would be beneficial for production as well as supply of better-quality crops. Due to a lack of resources and proper policy management, farmers in developing countries cannot implement such new technologies. Because of the existence of both reversible and irreversible benefits and costs, an assessment of the additional measures is needed of the hour for accelerated plant breeding on food security. Considering the present scenario, it is expected that phenotyping technology will be applied in farmers' fields to increase the integrated phenotype of the plant due to soil inputs, breeding, and management practices. Furthermore, the pattern of software development in this regard may be useful for the identification of integrated rhizosphere and shoot phenotypes with advantageous traits for further selection.

References

Abdel-Raouf, M.E., El-Saeed, S.M., Zaki, E.G., Al-Sabagh, A.M., 2018. Green chemistry approach for preparation of hydrogels for agriculture applications through modification of natural polymers and investigating their swelling properties. Egyptian Journal of Petroleum 27 (4), 1345–1355. Available from: https://doi.org/10.1016/j.ejpe.2018.09.002.

Adegboye, A.R.A., Bawa, M., Keith, R., Twefik, S., Tewfik, I., 2021. Edible insects: sustainable nutrient-rich foods to tackle food insecurity and malnutrition. World Nutrition 12 (4), 176–189.

Afzal, I., Javed, T., Amirkhani, M., Taylor, A.G., 2020. Modern seed technology: Seed coating delivery systems for enhancing seed and crop performance. Agriculture 10 (11), 526.

Ahmar, S., Gill, R.A., Jung, K.H., Faheem, A., Qasim, M.U., Mubeen, M., et al., 2020. Conventional and molecular techniques from simple breeding to speed breeding in crop plants: recent advances and future outlook. International Journal of Molecular Sciences 21 (7), 2590. Available from: https://doi.org/10.3390/ijms21072590.

Alonge, M., Wang, X., Benoit, M., Soyk, S., Pereira, L., Zhang, L., et al., 2020. Major impacts of widespread structural variation on gene expression and crop improvement in tomato. Cell 182 (1), 145–161. Available from: https://doi.org/10.1016/j.cell.2020.05.021. e23.

Baidakova, M., Sitnikova, V., Uspenskaya, M., Olekhnovich, R., Kremenevskaya, M., 2019. Polymer acrylic hydrogel with protein filler: synthesis and characterization. Agronomy Research 17 (Suppl. 1), 913–922.

Bejarano, A., Sauer, U., Mitter, B., Preininger, C., 2014. Encapsulation of plant-growth promoting bacteria in polymer matrix: Development of new biofertilizers. In: International Conference on Bioencapsulation, XXII.

Bhandari, S., Kasana, V., 2018. Applications of green chemistry principles in agriculture. Green Chemistry and Technology Letters 4 (2), 10–12. Available from: https://doi.org/10.18510/gctl.2018.422.

Bhatt, P., Nailwal, T.K., 2018. Crop improvement through microbial technology: a step toward sustainable agriculture. Crop Improvement Through Microbial Biotechnology. Elsevier, pp. 245–253.

Chen, J.-F., Adhelberg, J., 2020. Interspecific hybridization in Cucumis-progress, problems, and perspectives. Horticultural Science 35 (1).

Cobb, J.N., DeClerck, G., Greenberg, A., Clark, R., McCouch, S., 2013. Next-generation phenotyping: requirements and strategies for enhancing our understanding of genotype–phenotype relationships and its relevance to crop improvement. Theoretical and Applied Genetics 126 (4), 867–887.

Dar, S.B., Ram, H., 2016. Grain yield, nutrient uptake and water-use efficiency of wheat (Triticum aestivum) under different moisture regimes, nutrient and hydrogel levels. Indian Journal of Agronomy 61, 58–61.

Devaux, A., Goffart, J.P., Kromann, P., Andrade-Piedra, J., Polar, V., Hareau, G., 2021. The potato of the future: opportunities and challenges in sustainable agri-food systems. Potato Research 64 (4), 681–720.

Gaj, T., 2021. Next-generation CRISPR technologies and their applications in gene and cell therapy. Trends in Biotechnology 39 (7), 692–705.

Hasegawa, T., Sakurai, G., Fujimori, S., Takahashi, K., Hijioka, Y., Masui, T., 2021. Extreme climate events increase risk of global food insecurity and adaptation needs. Nature Food 2 (8), 587–595.

Henchion, M., Hayes, M., Mullen, A.M., Fenelon, M., Tiwari, B., 2017. Future protein supply and demand: strategies and factors influencing a sustainable equilibrium. Foods 6 (7), 53. Available from: https://doi.org/10.3390/foods6070053.

Katche, E., Quezada-Martinez, D., Katche, E.I., Vasquez-Teuber, P., Mason, A.S., 2019. Interspecific hybridization for Brassica crop improvement. Crop breeding. Journal of Genetics and Genomics 1 (1).

Khan, S.A.R., Razzaq, A., Yu, Z., Shah, A., Sharif, A., Janjua, L., 2022. Disruption in food supply chain and undernourishment challenges: an empirical study in the context of Asian countries. Socio-economic Planning Sciences 82, 101033.

Kolbert, Z., Szőllősi, R., Rónavári, A., Molnár, Á., 2022. Nanoforms of essential metals: From hormetic phytoeffects to agricultural potential. Journal of Experimental Botany 73 (6), 1825–1840. Available from: https://doi.org/10.1093/jxb/erab547.

Kumar, R., Kumawat, N., Sahu, Y.K., 2017. Role of biofertilizers in agriculture. Pop Kheti 5 (4), 63–66.

Lidder, P., Sonnino, A., 2012. Chapter 1 – Biotechnologies for the management of genetic resources for food and agriculture. Advances in Genetics 78, 1–167. Available from: https://doi.org/10.1016/B978-0-12-394394-1.00001-8.

Mir, R.R., Reynolds, M., Pinto, F., Khan, M.A., Bhat, M.A., 2019. High-throughput phenotyping for crop improvement in the genomics era. Plant Science 282, 60–72.

Moses, C., Kaur, P., 2019. Applications of CRISPR systems in respiratory health: Entering a new 'red pen' era in genome editing. Respirology (Carlton, Vic.) 24 (7), 628–637.

Neethu, T.M., Dubey, P.K., Kaswala, A.R., 2018. Prospects and applications of hydrogel technology in agriculture. International Journal of Current Microbiology and Applied Sciences 7 (5), 3155–3162. Available from: https://doi.org/10.20546/ijcmas.2018.705.369.

Paez-Garcia, A., Motes, C.M., Scheible, W.R., Chen, R., Blancaflor, E.B., Monteros, M.J., 2015. Root traits and phenotyping strategies for plant improvement. Plants 4 (2), 334–355.

Pedrini, S., Merritt, D.J., Stevens, J., Dixon, K., 2017. Seed coating: science or marketing spin? Trends in Plant Science 22 (2), 106–116.

Perlatti, B., Forim, M.R., Zuin, V.G., 2014. Green chemistry, sustainable agriculture and processing systems: a Brazilian overview. Chemical and Biological Technologies in Agriculture 1 (1), 5. Available from: https://doi.org/10.1186/s40538-014-0005-1.

Pramanik, D., Shelake, R.M., Kim, M.J., Kim, J.Y., 2021. CRISPR-mediated engineering across the central dogma in plant biology for basic research and crop improvement. Molecular Plant 14 (1), 127–150. Available from: https://doi.org/10.1016/j.molp.2020.11.002.

Qamruzzaman, M.D., Ahmed, F.O., Mondal, Md. A.H., 2021. An overview on starch-based sustainable hydrogels: potential application and aspects. Journal of Polymers and the Environment 30, 19–50.

Roy, A., Ghosh, A., Vashisht, D., 2022. The consumer perception and purchasing attitude towards organic food: a critical review. Nutrition & Food Science 53 (3), 578–599.

Scossa, F., Alseekh, S., Fernie, A.R., 2021. Integrating multi-omics data for crop improvement. Journal of Plant Physiology 257, 153352. Available from: https://doi.org/10.1016/j.jplph.2020.153352.

Seenivasagan, R., Babalola, O.O., 2021. Utilization of microbial consortia as biofertilizers and biopesticides for the production of feasible agricultural product. Biology. MDPI 10 (11). Available from: https://doi.org/10.3390/biology10111111.

Sohail, M., Pirzada, T., Opperman, C.H., Khan, S.A., 2022. Recent advances in seed coating technologies: transitioning toward sustainable agriculture. Green Chemistry: an International Journal and Green Chemistry Resource: GC 24, 6052–6085. Available from: https://doi.org/10.1039/D2GC02389J.

Strait, B.R., Bhandari, B., Bajracharya, J., Shrestha, P.K., Kumar, S., 2012. SGyawali, S. Biopesticides: a need for food and environmental safety. Biofertil Biopestici 3 (6202), 2155.

Swamy, K.R.M., 2017. Origin, distribution and systematics of culinary cucumber (Cucumis melo subsp. agrestis var. conomon). Journal of Horticultural Sciences 12 (1), 1–22.

Tan, J., Forner, J., Karcher, D., Bock, R., 2022. DNA base editing in nuclear and organellar genomes. Trends in Genetics 38 (11), 1147–1169.

Tanda, A.S., 2021. Why insect pollinators are important in crop improvement? Indian Journal of Entomology . Available from: https://doi.org/10.55446/IJE.2021.42.

Touchell, D.H., Palmer, I.E., Ranney, T.G., 2020. In vitro ploidy manipulation for crop improvement. Frontiers in Plant Science 11, 722. Available from: https://doi.org/10.3389/fpls.2020.00722.

Tracy, S.R., Nagel, K.A., Postma, J.A., Fassbender, H., Wasson, A., Watt, M., 2020. Crop improvement from phenotyping roots: highlights reveal expanding opportunities. Trends in Plant Science 25 (1), 105–118. Available from: https://doi.org/10.1016/j.tplants.2019.10.015.

Yuan, Y., Bayer, P.E., Batley, J., Edwards, D., 2021. Current status of structural variation studies in plants. Plant Biotechnology Journal 19 (11), 2153–2163.

Zhang, F., 2019. Development of CRISPR-Cas systems for genome editing and beyond. Quarterly Reviews of Biophysics 52.

Zhang, X., Liu, T., Li, X., Duan, M., Wang, J., Qiu, Y., et al., 2016. Interspecific hybridization, polyploidization, and backcross of Brassica oleracea var. alboglabra with B. rapa var. purpurea morphologically recapitulate the evolution of Brassica vegetables. Scientific Reports 6, 18618. Available from: https://doi.org/10.1038/srep18618.

Zhang, Y., Massel, K., Godwin, I.D., Gao, C., 2018. Applications and potential of genome editing in crop improvement. Genome Biology 19 (1), 210. Available from: https://doi.org/10.1186/s13059-018-1586-y.

Zhao, Q., Meng, Y., Wang, P., Qin, X., Cheng, C., Zhou, J., et al., 2021. Reconstruction of ancestral karyotype illuminates chromosome evolution in the genus Cucumis. The Plant Journal 107 (4), 1243–1259.

Index

Note: Page numbers followed by "*f*" and "*t*" refer to figures and tables, respectively.

D

H

I

J

K

N

O

Q

R

T